Extractables and Leachables

Extractables and Leachables

Characterization of Drug Products, Packaging, Manufacturing and Delivery Systems, and Medical Devices

Dennis Jenke

Registered Office
John Wiley & Sons, Inc., 111 River Street, Hoboken, NJ 07030, USA

Editorial Office
9600 Garsington Road, Oxford, OX4 2DQ, UK

For details of our global editorial offices, customer services, and more information about Wiley products visit us at www.wiley.com.

Wiley also publishes its books in a variety of electronic formats and by print-on-demand. Some content that appears in standard print versions of this book may not be available in other formats.

Library of Congress Cataloging-in-Publication Data

Names: Jenke, Dennis, author.
Title: Extractables and leachables : characterization of drug products,
    packaging, manufacturing and delivery systems, and medical devices /
    Dennis Jenke.
Description: Hoboken, NJ : Wiley, 2022. | Includes bibliographical
    references and index.
Identifiers: LCCN 2022015986 (print) | LCCN 2022015987 (ebook) | ISBN
    9781119605072 (cloth) | ISBN 9781119605089 (adobe pdf) | ISBN
    9781119605102 (epub)
Subjects: MESH: Drug Packaging–standards | Drug Delivery
    Systems–standards | Drug Contamination–prevention & control | Drug
    Stability | Consumer Product Safety
Classification: LCC RS159 (print) | LCC RS159 (ebook) | NLM QV 825 | DDC
    615.1/8–dc23/eng/20220525
LC record available at https://lccn.loc.gov/2022015986
LC ebook record available at https://lccn.loc.gov/2022015987

Cover Design: Wiley
Cover Image: © KJ_Photography/Shutterstock

Set in 9.5/12.5pt STIXTwoText by Straive, Chennai, India

Contents

Preface

I admit it, I watch reruns of the original Star Trak television series as a way to unwind from a hard day at work. Now I am not such a fanatic that I know all the episodes and that I immediately recognize dialogue and minor characters, but certain things have stuck with me. Pertinent to this book is a quote that is attributed to Scotty, the Enterprise's irascible Chief Engineer. Now I cannot remember the episode or the circumstances but in one of the seemingly endless impossible situations Scotty finds himself in, he notes "Fool me once, shame on you, fool me twice, shame on me."

So, what does this have to do with my second book on the subject of chemical characterization, popularly known and mis-represented as "extractable and leachables?" Well, in the preface of my first book on this subject, I indicated a certain trepidation at even starting that project, recognizing the difficulty in trying to bring a certain degree of order to a subject where there was little consensus, imperfect knowledge, and minimal meaningful guidance. In the end, I ultimately decided that while I could not bring the subject to closure, I could move the ball just a little closer to the end-zone.

As I suspected would be the case when I wrote the first book and as has been verified by ongoing and recent developments, the field has changed dramatically in the decade since the publication of that book. Facilitated by both scientific publication and more casual "word of mouth," many (but not all) questions of science have been engaged, addressed and, to a certain extent, answered. Based on the evolving understanding of good science practically applied, experimental and laboratory strategies, tactics, and practices have been improved so that better information is more routinely obtained in a more effective, efficient, and reproducible manner. A multitude of new case studies have opened our eyes to new threats and new opportunities while at the same time putting some of the old monsters back under the bed where they belong. Through the hard and largely thankless work of regulators, standard setting bodies, and professional organizations (and the volunteers who selflessly give of their time and talent), some progress has been made in terms of consensus-building that seeks to find the proper risk-based balance between the ivory tower of science and the practical realities of developing, registering, regulating, marketing, and supporting safe, effective, and appropriately inexpensive pharmaceutical and medical products that save, sustain, and improve lives. To a certain extent, this consensus-building has reached the point where actionable and appropriately conservative regulations, guidelines, and best practice recommendations have been developed, published, and executed. Unfortunately, in too many ways it is "still the wild west out there."

So, it is time to get back in the saddle again and write a new text considering the ongoing and ever-changing scientific and practical endeavor that we call "E&L." Not to throw out the old texts as invalid, irrelevant, imperfect, and incorrect but rather to say "given what we know today, we can improve on what we had in the past." Fool me twice, indeed!

Most of the readers will be familiar with the phrase "what happens in Vegas stays in Vegas," which gives people at least the impression that they can "walk a bit on the wild side" while visiting Vegas with no significant consequence. The corresponding expression in the E&L world go something like "What starts in the item should stay in the item." You see, as I have noted previously, drug products do not just magically appear to the user at time of use. Rather, drug products must be manufactured from their individual raw materials and ingredients. Once manufactured, drug products must be packaged for protection and preservation during distribution and storage. At time of use, drug products must be administered to the user, either by the action of the storage container or the action of a drug administration device.

Each of these activities, manufacturing, storage, and delivery, includes contact between the drug product (and/or its precursors) and one or more items. For example, process solutions encounter various items including mixing and storage containers, tubing, filters, gaskets, connectors, sensors, etc., as they are converted into the final drug product. During storage and distribution, the drug product is in contact with components of its packaging system. During administration, the drug product is in contact with that medical device that serves as the intermediary between the packaging and the patient.

Each contact described above provides an opportunity for the drug product (or its precursors) and the contacted item to chemically interact. If, during that interaction, "what starts in the item stays in the item," then life is good. The drug product is not altered by the contact and everyone goes away safe and happy.

Unfortunately, drug products and the items they contact are quite different chemically and thus some extent of drug product alteration will occur as a result of that contact. On the one hand, substances in the contact item will be driven by a concentration gradient to move from the item and into the drug product. In this way, substances leach from the item and into the drug product, thereby becoming foreign impurities in the drug product. As foreign impurities in the drug product, these substances may have an unwanted effect on the drug product, for example rendering it less stable. Alternatively, the foreign impurities may be toxic and thus they represent a safety hazard to drug product users. In any event, characterization of the drug product for foreign impurities (leachables) and/or characterization of the contact item for potential foreign impurities (extractables) becomes a pre-requisite for establishing that a drug product is suitable for its intended use.

Of course, "for every action there is an equal and opposite reaction." For leaching, this opposite action is sorption. While leaching involves the movement of substances from an item and into the drug product, sorption involves the movement of substances from the drug product and into the contact item. As it is rare that drug products are formulated with superfluous ingredients, loss of an ingredient can have a significant adverse effect on drug product potency, efficacy, and/or stability.

However, sorption is a topic for another author and another book. In this book we will focus strictly on those interactions that add substances to the drug products and/or their precursors. When one tests items to establish "what could come out," one is performing an extractables or extraction study. When one tests the drug product (or precursor solution) for foreign impurities derived from the items, one is performing a leachables or leaching study. Generally, through a combination of both extractables and leachables testing, one produces the data set with which the impact of the interaction on the patient, the drug product, and the item can be assessed.

A similar discussion can be held for a medical device, where chemicals present in the device can leach out into that medium by which the device directly or indirectly interacts (or "communicates") with the device's user.

Previous works in this field have focused on providing answers to two important questions: "What has to be done?" and "When does it have to be done?" While providing updated answers to these questions is still an important objective of this book, we will also consider and address two additional points: "How do we do what we need to do?" and "Why do we do it that way?" Although a fifth question is relevant, specifically, "Where does what needs to be done get done?" is pertinent, this question will not be addressed to any great extent as its answer is largely an individual organization's prerogative.

A few words about the construction of this book are appropriate. Actually, the hardest part about writing this book is not collecting the contents but rather organizing the concepts and contents in a logical and appropriate manner. It is challenging to construct a text that starts from the beginning, establishing essential concepts and principles, and then takes these basic building blocks and turns them into a finished structure. Nevertheless, the book is roughly constructed in this manner, starting with concepts and ending with tactics, strategies, and processes.

Chapter 1 serves as the introduction to the issue of chemical compatibility. In addition to providing a Problem Statement, Chapter 1 speaks to the evolution of chemical assessment as an area of interest, as a scientific discipline, and as a regulatory "topic of interest." Chapter 1 will delineate those significant issues facing the various stakeholders involved with designing, implementing, interpreting, reporting, and reviewing chemical characterizations, setting the stage for the answers that will be provided in the ensuing Chapters.

Chapters 2 through 4 are the places where we establish the building blocks upon which chemical characterization will be built. As I am convinced that one of the difficulties in addressing chemical characterization is a sloppy use of an evolving but disjointed nomenclature, definitions, derived from and consistent with emerging standards, will be provided for all relevant terms and concepts. Extraction and leaching studies will be described and discussed, in terms of both basic scientific principles and practical applications. Both types of studies will be discussed in terms of their two essential aspects, generating the test article (Chapter 2) and performing the actual chemical testing (Chapters 3 and 4). Concepts important for organic compounds, such as screening versus targeting, "how low do you go?" (the AET), identification, quantitation, and qualification (validation), are addressed in Chapter 3. The importance and use of extractables–leachables correlations will also be discussed and examples of the various levels of correlation will be provided. Extracted elements, anions, and general chemical testing are all addressed in Chapter 4.

The premise of my previous book was that there is a logical, effective, and efficient process performing the various activities involved in chemical characterizations and the book was constructed on the basis of that process. Chapters 5 and 6 build on that premise by focusing on the four major stages of a product's lifecycle, development, optimization, registration, and life cycle (post approval) management. Activities and actions that are relevant to and appropriate for each stage in the product's lifecycle are delineated and it will be established and demonstrated that chemical characterization is not a series of largely unrelated tasks but rather a logical sequence of tasks, each one serving a necessary and specific purpose, most appropriately performed at a specific time in the process. Specific attention will be given to the topic of managing change control once a drug product and its related items have been approved for and implemented in the commercial marketplace (Chapter 6).

The concept and reality of chemical assessment are relevant and appropriate for three major classes of pharmaceutical contact items, pharmaceutical packaging, components used in the manufacturing of pharmaceuticals, and medical devices, particularly those used to deliver a pharmaceutical product to the patient. While these three classes of items share common materials of construction and may have a similar lifecycle in terms of development, optimization, registration, and maintenance, the simple fact remains that a pharmaceutical packaging system is not the same as a pharmaceutical manufacturing component, which in turn is not the same as a medical device. Although oftentimes they are constructed from similar material classes (for example, polyethylene), compositional similarity is not always the case and, furthermore, they can differ greatly in configuration, they do differ greatly in terms of processing, and their conditions of use vary widely. Thus, it is logical, appropriate, and necessary that the process of chemical consideration differs, packaging versus manufacturing components versus medical devices. Therefore, Chapters 7 through 9 on Applications consider each of these items separately, starting with a review of the emerging regulatory landscape, considering particularly recent developments in compendial monographs and international standards. Characteristics that differentiate the various items are discussed, specifically in the content of how the differences translate into different chemical characterization strategies and tactics. The concept of a "risk-based approach," meaning that the amount and type of chemical characterizations should depend on the risk that a particular item could have an adverse effect, is introduced and applied to each item class, resulting in a risk classification of items within each class. For example, the process for classifying manufacturing components in terms of the chemical risk they represent will be described and the nature and extent of chemical characterization required in each risk category will be established. Lastly, chemical characterization strategies for each class of pharmaceutical items will be illustrated in selected case studies.

To that point, the book will have considered only one aspect of chemical assessment, the gathering and generation of chemical information. Although information gathering and generation is the foundation of chemical assessment, by itself it is a means without an end, an activity without an impact. Ultimately and appropriately, information gathering and generation serves the sole purpose of enabling information interpretation. For pharmaceutical items, the key concept that must be interpreted is impact; that is, what impact will the chemicals that have been identified and quantified during the testing component of chemical characterization have on a key quality attribute of the items undergoing contact? For packaging as an example, what impact will leachables have on the safety, efficacy, stability, purity, and compendial compliance of the packaged drug product? Furthermore, what impact will leaching have on the packaging system's ability to perform its essential function of containment, preservation, and protection? Establishing the impact of leaching on both the source item of the leachables as well as the item receiving the leachables is termed risk assessment. Chapter 10 provides an overview of risk assessment. First and foremost, the Chapter considers the adverse impact that leachables can have on the health and safety of the patient who is exposed to the leachables as a result of a medical therapy. The general concepts and principles of toxicological safety risk assessment are considered at a very high level and the means of toxicologically assessing extractables and leachables are discussed. The challenges associated with toxicologically assessing incompletely identified extractables or leachables are considered and the use of auxiliary information (such as compendial compliance) in toxicological risk assessment is addressed. Biological safety risk assessment, which is the process of performing and interpreting biological testing to address safety impact, is considered as an alternate to chemical characterization and the complementary nature of chemical and biological testing is established.

Although it is appropriate for patient safety to be of paramount concern, the direct and adverse impact of leachables on health is not the only concern a patient could have about leachables. For example, if a leachable reacts with the drug product's active ingredient, drug product efficacy, and therefore the positive therapeutic benefit, can be lessened. Furthermore, interactions between a leachable and a drug product ingredient can shorten the shelf-life by a number of means, increasing the cost of drug products. Additionally, the products of a reaction between a leachable and the drug substance may have adverse safety consequences.

Moreover, some leachables impact items other than the drug product itself. For example, leaching of silicone oil from the barrel of a pre-filled syringe can make it more difficult to move the syringe's plunger, preventing the syringe from performing its drug delivery function. Additionally, process-related leachables from manufacturing components can adversely impact the efficiency of upstream manufacturing steps. Thus, Chapter 10 also considers the risk assessment of leachables with respect to product quality attributes other than safety.

Lastly, the rise of biopharmaceuticals has put a spotlight on effects of leachables other than direct toxicity. Because of their chemical nature, biological drug substances are much more likely to chemically and physically interact with leachables than are their more traditional "small molecule" drug substance counterparts, and it is more likely that such an interaction would have a meaningful effect on the key quality attributes of the biopharmaceutical drug products. Thus, Chapter 10 includes a discussion of the special case of biopharmaceuticals.

I started this preface with the observation that much has happened in the chemical characterization world in the last 10 years or so. It is shortsighted to think that the evolution in the science and practice of chemical characterization is over or even that the rate of evolution has slowed. Thus, I spend time in Chapter 11 describing and advocating for the "future of E&L." First and foremost, the Chapter advocates for better science. In this vein, it addresses inherent "errors" that are a part of generally established chemical characterization practices, such as identification and quantitation, and advocates for a means of reducing the frequency of occurrence and the magnitude of such errors. Additionally, I consider how better science can result in better (more consistent, reproducible, and relevant) extractions. Lastly, I advocate for the more pervasive use of sound scientific principles (as opposed to highly generalized and frequently incorrect "rules of thumb" or pockets of "common knowledge") to design and justify chemical characterization studies.

The Chapter then moves on to the practical reality of the "silo approach" that is the current business model for chemical characterization. Generally speaking, the "silo approach" recognizes the fact that chemical characterization is a disjointed and insulated activity. It is disjointed as there are few standard methods, both in terms of how extractions are performed and how extracts (and drug products) are tested. Thus, it is possible for two companies to use the same item in the same manner but to support the item with chemical characterizations that differ in tactics and thus differ in outcomes. It is insulated because each company is responsible for the registration of its own products and because each company treats its chemical characterization as part of its intellectual property. In such an environment where there is little information sharing, one imagines that redundant (and therefore unnecessary) testing is the rule rather than the exception. Thus, I advocate for information sharing as the means of reducing redundant and unnecessary testing and consider means by which such information gathering can be accomplished.

Given future advances in science and practice and projecting that some of the items I have advocated for come to fruition, then the theory and practice of chemical characterization as it is practiced tomorrow will bear little resemblance to what we know it today. Chapter 11 closes with the prediction of "the end of E&L" and a description of a collaborative environment where the application of sound science, tempered with practical realities and the proper balance of risk and reward, has led a chemical characterization process that is consistent, credible, efficient, effective, and reliable.

The last section contains an abbreviations key and a glossary of the critical terms used throughout the book.

If you watch Star Trak re-runs like I do, then you have invariably seen those commercials that go "But wait, there's more." So just when we think that the book has reached its logical ending, there is this "but wait, there is more!" Here the "more" is an Appendix, devoted to a discussion of good identification practices for organic extractables and leachables. Here I consider the various categories of identification, define the categories in terms of the amount and number of "dimensions" of corroborating data and describe the means by which identities are "elevated" to higher confidence categories. Identification practices such as spectral matching and structure elucidation are discussed and illustrated by examples. Moreover, the use and application of "additional evidences" to support and secure high confidence identities are discussed.

I note in closing that the reader may find the text to be redundant to a certain point. The reason for this is that I want concepts and such to be understood not only conceptually but also in context. Thus, concepts that are relevant in several contexts (for example, the AET) are repeatedly addressed in those multiple situations and in multiple chapters. I personally find it frustrating to run across a concept in a text and then to have to go and search for its definition and explanation somewhere else. Thus, I have chosen to be redundant and apologize up-front to those who find it excessive.

Acknowledgments

I hope you find this book to be useful, well-designed, logically organized, understandable, and effective. Those who know me and my work well will understand that I could not produce such a book without considerable help. I thank Aruna Pragasam, Jonathon Rose, and the entire editorial staff at John Wiley & Sons Inc., not only for the considerable improvements they have made to this book but also for their help and encouragement throughout this process. Without Jonathon's gentle persistence, it is likely the book would have never gone past even the concept phase.

The expression that "an army travels on its stomach" has been attributed to no lessor individuals than Napoleon Bonaparte, Frederick the Great of Prussia, and Claudius Galen, chief physician to the Roman army. And yet, between you and me, it is rare that the cooks are the ones that get the medals and the promotions.

There is a similar concept in science that states "science advances on the backs of the individual scientists." Science advances only when individual scientists accept the challenge, responsibility, and burden of the scientific endeavor and shoulder their obligation to "the greater good" by sharing the outcome of their efforts, regardless of motivation and benefit. The pharmaceutical industry's ability to have a positive impact on the human conditions has progressed, in some small but discernible way, due to advances in the theory and practice of chemical characterization and these advances have been hard won by those scientists, academic, industry, and regulatory, who have accepted the challenge and embraced their responsibility. Although it is a small reward indeed, these scientists have my utmost respect and admiration and my sincere and heart-felt thanks. While I am tempted to mention individuals by name, I fear that I might inadvertently leave someone out, and in so doing damage the commitment and dedication that I am trying to support. You know who you are and you and I both know what it is that you have sacrificed and accomplished.

Likewise, I offer a special thanks to those who take up the challenge of applying their good judgment and technical expertise to convert theory into practice, to build consensus from chaos, and to find the common good among individual self-interest, for the purpose of establishing standards, creating regulatory guidance, and developing best demonstrated practices for chemical characterization. The task is difficult, the expectations are unattainable, the required effort is substantial, and the personal rewards are minimal. Despite these challenges, you have persevered and elevated the standard of regulatory practice to the point that it enables, as opposed to retards, the development, commercialization, and clinical application of safe and effective pharmaceutical and medical products.

I have been blessed to have worked with friends and colleagues who have been a constant source of support, encouragement, insight, knowledge, and inspiration. When you find the good that is in this book, I am sure you will see reflections of yourself and your contributions. It has been my good fortune to be associated with organizations such as Baxter Healthcare and Nelson Labs, which have provided me with the support, both tangible and intangible, that has made it possible to take the journey that I call my E&L experience.

The job is not done and yet my leg of the race is just about over. With high hopes and great expectations, I hand the baton over to the "next generation" of practitioners who will continue to "run the hard race" and "fight the good fight" until it can truly be said that chemical assessment is a strong enabler of the development, qualification, and commercialization of pharmaceutical and medical products that save and sustain lives. I wish you Godspeed and good luck in this endeavor.

Dennis R. Jenke
Chief Executive Scientist
Triad Scientific Solutions, LLC

1

Introduction and Essential Concepts

1.1 General Discussion

Materials such as plastics, elastomers, glasses, and metals are widely used in medical and pharmaceutical items such as containers, packaging systems, sets, transfer and transport systems, manufacturing systems/equipment, and medical devices. The physical and chemical nature of these materials provides medical and pharmaceutical products with their necessary and desirable performance characteristics. A number of medical products involve items, constructed in whole or in part from materials, whose primary purpose is the generation, production, protection, transport, storage, delivery, and/or administration of therapeutic products that are used either directly or indirectly by patients to produce a desirable therapeutic outcome. Additionally, such items may be used for the same purposes with precursors of the therapeutic product, such as manufacturing process streams. Lastly, such items themselves may provide the therapeutic benefit (e.g. medical devices).

While an important performance characteristic of materials used in medical and pharmaceutical applications is chemical inertness, it is an unfortunate reality that a completely inert material has yet to be developed. Whenever a material is put into contact with another entity that is chemically different than the material, an interaction must occur between the material and the contacting entity. Thus, interactions between a material and the pharmaceutical product it contacts are well documented. These interactions involve the transfer of substances between the material and the pharmaceutical product and include sorption (binding), the transfer of product components (such as the active ingredient) into the material, or leaching, the transfer of material components (such as plastic additives) into the pharmaceutical product (Figure 1.1).

Throughout this text, I will refer to the two items that come into contact as a material and a pharmaceutical product. I use these terms to avoid the awkwardness of terms like "item 1 and item 2" or "construct and therapeutic product." I point out that their intent is to be completely generic and to avoid exclusion. Thus, for example, the terms "material" and "pharmaceutical product" are meant to capture those contact situations which can occur between a drug product and its packaging, between a medical device and the body fluid or part it contacts, and between a manufacturing component and a manufacturing process solution.

If the transfer of substances between the material and the pharmaceutical product were merely of academic interest then there would be little practical need to study and quantify these transfer processes and establish their effects. However, both the material and the pharmaceutical product were formulated so that each and every ingredient in either the material or the product performs an essential function. If ingredients are lost from an item by transfer, it is logical to assume that the loss would have an undesirable effect on the item's ability to perform its intended function. Thus, for example, ingredients in the pharmaceutical drug product provide either the therapeutic benefit (the active substance), stabilize the product (pH buffers, solubilizing agents, metal chelating agents) or produce a cosmetic effect (flavors, dyes, etc.). Loss of these ingredients could seriously affect the pharmaceutical product's potency, efficacy, compositional compliance, and stability. In a similar way, ingredients are present in materials to perform a specific function within the material. For example, material additives provide the material with desirable physical properties (e.g. plasticizers and fillers), promote the material's stability (e.g. antioxidants, acid scavengers), give the materials certain desirable properties (e.g. colorants, flame retardants, antimicrobial agents), and improve the material's processing (e.g. anti-static agents, slip agents). Loss of these ingredients could alter the material's stability, processability, integrity, and/or functionality.

Furthermore, addition of unintentional substances (foreign impurities) to either a pharmaceutical product or a material could adversely affect a key quality attribute of the item and/or the item's suitability for intended use. Addition of an unintentional substance (foreign impurity) to a pharmaceutical product amounts to adulteration of that product and the effect of that adulteration must be ascertained. For

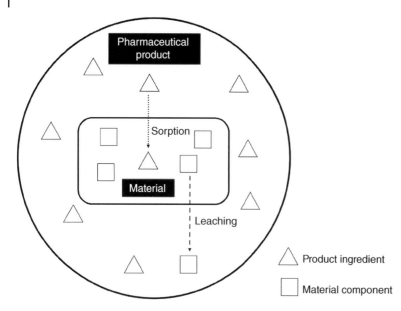

Figure 1.1 Interactions between a material and a pharmaceutical product. Such interactions include leaching, the transfer of material-related components to the pharmaceutical product and sorption, the transfer of pharmaceutical product ingredients to the material. Both processes impact the material's and the pharmaceutical product's key quality attributes and ability to perform their intended function (suitability for intended use).

example, addition of unintentional reactive substances to pharmaceutical products (foreign impurities) could alter the products' stability (reduced shelf-life), efficacy (active ingredient degradation or inactivation), and physical acceptability (e.g. precipitation, discoloration, flavoring). Moreover, the foreign impurity may be inherently toxic, and thus possesses the potential to adversely affect patient health and safety (direct safety impact). Furthermore, interactions between a reactive foreign impurity and a product's active ingredient could produce substances that themselves have an adverse effect on patient health and safety (for example, drug substance/leached substances reaction products acting as adjuvants, thereby triggering an immune response).

Although sorption of pharmaceutical product ingredients by a material could conceivably have an effect on the material's ability to perform its intended function, examples of this phenomenon have been rarely documented and it is likely that they are quite uncommon. However, adulteration of a material with pharmaceutical product ingredients certainly impacts the suitability of the material for reuse. Once a material is adulterated with a pharmaceutical product ingredient, it becomes a potential source of that ingredient should the material be used again. Thus, for example, an adulterated material that is reused with a second pharmaceutical product can be the means by which pharmaceutical products are cross-contaminated.

As was noted previously, if the interactions of sorption and leaching were sufficiently small that their effects were minor and inconsequential, then the need to study, quantify, and impact assess these processes would be purely academic. However, instances where sorption and leaching have resulted in serious and impactful effects

are well-documented. Considering leaching specifically, an often-cited case involved the leaching of a vulcanizing agent from uncoated stoppers used in pre-filled syringes, which has been proposed as a mechanism contributing to adverse clinical events associated with EPREX® [1]. Precipitation of phosphate salts in phosphate-buffered drug solutions due to the leaching of divalent cations (e.g. Ca^{+2}, Zn^{+2}) from rubber stoppers is well-documented [2]. Other examples of leachables exerting an undesirable influence on therapeutic biologic protein products include [3]:

- Leaching of a divalent metal cation from a vial's rubber stopper, which activated a metalloprotease process-related impurity in the drug product causing N-terminal degradation of the drug substance,
- Leaching of an organic solvent from the partially dried epoxy glue used to cement a needle to a syringe barrel, which caused an increase in protein oxidation followed by aggregation via disulfide switching,
- Leaching of residual aluminum oxide from a glass vial, which precipitated as aluminum phosphate in the phosphate-buffered drug product.

Leaching of tungsten from the barrels of syringes has been linked to aggregation of proteins in biopharmaceutical drug products by numerous authors [4, 5], while acrylic acid, leached from the acrylic adhesive used to bond a syringe needle to its glass barrel, has been demonstrated to interact with therapeutic proteins, modifying their structure and surface charge distribution [6]. Another documented case of leachables adversely affecting the stability of protein-based biopharmaceutical products involved the ability of leachables associated with rubber stoppers to facilitate the formation of soluble and insoluble high molecular weight aggregates of model proteins including

IgG and EPO [7]. As a last example, extractables from sterilizing filters used in biopharmaceutical manufacturing were found to be destabilizing, causing increases in (protein) aggregation, oxidation, and acidic species [8].

Cases of adverse effects of leachables are not limited to drug products and their associated container closure systems. For example, an antioxidant-related leachable associated with single-use bioprocess containers has been demonstrated to reduce cell growth rate, and thus drug substance yields, during cell culture [9]. A leachable from polycarbonate shaker flasks has been shown to inhibit the growth of CHO cells, interfering with biological testing results [10]. A similar ability of polypropylene-related leachables from microplates to adversely affect the results of biological test results has been reported [11].

These recent examples augment a long history of instances where the safety or efficacy of a pharmaceutical product has been compromised by its interaction with an item associated with its manufacturing, storage/distribution, and administration.

Although the previous citations have established the adverse effect of leachables, it is interesting to note that positive effects have also been ascribed to leachables. The ability of di-(2-ethylhexyl) phthalate (DEHP) leached from plasticized polyvinyl chloride (PVC) storage bags to stabilize stored blood and increase its storage shelf-life is well-documented [12–14]. Diphenylguanidine, leached from rubber closures, has been established to improve the stability of an oxytocin drug product formulated in an acetate-buffered solution by binding with the peptide and suppressing intermolecular degradation reactions involving oxytocin's disulfide bond [15].

Thus, contact between a material and pharmaceutical product (or its related precursors) is an opportunity for that material or product to be altered as a result of that contact. The purpose of a **compatibility assessment** is to assess the magnitude and impact, if any, of such an interaction and its associated alteration. Thus, the following definition is offered:

Compatibility assessment. This is the process by which the interaction between two contacting items (for example, a material and a pharmaceutical product) is investigated, quantified, and assessed for impact.

Logically, then, a **chemical compatibility assessment** is a compatibility assessment that focuses on the chemical interaction that occurs between two items in contact.

By convention, if the degree of alteration is small and has no meaningful adverse effect, then it is concluded that the material and the pharmaceutical product are compatible. To establish if there is the potential for a meaningful

adverse effect, a complete compatibility assessment considers numerous potential outcomes of the material/product interaction, as illustrated in Figure 1.2. In the most general sense, specific aspects of a compatibility assessment address either the issues of a pharmaceutical product's efficacy (does the product perform in a manner consistent with its labeling and indication and thus provide the expected therapeutic outcome), stability (does the pharmaceutical product provide the expected therapeutic outcome and/or remain within compositional specifications over its labeled shelf-life), compositional compliance (does its composition comply with compositional specifications including labeling and relevant compendial monographs), or safety (does the product produce an unanticipated and adverse patient response). Considering efficacy, while sorption (loss of ingredient from the drug product due to the ingredient's uptake by the material) is the most typically documented efficacy-impacting interaction, leaching interactions and effects are possible and significant. For example, a leachable and the drug substance in the pharmaceutical product could interact chemically, resulting in the ingredient's decomposition and/or inactivation, thereby reducing its potency and efficacy. Alternatively, the leachable may affect some chemical property of the pharmaceutical product to the extent that the drug substance becomes destabilized. For example, an acidic leachable could lower the pH of an unbuffered aqueous pharmaceutical product to an extent sufficient that the rate of acid hydrolysis of the drug product is increased. It is noted that efficacy does not solely reflect a substance's ability to deliver its specified therapeutic dose. If the active ingredient is deactivated, then surely a drug product's efficacy has been compromised.

To a certain extent, the aspects of efficacy, stability, and compositional compliance are related. As potency of the pharmaceutical product is a stability-indicating attribute, any interaction that affects efficacy also affects stability. However, efficacy (potency) is not the only stability-indicating attribute of a drug product, and thus leaching can affect product stability by other means. For example, a leachable may impact a chemical property of the pharmaceutical product (such as pH) whose change increases the rate of drug substance degradation, thereby reducing shelf-life. Alternatively, a leachable may react with a stabilizing agent in the pharmaceutical product, limiting the agent's stabilizing ability and shortening shelf-life. Lastly, as compositional compliance is a stability-indicating product attribute (that is, a product must comply with compositional specifications, such as those appearing in relevant compendial monographs over its entire shelf-life), any leachable that affects compositional and/or compendial compliance is also a leachable effect that impacts stability (shelf-life). For example, the accumulation of colored leachables may affect product stability by causing compendial appearance and color specification limits to be exceeded earlier than would be the case in the absence of leaching.

The aspect of compendial compliance has been largely addressed in the previous discussion. Since the level of the drug substance and limits on impurity levels are typically specified in a pharmaceutical product's compendial monograph, accelerated degradation affects the aspect of compendial compliance in the same way that it affects efficacy and stability. As compendial monographs typically contain specifications for general product attributes, such as pH, UV absorbance, and level of particulate matter, a leachable effect on these attributes can become an issue of compendial compliance.

The final aspect listed in Figure 1.2 is patient safety. When a patient either receives a pharmaceutical product or uses a medical device as part of a clinical therapy, the patient is exposed to leachables present in the pharmaceutical product or leachables leached from the medical device during its use. If the leachable is toxic, then patient health (safety) may be directly and adversely affected, with the magnitude of the adverse effect being proportional to the magnitude of exposure (amount of the leachable to which the patient is exposed), the frequency of exposure, the route of exposure, and the inherent toxicity risk posed by a specific leachable.

Additionally, a leachable may produce an adverse health effect indirectly, meaning that it is not the leachable itself that is the safety hazard but rather either a pharmaceutical product ingredient that has been altered via a chemical interaction with the leachable so that it is now toxic or a toxic degradation product that is formed via the chemical interaction of the leachable and a pharmaceutical product ingredient.

Lastly, it is also noted that the accumulation of leachables in a pharmaceutical product can have ramifications outside of the context of the pharmaceutical product's performance and key quality attributes. For example, leachables can complicate the chemical analysis of the product by producing analytical responses (for example, chromatographic peaks) that either directly interfere with targeted analyte quantitation or indirectly complicate the interpretation of the analytical information.

The previous discussion establishes those drug product aspects that are included in the compatibility assessment but do not address the process of performing the compatibility assessment. Clearly, compatibility assessment consists of two distinct parts: information collection and information interpretation (see Figure 1.3). The collected information may include existing information that is gathered (for example, a certificate of compendial compliance) or new information that is generated (for example, data from an extraction study). Processes and techniques for information interpretation vary depending on the effect being considered; however, generally speaking, the interpretation is either direct or indirect. For example, a drug product stability study performed in the product's

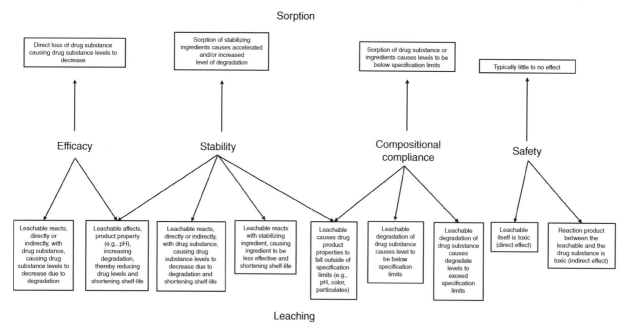

Figure 1.2 Dimensions of compatibility assessment. The complete compatibility assessment includes the aspects of efficacy, stability, compendial compliance, and safety, all of which can be influenced by either sorption or leaching.

Figure 1.3 The chemical compatibility assessment process. Collected information, obtained by either gathering or generation (or both), is interpreted to establish its effect on key product quality attributes, considering both interacting entities (drug product and material).

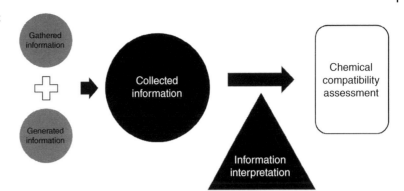

marketed packaging system will directly establish the effect of leachables on product stability. However, a leachable's probable effect on patient safety is inferred, indirectly, by the process of toxicological safety risk assessment.

In any circumstance, the processes of information collection and information interpretation are discussed in greater detail elsewhere in the book (for example, Chapters 2 and 3 for information generation and Chapter 10 for information interpretation).

1.2 Regulations, Guidelines, Standards, and Recommendations for Chemical Compatibility Assessments

The generation of safe and effective pharmaceutical products and medical devices is an obligation for companies that participate in the pharmaceutical and medical marketplaces. Nevertheless, it is one thing for a company to acknowledge that "we must generate a safe and effective product" and another thing altogether for that company to understand the process by which a product is demonstrated to be safe and effective. Considering safety specifically, it is one thing to say that "our product meets our own internally developed standards for safety" and another thing altogether to have the product declared to be safe based on an independent, scientifically rigorous and unbiased regulatory analysis of the safety assessment process and outcome.

Chemical compatibility assessment is a complex and challenging endeavor. The single reason this book and many other similar resources exist is to address and simplify this complexity and reduce the challenge. The aspect that makes chemical compatibility assessment so challenging is not a matter of companies recognizing and accepting their obligation to perform an assessment. Rather, the challenges associated with chemical safety assessment lie in the difficulty in establishing what represents an efficient and effective (and therefore acceptable) assessment to all the relevant stakeholders. That is to say that the largest

issue standing in the way of performing chemical compatibility assessment is not accepting the responsibility for performing the assessment but rather the lack of a standardized process for designing, implementing, reporting, and interpreting an assessment. The issue is not "Do I accept the responsibility for performing an assessment?" but rather "I do not know how to perform an efficient, effective and acceptable assessment."

This difficulty and uncertainty arise due to certain unavoidable truths associated with pharmaceutical and medical products:

a. The great diversity of pharmaceutical and medical products,
b. The great diversity in the conditions under which pharmaceutical and medical products are used to produce a positive clinical outcome,
c. The scientific complexity of chemical interactions,
d. The difficulty in attributing effect to a cause (that is, the difficulty in establishing the effect that a chemical will produce once it has been established that the chemical is relevant to the interaction being investigated).

Let us be clear. The ideal circumstance in any assessment is that:

1. The process for performing the assessment has been established and clearly communicated.
2. The process for performing the assessment has been standardized to the extent that a minimum number of process variations cover the maximum number of circumstances likely to require assessment.
3. The process for performing the assessment is relatively easily achieved with a reasonable degree of knowledge and skill.
4. The process for performing the assessment can be performed using universally available tools that are relatively inexpensive to procure, maintain, and operate.
5. The process for performing the assessment is logical and can be explained to individuals generally schooled in the art.

6. The assessment itself provides a tangible and recognizable outcome that is acceptable to and accepted by all stakeholders.

7. The assessment process and the outcome of the assessment is reproducible.

It is fine to understand the characteristics of the "ideal" chemical compatibility assessment. However, practically speaking, it is infinitely more important (and useful) that the ideal chemical safety assessment exists as a document (or documents) that contains and describes the process of chemical compatibility assessment. Such a document (or series of documents) would:

– Define those actions that need to be taken to perform the assessment,
– Establish, in a clear and unmistakable manner, how to accomplish those actions,
– Provide a means to verify that the actions were properly taken,
– Delineate how the results of the actions taken can be interpreted in the context of establishing effect and impact.

This would be a very short book indeed if all that was required at this point was to list existing chemical compatibility assessment documents that achieve the ideal case and say "here they are, go read them and do what they say." The circumstance that this book is as long as it is and that there is so much discussion and controversy around chemical compatibility assessment verifies the reality that ideal chemical compatibility assessment documents do not currently exist.

However, that is not to say that stakeholders are left to fend for themselves without aid and guidance. Resources that address chemical safety assessment to a certain extent exist and can provide an interested party with valuable insights, recommended practices and actionable processes. These resources include (see also Table 1.1):

- Regulatory Guidelines and Guidances. As a fundamental responsibility of a government is to protect its people, it is necessarily the case that the requirements for compatibility assessments are contained within the context of laws, legislation, and associated regulations and that such laws, legislation, and regulations are upheld and enforced by appropriate governmental authorities such as the US Food and Drug Administration (FDA), the European Medicines Agency (EMEA), Health Canada, China's National Medical Products Administration (NMPA), and other national Ministries of Health. To facilitate enforcement and compliance, such governmental authorities publish Guidelines or provide Guidance that essentially establish the authorities' expectations with respect to compatibility and, to a certain extent, delineate the means for fulfilling those expectations. Because pharmaceutical and medical products must be registered with the regulatory authorities before they can be marketed (meaning in essence that the products must be "approved" by the authority before they can be sold within the authority's jurisdiction), regulatory guidances and guidelines carry significant weight and are typically viewed as providing the ultimate authority on "what to do and how to do it."

- National and International Standards. Given the complexities involved with establishing guidelines and providing guidance, organizations whose sole purpose is the generation and maintenance of guidance and guidelines have evolved, such as, for example, the International Council for Harmonization of Technical Requirements for Pharmaceuticals for Human Use (ICH). These so-called "standards-setting organizations" have assembled appropriate teams of experts and stakeholders who are charged with the responsibility of generating standards. Additionally, these "standards-setting" organizations have developed processes by which the generated standards are reviewed and ultimately "approved" by the larger global community of experts and stakeholders. In this way, the standards represent the consensus opinion among experts and stakeholders. Equally important, these organizations have developed content, quality, and format requirements for their standards and have established processes for the periodic maintenance (review and revision) of the standards.

Generally speaking, these "standards-setting" organizations have no direct regulatory authority and thus no ability to enforce compliance to their standards. These standards become enforceable either via formal recognition by regulatory authorities (that is, the regulatory authorities will provide a statement that communicates their stance on the use and utility of the standard) or by the informal recognition by the community of stakeholders that "this is the right way to do things and thus this is the way we are going to do it." In this way, the standards are credible because the organizations generating the standards are credible.

- Compendial Monographs. To aid in the development, marketing, and control of safe and effective pharmaceuticals and medical devices, many nationalities have established a Pharmacopeia (a compendium for drug information) whose purpose is "to provide standards of identity, strength, quality, and purity that can help safeguard the global supply of medicines, dietary supplements, and food ingredients" (United States

Table 1.1 Sources of guidance and advice on chemical compatibility assessment.

Type	Example	Use	Applicability
Regulatory Guidance or Guideline	Guidance for Industry Container Closure Systems for Packaging Human Drugs and Biologics, US Department of Health and Human Services, Food and Drug Administration, Rockville, MD, May 1999	Direct legal and enforceable requirements for marketing approval	General concepts with wide applicability; focus is more on what to do and less on how to do it. May contain some specifics on methodology and acceptance criterion. Typically contains little explanation or justification of requirements.
National or International Standard	ISO 10993-1:2018. Biological evaluation of medical devices – Part 1: Evaluation and testing within a risk management process. Association for the Advancement of Medical Instrumentation (AAMI), Arlington, VA, 2018	Indirectly legal and enforceable requirements pending adoption by regulatory authorities. Recognized and adopted as consensus best practice	General concepts with wide applicability; focus is more on what to do and less on how to do it. May contain some specifics on methodology and acceptance criterion. Typically contains minimal explanation or justification of requirements.
Compendial Monographs	USP <661> Plastic Packaging Systems and Their Materials of Construction. USP42/NF37, pp. 6812–6818. Official May 1, 2019	Legal and enforceable requirements at the discretion of regulatory authorities; mandatory compliance for products so labeled	General concepts with wide applicability; provides a balance of what to do and how to do it. Typically contains both generalities and specifics on methodology and acceptance criteria. Typically contains little explanation or justification of methods and requirements.[a]
Best Demonstrated Practice Recommendations	Safety thresholds and best practices for extractables and leachables in orally inhaled and nasal drug products. PQRI Leachables and Extractables Working Group. September 8, 2006	Consensus advice provided by a panel of experts. May be incorporated into regulations, standards and monographs.	Contains both general concepts and specific recommended practices. Equal focus on both what to do, how to do it and why it should be done that way. Typically contains extensive justification and explanation.
Publications	D.R. Jenke and B.E. Rabinow. Proper Accounting for Surface Area to Solution Volume Ratios in Exaggerated Extractions. *PDA J. Pharm. Sci. Technol.* 71(3):225–233 (2017).	Advice provided by individual experts, which may be incorporated into regulations, standards, and monographs by reference.	Focused on a specific circumstance or aspect. Typically defines a set of circumstances and reflects the actions the author took under those circumstances.

a) Recent enforceable USP monographs are now supported by informational monographs that contain more explanation or justification of the contents of the associated enforceable monograph and which serve as "user manuals" for the associated enforceable monograph.

Pharmacopeia). The Pharmacopeia "establishes written (documentary) and physical (reference) standards for medicines, food ingredients, dietary supplement products, and ingredients." The generation, publication, and the maintenance of the Pharmacopeia is entrusted to an organization charged with these tasks, for example, the United States Pharmacopeial Convention (USP). To a certain extent, these organizations and their processes are similar to those of the standards-setting organizations discussed previously.

As was the case with Standards and "standards-setting" organizations, pharmacopeial organizations generally have no clear regulatory authority and thus little or no ability to enforce compliance to their Compendial standards. There is an "understanding," sometimes reinforced by law and legislation to a certain extent,

that the Compendial standards are enforced by the local regulatory authorities. Considering the USP as an example, the USP works closely with government agencies, ministries, and regulatory authorities around the world to help safeguard the global supply of medicines, dietary supplements, and food ingredients and USP standards are used by regulatory agencies and manufacturers to help to ensure that these products are of the appropriate identity, strength, quality, purity, and consistency.

- Best Practices Recommendations. It is a natural circumstance in any and all areas of expertise that experts routinely get together and "talk shop." Professional and trade organizations have been organized to facilitate and formalize "shop talk" and to harness the talk for productive purposes. To be honest, a fair portion of that "shop talk" revolves around deficiencies, real and perceived, in regulations, standards, and Compendial monographs. Although the topics and focus of these discussions vary, it is often the case that the expert's concern is that the regulations, standards, and monographs are insufficiently detailed to provide actionable guidance in the circumstance that the experts find themselves in. Certain organizations have embraced and accepted the responsibility for addressing and remediating such deficiencies so as to more generally improve the chemical compatibility assessment and related processes. They accomplish this, in the same way that the "standards-setting" and Compendial organizations do, by assembling a panel of experts, charging the experts to write a position paper, vetting the written paper with a wider audience or experts, stakeholders, and interested parties, and then publishing the position paper. Because these organizations have no regulatory authority and cannot claim even a tangential relationship with regulators, the position papers are generally known as "best demonstrated practice recommendations" that serve the purpose of providing "consensus expert advice." These recommendations derive their "authority" on the basis of their credibility and practicality and, in isolated cases, by informal recognition by regulatory authorities. Examples of such best demonstrated practice recommendations include:

– Safety thresholds and best practices for extractables and leachables in orally inhaled and nasal drug products, submitted to the Product Quality Research Institute (PQRI) Drug Product Technical Committee, PQRI Steering Committee, and US Food and Drug Administration by the PQRI Leachables and Extractables Working Group, September 8, 2006.
– Recommendations for Extractables and Leachables Testing, published by the Extractables and Leachables Subcommittee of the Bio-Process Systems Alliance (BPSA) in 2008 to address single-use manufacturing systems.
– Standardized Extractables Testing Protocol for Single-Use Systems in Biomanufacturing (Ding, W., Madsen, G., Mahajan, E., O'Connor, S., and Wong, K., *Pharm. Eng.* **34(6)**:1–6 (2014)), developed in conjunction with the Extractables Work Group of the BioPhorum Operations Group (BPOG).
– Assessing safety of extractables from materials and leachables in pharmaceuticals and biologics – Current challenges and approaches (Broschard, T., Glowiemke, S., Bruen, U.S., Nagao, L.N., Teasdale, A., Studlts, C.L.M, Li, K.L., Iceik, L.A., Erexson, G., Martin, E.A., and Ball, D.J. *Reg. Tox and Pharm.* **81**:201–211 (2016)) developed in conjunction with

the Extractables and Leachables Safety Information Exchange (ELSIE).

• Individual Publications. The generally employed means for individual authors (or small groups of authors) to share their experiences and insights, communicate (and defend) their opinions, advocate for particular strategies or tactics, and communicate scientific advances is via publication of scholarly documents. Whereas the previously discussed guidances, standards, monographs, and recommendations tend to be broad scope and strategic, individual publications tend to focus on a specific circumstance or concept from a tactical perspective. Thus, where guidances, standards, monographs, and recommendations tend to focus on the "what needs to be done," the focus in individual publications is more on the "how to do it." For example, an individual manuscript may describe a specific test procedure for performing a certain aspect of chemical compatibility assessment (such as chromatographically screening an extract for organic extractables). Other manuscripts might discuss and enumerate the scientific principles relevant to a specific aspect of chemical compatibility assessment (such as how to select and justify the appropriate extraction vehicle). Case study-type manuscripts describe a particular circumstance, chronicle the author's actions in those circumstances and present the results of those actions. Reviews collect and summarize the available relevant information and may use that collected information to advocate for a specific concept or action (for example, summarizing available information on extractable metals from plastics and using the information to support the proposal that extractable metals are low risk and therefore extensive testing is not required).

Regardless of the nature of the publication, it is clear that the publication itself is advice rather than guidance. Thus, the publication itself is not directly enforceable and implementation of any strategies and tactics presented in the publications may or may not address and fulfill regulatory requirements. Of course, it is a realistic expectation that guidances, standards, monographs, and recommendations embrace meritorious publications, either by reference or direct assimilation.

1.3 Why Chemical Assessment?

In the previous discussion we established that compatibility assessment is necessary as it can reasonably be inferred that a material and a pharmaceutical product are incompatible to a certain extent. That is, when the material and the pharmaceutical product come into contact, they will interact and it is possible, if not likely, that the interaction will have an adverse impact on a critical quality attribute of either the material or the pharmaceutical product (or both). Thus, an assessment is necessary to quantify the impact and to establish the acceptability of the impact.

However, we have yet to establish why the proper compatibility assessment is a chemical assessment. In fact, it is logical to suggest that a chemical compatibility assessment is not the assessment method of first choice. This is true for the simple reason that chemical compatibility assessment is an indirect assessment method where compatibility is generally inferred, not measured. Specifically, the chemical compatibility assessment involves the collection of chemical data relevant to the interaction followed by inferring the impact based on the chemical data (data generation followed by data interpretation). Thus, the "accuracy" of the chemical compatibility assessment is based on the quantity and quality of the chemical data and the certainty in the inference. An inaccurate assessment is produced when either insufficient or substandard data is the basis of the assessment and/or if the inference is irrelevant (e.g. an "apples to oranges" comparison) or based on questionable science. An accurate assessment is the outcome only if the chemical data is sufficient and "high quality" and the inference is based on relevant good science.

For these reasons, a direct determination of compatibility is the assessment method of choice. However, the practical issue is that many key quality attributes cannot be assessed directly or, if they can be assessed directly, the direct assessment is sub-optimal. For example, consider the key quality attribute of pharmaceutical product stability. Conceptually, it is relatively straightforward to directly establish the effect of a pharmaceutical product's packaging system on pharmaceutical product stability. Two sets of pharmaceutical product units are stored side by side over the course of a stability study, one set being the pharmaceutical product in its packaging system (the test system) and the second set being the pharmaceutical product stored in an inert container (the control system). Stability-indicating attributes of the pharmaceutical product are measured in both sets of units over the shelf-life and the results are compared, test versus control. Any differences in the stability-indicating attributes, test versus control, are attributed to pharmaceutical product/packaging system incompatibilities.

Although this experimental approach is straightforward enough in concept, there are practical issues to consider. For pharmaceutical products with long shelf-lives and/or numerous critical quality attributes, such a study would represent a considerable investment of resources. Furthermore, it is likely that such a study would be performed only after a considerable development process and effort for both the pharmaceutical product and its packaging.

It would be most unfortunate indeed to discover a meaningful incompatibility at this late stage of the product development cycle.

The problem associated with direct compatibility assessment is even clearer if the quality attribute being considered is patient safety. Let us use the same example of a pharmaceutical product and its packaging system. Certainly, one way to establish the safety of packaging-related leachables would be to perform the stability study noted above and dose patients with an aged product at certain times over its shelf-life using and then observing the related effect on patient health. Even as one reads this one understands the practical and ethical issues associated with such a direct approach of what amounts to a clinical trial.

Thus, one understands that the indirect approach of chemical compatibility assessment is often employed because more direct approaches have irreconcilable difficulties.

Furthermore, if a direct chemical approach to compatibility assessment is not possible, then what about alternate direct approaches to compatibility assessment? The simple truth is that for many critical quality attributes, there is no alternate, direct, and non-chemical approach or, if there is an alternate direct, non-chemical, approach, it too has its own issues. Reconsider the case of product stability. Product instability is generally chemistry-based; that is a pharmaceutical product's ingredients chemically degrade over its shelf-life, causing the product to either (i) become sub-potent, (ii) have excessive levels of degradation products, or (iii) exhibit an undesirable cosmetic property (e.g. discoloration). Because the mode of action of the incompatibility is chemical, there are no alternative assessment methods versus chemical assessment.

A slightly different situation exists for the critical quality attribute of patient safety. In this case, a viable alternative to a chemical assessment is a biological compatibility (biocompatibility) assessment, which is widely used in the safety impact assessment of medical devices. Unfortunately, a biocompatibility assessment suffers from its own shortcomings, both technical and practical. There is considerable ongoing discussion in the toxicology community with respect to the applicability, sensitivity, and morality of current methods for assessing biocompatibility and it is the consensus of this technical community that a biocompatibility assessment is not a substitute for all aspects of a chemical safety assessment, or conversely that a chemical safety assessment is not a suitable substitute for all biological endpoints. In fact, it is generally accepted that it is the combination of a biocompatibility assessment and a chemical safety assessment that provides the most rigorous assessment of the patient safety impact of pharmaceutical and medical products.

Thus, the simple answer to "why a chemical compatibility assessment?" is that there are few viable alternatives.

1.4 An Overview of the Chemical Compatibility Assessment Process

A chemical compatibility assessment consists of two unrelated activities, information collection and information interpretation (impact assessment). Each of these activities is discussed in greater detail elsewhere in this book. However, the following generalizations are offered as an introduction to these two activities.

At this point, the discussion of information collection will address what information should be collected as opposed to the mechanics of information collection. In considering what information should be collected, it is clear that there are two extreme conditions to assess in terms of compatibility; those conditions that produce the worst-case (greatest) incompatibility and those conditions that produce the best-case (or actual) incompatibility. If the mechanism of incompatibility involves the transfer of chemical substances from one item to another (for example, from a material to a pharmaceutical product), then it is clear that the worst-case incompatibility occurs when the transfer is the greatest, which takes place when the entire contents of one item (the material) is transferred to the second item (the pharmaceutical product). This circumstance is more figuratively referred to as "it all comes out."

In this circumstance, the answer to the question "what comes out?" is "everything that was in the item in the first place." One understands that performing a worst-case chemical compatibility assessment requires knowledge of a material's composition, as the material composition provides the answer to "what was in the item in the first place?" Thus, establishing the worst-case chemical incompatibility requires compositional profiling.

In addition to supporting a worst-case impact assessment, the material composition may provide an assessment outcome on another basis, compositional equivalence. The concept of chemical equivalence is as follows: if a material of interest is compared to a reference material used in an approved application and it is established that the material of interest and the reference material have the same composition, that the materials are processed in the same manner, and that the application of the material of interest is the same as the approved application of the reference material, then the compatibility of the reference material can be "transferred" to the material of interest with no additional consideration.

If a material's composition establishes the worst possible interaction, then the actual conditions of contact between the items (for example, a material and a pharmaceutical product) establishes the best case (or lowest) incompatibility, which is reflected in the actual transfer of substances between the items under their actual conditions of contact. This circumstance is more figuratively referred to as "what actually comes out" and involves, from the perspective of the pharmaceutical product, establishing the pharmaceutical product's leachables profile.

When the generation of a leachables profile is technically and practically feasible, then it is intuitive that generating a leachables profile provides the most meaningful and relevant information upon which to base an impact assessment. Thus, when screening leachables studies can be performed with the required level of rigor and a reasonable amount of effort, the sponsor is well-advised to perform the leachables testing and to skip everything else. However, there are numerous circumstances that impede, if not prevent, proper leachables profiling. For example, consider a medical implant. Assessing medical implant leachables would require that the medium of contact between the implant and the device would have to be sampled and tested over the entire time of implantation in order to establish both the magnitude and kinetics of leaching. The fact that the medium of contact is part of the body itself could impede sampling. If sampling is impeded, then clearly leachables profiling is confounded.

There may also be issues in "simpler" contact situations, for example a pharmaceutical product and its packaging. In such a situation, one could envisage several means of collecting a viable sample of the drug product for testing and thus sampling is not the issue. However, it may be that the pharmaceutical product is chemically complex to the extent that it is not analytically expedient or that the clinical use of the pharmaceutical product is such that the levels to which leachables would need to be detected is beyond the analytical method's capabilities. In either circumstance, the pharmaceutical product cannot be screened for leachables at their required levels and the compatibility assessment is thwarted.

It is possible, therefore, to end up in the situation where a worst-case chemical compatibility assessment based on composition suggests the possibility of an adverse impact and where the actual-case chemical compatibility assessment based on a leachables profile cannot be performed. In the absence of an alternate chemical assessment strategy, there is no escaping the assessment conclusion that the items under investigation are sufficiently incompatible that they are not viable as pharmaceutical or medical products. Since such a conclusion is not a desirable outcome, a third assessment option must be found. This third assessment option is to estimate the actual case by collecting information that establishes "what possible interaction could occur between the material and the pharmaceutical product." In essence, the concept behind this third and intermediate option is "if I can't directly measure leachables in the pharmaceutical product under clinical circumstances, can I estimate leachables under laboratory conditions?"

It is intuitively obvious, I suppose, that if the purpose of this intermediate process is to approximate the leaching process that occurs under clinical conditions, the process should involve or use leaching to accomplish its objective. However, to reserve the term leachables for an actual clinical circumstance, when a leaching process is performed in the laboratory it is referred to as an extraction. Thus, the substances found in an extract generated by an extraction performed in the laboratory are called extractables, and the process of producing, discovering, identifying, and quantifying extractables is termed extractables profiling.

It is obvious that the terms extractions and extractables are somewhat general terms as clearly there are many ways to perform an extraction in the laboratory. At one extreme, one can extract the material of interest with aggressive solvents under stressful conditions with the intent of "getting as much out of the material as is possible, thus providing an exaggerated worst-case approximation of the clinical circumstances." At the other extreme, one can extract the material of interest with solvents that have a "leaching power" similar to that of the pharmaceutical product and under conditions that approximate clinical use with the intent of producing an extractables profile that mirrors the leachables profile. Clearly, there are many extraction possibilities that fall somewhere between these two extremes.

The juxtaposition between ingredients (composition), extractables, and leachables is addressed in greater detail in Chapter 2. At this point suffice it to say that extractables profiling provides an intermediate chemical compatibility assessment option to compositional profiling and leachables profiling. Pictorially, the chemical compatibility assessment strategy that is based on these three possible assessment tactics is illustrated in Figure 1.4 and has been referred to in the literature as the Chemical Assessment Triad [16].

In considering the chemical compatibility tactics illustrated in the Triad, there is the implication that a complete chemical compatibility assessment is based on performing all three types of studies; that is, a complete chemical compatibility assessment requires compositional profiling, extractables profiling, and leachables profiling and thus the assessment starts with compositional profiling and ends after leachables profiling. Although there may be other reasons where one would generate all three profiles, strictly speaking they are not all necessary to establish compatibility. In fact, any one of these tactics is capable

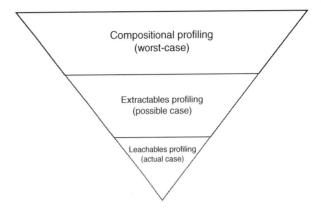

Figure 1.4 A revised chemical assessment triad. The chemical assessment process consists of some combination of three distinct actions; compositional profiling, extractables profiling, and leachables profiling.

of establishing compatibility and once one of the tactics has established compatibility then the other tactics are redundant and unnecessary. Thus, for example, if one can establish compatibility using compositional profiling ("it all comes out"), then what would be the relevance and necessity, from a compatibility assessment perspective, of extractables profiling? Similarly, if one could generate a leachable profile and establish compatibility on this basis, then what would be the relevance and necessity, from a compatibility assessment perspective, of generating composition or extractables profiles?

The important point to recognize from the previous discussion is that once compatibility has been definitively established by one means, reaffirming compatibility by another means becomes a luxury and not a necessity. Thus, an efficient chemical compatibility assessment strategy involves, as a first step, an evaluation that establishes which profiling tactic is most likely to be practically feasible and is most likely to produce the desired outcome that compatibility has been confirmed. An efficient chemical compatibility assessment then proceeds with this most likely tactic.

For example, consider the circumstance where one possesses sufficient knowledge that a material is intrinsically compatible because it is "relatively pure" (whatever that means). In this circumstance, compositional profiling is likely to be the most direct and definitive means of establishing compatibility. Alternatively, consider the circumstance where one possesses sufficient material knowledge that it is likely that only leachables profiling will establish compatibility and there are no impediments to performing leachables profiling. One might, for example, have the results of a generic extraction study performed by the material's vendor and may have concluded from those results that "there are too many extractables at too high levels." In this case, proceeding directly to leachables profiling is the most efficient compatibility assessment approach.

To reiterate, chemical compatibility assessment does not inherently require three profiling activities (compositional, extractable, and leachables) to be performed; rather, an efficient compatibility assessment is one that identifies and implements the profiling activity that is likely to definitively demonstrate compatibility. One identifies the profiling activity that is most likely to definitely establish compatibility based on available knowledge.

Now even though chemical compatibility assessment does not necessarily require all three profiling activities that does not mean there will not be circumstances where all three profiles are performed as part of the compatibility assessment. For example, let us say that there is insufficient available information to establish which is the tactic most likely to produce a credible outcome. In that case, one might choose to "start at the beginning" with compositional profiling to collect the necessary information either to complete the assessment (if one is lucky) or to decide whether extractable or leachables profiling is the appropriate next step. Alternatively, one understands that oftentimes it is the case that the identification of the optimal profiling activity is based on incomplete or imperfect information and thus it is possible that performing a particular profiling activity will produce an assessment that the items are incompatible. For example, let us say

that the analysis of the available data suggested that a material could be established to be compatible based on "it all coming out" (compositional profiling). However, to confirm this suggestion, a compositional assessment might be performed. Furthermore, it is possible that once the compositional profile is generated and assessed, the outcome of the assessment would be "potentially incompatible." Does that mean we throw up our hands, proclaim, "we're screwed," and go and update our resume? Well, I suppose that could be a viable path forward but a more rational response would be "OK, I guess I had better do an extraction study because the material might well be compatible if everything in it doesn't all come out."

This discussion leads to the generalized chemical compatibility assessment process illustrated in Figure 1.5. As per the previous discussion, the assessment process begins with the collection and review of relevant and available information, which is used to identify the profiling activity that is most likely to establish compatibility. Assuming that such an activity can be identified, then the assessment proceeds by performing that type of profiling. Once the profiling is complete, the profile is assessed for impact. If the impact assessment concludes that the material is compatible, the process of compatibility assessment is complete and the process flow is exited. If the impact assessment concludes that the material is potentially incompatible, this potential incompatibility is addressed by either rejecting

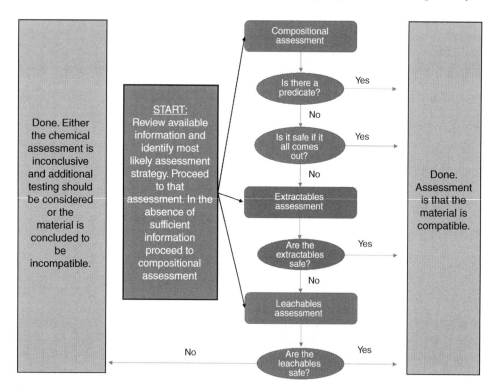

Figure 1.5 Process flow diagram, chemical compatibility assessment process for a material. (*See insert for colour representation of the figure.*)

the material as unusable or continuing the compatibility assessment by performing the next appropriate profiling activity.

If the available information is insufficient to allow one to identify the most likely profiling activity, one "starts at the top" with compositional profiling and proceeds through the process until an assessment outcome of "compatible" is obtained. One uses the information gained in a predicate profiling step to make a further profiling step more efficient. For example, if an assessment of an extractables profile indicates that an incompatibility is produced by a single extractable, then perhaps the leachables study required to demonstrate compatibility focuses solely on (or targets) the one potentially problematic extractable.

Should one complete the entire profiling process (compositional, extractables, and leachables) and fail to establish compatibility, then this is strong evidence indeed that the material is unsuited for this particular application. Although there are no remaining chemical means of establishing compatibility, the user may seek to employ other means to establish compatibility (for example, biological compatibility testing), fully recognizing that the burden of proof is likely to be considerable.

One of the promises I made to myself when I agreed to write this book is to try to be "fair" in terms of pointing out the challenges to strategies and tactics that I might propose. There are two "assumptions" in my previous discussion on the chemical compatibility assessment that are critical and relevant. The first is that the outcome of the compatibility assessment is definitive. The second is that the profiling activity, regardless of type, has been effectively carried out. Both of these "assumptions" are relevant in the concept that "you only have to establish compatibility once and then you are done." For example, what happens if the conclusion of compatibility is equivocal. In a chemical safety assessment, this might be that case where a margin of safety obtained by the assessment is 1.0, the minimal acceptable value. If one obtained this result in a chemical safety assessment, one can reasonably expect that reviewers of the assessment would question the qualitative (identification) and quantitative (quantitation) data upon which the assessment is based. Moreover, consider the case of an assessment based on a leachables profile. It is reasonable for a reviewer of that assessment to ask "how can I be sure that you have accounted for all the leachables?"

It is often the case that confirmatory information is obtained by performing a complementary profiling activity. Thus, for example, the proper response to the question of "how can I be sure that you accounted for all the leachables?" is to generate an extractables profile and perform an extractables/leachables correlation.

My point here is that there are many valid reasons, within and beyond chemical compatibility assessment, for generating profiles that are not rigorously necessary to establish compatibility. Although use in a compatibility assessment is a prime factor in terms of deciding what profiling activities to perform, it is not the only factor to consider.

1.5 Looking Ahead

Looking at the previous text one could realistically say that we have already jumped into the thick of chemical safety assessment without having established its foundations, such as providing critical definitions and discussing key concepts. However, by considering chemical compatibility assessment at a high level, we have established the context in which those definitions and concepts have to be understood and used. Furthermore, we have barely scratched the surface of the concept of "what must be done" and have not even begun to consider "how this should be done" and "why is that the right way to do them."

It is the current reality that while the need for chemical compatibility assessment is clear, the strategic and tactical means of establishing chemical compatibility are not. Given the great diversity in pharmaceutical and medical products and materials and the almost infinite set of clinical use circumstances, it should be obvious to both the novice and the expert, to the regulator and the regulated, and to the product's vendor and the product's user, that there is no "one size fits all" approach to chemical compatibility assessment. While one potential outcome of this reality is chaos, this need not be the case for a chemical compatibility assessment. This is true because the pharmaceutical/medical industry has amassed a wealth of experience in terms of the strategy and tactics of properly, efficiently, and effectively performed chemical compatibility assessments. Collaborators from industry, academia, and regulatory agencies have leveraged this experience to effectively reconcile the seemingly conflicting objectives of expeditiously bringing safe products to the market. The remaining chapters draw upon this experience and leverage the emerging regulations, standards, and recommendations to delineate and defend appropriate, effective, and compliant strategies and tactics for performing chemical compatibility assessments and establish the means for interpreting the results obtained from such assessments.

As noted previously, the chemical compatibility assessment process is an indirect one, involving the generation of relevant information and the interpretation of the information to infer effect and impact. Thus, a chemical compatibility assessment is often an exercise in risk management and it is important to understand that there is no chemical compatibility assessment process that is absolute in the sense that it reduces risk to zero. Rather, the degree of risk is minimized via the aggressive, systematic, extensive, and judicious utilization of "good" theoretical and practical science. Because the degree of acceptable risk may vary from situation to situation, a chemical compatibility assessment strategy or tactic that is wholly acceptable and accepted for one situation may be either hopelessly inadequate or grossly excessive in another situation. Thus, this book does not deal in absolutes and cannot (and does not) prescribe a strategy (or even tactics in many cases) that are universally applicable or "guaranteed to generate a safe, effective, and approvable product each and every

time or your money back." This book does not, cannot, and will not provide a chemical compatibility assessment process and solution for every circumstance encountered by every one of its readers. Rather, this book establishes the relevant scientific principles and practical considerations, enumerates and discusses options, makes and supports recommendations, considers pros and cons, and illustrates and clarifies these points through the use of relevant case studies, all intended to provide readers with the insight and knowledge they need to design, implement, report, and defend their own efficient, effective, justifiable, and compliant risk-based chemical compatibility assessments.

References

1 Sharma, B., Bader, F., Templeman, T. et al. (2004). Technical investigations into the cause of the increased incidence of antibody-mediated pure red cell aplasia associated with EPREX®. *Eur. J. Hosp. Pharm.* 5: 86–91.

2 Castner, J., Williams, N., and Bresnick, M. (2004). Leachables found in parenteral drug products. *Am. Pharm. Rev.* 7 (2): 70, 72–75.

3 Markovic, I. (2006). Challenges associated with extractable and/or leachable substances in therapeutic biologic protein products. *Am. Pharm. Rev.* 9 (5): 20, 22, 24–27.

4 Liu, W., Swift, R., Torraca, G. et al. (2010). Root cause analysis of tungsten-induced protein aggregation in pre-filled syringes. *PDA J. Pharm. Sci. Technol.* 64 (1): 11–19.

5 Seidl, A., Hainzl, O., Richter, M. et al. (2012). Tungsten-induced denaturation and aggregation of epoetin alfa during primary packaging as a cause of immunogenicity. *Pharm. Res.* 29: 1454–1467.

6 Liu, D., Nashed-Samuel, Y., Bondarenko, P.V. et al. (2012). Interactions between therapeutic proteins and acrylic acid leachable. *PDA J. Pharm. Sci. Technol.* 66 (1): 12–19.

7 Richter, C., Lipperheide, C., Lipke, U., and Lamprecht, A. (2018). Impact of extractables from rubber closures on protein stability under heat stress. *Eur. J. Pharm. Biopharm.* 130: 22–29.

8 Huang, M., Horwitz, T.S., Zweiben, C., and Singh, S.K. (2012). Impact of extractables/leachables from filters on stability of protein formulations. *J. Pharm. Sci.* 100 (1): 4617–4630.

9 Hammond, M., Nunn, H., Rogers, G. et al. (2013). Identification of a leachable compound detrimental to cell growth in single-use bioprocess containers. *PDA J. Pharm. Sci. Technol.* 67 (2): 123–134.

10 Peng, J., Zhao, Y., Hong, Y. et al. (2018). Chemical identity and mechanism of action and formation of a cell growth inhibitory compound from polycarbonate flasks. *Anal. Chem.* 90: 4603–4601.

11 Hill, E.J., Martin, S.J., and Weikart, C.M. (2018). Characterization of extractable species from polypropylene microplates. *SLAS Technol.* 23 (6): 560–565.

12 Estep, T.N., Pedersen, R.A., Miller, T.J., and Stupar, K.R. (1988). Characterization of erythrocyte quality during refrigerated storage of whole blood containing di-(2-ethylhexyl) phthalate. *Blood* 64: 1270–1276.

13 AuBuchon, J.P., Estep, T.N., and Davey, R.J. (1988). The effect of the plasticizer di-2-ehtylhexyl phthalate on the survival of stored RBCs. *Blood* 71: 448–452.

14 Hoorowitz, B., Stryker, M.H., Waldman, A.A. et al. (1985). Stabilization of red blood cells by the plasticizer, di-ethylhexyl phthalate. *Vox Sang.* 48: 150–155.

15 Zidan, A.S., Aqueel, S.M., Alayoubi, A. et al. (2017). Leachable diphenylquanidine from rubber closures used in pre-filled syringes: a case study to understand solid and solution interactions with oxytocin. *Int. J. Pharm.* 532: 491–501.

16 Jenke, D. (2012). A general strategy for the chemical aspects of the safety assessment of extractables and leachables in pharmaceutical drug products: The chemical assessment triad. *PDA J. Pharm. Sci. Technol.* 66 (2): 168–183.

2

General Principles of Chemical Compatibility Assessment: Extraction and Leaching

2.1 Key Definitions and Concepts, Extractables, and Leachables

If I had a dollar for every definition that has been proposed for the terms extractables and leachables, I could be sitting on a beach somewhere and not in my basement office staring at the computer screen. Furthermore, if I had an additional dollar when a practitioner, novice, or expert used the terms interchangeably and incorrectly, then I could probably buy the beach I could be sitting on. The considerable variety in the definitions are illustrated in Tables 2.1 and 2.2, which contain various definitions from well-recognized sources.

Perhaps the most curious aspect of the terms extractables and leachables is that these two terms describe what is essentially one process. The purpose of having two different terms for what are arguably similar concepts and processes is to distinguish and differentiate between the reality of the clinical environment and the potentiality of the laboratory. Thus, the term leachables was established to reflect the clinical situation and to provide a user of a pharmaceutical product or a medical device with the answer to the question "what substances am I being exposed to, when I use this product or device, that do not provide a therapeutic benefit or do not serve a necessary purpose?" The answer to this question is, at least partially, that "you are being exposed to leachables." For pharmaceutical products, it is understood that these leachables are foreign impurities present in the product because they have leached from either the product's packaging or manufacturing equipment. For medical devices, these are the substances that are leached from the medical device by the medium of contact between the device and its user under the conditions of use. If the device contacts the user indirectly (for example, a device used to administer a drug product), a user is exposed to the leachables that are present in the indirect contact medium because they leached from the medical device. If the device contacts the user directly (for example, an implanted device), a user is exposed to leachables that

have been leached from the device by whatever body item (organ, tissue, fluid) is contacted by the device.

It is the aspects of user exposure and clinical conditions of use that are essential to the concept of leachables and are captured in most definitions proposed for leachables. Where the definitions vary is when the terms are applied to different items (that is, the definition for a packaging system differs from the definition for a medical device which differs from the definition for a manufacturing component) and specifically when the terms are defined in the context of "what article does one test and how does one perform the test?"

Leveraging the common aspects of these leachables definitions and considering the patient exposure aspect, the following definition is proposed for the term leachable:

Leachable. A substance or chemical entity, leached from a packaging system, a manufacturing component, or a medical device by a pharmaceutical product, process stream, or a body fluid/tissue, that is present in the pharmaceutical product, process stream, or body fluid/tissue because these objects contacted the system, the component, or the device during their manufacturing, distribution, storage, or clinical use.

Note 1: If the leachable is derived from the pharmaceutical product's packaging system or a manufacturing component, the leachable is measured by testing a packaged drug product that has been manufactured by its commercial process and subjected to its typical storage and/or distribution conditions over its labeled shelf-life.

Note 2: If the leachable is derived from a medical device, the leachable is measured in the medium of contact between the patient and the device where such contact occurs during the device's labeled conditions of clinical use.

One of the complicating factors associated with the proper use of the term leachables is the circumstance where it is difficult to measure leachables. Consider, for example, an implanted medical device. The fact that the

Extractables and Leachables: Characterization of Drug Products, Packaging, Manufacturing and Delivery Systems, and Medical Devices, First Edition. Dennis Jenke.
© 2022 John Wiley & Sons, Inc. Published 2022 by John Wiley & Sons, Inc.

Table 2.1 Definitions of extractables and leachables, packaging applications.

Source	Definition of extractable	Definition of leachable
FDA 1999 Container Closure Guidance [1]	The term extractable is not defined and the term leachable is not used.	
EMEA Guideline for Plastic Immediate Packaging [2]	The terms extractables and leachables, although used in the text, are not formally defined. However, extractables are established to be the result of extraction studies and leachables are established to be the result of a migration study.	
PQRI Recommendations for OINDP [3]	Compounds that can be extracted from OINDP device components or surfaces of the OINDP container/closure system in the presence of an appropriate solvents (s) and/or condition(s)	Compounds which are present in the drug product due to leaching from container/closure system components
USP <1663> [4] and <1664> [5]	Organic and inorganic chemical entities that can be released from a pharmaceutical packaging/delivery system, packaging component, or packaging material of construction and into an extraction solvent under laboratory conditions	Foreign organic and inorganic chemical entities that are present in a packaged drug product because they have leached into the packaged drug product from a packaging/delivery system, packaging component, or packaging material of construction under normal conditions of storage and use or during accelerated drug product stability studies
Compatibility of Pharmaceutical Products and Contact Materials [6]	Substances that can be extracted from a material or construct using extraction solvents and/or extraction conditions that are expected to be at least as aggressive as the conditions of contact between a material and a primary therapeutic substance	Substances that are present in a primary therapeutic substance because of its interaction with a material or construct during its intended use (including production, storage, transport, and/or administration)
Leachables and Extractables Handbook [7]	Chemical entities that are derived from container/closure and/or device components under laboratory conditions	Chemical entities derived from a container/closure and/or a device component when they are part of the final drug product and under patient-use conditions

Table 2.2 Definitions of extractables and leachables, manufacturing and device applications.

Source	Definition of extractable	Definition of leachable
USP <1665> [8]	An organic or inorganic chemical entity that is released from a manufacturing component or system into an extraction solvent under laboratory conditions	None[a]
BPSA Recommendations for Testing and Evaluation of Extractables from Single-Use Process Equipment [9][b]	Chemical entities that migrate from process equipment under appropriate exaggerated conditions (e.g. solvent, temperature) that exceed "worst case" process conditions	Chemical entities that migrate into finished drug product dosage under actual product and process conditions
BPOG Extractables Protocol & Leachables Best Practices Guide [10]	Chemical entity that is extracted from a component of a process system into a solvent under controlled conditions that are usually more aggressive than normal operating conditions	Chemical entity that comes from single-use systems during normal use
ISO 10993-18 [11]	Substances that are released from a medical device or material of construction when the medical device or material is extracted using laboratory extraction conditions and vehicles	Substance that is released from a medical device or material during its clinical use

a) Although neither USP <665> [12] nor <1665> [8] define leachables, the term process equipment-related leachable is introduced and defined as "A foreign organic or inorganic chemical entity that is present in a process stream or manufactured API, DS, or DP because it has leached from a material or component used in the manufacturing system and has persisted through the entire manufacturing process."

b) In addition to extractables and leachables, this reference defines the term "migrant" as "chemical entities that migrate from process equipment into actual process fluids under 'worst case' conditions."

device is implanted into a living subject and that sampling the contacted part of the subject's body for testing might be difficult makes the concept of leachables awkward to apply and thus some authorities have "stretched" the definition to include experiments that simulate the clinical use. Although using the term leachables to describe simulated extractions may make the authorities feel more "comfortable," the fact remains that the definition of leachables should not be perverted when it is difficult or challenging to perform an actual leaching study. The inability to actually perform a leachables study is not an appropriate reason for altering or mis-using the definition of leachables.

Another complicating factor occurs in the case of pharmaceutical manufacturing. By definition, it is clear that a leachable derived from a manufacturing system is measured in the manufactured pharmaceutical product. However, this raises the question "what do we do in the case of process streams?" where the process stream is that solution that is actually being processed by or used in the manufacturing system. The point here is that the actual leaching during manufacturing likely does not occur between the pharmaceutical product and a manufacturing component but likely occurs between a component and a process stream, which is that solution which is further processed to produce the pharmaceutical product. A good case can be made to call substances leached into a process stream leachables, because these substances fulfill important parts of the leachable's definition (testing the actual contact medium under the actual conditions of contact). Moreover, substances in a process stream can have undesirable effects, for example, the previous cited cases where substances in the process stream inhibit cell growth, and thus reduce yields, in cell cultures. However, rigorously speaking the process stream is not the final product and substances in a process stream may be removed from the process stream prior to the generation of the final pharmaceutical product. If removal occurs, the substances in the process stream do not accumulate in the drug product and thus do not become true leachables.

If these substances in process streams are not called leachables then it follows that they must be called extractables by default. However, the definition of extractables does not quite fit for the same reasons that the definition of leachables does not fit. To reconcile this juxtaposition, the term process equipment-related leachable (PERL) was adopted. When a substance is referred to as a PERL, it is understood that the substance is present in a process stream (and not a finished drug product) due to contact between the process stream and a manufacturing component under actual manufacturing conditions.

Considering these difficulties, the following definitions are provided for leachables in packaged drug products and from medical devices.

Leachables in packaged drug products. Substances or chemical entities that are present in a packaged drug product and were derived, or leached, from the drug product's packaging system and/or manufacturing system.

Note 1: A patient is exposed to these leachables via the drug product when the patient is treated with the drug product as part of a clinical therapy.

Note 2: These leachables are foreign to the drug product in the sense that they are not associated with or derived from the drug product's ingredients or vehicle.

Note 3: A packaged drug product may contain foreign substances that are environmental contaminants, substances that are present in the packaged drug product's storage environment and have breached and migrated through the permeable packaging to become entrained in the packaged drug product. Although environmental contaminants can have important drug product quality implications and may be discovered during leachables screening of the packaged drug product, they are not, by definition, leachables.

Note 4: Leachables in a drug product may react with drug product ingredients, impurities, or degradation products to form additional impurities or degradation products. Although such impurities and/or degradation products can have important drug product quality implications and may be discovered during leachables screening of the packaged drug product, they are not, by definition, leachables.

Note 5: Manufacturing systems can be assessed for their ability to contribute leachables to packaged drug products in three ways:

1. The manufacturing system's output, either a drug product or drug substance, can be screened for manufacturing-related leachables.
2. The manufacturing system or its components can be subjected to a laboratory-based extraction study (e.g. USP <665>), thus establishing the system's or component's extractables profile.
3. The actual manufacturing process stream can be screened for extracted (or leached) chemicals. This process is such that the extracted (or leached) chemicals are neither truly extractables nor leachables. They are not true extractables because they are obtained by testing the actual process stream produced under actual process conditions. However, they are not true leachables because they are not discovered by testing the drug product. To address this juxtaposition, such chemicals are referred to as process equipment-related leachables (PERLs).

Leachables from medical devices. Substances or chemical entities that are leached from a medical device during clinical use of the medical device.

Note 1: If the clinical use of the medical device is such that it indirectly or externally "communicates" with the patient via a medium that first contacts the device and then is administered to a patient, then the medical device is leached by that medium, leachables are contained in that medium, and the leachables are established by testing that medium.

Note 2: If the clinical use of a medical device is such that there is direct contact between the medical device and the patient's body, then the medical device is leached by the body items (tissue, organs, and/or fluids) that the device contacts, the leachables are present in the body items, and leachables are established by testing the body items.

Note 3: Because of the challenge in screening body items for unspecified leachables, it may be uncommon for direct contact medical device leachables to be addressed in this manner. Rather, such medical devices may be screened for extractables using extraction conditions that provide an extractables profile that reflects the worst-case leachables profile.

The discussion of PERLs is a good transition from leachables to extractables. As was indicated for a PERL, the simplest definition of an extractable is that an extractable is "everything that is not a leachable." Although this may be a workable definition, it is not a proper definition as it does not define extractables independently from leachables. However, it is clear that extractables must be defined in such a way that extractables are very clearly differentiated from leachables. To this end, the following definition is provided:

Extractable. A substance or chemical entity, extracted from a test article by an extraction medium under specified laboratory test conditions, including temperature, duration, extraction process, and dimensions of contact

(e.g. ratio of test article weight or surface area to extraction medium volume).

Considering this definition of extractables, it is not uncommon for extractables to be defined in terms that refer to a particular extraction solvent or vehicle. However, thermal extraction of test articles is often an important part of an extractables study and the substances released due to thermal extraction are called extractables, the same term applied to substances released during solvent extraction. Clearly, a definition of an extractable that refers only to solvent extraction is not correct.

In comparing and contrasting the definitions for extractables and leachables (see Table 2.3), the similarities and differences between the terms become clear. Considering similarities, both extractables and leachables are substances that are transferred from a source to a receiver because these items have come into contact. Considering differences, extractables and leachables differ in at least one of the following ways:

- The conditions of contact,
- The identity of their source,
- The identity of the receiver.

It is clear, both from their definitions and their practical applications, that extractables and leachables derived from the same source should be related. In fact, best practice recommendations and compendial standards for extractables and leachables note that a leachables–extractables correlation (either qualitative or quantitative) should be established by linking, for example, actual drug product leachables with extractables from corresponding controlled extraction studies performed on either the packaging system itself, individual packaging components, or individual materials of construction [3–5]. These same recommendations and standards note that "leachables–extractables correlations are important for several reasons, including justifying the use of routine extractables release tests of packaging components as an alternative to leachables

Table 2.3 Definitions for extractables and leachables.

Term	Definition
Extractable	A substance or chemical entity, extracted from a test article by an extraction vehicle, that is present in that vehicle because the vehicle contacted the test article under specified laboratory test conditions, including temperature, duration, process (e.g. refluxing, sonication, etc.), and dimensions of contact (e.g. ratio of test article weight to extraction vehicle volume).
Leachable	A substance or chemical entity, leached from a packaging system, a manufacturing component, or a medical device by a pharmaceutical product, process stream, or a body fluid/tissue, that is present in the pharmaceutical product or body fluid/tissue because these items contacted the system, the component or the device under the item's manufacturing, distribution, storage, or clinical use conditions.

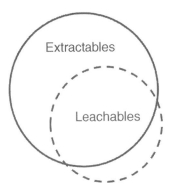

Figure 2.1 Venn diagram reflecting the generalization that "leachables are a subset of extractables but not all leachables are necessarily extractables." The degree of overlap between extractables and leachables depends on the purpose and design of the extraction study.

testing during stability studies for high-risk drug products, establishing the source of a leachable producing an Out-of-Specification (OOS) result for a low-risk drug product, change control, and ongoing quality control, etc."

Acceptance, adoption, and implementation of the recommendations for generating extractables–leachables correlations have led to inevitable but unfortunate generalizations about how the two are related. Phrases such as "leachables are a subset of extractables" or "extractables are potential (or probable, or worst-case) leachables," or pictures such as the infamous extractables–leachables Venn Diagram (Figure 2.1) are staples in the E&L knowledge space. In fact, it is generally accepted that one cannot consider oneself to be an "E&L expert" unless one has their own version of the E&L Venn Diagram. However, as is the case with many generalizations, these generalizations about the relationship between extractables and leachables have created confusion and discord among practitioners of the art and science of extractables and leachables assessment rather than fostering understanding and harmonization. If one wants to start a contentious debate among E&L experts, showing Figure 2.1 and asking whether it accurately reflects reality or not is a sure way to get the conversation going.

The root of this confusion and discord is no more complex than the observation that the terms extractables and leachables are not concise. For example, since the definition of an extractable does not specify a specific set of contact conditions (it just says the conditions are to be specified), one intuitively understands that an extractables profile (a list of extractables and their concentrations) will differ, depending on the extraction conditions used, which are linked to the purpose for which the extractables profile will be used. For example, an extraction study whose purpose is to obtain a test article's composition is likely to have

different extraction conditions than a study whose purpose is to mimic leaching. These different extractions will undoubtedly produce different extractables profiles and it is clear that the different profiles will be linked to leachables in different ways.

Moreover, since the definition of a leachable does not specify a specific drug product or a specific set of clinical circumstances, one understands that chemically different drug products packaged in the same packaging system could have different leachables profiles, and that even a single drug product in a specified packaging system could have a different leachables profile depending on the clinical circumstances (for example, beginning and end of shelf life).

Thus, it is clear why generalizations such as that shown in Figure 2.1 lead to discord and confusion, since the relationship between extractables and leachables differs depending on the details of the extraction and leaching studies. For example, if the purpose of the extraction study is to establish a test article's composition, it is likely that the extraction conditions used will be quite different from (and generally more aggressive than) the clinical conditions of contact, for example, between a pharmaceutical product and its packaging system or a target body fluid or tissue and a medical device. In this case, the relationship between extractables and leachables is likely to be accurately reflected by Figure 2.1, with the degree to which the extractables and leachable profiles (circles) overlap reflecting the difference in extraction conditions versus leaching conditions.

The other general purpose for generating an extractables profile is to mimic the leachables profile. Extraction studies designed to mimic leaching generally come in two "flavors," those studies designed so that the extractables profile reflects the worst possible leachables profile (which means that all leachables are extractables, although not all extractables are leachables, and the levels of the extractables are higher than the levels of the leachables) and those studies designed so that the extractables profile reflects, as closely as possible, the leachables profile (which means that all extractables are leachables, all leachables are extractables, and the levels of extractables and leachables are comparable). These circumstances are illustrated in Figure 2.2.

There are three final points to consider before concluding the discussion of the relationship (correlation) between extractables and leachables. The first point focuses on the term correlation and its interpretation. As an example, consider the case where Irganox 1076 has been established to be an extractable. The most extreme interpretation of correlation is that Irganox 1076 as an extractable is only correlated to a leachable if that leachable is Irganox 1076. This extreme interpretation is not consistent with the concept of correlating extractables and leachables. Rather,

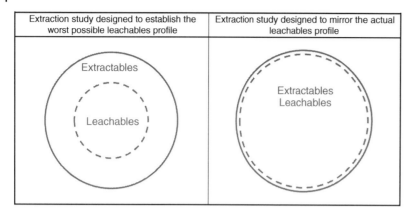

Figure 2.2 Venn diagrams reflecting the two possible situations of an extraction study designed to produce an extractables profile that mimics the leachables profile: the extraction study designed to establish the worst possible leachables profile and the extraction study designed to mirror the actual leachables profile.

a correlation between Irganox 1076 as an extractable and a leachable (that is not Irganox 1076) can be made if a chemical relationship between Irganox 1076 and the leachable can be established. For example, the compound 2,4-di-tert-butyl phenol is routinely reported as a leachable in drug products stored in containers that consist of polymers that are stabilized with Irganox 1076. As shown in Figure 2.3, there is a clear structural relationship between Irganox 1076 and 2,4-di-tert-butylphenol as 2,4-di-tert-butyl phenol is a well-known degradation product of Irganox 1076. Thus, an acceptable correlation can be drawn between the extractable Irganox 1076 and the leachable 2,4-di-tert-butyl phenol.

The second point to consider is the "nature," "level," or "degree" of correlation. There are two aspects to this concept. Considering the nature of the correlation, a correlation can be either qualitative or quantitative. The correlation between Irganox 1076 and 2,4-di-tert-butyl phenol illustrated in Figure 2.3 is an example of a qualitative correlation, where the correlation is based on similar or relatable structures. In a quantitative correlation, for example, the amount of the extractable in the extract of a packaging system is compared to the amount of the leachable in the packaged drug product. Using Irganox 1076 and 2,4-di-tert-butyl phenol as examples, let us say that the level of Irganox 1076 in the extract was 70 mg/l and the level of 2,4-di-tert-butyl phenol in the drug product was 25 mg/l. Examination of the structures of these two

substances establishes that they are related on an eqi-molar basis, meaning that the oxidation of one mole of Irganox 1076 produces one mole of 2,4-di-tert-butyl phenol. The ratio of the concentrations, extractables versus leachable, is 2.8 and the ratio of molecular weights is 2.6 (530.878 for Irganox 1076, 206.329 for 2,4-di-tert-butyl phenol). As these two ratios agree to within the accuracy of the analytical methods used to quantify these two substances, a quantitative correlation between the extractable and leachable has been established.

The second aspect to consider is the "degree" or "level" of the correlation; in other words, how "good" or "strong" (or scientifically justified) is the correlation. This aspect is important largely from the perspective that the required "strength" of a correlation may depend on the purpose for which the correlation is being used. For example, during product development and for internal use only, an organization might justify the correlation of Irganox 1076 as an extractable to 2,4-di-tert-butyl phenol as a leachable because one can examine their structures and conclude that their relationship is "obvious." However, during registration of the drug product, telling the reviewer "everyone can see that they are related" may not be a sufficient justification of a correlation. In fact, it is often the case that a relationship based on "visual similarities" must be fortified with a discussion of mechanism and pathway in order for the correlation to be accepted as valid and confirmed.

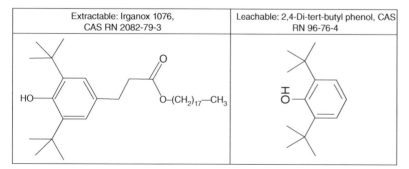

Figure 2.3 Correlation between an extractable and a leachable. A clear structural relationship can be established between these two substances, resulting in an acceptable extractable–leachables correlation. Extractable: Irganox 1076, CAS RN 2082-79-3. Leachable: 2,4-di-tert-butyl phenol, CAS RN 96-76-4.

Lastly, the third point is the circumstance of the so-called "magically appearing" or "mystery leachable." This is a leachable that is found during a leachables screening study unexpectedly and with no obviously correlated extractable, that falls in the portion of the leachable circle of Figure 2.1, but not the extractable circle, and that supports the assertion that "one must always do both extractables and leachables screening studies because you can never be sure when a 'magically appearing' leachable will show up."

There are two obvious reasons why leachables can "magically" show up. The first is a disconnect between the test article in the extraction study and the test article in the leaching study. It stands to reason that the best correlation between extractables and leachables is obtained when the leachables study is performed on a system that is manufactured from the same materials (and/or components) that were tested for extractables; in fact, this is one of the chemistry-based recommendations from the PQRI OINDP document. While simple enough in concept, accomplishing this objective can be quite challenging due to many complicating practical factors.

An obvious example of such a disconnect is that a test article, let us say a stopper, is treated (for example, sterilized by irradiation) prior to its use in a packaging system. If the stopper is tested non-irradiated in the extraction study and after irradiation as part of the packaging system in the leachables study, then it is almost a forgone conclusion that there will be "magically appearing leachables." However, you may protest, this is such an obvious example. Then let us take a not-so-obvious example, the case of tungsten leaching from a pre-filled syringe. I am sure that the first investigator who found tungsten causing protein degradation in biopharmaceutical drug products stored in pre-filled syringes was astonished and could not account for where the tungsten was coming from. Certainly, testing of the glass syringe barrel itself in an extractables study would not reveal tungsten. Thus, tungsten is an example of a "mystery" or "magically appearing" leachable.

Then somebody got smart (or lucky) and looked into the process of manufacturing a pre-filled syringe. It turns out that a tungsten filament is inserted into the hot glass barrel to form the cavity in which the needle is placed. As a result of this contact, tungsten residues are left in the cavity. It is these tungsten residues that are leached into the drug product and which cause the protein to degrade. When one tests a glass barrel with a needle in place, and if the extraction is such that the needle/glass interface is extracted, then tungsten will be found (although correlating it to the drug degradation might not be obvious at first). Mystery solved!

Therefore, in both cases (irradiation and tungsten), the extraction study and the leaching studies were not performed on the same test article. Properly, the leachables

study was performed with the final packaging system. However, the extraction study was performed on a component of the packaging system. Now doing an extraction study on components (or raw materials) is not necessarily a bad practice and there are any number of reasons why one would want to generate a material's extractable profile (change control, for example). However, if one wants to correlate extractables and leachables (and avoid "mystery leachables"), then surely it is intuitively obvious that the extractables study needs to be performed on the same packaging system as the leachables study.

Now there is a second type of "mystery leachable." The reason that the second type of "mystery" leachable shows up in the leachables profile is the nature by which leachables screening is accomplished. In general, screening is accomplished in a differential manner; that is, the analytical outcome obtained for a test sample (for example, the packaged drug product) is compared to the analytical outcome obtained for a matched blank (for example, the drug product stored in an inert container). In the case of chromatographic screening for organic leachables, for example, a leachable is "found" as either a peak in the test article chromatogram that is not in the blank chromatogram or a peak that is larger in the test article chromatogram than it is in the blank chromatogram. Now in many cases, this finding is accurate and in fact the substance "found" as a leachable and reported to be a leachable is truly a substance that has leached into the drug product from its packaging system. However, there are two situations where a difference compound is not an actual compound that leached from the packaging system. That is, the compounds are reported as leachables but they did not actually leach. In both cases, while the substances are called leachables by the testing laboratory that "found" them, they are not real leachables and thus their correlation to extractables may be "complicated."

The first situation is the case of unstable leachables; that is, a substance that leaches as one compound but converts to another compound in the drug product. For example, consider the case of the oxidation and/or hydrolysis products of Irganox-type antioxidants. When one performs an extraction on the test article with a non-polar solvent, it is frequently the case that the antioxidant itself is reported in abundant quantities but there are few "antioxidant-related substances" that are discovered and reported. However, "antioxidant-related substances," such as hydrolysis or oxidation products, and not the antioxidant, are frequently reported in an aqueous drug product packaged in the test article screened for leachables, especially after a thermal stress (e.g. autoclaving or long-term storage at elevated temperatures). So here we have a "mystery leachable" that is not strictly and directly correlated with an extractable.

Now it is true that with today's knowledge, this seems like a trivial case, as "everybody" who is familiar with the subject surely knows about antioxidant degradates. While that may be true to a certain extent, I am sure that the first time that these types of leachables were "found" the investigators scratched their heads and wondered "OK, now where in the heck did this compound come from?" and "what am I going to do with this mystery?"

This second situation is the "mystery leachable" that arises due to the interaction between a real leachable and an ingredient in the drug product. When the drug product is screened for leachables, it is the interaction product that is found. So now again we have a "mystery leachable."

First, I am going to invoke a technicality. The interaction product that was found during leachable screening is not a leachable for the simple reason that the interaction product did not leach from the packaging system. Thus, calling the interaction product a "mystery leachable" is a mis-use of the term leachable. Second, I would make the point that eventually a correlation between the "mystery" substance and an extractable will be worked out and then the substance will no longer be a mystery. This is truly splitting hairs and the fact of the matter is that the interaction product is a new and, most importantly, an unexpected

foreign impurity whose impact on the product's key quality attributes must be established. If these extractables/drug product ingredient interaction products are common, then this is certainly a reason to perform both screening extractables and leachables studies.

Turning away from the topic of "mystery" leachables, one of the best demonstrated practice recommendations proposed by the PQRI in their OINDP manuscript was that "during the controlled extraction study process, sponsors should revisit supplier information describing component formulation" [3]. This statement is true because there is another correlation that is relevant to chemical compatibility assessment, the correlation between formulation ingredients and extractables. In the same way that it is intuitive that extractables and leachables should be related, it is also intuitive that a test article's ingredients and its extractables should be related. In fact, much of the previous discussion on extractables–leachables correlations is also relevant to ingredients–extractables correlations.

The power of these correlations lies in their predictive capacity. If ingredients can be correlated to extractables and extractables can be correlated to leachables (see Figure 2.4), then ingredients can be correlated to leachables. Practically speaking, this means that if I know the

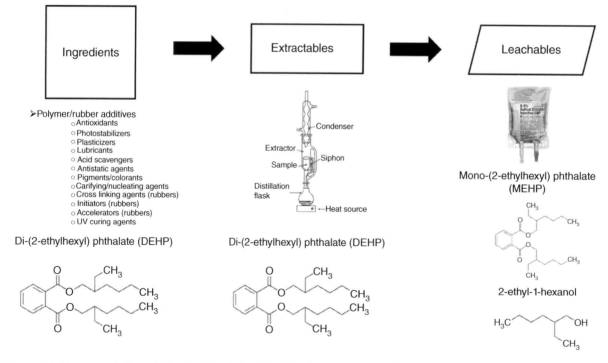

Figure 2.4 The concept of correlations. The logical relationships between a test article's ingredients, extractables, and associated leachables is that ingredients become extractables and extractables become leachables. In this illustrated example of an aqueous drug product in a flexible plasticized PVC bag, DEHP is the plasticizer (ingredient) in the bag. An extraction study, which may use an organic solvent as an extraction medium, reveals DEHP as an extractable. However, DEHP is not a leachable in the aqueous drug product because DEHP is quite insoluble in water. Rather, the leachables that are observed are water-soluble substances that are clearly structurally related to DEHP.

ingredients in a component of a packaging system, I should be able to forecast, to a certain extent, the leachables' profile of the drug substance packaged in that system. Examples of this forecasting process have been published for several materials including plasticized poly (vinyl chloride), polyethylene, and a bromobutyl rubber [13, 14].

2.2 Extraction Studies

2.2.1 Key Definitions and Concepts

Information generation in the chemical compatibility assessment is typically a two-step process; first the test article is extracted and then the extract is subsequently chemically analyzed to establish the extracted substances. Focusing on extraction, it is interesting that when one reviews the "definitive" literature on chemical compatibility assessment, one does not find a definition of the term "extraction." Oh, there are definitions for extractables, for extraction studies, for extractables profiles, and other related terms, but it is difficult to find a definition for the specific term extraction. Turning to the dictionary, one finds that definitions such as "the process of extracting something" are circular and essentially useless. In chemistry, the most common usage of the noun extraction is "the process of separating a substance from a mixture of substances." More generally applicable definitions include "the process of removing something, especially by force" and "to get, pull, or draw out, usually with extra effort, skill, or force."

In considering these definitions in total, it is clear that there are certain central themes associated with the term extraction. First, there are two items involved in an extraction, the item being extracted and the item doing the extraction. Second, there is a substance, present in the item being extracted, which will be transferred from the item being extracted to the item doing the extraction. Third, the substance is being transferred due to some action performed by the extracting item, on the item being extracted. Thus, we can begin to flesh out a workable definition for extraction that sounds something like "the process of transferring a substance from an extracted item to an extracting item due to the action of the extracting item on the extracted item." Such a definition makes a certain degree of sense when applied to the general world of pharmaceutical products and medical devices, where the extracted item typically is a packaging system, a manufacturing component or a medical device and the extracting medium is some kind of solvent or solution. However, there is a problem with the concept of "due to the action of the extracting item on the extracted item"

when this concept is applied to a chemical compatibility assessment. While it is certainly true that some extractions are accomplished by an action of the extracting item that is performed on the extracted item (for example, an extraction that swells a polymeric material being extracted), in most cases extractions that materially change the item being extracted are avoided in chemical safety assessment, as it is usually the case that in clinical practice packaging systems, manufacturing components and medical devices remain intact during clinical use. Thus, the transfer of a substance from an extracted item to an extracting item is generally a "natural" consequence of the concentration gradient that exists between the extracted item and the extracting item for the substance. That is, substances will diffuse from the extracted item to the extracting item as a natural means of equalizing the concentrations without any action performed by the extracting item.

This picture of passive diffusion versus active extraction is quite useful in debunking one of the great myths in E&L, which is "leachables do not accumulate in solid dosage forms because the solid dosage form has essentially no "extracting power" when compared to a liquid dosage form". In fact, diffusion of substances between two solids (for example, a packaging system and a lyophilized powder) is just as likely and just as common as diffusion of substances between a solid and a liquid (for example, a packaging system and a liquid dosage form) and the presence of leachables in solid dosage forms is confirmed by the published literature [15–18].

One last factor to consider in defining extraction is that we require that it be differentiated from leaching, even though the physiochemical phenomena occurring in both processes are likely to be similar. Thus, an extraction in a chemical compatibility assessment is an action that occurs in the laboratory or under laboratory conditions while leaching is a process that occurs under clinical conditions of contact.

Taking all these factors into account, the following definition is proposed for the term extraction as it is applied in chemical compatibility assessment:

Extraction. The transfer of a substance (or substances) from a donor item (the item being extracted) to a receptor item (the extracting medium) when contact between the two items is initiated and maintained under laboratory conditions.

Building from this definition provides definitions for other related terms:

Extraction Study. A laboratory study that examines the transfer of a substance (or substances) from a donor item to a receptor item when contact is initiated and maintained between the two items under laboratory conditions.

Extractable. A substance that is transferred from a donor item to a receptor item when contact is initiated and maintained between the two items under laboratory conditions.

Extractables profile. All extractables, identified and quantified.

2.2.2 Types of Extractions

As has been mentioned previously, the definition of extraction virtually ensures that there will be some confusion when designing, implementing, and interpreting the results of an extraction study, as the outcome of the extraction study (the extractables profile) will differ depending on the laboratory conditions under which the study is performed. The more "aggressive" the extraction conditions, the more likely the extractables profile will consist of more extractables at higher concentrations. The more "gentle" the extraction conditions, the more likely the extractables profile will consist of fewer extractables at lower concentrations.

In any event, an ideal extraction, regardless of how it is accomplished, possesses the following characteristics:

- The extraction should release extractables from the test article (including intentional and unintentional additives, residuals, and impurities) in quantities sufficient for detection, identification, and quantitation.
- The extraction should release extractables from the test article in quantities that bear some relationship to a leachables profile.
- The extraction should be simple, straightforward, and safe to perform using routinely available laboratory instrumentation.
- The extraction should preserve the chemical and structural integrity of all extractables.
- The extraction should produce an analytically expedient extract.
- The extraction should be performed in a manner that is justifiable based on sound scientific principles.
- The extraction should be reproducible.

Many different types of extraction studies have been used as the basis of a chemical compatibility assessment, from decidedly "aggressive" to appropriately "gentle." The circumstance that extraction studies can be performed in any number of different ways has created quite a problem in chemical compatibility assessment, where the problem is linking the conditions of the extraction study to the clinical circumstances being assessed. Now there are many reasons outside of chemical compatibility assessment for performing an extraction study. However, within the confines of chemical compatibility assessment, the only purpose for performing an extraction study is to facilitate the assessment. Since a proper and definitive chemical compatibility assessment is based on leachable profiling, the only purpose for performing an extraction study is to produce an extractables profile that is somehow relatable to the leachables profile. The desired relationship is that the extractables equal or exceed the leachables profile because an extractable profile that is less than the leachables profile will lead to a faulty outcome in the chemical compatibility assessment. Nevertheless, the nature of this relationship can vary. For example, one investigator might want to establish the worst possible leachables profile, which is the situation that "it all leaches out." There is a certain logic to this, for if it can be established that two items are compatible based on the worst possible leachables profile then the items will be compatible under any and all conditions of contact. Performing a compatibility assessment using such information is desirable in certain situations, such as in a regulatory review of product submissions, as the worst possible case aspect of the study provides the assessment with a margin for error. However, performing a compatibility assessment based on the worst possible case can be undesirable in the circumstance that the assessment concludes that the items are incompatible as it could potentially lead to the rejection of an item whose only problem was that it was tested too aggressively.

Thus, another investigator might say "I want to overestimate the interaction but not to the extent of the worst possible case so I will exaggerate things slightly." Yet another investigator will say "my product's shelf-life is three years and I do not want to wait three years for extractables results so I will accelerate things somewhat." A final investigator might say "I want to do an extraction study that produces an extractables profile that is actually the same as the leachables profile." All these investigators will design an extraction study and the designs will be different. But they will all be called extraction studies and they will all produce an extractables profile. Thus, the generic terms "extraction study," extractables, and extractables profile are not sufficiently descriptive of the conditions and intent of the study to facilitate a discussion of the study. Therefore, a descriptive nomenclature has been developed that establishes a spectrum of different types of extraction studies and extractables (see Table 2.4).

In considering and contrasting these various types of extraction studies, it is noted that the minimally acceptable or baseline extraction study uses the clinical conditions of use as extraction conditions. Extraction conditions that are more "gentle" or less "harsh" than the clinical conditions of use are of no value in chemical compatibility assessment as they likely underestimate patient exposure to leachables and thus result in a flawed assessment. Of course, an

Table 2.4 Types of extractions.

Term	Definition	Purpose
Clinical conditions of use	Conditions under which the clinical contact occurs	Produce an extractables profile that matches the leachables profile
Simulated	An extraction using a method that simulates clinical use[a]	Produce an extractables profile that closely matches the leachables profile[b]
Accelerated	An extraction whose duration is shorter than the duration of clinical use but whose conditions do not result in a chemical change to the substances being extracted[a]	Produce an extractables profile that closely matches the leachables profile[b] but in less time than the actual clinical duration of the contact
Exaggerated	Extraction that is intended to result in a greater number and/or amount of chemical constituents being released as compared to the amount generated under the clinical conditions of use[a]	Produce an extractables profile that is "worse" than the leachables profile[c],[d]
Aggressive	Extraction that is intended to result in a greater number and likely greater amount of chemical constituents being released as compared to the amount generated under the clinical conditions of use	Produce an extractables profile that is "worse" than the leachables profile[c],[d]
Exhaustive	An extraction that removes the entire amount of all ingredients and constituents from the test article Multi-step extraction conducted until the amount of material extracted in a subsequent extraction step is less than 10% by gravimetric analysis (or achieved by other means) of that determined in the initial extraction step	Produce an extractables profile that reflects the ingredients of and constituents in the test article, establishing the worst possible leaching case that "it all comes out"
Dissolution	Process of completely solubilizing a test article, generally preserving the molecular structures of its constituents[a]	Produce an "Extractables" profile for the organic ingredients and constituents in the test article, establishing the worst possible leaching case that "it all comes out"
Digestion	Process of completely solubilizing a test article by breaking it down into its fundamental structural units, such as its elemental constituents or monomeric units[a]	Produce an "Extractables" profile for the elemental or inorganic ingredients and constituents in the test article, establishing the worst possible leaching case that "it all comes out"

a) From ISO 10993:18(2019) [19].
b) All the leachables are extractables and the levels of extractables are comparable to the level of leachables.
c) There are likely more extractables than leachables and the levels of extractables are greater than the levels of leachables.
d) The difference between an exaggerated and aggressive extraction is a matter of degree. An exaggerated extraction uses extraction conditions closer to clinical use while an aggressive extraction uses extractions closer to an exhaustive extraction. Thus, an aggressive extraction is more "exaggerated" than an exaggerated extraction.

extraction study performed under clinical use conditions is a leachables study if the test article is a packaging system, manufacturing component, or medical device and the extracting solution is the clinical patient contact medium. An extraction study performed under clinical use conditions is not a leachables study if the test article is a material or part of a packaging system, manufacturing component, or medical device and/or if the extracting solution is not the clinical patient contact medium.

There is the perception that an extraction performed using the clinical conditions of contact is a "mild" extraction because, after all, these are the baseline extraction conditions. However, clinical conditions of use are not necessarily "mild" conditions. For example, a packaged

parenteral i.v. drug solution could be sterilized by autoclaving (heating at 121 °C for a period of more than an hour) and then stored at ambient temperature for up to three years. Some of these i.v. solutions contain high proportions of lipids, blood components, or solubilizing agents, and thus are moderately "aggressive" extraction vehicles. Permanent medical implants are exposed to body fluids or tissue at body temperature for periods measured in years, if not tens of years. In these and other cases, extraction performed under clinical use conditions can be quite aggressive.

There are many reasons that prevent an investigator from actually or practically carrying out a leachables study (or an extractables study that uses clinical use as its extraction condition), most frequently related, one way or another, to the inability to test drug products, process streams, and/or body fluids/tissues with the sensitivity required to perform a credible chemical compatibility assessment. For example, the drug product, process stream or body fluid/tissue may be so compositionally complex that it cannot be effectively tested. This occurs when either the item to be tested is physically incompatible with the testing apparatus or when the item to be tested produces a test response that obscures the test responses related to the extracted or leached substances. Additionally, it may be that the clinical use of the item involves so much of an item that the patient exposure to the item itself is large. An example of this is a large-volume aqueous parenteral drug product, where the daily patient dose volume can be so large that leachables must be accounted for at levels lower than the analytical methods' detection or quantitation capabilities. Lastly, there may be practical issues that make leachables profiling resource prohibitive. For example, consider the case of a packaging system that will be used to package tens of drug products. It may be more practical to perform a single simulated extraction study, using extraction vehicles that bracket the "leaching power" of the drug products, than it is to perform a leachables study on each and every drug product.

If one cannot obtain a leachables profile by testing a drug product, process stream, or body fluid/tissue, then surely the next best experimental approach to address the clinical reality is to perform an extraction study that uses extraction conditions that simulate the clinical conditions of use, as presumably the extractables profile thus obtained would mirror the leachables profile. This means that (i) all extractables would also be leachables, (ii) all leachables would also be extractables, and (iii) that the level of extractables would be comparable to the level of leachables.

Although the simulation study is the proper extraction alternative to leachables studies in concept, there are certain practical issues associated with performing simulation studies that have retarded their regulatory adoption. The first practical issue is establishing and justifying the proper simulating extraction conditions. In circumstances when the factors that influence leaching (and/or extraction) are many and complex, it may be that the science is not sufficiently advanced (or well understood) to establish and justify these proper extraction conditions. The fear is that a simulating extraction would be less "intense" or "aggressive" than the clinical conditions of use and thus that the simulated extractables profile would under-represent the leachables profile. Even worse than actually underestimating the profile is the possibility that one would not be able to recognize when an underestimation has occurred.

Additionally, it is well understood that there is uncertainty in the chemical compatibility assessment process. Extraction, analytical testing of the resultant extracts, and impact interpretation of the test results are processes that are generally recognized as having some variation, uncertainty, or "estimation errors" in them. Thus, even if a simulated extraction study was appropriately designed and executed, it is possible that the resultant extractables profile, and thus the compatibility assessment based on the profile, would underestimate clinical patient exposure to leachables. Because underestimation is unacceptable, regulators have generally concluded that proper simulation studies do not typically have a "margin for error" or a "safety factor" built into them and thus the regulatory preference is an exaggerated extraction study, if not the worst case. Thus, for example, an extraction study is defined in the EMEA Plastic Immediate Packaging Guidelines [2] as "testing that specifically involves exposing a sample of a component to an appropriate solvent system at extreme conditions." In Attachment C of the FDA Container-Closure Guidance, the extraction study is described as having exaggeration factors such as "elevated temperature to increase the rate of extraction … to maximize the amount of extractables obtained from a sample" and a test sample that is "subdivided to increase surface area." Considering medical devices, ISO 10993:18 [11] notes that "extraction conditions are expected to be at least as aggressive as the conditions of clinical use."

Before we consider exaggerated extraction studies in detail, there is a certain type of exaggerated extraction study that differs from the others of that class, the accelerated extraction study. The purpose of the accelerated extraction study is to accomplish a simulation study (that is, produce an extractables profile that closely mimics the leachables profile) in a shorter time than clinical use. While acceleration has little benefit in and is typically not justifiable in short-term use conditions (for example, limited duration medical devices), acceleration can have a significant practical benefit in long-term use conditions (for example, a packaged drug product with a three-year shelf-life) and acceleration is generally a well-accepted extraction practice.

Although an extraction can be accelerated by altering several extraction conditions, it is almost always the case that acceleration is achieved by using elevated extraction temperatures, which leads to shorter extraction durations as a result of the effect of temperature on rates of diffusion.

Considering temperature further, it is noted that elevating temperature can be both an accelerating and an exaggerating factor. When temperature is used merely to accelerate (as previously noted) then the extraction study that uses the elevated temperature is an accelerated extraction study. When temperature is used primarily to exaggerate the extraction (although acceleration may be a secondary effect), then the extraction study that uses the elevated temperature is called an exaggerated extraction study.

Truly there is no hidden meaning in the term exaggerated extraction study. An exaggerated extraction study is an extraction study where one or more of the extractions conditions have been "exaggerated," or made worse, than the clinical conditions of use. The purpose of the exaggerated extraction is equally obvious and is to produce an extractables profile that is worse than the leachables profile. This worse extractables profile may have extractables that are not leachables, extractables that are at higher levels than the leachables, or a combination of both.

The example of using temperature as an exaggerating factor was noted previously; however, virtually every factor important to performing an extraction has been and will continue to be exaggerated. These include:

- Using an extraction vehicle (or solvent) that is more "aggressive" at leaching than is the clinical contact item,
- Using the combination of an elevated temperature and/or a longer duration to increase the "leaching driving force" (e.g. increasing the rate of diffusion),
- Increasing the stoichiometry of the extraction (e.g. increasing the ratio of contact surface area to either the mass or volume of the extraction vehicle) to increase the extent or rate of extraction versus leaching,
- Increasing the number of items extracted per unit measure of the extracting item versus the number of items associated with clinical use,
- Increasing the contact surface area per unit amount of extracting medium by increasing the surface area of the item being extracted (e.g. cutting, grinding, etc.),
- Using more aggressive extraction processes (e.g. refluxing versus soaking).

It is generally the case that the issue with designing a proper exaggerated study is to design a study that is not too exaggerated. While the intent of the exaggerated extraction is to produce an extractables profile that is worse than the leachables profile, it is not the intent of the exaggerated extraction study to distort the leachables profile. Thus, the exaggerated extraction conditions cannot materially change the make-up of the leachables profile. This requirement means that while the exaggerated extraction process should release extractables from the test article, it should neither generate new extractables from the test article nor transform existing extractables into new extractables.

In moving from an exaggerated to an aggressive extraction, one crosses the subtle line separating an inflated actual case from the worst possible case. Although an exaggerated extraction is geared toward establishing "this is likely to be the worse it can get," once one moves to an aggressive extraction one enters the territory of "this is the absolute worst it can get." In moving into an aggressive extraction, one moves away from answering the question "what could leach from the test article?" to "what is in the test article (regardless of whether it can actually leach or not)?"

Unfortunately, there is no hard and fast line that separates an exaggerated extraction and an aggressive extraction. Where an exaggerated extraction might exaggerate one or two aspects of the extraction to a limited extent, an aggressive extraction exaggerates multiple aspects to a greater extent. Nevertheless, an aggressive extraction is still an extraction and thus it must abide by the rules that say an extraction should not alter either the item being extracted or the substances that were extracted.

In theory, the concept of an exhaustive extraction is clear and objective, as its implied intent is to extract everything that is in a test article completely out of the test article. This objective separates the exhaustive extraction from all other extractions in the sense that the claim of having exhausted the test article must be verified. Realistically, the claim of exhaustive can only be substantiated by extracting the supposedly exhausted test article one more time and demonstrating that in fact "there is nothing more in the last extract."

It is this requirement to verify exhaustion that makes the practical implementation of the exhaustive extraction so difficult. In a practical sense, what does it mean that "there is nothing present in this last extract?" Does this mean there are no detectable extractables in the last extract or that the levels of extractables in the last extract is some fraction of the extractables from the previous extraction steps? If it is the latter, what is the fraction and is it a fraction of all the previous extraction steps, or one of the extraction steps (and, if so, which one)? Furthermore, how is exhaustion established? Is exhaustion established for one extractable or for all extractables? If one, which one? Or can exhaustion be established by some property shared by all extractables such as total organic carbon or non-volatile residue?

ISO 10993:18 for medical devices attempts to answer these questions by its definition of an exhaustive extraction: a multi-step extraction conducted until the amount of material extracted in a subsequent extraction step is less than 10% by gravimetric analysis (or achieved by other means) of that determined in the initial extraction step [11]. Thus, we see that the preferred means of establishing exhaustion is gravimetric analysis (e.g. NVR) and the criterion for exhaustion is that the last step is 10% or less of the first step. Now it is clear that this approach, especially involving the use of NVR is not perfect as NVR, after all, only captures non-volatile extractables. This is why the definition includes the phrase "or achieved by other means." While this phrase does not provide much in the way of actionable guidance (for example, it does not explain what these other means might be), it does provide investigators with the opportunity and flexibility to establish their own means of quantifying exhaustion with the caveat that the means will require justification. For example, documenting the extraction of the most non-polar ingredient in a plastic could be justifiable as a means of establishing exhaustion, as conceptually it would be this most non-polar ingredient that would be the most difficult to extract.

There are two major practical issues with performing an exhaustive extraction. First, sequential extraction is efficient when the extraction sequence consists of the fewest possible numbers of extraction steps while not degrading a test article's additives and ingredients. A poorly designed exhaustive extraction that requires too many steps is not efficient and places great pressure on the analytical methods (as analyte levels in individual extracts may be low). This problem is typically encountered when the extraction conditions used are not aggressive enough and thus only a small part of the extractable's pool in the test article is released in each extraction step.

The second problem involves the exhaustive extraction of test articles that are intrinsically low in extractables. The problem encountered is that the analytical methods employed are not sensitive enough to confirm the "less than 10%." For example, consider the example of an analytical method with a limit of detection of 10 ppb. If the amount of extractables in the first extract was 90 ppb, then the method would have to be able to detect 9 ppb of the extractables in a subsequent extract in order to confirm 10%, which of course it cannot do.

There are several potential solutions to this problem. The first potential solution is to use extraction conditions that produce a more concentrated first extract. The second would be to revise the definition of exhaustion to read something like "until the amount of material extracted in a subsequent extraction step is either undetectable or less than 10% ...." A third potential solution is to mathematically project the extraction sequence. That is to say, if each extraction step is such that equilibrium is established, then the entire extraction sequence can be mathematically calculated, via application of partitioning theory, on the basis of two consecutive extraction steps [19].

The transition from exhaustive extraction to dissolution (generally relevant for organic extractables) or digestion (generally relevant for inorganic or elemental extractables) is not a transition of purpose but rather a transition of process. As was the case with an exhaustive extraction, the purpose of either dissolution or digestion is to establish the total amount of constituents (intentional and unintentional additives and residuals and impurities) in a test article. However, while an exhaustive extraction accomplishes this purpose in a multi-step process that essentially leaves the test article intact, both dissolution and digestion accomplish the same purpose in a single step in which the test article is completely solubilized. The difference between dissolution and digestion is what types of extractables they are applied to, the chemical state of the extractables after solubilization, the chemical nature of the test article, and the means by which the solubilization is accomplished.

When the impact of an organic extractable is assessed, the assessment is generally based on properties or characteristics of the specific organic compound that was extracted. For example, the safety impact assessment for an organic extractable involves the collection and evaluation of safety-related data specific to that compound. Thus, if complete solubilization is a means used to assess organic extractables, the solubilization process must preserve the chemical structure of the extracted organic compounds. This is why dissolution is typically applied to establish the organic constituents (and thereby potential organic extractables) in a test article.

Additionally, the solubilization process used depends on the chemical nature of the test article. Organic test articles, such as plastics, polymers, and elastomers, can typically be dissolved by appropriate organic solvents. Thus, dissolution with organic solvents is an effective means of addressing organic constituents in organic test articles. However, inorganic tests articles, such as glass, metals, and ceramics, are not dissolved by organic solvents. Thus, treatment with organic solvents may not be an effective means of establishing inorganic and/or elemental constituents in inorganic test articles.

Alternatively, inorganic or elemental extractables are frequently assessed for impact based strictly on the elements that make up the extractable. Thus, extracted metals are assessed for impact based on the elemental identity of the metal is and not necessarily on the chemical form the metal is in. This is not to suggest necessarily that a metal's speciation is not relevant to an element's impact assessment, but rather that speciation is taken into account by considering the most impactful form of an element. For example, in the ICH Q3D Guideline for Elemental Impurities, used to assess the patient health impact of elements contained in drug products, ICH toxicologists note that "when the toxicities of species of the same element are known, the PDE (permissible daily exposure) has been established using toxicity information on the species expected to be in the drug product" [20].

If an element exists in a test article in an inorganic form, it is unlikely to be solubilized by dissolution. Once it has been solubilized, it is measured in its elemental form. For both these reasons, profiling a test article for elemental constituents generally involves digestion of the test article, which is typically accomplished by contacting the test article with strong mineral acids at high temperature (and possibly under high pressure).

2.2.3 Examples of Extraction Sequences

Given a specified clinical circumstance, one can devise appropriate extractions processes to cover each of the extraction types considered previously. An extraction sequence is used in this document to mean a set of proposed extraction processes, linked to a specified clinical circumstance, that cover all extraction types. Examples of extraction sequences are provided in Tables 2.5–2.8 for a specified packaged drug product, indirect contact medical device, a direct and permanent medical device, and a component used in pharmaceutical manufacturing.

Table 2.5 addresses the case of an aqueous drug product packaged in a flexible container (bag). The drug product formulation includes the drug substances, at a level of 1 mg/ml, normal saline (0.9% by weight sodium chloride) as the diluent, and a phosphate buffer to control pH within the product's compendial range of 6–8. The fill volume of the bag is 50 ml, the product is steam sterilized in the bag (121 °C for one hour), and has a post-sterilization shelf-life of three years at an ambient temperature (taken as 25 °C).

Table 2.5 Extraction sequence, pharmaceutical packaging.

Extraction type	Description of extraction conditions used
Clinical conditions of use	Drug product (1 mg/ml drug in phosphate buffered saline, PBS, pH 6–8), 50 ml fill of a DEHP-plasticized PVC bag, terminally sterilized (121 °C for 1 hr), 3-year (36 month) ambient temperature (25 °C) shelf life
Simulated extraction	PBS (pH 6–8), 50 ml fill into bag, terminally sterilized (121 °C for 1 hr), 3-year ambient temperature (25 °C) shelf-life
Accelerated extraction	PBS (pH 6–8), 50 ml fill into bag, terminally sterilized (121 °C for 1 hr), store at 40 °C for 12 months
Exaggerated extraction	PBS (pH 6–8), 25 ml fill into bag, terminally sterilized (121 °C for 1 hr), store at 55 °C for 12 months
Aggressive extraction 1[a]	PBS (pH 4 and 10), 25 ml fill into bag, terminally sterilized (121 °C for 1 hr), store at 70 °C for 9 months
Aggressive extraction 2[a]	Cut bag. Reflux in isopropyl alcohol for 8 hours
Exhaustive extraction 1[a]	Repeat the following extraction process for multiple steps: PBS (pH 4 and 10), 25 ml fill into bag, 121 °C for 2 hr
Exhaustive extraction 2[a]	Repeat the following extraction process for multiple steps: cut bag, reflux in isopropyl alcohol for 8 hr
Dissolution	Dissolve bag in an appropriate organic solvent
Digestion	Digest bag with strong mineral acids at elevated temperature (e.g. microwave-assisted digestion)

a) Two alternate extraction processes are illustrated. Either one or the other should be performed, not both.

Table 2.6 Extraction sequence, indirect contact medical device; blood set.

Extraction type	Description of extraction conditions used
Clinical conditions of use	Administration set used to transfer whole blood from storage bag to patient; single-use, duration of use not to exceed 12 hr at a flow rate not to exceed 200 ml/hr
Simulated extraction	Recirculate 1200 ml (100 ml/hr × 12 hr) of a blood-simulating solvent (e.g. 46% ethanol) through the administration set for 12 hours at 25 °C
Accelerated extraction	Recirculate 1200 ml (100 ml/hr × 12 hr) of a blood-simulating solvent (e.g. 46% ethanol) through the administration set for 4 hr at 40 °C
Exaggerated extraction	Recirculate 1200 ml (100 ml/hr × 12 hr) of a solvent (55% ethanol) through two tandem administration sets for 4 hours at 40 °C
Aggressive extraction	Recirculate 600 ml of ethanol through the tandem administration sets for 8 hr at 40 °C
Exhaustive extraction	Repeat the aggressive extraction for multiple steps
Dissolution	Dissolve the administration set in an appropriate organic solvent
Digestion	Digest the administration set with strong mineral acids at elevated temperature and pressure (e.g. microwave-assisted digestion)

Table 2.7 Extraction sequence, direct contact medical device, and non-dissolving Class III permanent implant.

Extraction type	Description of extraction conditions used
Clinical conditions of use	Device, implanted in bladder, for sustained or controlled drug delivery. Removed after 6-month duration of therapy. The clinical temperature of contact is taken as the normal body temperature (37 °C)
Simulated extraction[a]	Suspend device in artificial urine,[b] store at 37 °C for 6 months with constant agitation
Accelerated extraction[a]	Suspend device in artificial urine,[b] store at 55 °C for 2 months with constant agitation
Exaggerated extraction	Suspend device in a 1/1 mixture of ethanol/artificial urine,[b] store at 60 °C for 3 months with constant agitation
Aggressive extraction	Reflux device in hexane for 8 hr
Exhaustive extraction	Repeat the aggressive extraction for multiple steps
Dissolution	Dissolve the device in an appropriate organic solvent
Digestion	Digest the device with strong mineral acids at elevated temperature and pressure (e.g. microwave-assisted digestion)

a) Simulated and accelerated extractions are generally not considered to be adequate for registration of permanent Class III medical devices per ISO 10993-18. However, in this case a reasonable simulation process can be envisioned, justified, and implemented.
b) For example, a mixture containing: sodium chloride, 6.773 g/l; sodium phosphate monobasic dihydrate, 1.33 g/l; potassium chloride, 6.065 g/l; calcium chloride dihydrate, 0.883 g/l; sodium citrate dihydrate, 0.584 g/l; sodium phosphate dibasic, 0.435 g/l; sodium sulfate, 2.431 g/l; magnesium sulfate heptahydrate, 0.731 g/l; ammonium chloride, 2.322 g/l; urea, 25 g/l; pH 6.0.

Table 2.8 Extraction sequence, manufacturing component.

Extraction type	Description of extraction conditions used[a]
Clinical conditions of use	Sterilizing filter used prior to final fill. Used to filter a proteinaceous biotherapeutic drug product formulated at neutral pH (6–8) containing appropriate organic solubilizing/stabilizing agents. Worst-case use is 8 hr at elevated temperature (30 °C) at a flow rate of 200 ml/hr
Simulated extraction	Recirculating extraction with 1600 ml (200 ml/hr) of a 30% ethanol/water vehicle (justified as an appropriate simulant for the drug product). Recirculation performed at 30 °C for 8 hr
Accelerated extraction	Recirculating extraction with 1600 ml (200 ml/hr) of a 30% ethanol/water vehicle (justified as an appropriate simulant for the drug product). Recirculation performed at 50 °C for 2 hr
Exaggerated extraction	Recirculating extraction with 800 ml (200 ml/hr) of a 40% ethanol/water vehicle. Recirculation performed at 50 °C for 4 hr
Aggressive extraction	Recirculating extraction with a vehicle volume calculated to provide the appropriate surface area/vehicle volume ratio. Perform three separate extractions, using different vehicles (pH 3, pH 10 and 1/1 ethanol/water). Recirculate at 40 °C for 24 hr at a proper flow rate
Exhaustive extraction	Repeat the aggressive extraction multiple times
Dissolution	Dissolve the entire filter in an appropriate organic solvent
Digestion	Digest the entire filter with strong mineral acids at an elevated temperature and pressure (e.g. microwave-assisted digestion)

a) Extraction conditions recommended by USP <665> if the filter is classified as higher risk [12].

In this case, the extraction study that exactly reflects these product's clinical aspects (essentially mimicking the leaching study) would involve filling bags with the product, sterilizing the filled bags and then storing the filled bags for three years at 25 °C. At various points after sterilization and during storage, the contents of the bags are sampled and the samples are screened for extractables. While such a study is the equivalent of a leachables study in terms of its design, it might differ from a true leachables study as the product units might be the equivalent of a

commercial product (e.g. a product not produced under GMP conditions, bags not fully representative of the final container design, etc.).

Unless the daily dose volume of this drug product is large (driving the analytical evaluation threshold, AET, lower) or the drug substance is chemically complex (likely causing analytical interferences that would reduce the screening method's LoD), it is possible that the drug product could be screened for leachables at sufficiently low levels to support an appropriate and rigorous impact assessment, thereby eliminating the need to perform an extraction study (at least from the perspective of impact assessment for this one drug product). However, in the circumstance that extractable profiling was desired, one can establish what proper extraction conditions would look like for this drug product over the spectrum of extraction types.

In this relatively simple case, the most appropriate simulating extraction is to use the drug product placebo (drug product without the drug substance) as the extraction vehicle. By removing the drug substance from the extraction vehicle, one is potentially reducing the possibility and/or magnitude of analytical interferences, presumably without decreasing the "leaching power" of the drug product (which is likely the case with such a low level of active ingredient in the drug product).

With a three-year shelf-life, there is a significant opportunity to reduce the study's duration via an accelerated extraction. Raising the extraction temperature from 15 °C to 40 °C allows the accelerated extraction to be completed in one year (estimated). It is noted that the accelerated study would still include testing at multiple timepoints during the higher temperature aging.

Because the drug product is sterilized by autoclaving at a high temperature, it is likely that the packaged drug product can be stored at a temperature higher than 40 °C (to exaggerate the thermal driving force for diffusion) without changing fundamental physical properties of the packaging system (for example, moving through its glass transition temperature). Thus, the exaggerated extraction is performed at the elevated temperature of 55 °C. Because the primary purpose of the elevated temperature is to exaggerate (and not necessarily to accelerate, although acceleration may be occurring), the duration of the aging is not shortened to account for acceleration. To provide even a greater exaggeration, the fill volume is decreased, which has the net effect of increasing the ratio of either the bag weight to fill solution volume or the exposed bag surface area to fill solution volume.

In this example case, the extraction transitions from exaggerated to aggressive when a third variable is modified, in this case the pH of the extraction vehicle. When the pH of the extraction vehicle is lowered, the levels of basic extractables should increase in response to the extractable's increasing solubility in the extraction vehicle. Furthermore, if the test article's ingredients are degraded by an acid-catalyzed process (e.g. hydrolysis), lowering the pH of the extraction vehicle will increase the amount of degradation. Conversely, when the pH of the extraction vehicle is raised, the levels of acidic extractables should increase in response to the extractable's increasing solubility in the extraction vehicle. Moreover, if the test article's ingredients are degraded by a base-catalyzed process (e.g.

hydrolysis), raising the pH of the extraction vehicle will increase the amount of degradation. Thus, by changing the pH of the extraction vehicle, one has potentially and aggressively changed the resulting extractable profile.

In addition to the pH change, the aggressive extraction is made more time efficient by replacing elevated temperature aging with an extreme temperature aging that uses an autoclave cycle of longer duration.

Now the previous example of an aggressive extraction followed a progression of altering the extraction conditions versus the clinical conditions of use by changing those conditions of use one variable at a time. As each variable was changed to increase the "extracting power," the extraction became progressively more aggressive. However, it is also possible to envisage a discontinuous aggressive extraction where the conditions of extraction bear little or no resemblance to the clinical conditions of use. Thus, in the second example of a discontinuous aggressive extraction, the entire extraction process is changed. Aging at an elevated temperature is replaced by the more aggressive process of refluxing. The extracting power of the extraction vehicle is increased (at least for the non-polar constituents generally present in the plastic used in flexible bags) by replacing the aqueous polar drug product with a non-polar extraction vehicle.

The conditions of both examples of an aggressive extraction were chosen so that each extraction was likely to achieve equilibrium (that is, the extraction proceeded as far as thermodynamic considerations allowed it to). Based on the author's experience, it is likely that the longer autoclave cycle used in the aqueous aggressive extraction is sufficient to drive diffusive leaching to completion (that is, leaching stops because the concentration gradient between the extraction vehicle and the test article that drives leaching is eliminated). The same outcome is expected to be achieved when refluxing is continued for 8 hr or longer. The attainment of maximum extraction (equilibrium) in a single extraction step is the ideal situation for an exhaustive extraction as it means that each successive extraction step in the exhaustive process will remove as much of the extractable from the test sample as is thermodynamically possible. In such a circumstance, the exhaustive extraction would be accomplished in the minimum number of extraction steps. Therefore, the two examples of an exhaustive extraction are simply "repeat the aggressive extraction process sequentially until the requirements for an exhaustive extraction have been met."

The DEHP-plasticized PVC bag material considered previously is challenging for an exhaustive extraction because it differs greatly from other plastics in terms of its levels of additives. While most plastics contain additives in the parts-per-thousand range (antioxidants, stabilizers, colorants, etc.), the level of the DEHP in plasticized PVC can range from 25 to 40% by weight. Additional additives, such as a secondary plasticizer, may also be present in amounts greater than 1%. Thus, an exhaustive extraction for plasticized PVC typically would require much more aggressive extraction conditions and many more sequential extraction steps than a properly designed and successful exhaustive extraction for other types of plastics.

Dissolution and digestion are self-explanatory and the proper conditions for dissolution and digestion will differ depending on the test article. In the case of the plasticized PVC used in this example, tetrahydrofuran has been identified as an appropriate solvent for PVC [21] while microwave-assisted digestion using a mixture of nitric acid and hydrogen peroxide has been demonstrated to be effective for the purpose of trace metal profiling [22].

Table 2.6 addresses the case of an indirect contact medical device, using a blood transfer administration set as an example. This administration set used to transfer whole blood from a storage bag to the patient; the set is single-use and the duration of use is no more than 12 hr at a flow rate not to exceed 200 ml/hr. This device is thus classified as externally communicating, blood path indirect, limited contact duration per ISO 10993-1.

For this specific medical device and clinical use circumstances, a simulated extraction would involve a laboratory simulation of the clinical use and a substitution of a more analytically expedient extraction vehicle (46% ethanol; see reference [23] for a justification of this vehicle) for blood, which could be difficult to analytically screen for extractables at the required levels. Because the clinical conditions of contact involve flowing blood, the simulated extraction uses a flowing extraction vehicle. The only difference between clinical use and the simulated extraction is that the extraction vehicle is recirculated through the device, with the aim of creating a more concentrated extract.

As noted in Table 2.6, acceleration by a factor of 3 (duration reduced from 12 to 4 hr) is achieved by increasing the extraction temperature to 40 °C while keeping all other simulated conditions unchanged. In the exaggerated extraction, exaggeration is accomplished by increasing the % ethanol in the extraction vehicle (presumably increasing the "leaching power" of the vehicle for semi-polar and non-polar extractables) and increasing the extraction vehicle's exposure to the device by placing two devices in the vehicle's flow path. In the more extreme aggressive extraction, the semi-polar extraction vehicle used in the exaggerated extraction has been replaced with a non-polar vehicle and the duration of the extraction has doubled. As was the case with the other examples discussed herein, the aggressive extraction is designed with the intent of achieving equilibrium (otherwise known as asymptotic

extraction) and thus the exhaustive extraction simply becomes a matter of repeating the aggressive extraction until the requirements for exhaustion are met. Lastly, conditions for the dissolution and digestion of the blood set would have to be established.

If the blood set example can be considered to be a simpler case for a medical device where the contact is indirect and the conditions of contact are relatively mild, then a more complex, harsher contact medical device is a non-dissolving permanent implant. Because the contact is direct and long-term and because the contacting medium is likely to have a considerable "leaching power," there is the reasonable expectation that leaching will occur to the fullest extent possible, which generally is the "it all comes out" scenario of an exhaustive extraction. Furthermore, exhaustive extractions are typically required for certain permanent implants (such as those that contact tissue) because it is difficult to design a justifiable simulation or accelerated extraction that can be easily and readily implemented.

The example in Table 2.7 is a drug-releasing device that is placed in the bladder, presumably for the treatment of bladder-related conditions. This case is chosen because one can envision a means of simulating clinical use and thus it provides an opportunity to establish the entire extraction sequence. Moreover, the ability to simulate clinical use is valuable as the simulated study may allow one to establish the extractable's release kinetics and could be relevant for impact assessment.

In this case, envisioning and implementing a simulated extraction is relatively straightforward. In essence, an extraction vessel is used to simulate the bladder and artificial urine is used as the extraction medium. The conditions of extraction mirror those of the clinical contact in terms of temperature and duration and one can envision a process where small aliquots of the extracting vehicle are retrieved over time to provide kinetic release data. This extraction is accelerated by increasing the clinical use temperature (37 °C) to an extraction temperature of 55 °C. The accelerated extraction is further exaggerated by increasing the extraction temperature even more, increasing the extraction duration and replacing the largely polar extracting vehicle with a semi-polar mixture of an alcohol and the simulated aqueous vehicle.

There is a complicating factor in this case that should be considered in the simulated, accelerated and exaggerated extractions. In the clinical use of the device, the contacting medium (urine) is periodically replenished in the clinical circumstances, essentially providing a fresh leaching medium. In the simulating extraction shown in Table 2.7, a single portion of extraction vehicle is in contact with the device for the entire duration of the contact, which clearly is an imperfect representation of clinical use. However, the simulation study could easily be modified to allow for periodic replenishing of the extraction vehicle, and, if the replenishing were frequent enough, analysis of the replenished extraction vehicle sample could provide the desired kinetic information. In this circumstance, the simulated extraction is similar to an exhaustive extraction in the sense that both involve multiple extraction steps (replenishments of the extracting solvent). However, they likely to differ in terms of the number of replenishing steps employed.

In this particular case, the exaggerated extraction is already a fairly aggressive extraction and thus the aggressive extraction shown in the sequence reflects the discontinuous process (that is, a process that has not evolved from previous processes for other extractions). Of course, the use of this aggressive extraction process is predicated on the requirement that the device is not degraded or dissolved to any appreciable extent due to the aggressive extraction.

As was the case with the other examples discussed herein, this aggressive extraction is designed with the intent of achieving equilibrium (otherwise known as asymptotic extraction) and thus the exhaustive extraction simply becomes a matter of repeating the aggressive extraction until the requirements for exhaustion are met. Lastly, conditions for the dissolution and digestion of the implanted device would have to be established.

It is important to be sure that I have not created a false impression. It is reasonable to devise a complete extraction sequence for this case because one can easily design, implement, and justify a simulated, accelerated, and exaggerated extraction. However, there will be many permanent implanted devises where the clinical conditions of use cannot be simulated in the laboratory (either because it is not practically possible or it cannot be scientifically justified) and in these circumstances aggressive or exhaustive extractions are the only viable approach.

The last example is for a component used in pharmaceutical manufacturing. Before we consider this example in greater detail, we would do well to remember the unique situation of manufacturing components with respect to the relationship between extractables and leachables. This means that, in order for manufacturing component-related extractables to become leachables, two actions must occur, the extractables must be extracted by the process stream (producing the corresponding process equipment-related leachable, PERL) and the PERL must persist in the process stream through the entire manufacturing process (at which point the PERL becomes an actual leachable in the finished drug product). An extraction study performed on a manufacturing component addresses this first action only, and thus the purpose of an extraction study with manufacturing components is to forecast PERLs.

With this thought in mind, the manufacturing component considered in Table 2.8 is a sterilizing filter, used just prior to the final filling of the drug product into its packaging. In this example, the sterilizing filter is used to filter a proteinaceous biotherapeutic drug product formulated at neutral pH (pH between 6 and 8) and containing appropriate organic solubilizing/stabilizing agents. Under the worst possible manufacturing conditions, the filter can be used for eight hours at an elevated temperature (30 °C) and with a flow rate of 200 ml/hr.

As was the case with the blood set medical device, the filter is used under flow conditions and thus the simulating extraction must use flow contact as well. The simulating extraction utilizes a recirculating loop in the interest of producing a more concentrated extract. The simulating extraction vehicle to be used, 30% ethanol/water, will have been justified.

The accelerated extraction is accomplished by increasing the extraction temperature from 30 to 50 °C and decreasing the extraction duration fourfold (this decrease must be justified). In the exaggerated extraction, the exaggeration is accomplished by decreasing the extraction volume, increasing the "leaching power" of the extraction vehicle by increasing its ethanol content, and increasing the extraction duration.

One could use the concept of increasing aggressiveness that was used in previous examples to establish the appropriate aggressive extraction. Thus, for example, the aggressive extraction could have used a 50% ethanol/water extracting vehicle as one of its exaggerating factors. However, in the case of manufacturing components, a standardized extraction protocol has been developed for performing aggressive extractions for "higher risk" components, USP <665> [12]. It is likely that the sterilizing filter will be classified as a higher risk component, simply on the basis of its use in the last manufacturing process step prior to the final fill (which means there is little to no possibility for PERLs to be removed from the process stream prior to the final fill). Thus, in this case, the USP <665> standard extraction protocol conditions are used to generate the aggressive extract.

As was the case with the other examples discussed herein, the USP standard extraction protocol (aggressive extraction) was designed so that it is likely that the extraction achieves equilibrium (otherwise known as asymptotic extraction); thus, the exhaustive extraction simply becomes a matter of repeating the aggressive extraction until the requirements for exhaustion are met.

Lastly, conditions for the dissolution and digestion of the sterilizing filter would have to be established.

2.2.4 Required/Recommended Extractions

Although the guidelines, standards, monographs, and recommendation documents typically lack detail in terms of the specific conditions with which an extraction must be performed, they do contain clear and specific guidance in terms of what type of extraction studies are most appropriate under certain specified circumstances.

For packaging systems. As was noted previously, published regulatory guidelines for extraction tend to favor an exaggerated or aggressive extraction. For example, the EMEA notes in its Plastic Immediate Packaging Guidelines [2] that extractions should be performed under conditions "that specifically involves exposing a sample of a component to an appropriate solvent system at extreme conditions." Similarly, the FDA notes in its Container-Closure Guidance [1] that an extraction should be performed with exaggerating factors such as "elevated temperature to increase the rate of extraction … to maximize the amount of extractables obtained from a sample" and use of a test sample that is "subdivided to increase surface area." Of course, these references are somewhat dated and reflect a time when impact assessment was inordinately obsessed with the concept of worst-case.

USP <1663>, an informational chapter that deals with the assessment of extractables associated with packaging and delivery systems [4], is less clear in terms of specifying the type(s) of extractions necessary for packaging systems. However, it makes reference to and endorses the use of simulation studies (described in greater detail in USP <1664> [5]) for generating extractables profiles that are similar to leachables profiles.

Specific requirements and recommendations for extraction conditions of packaging systems are discussed in more detail in Chapter 7.

For medical devices. The following requirements for the extraction of medical devices (Table 2.9) are contained in the normative text of Section 5.6 in ISO 10993:18 [11].

Specific requirements and recommendations for extraction conditions contained in ISO 10993:18 are discussed in more detail in Chapter 9.

For manufacturing components. The appropriate reference for the extraction required for manufacturing components can be found in USP <665> [12]. Although this monograph specifies specific extraction conditions depending on the particular manufacturing component and the level of risk associated with the use of the component, in general the standard extraction protocol required for higher risk components is an aggressive extraction, requiring an elevated extraction temperature, generally aggressive extraction vehicles (low pH, high pH, and semi-polar), and worse than typical use of surface area to solution volume ratios. Extraction conditions that comply with the USP <665> standard extraction protocol are discussed in more detail in Chapter 8.

2.2.5 Principles of Extraction

2.2.5.1 Thermodynamics and Kinetics of Extraction and Leaching

Regardless of the type of extraction performed, extraction is a complex process influenced by factors including time, temperature, surface area-to-volume ratio, extraction vehicle, and the partitioning behavior of the substances in the test article relative to the extraction vehicle [11]. Considering partitioning, the partition coefficient, K, is a unitless

Table 2.9 Recommended extraction conditions for medical devices per ISO 10993:18 [11].

Contact category	Recommended extraction conditions	Credible alternatives
Limited contact devices	Simulated use conditions[a]	Exaggerated conditions
Prolonged contact devices	Exhaustive conditions	Exaggerated conditions[b),c]
Long-term contact devices	Exhaustive conditions	Exaggerated conditions[b),c),d]

a) With suitable justification.
b) Examples of instances where exhaustive extraction would not typically be required include:
   - Single-use devices used for less than 24 hr, where repeat use of a new device each day would result in categorization as prolonged or long-term contact;
   - Single-use devices used for several days, where repeat use of new devices would result in categorization as prolonged or long-term contact; and
   - Reusable devices, where a patient may be exposed to repeated use of the same device, resulting in categorization as prolonged or long-term contact. When an exaggerated extraction is used for a reusable device, the extraction should properly account for the duration of each individual use.
c) Exaggerated conditions can be appropriate for external communicating or non-absorbable surface contact devices, with justification.
d) An example is a device comprised entirely of non-absorbable metal (e.g. a vascular stent), because migration of constituents from within the material is not possible and the constituents of interest are related to the surface only and exaggerated extraction can be adequate to generate a complete extractables profile.

fundamental physicochemical (thermodynamic) parameter that describes the distribution of a solute between two contacting and immiscible phases at equilibrium. For our purposes, the solute (s) is leachable (or extractable), one phase is the receiving phase (taken as a liquid L, into which the solute is transported), and the other phase is the donor phase (taken as a plastic P, from which the solute is transported). In essence, the partition coefficient is the concentration (C) ratio of the solute in the two contacting phases at equilibrium:

$$K_{P/L} = C_{P,s}/C_{L,s} \tag{2.1}$$

For example, if a plastic and a liquid are in contact and after they reach thermodynamic equilibrium, $C_{P,s} = 10\,\text{mg/g}$ and $C_{L,s} = 5\,\text{mg/g}$, then K_P/L becomes

$$K_{P/L} = C_{P,s}/C_{L,s} = 10\,\text{mg/g} \div 5\,\text{mg/g} = 2$$

Although this is a useful example, it is rare that $C_{P,s}$ and $C_{L,s}$ are measured directly. Rather, it is typically the case that $C_{P/s}$ is known (or is determined), $K_{P/L}$ is known (or determined), and the desired variable to calculate is $C_{L,s}$: that is, what is the equilibrium (maximum) concentration that a leachable will achieve in a liquid drug product in contact with a plastic packaging system?

To derive an equation that is more useful in this circumstance, let us examine the situation at the point of time of first contact between the liquid drug product and its packaging. At the point of initial contact, the total amount of the solute is in the packaging system and has the value $m_{P,i}$. Clearly, the amount of the solute in the liquid drug product at the point of initial contact is 0 and thus $m_{L,i} = 0$. At equilibrium, the mass of the solute in the drug product is denoted by $m_{L,e}$ and the mass of the solute in the packaging system $m_{P,e} = m_{P,i} - m_{L,e}$. If the volume of the polymer is V_P and the volume of the drug product is V_L, then the partition coefficient takes the form

$$K_{P/L} = (m_{P,i} - m_{L,e})/V_P \div (m_{L,e})/V_L \tag{2.2}$$

Rearranging Equation (2.2) to isolate $m_{L,e}$ produces

$$m_{L,e} = (m_{P,i})/[(K_{P/L} \times V_P) + 1] \tag{2.3}$$

For example, consider a leachable that is present in a packaging with a total pool ($m_{P,i}$) of 100 mg. If $K_{P/L} = 10$ (the leachable is more "plastic-like" than "liquid-like"), $V_p = 100\,\text{cm}^2$ and $V_L = 250\,\text{cm}^2$, then the amount of the leachable in the drug product at equilibrium ($m_{L,e}$) becomes (from Equation (2.3))

$$\begin{aligned} m_{L,e} &= (m_{P,i})/[(K_{P/L} \times V_P)/V_L + 1] \\ &= (100\,\text{mg})/[(10 \times 100\,\text{cm}^2)/250\,\text{cm}^2 + 1] \\ &= 20\,\text{mg} \end{aligned}$$

and the maximum equilibrium concentration of the leachable in the drug product is

$$C_{L,e} = m_{L,e}/V_L = 20\,\text{mg}/250\,\text{cm}^2 = 0.08\,\text{mg/cm}^2$$

So we see that if we are able to obtain the proper partition coefficient ($K_{P/L}$) and the total amount of the substance in a donor item (such as plastic packaging or a medical device, $m_{P,i}$), we would be able to calculate the maximum equilibrium level of the substance as an extractable in an extract or as a leachable in a drug product. Obtaining $m_{P,I}$ is straightforward and is accomplished by performing an exhaustive extraction of the donor item. However, the issue is obtaining $K_{P/L}$ as there are no published sources of even a partial list of $K_{P/L}$ values for all possible extractables and leachables and all possible donor items such as plastics and elastomers. Of course, individual researchers could compile their own lists of $K_{P/L}$ values by obtaining these results experimentally; however, the effort required to do so would be considerable even for a single pharmaceutical product.

Although there are numerous means of mathematically estimating polymer/liquid partition coefficients (see reference [24] for a review of these various means), they typically require information that may be as elusive to find as the partition coefficients themselves. An empirical approach to obtaining K_p/L values was proposed and developed by Jenke and co-workers, who observed that octanol/water partition coefficients ($P_{o/w}$) are routinely available for many organic compounds as a compound's log $P_{o/w}$ has been related to numerous compound characteristics such as solubility and bioavailability. These researchers noted that both $K_{P/L}$ and $P_{o/w}$ could be mathematically related to the extent that (a) the polymer P "behaves" in the same way as octanol and (b) the liquid L "behaves" in the same way as water. This relationship takes the form:

$$\log K_{P/L} = \text{slope} \times \log P_{o/w} + \text{intercept} \tag{2.4}$$

While $P_{o/w}$ values for many organic compounds are well-published and readily available, it is clear that the slope and intercept of Equation (2.4) will be different for different plastics. Values for the slope and intercept can be obtained by experimentally measuring $K_{P/L}$ for compounds of known log $P_{o/w}$ within the context of a given plastic/liquid system and then establishing the linear relationship between these two quantities. For example, see Figure 2.5, which illustrates the relationship between $K_{P/L}$ (which these authors referred to as the equilibrium binding constant E_b) and log $P_{o/w}$ for a number of plastic and elastomeric materials.

Equilibrium binding models for specific plastic and elastomeric materials have been published, for example, in

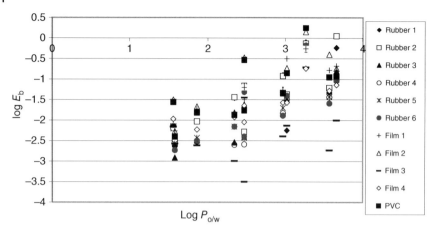

Figure 2.5 Equilibrium binding models for various plastic films and rubber materials. The equilibrium binding models establish the linear relationship between $K_{P/L}$ for a given material (here called the equilibrium binding constant, E_b) and $P_{o/w}$ for a particular substance. The best-fit line obtained for each material defines a relationship of $\log P_{K/L} = \text{slope} \times (\log P_{o/w}) + \text{intercept}$, which can be used to calculate a compound's $K_{P/L}$ from its $P_{o/w}$ value for each individual material.

references [25–31]. Feenstra and associates have recently studied the partitioning of migrant molecules between a solvent and polymeric packaging material and report an interaction model, using the linear interaction energy algorithm, which reportedly yields better results than the $P_{o/w}$-based models [32]. During the time this book is created, its author is aware of other researchers who are addressing this issue but whose work has yet to be published.

The fact that the partition coefficient is associated with a thermodynamic (equilibrium) process does not mean that it is unaffected by temperature. Rather, the temperature-dependence of the partition coefficient is given by the van't Hoff equation (in its integrated form):

$$\ln(K_2/K_1) = (-\Delta H/R) \times (1/T_2 - 1/T_1) \qquad (2.5)$$

One notes from Equation (2.5) that the higher the temperature, the lower will be the partition coefficient. Thus, as the temperature of an extraction increase, so too will the concentration of an extractable in an extracting medium.

This effect is illustrated in Figure 2.6, which considers the effect of temperature on the leaching of dipropyl

phthalate (DPP) from a bromobutyl rubber disk to isopropyl alcohol (IPA). The increase in the equilibrium concentration achieved by this extractable is a direct reflection of the effect of temperature on the $K_{P/L}$.

In many circumstances, a donor item and its contacting solution reach equilibrium during contact and thus the worst-case transfer of a leachable from the donor to the solution is expressed by Equation (2.3). For example, an aqueous drug product that is steam sterilized in a plastic container (bag) and then stored in the same bag for a two-year shelf-life at ambient temperature is likely to achieve equilibrium at some point during its shelf-life. However, this equilibrium will not have been obtained at all times during its shelf-life and thus the actual concentration of a leachable may be lower than this maximum equilibrium value at many points over the shelf-life. As another example, consider a drug administration transfer set whose maximum time of use is 24 hr at ambient temperature. It is unlikely that equilibrium between the administration set and the delivered drug product is achieved during contact and thus the levels in leachables in

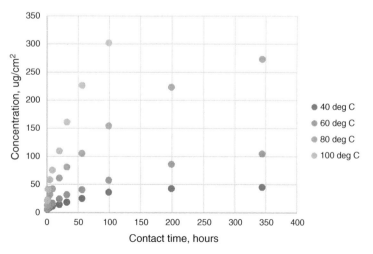

Figure 2.6 Leaching of dipropyl phthalate (DPP) into isopropyl alcohol (IPA) from a bromobutyl rubber (BIIR) disk. The effect of extraction temperature on the equilibrium concentration of this extractable is clearly evident, as the equilibrium concentration increases as the temperature increases. (*See insert for colour representation of the figure.*)

the delivered drug product will be less than the calculated equilibrium value.

This discontinuity between the maximum possible concentration of a leachable and its actual concentration arises from the simple fact that the transfer of a leachable from a donor item to a receiving liquid is not instantaneous. This is the case because the donor item is three-dimensional and the leachable is typically distributed in the item across its entire cross-section. This means, simply, that all of the leachable is not present at the donor item/liquid interface at the time of contact and thus cannot transfer instantaneously from the donor item to the liquid. Rather, the transfer proceeds via a two-step process:

1. That amount of a leachable that is at the donor item/liquid interface at the time of contact is immediately transferred from the donor item to the liquid and
2. That amount of the leachable that is dispersed in the donor item (but not at the interface) must move to the interface before it can be transferred from the donor item to the liquid.

The movement of a substance through a medium in response to a concentration gradient is termed diffusion. The rate of diffusion will affect the time it takes a donor item and its contacting liquid to reach equilibrium. Thus, while maximum leachable levels over the entire duration of contact are dictated by thermodynamics (partitioning), the actual leachable levels during contact are dictated by kinetics (diffusion).

I can assure you that the derivation of the equations that express the rate at which substances leach from a donor item and into a solution is an exercise of futility for most scientists with even a passing interest and capability in mathematics. Skipping the heavy mathematics, the diffusion equations have been solved to provide expressions that can be related to the levels of leachables in a liquid over time [27, 33, 34], based on the following assumptions:

1. Diffusion occurs in only one direction (out of the donor and into the liquid),
2. The diffusion coefficient D is concentration independent,
3. The diffusing substance is evenly distributed in the donor item,
4. The concentration of the diffusing substance at the item/liquid interface is instantaneously brought to zero.

For short contact times (where $M_t/M_\infty < 0.5$):

$$M_t/M_\infty = (4Dt/\pi\delta^2)^{1/2} \qquad (2.6)$$

For long contact times (where $M_t/M_\infty > 0.5$):

$$M_t/M_\infty = 1 - [(8/\pi^2)\exp(-\pi^2 Dt/\delta^2)] \qquad (2.7)$$

where

M_t = amount of the leachables present in the liquid at contact time t

M_∞ = amount of the leachables present in the liquid at infinite contact time (equilibrium amount)

D = diffusion coefficient

δ = thickness of the donor item

t = contact time

Presumably, these equations provide the means of calculating leachable profiles over contact time. However, the practical reality is that obtaining the required input data is difficult and mathematical modeling of leaching is rarely considered to be acceptable by regulators in product registration dossiers (reflecting the preference for data as opposed to theory). However, where these concepts are relevant is in the design of accelerated extraction of leaching studies, which is considered in greater detail in Section 2.2.5.3.

Several recent articles have been published on the modeling of substances migrating out of and through plastic materials. Lei Yu et al. investigated the release of ethylene oxide (a sterilizing agent) from plastic containers and its effect on monoclonal antibodies [35]. These authors note that the kinetic behavior of this compound could be modeled by considering the combined effects of Fickean diffusion and first-order chemical reaction kinetics (to account for the hydrolysis of ethylene oxide in the aqueous solutions). Fang and Zhao studied the permeation rates of six solvents through a fluoropolymer film, typical of a barrier layer used in some multi-layered pharmaceutical packages [36]. The authors developed a mathematical migration model, based on Hansen solubility parameters, which reportedly facilitates expedited screening for potential leachables, allowing investigators to focus on higher-risk leachables. Parthasarathy and co-workers studied the leaching of unpolymerized monomers from methacrylate-based cross-linked adhesives, and reported diffusion coefficients for several of the monomers [37]. Jenke used experimental data from 10 case studies to evaluate commonly used, Arrhenius-based, approaches to acceleration and recommendations were made in terms of the proper approaches to be used for concentration and duration extrapolations [38] (see also Section 2.2.5.4).

2.2.5.2 Extraction Solvents, Polarity

It is intuitively obvious that one of the predominant factors driving the transfer of a substance from a donor phase (e.g. plastic item) to a receiver phase (liquid drug product) is the chemical characteristics of the receiver phase, as the chemical characteristics of the receiver phase impact the partitioning of the substance between the two phases and the solubility of the substance in the receiver phase.

One of the critical properties of the receiving phase (extraction medium or drug product) is its polarity. This is especially true for aqueous parenteral drug products that contain agents for solubilizing drug substances and for drug products (or device-contacting media) that are lipophilic, such as blood, blood-derived substances, tissues, topical dosage forms, and lipid emulsions. Because extractables and leachables may vary significantly in terms of their lipophilic nature, the polarity of the extraction solvent and the drug product is a critical extraction design parameter, as the solubility and the accumulation level of a leachable (or extractable) will be polarity-dependent.

The concept of polarity and its effect on solubility is simplistically explained as follows. A molecule is composed of one or more chemical bonds between molecular orbitals of different atoms. Not all atoms attract electrons with the same force. The amount of "pull" an atom exerts on its electrons is called its electronegativity. Atoms with high electronegativities, such as fluorine, oxygen, and nitrogen, exert a greater pull on electrons than atoms with lower electronegativities. In a bond, this leads to unequal sharing of electrons between the atoms, as electrons will be drawn closer to the atom with the higher electronegativity. Because electrons have a negative charge, the unequal sharing of electrons within a bond leads to the formation of an electric dipole: a separation of positive and negative electric charge.

Bonds can fall between one of two extremes of being completely nonpolar or completely polar. A completely nonpolar bond occurs when the electronegativities are identical and therefore the bond's net polarity is zero. A completely polar bond is more correctly called an ionic bond and occurs when the difference between electronegativities is large enough that one atom actually takes an electron from the other.

A molecule may be polar either as a result of polar bonds due to differences in electronegativity, as described above, or as a result of an asymmetric arrangement of nonpolar covalent bonds and non-bonding pairs of electrons known as a full molecular orbital. Water (H_2O) is an example of a polar molecule since it has a slight positive charge on one side and a slight negative charge on the other. The dipoles do not cancel out, resulting in a net dipole. Due to the polar nature of the water molecule itself, polar molecules are generally able to dissolve in water ("like dissolves like").

A molecule may be nonpolar either when there is an equal sharing of electrons between the two atoms of a diatomic molecule or because of the symmetrical arrangement of polar bonds in a more complex molecule. Examples of nonpolar compounds include fats, oil, gasoline, antioxidants, plasticizers, and other intentional plastic additives. Therefore, most nonpolar molecules are water-insoluble (hydrophobic) at room temperature. Many nonpolar organic solvents are able to dissolve nonpolar substances.

Whether or not a molecule is polar has profound effects on the physical properties of the substance, like solubility, boiling point, and melting point. Molecular polarity determines the strength and types of intermolecular forces of attraction at work in a sample of the substance. Molecular polarity affects solubility in that polar molecules are best solvated by polar solvent molecules and nonpolar molecules are best solvated by nonpolar solvent molecules; i.e. "like dissolves like."

Binary mixtures of a short-chained alcohol (such as ethanol and isopropanol) and water are commonly used as extraction simulants for aqueous drug products and alcohol/water mixtures are well established as stimulants for foods and pharmaceuticals in extraction studies. Considering pharmaceutical applications specifically, both ethanol/water and isopropanol/water mixtures have been used as simulating solvent systems [39–44]. In such cases, the proportion of the alcohol and water in the simulating solvent is adjusted so that the polarity (extracting power) of the stimulant matches the polarity (leaching power) of the drug product. Thus, the key issue in establishing and justifying the proper extraction solvent is determining the proportion of the alcohol that is present in the extraction solvent.

Examples of how the extracted levels of extractables of varying chemical nature are affected by the polarity and ethanol/water ratio of simulating solvents are contained in the case studies that appear at the end of this section.

Although there are several means of establishing (calculating) the polarity of solvents (for example, by considering their dielectric constant) and numerous polarity indices exist for such solvents, it is more difficult to directly measure or calculate the polarity of a drug product, particularly an aqueous one. Therefore, it is challenging to link a simulating solvent to an aqueous drug product on the basis of known or calculated polarities.

However, empirical means can be used to estimate a drug product's polarity via analytical studies. For example, a representative measure of polarity in relation to solubilization power may be obtained by the use of polarity-sensitive molecular probes. One such well-established solvatochromlc polarity probe is the so-called Reichardt's dye [44, 45]. The principle of the polarity measurement using the Reichardt dye is as follows. In the structure of Reichardt's dye, positive and negative charges are separated, and such neutral/dipolar molecules are called zwitterions. This is the ground state of the molecule. When

Figure 2.7 Solvent strength model. Levels of DEHP extracted from the PVC standard source material by various simulating solvents containing ethanol/water. If the equilibrium level of DEHP was measured in a drug product contacted by the PVC material, then this level could be used to establish the proper simulating solvent (in terms of ethanol content) for the drug product.

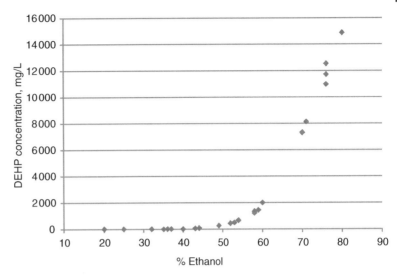

the zwitterionic dye absorbs light, an electron is transferred from the HOMO (Highest Occupied Molecular Orbital) to the LUMO (Lowest Unoccupied Molecular Orbital). HOMO and LUMO are over the phenoxide and pyridinium moieties respectively, so there is a charge delocalization induced by light. This electronic transfer (or charge transfer, CT) from the negative end of the molecule (from the oxygen atom) to the positive end (the nitrogen atom) is the basis of the polarity determination. In theory, polar solvents would stabilize the polar ground state relative to non-polar solvents while the opposite occurs for the excited nonpolar state. Consequently, less energy is required to excite the dye in nonpolar solvents, i.e. light of a higher wavelength (lower energy light), which belongs to the red part of the visible spectrum, is absorbed in nonpolar solvents whereas the opposite is true for polar solvents. Accordingly, by measuring the maximum absorption wavelength of the CT band, a polarity scale can be constructed.

Another example of an empirical means of linking simulating solvents and drug products by polarity is a solvent strength model that was developed by correlating the levels of a model compound, di-(2-ethylhexyl) phthalate (DEHP), extracted from a reference source material, plasticized poly-(vinyl chloride) (PVC) resin, and the proportion of ethanol in ethanol/water extractions solvents [46]. This model was established by experimentally investigating DEHP leaching (see Figure 2.7) and takes the form:

Ethanol in the simulating solvent

$$= 28.84 \times (\text{concentration of DEHP})^{0.10} \qquad (2.8)$$

If the level of DEHP extracted from the standard source PVC resin by a drug product is measured, then the level can be input into the above equation and the proper ethanol

content of the appropriate simulating solvent can be determined.

The levels of DEHP extracted from the reference source material by various formulations and drug products were measured and input into the solvent strength model and the ethanol/water proportions in the corresponding simulating solvents were calculated. These results are tabulated in Table 2.10.

Additionally, other studies reported in the literature are relevant to the concept of establishing the proper ethanol/water proportion to mimic a specific drug product or drug product formulation component. For example, Pearson and Trissel studied the leaching of DEHP from PVC bags into selected drugs and formulation components [47]. These findings established that glycols (such as polyethylene glycol and propylene glycol) are weak solubilizing agents and can be simulated by ethanol/water mixtures containing 25% ethanol or less (see Table 2.11).

Cremophor EL (polyoxyethylated castor oil) is a commonly used solubilizing agent for chemotherapeutic drug products. Cremophor's ability to extract DEHP from PVC bags has been investigated in the context of establishing the compatibility of reconstituted chemotherapeutics in diluents stored in PVC bags. For example, Faouzi et al. studied the leaching of DEHP from PVC bags by diluted drug formulations containing approximately 1–2% Cremophor (after dilution) [48, 49]. The levels of DEHP leached after 24 hours of room temperature extraction were 60–200 mg/l. Application of the simulating solvent strength model to these concentrations would establish 50/50 ethanol/water as an appropriate simulant for drug formulations containing 1–2% Cremophor EL.

There are certain confounding factors, both thermodynamic and kinetic, that affect the ability of an alcohol/water mixture to simulate the "leaching power" of receptor

Table 2.10 Forecasted ethanol/water simulants and polarities of drug products.

Drug matrix	Proportion	Measured DEHP level (mg/l)	Ethanol/water simulant	Polarity
Lipid emulsion	5%	1090	58/42	21.8
	10%	2493	63/37	21.2
	20%	2697	64/36	21.1
Albumin[a]	5%	16.3	38/62	23.8
	20%	35.3	41/59	23.6
	25%	9.6	36/64	24.0
Polysorbate 80	0.2%	9.6	36/64	24.0
	1%	210	49/51	22.8
	5%	1910	61/39	21.4
Blood and blood fractions	N/A	100	46/54	23.2

a) The mean ethanol/water stimulant for albumin in the range of 5–25% is 39/61 and the mean polarity is 23.8.
Source: From reference [46] with permission.

Table 2.11 Leaching of DEHP from PVC bags by propylene glycol.

Formulation components	DEHP concentration (mg/l) after 24 hours at ambient temperature
25% ethanol	<0.12
Polyethylene glycol 25%	<0.12
Propylene glycol 25%	<0.12

Source: Information from reference [47].

phases. Considering thermodynamics, the fundamental assumption in a solvent selection and justification strategy is that solvent and substance polarity is the predominant factor is establishing the "leaching power" of a drug product and a simulating solvent. There are three circumstances where this may not be the case:

1. Selective interactions,
2. Reactive interactions, and
3. Non-partitioning interactions.

Considering the first of these, it is noted that a polarity-based selection and justification of a simulating solvent is based on the assumption that the plastic test article, the drug product, and the simulating solvent all have no unique affinity for extractables or leachables and thus that the distribution of the extractables between the plastic and the drug product (and the plastic and the simulating solvent) is determined solely by their relative polarities. If this assumption is incorrect and the extractable has an affinity for any of these phases, then the extractable's leaching properties in the drug product could be different from the extractable's leaching properties in the simulating solvent. Consider, for instance, the case of a biopharmaceutical drug product whose active ingredient is a protein that exhibits selective solute binding properties. Extractables whose chemical nature is such that they can be bound by the protein could be leached into the drug product at higher levels than might be anticipated based on polarity alone. A model solvent chosen based on polarity would thus underestimate the "leaching power" of the drug product.

Considering the second case, it is further noted that a polarity-based selection and justification of a simulating solvent is based on the assumption that the extractable is unaltered or non-reactive in either the drug product or the simulating solvent. If in fact the extractable is transformed in either the drug product or the simulating solvent but not equally in both, the accumulation of that extractable will be different, drug product versus simulating solvent. For example, it has been reported that under more aggressive extraction conditions (higher temperatures, longer durations) ethyl esters of acidic extractables are observed in ethanol/water extracts but not in the drug product. Clearly such ethyl esters are reaction products arising from a reaction between the extractable (acid) and the extraction solvent (ethanol). Due to the formation of the ethyl esters, the extractables profile could be different from the leachables profile as (i) the ethyl esters will not be present in the drug product as leachables and (ii) the total level of extracted acid will be greater, simulating solvent versus drug product.

Considering the last case, there are certain situations where leaching is not accomplished by polarity-driven processes. For example, at high pH the stearate ion is leached from plastics containing stearate salts at levels much higher than would be predicted based on the salt's solubility or partitioning because the stearate ion undergoes ion exchange with the excess hydroxide available in the contact solution. Thus, the level of leached stearate will be greater in a drug product at high pH (pH 10 or higher) than it would be in a simulating solvent chosen and justified based strictly on polarity alone.

Although exceptions such as those listed above can be envisioned, it should be noted that the exceptions, although theoretically possible, are not routinely encountered in pharmaceutical applications.

It is important to note that the developed model is a thermodynamic model in that it is valid when equilibrium is established between the plastic material and the contact solution. Such a situation may not be achieved in certain pharmaceutical applications. For example, consider the case of a multi-layered film that is used as the container body of a packaging system for a liquid dosage form. If the extractable of interest is present in a non-solution contact layer of the film, its accumulation in solution may be limited by its rate of diffusion through the film. This is important as the nature of the simulating solvent (ethanol) and the dosage form that it is simulating may affect the diffusion rate. For example, when one contrasts an ethanol/water mixture with a lipid-containing drug product, one notes that they are fundamentally different in terms of their ability to penetrate the film. Specifically, the diffusion rate in the ethanol/water solvent system will be faster than it is for the lipid-containing product because ethanol is a penetrating solvent while the lipid is not.

In such a circumstance (penetrating simulant, non-penetrating drug product), the ethanol/water simulant will overestimate an extractable's accumulation level in the drug product until the contact duration is long enough for equilibrium to be achieved, at which point the accumulation levels in the model simulant and in the drug product will be the same.

Considering kinetics effects further, a simulant can also affect the migration of substances through a plastic by changing the structure of the plastic. For example, swelling of a polymer by an extraction solvent will surely increase the speed with which compounds will migrate (diffuse) through the polymer and thus will increase the speed with which an equilibrium between the solvent and the plastic is attained. Thus, an extractables profile obtained with a swelling solvent will exceed the leachables profile of a non-swelling drug product until the contact time is sufficiently long that equilibrium is achieved.

There is an ongoing discussion that considers the question "what is the best organic solvent to use, in combination with water, as a simulating solvent?" It is possible to make a case, based on certain extracted materials, certain extractables, and certain extraction conditions that "isopropanol (IPA) is a better solvent to use than ethanol because it extracts greater quantities of extractables." The use of IPA/water mixtures is strongly advocated by researchers at the National Institute for Bioprocessing

Research and Training (NIBRT) [50]. In other circumstances, the opposite conclusion can be drawn. Practical arguments such as "ethanol is a more reactive solvent" or "ethanol penetrates to a greater extent than isopropanol" or "ethanol provides a link to previous research which was performed with predominately ethanol-based solvents" or "ethanol-based solvents are specifically recommended in related global standards" can be made to support each case. Other authors have even suggested the use of mixed solvents that use organic phases that are not alcohols (such as acetonitrile), citing practical considerations such as ease of analysis.

In any event, it is not possible, in a general sense, to justify one organic solvent as being more universally acceptable than another as the organic component of a simulating solvent. The truth of the matter is that any number of solvent/water mixtures can be justified as appropriate simulating solvents for pharmaceutical drug products based solely on polarity considerations. If a particular organization has the valid technical justification for supporting, for example, a 40/60 mixture of IPA/water and another organization has a valid technical justification for supporting a 50/50 mixture of ethanol/water, it is possible that both organizations are right and that neither organization is wrong.

Of course, it should be obvious that the potential organic solvents are not equivalent in their "leaching power" as they do not have equivalent polarities. For example, the Hildebrand solubility parameters for the various solvents are (in units if $MPa^{1/2}$):

- Water = 48.0 (highly polar)
- Ethanol = 26.5
- Isopropanol = 23.5
- n-Hexane = 14.9 (highly nonpolar)

Given that ethanol is slightly more polar than IPA, simulating solvents having the same polarity would have a lower proportion of IPA than of ethanol.

In conclusion, the levels to which extractables will be leached into extraction solvents and leachables will be leached into drug products is affected by the polarity of the extraction solvent or drug product. The more non-polar the extraction solvent or drug product, the higher will be the leached level of a non-polar extractable. The opposite is true for a polar extractable.

Various means, including the solvent strength model discussed previously, are appropriate and effective ways of establishing and justifying an alcohol/water simulating solvent for drug products. For some classes of drug products, simulating solvents have been established and justified and should be used in controlled extraction simulation studies.

Care must be exercised in selecting simulating solvents where product use conditions are such that equilibrium might not be established between the drug product (or its simulating solvent) and the contact plastic, as polarity-matching the simulating solvent and the drug product might not account for properties of either phase that alter the speed with which leaching occurs. The composition of a simulating solvent should therefore be adjusted to account for kinetic effects.

The choice of the specific organic solvent that is used in solvent/water mixtures is driven more by practical considerations and there is no organic solvent that is universally preferred. In any case, solvent selection must be justified both scientifically and practically.

Several case studies, illustrating the effect of solvent polarity (ethanol/water proportion) of the levels of extractables are discussed as follows.

Leaching of Organic Acids from Gamma-irradiated EVA Film [51]:

In this study, the levels of organic acids extracted from an ethylene vinyl acetate (EVA) film were measured as a function of the polarity of several aqueous extracting solutions. The behavior of myristic acid, shown in Table 2.12, was representative of the behavior of poorly soluble fatty acid extractables and clearly illustrates the increase in solubility as a function of increasing polarity (ethanol proportion).

Leaching of Extractables from a Polyolefin Film [52]:

In this study, a polyolefin material was loaded to contain known quantities of chemically diverse model extractables. Portions of the loaded material were extracted with solvents of varying compositions and the levels of extracted substances were measured. Aqueous solutions at various pH values (as well as unbuffered water) were among the various solvents used in this study. The model extractables included both neutral and acidic compounds. The results of this study are summarized in Table 2.13, which clearly illustrates the increase in solubility as a function of increasing polarity (ethanol proportion). Additionally, the effect is more pronounced the more nonpolar the substance is (higher log $P_{o/w}$).

Leaching of Extractables from a Polyolefin film [41]:

In this study, a polyolefin material was extracted with solvents of varying compositions and the levels of extracted substances were measured.

The results of this study are summarized in Table 2.14, which clearly illustrates the increase in solubility as a function of increasing polarity (ethanol proportion). Trends as a function of the polarity of the individual extractables are more complicated to interpret as the total pools of each of the substances are different. However, in general the effect is more pronounced for the more nonpolar, higher-log $P_{o/w}$ substances.

Leaching of Extractables from Plastic Materials by Isopropanol/Water Mixtures [31]:

In this study, three pharmaceutical systems were leached and the levels of a specific extractable was determined as a function of the proportion of isopropanol (IPA)/water as the extracting solvent. Systems and extractables examined included:

- PVC i.v. bag, di-(2-ethylhexyl) phthalate (DEHP)
- Single-use bioprocessing bag, butylated hydroxytoluene (BHT)
- Polypropylene bottle, Irganox 1010, Irgafos 168

Table 2.12 Leaching of fatty acids from a gamma-irradiated EVA material.

Ethanol/water ratio	Polarity	Extracted fatty acid concentration, μg/l (ppb)		
		Myristic acid	Palmitic acid	Stearic acid
0 (water)	25.52	<10	<10	<10
1/3	22.43	100	10	10
1/1	189.34	400	700	300

Table 2.13 Leaching of organic substances from a polyolefin material.

Compound	Log $P_{o/w}$	Extracted concentration, mg/l			
		Water, polarity = 25.5	20% Ethanol, polarity = 24.7	40% Ethanol, polarity = 23.4	60% Ethanol, polarity = 21.7
Benzyl alcohol	1.05	0.28	0.31	0.31	0.32
Dimethyl phthalate	1.56	1.3	1.5	1.5	1.6
Diethyl phthalate	2.47	1.5	3.0	4.1	4.9
Dipropyl phthalate	3.66	0.35	1.6	5.7	8.3
Dibutyl phthalate	4.72	0.05	0.33	3.9	10

Table 2.14 Leaching of organic substances from a polyolefin material.

Compound	Log $P_{o/w}$	Extracted concentration, µg/l			
		Water, polarity, 25.5	20% Ethanol, polarity, 24.7	40% Ethanol, polarity, 23.4	60% Ethanol, polarity, 21.7
Caprolactam	−0.32	<10	<10	<10	12
1,4-Dioxacyclotetra-decane-5,14-dione	1.20	<10	<10	22	60
3,3-Dimethyl-1,5-dioxa-cycloundecane-6,11-dione	2.62	<10	27	79	129
Irganox degradant	4.48	<10	47	89	145
Stearic acid	8.22	<10	24	58	152

Instead of reporting the results in terms of the concentration of the extractable in the leaching solution, the author reported a quantity referred to as the equilibrium binding constant, E_b, which is equivalent mathematically to $K_{P/L}$. Without going into the details of the mathematical derivation of E_b, it is sufficient to note that E_b and the leached concentration are inversely proportional; that is, the smaller the E_b, the larger the leached concentration.

E_b versus the proportion of IPA/water data are summarized in Table 2.15, which clearly illustrates the increase in the extractables level (solubility) as a function of increasing polarity (increased proportion of IPA).

2.2.5.3 Extraction Solvents, pH

Another critical property of an aqueous extraction medium or drug product is its pH, especially for aqueous drug products that are generally formulated with a pH that can range from slightly acidic (pH 3 or lower) to slightly basic (pH as high as 10). This is also the case for manufacturing process streams, although they can exhibit an even wider range of pH values. The pH of the extraction solvent and the drug product is a critical experimental design parameter for an extraction study, as the solubility and the accumulation level of an acidic or basic leachable (or extractable) will be pH-dependent. This phenomenon occurs since the solubility of the dissociated form of an acid

or base is greater than the solubility of the undissociated acid or base itself; thus, the solubility of an acidic or basic substance is pH-dependent. The effect of drug product or extraction solvent pH on the accumulation of leachables and extractables is well-documented [6, 51–57].

The equilibrium distribution of a leachable (or extractable) between its source article and the drug product (or extraction medium) is reflected in a source/solution partition coefficient. To a first approximation, such a partition coefficient is related to the solubility of a leachable in the drug product or of an extractable in the extraction medium. Because solubility data are more readily available than source/solution partition coefficients, the following discussion focuses on solubility.

Considering the effect of pH on the accumulation of acidic and basic extractables further, the relationship between the solution pH, the extractable's acid dissociation constant (pK_a), its effective solubility (S_e) at a particular pH, and its intrinsic solubility (S_0) are expressed as follows:

$$\text{For an acid: } \log S_e = \log S_0 + \log \left[1 + 10^{(pH - pK_a)} \right] \tag{2.9}$$

$$\text{For a base: } \log S_e = \log S_0 + \log \left[1 + 10^{(pK_a - pH)} \right] \tag{2.10}$$

Table 2.15 Leached amounts of extractables as a function of the IPA/water proportion.

	E_b, equilibrium constant[a]			
	PVC bag	Single-use biocontainer	Polypropylene bottle	
% IPA	DEHP	BHT	Irgafos 168	Irganox 1010
0	ND[b]	ND	ND	ND
10%	ND	ND	ND	ND
25%	4.82	ND	ND	ND
40%	2.03	2.31	3.28	2.30
55%	0.47	1.46	1.21	1.31
70%	0.17	1.15	0.90	0.55

a) E_b is inversely proportional to the extracted concentration. The smaller the E_b, the larger the concentration.
b) ND = Not detected. The compound was not present in this extract at detectable levels.

This relationship between solubility and pH is illustrated in Figure 2.8 for an acid and base that have a pK_a of 5.0 and an arbitrary S_0 of 100. In essence, Figure 2.8 suggests that as the pH changes, the solubility will change without limit. In practical reality, the dissociated ion itself will not be infinitely soluble and thus the solubility versus pH plot eventually reaches a plateau whose value reflects the intrinsic solubility of the dissociated ion. Additionally, if the total pool of the extractable is limited, the calculated solubility might exceed the total pool. In this circumstance, further increasing the pH of the extracting medium will not produce a further increase in the concentration of the acidic extractable in the extract because its pool has been depleted.

Clearly, pH is an important property of an extracting medium that will affect the extractables profile obtained, both qualitatively (what extractables are present in the profile) and quantitatively (at what level is the extractable present in the extract). Thus, when performing a controlled extraction study to forecast the leachables in a single drug product, the pH of the extraction solvent must match the pH of the drug product. However, in many cases the purpose of the controlled extraction study is to forecast the behavior of a number of drug products in a given packaging system or to forecast the behavior of a material in numerous applications. In this case, the controlled extraction study must include multiple extracting solvents whose pH values span the application range. Then again, when the application range in pH is wide, or when it includes extreme pH values, there are several practical issues that must be addressed to ensure that the proper number and type of extraction solvents are used. Specifically, two questions are relevant:

1. What is the proper range in pH values that provides the most complete extractables profile that is applicable to the most common marketed aqueous drug products?
2. Is there any value in including extraction solvents with intermediate (e.g. neutral) pH values?

Knowing the acid/base properties of commonly encountered extractables is essential in assessing the impact of extraction solvent pH on a qualitative or quantitative extractables profile. Thus, Tables 2.16 and 2.17 were compiled to include acidic or basic extractables that have been reported in the literature. Considering acidic extractables (Table 2.16), the more commonly encountered acidic organic extractables include weak organic acids such as acetic acid, formic acid, benzoic acid, and fatty acids of variable molecular weight. In general, these acidic extractables possess acid dissociation constants (pK_a) in the relatively narrow range of 3.2 to 5.5. A somewhat shorter list of basic organic extractables is compiled in Table 2.17. Unlike the organic acid extractables, the basic organic extractables exhibit a fairly wide range in pK_a values.

Rather than considering the behavior of each individual extractable listed in Table 2.16, the effect of the extraction solvent pH on the accumulation of acidic or basic extractables can be established by focusing on representative acidic and basic extractables that meet two requirements:

1. They bracket the behavior of the two groups of extractables listed in Tables 2.16 and 2.17 and
2. They have published solubility data over a wide range in pH.

Figure 2.9 plots the mass solubility for the representative extractables as a function of solvent pH (listed in Table 2.18). The mass solubility axis has been standardized by expressing the solubility in proportion to the maximum solubility reported for the individual extractable. This allows the data for the individual extractables, which exhibit a wide range of absolute solubilities, to be presented in a single plot.

The following observations about Figure 2.8 are generally pertinent:

1. As the solubilities of the representative extractables are strongly impacted by the pH of the extraction solvent, the extractables profile at two individual pH values could be different, both in terms of the specific extractables that are detected and their measured concentrations.

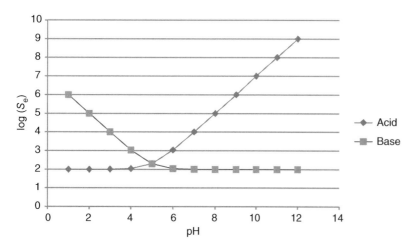

Figure 2.8 The effect of pH on the conditional solubility (S_e) of an acidic or basic extractable. The figure considers the case of an acidic or basic extractable with a pK_a of 5.0 and an intrinsic solubility (S_0) of 100 (in arbitrary units). As the pH of the extracting medium increases, the solubility of the acidic extractable increases. Similarly, as the pH of the extracting medium decreases, the solubility of a basic extractable increases. The concentration of the extractable in an extract will increase in response to pH up to the point where (a) the total pool of the extractable is exhausted or (b) the chemistry of the extract is such that the solubility of the dissociated form of the extractable is exceeded.

Table 2.16 Acidic extractables and their associated acid dissociation constants (pK_a).

Extractable	CAS RN	pK_a	Extractable	CAS RN	pK_a
Oxalic acid	144-62-7	1.23	Ricinoleic acid	141-62-8	4.74
2-Furancarboxylic acid	88-14-2	3.16	Palmitic acid	57-10-3	4.75
Mono-(2-ethylhexyl)phthalate (MEHP)	4376-20-9	3.37	Acetic acid (AA)	64-19-7	4.76
Terephthalic acid	100-21-0	3.51	Linoleic aid	60-33-3	4.77
m-Phthalic acid	88-99-3	3.54	9-Oxononanoic acid	2553-17-5	4.77
Formic acid	64-18-6	3.75	9-Oxononanoic acid	2553-17-5	4.77
3-(Hydroxyphenyl)lactic acid	23508-35-2	3.80	9.10-Epoxystaric acid	2443-39-2	4.78
Hydroxy acetic acid	79-14-1	3.83	Hydroxystearic acid	106-14-9	4.78
Dihydroxymyristic acid	103963-71-9	3.85	Pentadecanoic acid	1002-84-2	4.78
2-Hydroxyheptanoic acid	636-69-1	3.86	9-Hexadecenoic acid	2091-29-4	4.78
2-Hydroxypropanoic acid	79-33-4	3.91	Eicosanoic acid	506-30-9	4.78
4-Chlorobenzoic acid	74-11-3	3.98	Erythro-9,10-dihydroxystearic acid	3639-32-5	4.78
Succinic acid	110-15-6	4.16	Undecanoic acid	112-37-8	4.79
Benzoic acid	65-85-0	4.19	Cyclohexaneacetic acid	5292-21-7	4.80
Acrylic acid	79-10-7	4.25	2-Hexenoic acid	1191-04-4	4.80
Pentanedioic acid	110-94-1	4.31	3-Ethylheptanoic acid	14272-47-0	4.80
Adipic acid	124-04-9	4.43	Butyric acid	107-92-6	4.83
1,8-Octadecanoic acid	505-48-6	4.46	Pentanoic acid	109-52-4	4.84
Sebacic acid	111-20-6	4.48	Hexanoic acid	142-07-7	4.85
4-Ethoxybenzoic acid	619-86-3	4.49	Propionic acid	79-09-4	4.86
Azelaic acid	123-99-9	4.55	Heptanoic acid	111-14-8	4.89
2-Propylvaleric acid	99-66-1	4.60	Octanoic acid	124-07-2	4.89
4-Hydroxyhexanoic acid	13532-38-2	4.62	Myristic acid	544-63-8	4.90
Abietic acid	514-10-3	4.64	Nonanoic acid	112-05-0	4.96
Dehydroabietic acid	1740-19-8	4.66	Oleic acid	112-80-1	5.02
Pimaric acid	79-54-9	4.68	2,2-Dimethylpropanoic acid	75-98-9	5.03
Caproic acid	149-57-5	4.72	Dodecanoic acid	143-07-7	5.30
6-Chlorohexanoic acid	4224-62-8	4.73	Stearic acid (SA)	57-11-4	5.50

For multi-protic acids, the pK_a listed is for the most acidic proton. The highlighted entries are the representative extractables chosen for inclusion in Figure 2.9.

2. The pH range of approximately 2–10 defines a "zone of divergence," which is the pH range in which the solubilities of the representative extractables are changing and within which the maximum solubility for the extractables is achieved. Use of extraction solvents at the zone's pH extremes (pH 2 and pH 10) would produce extractables profiles that include the greatest number of acidic or basic extractables at their highest concentrations. Furthermore, aqueous parenteral drug products are generally formulated with a pH value in this range. Thus, if the purpose of the controlled extraction study was to establish a material's "worst case" extractables profile (all possible extractables at their highest levels), this purpose would be achieved by performing the extractions at both pH 2 and 10. Generating extracts at pH values outside the zone of divergence would not materially change the extractables profiles versus what they were at the endpoints of the zone. Nevertheless, if the purpose of the extraction study is to cover a set of drug products whose pH range is smaller than the zone of divergence, then the proper pH range to use is the product range (consistent with point 1).

3. None of the targeted extractables reach and then retreat from a solubility maximum in the pH range of roughly 6–8. Thus, if the purpose of a controlled extraction study was to establish a material's "worst case" extractables profile (all possible extractables at their highest levels) and the study included extraction solvents whose pH values were beyond this range, then performing an extraction with a solvent whose pH is within this range would add no value in terms of revealing "new" extractables or producing higher concentrations for the already-measured extractables.

Table 2.17 Basic extractables and their associated acid dissociation constants (pK_a).

Extractable	CAS RN	pK_a	Extractable	CAS RN	pK_a
Dibutylamine	111-92-2	11.25	Dibenzylamine (DBA)	103-49-1	8.76
Piperidine	110-89-4	11.1	Drometrizole	2440-22-4	8.15
Diethylamine	109-89-7	10.76	Triisopropanolamine	122-20-3	8.06
Dimethylamine	124-40-3	10.73	4,4'-Methylenedianiline	101-77-9	5.32
N,n-Dimethylcyclohexylamine	98-94-2	10.72	Pyridine	110-86-1	5.23
Stearylamine	124-30-1	10.7	2,4-Toluenediamine (TDA)	95-80-7	5.02
N-Ethylamine	75-04-7	10.64	Melamine	108-78-1	5.00
Cyclohexylamine	108-91-8	10.57	2,4-Dimethylaniline	95-68-1	4.88
2(3H)Benzothiazole	934-34-9	10.4	2,6-Toluenediamine	823-40-5	4.74
N-Ethyl-1-phenyl-cyclohexamine	2201-15-2	10.17	Aniline	62-53-3	4.63
Diphenylguanidine	102-06-7	10.12	2-Benzothiazolamine (BTA)	136-95-8	3.94
2-Mercaptobenzothiazole	149-30-4	9.80	1,1'-Carbonothiobispiperdine	1013-92-9	1.80
Benzylamine	100-46-9	9.33	Benzothiazole	95-16-9	0.85
N-Benzyl-N,N'-Dimethylamine	103-83-3	8.80	Diphenylamine	122-39-4	0.78

The highlighted entries are the representative extractables chosen for inclusion in Figure 2.9.

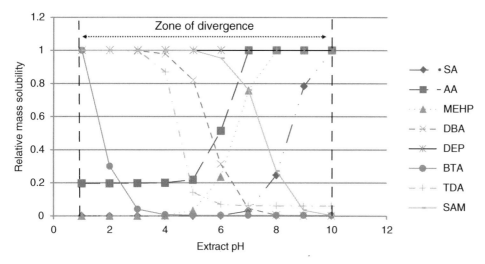

Figure 2.9 The Effect of pH on the reported solubility of extractables. Because DEP is non-ionic, its solubility is unaffected by the pH of the extract. The solubility of the acidic extractables (AA, SA, and MEHP) increases with increasing pH, depending on their specific pK_a values. The solubility of the basic extractables (SAM, DBA, TDA, BTA) increases with decreasing pH, again consistent with their pK_a values. A Zone of Divergence exists between the pH values where the weakest acid (SA) and the weakest base (BTA) achieve their maximum solubilities. If a set of extraction solvents is designed to capture essentially all possible acidic or basic extractables at their likely highest concentration in the extract, then the set must have a pH that spans the values that define the Zone of Divergence. Source: From reference [58] with permission. (*See insert for colour representation of the figure.*)

4. Generally, there is no single pH at which all the potential extractables would be readily measurable. At pH values less than 6, the levels of stearic acid (and other acidic extractables with similar pK_a values) would likely be sufficiently low that the stearic acid would be present in the extract at trace levels. Similarly, at a pH greater than 6, the levels of 2,4-toluenediamine and 2-benzothazolamine would likely be sufficiently low that they would be present in the extract at trace levels. Thus, a controlled extraction study designed to qualitatively establish all of a material's potential extractables would need to include multiple extraction solvents at different pH values. In a specific extraction study, it may be the case that the members of an extractables profile are such that the profile does not consist of chemically diverse compounds and therefore that a single extraction pH is adequate for establishing the qualitative profile.

5. The accumulation of a non-ionic extractable, represented by diethyl phthalate (DEP) in Figure 2.9, is unaffected by pH and thus an extraction solvent at any pH is appropriate for qualitatively and quantitatively delineating the non-ionic members of an extractables profile.

Table 2.18 Extractables included in Figure 2.9, the effect of pH on solubility.

Extractable	CAS RN	Formula	Structure	pK_a
			Neutral extractable	
Diethyl phthalate (DEP)	84-66-2	$C_{12}H_{14}O_4$		—
			Acidic extractables	
Acetic acid (AA)	64-19-7	$C_2H_4O_2$		4.76
Stearic acid (SA)	57-11-4	$C_{18}H_{36}O_2$		5.50
Mono-(2-ethylhexyl) phthalate (MEHP)	4376-20-9	$C_{16}H_{22}O_4$		3.37
			Basic extractables	
Stearylamine (SAM)	124-30-1	$C_{18}H_{39}N$		10.67
Dibenzylamine (DBA)	103-49-1	$C_{14}H_{15}N$		8.76
2,4-Toluenediamine (TDA)	95-80-7	$C_7H_{10}N_2$		5.02
2-Benzothiazolamine (BTA)	136-95-8	$C_7H_6N_2S$		3.94

While these conclusions may be valid for many pharmaceutical drug products, they are especially meaningful to those aqueous drug products whose pH falls between 3 and 10.

There are certain confounding factors associated with establishing the effect of pH on leachables accumulation. For example, the previous discussion focuses on the thermodynamic effect of pH on a compound's solubility. However, in certain cases an extractables or leachables profile can be the result of factors other than thermodynamic solubility. For example, the extractables profile can be affected by the acid- or base-mediated reaction of the extractables themselves or of their source materials. For instance, the leaching of extractables known as "cyclic esters" is pH-dependent as the cyclic esters are prone to acid- and base-hydrolysis [59]. Similarly, it is well-documented that leaching of Bisphenol A from polycarbonate is pH-dependent, even though Bisphenol A is neither an acid nor a base. Rather, the polycarbonate itself is degraded as the pH increases; thus, as solution pH increases, more polymer is degraded and more of its monomer (Bisphenol A) is leached. A similar effect is noted for mono-(2-ethylhexyl) phthalate, whose leached levels from di-(2-ethylhexyl) phthalate (DEHP)-plasticized PVC materials increases with increasing pH due to the base-mediated decomposition of DEHP. Lastly, the leaching of stearic acid at high pH (pH values greater than 10) is increased well beyond its intrinsic solubility as ion exchange occurs between the hydroxide ion in the leaching solution and the stearate ion in the leached sample [57].

Additionally, it is possible that the kinetics of extraction and leaching could be changed at high or low pH due to a pH-mediated change in the physical characteristics of the material or system being tested.

If the primary process involved in producing extractables or leachables is extraction or leaching itself, then the generalizations noted previously are relevant and applicable. If the process of producing extractables or leachables includes pH-dependent reactions and diffusion kinetics, then the generalizations noted previously are less relevant, depending on the relative importance and pH-dependence of the individual processes (solubility/partitioning, reactivity, or diffusion).

One of the most commonly encountered confounding factors associated with the extraction solution pH is the decision whether to use a buffered extraction solvent or not. The major advantage of using a buffered extracting solvent is clear, the pH of the extracting solvent is unchanged as a result of the extraction, and thus the "leaching strength" of the extraction solvent is preserved throughout the extraction process, thereby maximizing the amount of substance that is extracted. The use of a buffered extraction solvent is generally proper when the drug product being simulated is itself buffered. The major disadvantage of using a

buffered extraction solvent is that it does not mirror the effect that leaching of acidic or basic extractables will have on the extraction solvent. For example, when an acidic extractable is extracted, the pH of an unbuffered extracting solution will go down. Generally, this reduction in extraction solvent pH will reduce any further leaching of acidic extractables. Additionally, such a pH decrease is a useful indicator that the extractables profile will contain one or more acidic extractables and thus could enable complete extractable screening. The use of an unbuffered extraction solvent is generally proper when the drug product being simulated is not buffered (e.g. 0.9% sodium chloride, water for injection, 5% dextrose) or when delta pH is used as a diagnostic tool to enable complete extractables profiling.

Several case studies, illustrating the effect of pH on the levels of extractables, are discussed as follows.

Leaching of Organic Acids from Gamma-irradiated EVA Film [51]

In this study, the levels of organic acids extracted from an EVA film were measured as a function of the pH of several aqueous extracting solutions. The behavior of myristic acid, shown in Table 2.19, was representative of the behavior of poorly soluble fatty acid extractables and clearly illustrates the increase in solubility as a function of increasing pH.

Leaching of Extractables from a Polyolefin film [52]:

In this study, a polyolefin material was experimentally loaded to contain known quantities of chemically diverse model extractables. Portions of the loaded material were extracted with solvents of varying compositions and the levels of extracted substances were measured. Aqueous solutions at various pH values (as well as unbuffered water) were among the various solvents used in this study. The model extractables included both neutral and acidic compounds.

Table 2.20 summarizes the measured levels of several of the targeted extractables as a function of solution pH. This data corroborates the observations made previously, specifically:

1. The accumulation of the neutral extractables is generally unaffected by solution pH,
2. The accumulation of the acidic extractables is noticeably affected by solution pH,

Table 2.19 Leaching of myristic acid ($pK_a \sim 6$) from a Gamma-irradiated EVA material.

pH of extracting solution	Myristic acid concentration, µg/l (ppb)
3	<10
5.5	10
8	50
9.5	300
11	300

Table 2.20 Accumulation levels of targeted extractables as a function of solvent pH.

Extractable	Accumulation level, mg/l				
	pH 3	Water[a)]	pH 4	pH 5	pH 7
4-Methylbenzoic acid[b)]	1.3	2.0	2.0	2.4	2.6
2-Ethylbenzoic acid[b)]	1.4	2.4	2.4	3.5	3.8
Benzyl alcohol[c)]	0.29	0.28	0.30	0.30	0.29
Diethyl phthalate[c)]	1.3	1.5	1.5	1.4	1.3
Dipropyl phthalate[c)]	0.30	0.35	0.36	0.33	0.29

a) The pH of this unbuffered extraction solution was 3.9 after the extraction was performed.
b) Acidic extractables with pK_a values of approximately 4.3.
c) Neutral extractables.

3. The accumulation levels in water are effectively bracketed by the accumulation levels in the extracting solvents with the extreme pH values.

Leaching of Organic Fatty Acids from Polymeric Materials [57]:

In this study, the levels of organic fatty acids extracted from various polymeric materials were measured as a function of the pH of several aqueous extracting solutions. The behavior of palmitic acid, shown in Table 2.21, was representative of the behavior of poorly soluble fatty acid extractables and clearly illustrates the increase in solubility as a function of increasing pH.

Various compendial, regulatory, and scientific publications address the topic of extraction and thus are logical sources of information with respect to setting the proper pH of an extraction solvent. Considering pharmaceutical container closure systems, although certain documents may discuss extraction solvent composition and pH in general terms, references such as the FDA Container Closure Guidance [1], the EMEA Guidelines for Immediate Packaging Materials [2], the PQRI Parenteral and Ophthalmic Drug Product (PODP) Recommendations [60] and the USP <1664> monograph on leachables [5] do not contain specific recommendations for extraction solvent pH. Although the PQRI PODP document does not specify a pH range for its extraction solvents, when the PQRI chemistry team performed a controlled extraction on several plastic materials, the extracting solvents used included a pH 2.5 salt solution and a pH 9.5 phosphate buffer. Additionally, the USP monograph on extractables (USP <1663>, Table 1) [4] suggests the use of solvents at pH 5.2 and 9.5 to simulate an aqueous drug product whose nominal pH is 7, recognizing the situation that the solubility of an acidic or basic extractable is affected most strongly by extraction solvent pH in the pH range that is ±2 pH units from the extractable pK_a.

Considering medical devices, while informative Annex C ISO10993-18 (Biological evaluation of medical devices – Part 18: Chemical characterization of materials) [11] discusses extraction solvent pH as an exaggerating factor for acidic/basic extractables and notes that using extraction solvents whose pH is above and below the product's pH exaggerates the product's profile of acidic and basic leachables, Part 18 does not provide any guidance in terms of the selection of proper pH values for extracting solvents.

Considering systems used to manufacture drug substances and/or drug products (and the polymeric materials and components from which they are constructed), standard extraction protocols, including the use of low and high pH extraction solvents, have been developed by a system user consortium (BPOG) [61] and the USP (<665>) [12]. Among the six extraction solvents that were originally included in BPOG's standard extraction protocol, one is

Table 2.21 Leaching of palmitic acid ($pK_a \sim 6$) from various plastic materials.

pH of extracting solution	Concentration in the extract, mg/l (ppm)			
	Polypropylene	Polyethylene	Plasticized PVC	Elastomer (µg/g)
<6.5	0.04	0.04	0.04	5
6.5	0.04	0.03	0.04	6
7.0	0.05	0.04	0.04	6
7.5	0.06	0.04	0.05	12
8.0	0.35	0.07	0.10	51
8.5	1.98	0.27	0.52	160
9.0	2.46	0.69	1.71	450
9.5	5.61	0.74	3.94	680
10.0	8.37	1.15	4.25	920
10.5	8.34	2.04	4.78	1550
11.0	7.75	2.17	9.41	1890

a high pH caustic solution (0.5 N NaOH, pH ~ 13.6) and another is a low pH acidic solution (0.1 M phosphoric acid, pH ~ 1.5). Among the three extraction solvents mentioned in the USP <665> Standard Extraction Protocol, one has a low pH (potassium chloride, hydrochloric acid mixture, pH 3.0) and another has a high pH (phosphate buffer at pH 10.0). The difference in the pH values targeted by BPOG and the USP reflect their different approaches to selecting a proper extraction solvent. While both the USP and BPOG defined extraction solvents to cover a range of solutions encountered in pharmaceutical manufacturing, their criterion for establishing the extraction solvent's pH was different. Whereas the USP established a pH range that was meant to encompass a majority of the process solutions encountered in manufacturing, BPOG established its range to reflect the worst-case (lowest and highest pH) process solutions encountered in typical manufacturing operations.

Based on scientific principles and available extraction study data, the differences between the USP and BPOG low- and high-pH extraction solvents were reconciled as follows:

- At low pH, the "leaching power" of the USP pH 3 and the BPOG phosphoric acid solvents are sufficiently similar that these two solvents are considered to be "interchangeable" by the USP. Thus, an extraction performed with the BPOG solvent and tested by the USP criteria would fulfill the USP <665> requirements at low pH.
- At high pH, the apparent "leaching power" of the BPOG caustic solvent (pH 13.6) can be greater than that of the USP phosphate buffer (pH 10) as the higher pH changes the nature of the leaching process. For example, as noted earlier, actual decomposition of the plastic (e.g. polycarbonate to Bisphenol A) is exacerbated by the higher pH of the BPOG solvent. In fact, it is noted that it is possible that the pH of the BPOG solvent is sufficiently high that it is actually out of the compatibility range for certain plastics (such as polycarbonate). Additionally, dissociation of fatty acids at lower pH (pH 10) is augmented by ion-exchange at higher pH (pH 13.6), increasing the "driving force" for the leaching of these substances.

To handle the lack of alignment at high pH, the USP noted that circumstances where the pH of a process solution was greater than 10 were sufficiently rare and sufficiently isolated that the higher pH situations should be dealt with as exceptions rather than as the routinely-applied standard. Nevertheless, when use of the higher pH extraction solvent is consistent with manufacturing operations, the USP allows its pH 10 extraction solvent to be replaced by a more appropriate solvent, which could be the BPOG caustic extraction solvent.

Considering the various questions posed earlier, the following answers and conclusions are appropriate:

1. The proper pH range for simulating extracting solvents used in controlled extraction studies designed to forecast drug product leachables spans the pH range of the relevant drug products. Nevertheless, relatively complete extractables profiles, reflecting all relevant extractables and establishing their highest possible levels as drug product leachables, are obtained at pH extremes of 2 (or 3 if more consistent with the drug products in question) and 10. A pH range smaller than pH 2 to pH 10 can be used if the population of drug products (or process solutions) has pH values that fall within that smaller range.

2. Use of an extraction solvent with a neutral pH between 6 and 8 adds little or no value in terms of establishing the worst-case extractables profile (greatest number of extractables at their highest levels). This is the case as it is rare that an extractable achieves a concentration in an extract in the pH region of 6 to 8 that is higher than the concentration achieved at lower and higher pH.

3. When choosing the proper pH for a simulating solvent for a single drug product (or process solution), the nominal product limits for pH should be considered. If the intent of the study is to accurately simulate the drug product, then one should consider using two extracting solutions, one each with a pH at either extreme of the product's pH limits. However, if the range in the product limits for pH is small (2 pH units or less) or if the range is in the pH region of 6 to 8, then a single extraction solvent at either the low or high pH extreme might be justifiable (as the effect of pH in this range is small). Such a justification would require some knowledge of the possible extractables, as the possibility of acidic extractables would dictate the use of a higher pH solvent while the possibility of basic extractables would dictate the use of a lower pH solvent.

4. Simulating the "leaching power" of drug products that span a specific pH range requires the use of extraction solvents whose pH values closely match the extreme pH values of the drug products. Expanding the pH range of the simulating solvents much beyond the pH range of the drug products produces an extractables profile that unnecessarily over-exaggerates the drug products' leachables profile. Thus, the use of extracting solvents with excessively broad pH ranges is not recommended.

5. Once the proper pH range has been established, and extracting solvents have been generated at the pH extremes, it is rarely the case where use of a third extraction solvent, at an intermediate pH value, provides either additional insight into the individual extractables that make up a test article's extractables profile (i.e. reveals new extractables) or higher concentration

estimates. Thus, use of a third extraction solvent at intermediate pH is not recommended.

6. If the objective of the extraction study is to simulate a drug product, the buffering of the extraction solvent should match the buffering of the drug product. If the drug product is buffered, the extraction solvent should be buffered as well. If the drug product is unbuffered, then the extraction solvent should be unbuffered.

7. Point 6 notwithstanding, if it is desirable that the pH of an extraction solvent remains constant throughout the extraction, then a properly buffered extraction solvent should be used. If, however, the delta pH of the extracting solvent is to be used to assess the presence of acidic or basic extractables, then an unbuffered extraction solvent should be used.

8. The discussion of the pH effects is relevant for ionizable extractables. The accumulation levels of neutral extractables in extracting solvents are largely unaffected by the extraction solvent pH.

9. Exceptions to point 5 can occur at extreme pH values, generally greater than 10 and less than 2, when the extreme pH causes either decomposition of the plastic or alteration of the extraction mechanism. Thus, extractions should not be performed with solvents whose pH is outside this range, unless there is sufficient justification to do so.

2.2.5.4 Temperature and Duration

Contact temperature and duration are frequently chosen in extractables and/or leachables studies to accelerate the interaction that occurs between a plastic material, component, or system (e.g. container closure, medical device, production equipment) and a solution, extraction solvent, drug product, body fluid, or tissue under typical conditions of use. For the acceleration to be valid, a means must be established to identify and justify the proper accelerated contact conditions based on sound scientific principles.

ASTM standard F1980-16 [62] is often used as a means of calculating and justifying accelerated contact conditions. The Scope of this standard is "to rapidly determine the effects, if any, due to the passage of time on the sterile integrity of the sterile barrier system … and the physical properties of their component packaging materials." In its Scope, the standard notes that it "does not purport to address all of the safety concerns, if any, associated with its use." Although the Scope of ASTM F1980-16 does not include packaging systems for drug products and does not include the acceleration of chemical processes per say, it has typically been used to establish and justify accelerated contact conditions for studies that involve the migration of substances (either extractables or leachables) from plastics and into a contact solution (such as extraction solvent or drug products).

The accelerated aging techniques noted in ASTM F1980-16 are based on the assumption that the chemical reactions being considered follow the Arrhenius reaction rate function, which essentially states that a 10 °C increase or decrease in temperature of a homogeneous process results in an approximately twofold change in the rate of a chemical reaction. The technique is based on an accelerated aging factor (AAF), which mathematically is expressed as

$$AAF = Q_{10}^{[(TAA-TREF)/10]} \qquad (2.11)$$

where

Q_{10} = aging factor, which has a conventionally accepted value of 2.0 for a first-order chemical reaction,

TAA = accelerated temperature of contact, and

TREF = reference temperature of typical use.

The AAF term is used to calculate the accelerated aging time (AAT) as expressed as

$$AAT = tref/AAF \qquad (2.12)$$

where tref is the reference time of typical use.

To illustrate the use of Equations (2.11) and (2.12), consider the circumstance where one seeks to determine how long a test article must be aged at 40 °C to accelerate a typical use contact of two years (taken as 730 days) at room temperature (taken as 20 °C). Using this information, Equations (2.11) and (2.12) become:

$$AFF = 2.0^{[(40-20)/10]} = 4.0$$
$$ATT = 730/4 = 183 \text{ days}$$

This calculation is the basis for the convention that storage at 40 °C for six months is an appropriate acceleration of a two-year ambient temperature shelf-life.

Extracting or leaching substances out of plastic entities is a diffusion-mediated process. The extraction or leaching of a material's constituent by a contacting solution requires the movement of the constituent through the material via diffusion. The diffusion of constituents through polymers has been extensively studied; conditions of contact that strongly influence the magnitude of migration (diffusion) include the contact time, the contact temperature (as temperature affects the diffusion coefficient), and certain dimensional aspects such as the length of the diffusion path (relatable to the thickness of the contacted material) and the contact surface area.

Considering the contact time, it is well-established that the magnitude of migration is proportional to the square root of migration time; see, for example, the following equation [59]:

$$M_t/M_\infty = 4(Dt/\pi\delta^2)^{1/2} \qquad (2.13)$$

where M_t is the amount leached into solution at time t, M_∞ is the equilibrium accumulation at infinite time, D is the diffusion coefficient, and δ is the polymer's thickness.

Considering temperature, the mathematical relationship between the diffusion coefficient and temperature is well-defined; however, obtaining relevant input values to quantify temperature effects, such as activation energies, may be challenging. Thus, a more qualitative relationship between the diffusion coefficient (*D*) and contact temperature may be adequate for the purpose of risk assessment and management. In such a circumstance, a qualitative generalization, known as the "factor 10 rule," reference [63] may be applicable. This factor 10 rule, which was developed for and is extensively used for food packaging, is based on the observation that activation energies for migrating substances in polymers relevant to packaging are typically in the range of 80 to 100 kJ/mol. In such a circumstance, the diffusion coefficient increases by roughly an order of magnitude for every 20 °C increase in contact temperature. Thus, for example, the migration rate at 40 °C is ten times faster than the migration rate at 20 °C:

$$D1 \text{ (at T1} = 20°C)/D2 \text{ (at T2} = 40°C) = 1/10 \quad (2.14)$$

Although Equation (2.14) can be interpreted in many ways, one interpretation is that the migration time needed at 40 °C to reach the same concentration as after 730 days at 20° is 10 times shorter, (730/10) or 73 days. This is significantly shorter than the 183 days established by ASTM F1980-16.

If one were to convert the concepts in the "factor ten rule" to the equation format used in ASTM F1980-16, then Equation (2.11) would become

$$AAF = Q_d^{[(TAA-TREF)/20]} \quad (2.15)$$

and Q_d would have a value of 10. In its practical application, the AAF determined in Equation (2.15) would be used in Equation (2.12) to calculate an AAT in the same manner as the AAF determined from Equation (2.11) was used for the same purpose.

A comparison of accelerated conditions calculated using either the ASTM F1980-16 or the "factor ten rule" is shown in Figure 2.10 and Table 2.22.

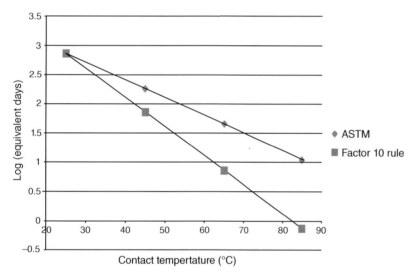

Figure 2.10 Acceleration of a two-year (730 days) ambient temperature shelf-life.

Table 2.22 Equivalent storage time for acceleration of a two-year (730-day) shelf-life at room temperature (25 °C).

Storage temperature (°C)	Equivalent storage time (days)		
	Per ASTM F1980-16	Per factor 10 rule	Ratio, ASTM vs factor 10
25 (based value)	730	730	1.0
40	258	130	2.0
45	183	73	2.5
50	129	41	3.1
55	91	23	4.0
65	45	7.3	6.2
75	22	2.3	9.6
85	11	0.73	15

Table 2.23 Extracted acetic acid levels from two multi-layered films.

Storage duration (days)	Data for film C1, buried EVA			Data for film C2, solution contact EVA		
	Concentration, mg/l		Ratio	Concentration, mg/l		Ratio
	45 °C	25 °C	45 °C/25 °C	45 °C	25 °C	45 °C/25 °C
1	2.2	0.7	3.14	26.1	21.7	1.20
7	3.3	1.4	2.36	23.1	21.2	1.09
14	3.9	1.7	2.29	23.9	21.2	1.13
30	5.6	2.1	2.67	26.1	21.9	1.19
60	7.5	2.4	3.13	29.1	22.2	1.31
90	9.1	2.5	3.64	33.7	24.1	1.40
120	10.4	2.8	3.71	38.8	25.2	1.54

Since it is clear that the two approaches produce different outcomes, there is some question as to which approach is appropriate in which circumstances. To address this point, Jenke used experimental data from 10 published migration studies to evaluate these two Arrhenius-based approaches to acceleration and made recommendations in terms of the proper approaches to be used for concentration and duration extrapolations [38]. Based on his analysis of the data, the Arrhenius model formulated by ASTM for the accelerated aging of medical devices, using a Q10 value of 1.5, most frequently provides the best fit to experimental leaching data when accumulation levels are projected from a clinical temperature to an elevated temperature. Alternatively, when contact durations are projected from a clinical to an elevated temperature, the Arrhenius model reflected in the empirically-derived "factor 10 rule," developed for and applied to food contact packaging, most frequently provides the best fit to experimental leaching data.

Several case studies, illustrating how the temperature and duration of contact affects the levels of extractables, are discussed as follows.

Case 1: Acetic Acid Migration from Multi-layered Plastic Films [64]

In this study, the migration of acetic acid out of two multi-layered films was studied. One film, C1, consisted of four layers: an outer layer of ultra-low-density polyethylene (ULDPE), a core layer of EVA, a gas barrier layer of a polyethylene vinyl alcohol copolymer (EVOH), and a solution contact layer of ULDPE. A second film, C2, consisted of three layers: linear low-density polyethylene (LLDPE) as the outer layer, EVOH as a gas barrier layer, and EVA as the solution contact layer. Both bags were irradiated, producing acetic acid from the EVA (primary source) and EVOH layers. This is important because the EVA layer in film C1 is buried in the structure while the EVA is the solution contact layer in film C2. Thus, acetic acid from C1

has to migrate through a PE layer to get into solution while acetic acid from C2 is directly available to the solution.

Bags were filled with water as the extracting solvent and aged at either room temperature (taken as 25 °C) or 45 °C for 1, 7, 14, 30, 60, 90, and 120 days. Although TOC and acetic acid levels were measured in the resulting extracts, only the acetic acid data will be discussed.

From Equations (2.11) and (2.15), one expects a four-fold difference in extracted concentration from the ASTM method and a ten-fold difference in extracted concentration from the factor rule of 10 for a 20 °C difference in temperature. Table 2.23 summarizes the ratio of the measured acetic acid levels at each test interval, 45 °C versus 25 °C.

In examining Table 2.23, one immediately sees the difference in behavior, C1 versus C2. With the primary source of acetic acid in the solution contact layer, release of acetic acid from film C2 is immediate and there is very little diffusion involved (note the higher amounts of acetic acid at one day, C2 versus C1, and the concentration ratio, 45 °C versus 25 °C, for C2 at day 1 of nearly 1). For film C1, the value of the ratio, 45 °C versus 25 °C is nearly 4, the value suggested by the ASTM method. In this film, the acetic acid must migrate through one layer before it accumulates in the extracting solution.

One also notes from Table 2.23 that as the time increases past 14 days, the ratio increases as well. One envisions that this is due to a two-step mechanism of acetic acid leaching. Up to a time of roughly 14 days, acetic acid that is readily available is leaching from the film. After 14 days, acetic acid that is diffusing is being released from the film and the ratios begin to approach a diffusion-mediated value.

This study suggests that the ASTM equations more appropriately model this situation. This is not a surprising outcome as (i) acetic acid is a small molecule with a relatively large diffusion coefficient and (ii) the acetic acid migrates through a PE film, which is a material in which solutes generally have relatively high diffusion coefficients.

Case 2: Di-(2-ethylhexyl) phthalate (DEHP) migration into solutions stored in a DEHP-plasticized bag.

The release of DEHP from plasticized PVC bags, tubing, and other medical devices has been extensively studied. In one case, the leaching of DEHP from DEHP-plasticized intravenous bags into an intravenous teniposide drug product was examined at refrigerated (4 °C) and room temperature (22 °C) for a period of time up to 50 hours [48]. The ratios of DEHP leached at 22 °C versus DEHP leached at 4 °C after 24 and 50 hours of contact were 3.75 and 3.33 respectively. Projected ratios based on the ASTM and factor rule of 10 methods were 3.48 and 7.94 respectively. As was the case with the previous case study, this data would support use of the ASTM method.

In another study, levels of DEHP in aqueous solutions (water 0.9% saline and 5% dextrose) stored in PVC pouches were measured under various storage conditions [65]. For example, the DEHP level in these solutions was reported to be approximately 454 µg/500 ml after storage for 10 days at 60 °C and approximately 625 µg/500 ml after 6 months (taken as 180 days) at room temperature (taken as 25 °C). Projecting the 60 °C data to 6 months at room temperature via the ASTM method produces a projected DEHP level of 720 µg/500 ml. The same projection via the factor rule of the 10 method produces a projected DEHP level greater than 1000 µg/500 ml. This analysis also suggests that the ASTM method produces the more accurate projection.

It is noted, however, that since plasticized PVC is generally 25–50% by weight DEHP and that the DEHP is more or less evenly distributed through the PVC matrix, DEHP migration for short times is typically not diffusion constrained.

Considering the previous discussion and case studies, the following conclusions are applicable. The extraction or leaching of entities from plastic articles into contact solutions is a diffusion-mediated process. Thus, the proper means of establishing and justifying accelerating conditions of contact that are used in extractables and leachables studies, such as temperature and duration, is the use of diffusion-based scientific principles. Although the "factor 10 rule" is empirically derived, it is supported by scientific analysis and its use is well-established. Thus, the combination of Equations (2.15) and (2.12) represent the most scientifically appropriate means of establishing accelerated contact conditions.

Nevertheless, use of the ASTM F1980-16 approach to calculating accelerated contact conditions is appropriate from a risk management perspective as this approach produces contact durations that are longer and temperatures that are higher than those calculated via the "factor 10 rule" approach. In essence, calculations performed using the ASTM F1980-16 approach produce a worst-case exaggeration versus calculations performed using the "factor 10 rule" approach.

The following two approaches are recommended to address the indicated situations:

- To answer the question "If I know the amount of time it takes for an extractable to reach a certain concentration at a certain temperature, can I estimate the amount of time it would take to reach the same concentration at a different temperature?", the most generally applicable model for duration projections is the "factor 10 rule." However, deviations from this model can be anticipated. In general, and using the factor 10 rule, the time (t2) required for an extractable to reach a certain concentration at a temperature T_2 can be estimated from the time (t1) required for the same extractable to reach the same concentration at a reference temperature T_1 using the following equation, although exceptions will occur:

$$t2 = t1 \div 10^{[(T_2 - T_1)/20]} \tag{2.16}$$

- To answer the question "If I know what the concentration of an extractable is at a certain time and temperature, can I estimate what the extractable's concentration will be at the same time but a different temperature?", the most generally applicable model for concentration projections is the ASTM with a Q10 = 1.5. However, deviations from this model can be anticipated. In general, the concentration (C2) of an extractable at a certain duration of contact at temperature T_2 can be estimated from the concentration of the same extractable (C1) at the same duration of contact at a reference temperature T_1 using the following equation , although exceptions will occur:

$$C2 = C1 \times 1.5^{[(T_2 - T_1)/10]} \tag{2.17}$$

2.2.5.5 Stoichiometry

The surface area of a test article extracted or leached per unit volume of extracting solution is frequently used in extractables studies as a means to exaggerate the contact that occurs between the test article and a contacting solution under the test article's typical conditions of use. In some situations, the exaggeration is used to produce a worst-case extraction. For example, when a drug product is packaged in containers that can vary in size, it is the smallest available sized container that is typically tested, as the container surface area used to fill the solution volume ratio is largest for the smallest available size. In other situations, exaggeration factors are used to forecast what extractables levels would be when the conditions of the extraction differ from the conditions of use. For example, a 50-ml i.v. bag may include an injection septum. In an extractables study one might extract five septa per 50 ml of extraction solvent (potentially a five-fold exaggeration) in the belief that so doing will produce a more concentrated extract, thus facilitating the identification of extractables. To establish the amount of the extractable one might anticipate to be present in the bag's fill solution as a leachable, one might divide the concentration of an extractable in the extract by the fivefold exaggeration factor. In both situations, the underlying assumption is that the level of extractables in an extracting solution will proportionally increase with an increasing surface area to solution volume ratio.

This ratio of the test article's extracted surface area to the extraction solution's volume ratio can affect both the thermodynamics and kinetics involved with extraction or leaching. Thermodynamically, the greater the surface area to solution volume ratio, the greater is the total pool of an extractable or leachable (at constant test article thickness). Kinetically, the increased surface area to solution volume ratio will increase the speed of extraction or leaching as increasing the contact surface area increases the rate of leaching.

If the extraction or leaching conditions (e.g. duration and temperature) are such that equilibrium is attained between the test article and the extracting solution, then the kinetics effect is irrelevant and only the thermodynamic aspects are important to consider. For example, terminally sterilized, aqueous packaged drug products may have fairly long ambient temperature shelf lives (multiple years). As a result of the high heat stress of terminal sterilization and the long post-autoclave contact duration over shelf-life, the packaged drug product and the packaging system reach an equilibrium with respect to leachables.

The effect of contact surface area on the equilibrium level of extractables and leachables is addressed as follows. The equilibrium partitioning of a substance between a plastic phase (the test article; for example, a container or material) and a liquid solution phase (the fill solution or extracting solution) is described by a plastic/liquid partition coefficient, $K_{P/L}$, as follows:

$$K_{P/L} = (m_{p,e}/V_p)/(m_{l,e}/V_l) \qquad (2.18)$$

where m = the mass of a substance in a particular phase, V is the volume of either phase and the subscripts p, l, and e refer to the plastic phase, the liquid phase, and equilibrium respectively.

Equation (2.18) can be rewritten to express the relationship between the concentration of a substance in the liquid phase at equilibrium ($C_{l,e}$) and the various other relevant parameters (obtained from reference [66]):

$$C_{l,e} = m_{l,e}/V_l = m_{p,o}/[V_l + (K_{P/L} \times V_p)] \qquad (2.19)$$

where the subscript o refers to the initial condition prior to contact. As the volume of a test article shaped as a sheet is the product of the sheet's surface area (SA_p) and its thickness (t_p), Equation (2.19) can be adjusted to produce a relationship that expresses the substance's solution level at equilibrium in terms of the material's surface area and the solution's volume:

$$C_{l,e} = m_{l,e}/V_l = m_{p,o}/[V_l + (K_{P/L} \times SA_p \times t_p)] \qquad (2.20)$$

As noted from Equation (2.20), a substance's plastic/solution partition coefficient ($k_{P/L}$) will significantly impact the effect that changing the extracted surface area to the extraction solution volume ratio has on the substance's equilibrium concentration in solution ($C_{l,e}$). When $K_{P/L}$ is large (the substance partitions favorably into the plastic phase), the $K_{P/L}$ term is dominant and the effect of the surface area is minimized. When $K_{P/L}$ is small (the substance partitions favorably into the solution phase), the $K_{P/L}$ term is small and the effect of the surface area is maximized.

To illustrate this effect more clearly, the following situation is considered: $m_{p,o} = 10$ mg/cm^2, $V_l = 100$ ml $= 100$ cm^3, $t_p = 1$ cm, and $K_{P/L}$ takes values ranging from 0.1 (the substance is highly soluble in the solution) to 1000 (the substance is poorly soluble in the solution).

Under these conditions, the effect of increasing the extracted surface area of the test article (SA_p) on the equilibrium concentration of the substance in the solution ($C_{l,e}$) is shown in Figures 2.11 and 2.12. As anticipated, the effect of surface area is greatly reduced as the substance's $K_{P/L}$ increases. For a substance that is highly soluble in the solution, an increase in the material surface area produces nearly a proportional increase in the concentration of the substance in the solution. For example, when the surface area is increased by a factor of 100 for a substance with a $K_{P/L}$ of 0.1, the increase in the substance's concentration in solution is also nearly a factor of 100. For a substance that is poorly soluble in the solution ($K_{P/L} = 100$) a 100-fold increase in the surface area produces barely a doubling of the substance's concentration in solution.

Previously, two scenarios were considered in terms of the use of exaggerated surface area to solution volume ratios. The first scenario addressed the situation where a family of packaged products, differing in size, where deemed to be represented by the one product that had the highest packaging surface area to solution volume ratio. For a package, the highest surface area to solution volume ratio is attained by the smallest sized package in the family as it is usually the case that the size of the package does not increase in direct proportion to its fill volume. The rationale for choosing the size with the largest surface area to solution volume ratio was that this would be the product size that had the highest concentration of substances in solution and thus was

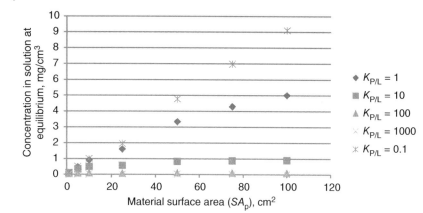

Figure 2.11 Relationship between the test article's surface area and the concentration of an extractable in an extracting solution at a constant extracting solution volume. The relationship is shown for extractables with $K_{P/L}$ ranging from 0.1 (extractable is highly soluble in the solution) to 1000 (extractable is poorly soluble in the solution).

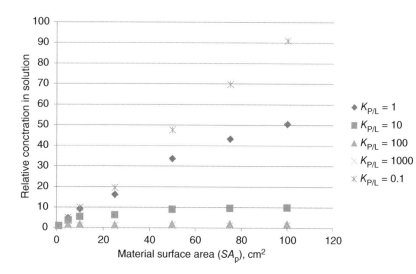

Figure 2.12 Relationship between the test article's surface area and the relative concentration of an extractable in an extracting solution at a constant extracting solution volume. The relationship is shown for extractables with $K_{P/L}$ ranging from 0.1 (extractable is highly soluble in the solution) to 1000 (extractable is poorly soluble in the solution).

worst case in terms of substance concentration. The analysis shown in Figures 2.11 and 2.12 confirms this rationale, as it is always the case that the substance's concentration in solution is highest at the highest surface area to solution volume ratio.

The second scenario addressed the situation where one would use an exaggeration factor to extrapolate an extractable's concentration in solution obtained at one surface area to volume ratio to the extractable's concentration in solution at another surface area to solution volume ratio. In this situation, it was supposed that the exaggeration factor would be equal to the surface area to solution volume ratios. More specifically, if the surface area to solution volume ratio in case 2 was twice the surface area to solution volume ratio in case 1, then the solution concentration of a substance in case 2 would be twice the solution concentration in case 1.

Figures 2.11 and 2.12 clearly show that this practice is not always proper and correct. While a 100-fold increase in surface area to solution volume ratio produces nearly a 100-fold increase in the solution concentration for a substance that is highly soluble in the solution ($K_{P/L} = 0.1$), the increase becomes much less as the substance's solution solubility decreases. For example, the increase in the substance's concentration in solution is roughly 50-fold when the $K_{P/L}$ is 1. For poorly soluble substances ($K_{P/L} = 1000$), a 100-fold increase in surface area to solution volume ratio produces less than a 10% increase in the substance's solution concentration.

Thus, while it is clear that 100-fold extrapolations of surface area to volume ratios are not proper or viable in many cases, it may be the case that smaller extrapolations would be acceptable. Table 2.24 addresses this possibility by collating the concentration exaggeration factors when the surface area to solution ratio is changed by a factor of 5 or less.

One notes from this table that when the surface area is increased from 1 to 3 (a potential exaggeration factor of 3), even in the case of a poorly soluble compound ($K_{P/L} = 100$), the actual exaggeration factor is close to the potential factor (1.5 versus 3, or approximately 50%). However, at a surface area exaggeration factor of 5, the actual exaggeration factor

Table 2.24 Relative substance concentrations in solution at various surface area to volume ratios.

Surface area (SA_p, cm^2)	Relative concentration in solution, $C_{e,l}$				
	$K_{P/L} = 0.1$	$K_{P/L} = 1$	$K_{P/L} = 10$	$K_{P/L} = 100$	$K_{P/L} = 1000$
1	1	1	1	1	1
2	2.00	1.98	1.83	1.33	1.05
3	2.99	2.94	2.54	1.50	1.07
5	4.98	4.81	3.67	1.67	1.08

for a substance with $K_{P/L} = 100$ is only 33% of its potential value. This analysis suggests that a threefold exaggeration of the surface area to solution volume ratio would be acceptable across a wide population of extractables with a wide range of $K_{P/L}$ values. However, once the exaggeration factor exceeds 3, the exaggeration is no longer accurate for extractables with higher $K_{P/L}$ values, with the inaccuracy of the exaggeration increasing as $K_{P/L}$ increases.

A case study that examines these effects is as follows [67]. In one of the examples previously considered was the situation of a drug product that is commercially available in multiple package sizes (e.g. multiple fill volumes), which can be the case for certain drug products used as parenteral injections. The measurement of targeted leachables over storage time across multiple package sizes for one or more parenteral solutions of similar composition produces a set of data relevant to a consideration of the effect of the ratio of surface area to solution volume (SA/V) on the leachable's concentration in the drug product.

The test articles for this case study were three different packaging systems that are used to package aqueous parenteral injections such as Sterile Water, Normal Saline and Dextrose for Injections. Filled packages of multiple sizes (25–3000 ml) were treated in a manner appropriate for such products, which means that the filled units were terminally sterilized (autoclaved) and then stored for a period of time reflecting multiple year, ambient temperature shelf lives. Over the course of such storage, the fill solutions were sampled and the samples were tested for their levels of targeted leachables. Leachables were targeted based on their known association with the three container systems studied and included di-(2-ethylhexyl) phthalate (DEHP) and Zn from a plasticized poly(vinyl chloride) packaging, caprolactam from a plastic package whose materials of construction included Nylon 6, and a class of compounds termed cyclic esters from a laminated plastic structure that included polyurethane-based tie layers. Validated analytical methods for quantifying the targeted leachables included inductively coupled plasma atomic emission spectroscopy (ICP-AES) for Zn, liquid chromatography with UV detection (LC/UV) for DEHP,

liquid chromatography with mass spectrometric detection (LC/MS) for caprolactam, and gas chromatography with flame ionization detection (GC/FID) for the cyclic esters. Polymer/solution partition coefficients ($K_{P/L}$) for these targeted leachables are contained in Table 2.25.

A typical use of exaggeration factors is extrapolating the higher concentrations measured in the smallest package to lower concentrations relevant to larger packages. To illustrate this situation, Figure 2.13 is constructed from the experimental data, where normalizing is the data using the smallest package as the point of reference.

Figure 2.13 shows a divergence between the experimental behavior and the ideal outcome. The significance of the divergence is illustrated in the following situation. The equilibrium concentration of a targeted leachable has been measured in the smallest sized package and the investigator seeks to extrapolate that outcome to a larger package that has a five times smaller SA/V ratio. Were the "ideal" situation to be encountered, the investigator would project the leachable's concentration in the larger package to be one-fifth of its level in the smaller package. As shown in Figure 2.13, however, the actual "extrapolation factor" for the leachables with low partition coefficients would be closer to a factor of one-fourth, yielding a higher concentration estimate.

The under-exaggeration is far worse for a higher partition coefficient extractable such as DEHP. As was the case for the other leachables, the ideal extrapolation would indicate that the DEHP level in the smaller package should be one-fifth that measured in the larger package. In fact, the experimental data shows that the concentrations of DEHP in the two packages are essentially equal.

Thus, extrapolating small package data to larger packages based on the "ideal" circumstance underestimates the leachable's concentration and underestimates its potential safety effect. The higher the leachable's polymer/solution partition coefficient $k_{p/l}$, the larger the underestimation.

To this point in our discussion, the phenomenon that is being investigated has been expressed in terms of a test article's surface area to solution volume ratio. So doing implies a cause-and-effect relationship; that is, it is the

Table 2.25 Frequently encountered extractables and their $P_{o/w}$ and $K_{P/L}$ values.

Extractable	CAS RN	Log $P_{o/w}$	Approximate $K_{P/L}$		
			PVC	**Polyolefin**	**Rubber**
Stearic acid, pH 3	57-11-4	7.82	>1 000 000	419 000	13 200
Stearic acid, pH 9	57-11-4	4.21	1830	221	141
2-Ethylhexanoic acid, pH 3	149-57-5	2.57	62.9	7.17	17.9
2-Ethylhexanoic acid, pH 9	149-57-5	−1.03	0.039	0.0039	0.194
Caprolactam	105-60-2	−0.04	0.295	0.031	0.673
Bisphenol A	80-05-7	3.64	566	67.2	68.8
Cyclic ester CE 228 (I)	5578-82-5	1.19	3.69	0.400	3.16
Cyclic ester CE270	94113-50-5	2.49	53.3	6.07	16.2
Butylated hydroxytoluene (BHT)	128-37-0	5.17	13 100	1650	471
Irganox degradate, pH 3	20170-32-5	3.54	461	54.5	60.7
Irganox degradate, pH 9	20170-32-5	−0.12	0.251	0.026	0.609
Irganox degradate	82304-66-3	2.47	51.2	5.82	15.8
Mono-(2-ethylhexyl) phthalate, pH 3	4376-20-9	4.39	2640	322	177
Mono-(2-ethylhexyl) phthalate, pH 9	4376-20-9	1.39	5.57	0.608	4.06
Di-(2-ethylhexyl) phthalate (DEHP)	117-81-7	8.52	>1 000 000	>1 000 000	31 763
Irganox 1010	6683-19-8	18.8	>1 000 000	>1 000 000	>1 000 000
Irgafos 168	31570-04-4	13.7	>1 000 000	>1 000 000	>1 000 000

Bolded substances are illustrated in Figure 2.13.

changed SA/V that causes the change in the extractable's concentration. However, in most situations, the increase in SA/V is also reflected in an increase in the mass of material extracted per volume of extracting solution and it may be the case that it is the change in extracted mass, and not the change in surface area, that drives the change in the extractable's concentration. For example, the surface area and the material mass are proportionally related for a container whose composition and thickness are constant over a given range of sizes.

The following conclusions and recommendations can be drawn about the proper use of an exaggerated test article surface area to extraction solution volume ratios in exaggerated extractables studies:

1. Extractable concentrations can be extrapolated as a function of surface area to solution volume ratios so long as the exaggeration factor is no more than three.
2. Exaggeration factors larger than three can be justified on a case-by-case basis if the nature of the extractables can be established (specifically their $K_{P/L}$ values).

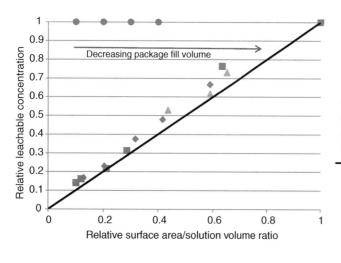

Figure 2.13 Normalized plot showing the experimental effect of a package's surface area to solution volume ratio (SV/A) on the equilibrium concentration of leachables in the contained solution. As the package's size (fill volume) decreases, its surface area to solution volume increases, resulting in an increased extractable concentration in the contained solution. Concentrations and SA/V ratios have been normalized to the corresponding values for the smallest package.

3. Exaggeration factors larger than 10 should rarely be used and only in those circumstances were the $K_{P/L}$ values for all extractables are less than 10. Use of exaggeration factors greater than 10 would require justification on a case-by-case basis.
4. If one proportionally extrapolates concentrations from a lower surface area to solution volume ratio to a higher surface area to solution volume ratio, the extrapolated concentration will always be greater than or equal to the actual concentration (making the extrapolated concentration an overestimated worst case). The larger a substance's $K_{P/L}$, the greater the relative difference between the extrapolated and actual concentrations.
5. If one proportionally extrapolates concentrations from a higher surface area to solution volume ratio to a lower surface area to solution volume ratio, the extrapolated concentration will not be greater than the actual concentration (and thus the extrapolated concentration will be an underestimation and not a worst case). The larger a substance's $K_{P/L}$, the greater the relative difference between the extrapolated and actual concentrations.

Since knowing $K_{P/L}$ values for extractables is essential in the application of the above equations, but published $K_{P/L}$ values are hard to find, the following discussion provides a means of calculating $K_{P/L}$ values from the more commonly available octanol-water partition coefficient, $P_{o/w}$.

A polymer/solution partition coefficient ($K_{P/L}$) is equivalent to a solvent/water partition coefficient ($P_{s/w}$) in terms of its mathematical derivation and its application. If one could identify an organic solvent that behaved in the same manner as a particular polymer, then a substance's polymer/solution and solvent/water partition coefficients would be equivalent.

Octanol/water partition coefficients ($P_{o/w}$) are commonly available for a large number of organic substances (including extractables), thus making $P_{o/w}$ an excellent starting point for determining $K_{P/L}$ values. While octanol is not an exact model solvent for polymers used in pharmaceutical applications, relationships between $K_{P/L}$ and $P_{o/w}$ have been documented in the literature, for example, references [25] and [68]. Such relationships for three types of polymers, plasticized PVC, polyolefin, and rubber, are as follows:

For plasticized *PVC*: $\log K_{P/L}$
$$= (0.892 \times \log P_{o/w}) - 0.494 (*)$$

For polyolefin: $\log K_{P/L1}$
$$= (0.908 \times \log P_{o/w}) - 1.48 (*)$$

For rubber: $\log K_{P/L} = (0.546 \times \log P_{o/w}) - 0.15 (\#)$

(*) = obtained from data in reference [68]
(#) = adjusted from data in reference [25]

Although these correlations are neither particularly strong for a given polymer and certainly not applicable to all polymers of a given class (for example, different plasticized PVC materials may have different models as a function of the amount of plasticizer in the material), they are sufficiently applicable in a general sense that they can be used to estimate $K_{P/L}$ values for the purpose of designing extractables or leachables studies in terms of the surface area to solution volume ratio. Furthermore, other thermodynamic modeling approaches for estimating $K_{P/L}$ are being investigated and reported.

2.2.5.6 Additional Factors to Consider

The previous discussions about establishing and justifying extraction conditions are predicated on the premise that there is freedom to choose these conditions. It is certainly the case that USP <665> establishes the required extraction conditions for a manufacturing component and although these conditions may be based on the principles previously discussed, they cannot be altered based on those principles. The same situation is true, to a certain extent, for medical devices, where ISO 10993-18 provides recommendations for all critical dimensions of the extraction.

Although the application of the previously-discussed principles may establish certain extraction conditions as being technically justified, there are a few practical issues to consider. In general, the extraction conditions should not alter the test article as alteration of the test article could change the amount and/or type of extractables released from the test article. For example, consider the case where a plastic packaging bag becomes cloudy (or sticky) during an extraction. Unless the bag also becomes cloudy (or sticky) during its clinical use (an unlikely circumstance), the fact that it turns cloudy or becomes sticky during the extraction suggests that it is being altered by the extraction. Thus, a test article's chemical properties may also need to be considered when selecting extraction conditions, for example, to avoid solubilization or alteration of the test article's material(s) of construction.

Another practical issue to consider is whether the extraction system is open or closed. The ideal situation is that the extraction system is closed. This is the case for several reasons. First, a closed system is isolated from environmental contaminants during the extraction process. Second, a closed system minimizes the loss of extraction solvent. When loss of extraction solvent occurs, the practical issues to address are how to minimize solvent loss and how to account for that solvent which is lost. In general, it is not recommended to compensate for solvent loss by adding additional solvent after extraction is complete; rather, steps should be taken to reduce evaporative loss (e.g. by covering sample containers) or the final extract volume should be measured for use in later calculations concerning extractable's levels. Measurement of solvent volumes in order to compensate for solvent loss due to swelling should be done cautiously, given that the amount of solvent that swells a test article may be unknown and difficult to measure. In either case, the final extract volume should be measured and reported for later calculations of, for example, extractables per device. Moreover, excessive solvent loss during extraction (for example, loss of greater than 20%) could be the basis for rejecting an extraction process as being appropriate, as loss of that much solvent during extraction raises serious questions about the integrity and validity of the extract. For example, if one starts out with an extraction solvent that is 50% ethanol and 20% of the solvent's volume is lost during extraction, it is reasonable to conclude that what was lost was the ethanol, changing the "extracting power" of the extraction solvent. Moreover, if solvent is lost during extraction by volatilization, it is also likely that volatile and semi-volatile extractables may be lost if the test system is not closed.

Another question that frequently arises in implementing an extraction is whether the extraction should be static or dynamic. The major concern here is that extraction may be retarded (diffusion is slowed) if the extraction medium is not mixed well. Such a concern is especially relevant for extractions performed at lower temperatures, for shorter durations, and with "weaker" extraction vehicles. Generally speaking, it is always better to perform a dynamic extraction unless there are physical constraints that prevent this.

Once a dynamic extraction is chosen, then the means of performing the extraction must be established. There are typically only two means of performing a dynamic extraction, agitation (for example, shaking) or circulation. In some cases, practical realities make the choice obvious. For example, extracting a bag used as a packaging system is usually accomplished by filling the bag and thus agitation is appropriate. On the other hand, a filter used in a manufacturing step is used to filter a flowing solution and thus the appropriate extraction that simulates use is an extraction that involves the circulation of the extraction vehicle through the filter.

When circulation is used as the extraction process, there are two options; continuous circulation, where fresh extracting solvent is circulated through the test item throughout the extraction and is collected for testing after it leaves the test item, and recirculation, where the same extraction solvent is re-directed through the test article throughout the extraction. If the clinical application of a test item involves recirculation, then surely recirculation is the proper extraction strategy. However, if the clinical application does not involve recirculation, then the choice of circulation type is not clear. When the extraction involves high flow rates for extended periods of time, constant circulation of fresh extracting solvent has the obvious disadvantage that the total volume of the extracting solution will be large. Not only are there practical issues with this (one has to make a large volume of extracting solvent and have a suitable vessel to collect the volume) but it is possible that the concentration of extractables in a continuous extract could be low, confounding extract analysis.

That is not to say that recirculation is not without its issues. I think there is a misconception out there that somehow recirculation allows one to violate the laws of thermodynamics. It does not. Once an extraction solvent and the test article reach equilibrium, then the extraction will stop, recirculation or not. Although recirculation might speed up the attainment of equilibrium (by preserving the concentration gradient that drives diffusion), it will not shift the equilibrium.

Another factor to consider is the generation and testing of replicates. For an extraction study, this could mean performing multiple extractions, analyzing multiple portions of the same extract, or a combination of both.

The issue of replication in extraction studies is currently receiving considerable attention as there is largely circumstantial "evidence" to suggest that extractables profiles are not very reproducible, both in terms of the extractables reported and their reported concentrations.

To address reproducibility objectively, one must understand that there is interlaboratory variation, which cannot be addressed by replicates, and intralaboratory variation, that can be addressed by replicates. Interlaboratory variation occurs and is likely to be large as there is little or no standardization between testing laboratories in terms of extraction conditions and analytical procedures. Even when the extraction conditions are specified, as in the standard extraction protocol of USP <665>, not every detail is specified and subtle but impactful differences can be present in the individual laboratory's extraction process. When the extraction conditions are not specified, it is not uncommon for two different testing laboratories to devise two different extraction processes to accomplish the same objective.

Considering analytical testing, testing laboratories have standardized the testing methodologies to some extent. Thus, for example, every laboratory uses some combination of HPLC, coupled with some form of mass spectrometric (MS) detection, to address non-volatile organic substances. However, there is little to no standardization on instruments, instrumental operating parameters, consumables (columns and stationary phases), and data processing methods.

Thus, it would not surprise me at all that the same test article, submitted to three different testing laboratories, would be reported to have three different extractables profiles. However, standardization of methods, and not replication, is the solution to this reproducibility problem.

Intralaboratory reproducibility is also an issue but this issue can at least be quantified by replication. In my mind intralaboratory variation exists because little or no attempt has been made to optimize extraction and analysis processes for reproducibility. For example, consider two extraction processes, one that achieves equilibrium during extraction and one that does not. It is intuitive that an extraction process that achieves equilibrium would be highly reproducible, as equilibrium is the same, extraction to extraction. However, if the extraction process does not achieve equilibrium that means more extraction is still occurring when the extraction has stopped. Again, it is intuitive that small variations in incomplete extraction conditions could lead to one extract being more "mature" than its corresponding replicate.

Another source of variation is test article inhomogeneity, which surely will depend on the nature of the test article and the sample size used in the extraction.

In any event, the imprecision of the extraction process can be established by performing replicate extractions. The imprecision of the extract analysis process can be established by analyzing replicates of each extract. However, practically speaking, the number of samples that must be tested grows with each level of replication. For example, producing three replicate extracts and then testing three aliquots of each extract produces a total of nine tested samples for each test article, which is a significant analytical burden. Thus, the desire to have some idea of the magnitude of test method imprecision (via replication) stands in direct conflict with the practical reality that "each replicate costs time and money."

Therefore, although most guidelines, monographs, and recommendation documents broach the subject of replicates, few contain definitive requirements. The following language, contained with the normative contents of ISO 10993:18, is typical:

> "Considering replication of extractions, a single extraction replicate for each vehicle shall be sufficient in those circumstances where it can be established that the variation in the test article's composition and/or the variation in the extraction process is low, establishing that the single extraction is properly representative of the test article and the extraction process. In cases where other information (e.g. engineering testing) indicates higher variability either within or across test article units or lots or inherent

to the extraction process, multiple (e.g. duplicate or triplicate) extractions may be necessary. Multiple extractions should also be performed in those circumstances where the test article and/or extraction variability is unknown. Regardless of the number of replicate extractions performed, the number of extracts generated should be justified."

Truly, it is difficult to understand precisely what the exact acceptable practice is from this text. Presumably if one had data that established that the extraction variation was small, a single replicate extract would be adequate. In the absence of such data, or if the data suggested that variability was high, then multiple extractions, up to three, are necessary.

Lastly, I want to address the issue of "who should be doing extraction studies and be providing extraction study data?" This question is one that is debated between users and vendors of items such as manufacturing components and materials used in medical devices and packaging systems and is based on the user's desire that the vendors supply extractables data.

It is easier to start working through an answer to this question if we first consider a related question, which is "who must possess extractables data to support a regulatory submission?" The answer of course is the organization which owns the product being submitted, which is the user. Thus, ultimately the user must have relevant extractables data, no matter who actually performed the extraction study. If the user cannot get relevant extraction data from a vendor then the user has little choice but to generate that data themselves.

However, to understand this question fully, one has to understand that extractables data could serve two purposes, selection of items and qualification of items. That is, when a user decides to use an item, that decision should be based on data, some of which should be extractables data. Furthermore, when the user submits a product for regulatory approval, the submission will include extractables data that qualifies the item as being compatible and suited for its intended use. However, it is not necessarily the case that the extractables data used for selection is the same extractables data that is appropriate for qualification. There is a simple reason for this. The data used for selection supports the selection process most efficiently when the same data is obtained from all possible vendors of an item. In this way the comparison of the extractables data from the various vendors can be compared in an "apples to apples" manner. Obtaining the same information (from extraction studies performed in the same way) from all potential vendors would mean that there is an agreement up-front as to the design, implantation, and reporting of the extraction study. All vendors would agree to perform the same study and all vendors would agree to report the same type of results to all possible users.

It is reasonable to expect that vendors of items would supply this type of extractables data based on the logic of simple economics. Economically, it makes sense that the vendor would do the standardized study once and provide the results to all possible users. It makes no economic sense that each potential user would individually and repetitively test the same item by the same methodology, producing essentially the same results over and over again. Of course, this repetitive testing occurs quite frequently because in the real world users do not share extractables data.

A certain level of standardization has been achieved for manufacturing components in the USP <665> monograph [12], which clearly specifies a standard extraction process. While <665> does not standardize the analytical testing, it does standardize the extraction process. The BPOG recommendation for performing extractables studies for manufacturing components [10] actually does go this extra step and specifies specific analytical methodologies. In either case, however, it is clear that the most efficient means of producing either USP <665> or BPOG data is for the item's vendor to perform the extraction test once and to provide the test results to potential users. In fact, vendors of single-use manufacturing components have largely supported and adopted this strategy.

However, the trade-off is that these extraction studies which support selection are performed with standardized extraction conditions that may not be an exact fit to a particular user's product's clinical conditions of use. Thus, when the vendor's extractables study is submitted by the user for the purpose of product qualification, the typical regulatory response is that the extractables study design was not adequate for qualification and thus that another extractables study is required for submission. This is not a surprising outcome for two reasons. First, it is impossible to design a standard extraction that is relevant to all possible clinical use or clinical contact conditions because there are too many variations of these conditions. Second, the vendor tests an item in the form the vendor supplies the item to its user. In many cases, the user will alter, modify, or process the vendor's item to make the user's product. For example, a purchased plastic resin might be extruded by the user to form a part and then the part might be sterilized by irradiation. It is clear that this process of alternation, modification, and processing could have a profound effect on the extractables profile and thus that the extractables profile of the resin, supplied by its vendor, would be different than the extractables profile of the part used by the user.

In this case, someone has to test the part to provide qualification data for submission. Reasonably speaking, it is "unfair" and "uneconomic" (from the vendor's perspective) to have to test their item in all the many different ways that users might use the item. It makes more sense that testing that is so specific to the user's use of the item be performed by the user.

Thus, we have logically developed an understanding that an item's vendor would supply potential users with standardized extractables data to support the user's selection process. In those circumstances where the conditions of the vendor's test are a good match to the user's clinical conditions or use and contact, then it might be possible that the vendor data could be used to qualify the user's product for the purpose of securing regulatory approval. However, vendors, users, and regulators all understand that this outcome will not always be the case and that many times a vendor's extraction study will not be performed under conditions that match the final product's clinical conditions of use or contact. Under these circumstances, it is understood that the user must perform their own extraction study to cover their product's conditions of clinical use and contact.

Now this does not mean that the vendor's extractables study results are useless to the user. It is certainly the case that the vendor's extractables data provides the user with useful data about potential extractables and their potential levels and thus the vendor's extractables data may help the user to more efficiently and effectively devise, justify, implement, and interpret their own extractables study.

2.2.6 Compromised Extracts

As one performs more and more extraction studies, they will invariably run into the circumstance of an extract that does not look right and/or a test article that has been markedly altered, either physically or functionally, by the extraction process. Considering the circumstance of an extract not looking right, the essential assumption in an extraction is that the extraction medium should have the same physical properties before and after extraction. This essential assumption is valid as it is the case that, during leaching, the contact medium is generally unaltered by the leaching process. For example, when a drug product is stored in a container, one expects that the drug product is not physically altered during storage. Similarly, during manufacturing, a process stream contacted by a manufacturing component should not be physically altered by the contact. Lastly, a tissue contacted by a medical device during its clinical use should not be altered as a result of the contact.

Thus, when an extraction is designed to simulate leaching, the extraction medium should be physically unaltered by the extraction. For example, an extraction medium that started out clear and colorless should not contain particulate matter nor be colored after the extraction has been completed.

Furthermore, there is the expectation that when an item is used in a clinical setting it retains its form and function throughout use. For example, a drug product container is not suitable for use if it starts leaking the drug product halfway through the drug product's shelf-life. Likewise, a medical device should not become so deformed during clinical use that it becomes non-functional. Thus, the test article should not be deformed or lose its functionality when it is extracted.

An extract is established to be compromised in one of three circumstances:

1. The extraction medium is physically altered by the extraction process,
2. The extracted item is changed, in form and/or function, by the extraction,
3. Both 1 and 2.

The fundamental issue associated with a compromised extract is whether the extract should be analytically tested. This issue has both practical and philosophical aspects. Practically speaking, testing becomes a moot point if the physical nature of the extract is such that it cannot be directly tested. For example, a cloudy or particulate-containing extract typically cannot be tested directly without severe analytical consequences. In extreme cases, it is not be possible to make the extract analytically viable by eliminating the complicating factor (e.g. removing the particulate by filtration). It is obvious that if an extract cannot be made analytically viable, the extract cannot be tested.

However, what about cases where the complicating factor can be eliminated by processing the extract? In cases where the complicating factor can be eliminated, the question exists as to whether the extract has been altered by the eliminating process. For example, consider the case of an extract containing particulate matter. Logically, the particulate matter arises due to the extraction. Thus, eliminating the particulate by filtration clearly eliminates some of the extracted substances, specifically those present in the particulate. Moreover, the extract could be chemically altered during filtration via both the leaching of substances from the filter and by sorption of extractables by the filter.

It is obvious that a compromised extract cannot be tested if processing of the extract cannot make it analytically viable. However, the question of whether to test an extract becomes more complicated if either the compromising factor does not prevent direct testing (for example, a discolored extract can readily be tested) or if the compromised extract can be made analytically viable by processing. In these circumstances the issue is not whether the extract could be tested but rather whether should the extract be tested.

To this author, the clear and obvious answer is that a compromised extract should not be tested; rather, the extraction process must be altered and repeated so that an uncompromised extract is generated. The reason for this was stated previously; if the contact medium during leaching is not compromised, then the extract generated to address leaching should also not be compromised.

To put the concept of compromised extracts in the proper context, consider a similar circumstance in another situation. A baker has baked some cookies. The cookies are burnt. What do you do with the burnt cookies? Well, if they are not too badly burned, I suppose you could eat them as they are. Or, you could scrape off the burnt part of the cookies and eat the rest. However, I think I would just say "throw them away and make another batch" as the consequences of eating a burnt cookie may be more than just "the burnt part does not taste good."

The other aspect of a compromised extract is the impact of the extraction on the extracted item. As I have stated earlier, in most clinical situations an item remains intact, physically unaltered, and functional during use. Thus, an extraction that causes the extracted item to become deformed, physically altered (e.g. discolored), or non-functional is an invalid extraction from the perspective of addressing leaching. The argument that the extraction must be more aggressive (and thus a justifiable worse case) than leaching if the test article is deformed is not viable as it ignores the possibility that the extractables have been altered to the extent that they are not viable leachables.

An extraction that results in an altered test article can produce an extract that is either analytically viable without processing, analytically viable with processing, and not analytically viable even with processing. Regardless of these outcomes, it is this author's recommendation that an extract that is compromised because of physical or functional alternation of the extracted item should not be tested; rather, the extract should be discarded and a new extraction be designed and implemented.

Now perhaps the recommendation of "discard and re-extract" is too harsh a judgment. However, while it is true that some useful information (data) could be generated by testing a compromised extract, the information thus generated cannot be used in impact assessment because the link between the process of generating the extract and clinical use of the test article cannot be established.

Having said this, there are two circumstances where testing of a compromised extract arising from complete destruction of a test article could be appropriate. These circumstances are (i) when the purpose of the testing is compositional profiling (deformulation) or (ii) when the extracted item actually dissolves during clinical use. In both circumstances, testing of the extract produced by destructing the test article fulfills the goals of the testing. However, just because both clinical use and extraction result in the destruction of the extracted item does not mean that the extraction conditions are an acceptable mimic of the clinical use. For example, although a medical implant may dissolve after both a certain period of implantation or after digestion with a strong acid, this same outcome does not mean that the extractables in the digest are indicative of the device-related chemicals to which the patient is exposed as the device dissolves.

2.3 Leaching Studies

A leaching study, also known as a migration study, is generally self-explanatory. Considering packaged drug products, for example, leachables studies are defined in USP <1664> as "laboratory investigations into the qualitative and quantitative nature of a particular leachables profile over the proposed shelf-life of a particular (packaged) drug product" [5]. It is obvious, then, that a leachables study performed on a packaged drug product involves testing the packaged drug product at multiple points over its shelf-life to generate a leachables profile (list of all leachables above a defined threshold, including both the identities and concentrations of the leachables) at each time point. Considering medical devices, although ISO 10993-18 does not define the term "leachables study" per say, it is clear from other definitions and the normative text that a leachables study is a study whose purpose is to quantify leachables (substances) that are released from a medical device and into the device-contacting medium under the device's and medium's clinical conditions of contact. For components used in pharmaceutical manufacturing, the situation is a bit more complicated due to the two flavors of manufacturing leachables: "true" manufacturing leachables that are present in the drug product as it exits the manufacturing process and PERLs, process equipment-related leachables that are present in a manufacturing process stream. Clearly, a "true" leachables study involves establishing the leachables profile of a manufactured drug after the last manufacturing step has been completed (but before the drug product is transferred to its packaging system) while a PERL study involves characterizing a process stream after it has contacted the manufacturing component in question.

There is an important point to raise here about the testing of manufactured drug products for manufacturing system-related leachables and concerns the use of multiple timepoints during testing. One remembers in the discussion of leachables testing of packaged drug products that the drug product is tested for leachables at multiple timepoints during its shelf-life to establish trends that define the highest attained leachables levels. I sometimes hear people talking about multiple timepoints in terms of testing drug products for manufacturing related leachables. What one is talking about here is testing various drug product units over the course of a manufacturing campaign on the premise that over the course of the manufacturing campaign the leachables levels may change. Thus, for example, one might test a product manufactured at the beginning of the batch, middle of the batch, and end of the batch. However, this does not mean testing drug products over the shelf-life with the expectation that the levels of manufacturing-related leachables would change. The level of manufacturing-related leachables will not change over the shelf-life because contact between the drug product and the manufacturing system ends once the manufacturing process is completed and the manufactured drug product leaves the manufacturing system and is packaged. The only exception to this statement is reactive leachables that degrade over time.

Considering the complexities involved with designing and justifying an extraction study, devising and rationalizing a leaching study is straightforward. By definition and context, the leaching solvent is specified (the contacting medium), the temperature/duration of the contact is specified (clinical conditions of use), the stoichiometry is known (actual commercial situation), and the other details (e.g. static or dynamic) are also dictated by the clinical conditions of the contact.

Furthermore, there are clear regulatory and recommended practices that should or could be applied to enable the design of a proper leaching study. First, lets us consider one such recommendation/practice that challenges one of the "indisputable truths" of a leaching study – that it uses the actual contacting medium. In almost all cases, guidelines, monographs, and recommended practices, talk about a placebo being an appropriate substitute for the contact medium itself. When the contacting medium is a drug product, the term placebo means a solution that contains all the ingredients of the drug product except its active ingredient. The suitability of this substitution, placebo for drug product, has its roots in the observation that for most drug products the active ingredient (drug substance) is not itself an aggressive leaching agent and the concentration of the active substance in the drug product is low relative to its other ingredients. The value in using a placebo that does not contain the active ingredient is largely practical as (i) the drug may complicate the analytical process because it interferes with the leachables screening assays, (ii) the drug may be expensive or in limited supply, or (iii) the drug may be difficult to handle.

Applying the concept of a placebo to a medical device can be a bit more complicated. Clearly, if the contact medium with a device is a drug product, then the concept of a placebo is readily applicable. However, if the contact medium is either blood (or a related material) or a tissue, then the concept of the placebo fails for lack of an appropriate placebo. Remember, in a leaching study a placebo is not a substitute for the contact medium; rather, it is the contact medium from which typically one constituent has been removed.

Applying the concept of a placebo to manufacturing situations is equally complicated. In the case of a true leachable that is measured in the manufactured drug product, it is difficult to imagine using the manufacturing process to produce a placebo that does not contain the drug substance, as typically the manufacturing process includes both the generation, purification, isolation, and handling of the drug substance. In the case of a PERL, it is not always clear that a placebo is needed to replace a process stream (depending on the composition of the process stream) and when a placebo might be useful, it is not clear what "active ingredients" can be removed from the process stream without affecting its "leaching power." Nevertheless, when a proper placebo can be identified and justified, then its use in a manufacturing leaching study is consistent with what is regarded as acceptable practice.

The concept of replication was addressed previously for extraction studies. The concept of replication in leaching studies is a little more clear-cut. If the test article of a leaching study is a packaged drug product, the regulations and recommendation are consistent and universally clear. A leaching study includes three manufacturing batches of drug product, with three replicate product units being tested per batch at each test interval. Although less clearly stated, it is generally understood that a single aliquot of drug product is tested per individual unit.

The question that is frequently raised with drug product leachables testing is "what is meant by three batches?" This question is raised because three production batches do not necessarily conform to three batches that use different campaigns of packaging. In fact, it is not uncommon that several production batches of drug product are packaged in the same batch of packaging materials. This circumstance is unfortunate because it is likely that the greatest batch-to-batch variation in leachables is not the result of different drug product batches being generated in the same packaging batch but rather with the same drug batch being generated in different batches of packaging. However, the clear logic that "three batches means three packaging batches" is counterbalanced by the practical difficulty of obtaining samples from three separate packaging batches as separate packaging batches may be separated by years in terms of their production; that is, a single batch of packaging may be used in the generation of multiple batches of the manufactured drug product.

The same issue is encountered with medical devices and manufacturing components. Replicate batches of medical devices and manufacturing components truly reflect the inherent batch-to-batch variation in these items only when the replicate batches include different batches of starting raw materials and separate manufacturing campaigns.

Another design feature to consider with respect to leaching studies is the frequency of testing. The frequency of testing decision is driven by the regulatory necessity to perform a compatibility assessment based on the worst possible leachables profile that will be obtained over the entire course of an item's clinical use and/or shelf life. Intuitively, one would propose the argument that "the worst-case leachables profile will be obtained under the longest conditions of contact" and thus that the only relevant testing interval for a leaching study is "end of contact." However, it is well-established that leaching is not such a predictable phenomenon and that leachable's levels can exhibit various trends as a function of contact time. Thus, it is almost always the case that a leaching study consists of multiple time points (test intervals), unless the clinical conditions of contact or use are so short that leachables trending versus time is impractical.

When a leaching study requires multiple test intervals, it is clear that the number of test intervals must be sufficient to establish the leachable's accumulation trends. This is necessary as an understanding of the accumulation trends allows one to establish the worst-case leachables profile, which is basis of the compatibility assessment. It is also clear that, minimally, the leaching study must consist of two test intervals, the beginning of the contact and the end of the contact. However, it is also clear that since "two points determine a line," two timepoints do not effectively establish a leachable's accumulation trend if that trend is not a line and thus does not satisfy the requirement that the trend has been established.

If one does not know what the leachable's accumulation profile is likely to be, then it is impossible to establish the minimum number of time points and the frequency of time points required to fully define the accumulation profile. Certain information, such as a multiple timepoint extractables study or relevant historical data on similar products in similar circumstances, may suggest the nature of the accumulation profile. Computational tools such as computerized mathematical migration modeling may also be useful when the relevant input information is available.

I do not know if it is necessary to state the obvious, but there are two points that are relevant to the aspects of the number and timing of test intervals in a leachables study. The first point is that the longer the conditions of clinical contact and use, the more test intervals are required. The second is that since it is the worst-case leachables profile that will be considered in the compatibility assessment, there is no need for additional test intervals once the worst-case leachables profile has been obtained by testing. For example, consider the situation where equilibrium is established between the donor item and the receptor product (for example, the package and the packaged drug product). This will produce an accumulation profile that exhibits a plateau (which is referred to as asymptotic leaching). Once sufficient testing has been done to establish that the plateau has been reached, then further testing serves no useful purpose. Thus, it is possible that one might design a leaching study with many testing timepoints and then cancel certain later timepoints when the available data from the earlier timepoints makes it clear that such testing has no intrinsic value.

In the specific case of packaged drug products, leaching studies are typically performed as part of formal stability studies that serve as the basis for justifying the packaged product's shelf-life. Depending upon the length of the shelf-life, stability testing intervals may be established by relevant regulations and standards. These stability testing schedules also include accelerated aging. The operational question for the chemical compatibility assessment then becomes "do I need to perform leachables testing at each stability timepoint? If not, what intervals do I need?"

I know of no scientific basis that can be used to provide the right answer to these questions in every circumstance. The best advice I can offer is based less on science and more on practical experience. First, I would perform all test intervals up to the end of the accelerated aging part of a stability study. This includes all timepoints at any and all storage conditions. The reason for this recommendation is simple enough. The last timepoint of the accelerated aging conditions is typically used to approximate the end of shelf-life under normal aging. Thus, all the information gained from the test intervals performed up to the end of accelerated aging may be sufficient to establish the worst-case leachables profile (asymptotic leaching), in which case the leachables profiling activity has been completed. Furthermore, it is possible to submit leachables data up to the end of accelerated aging as a means of securing a timely regulatory review of the available data. Providing the leachables data at the later testing intervals then becomes a post-approval commitment.

Second, if the worst-case leachables profile has not been established by the end of accelerated aging, then I would continue the "normal use" conditions testing only to the point that it is required to establish the worst-case profile. Thus, I would plan on performing leachables testing at all the later timepoints and only actually do the testing at the timepoints necessary to establish the worst-case profile.

Third, I would say that it will probably be the case that the pre-established stability testing time points will not coincide exactly with the time at which asymptotic leaching is achieved. This situation is not problematic as it is the asymptotic level of leaching that is important to the compatibility assessment. While the time at which asymptotic leaching is achieved may be interesting, it is not a relevant point to consider in the compatibility assessment (unless it is the compatibility assessment that is driving the product's shelf-life).

Lastly, I have argued that the end of shelf-life testing under actual storage conditions might not be relevant if asymptotic leaching levels are achieved before the end of shelf-life. However, practically speaking it would be a rare circumstance indeed if a regulatory file was considered to be complete without the end of shelf-life leachables testing. For this reason, skipping the end of shelf-life leachables testing is not recommended.

2.4 Variation in Extraction and Leaching Studies

Previously in Section 2.2 the topic of variation in extractions was discussed in the context of generating replicate extracts. Considering this topic, I note upfront that the possibility that extraction study results are highly variable is a topic of considerable interest and debate in the E&L community. Although there is no definitive published data that has been intentionally and systematically collected to address this matter in a rigorous and methodical manner, there is sufficient published and unpublished data to empirically support the observation that extraction studies can produce variable results and even a superficial review of typical extraction practices reveals that the practices are such that the opportunity for variation exists.

The topic is significant as variation in extraction results can produce flawed compatibility assessments, with the greatest concern being that patient exposure is underestimated if, due to variation, not all the extractables are reported, the reported concentration of the extractables is lower than the actual concentration, and the reported identities are incorrect. Here I want to address variation in terms of identifying sources of variation and suggesting ways to reduce variation.

First, I want to address the topic of intralaboratory variation. By this I mean that two testing laboratories are provided with the same test article and are asked to perform an extraction study. Unless the testing laboratories are given specific instructions as to how to perform the extraction, it is possible that they could adopt completely different extraction processes, including different extraction vehicles, different extraction times and durations, different extraction methods, etc. In such a circumstance it is clear that there is a very high probability that the extractables profile generated by the two testing laboratories would be quite different and the extraction judged to be highly variable and non-reproducible.

However, even if the two laboratories were given the same instructions, it is possible that the extractions performed would be somewhat variable. First, the extraction procedure itself might have intrinsic sources of variation. Second, the test article itself may be non-homogenous, adding another dimension of variation. These two items are largely beyond the laboratory's controls and would be experienced even if the testing were performed in the same laboratory. However, there is a third source of variation in the sense that even well-specified extraction processes have some "wiggle room" when they are implemented. For example, consider the case where the extraction vehicle is specified as "low pH buffer, pH between 2.9 and 3.1," which would be a reasonable specification of an extraction vehicle with a nominal pH of 3.0. If one lab prepared the extraction vehicle at one extreme and the other lab prepared the vehicle at the second extreme, one could envisage that the extraction of acidic and basic extractables might be different between the labs to a certain degree.

This discussion leads to Recommendation 1:

Recommendation 1. The sponsor and the testing laboratory should agree on the technical details of the extraction to be performed and those details should be clearly and specifically established in the testing protocol.

Now lets us consider the case of interlaboratory variation, a situation where the intralaboratory sources of variation have largely been eliminated or mitigated. In this case, the intrinsic inhomogeneity of the test article is still a contributor to variation. It is obvious that when a test article is inherently inhomogeneous that the smaller the amount of test article extracted, the larger will be the extract-to-extract variation. It is also obvious that the more structurally complex is the test article, the greater the potential variation will be in the extraction, especially with regard to indirect contact surfaces (indirect contact between the surface and the extraction vehicle). This leads to the second and third Recommendations:

Recommendation 2. An extraction should be designed so that the maximum amount of test article can be extracted, consistent with the nature of the test article and the nature of the extraction process.

Recommendation 3. When a test article is structurally complex, the extraction should be designed to account for and minimize the impact of the design on extraction variation.

Although Recommendation 2 is straightforward, Recommendation 3 may be more difficult to understand. To facilitate understanding, consider the circumstance that the test article is a multi-layered film to be used as a container of parenteral drug products. In this case the film can be tested in three ways:

1. The film is formed into a pouch, which is subsequently filled with the extracting vehicle,
2. The film is immersed, intact, in the extracting vehicle,
3. The film is cut up and the pieces are immersed in the extraction vehicle.

At first glance, a compelling argument can be made to perform the extraction in the first way, as so doing tests the film in the manner that it is being used. However, unless the extraction is such that equilibrium is achieved during extraction, the first way leads to the greatest potential variation in the extraction. In fact, it is the third option that will produce the lowest extraction-to-extraction variation.

Now it is the practical reality that the choice of any extraction process considers multiple parameters and it still may be that in the previous example that the proper extraction to perform is "pouch and fill." Nevertheless, one must understand and accept the effect that this decision could have on extraction variation.

Continuing our discussion further, one notes that the extraction generation process consists of two distinct actions, the actual generation of the extract and what is done with the extract between the point where it is generated and the point that it is tested (extract storage). Considering the simpler of these two, extract storage, there are several good practice recommendations that are relevant not only in terms of reducing variation but also ensuring extract integrity. These Recommendations are as follows:

Recommendation 4. Extracts must be handled and stored carefully between the time that they are generated and the time they are tested. Careful handling means that:

1. If extraction is accomplished at elevated temperatures, the extracts must be allowed to completely equilibrate to ambient temperature before they are further manipulated.
2. After the extracts have been equilibrated, the extraction vehicle should carefully be separated from the extracted test article.
3. Sample handling processes should be kept to a minimum and when they are necessary, they should be

designed to minimize the adverse effects of contamination and/or analytic loss. For example, unless filtration is obviously necessary, it should be avoided. If samples are to be stored in vessels prior to analysis, the vessels must be inert and the empty headspace should be minimized.

4. Care must be exercised in the storage of extracts prior to testing. Extracts should be protected from light, heat, etc. The storage duration should be as short as possible and storage durations longer than two weeks at ambient temperature are discouraged. Although refrigerated storage of extracts may be acceptable, it is generally recommended that extracts are not frozen, as the effect of freezing on the extract may be complicated and unpredictable.

5. Obvious changes to an extract during storage (e.g. discoloration, particulate formation, etc.) must lead to the rejection of the extracts as viable test subjects, as alternations to the extract to account for these effects (for example, filtration, pH adjustment, dilution) compromises sample integrity and potentially removes analytes of interest from the extracts.

Two of the factors relevant to the storage of the extract, prevention of contamination and protection from solvent loss, are also relevant to the extraction itself. In either case, it is clear that the most reproducible extraction is one where there is no communication between the extraction (vehicle + test article) and the external environment as the communication processes are not only a means for corrupting the extract but are generally highly non-reproducible and therefore add variation to the extraction. Thus, Recommendation 5 becomes:

Recommendation 5. Extraction should be performed in a closed environment or, if not practically achievable, an environment that minimizes the possibility that substances from the environment can transfer to the extract or that the extraction vehicle is lost, in all or in part, from the extraction vessel. Considering the extraction process itself, I have previously noted in the discussion of replicates that an extraction that proceeds to equilibrium is a more reproducible extraction than one that ends while the transfer (diffusion) of substances between the test article and the extraction vehicle is still occurring. An extension of this point is that an extraction that ends while an ongoing transfer is slow is more reproducible than an extraction that ends while the ongoing transfer is fast or when the rate of transfer is rapidly changing. These observations lead to Recommendation 6:

Recommendation 6. When it is consistent with the purpose of the extraction study, an extraction should be designed so that equilibrium is achieved. If this cannot be justified, extractions should be designed to avoid circumstances where the extraction is ended while it is proceeding rapidly or where the rate of extraction is rapidly changing.

There are several aspects of an extraction whose control provides an obvious means for reducing variations as it is clear that the means by which the extraction is accomplished can reduce variations. These aspects are captured in Recommendation 7.

Recommendation 7. An extraction that minimizes variation is one:

1. Which is accomplished in a minimum number of steps,
2. Whose individual steps consist of the minimum number of simple, easy to reproduce actions,
3. Which can be automated,
4. Wherein contact between the test article and extraction vehicle is constant, continuous and complete.

While items 1 to 3 in Recommendation 7 are obvious, I would like to explain, through examples, what is meant by item 4. Most researchers who routinely perform extractions have encountered the situation (likely more than once) were the relative density of the test article and the extraction vehicle are such that the test article "floats" on top of the extraction vehicle. It should be clear that such circumstances are highly unreproducible and thus means must be taken to ensure more complete and more reproducible contact between test article and extraction vehicle. Another example is static versus dynamic extraction. Especially for low temperature, short-durations extractions with viscous extraction vehicles, it can be the case that "natural" diffusion of an extractable from the test article to the extraction vehicle is thwarted by the lack of good mixing of the extraction vehicle, as an extractable-rich boundary layer may form between the bulk of the extraction vehicle and the test article. One anticipates that the formation and extent of such a boundary layer may vary across different extracts, leading to variations. Another example is the case of difficult to wet test article surfaces. For example, a circulating extraction of a manufacturing filter facilitates complete and effective contact between the entire cross-section of the filter and the extraction vehicle, leading to more reproducible extractions. An extraction in which the filter is immersed in the extraction vehicle may not accomplish complete contact between the filter's entire cross-section and the vehicle, even if the extraction unit is agitated.

It is clear that specific extraction processes can be further optimized, with respect to minimizing variation, in ways that are specific to the individual extraction processes. As it is not possible to address all possible extractions and thereby propose proper variation-reducing modifications, it is nevertheless important to stress that the reduction of variations is a characteristic of a well-designed extraction and not an accidental outcome of an optimized extraction.

If an extraction cannot be optimized to reduce variations or if an extraction has already been optimized with respect to variations, it is still possible that the variations can be further reduced via means of correcting for them. This concept is manifested in the addition of an internal standard to the test article prior to or during the extraction process. In theory, the internal standard experiences the same sources of variation as the extractables. Thus, if the extractables results are established relative to those of the internal standard, then the extractables results from replicate to replicate have been "corrected" for extraction variations between replicates.

In the case of extractables screening, one immediately understands the difficulties in implementing internal standards. First, the internal standard is most applicable if it "experiences" every process occurring during extraction that is experienced by the extractables and that it experiences these processes in the same way as the extractables. Taken to its logical extreme, this would mean that the internal standard has to be taken up by or added to the test article as if the internal standard was an actual ingredient in the test article, which practically speaking may be difficult to accomplish. When one spikes an aqueous extract with a small portion of an aqueous solution containing a dissolved internal standard, it is clear that the internal surrogate standard does not provide a perfect "correction" as it is, after all, already dissolved in the extraction vehicle and thus is not extracted from the test article. Thus, it is often the case that an internal standard may compensate for variations in certain, but not all, steps of the extraction process. Second is the quite variable chemical properties

of all potential extractables. Because potential extractables are chemically diverse, it is most unlikely that a single internal standard would properly mirror the behavior of all the potential extractables. So now the challenge becomes one of identifying a set of internal standards that (i) are not extractables themselves, (ii) do not chromatographically interfere with extractables, and (iii) behave as if they are extractables. At the current time, there is no set of internal standards that have been proposed, justified, and "validated" for use in controlling extraction-to-extraction variations.

My point in making this last statement is not to have analysts throw up their hands in despair and proclaim "I am cursed." Rather there are two points that I want to make:

1. It is unlikely that the use of an internal standard (or internal standards) will reduce variation to its lowest possible level (therefore expectations for improved reproducibility should be tempered with reality) and
2. A poorly devised and/or implemented internal standard approach is just as likely to increase variation as it to decrease it.

Variation in leachables studies is a much easier topic to handle, as in a leachables study the test article is the finished drug product. Since one is not generating the test article per say, the only sources of variation in samples for leachables studies is the natural variation in the drug product. This would include unit-to-unit variation (assuming the leachables study is designed to test multiple product units per test interval) and batch-to-batch (or lot-to-lot) variation (assuming the leachables study is designed to test multiple produce batches per test interval). In fact, while one seeks to reduce variation in the extraction process, establishing unit-to-unit and batch-to-batch variation is a critical aspect to address in the leaching study and thus "natural" variation in these parameters is not a property to reduce but rather a property to establish.

References

1 *Guidance for Industry* (1999). Container closure systems for packaging human drugs and biologics. US Department of Health and Human Services, Food and Drug Administration, Rockville, MD

2 *Guideline on Plastic Immediate Packaging Materials.* (2005) European Medicines Agency, CPMP/QWP/43359/03, EMEA/CVMP/205/04.

3 Safety thresholds and best practices for extractables and leachables in orally inhaled and nasal drug products. (2006). PQRI Leachables and Extractables Working Group.

4 USP <1663>. Assessment of extractables associated with pharmaceutical packaging/delivery systems. USP 43/NF38, pp. 8442. Official as of 1-Dec-2020.

5 USP <1664>. Assessment of drug product leachables associated with pharmaceutical packaging/delivery systems. USP43/NF38, pp. 8455. Official as of 1-Dec-2020.

6 Jenke, D. (2009). *Compatibility of Pharmaceutical Products and Contact Materials: Safety Considerations Associated with Extractables and Leachables*, 8. Hoboken, NJ: Wiley.

7 Ball, D.J., Norwood, D.L., and Nagao, L.M. (2012). Chapter 1: Overview of leachables and extractables in orally inhaled and nasal drug products. In: *Leachables and Extractables Handbook: Safety Evaluation, Qualification and Best Practices Applied to Inhalation Drug Products* (ed. D.J. Ball, D.L. Norwood, C.L.M. Stults and L.M. Nagao), 6. Hoboken, NJ: Wiley.

8 USP <1665>. Characterization and qualification of plastic components and systems used to manufacture pharmaceutical drug products and biopharmaceutical drug substances and products. USPNF 2022, Issue 1. Official as of 1-May-2022.

9 Recommendation for testing and evaluation of extractables form single-use process equipment. Bio-Process Systems Alliance (BPSA), Washington, DC (2010).

10 BioPhorum. BioPhorum best practices guide for extractables testing of polymeric single-use components used in biopharmaceutical manufacturing. 22-Apr-2020.

11 ISO 10993:18(2019) *Biological Evaluation of Medical Devices – Part 18: Evaluation and Testing within a Risk Management Process.* Arlington, VA: Association for the Advancement of Medical Instrumentation (AAMI).

12 USP <665>. Plastic components and systems used to manufacture pharmaceutical drug products and biopharmaceutical drug substances and products. USPNF 2022, Issue 1. Official as of 1-May-2022.

13 Jenke, D. and Ruberto, M. (2014). Using the correlation between material composition and extractables and leachables to forecast extractables and/or leachables profiles. Part 1: Concepts and plasticized poly-(vinylchlorde) case study. *Pharm. Outsourcing* 15 (1): 32, 34, 36, 38–41.

14 Jenke, D. and Ruberto, M. (2014). Using the correlation between material composition and extractables and leachables to forecast extractables and/or leachables profiles. Part 2: Polyethylene and Bromobutyl rubber case studies. *Pharm. Outsourcing* 15 (2): 22, 24, 26–31.

15 Fang, X., Cherico, N., Barbacci, D. et al. (2006). Leachables study on solid dosage form. *Am. Pharm. Rev.* 9 (6): 58–63.

16 Connolly, C. and Miron, N. (2014). Leachables studies – Investigating leachables in oral solid dosage forms. *Drug Dev Deliv.* 14 (7): 30–35.

17 Zdravkovic, S.A. (2018). Assessment of patient exposure to leachables from lyophilized drug formulations following reconstitution, storage, and administration via polymeric packaging/delivery systems. *J. Pharm. Sci.* 107 (11): 2837–2846.

18 Zdravkovic, S.A. (2017). Comparison of a lyophilized drug product to other solid and liquid media for the extraction of elastomeric oligomers from a butyl rubber stopper. *PDA J. Pharm. Sci. Tech.* 71 (6): 488–501.

19 Jenke, D.R. (1994). Determination of solute-polymer interaction properties via two-stage extraction. *Pharm. Res.* 11 (5): 774–775.

20 ICH Q3D (2018). *Elemental Impurities in Drug Products, Guidance for Industry.* Center for Drug Evaluation and Research (CDER) and Center for Biologics Evaluation and Research (CBER), Food and Drug Administration.

21 Grause, G., Hirahashi, S., Toyada, H., and Kameda, T. (2017). Solubility parameters for determining optimal solvents for separating PVC from PVC-coated PET fibers. *J. Mater. Cycles Waste Manag.* 19 (2): 612–622.

22 Lehtimaki, E. and Vaisanen, A. (2017). Determination of metals concentrations in certified plastic reference materials after small-size autoclave and microwave-assisted digestion followed with inductively coupled plasma optical emission spectrometry. *Spectrochim. Acta B* 127: 64–69.

23 Jenke, D., Liu, N., Hua, Y. et al. (2015). A means of establishing and justifying binary ethanol/water mixtures as simulating solvents in extractables studies. *PDA J. Pharm. Sci. Technol.* 69 (3): 366–338.

24 Baner, A. and Pronger, O. (2008). Partition coefficients. In: *Plastic Packaging* (ed. O.G. Pringer and A.L. Baner), 89–122. Wiley-VCH Verlag GmbH & Co.

25 Hayward, D.S., Kenley, R.A., and Jenke, D.R. (1990). Interactions between polymer containers and parenteral solutions: The correlation of equilibrium constants for polymer-water partitioning with octanol-water partition coefficients. *Int. J. Pharm.* 59: 245–253.

26 Kenley, R.A. and Jenke, D.R. (1990). Determination of solute-polymer interaction properties and their application to parenteral product container compatibility evaluations. *Pharm. Res.* 7 (9): 911–918.

27 Jenke, D.R., Hayward, D.S., and Kenley, R.A. (1990). Liquid chromatographic measurement of solute solvent/solvent partition coefficients: Application to solute/container interactions. *J. Chromatogr. Sci.* 28 (12): 609–612.

28 Hayward, D.S. and Jenke, D.R. (1990). Interactions between polymer containers and parenteral solutions: Correlating equilibrium polymer-water interaction constants with solute partition coefficients. *Int. J. Pharm.* 66 (1–3): 87–96.

29 Jenke, D.R., Kenley, R.A., and Hayward, D.S. (1991). Interactions between polymeric containers and their contained solution: Modeling of polymer-water solute partitioning via coupled solvent-water partition coefficients. *J. Appl. Polymer Sci.* 43: 1475–1482.

30 Jenke, D.R. (1994). Drug binding by reservoirs in elastomeric infusion devices. *Pharm. Res.* 11: 984–989.

31 Zdravkovic, S.A. (2016). Comparison of the solubilization properties of polysorbate 80 and isopropanol/water solvent systems for organic compounds extracted from three pharmaceutical packaging configurations. *Eur. J. Pharm. Sci.* 93: 475–483.

32 Feenstra, P., Brunsteiner, M., and Khinast, J. (2014). Investigation of migrant-polymer interaction in pharmaceutical packaging material using the linear interaction energy algorithm. *J. Pharm. Sci.* 103: 3197–3204.

33 Peppas, N.S. (1983). A model of dissolution-controlled solute release from porous drug delivery polymeric systems. *J. Biomed. Mater. Res.* 17 (6): 1079–1087.

34 Ayres, J.L., Osborne, J.L., Hopfenburg, H.B., and Koross, W.J. (1983). Effect of variable storage times on the calculation of diffusion coefficients characterizing small molecule migration in polymers. *Int. Eng. Chem. Prod. Res. Dev.* 22 (1): 86–89.

35 Yu, B.L., Han, J., Hammond, M. et al. (2017). Kinetic modeling of the release of ethylene oxide from sterilized plastic containers and its interaction with monoclonal antibodies. *PDA J. Pharm. Sci. Technol.* 71 (1): 11–19.

36 Fang, L. and Zhao, C. (2019). Modeling the permeation rates of organic migrants through a fluoropolymer film. *PDA J. Pharm. Sci. Technol.* 73 (1): 70–82.

37 Parthasarathy, R., Misra, A., Park, J., and Spencer, P. (2012). Diffusion coefficients of water and leachables in methacrylate-based crosslinked polymers using absorption experiments. *J. Mater. Sci.: Mater. Med.* 23: 1157–1172.

38 Jenke, D. (2017). Application of Arrhenius kinetics to acceleration of controlled extraction studies. *PDA J. Pharm. Sci. Technol.* 71 (1): 11–19.

39 Weltzmann, C.J. (1997). The use of model solvents for evaluating extractables from filters used to process pharmaceutical products. *Pharm. Technol.* 21 (4): 72–99.

40 Jenke, D.R., Brennan, J., Dotty, M., and Poss, M. (2003). Use of binary ethanol/water model solutions to mimic the interaction between a plastic material and pharmaceutical formulations. *J. Appl. Polym. Sci.* 89 (4): 1049–1057.

41 Jenke, D., Odufu, A., and Poss, M. (2006). The effect of solvent polarity on the accumulation of leachables from pharmaceutical product containers. *Eur. J. Pharm. Sci.* 27 (2): 133–142.

42 Roark, J.M., Jacques, M.N., Tullo, E.J. et al. (2013). Alternative solvents for extractables and leachables evaluation. *Pharm. Technol.* 37 (8): 51–53.

43 Ding, W. (2013). Determination of extractables and leachables from single-use systems. *Chem. Ing. Tech.* 85 (1–2): 186–196.

44 Reichardt, C. (1994). Solvatochromic dyes as solvent polarity indicators. *Chem. Rev.* 94: 2319–2358.

45 Reichardt, C. (2005). Polarity of ionic liquids determined empirically by means of solvatochromic pyridinium N-phenolate betaine dyes. *Green Chem.* 7: 339–351.

46 Jenke, D., Liu, N., Hua, Y. et al. (2015). A means of establishing and justifying binary ethanol/water mixtures as simulating solvents in extractables studies. *PDA J. Pharm. Sci. Technol.* 15 (4): 20, 22, 24–27.

47 Pearson, S.D. and Trissel, L.A. (1993). Leaching of diethylhexyl phthalate from polyvinyl chloride containers by selected drugs and formulation components. *Am. J. Health-Syst. Pharm.* 50 (7): 1405–1409.

48 Faouzi, M.A., Dine, T., Luyckx, M. et al. (1994). Leaching of diethylhexyl phthalate from PVC bags into intravenous teniposide solution. *Int. J. Pharm.* 105 (1): 89–93.

49 Faouzi, M.A., Dine, T., Luyckx, M. et al. (1995). Stability, compatibility and plasticizer extraction of miconazole injection added to infusion solutions and stored in PVC containers. *J. Pharm. Biomed. Anal.* 13 (11): 1363–1372.

50 Dorival, N., Larsson, I., and Bones, J. (2017). Non-volatile extractable analysis of prefilled syringes for parenteral administration of drug products. *J. Pharm. Biomed. Anal.* 142: 337–342.

51 Jenke, D., Zietlow, D., and Sadain, S. (2004). Leaching of organic acids from irradiated EVA plastics as a function of solution pH and polarity. *PDA J. Pharm. Sci. Technol.* 58 (1): 24–31.

52 Jenke, D. and Couch, T. (2006). A consideration of the impact of solution composition on the accumulation of organic substances leached from plastics used in container/closure systems. *PDA J. Pharm. Sci. Technol.* 60 (1): 60–71.

53 Jenke, D.R., Chess, E.K., Zietlow, D.C., and Rabinow, B.E. (1992). Model for estimating the accumulation of solutes leaching from polymeric containers into parenteral solutions. *Int. J. Pharm.* 78: 115–122.

54 Jenke, D. (2011). A general assessment of the physiochemical factors that affect the leaching of organic substances from plastic contact materials into aqueous pharmaceutical solutions. *PDA J. Pharm. Sci. Technol.* 65 (2): 166–176.

55 Jenke, D., Castner, J., Egert, T. et al. (2013). Extractables characterization of five materials of construction representative of packaging systems used for parenteral and ophthalmic drug products. *PDA J. Pharm. Sci. Technol.* 76 (5): 448–511.

56 Teasdale, A., Jahn, M., Bailey, S. et al. (2015). Controlled extraction studies applied to polyvinylchloride and polyethylene materials: conclusions from the ELSIE controlled extraction pilot study. *AAPS PharmSciTech.* 16 (3): 664–674.

57 Jenke, D. (2010). Extraction of stearate salts from plastic materials used in pharmaceutical applications. *PDA J. Pharm. Sci. Technol.* 64 (3): 200–210.

58 Jenke, D. (2014). Establishing the proper pH of simulating solvents used in organic extractables assessments for packaging systems and their materials of construction used with aqueous parenteral drug products. *Pharm Outsourcing* 15 (4): 20, 22, 24–27.

59 Cruz, L.A., Jenke, M.P., Kenley, R.A. et al. (1990). Influence of solute degradation on the accumulation of solutes migrating into solution from polymeric parenteral containers. *Pharm. Res.* 7 (9): 967–972.

60 Paskiet, D., Jenke, D., Ball, D. et al. (2013). The Product Quality Research Institute (PQRI) leachables and extractables working group initiatives for parenteral and ophthalmic drug product (PODP). *PDA J. Pharm. Sci. Technol.* 67 (5): 430–447.

61 Ding, W., Madsen, G., Mahajan, E. et al. (2014). Standardized extractables testing protocol for single-use systems in biomanufacturing. *Pharm. Eng.* 34 (6): 1–11.

62 ASTM F1980-16 Reapproved(2016). *Standard Guide for Accelerated Aging of Sterile Barrier Systems for Medical Devices*. West Conshohocken, PA: ASTM International Approved 15 September 2016.

63 Franz, R. and Stormer, A. (2008). Migration of plastic constituents. In: *Plastic Packaging: Interactions with Foods and Pharmaceuticals*, 2e (ed. O.G. Piringer and A.L. Baner), 368. Wiley-VCH.

64 Jenke, D. and Barge, V.J. (2014). Factors affecting the release of extractable acetic acid from multi-layered plastic films containing ethylene vinyl acetate (EVA) and polyethylene (PE) layers. *Pharm. Outsourcing* 15 (3): 52, 54, 56–59.

65 Srivastava, S.P., Saxena, A.K., and Seth, P.K. (1985). Migration of di (2-ethylhexyl) phthalate (DEHP) from plastic containers used for packaging infusion solutions. *Indian J. Pharm. Sci.* 1985: 156–158.

66 Franz, R. and Stormer, A. (2008). Migration of plastic constituents. In: *Plastic Packaging: Interactions with Foods and Pharmaceuticals*, vol. 2 (ed. O.G. Piringer and A.L. Baner), 370. Wiley-VCH.

67 Jenke, D.R. and Rabinow, B.E. (2017). Proper accounting for surface area to solution volume ratios in exaggerated extractions. *PDA J. Pharm. Sci. Technol.* 71 (3): 225–233.

68 Jenke, D., Couch, T., and Gillum, A. (2010). Comparison of solute-binding properties of plastic materials used as pharmaceutical product containers. *PDA J. Pharm. Sci. Technol.* 64 (1): 37–43.

3

General Principles of Chemical Assessment: Analytical Testing for Extractables and Leachables; Organic Compounds

3.1 Key Definitions and Concepts

We enter this part of the chemical compatibility assessment with a sample in hand, whether that sample be an extract, a drug product, or another type of patient contact medium. In any case, these samples have been generated by contacting the sample with an item for the purpose (or outcome) of extracting (or leaching) substances from that item. Clearly, in order to perform the chemical compatibility assessment these extracted or leached substances must be discovered, identified, and quantified (see Table 3.1). To be sure that there is clarity and consistency in usage, these terms are defined as follows:

Discovery. The process by which analytes present in a sample are revealed by an analytical method, typically manifested by the analyte producing a response that is recognized by the method.

Identification. The process of assigning a molecular structure, a chemical name, or other identifying information to an organic compound, or assigning constituent elements or molecular structure as appropriate, and a chemical name to an inorganic compound [1].

Quantification (quantitation). The process of establishing the concentration of an analyte present in a sample [1].

A fourth process, scouting, is the use of analytical methods that address general properties of an extract for the purpose of establishing certain general properties of individual extractables (for example, extract pH can be used to establish if there are acidic or basic extractables). Scouting is considered in greater detail in Chapter 4, Section 4.3.

Extractables and leachables can "naturally" be differentiated into three classes, depending on the chemical nature of these substances and the analytical method by which they are addressed. One major class of extractables and leachables are organic substances, which are typically addressed analytically by chromatographic means. Now while it might seem obvious that a second class

of extractables and leachables is inorganic substances, this is not necessarily the case, primarily as a matter of the analytical realities involved. In fact, the second class is termed **elemental extractables and leachables** and are essentially extractables and leachables that are associated with constituent elements other than carbon, hydrogen, oxygen, and nitrogen. They are termed elemental extractables and leachables because the test methods employed in addressing these analytes are elemental methods (e.g. atomic spectroscopy) and the compatibility impact of these extractables and leachables is based on the element's properties. Because these extractables and leachables are measured and assessed as elements, their actual chemical form may not be known or specified. It is for this reason that it is not proper to characterize these elemental extractables and leachables as being inorganic, as some extracted or leached elements may have been extracted as element-containing organic substances such as organometallic compounds.

A third, and relatively more minor, class of extractables and leachables are called ionic extractables and leachables because they exist as ions in the test sample and are measured as ions by the test methods (for example, ion chromatography). Examples of ionic extractables and leachables include the counterions of strong and weak acids such as chloride (Cl^-), bromide (Br^-), nitrate (NO_3^-), phosphate (PO_4^{3-}), sulfate (SO_4^{2-}), acetate ($C_2H_3O_2^-$), and formate (CHO_2^-).

Analytical methods applicable to organic extractables are covered in this Chapter 3. Analytical methods for elemental extractables and ionic extractables are considered in greater detail in Chapter 4, Sections 4.1 and 4.2 respectively.

Whether the desired outcome of analyzing a sample is to account for all extractables uniquely present in an extract (versus an extraction blank) above an established threshold or to establish all leachables uniquely present in a drug product (versus some sort of control or blank) above an established threshold, the desired outcome is obtained by analyzing the extract, drug product, or contact medium

Extractables and Leachables: Characterization of Drug Products, Packaging, Manufacturing and Delivery Systems, and Medical Devices, First Edition. Dennis Jenke.
© 2022 John Wiley & Sons, Inc. Published 2022 by John Wiley & Sons, Inc.

Table 3.1 Analytical methods for extractables survey, material evaluation.

Nature	Purpose	Desirable characteristics	Examples
Scouting	Establish general chemical properties of the extractables	Responsive to bulk chemical properties of the extractables	TOC, NVR, pH, UV absorbance, IR, conductivity
Discovery	Find the extractables	Sensitive, specific, broad in scope and range, "universal" response	GC/FID, GC/MS, LC/PDA, LC/MS, LC/ELSD, Headspace GC/FID, IC, atomic spectroscopy (scanning)
Identification	Identify the extractables that have been found	Sensitive, specific, "information rich"	LC/MS/MS, LC/MS (accurate mass), LC/NMR, GC/MS, GC/IR, IC/MS, Headspace GC/MS, "off-line" spectroscopy coupled to preparative techniques, specialty detection methods, atomic spectroscopy, (scanning)
Quantification	Measure what has been identified (and estimate what remains unidentified)	Accurate, precise, specific, sensitive, "universal" response, known response function	GC/FID, LC/UV, LC/MS, LC/ELSD, IC, Headspace GC, atomic spectroscopy (analyte specific)

(and its associated blank or control) with analytical methods that are able to produce useful and interpretable responses for each and every potential extractable or leachable. The analytical methods employed for this purpose are generally of three types: target, screening, or targeted screening. Definitions for the first two analytical processes are as follows:

Analytical screening method. A method whose purpose is to discover, identify, and semi-quantitatively estimate the concentration of all relevant, but unspecified, analytes in a test sample above an established reporting threshold (such as the AET) [1].

Analytical targeting method. A method whose purpose is to quantify, with an appropriately high degree of accuracy and precision, specified analytes in a specified test sample over a specified concentration range [1].

In targeted analyses, the extractables or leachables that the method is intended to address are typically a smaller set of substances of known or suspected significance. That is, specific extractables or leachables have been established as targets for the analysis. As the targets are known, the sole purpose of the target analysis is to quantify the targets in either the extracts or the drug product. As quantitation is the sole purpose and the suite of targeted substances is generally small, targeted methods are optimized (and validated) in terms of attributes relevant to quantitation, such as accuracy, precision, sensitivity, and selectivity. A secondary purpose of target analysis is to ensure that the compound that has been identified as the target compound in the test article is in fact the target compound and not an imposter.

It is often the case, however, that the extractables in an extract or leachables in a drug product cannot be specified up-front and thus they must be discovered, after which the discovered substances must be identified, as they were not specified up-front. Lastly, the discovered substances must be quantified. The analytical process by which unspecified extractables or leachables are discovered, identified, and quantified is termed screening (see Figure 3.1).

The difference between screening and targeting is sometimes explained via an analogy to fishing. For example, when one fishes with a net, one is essentially indiscriminately fishing for whatever fish are caught in the net, which is akin to screening. However, when one is fishing with a rod that is baited with a specific lure and bait, one generally optimizes the fishing conditions so as to catch only a certain type of fish (targeting). This concept is illustrated in Figure 3.2.

As may be clear from the previous definitions, targeted screening is a hybrid between the target and screening methods. A targeted screening method retains the aspects of a screening method in that the intent of the analysis is to screen a sample for generally a larger population of extractables; however, in this case the individual extractables of interest are specified upfront (targeted). Moreover, like targeted analysis (and unlike screening analysis), the method employed to estimate the concentration of the analytes is quantitative. Nevertheless, while the target screening method is qualified (like a screening method), it is likely not to be validated to the extent of a targeted method.

The distinction between screening targeting and targeted screening is illustrated in Table 3.2.

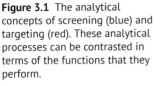

Figure 3.1 The analytical concepts of screening (blue) and targeting (red). These analytical processes can be contrasted in terms of the functions that they perform.

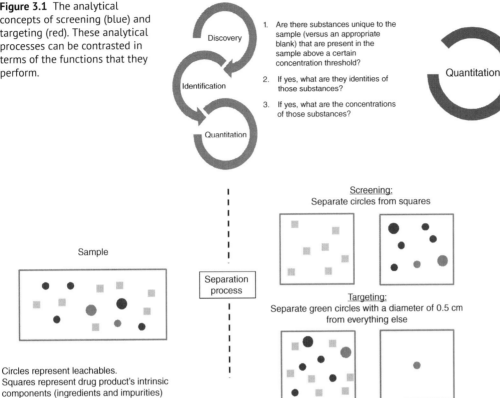

Figure 3.2 Comparison of screening versus targeting. In screening, the separation is at a high level, with the screening method separating the analytes (circles) from the other constituents in the sample (squares). However, in targeting the separation is much more specific, focusing on a specific analyte. (*See insert for colour representation of the figure.*)

Table 3.2 Test methods for extractables and leachables.

	Type of analytes	
Type of quantitation	**Unspecified**	**Specified**
Estimate	Screening	Not likely[a]
Semi-quantitative	Not likely[b]	Target screening[c]
Quantitative	Not possible[d]	Targeting

a) There is little to be gained by targeting an analyte and only estimating its concentration.
b) Given the response variation between analytes, it is difficult to quantify unspecified extractables in a semi-quantitative manner.
c) Targeted screening is semi-quantitative, versus estimate for screening methods, because it is supported by a database of response factors.
d) Given the response variation between analytes, it is virtually impossible to quantify unspecified extractables in a quantitative manner.

Screening as an analytical process for organic extractables and leachables will be considered in greater detail in Section 3.2, while target analysis will be discussed in Section 3.3 and targeted screening is discussed in Section 3.4 .

Desirable characteristics of any analytical method used to characterize a sample for its contents include the following:

- Sensitivity
- Selective (specificity)
- Breadth (broad scope, "all encompassing")
- Uniform in response across compounds and concentrations
- Accurate
- Precise
- Reproducible (rugged and robust)

When the purpose of an analytical method is established as either screening or targeting, one understands that there will be a trade-off in terms of obtaining these desirable characteristics. For example, perhaps the most important characteristic for a screening method is that it should be broad scope, in the sense that it produces a response for the largest possible number of potential analytes. However, a method that is broad scope will sacrifice certain other desirable performance characteristics, most notably selectivity and most likely sensitivity. Furthermore, since the characteristic of a uniform response is a goal that is not so easily achieved, accuracy suffers to a certain extent.

Alternatively, a target method places a premium on characteristics including selectivity, accuracy, and precision, while sacrificing breadth of scope.

Lastly, targeted screening attempts to marry the breadth of scope aspect of a screening method with the more quantitative aspect of a target method.

3.2 Organic Substance Analysis, Screening

3.2.1 Relevant Analytical Methods

The universe of potential organic extractables and leachables is large and diverse and thus it is the case that no single analytical method has the ability to screen extracts or drug products for all possible extractables and leachables. Therefore, a set of complementary, overlapping, and orthogonal analytical methods are employed for the purpose of extractables and leachables screening.

As chromatographic methods are uniquely appropriate for analyzing test samples for organic constituents, it has become standard industry practice to accomplish E&L screening via chromatographic methods as the chromatographic process is a necessary and effective means of separating individual analytes from one another. This separation of analytes is an essential aspect in terms of securing identities and concentration estimates. Moreover, the various types of chromatography lend themselves well to the concepts of complementary, overlapping, and orthogonal.

For example, it is widely understood that gas chromatography (GC) coupled with headspace sampling is an effective means of analyzing a sample for volatile organic substances. Additionally, so-called direct injection gas chromatographic methods are suited for semi-volatile organic analytes while liquid chromatography is well-suited for non-volatile organic analytes (see Figure 3.3). As they rely on different mechanisms of separation, these methods are orthogonal; because the distinction between volatile, semi-volatile, and non-volatile is imprecise, the methods are complementary in the sense that analytes can be accounted for by two of the three methods.

One understands in examining Figure 3.3 that the classification of organic extractables in terms of their volatility and then linking specific chromatographic methods to the volatility classes is an obvious over-simplification, as the chromatographic methods themselves do not primarily differentiate between analytes based solely on volatility. For example, analyte polarity is another differentiating factor for the organic extractables and their associated chromatographic method. Nevertheless, Figure 3.3 effectively establishes the concept of effective screening through the use of multiple orthogonal and overlapping analytical techniques.

3.2.2 Survey of Chromatographic Screening Methods

The screening of samples for extractables and leachables presents unique challenges to analytical scientists for the following reasons:

1. **Structural diversity.** Additives to rubber and plastic materials serve a variety of important functions within these materials during their manufacturing processes. These functions include: antioxidant, antistatic, anti-slip, curing agent, curing accelerator, ultraviolet

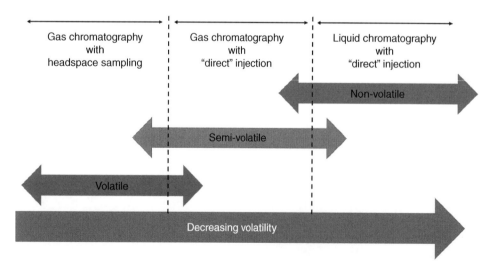

Figure 3.3 Application of chromatographic methods in extractables and leachables screening. Various overlapping but orthogonal chromatographic separation methods are used to encompass the widest possible range of organic substances, specifically considering the substances' volatility. The orthogonal aspect of the methods provides the widest breadth while the overlap reduces the possibility of there being "gaps" between the methods.

(UV) stabilizer, filler, lubricant, pigment, etc. Substances performing an individual function can have significantly diverse chemical structures and therefore different chemical and physical properties. For example, antioxidants can include phosphites, hindered phenols, aromatic amines, phenolic sulfides, thiophenols, sulphoxides, sulphones, metal dithiolates, organoboranes, etc.

2. **Sample diversity.** The materials used in packaging, delivery, and/or manufacturing systems are quite diverse. Considering rubber materials specifically, there are two main categories of rubber used for pharmaceutical container closure systems, sulfur cured and peroxide cured. Each category includes several different polymeric types with different applications. Also of interest are thermoplastic elastomers, which include a plastic such as low-density polyethylene as a reinforcing agent in the rubber matrix. Each elastomer has different swelling properties during laboratory extraction and drug product leaching. Plastics behave differently than rubber during laboratory extraction studies. Extractables associated with metal and glass components are usually surface residues, and are often complex mixtures. Thus, the universe of potential extractables and leachables is large and diverse. Moreover, the test samples themselves can be chemically diverse. Drug products are widely variable in composition and thus present a considerable analytical challenge for leachables' assessments, particularly non-aqueous formulations. Additionally, numerous compositionally unique extracting media have been reported and recommended for use in extraction studies.

3. **Concentration levels.** Leachables can be present in drug products over a wide concentration range. For example, in MDI drug products, primary rubber gasket additives can appear as leachables at the tens to several hundreds of μg/canister levels, polyaromatic hydrocarbons at the ng to low μg/canister levels, and N-nitrosamines at low ng/canister levels. The specific levels notwithstanding, these levels are all considered "trace" and require special considerations for analytical method development.

Given the complexity of the analytical challenge faced by those performing extractables and leachables testing, it is not surprising that success in such an endeavor is achieved via the use of a multitude of complementary, yet orthogonal, analytical methodologies. As noted in the Product Quality Research Institute's (PQRI) Orally Inhaled and Nasal Drug Products (OINDP) document proposing "Best Practice" recommendations for extractables and leachables studies in inhalation drug product development [2]: "No single analytical technique will be sufficient to detect and/or identify all possible extractables (… or leachables) from any particular container/closure system component

(… or drug product matrix), therefore, multiple broad spectrum techniques should be used to ensure complete evaluation of an extractables (… or leachables) profile."

In the case of organic extractables and leachables, many authorities have specifically recommended the use of Gas Chromatography (GC)/Flame Ionization Detection (FID), Gas Chromatography/Mass Spectrometry (GC/MS), and HPLC combined with both Mass Spectrometry (LC/MS) or UV detection (LC/UV). The separating power of capillary GC and ultrahigh-performance liquid chromatography (UPLC), combined with the universality, sensitivity, and information content of their associated detection methods (e.g. FID and MS), makes this family of techniques highly applicable to extractables/leachables screening.

The application of the various chromatographic methods has essentially mirrored their scientific and practical development. Thus, GC was the primary analytical method employed in early (ca. 1990 and before) screening studies and several early applications of GC for extractables screening are listed in a recent review by Jenke [3]. GC screening methods based on MS or tandem FID/MS detection are enumerated, in varying degrees of detail, in virtually every E&L publication that speaks to screening. Although it is beyond the scope of this book to provide a comprehensive review of all published GC screening methods, Table 3.3 is a listing of notable recent publications.

Despite its status as a "must have" methodology for organic E&L screening, many extractables and leachables are not totally amenable to GC analysis, because of the lack of volatility, thermal instability, or chemical/physical interactions with the GC column or system. Under such circumstances, HPLC with different detection possibilities becomes the method of choice for a wide variety of leachables/extractables.

In a somewhat philosophical treatment of the fundamental principles of analytical chemistry, Valcarce [4] distinguishes between three types of analysis:

- **Qualitative analysis.** "A type of chemical analysis by which the analyte or analytes in a sample are identified. The result is a YES/NO binary response."
- **Quantitative analysis.** "A type of chemical analysis by which the proportion or amount of each analyte in a sample is determined. The result is a numerical response."
- **Structural analysis.** "A type of chemical analysis by which the structure of a sample (viz. the spatial distribution of its constituents) or a pure analyte is established."

Given these classifications in analytical chemistry, one can differentiate between the two detection methods used with HPLC separations for extractables and leachables screening, UV absorbance, and mass spectrometry (MS). LC/UV is clearly capable of qualitative and quantitative analysis but is limited in its capability for structural

Table 3.3 Relevant references for the application of gas chromatography (GC) to E&L screening.

Authors	Title	Source	Contents
Zweiben, C., and Shaw, A.J.	Use of thermal desorption GC-MS to characterize packaging materials for potential extractables	*PDA J. Pharm. Sci. Technol.* 63(4): 353–359 (2009)	Use of thermal desorption allows for the characterization of polymers for ingredients that could become extractables. Solid sampling eliminates the need for extraction.
Armstrong, B.L., Senyurt, A.F., Narayan, V., Wang, X., Alquier, L., and Vas, G.	Stir bar absorptive extraction combined with GC-MS/MS for determination of low-level leachable components from implantable medical devices	*J. Pharm. Biomed. Anal.* 74:162–170 (2013)	This combination of sample preparation and GC analysis was capable of screening extracts from implantable medical devices to levels as low as 150 ng/device and a method was validated for antioxidant-related leachables from orthopedic knee-inserts made of polyethylene.
Ye, X., Ma, J., Yang, Y., and Jin, F.	Analysis of 21 phthalate leachables in metered dose inhalers by gas chromatography tandem mass spectroscopy	*Anal. Methods* 6:4083–4089 (2014)	The optimized and validated method had limits of detection of approximately 10 ng/ml for the targeted analytes.
Zdravkovic, S.	Solid-phase extraction in tandem with GC/MS for the determination of semi-volatile organic substances extracted from pharmaceutical packaging/delivery systems via aqueous solvent systems	*J. Pharm. Biomed. Anal.* 112:126–138 (2015)	The applicability of the combination of sample preparation and GC analysis to screening was considered in the context of 31 commonly observed extractables as model compounds. The method was able to quantify extractables in the range of 20 ng/ml to 50 µg/ml with recoveries in the range of 70–130% and precision of 15% or less.
Hoffmann, A., Albinus, T., Almasi, E., and Thaxton, K.	Extractables and leachables analysis of IV bag systems using direct thermal extraction of materials and stir bar sorptive extraction of aqueous solutions coupled with thermal desorption gas-chromatography with unit mass and high-resolution mass spectrometric detection	Gerstal Inc. (Linthicum, MD) AppNote 2/2015	The authors concluded that direct thermal desorption analysis of the packaging components in tandem with stir bar sortive preparation of aqueous extracts provided an efficient means of creating a comprehensive target list for leachables studies. The ability of the high-resolution, accurate mass MS for the confirmation and exclusion of compounds present in the sample or the extract was discussed.
Zdravkovic, S.A, Duong, C.T., Hellenbrand, A.A., Duff, S.R., and Dreger, A.L.	Establishment of a reference standard database for use in the qualitative and semi-quantitative analysis of pharmaceutical contact materials within an extractables survey by GC-MS	*J. Pharm. Biomed. Anal.* 151:49–60 (2018)	A set of 154 compounds representing a diverse range of chemical functionalities and properties was analyzed on multiple instruments. The authors concluded that relative response factors and retention times were highly constant across instruments.
Cavagnino, D.	Software tools that enable confident GC-MS analysis of extractables in pharmaceutical products	*Eur. Pharm. Rev.* 23(4):26–228 (2018)	The author has demonstrated how the use of powerful deconvolution software can be used to reveal individual extractables in complex mixtures and how chromatographic libraries can produce and support higher quality identifications.
Scherer, N., Marcsekova, K., Posset, T., and Winter, G.	Evaluation of stir-bar sorptive extraction coupled with thermal desorption GC-MS for the detection of leachables from single-use systems to drugs	*J. Pharm. Biomed. Anal.* 152:66–73 (2018)	Using the described methods, the authors report that the limit of detection of targeted leachables was 3 µg/l or less, even if they were analytically challenging samples, and claim that a broad range of leachables can be detected by this approach.
Scherer, N., Marcsekova, K., Posset, T., and Winter, G.	New studies on leachables in commercial-scale protein drug filling lines using stir bar sorptive extraction coupled with TD-GC-MS and UPLC/QTOF-MS/MS analytics	*J. Pharm. Biomed. Anal.* 555:404–419 (2019)	This publication describes the use of the indicated analytical methods to screen actual manufacturing process streams for process equipment-related leachables. The method is reportedly capable of quantifying leachables at levels of 0.27 µg/ml.

analysis. UV (and visible) spectrophotometry depends on an analyte molecule having the ability to absorb light at specific wavelengths in amounts directly proportional to the concentration of the analyte (i.e. Beer's Law). Certain structural features of analyte molecules (termed "chromophores") impart this light absorbing ability. Typical chromaphores include aromatic and heteroaromatic rings, unsaturated carbon–carbon bonds, and various functional groups such as aldehyde, azo, carboxyl, ester, ketone, nitrile, nitro, sulfone, thiol, thioether, thioketone, etc. Many of the chemical entities that are additives to rubber and plastic, and can therefore appear as extractables and leachables, have chromophoric structural features. The variety and multitude of entities and structures dictates that the peak capacity of an HPLC system should be maximized by maximizing the number of theoretical plates and consequently maximizing the kinetics of analytes mass transfer. The latest column technologies, such as fused core shell stationary phases, allow for fast, selective, and efficient separation of most leachables/extractables.

The ability of photodiode array detectors (PDAs) to acquire UV spectra over defined wavelength ranges has given LC/UV the ability to generate more structural information. However, this information is for the most part limited to compound class identification and not individual analyte structure elucidation. LC/UV is not, therefore, compound specific, in the sense that it is capable of providing information unique to and characteristic of a particular chemical entity.

Unlike spectrophotometry, mass spectrometry is a compound specific detection method. In general, mass spectrometers can produce the following information for organic analytes (down to trace levels):

- Molecular weight.
- Molecular formula (i.e. elemental composition).
- Fragmentation behavior according to defined chemical rules.

This compound-specific information often (but not always) allows for the structural analysis of trace level organic chemical entities, including extractables and leachables. However, mass spectrometry is limited in its ability to distinguish between isomers of various kinds, including geometric and structural isomers. It is also limited in its ability to elucidate aromatic substitution patterns. The capability to elucidate these structural features for extractables/leachables must await the fulfillment of the promise of LC/NMR (Nuclear Magnetic Resonance) spectroscopy. However, in spite of these limitations, mass spectrometry has become a primary technique for both structural and qualitative analysis of extractables and leachables. Further, due to its high selectivity and specificity relative to other detection techniques, mass spectrometry has also become a primary tool for high-sensitivity semi-quantitative analysis.

Over the past decade, a number of comprehensive reviews on the application of HPLC to polymer characterization in general, and extractables and leachables in particular, have been published. For example, Norwood and associates published a comprehensive review in 2009 [5]. A more recent review by Jenke captured several more recent applications of HPLC (and GC) to E&L assessment [2]. Even more recently, Legrand and associates published an HPLC/UV method for screening extracts containing calcium glucoheptonate for 11 additives-associated with PP and PE containers [6]. Additionally, Scherer et al. documented the use of stir bar sorptive extraction coupled with a UPLC/QTOF-MS/MS method to screen filling line solutions (from single-use manufacturing systems) for process equipment-related leachables [7].

3.2.3 Derivation and Use of the Analytical Evaluation Threshold (AET)

3.2.3.1 Definitions and Concepts

Perhaps we are jumping the gun a bit here and getting into the specific facet of screening before we address the general actions of screening, but the concept of a reporting threshold is so important to chemical compatibility assessment in general and screening in particular that it is best to get this topic behind us.

Analytical methods used to screen an extract for extracted substances should perform four functions:

1. They should detect the extractables.
2. They should distinguish between the extractables so that each extractable provides a unique response.
3. They should provide information from which the extractable's identity can be elucidated.
4. They should provide information from which the extractable's concentration can be established.

Considering chromatographic methods used to screen extracts for organic extractables, the methods can be better at detecting extractables than they are at providing the information necessary to correctly identify or accurately quantify extractables.

When an extractable has been detected, it is necessary to consider the compatibility impact that the extractable might have as a leachable. However, if the extractable's identity cannot be established, a safety risk assessment of this extractable, as described in ISO 10993:17 [8] cannot be performed. Furthermore, if the extractable is inaccurately quantified, the outcome of any compatibility risk assessment may be incorrect.

Considering the safety aspect of compatibility specifically, perhaps the most pressing question concerning the chromatographic method for screening samples is a matter of sensitivity, expressed somewhat figuratively as "how low do you go?" In more scientific terms, how sensitive must a chromatographic method be so that it is capable of performing the four critical functions noted for all extractables or leachables that potentially present a risk to patient safety?

In the very recent past, the answer to this question was "go as low as you can go," meaning that "every peak that you see in a chromatogram needs to be identified, quantified and reported for toxicological safety assessment." Implementing this approach is problematic in two situations:

1. When the analytical methods are capable of exceptional sensitivity, a large number of peaks may be revealed. Not only is identifying and quantifying the multitude of peaks resource-intensive, but as the peaks get smaller, they become much more difficult to identify and quantify. Furthermore, assessing the impact of so many chromatographic peaks also has its challenges.
2. When the analytical methods are capable of only limited sensitivity, there is the possibility that potentially impactful substances will be missed as they fail to produce a chromatographic response of sufficient size to be recognized.

Thus, what was needed was a threshold, or measure, that establishes that sensitivity which a method must possess to reveal all extractables and leachables that could be likely to pose a safety risk. Now it is clear that such a threshold could be established in one of two ways. The most conservative approach would be to establish the extractable or leachable that is unsafe in the lowest amount and use that amount to establish the threshold. However, it is likely that such an overly conservative approach would provide thresholds that could not readily be achieved and which would result in the assessment of a multitude of substances that are inherently safe.

Another approach would be statistically-based. That is, one would statistically evaluate a large database of compounds and their associated safe toxicological doses (for example, permissible daily exposure, PDE), establish a required level of safety assurance (for example, one event in 100 000 events with an adverse safety affect), set a desired level of statistical confidence (for example, 95% level of confidence), and then establish the statistically safe dose. In fact, something similar to this approach, but infinitely more sophisticated than this rather simple representation, was performed by a team of toxicologists, working within the purview of the PQRI to establish what it termed the Safety Concern Threshold (SCT) and the Qualification Threshold (QT) for a class of drug products termed OINDP [2]. These terms were defined as follows:

Safety Concern Threshold (SCT). The threshold below which a leachable would have a dose so low as to present negligible safety concerns from carcinogenic and non-carcinogenic toxic effects.

Qualification Threshold (QT). The threshold below which a given non-carcinogenic leachable is not considered for safety qualification (toxicological assessment) unless the leachable presents structure-activity (SAR) concerns.

While these definitions were developed for leachables, the threshold concept can be applied to extractables in the circumstance that extractables are used to project the worst-case release of leachables.

Although these dose-based thresholds (DBTs) are useful from a toxicological assessment perspective, they have limited direct use to an analytical chemist as their units (µg/day) are not the concentration related units (for example, µg/ml) that are analytically measured. The application of the threshold concept requires that a DBT be converted to a concentration-based threshold, as such a conversion would facilitate leachables and extractables assessment decisions based on the concentration of the leachable or extractable in a test sample.

Such an analytical threshold has been termed the Analytical Evaluation Threshold (AET) and is defined as follows:

Analytical Evaluation Threshold (AET). The concentration threshold below which the analyst need not identify or quantify leachables or extractables or report them for potential toxicological assessment.

By definition, then, the AET establishes a threshold for the toxicological risk assessment of extractables or leachables. Extractables whose concentrations are at or above the AET should be identified and quantified as a prerequisite for their toxicological risk assessment, as there is a sufficient possibility that the extractables could produce an adverse safety effect. On the other hand, extractables whose concentrations are below the AET do not need to be identified or quantified for toxicological risk assessment, as it is unlikely that they would produce an adverse safety effect.

In essence, the AET is a "protective" threshold in the sense that compounds at or above the AET must be toxicologically assessed, thereby "protecting" patient safety.

The concept of the AET is unique to organic extractables and leachables. Although PDEs for individual metals have been established [9], an analysis of this data to produce quantities for metals that are equivalent to the SCT or QT for organics has not been performed and thus these quantities have not been established for extracted elements.

The application of the AET to chromatographic analyses is straightforward. First, it is noted that chromatographic

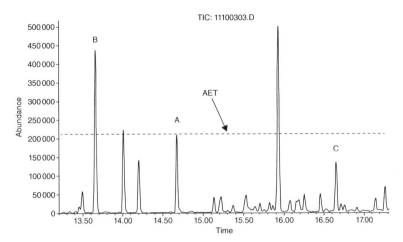

Figure 3.4 Representation of the AET. Peak A corresponds to an internal standard (IS) present in the test sample at a concentration equal to the AET. Thus, a horizontal AET line can be drawn across the chromatogram using the apex of A as the reference point. Peaks whose response apex falls above such a line (for example, peak B) are present in the sample at levels above the AET and the compound responsible for peak B should be identified and quantified to enable its toxicological assessment. Peaks whose response falls below the line (for example, peak C) are present in the sample at levels below the AET and can be toxicologically assessed as safe without identification or quantification. Although the concept of the AET is illustrated via peak heights, it is also applicable for peak areas as well.

methods do not produce concentrations directly but rather they produce a response measurable in units that should be convertable to concentrations. For example, the output of chromatographic analysis of a sample is a chromatogram in which analytes appear as peaks in the chromatogram (see Figure 3.4). In the case shown in this figure, the AET has been calculated and a compound A has been added to the sample (as an internal standard, IS) at a concentration equal to the AET. Thus, a horizontal line representing the AET can be drawn across the chromatogram using the apex of A as the reference point. Peaks whose responses fall at or above such a line (for example, peak B) are present in the sample at levels at or above the AET and the substance responsible for peak B should be identified, quantified, and reported for toxicological risk assessment. Peaks whose responses fall below the line (for example, peak C) are present in the sample at levels below the AET and do not need to be identified or further quantified for toxicological risk assessment.

Although Figure 3.4 illustrates the application of the AET in terms of peak height, peak area may also be used to compare individual extractables peaks with the AET and may be more appropriate. Furthermore, while this example and illustration specifically relate to chromatographic analysis, the concept of the AET is widely applicable to many other analytical techniques.

The conversion from a dose-based threshold (e.g. DBT, µg/day) to a concentration-based threshold (AET) for leachables requires input that establishes a patient's daily exposure to leachables. For example, consider a packaging-related leachable present in an administered drug product. If the therapeutic application of the drug product is such that a patient is exposed to X doses per day, the volume of each product unit is Y ml and the number of doses per unit is Z, then the AET becomes:

$$\text{AET (µg/ml)} = \text{DBT (µg/day)} \div [(X \text{ (doses/day)}$$
$$\times Y \text{ (ml/unit)})/Z \text{ (doses/unit)}]$$
(3.1)

For example, consider two cases. The first case is for a drug product, such as a small-volume parenteral, packaged in a single-dose container (flexible bag). In this case, the drug product is used in a therapy where the daily exposure is six doses, the fill volume of a single unit container is 50 ml, the number of doses per container is one and the DBT = 1.5 µg/day (chronic therapy, mutagenic impurities). In this case, the AET becomes:

$$\text{AET (µg/ml)} = \text{DBT (µg/day)} \div [(X \text{ (doses/day)}$$
$$\times Y \text{ (ml/unit)})/Z \text{ (doses/unit)}]$$
$$= 1.5 \text{ µg/day} \div [(6 \text{ containers/day}$$
$$\times 50 \text{ ml/container})/1 \text{ dose/unit}]$$
$$= 0.005 \text{ µg/ml}$$

which is a relatively low value that may be difficult to achieve analytically in certain SVP drug products with chemically complex formulations. It is noted that although the duration of therapy is relevant to the calculation of the AET, duration comes into the equation as part of establishing the proper DBT to use (see Section 3.2.3.4).

The second case is for a drug product, such as an inhaled drug product, packaged in a multi-dose container. In this case, the drug product is used in a therapy where the daily exposure is one dose (coming from a single inhaler), the fill volume of a single container is 5 ml, the number of doses per container is 100, and the DBT = 1.5 μg/day (chronic therapy, mutagenic impurities), and the AET becomes:

$$\text{AET (μg/ml)} = \text{DBT (μg/day)} \div [(X \text{ (doses/day)}$$
$$\times Y \text{ (ml/unit))}/Z \text{ (doses/unit)}]$$
$$= 1.5 \text{ μg/day} \div [(1 \text{ dose/day}$$
$$\times 5 \text{ ml/inhaler})/100 \text{ doses/inhaler}]$$
$$= 30 \text{ μg/ml}$$

which is a relatively high value that can typically be achieved in most inhaled drug products.

The third case is for a medical device that indirectly communicates with a patient via a solution that contacts the device and then is dosed to a patient, for example, a drug administration set. In this case, Equation (3.1) is simplified somewhat as the number of doses per set (*Z*) simplifies to 1. In this case, an individual set is used to deliver a drug at an infusion rate of 10 ml/hour. Although the therapy requires that the drug be administered over the course of an entire day, the set is switched out after 8 hours of use. Thus, $X = 3$ sets/day and $Y = 8$ hours $\times$ 10 ml/hour = 80 ml/unit. If a DBT = 5 μg/day is adopted (less than chronic therapy as it is unlikely that a therapy would last for a lifetime, taken as greater than 10 years), the AET becomes:

$$\text{AET (μg/ml)} = \text{DBT (μg/day)} \div [(X \text{ (sets/day)} \times Y \text{ (ml/set)}]$$
$$= 5 \text{ μg/day} \div [3 \text{ sets/day} \times 80 \text{ ml/set}]$$
$$= 0.021 \text{ μg/ml}$$

which is a relatively low value that may be difficult to achieve in certain drug products with chemically complex formulations.

The conversion from a dose-based threshold (e.g. DBT) to a concentration-based threshold (AET) for extractables requires inputs including:

– The frequency and duration of the test item's clinical use;
– The various extraction conditions used to produce the extractables profile;
– The uncertainty of the analytical method.

The AET in concentration units (μg/ml) can be calculated as given below :

$$\text{AET} = \frac{\text{DBT} \times \frac{A}{BC}}{\text{UF}} \qquad (3.2)$$

where

A	is the number of test items that were extracted to generate the extract;
B	is the volume of the extract (measured in ml);
C	is the clinical exposure to the test items (number of items a patient would be exposed to in a day under normal clinical practice);
DBT	is the Dose-Based Threshold (e.g. SCT or QT) in μg/day (note that a toxicologist should be consulted in selecting the specific threshold that can support a proper risk assessment);
UF	is an uncertainty factor that could be applied to account for the analytical uncertainty of the screening methods used to estimate an extractable's concentrations in an extract.

Any extract processing (e.g. any dilution or concentration steps) must be considered and the calculation of the AET value adjusted accordingly.

3.2.3.2 Sample AET Calculations

Four examples of calculating the extractables AET via Equation (3.2) are provided, one for a packaged drug product, two for a medical device (limited and permanent contact), and one for a manufacturing system component (both extractables and leachables AET).

Packaged Drug Product. A drug product is packaged in a flexible container with a fill volume of 10 ml. The drug product is dosed to the patient at a volume of 1 ml/day and therefore the number of items to which the patient is exposed is 0.1 (1 ml/day ÷ 10 ml/item = 0.1 item/day). Because the drug product is used in a chronic therapy, the AET is calculated based on the SCT of 1.5 μg/day as the Point of Departure (PoD). The container was extracted by adding 10 ml of extraction vehicle to it. The extract was processed for testing by evaporative concentration, where the concentration factor was 5. In this example, an uncertainty factor (UF) (which will be discussed in greater detail later) of 2 was used.

In this case,

$$A = 1 \text{ container extracted}$$
$$B = 10 \text{ ml extraction volume}$$
$$C = 1 \text{ container per day patient exposure}$$
$$\text{UF} = 2$$

The AET is calculated as:

$$\text{AET (μg/ml)} = [\text{DBT (μg/day)} \times (A/(B \times C))]/\text{UF}$$
$$= [1.5 \text{ μg/day} \times (1/10 \text{ ml/day} \times 0.1)]/2$$
$$= 0.75 \text{ μg/ml}$$

However, this AET does not take into account that the extract was concentrated by a factor of 5 prior to analysis. Thus, the working AET, accounting for sample preparation is:

Working AET (µg/ml) = AET (µg/ml)

$$\times \text{analytical concentration factor}$$
$$= 0.75\,\mu g/ml \times 5$$
$$= 3.75\,\mu g/g$$

Limited Contact Medical Device (e.g. a balloon catheter). In this example, a single device is used clinically and therapy is completed in less than one day. In this case, the value of the DBT used as a PoD was set to the ICH M7 TTC for potentially mutagenic impurities [10], 120 µg/day (duration of treatment 24 hours). In the extraction study, a single device was extracted in 9.0 ml of extracting vehicle. The resulting extract was neither diluted nor concentrated. GC/FID was used as the analytical method and the analytical method was supported by a response factor database, which established that the response factors were acceptably consistent across all possible extractables and thus that a UF = 1 was appropriate.

In this case,

$A = 1$ device extracted

$B = 9$ ml extraction volume

$C = 1$ device/day patient exposure

$UF = 1$

The AET is calculated as follows:

$$AET\,(\mu g/ml) = [DBT\,(\mu g/day) \times (A/(B \times C))]/UF$$
$$= [120\,\mu g/day \times (1/1 \times 9\,ml/day \times 0.1)]/1$$
$$= 13.3\,\mu g/ml$$

Permanently Implanted Medical Device (e.g. a cardiovascular stent). As a prelude to the calculation of the AET in this case, it is noted that since this is a permanently implanted device, it is required that the extraction study be exhaustive. In the extraction study, 20 devices were extracted in 100 ml of extracting vehicle. The exhaustive extraction was accomplished in two sequential extracts, meaning that the levels of extractables present in the second extract were less than 10% of the levels present in the first extract. The resulting extract was neither diluted nor concentrated. The analytical method had a response factor database which established that the %RSD of response factors was 25%, suggesting that a UF value of 2 is appropriate. The typical clinical circumstance is that a single stent will be used at any given time.

A critical issue is establishing the proper AET for a permanent implant is using the proper value for the PoD. Because the device is a permanent implant, the most likely leaching scenario is that all extractables present in the medical device will leach out of the device during the entire duration of the device/patient contact. This is why the proper extraction study for a permanent implant is an exhaustive extraction. Considering potentially mutagenic extractables, a DBT of 120 µg/day is appropriate, regardless of leaching kinetics, as illustrated below.

Consider a mutagenic substance, revealed after exhaustive extraction with a level of 120 µg/day, which corresponds to 120 µg/device in the example above based on a single device:

- If the 120 µg/device is leached in one day, the amount leached is equal to 120 µg/day, which is the TTC for this duration category per ICH M7 [10].
- If the 120 µg/device is leached in 31 days (one month), the amount leached is 120/31 = 3.9 µg/day, which is lower than 20 µg/day, the TTC for this duration category per ICH M7.
- If the 120 µg/device is leached in 365 days (one year), the amount leached is 120/365 = 0.33 µg/day, which is lower than 10 µg/day, the TTC for this duration category per ICH M7.
- If the 120 µg/device is leached in 3650 days (10 years), the amount leached is 120/3650 = 0.033 µg/day, which is lower than 1.5 µg/day, the TTC for this duration category per ICH M7.

Note that 20 µg/day for 31 days would be an exposure of 620 µg, 10 µg/day for 365 days would be an exposure of 3650 µg, and 1.5 µg/day for 3650 days would be an exposure of 5.475 µg. Each of these theoretical extreme approaches would therefore be less conservative.

In this case, the calculation of the AET proceeds as follows:

$$DBT = TTC = 120\,\mu g/day$$

(Note, however, that this DBT is "distributed" over both extraction steps. Thus, the DBT for each extraction step is 120 µg/day ÷ 2 extracts = 60 µg/day.)

$A = 20$ medical devices extracted

$B = 100$ ml extraction volume

$C = 1$ medical devices/day patient exposure

$UF = 2$

The AET is calculated as follows:

$$AET\,(\mu g/ml) = \{60\,\mu g/day$$
$$\times [20\,devices/(1\,device/day \times 100\,ml)]\} \div 2$$
$$= 6\,\mu g/ml$$

Bioprocessing bag used to store a bulk drug substance prior to final processing and filling of the drug product (e.g. vaccine). This case addresses the production of a biopharmaceutical product. After the bulk drug substance has been generated by any number of possible upstream manufacturing steps, it is stored refrigerated for up to 30 days in a bioprocessing bag before it is formulated, sterile filtered, and filled into its packaging system.

The concentration of the drug substance in the storage medium is 2 g/ml. The final drug product contains 20 mg/ml of the drug substance. The final drug product is administered to a patient via injection and up to two injections per day are possible. The drug product is stored in a 2-ml vial and each injection corresponds to a 2-ml delivered dose. The drug product therapy can be repeated for an additional two consecutive days and the three-day regimen can be repeated once more, three weeks after the initial therapy was initiated.

The bioprocessing bag was extracted by filling it to its nominal volume with a simulating solvent. The resulting extract was neither diluted nor concentrated. LC/MS was used as the analytical method and was supported by a response factor database, which established that the response factors were somewhat inconsistent across all possible extractables and that a UF = 5 be applied.

In this case,

$$DBT = TTC$$
$$= 120 \,\mu g/day \text{ as the drug product is used in}$$
$$\text{an acute therapy}$$
$$A = 1$$
$$B = 2 \text{ ml per injection}$$
$$C = 2 \text{ injections/day}$$
$$UF = 5$$

The leachables AET is calculated as follows:

$$AET_{leachables} \,(\mu g/ml)$$
$$= \{120 \,\mu g/day \times [1/(2 \text{ injections/day}$$
$$\times 2 \text{ ml/injection})]\} \div 5$$
$$= 6 \,\mu g/ml$$

To calculate the AET for extractables from the bioprocessing bag, it is noted that there is a factor of 100 dilution (2 g/ml to 20 mg/ml) going from the drug substance to the drug product. Furthermore, the bioprocessing bag was extracted with the same volume of extracting solution as the volume of the stored drug substance. Thus, the AET for extractables is calculated as:

$$AET_{extractables} = AET_{leachables} \times \text{dilution factor}$$
$$= 6 \,\mu g/ml \times 100 = 600 \,\mu g/ml$$

This calculation illustrates the reason why the patient safety risk for extractables from manufacturing components can, under certain circumstances, be low (that is, the extractables are typically distributed over a large number of doses). Given the high AET calculated in this example, the bioprocessing bag would have to be aggressively extracted, typically by means well beyond its conditions of use, for an extractable to achieve an extracted concentration greater than the AET.

3.2.3.3 Uncertainty Adjustments to the AET

Quantitation in extractables screening is achieved by means that differ with respect to the degree of uncertainty in the estimated concentration, which depends on the quantification strategy employed. For example, quantification could involve the use of a surrogate standard (SS) to normalize the responses obtained for all analytes and estimate the concentration of each analyte based on the simplifying assumption that all analytes respond similarly, among themselves, and with respect to the SS. Depending on the validity of this simplifying assumption, the concentration estimates thus obtained could have widely differing uncertainties and degrees of accuracy. If the simplifying assumption is true and response factors are constant, then the resulting concentration estimates for all analytes will be highly accurate. If the simplifying assumption is false and the response factors vary widely, then the accuracy of the resulting concentration estimates for the analytes will vary widely.

In other cases, the degree of certainty can be high. For example, if quantification is achieved through the use of authentic standards employed in qualified analytical methods, the concentration estimates obtained for the qualified analytes will be highly accurate. Considering quantification via an SS, if the simplifying assumption noted previously is true and response factors are constant, then the resulting concentration estimates for all analytes will also be sufficiently accurate for toxicological risk assessment.

Other quantification strategies could produce concentration estimates whose uncertainty is somewhere between these two extremes: lower uncertainty than use of an SS response factor but higher uncertainty than use of a calibration curve generated with an authentic reference standard. For example, relative response factors (RRFs) can be obtained for extractables, where the RRF is the ratio of the response of the extractable versus that of the SS at equal concentrations of extractable and the SS. Use of RRFs in quantification essentially accounts for differences in response factors, extractable versus SS, and produces more accurate, semi-quantitative concentration estimates.

When the concept of the AET was developed, these sources of uncertainty in the quantitation methods used by chromatographic screening methods were recognized. That is, while it is desired that the response factors of the chromatographic methods be the same for the entire population of organic extractables/leachables, the practical reality is that the analytical response factor varies across compounds. The implication of this variation is that the AET becomes potentially less "protective." For example, compounds with low response factors could appear to be below the AET based on response, when in fact their concentration is greater than the AET. This means that these low-responding compounds would not be reported for toxicological safety risk assessment, even though there is a chance that they might be safety-impacting. In such cases, the AET is not "protective" for these potentially harmful, low-responding substances.

This "loss of protection" suggests that the AET should be adjusted so that it can be "protective" for low-responding substances. This adjustment can be accomplished by use of a UF, which is added to the calculation of the AET to account for the analytical uncertainty that arises due to the variable responses.

In cases where the analytical uncertainty is known to be acceptably low, a UF value of 1 can be justified. Examples of these cases are methods with comparable response factors between expected extractables and applied SSs in qualified methods for targeted extractables. Otherwise, the value of the UF is based on an assessment of the analytical methodology to which the AET is applied. For example, a UF value of 2 has been proposed as being appropriate for the screening of extracts for semi-volatile extractables via GC-FID or GC-MS, as analytical FID or MS response factors for extractables are somewhat consistent, extractable to extractable [11, 12]. Alternatively, response factors for other analytical methods used for extractables screening, such as HPLC-UV and HPLC-MS (which are typically applied to non-volatile extractables), may be higher given the frequently wide variation in response factors among extractables by this methodology. At the current time, there is no available general guidance that recommends a specific value for the UF for these methods, although values as high as UF = 10 (or higher) are generally considered to be possible.

A statistical approach to establishing and justifying a particular UF has been proposed for medical devices [1] and is based on statistical analysis of a database of response factors specific to the analytical method being considered and the population of extractables for which that method is applicable. In one possible approach, the value of the UF would be linked to the relative standard deviation of the response factors according to :

$$UF = 1/(1 - RSD) \qquad (3.3)$$

where

RSD = the relative standard deviation in the response factor from the reference database

When the variation in response factors is large relative to the mean response factor (for example, RSD = 0.9 × mean), the variation in response factors is so large that although a UF can be calculated, its scientific validity becomes questionable. For example, while a UF > 10 can be calculated, the fact that the UF is as large as 10 (or larger) suggests that the quantification method being used is inherently inaccurate and thus may not be appropriate for the purpose of toxicological safety risk assessment. In this case, an adjusted AET should not be established and the concept of an AET should not be applied to the method. In cases where RSD > 1, a UF cannot be calculated, as the result is either infinity or a negative number. Clearly an analytical method with this much variation in response factors is not suitable for the purpose of producing concentration estimates that are the basis of a proper toxicological safety risk assessment.

The implications of using an UF can be illustrated by considering a simplified distribution of response factors for IS and SS candidates and extractables/leachables (see reference [13]). Consider the situation shown in Figure 3.5, where the response factors for all extractables or leachables exhibits a normal distribution and the AET is established based on an IS whose response factor falls somewhere within the distribution of the response factors for all extractables and leachables. Depending on where the response factor of the IS falls, the AET either becomes more or less protective. For example, if the IS has a response factor exactly at the distribution's median (and mean), then the initial or estimated AET will be protective for the 50% of the population whose response factor is equal to or greater than the response factor of the IS. Conversely, the AET may not be protective for the 50% of the population whose response factors are less than the mean. As a "50% level of protection" is not likely to be adequate, this establishes either the need to:

1. Properly choose the IS or
2. Perform a final AET adjustment to address variation in response factors.

Generally, the IS used in chromatographic screening is chosen based on chromatographic and practical factors (sensitivity, selectivity, availability in highly purity, stability, use in other fields such as environmental) and not in terms of its position in the distribution of the universe of response factors. Thus, its position in the universe of response factors is almost an accidental afterthought. As shown in Figure 3.5, however, failing to account for the IS response factor's position on the population can have consequences in terms of the AET's "protective power." When the AET is based on an IS with a response factor equal to the population mean, the AET is only "50% protective."

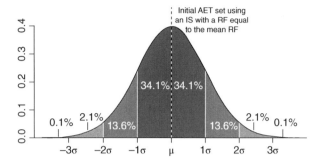

Figure 3.5 Illustration of the "Protective Power" of the AET. If the AET is established by an internal standard whose response factor is equal to the mean of the response factor distribution among all extractables and leachables, then the AET is "protective" only for those substances whose response factors are greater than or equal to the mean (50% protective).

We see, therefore, that we can affect the AET's "protective power" by our choice of IS. This concept is illustrated in Figure 3.6. If we pick, for example, an IS with a lower response factor (specifically, mean of − 1 STD), we can increase the AET's "protective power," in this case to 84.1% (that is, 84.1% of the substances are covered by an AET set with an IS whose response factor is mean − 1 STD). However, if the IS were such that it has a higher response factor (specifically mean + 1 STD), then we have decreased the AET's "protective power," in this case to 15.8% (that is, 15.8% of the substances are covered by an AET set with an IS whose response factor is mean + 1 STD).

The practical implication of this analysis is that if one has a database of response factors, then one can use that database to establish and justify an initial AET that potentially needs no uncertainty adjustment. Specifically, if the

AET is set with an IS whose response factor is mean − 1 STD, then the AET is sufficiently protective without adjustment as it accounts for 84.1% of all compounds in the database.

In this way, the proper value of the response factor of the IS is established. The identity of the proper IS is established by querying the database and finding that compound that has the right response factor and has the other characteristics of a good IS, as noted previously. Thus, the extractables and leachables database should also include response factor data for potential ISs.

The conclusion of the previous paragraph notwithstanding, the "protective power" of the AET implemented with an IS with any response factor can be increased by adjusting the AET to account for the variation in response factors via Equation (3.3). Figures 3.7–3.9 illustrate the effect that using a UF calculated using Equation (3.3) to adjust the initial AET will have on the "protective power" of the adjusted final AET [14].

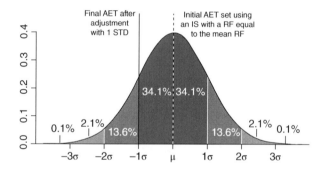

Figure 3.7 Adjusting an Initial AET set with a response factor equal to the population mean. Adjustment via a UF equal to 1/(1 − RSD) increases the "protection power" of the Final AET from 50 to 84.1% of the population of extractables and leachables.

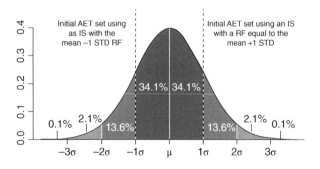

Figure 3.6 Illustration of the "Protective Power" of the AET based on various IS response factors. If the AET is established by an internal standard whose response factor is equal to the mean + 1 STD of the response factor distribution among all extractables and leachables, then the AET is "protective" only for 15.8% of the compounds in the database. However, if the AET is established by an internal standard whose response factor is equal to the mean − 1 STD of the response factor distribution among all extractables and leachables, then the AET is "protective" for 15.8% of the compounds in the database.

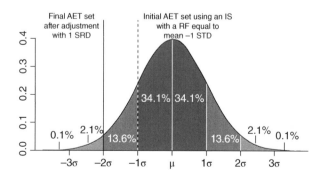

Figure 3.8 Adjusting an Initial AET set with a response factor equal to the population mean − 1 STD. Adjustment via a UF equal to 1/(1 − RSD) increases the "protection power" of the Final AET from 84.1 to 97.7% of the population of extractables and leachables.

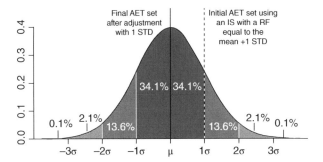

Figure 3.9 Adjusting an Initial AET set with a response factor equal to the population mean + 1 STD. Adjustment via a UF equal to 1/(1 − RSD) increases the "protection power" of the Final AET from 15.7 to 50% of the population of extractables and leachables.

This analysis of the AET UF can be illustrated by considering an actual database of response data. This database is a collection of relative response factors (RRFs), where the RRF is the ratio of response factors (RFs), compounds versus IS:

$$RRF_{compound} = RF_{compound}/RF_{IS} \qquad (3.4)$$

Thus, the RRF is essentially a compound RF "normalized" versus the IS RF. Trends exhibited in RF will also be exhibited in RRF.

Table 3.4 represents the contents of a recently published [13] database of RRF values for extractables and leachables screening by GC with mass spectrometric detection (GC/MS), produced by Nelson Labs (hereafter referred to as "the Database"). Considering this information, the RSD for the database is calculated as the STD/mean = 0.425/0.645 = 0.659 (all data). For this data, 1 − RSD = 1 − 0.659 = 0.341 = UF. This would mean that one would divide the AET by a factor of 1/0.341 = 2.93 to get the final AET and that all peaks above this final AET would

Table 3.4 Distribution of relative response factors related to semi-volatile compounds in the Nelson database for GC/MS. Bracketed values correspond to the subset of the database, which includes "quantifiable" compounds with RRF values between 0.1 and 5.0. It is noted that the RRFs are calculated versus 2-flurobiphenyl as the IS.

Parameter	GC-MS
Total number of entries	3076
Total number of entries with RRFs:	2247
RRFs <0.1	7%
Median RRF	0.590 (0.628)
Mean RRF	0.645 (0.687)
Standard deviation	0.425 (0.407)
Minimum RRF	<0.001
Maximum RRF	4.611

have to be identified and reported for toxicological safety assessment. It is noted that this distribution is specific to the analytical methods and processes used by this laboratory.

The UF is somewhat smaller if one considers the database minus the compounds with the really low and likely insufficient response factors, taken as UF < 0.1 (the numbers in ( ) in Table 3.4). This is justified if one takes the position that compounds with a really low response factor (0.1 or less) should not be reported by the method and should be addressed by another method where the compounds have a higher response factor. This is the case here as there is likely to be greater uncertainty in the response factors if the response factors are low to begin with. Using this data, RSD = 0.407/0.687 = 0.592, 1 − RSD = 0.408, and UF becomes 2.45.

It is noted that in both cases the UF is larger than "a factor of 2," which is the "default UF" recommended by the PQRI. This means that a UF of 2 is generally not sufficiently protective.

Figures 3.5–3.9 addressed the case where the RF (or RRF) values were characterized by a normal distribution. As noted in Figure 3.10, the distribution of RRF values in the database is not strictly normal and is skewed toward lower RRF values. Nevertheless, the normal distribution is an acceptable approximation of the character of the database and the observations made previously for the normally distributed data would be qualitatively relevant to the compounds and information in the database.

In summary, making the AET sufficiently protective to account for potentially harmful extractables or leachables requires that the derivation of the AET be carefully considered. The AET's "protective power" can be increased by either selecting an IS to represent the AET whose response factor is lower than the mean response factor for a population of extractables, leachables, and potential ISs or by adjusting the AET lower with an UF that is statistically derived from the response factor database. Selecting an IS whose response factor is the population mean minus the population standard deviation will be sufficiently protective as it will account for 84.1% of the extractables and leachables in the database. While adjusting the AET down by a UF factor of 1/(1 − RSD) will make the AET more protective, the degree of protection provided by the AET will depend on the position of the response of the IS used to represent the AET in the response factor distribution. The combination of choosing the IS as a compound whose response factor equals mean − 1 standard deviation and adjusting the AET down by a UF equal to 1/(1 − RSD) will produce an AET that is protective for 97.7% of all compounds in the response factor database. It is noted that this discussion is based on a database whose distribution of RF values is normal.

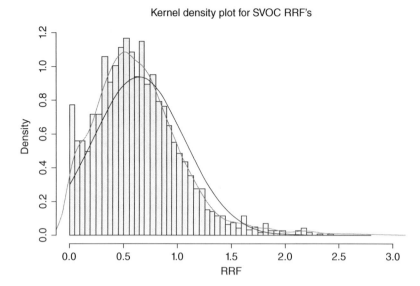

Kernel density plot for SVOC RRF's

Figure 3.10 Frequency distribution of relative response factors (RRFs) from the database. The individual histograms and the red curve represent the actual experimental data, while the blue shapes represent a normal distribution with the population mean and standard deviation. The distribution is skewed low, meaning that there are more compounds with lower RRFs than there are compounds with higher RRFs. The RRF values are obtained with 2-fluorobiphenyl as the IS.

To illustrate this point, consider the following example of a small (and manageable) database of response factors for GC/MS, originally published by Mullis et al. [11]. Specifically, this GC/MS response factor database was generated for 32 organic extractables, using either 2-fluorobiphenyl (2-FB) or p-terphenyl-d14 (PT14) as ISs. Statistical analysis of this database is shown in Table 3.5 and the distribution of response factors is illustrated in Figure 3.11.

In addition to the individual response factors, Figure 3.11 illustrates AET values adjusted for response variation, either using UF = 1 standard deviation or UF = 2 as the adjustment factor. The important point to note is that even with the adjustment, many of the individual compounds have response factors below the adjusted AETs. Thus, adjustment of the AET for response factor variation will not be sufficient to move these compounds "above the line." Therefore, they will not be captured as a reportable compound and will not be toxicologically safety assessed. In this case, the adjusted AET is therefore deemed to be insufficiently protective. This outcome is to be expected, as the mean value of the response factor distribution is less than 1.

However, this situation is mitigated if the AET is adjusted for both the magnitude of the mean of the response factors and the variation in response factors [14]. This two-step

process manifests itself as follows and is illustrated in Figure 3.12. First, the AET is adjusted for the mean value of the response factors, which for 2-FB is 0.642 (dashed AET line in Figure 3.12). Next, the AET adjusted for the mean is further adjusted for the response factor variation using UF = 2. This has the outcome of moving the AET down to a value of mean × UF = 0.642 ÷ 2 = 0.321. This is illustrated in Figure 3.12 by the dotted adjusted line. One notes that the net outcome of this two-step adjustment is that virtually all the 2-FB response factors and a majority of the PT_{14} response factors fall "above the line," meaning that they would be reported for and captured by a toxicological safety assessment. Thus, the AET is made sufficiently protective by adjustment for both the magnitude of the mean response factor and the variation (RSD) in the response factors.

Lastly, it is noted that this assessment considers GC/MS data, whose response factors are generally recognized as being relatively constant, compound to compound. It is anticipated that the value of the UF, and therefore the lowering of the AET resulting from application of the UF, is more significant for an analytical method such as LC/MS, where it is generally accepted that the response variation between compounds is much greater.

As one observes from the previous discussion, the selection of a properly protective UF is challenging, based solely on statistical analysis alone. A simpler approach, which does not directly involve statistical analysis of the distribution of response factors, reflects the concept of "diminishing returns." By this concept, the proper value of the UF is established as that value above which a further increase in the UF produces a relatively minor and inconsequential increase in the AET's coverage.

For example, consider an extraction that is taken to completion, that is, the point at which equilibrium

Table 3.5 Statistical analysis of a database of GC/MS response factors.

Item	Result with 2-FB as the IS	Result with PT_{14} as the IS
Number of compounds	32	32
Mean	0.642	0.487
Standard deviation	0.254	0.184
RSD	0.395	0.184
% RSD	39.5%	37.7%

Figure 3.11 Distribution of GC/MS response factors for the database summarized in Table 3.5. (*See insert for colour representation of the figure.*)

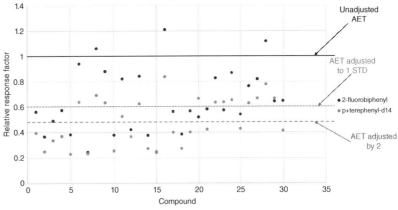

Figure 3.12 Adjustment of the AET for both the magnitude of the mean response factor and the variation in the response factor. The AET adjustment is based on 2-FB as the IS. (*See insert for colour representation of the figure.*)

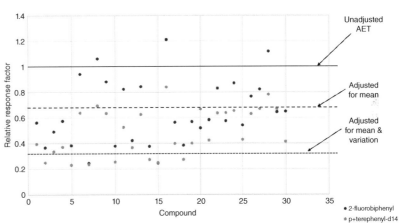

between the extracted item and the extracting solution is achieved (so-called asymptotic extraction), illustrated in Figure 3.13. One notes that the concentration of the extracted substances increases rapidly during the early duration of extraction and more slowly as equilibrium is approached. If one examines Figure 3.13 to answer the question "what is the proper extraction duration?" it is noted that there are two possible answers. The most rigorous answer is that the proper extraction duration is a duration at or after equilibrium has been definitively established and extraction is conclusively 100% complete, which, in Figure 3.13, is 1000 hours.

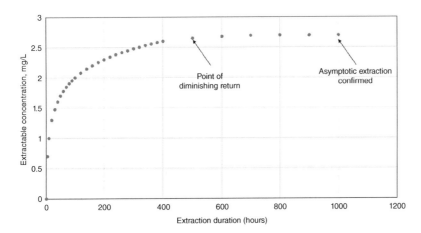

Figure 3.13 Extraction profile, extracted concentration versus extraction duration. It is clear that the extraction has reached equilibrium (asymptotically reached its maximum value) after 1000 hours of extraction. However, the extraction is very nearly complete after 500 hours and the incremental increase in the extractable's concentration between 500 and 1000 hours is small. Although rigorously speaking the extraction should proceed for 1000 hours to achieve full extraction, practically speaking the "point of diminishing returns" has been achieved at 500 hours and ending the extraction at 500 hours would only marginally underestimate the extractable's maximum extracted amount while reducing the required extraction time by half.

However, a more practical answer would be that the proper extraction duration is the duration at which continued further extraction produces a minimal further increase in the extractable's concentration. Such a "point of diminishing return" duration, 500 hours as noted in Figure 3.13, essentially reduces the recommended extraction duration in half without materially under-reporting the extractable's highest (worst-case) concentration.

Such a "point of diminishing returns" discussion can be used to establish an appropriate value of the UF for AET adjustment. Specifically, Figure 3.14 expresses the AET coverage as a function of UF values for various individual and combined chromatographic screening methods used for a population of over 200 organic extractables. In Figure 3.14, AET coverage is established as the ability of the AET, adjusted by the UF, to properly flag extractables

as being at or above the AET concentration when the extractable's concentration is equal to the AET. This ability to properly flag compounds depends on the compound's RRF; poorly responding extractables will have peaks that appear to be lower than the AET and thus will not be flagged or "covered" unless the AET is adjusted downward by the UF. The higher the UF, the more compounds that are flagged and the greater the coverage of the AET.

Figure 3.14 is interpreted to establish the proper value of the UF using GC/MS as an example as follows. As was the case for the extraction profile (Figure 3.13), the % coverage versus UF plot for GC/MS (and all methods, individual and combined) exhibits asymptotic behavior; that is, as the UF increases, the incremental increase in coverage decreases until a plateau is reached. At that point, further increases in the UF produce small increases in % coverage. For GC/MS,

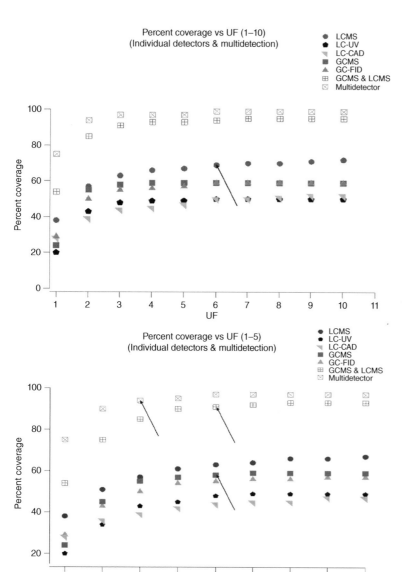

Figure 3.14 Coverage of organic extractables by various chromatographic screening methods. An extractable is "covered" by the AET if its response is greater than or equal to the AET when its concentration is equal to the AET. When this happens, the extractable is "flagged" for toxicological risk assessment. If the extractable has a poorer response than the internal standard used to set the AET, it will only be flagged if the AET is adjusted down via an uncertainty factor. The % coverage is the percent of compounds in a response factor database that are properly flagged as being at or above the AET when they are present in the extract at the AET concentration. As the UF increases in value, necessarily the % coverage increases. The upper plot illustrates the effect of a UF ranging from 1 to 10 while the lower plot expands the x-axis so that the points of diminishing return (designated by the arrow) are more obvious. Data provided by Jordi Labs.

a UF of 3 reflects the "point of diminishing return" and thus 3 is the recommended UF to be applied generally for this technique. LC/MS, characterized by more highly variable response factors, reaches the point of diminishing return at UF = 6 and this value is the recommended UF to be applied generally for this technique.

It is noted that these suggestions for UF values are slightly lower than values that have been communicated by the FDA staff members as being appropriate for these two techniques (i.e. UF = 4 for GC/MS and UF = 10 for LC/MS) [15].

It is noteworthy that the % coverage for any single analytical technique is not 100%. This is the case given the diverse chemical nature of extractables. That is, the universe of extractables invariably and expectedly consists of compounds that cannot be addressed by all screening methods. For example, it is clear that highly non-volatile extractables will not be amendable to GC/MS for the simple fact that they cannot be eluted in the GC portion of this technique. Thus, the % coverage of GC/MS for a diverse population of extractables cannot be 100% as some compounds will not be amendable to GC analysis.

Figure 3.14 also illustrates the advantages of an analytical strategy that is based on multiple, complementary, and orthogonal chromatographic methods. Not only does the % coverage increase for multiple methods (and essentially approaches 100%), but the point of diminishing return, which establishes the general UF, also decreases. Thus, for the combination of GC/MS and LC/MS, which is the current industry standard, the general UF, applied to both methods, is 3. For the combination of all five techniques illustrated in Figure 3.14 (six techniques if one considers LC/MS + and − ions), the point of diminishing return is reached at a UF = 2, which becomes the general UF to be applied to all techniques in this so-called multi-detector approach.

In any event, the use of the UF, and the value of the UF that is used, should always be justified. In some cases where the variation in response factors among extractables cannot be established or where the variation is established to be large, the value of the UF may be so large (e.g. UF values of 10 or greater) that the AET becomes so low that the AET concept has little practical value (e.g. the analytical method's limit of detection (LOD) or limit of quantification (LOQ) are greater than the AET). In such circumstances, use of the AET cannot be justified and thus the AET should not be applied. In such cases, it may be necessary to identify and quantify all the compounds associated with all observed analytical responses obtained by the screening analyses.

3.2.3.4 Selecting the Basis for the AET

As originally published in the PQRI recommendations for both OINDP and PDP drug products, the SCT and the QT were based on therapies that were chronic. However, not all therapies are chronic and in the previous examples for the AET, various values were used for the DBT, depending on the duration of the therapeutic use of the pharmaceutical item. This staged approach for the DBT was adopted from the ICII M7 guidelines for the assessment of mutagenic impurities in pharmaceuticals in order to limit potential carcinogenic risk [10]. In the M7 guidance, the TTC for lifetime exposure was established at 1.5 μg/day, corresponding to a theoretical 10^{-5} excess risk of cancer. TTCs for less-than-lifetime exposures are adjusted at a higher level, based on the premise that such exposures can have higher acceptable intakes of mutagenic impurities and still maintain comparable risk levels. This premise is based on the observation that mutagenicity is characterized by a dose–response relationship.

When the M7 guidance was first published, its less-than-lifetime TTCs were adopted as the basis for AET calculations. However, so doing violated an essential concept of safety assessment, which is that the assessment should be based on the most conservative approach applied to the most conservative toxicological endpoint. Basing the AET on TTC derived for mutagenicity assumes that mutagenicity provides the lowest acceptable threshold for adverse effects. However, there are other toxicological endpoints to consider when establishing which DBT can be the basis of the calculation of the AET. For example, a TTC of 5 μg/day has been established for impurities that can be sensitizers or irritants. In the United States, the FDA has supported the use of 5 μg/day as the proper TTC for general toxicity [16]. What is particularly noteworthy about these endpoints is that they do not exhibit a dose–response behavior and thus their value does not change as a function of the duration of therapy.

This juxtaposition of the TTCs for mutagenicity, sensitization – irritation – and general toxicity is illustrated in Table 3.6. One observes that the proper value to use as the basis of the AET for each indicated duration of therapy is the lowest value in each respective duration column. Thus, for chronic therapies (duration >10 years), the mutagenicity TTC value of 1.5 μg/day is the proper basis for the AET. For all other therapeutic durations, the proper, most conservative, basis for the AET is the sensitization–irritation and general toxicity TTC of 5 μg/day.

3.2.3.5 The Relationship Between the AET and Other Measures of Analytical Sensitivity

The proper relationship between the AET and analytical sensitivity limits, such as the LOD and LOQ, is as follows (see Figure 3.15). As the AET is a threshold that requires

Table 3.6 Reference DBT values for the AET calculation.

Toxicological endpoint	Duration of therapy			
	≤1 month	1–12 months	1–10 years	>10 years
Mutagenicity (TTC, SCT)	120	20	10	1.5
Sensitization–Irritation[a]	5	5	5	5
General toxicity,[a] FDA, PQRI[b]	5	5	5	5

Shaded = value to pick as the DBT to use for the AET.
a) Toxicological endpoint that does not exhibit a dose–response behavior.
b) Value of the QT proposed for parenteral drug products by the PQRI for PDP drug products.

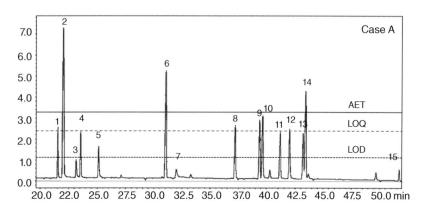

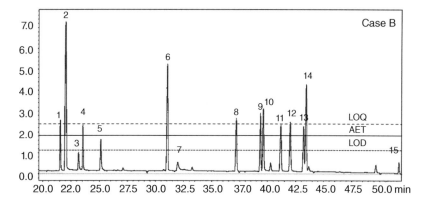

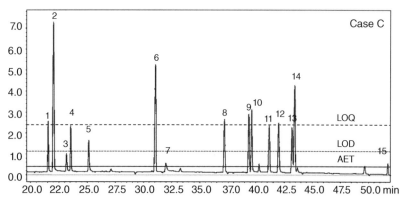

Figure 3.15 Examples of the possible relationships between the AET, the LOQ, and the LOD. In the ideal case, Case A, the LOD and the LOQ are both less than the AET, meaning that all compounds that must be reported (those above the AET) can be detected and quantified. In Case B, the AET is between the LOQ and the LOD. This means that while all compounds that must be reported can be detected, compounds that fall between the AET and the LOQ cannot be quantified. In the worst possible case, Case C, the AET is below both the LOQ and the LOD, meaning that there are compounds that must be reported but will neither be detected nor quantified.

the compound responsible for an analytical response to be identified and quantified, clearly the analytical response must be discernible above the analytical noise (detected) before its source compound can be identified. Thus, the AET should be greater than or equal to the LOD as an AET lower than the LOD would indicate that the analytical method cannot produce analytical responses at the necessary concentration levels for relevant compounds. Although the LOD might not be determinable for the specific individual compounds detected during the screening process, the method's LOD can be established and taken as the mean of the LODs of one or more relevant surrogates or SSs.

It is also clear that if the purpose of the analytical testing is quantitation, the AET should be higher than or equal to the LOQ. However, it is understood that semi-quantitative concentration estimates obtained in screening may not meet the rigorous accuracy and precision expectations inherent in an LOQ and thus there may be cases where screening studies provide concentration estimates when the AET is lower than the rigorously determined LOQ. Concentration estimates below a method's established LOQ may not be sufficiently accurate to support a valid toxicological risk assessment.

Lastly, it is observed that the AET is also an identification threshold and that the process of identification requires that the response contain more complex and/or advanced information than does the process of quantification (that is, quantification can typically be accomplished at concentrations lower than those required for identification). This being the case, it is possible that the AET could be above the LOQ but it would still not be possible to secure an identification for an analyte present in the sample at the AET.

3.2.4 Discovery

Before a compound can be identified, it must be discovered. A compound is discovered when it produces an analytical response that can be recognized above the background response (noise) of an analytical method. It is clear, then, that an ideal analytical method for discovery would be one that produces a response for a large number of compounds (broad scope) with a high degree of sensitivity. While the ideal discovery method should respond to a large number of compounds, it must produce a unique response for each individual compound (high specificity). Furthermore, it would be ideal if the analytical method produced a similar response (either in nature or magnitude) for all compounds (so that one could make statements as to the relative levels of discovered compounds), although this is a performance characteristic that is more relevant to quantitation than to

discovery. Lastly, it would be ideal if the analytical method produces information from which the identity of the discovered compound could be established (as the identity of a compound is the key to assessing its impact), although this is a performance characteristic that is more relevant to identification than to discovery.

As the primary desirable performance characteristics of discovery methods are conflicting, as opposed to complementary, the methods used in discovery are generally composite or hyphenated, where one part of the method provides the broad scope and selectivity while the other part of the method provides the sensitivity and "universal" response. For example, chromatographic methods are used to discover organic extractables in extracts (and organic leachables in a drug product) with a high level of success. In such a method, compound specificity is facilitated at the "front" end of the chromatographic analysis, typically by the combination of the sample preparation process and the chromatographic separation. The sample preparation and separation processes are designed so that they fulfill the requirement that chromatographic screening methods be broad-scope. The universal response is obtained at the "back end" of the chromatographic method via the detection methodology (for example, mass spectrometry practiced in various forms). Considering inorganic entities, certain atomic spectroscopy methods, such as inductively coupled plasma spectroscopy (ICPS), are effective at both discovery and identification as the universally applicable sample atomization and excitation processes are coupled to an element-specific detection process (for example, either light emission at characteristics wavelengths or mass spectrometry).

When the purpose of a method is to discover compounds, the relevant sensitivity measure is the LOD. Chromatographic screening methods employed for organic extractables and leachables typically have LOD values as low as 10–50 ng/ml, although the actual required LOD will depend on the AET established for a specific application. Depending on the sample matrix, ICPS methods for elemental extractables and leachables typically have LOD values in the 1–10 ng/ml range, although the LOD varies significantly in terms of the detection method and the specific element.

Although this topic will be discussed in greater detail in other sections of this book, it is proper to address factors that impact the ability of screening methods to perform their discovery function. It is clear that the ability to discern an extractable response from the background response depends significantly on the nature of the background response. It is obvious that the more complex the background response, the more difficult it is to discern a response above the background. This is the driving force, for example, for good practice recommendations such as "solvents used in extractions should be as compositionally 'simple' and pure as possible," that "extractions should be designed to reduce the possibility of extract contamination," and that "analytical instrumentation should be properly cleaned and maintained." It also illustrates the discovery challenges faced when screening complex drug products for leachables; that is, it is hard to differentiate small peaks due to leachables among a multitude of large peaks due to drug product ingredients.

The ability of screening methods to discern a "real" response from the background is complicated as screening is often a comparative process. Consider chromatographic screening methods applied to packaged drug products, for example. When screening an extract for extractables, one typically differentiates peaks for extractables from "background" peaks of varying origin by analyzing an extraction blank. Those peaks that are common to both the extract and extraction blank are "background" peaks that are not related to extractables and thus are ignored. When screening a drug product for leachables, one typically differentiates peaks for leachables from "background" peaks due to the drug product itself by analyzing the drug product that has been stored in an inert container. Those peaks that are common to both the packaged drug product and drug product stored in the inert container are "background" peaks that are not related to extractables and thus are not addressed in the toxicological safety assessment.

The purpose of this discussion is to point out the need for careful matching of the blank and the sample. For compositionally simple and stable extraction solvents, properly matching the blank and the sample is typically trivial. For compositionally complex and unstable drug products, properly matching the blank and the sample is essential but difficult to accomplish. For example, consider a migration study that consists of three lots of drug product tested over multiple time points over their three-year shelf-life. Obviously, the most appropriate blanks are the three lots of the drug product, also tested over the course of three years of storage. Thus, the study must be designed so that the proper blanks are available for use.

The matter of having a proper blank typically happens during an issue investigation and resolution, when proper blanks may be difficult or impossible to obtain. In certain circumstances, one may be inclined to believe that "any blank is better than no blank" only to find out the comparison of the sample to the imperfect blanks leads to issues that are numerous and unresolvable. The point here is that great care and clear thinking must be applied to the study design so that proper blanks are generated, available, and used.

3.2.5 Identification and Identification Categories

Two characteristics of extractables and leachables are relevant and necessary in assessing compatibility:

1. The substance's identity, as the identity is used to establish the substance's behavior, characteristics, or properties, and
2. The substance's concentration (in either the extract or the drug product), as the magnitude of an extractable's or leachable's effect on a product's suitability for use is directly related to the amount of the substance that is, or could be, present in the drug product.

The term "identification" was defined previously in Section 3.1. It is clear that the analytical methods used to secure an identification do not directly provide an identity; rather, a substance's identity is inferred based on the interpretation of the analytical data that is available for the substance. The amount of detail contained within the identity and the likelihood that the identity is correct increases in direct proportion to the amount of collaborating data and the information content of the data. In the absence of collaborating data or in cases where the information content of the test method is poor (for example, a UV absorption spectrum), an established identity may not be specific (for example, a compound class may be established but the exact member of the class cannot be ascertained) and/or the likelihood that the identity is correct is low. On the other hand, having multiple pieces of collaborating data or in cases where the information content is high (for example, a chemical ionization (CI) mass spectrum illustrating a significant fragmentation pattern) produces an established identity that may be highly specific (specifying an exact compound) and that is likely to be correct.

The proper use of an extractable's or leachable's identity requires that the extractable's or leachable's identity be reported in a manner that conveys the status of the identity with respect to its level of detail and accuracy. This is true because the validity of an assessment based on the identity is only as strong as the validity of the identity upon which it is based. If the identification is tentative, and thus of lower certainty and validity, then assessment based on the identity will be tentative and of lower certainty. However, if one is certain that the identity is correct because the inferred identity has been confirmed, then the assessment based on that identity will more likely be correct.

To this end, identification classes have been proposed and justified. For example, in their Recommendations Document for Safety Thresholds and Best Practices for Extractables and Leachables in OINDP, the PQRI Leachables and Extractables Working Group proposed three "levels" of identification: tentative, confident, and confirmed and provided guidelines for establishing what analytical data was required to achieve and justify each identification level [1]. A fourth level, unidentified or unknown, can readily be envisioned [17].

Though useful, the PQRI classification is awkward to use. For example, in most people's minds the term "tentative" implies that a compound has been fully identified (name and CAS RN provided) but we are just not very confident that this is the right identity. The way it is defined by PQRI, the term tentative is used to mean identities that are not "firm" enough to establish the exact compound but are only sufficient to establish a compound's general functionalities. To be more grammatically correct, such an identity is a partial identification. Moreover, the category that is now referred to as tentative (meaning a name and CAS RN is proposed) can be broken into two categories, interpretative and matching, depending on the nature of the process used to secure the identity. Thus, an interpretative identification is made on the basis of independent and subjective scientific interpretation of data (for example, the infamous accounting for all the peaks in a mass spectrum by drawing structures on a napkin, more rigorously referred to as structure elucidation) whereas a matching identification is made on the basis of objective comparative data such as a mass spectral match to a database. Although either means may be appropriate to secure a tentative identity and it is likely that both means are equally valid, they are quite different means and processes to get to the same point.

In reviewing the criteria for a tentative identification, we understand that it is a one-dimensional outcome; that is, it is an identification based on one item of confirming information. It is intuitive that the more confirming information one has, the greater confidence one has, and the greater the likelihood that the identification is correct. So, for example, a two-dimensional identification, that is, an identification based on two independent pieces of information, becomes a confident identification. For example, a tentative identification can be secured independently

by both means, marching and interpretation. These two pieces of information, independent and confirmatory, together support a confident identification.

At some point, one can collect so much confirmatory information that they become highly confident that their identification is correct and it is the case that the chances that the identification is wrong become vanishingly small. For example, a three-dimensional identity (an identity based on three pieces of confirmatory information) is surely an identification in which one has high confidence. In the circumstance of compelling and overwhelming confirmatory information, surely one can envisage that the identification has been confirmed by the preponderance of data.

There is also the more traditional understanding of a confirmed identity, that is, a confident identity that has been confirmed by comparison to an authentic reference standard. In this case, comparison to an authentic reference standard can be considered to be the third dimension of identification. Thus, in Figure 3.16 there are two flavors of a confirmed identity, data-based and standard-based.

These identification categories establish an identification hierarchy consisting of the five categories, as listed in Figure 3.17 [18].

While it is fine to have established a hierarchy, it is infinitely more useful to have established the criteria upon which placement in the hierarchy is based. To this end, a hierarchy has been established in Table 3.7 for a specific analyte, ultimately established to be di-(2-ethylhexyl phthalate). Based on the available information, the identification of this analyte is tracked from unknown to confirmed as more and different information is obtained.

Identification categories for extractables and leachables have been established in the USP extractables chapter <1663> [19]. The fit of the five classification categories in Table 3.7 and Figure 3.16 with the four categories in USP <1663> is established in Table 3.8. It is concluded that the Identification Hierarchy discussed herein is consistent

Figure 3.16 The Identification hierarchy, which establishes identification categories. As the amount and rigor of confirming information increase, the information content of the identity increases and the likelihood that the identity is correct increases. (*See insert for colour representation of the figure.*)

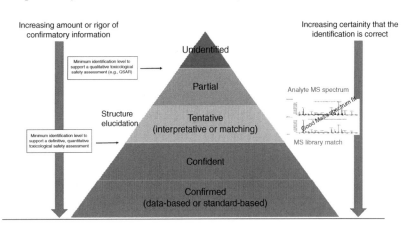

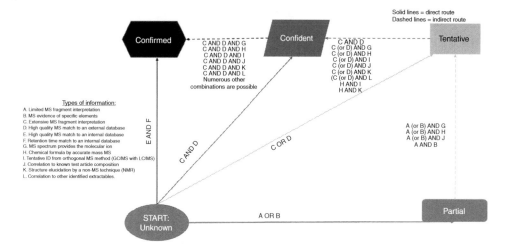

Figure 3.17 The process of securing an identification. When a chromatographic response has been obtained for an analyte, the analyte is classified as an unknown until the information contained within that response is evaluated. Depending on the nature and amount of available information, an identification within a certain class can be secured. As more information is secured, one can "move up" in the identification class, as the additional information corroborates the initial classification and provides more confidence that the initial identification is correct.

with, and is an embellishment of, and is an improvement over, the USP <1663> identification categories.

In reviewing the hierarchy, one notes that the terms tentative and speculative occupy the same position in the hierarchy, which is the lowest position (least amount of information) where one is able to make an "educated guess" at the compound's specific identity. These two classes differ, not necessary in terms of the information from which the identity is deduced but rather by the means the identity is deduced. For example, consider a compound revealed by GC/MS with its associated mass spectrum. If the mass spectrum is interpreted by an expert to elucidate the compound's identity, then the identity is the result of an interpretative process. However, if the mass spectrum is matched to a mass spectral database, then the inferred identity is the result of a matching process.

An important question concerning the use of identities in compatibility assessment frequently comes up, although it is phrased in different ways. This question is "what level of identification is required to confidently use the identification as a basis of the compatibility assessment?"

An extractable's identity is typically used to infer or assess the effect of that extractable, typically, but not solely, its safety impact. As an identification of "unknown" has no significance except that an extractable has been discovered at a level above a relevant threshold, it is clear that an "unknown" cannot be assessed, in terms of suitability for use impacts, in a direct manner. Alternatively, confident and confirmed identities can be assessed for their potential suitability impact directly as the degree of confidence in the identity is sufficiently high that it is likely the assessment will be proper and the information content of the identity

is sufficiently high that relevant compatibility information can be found (assuming that such data exists). Nevertheless, in performing the assessment it is pertinent to consider whether the uncertainty of a speculative tentative identification needs to be accounted for in that assessment.

A partial identification provides structural information but not a specific identity. Thus, a partial identification cannot be assessed in the same manner as a higher-level identification. However, a partial identity can be assessed by means that are structure driven; for example, in silico means of inferring safety, carcinogenicity, and mutagenicity. The value of the partial identity is that its in silico assessment may reveal the absence of structural alerts for incompatibilities, possibility allowing a higher threshold to be used. For example, if the in silico assessment of a partially identified extractable indicates that it has no structural alerts for mutagenicity, then perhaps the calculation of the AET can be based in the higher value of the QT versus the SCT.

Mechanistically, a tentative identification can be assessed for suitability for used considerations as the tentative or speculative identification typically assigns a single and specific identity to an extractable or leachable. However, given the level of uncertainty in a tentative identification, it is appropriate to question whether these types of identification are sufficiently robust to be the basis of a rigorous compatibility assessment. Thus, it is prudent to re-visit a tentative identification depending on the outcome of the compatibility assessment. If the outcome of the assessment is definitive (ether compatible or non-compatible), then the tentative identity is judged to be sufficiently robust to

Table 3.7 The proposed identification hierarchy. The type of data required for an identity to fall within a given category is given and the use of that information to establish an identity is illustrated using an example compound, di-(2-ethylhexyl) phthalate.

Level	Information content	Information type	Information example	Title	Reporting example
1	Insufficient information to propose a structure or identity	Chromatographic and/or spectroscopic data that provides little or no insight into the possible structure or identity	1. GC/FID peak with no GC/MS peak. 2. GC/MS peak with no library hits and uninterpretable spectra. 3. LC/UV peak with no MS signal. 4. LC/MS peak with uninterpretable spectra.	Unknown	Unknown
2	Sufficient information to propose a structure but insufficient information to propose a specific identity	Chromatographic and/or spectroscopic data that is sufficient to infer a general structure but not sufficient to propose a specific identity	1. Peak with GC/MS or LC/MS spectrum that can be interpreted sufficiently that a structure can be inferred. 2. Structure can be inferred by reference to internal or external literature. 3. Accurate mass MS that infers the empirical formula only.	Partial	A phthalate (for example, di-2-ethylhexyl phthalate)
3	Sufficient information to propose a specific identity	Chromatographic and/or spectroscopic data that is sufficient to infer a specific identity; however, inference is not supported by additional data	1. GC/MS peak with library spectral match. 2. LC/MS peak with definitive molecular weight and match to an equivalent molecular weight in a GC/MS peak. 3. GC/MS or LC/MS peak whose structure has been elucidated by expert interpretation of the mass spectrum.	Tentative (interpretative or matching)	Di-(2-ethylhexyl phthalate), tentative
4	Sufficient information to support a proposed specific identity	Chromatographic and/or spectroscopic data that is sufficient to infer a specific identity; inference is supported by additional data	1. GC/MS peak with library spectral match or LC/MS peak with definitive molecular weight and match to an equivalent molecular weight in a GC/MS plus retention time match. 2. GC/MS peak with library spectral match or LC/MS peak with definitive molecular weight and match to an equivalent molecular weight in a GC/MS plus accurate mass. 3. GC/MS peak with library spectral match or LC/MS peak with definitive molecular weight and match to an equivalent molecular weight in a GC/MS plus MS/MS fragment pattern interpretation. 4. GC/MS peak with library spectral match or LC/MS peak with definitive molecular weight and match to an equivalent molecular weight in a GC/MS plus inference from internal or external literature. 5. GC/MS peak with library spectral match or LC/MS peak with definitive molecular weight and match to an equivalent molecular weight in a GC/MS plus inference from NMR. 6. The same identity is proposed by orthogonal analytical methods (e.g. both GC/MS and LC/MS) independently.	Confident	Di-(2-ethylhexyl phthalate), confident
5	Sufficient information to confirm a specific identity	Chromatographic and/or spectroscopic data that is sufficient to infer a specific identity; inference is supported by comparison to an authentic reference standard of the inferred substance	Retention time and spectral match to an authentic reference standard	Confirmed (data-based or standard-based)	Di-(2-ethylhexyl phthalate), confirmed

Table 3.8 Comparison of the identification hierarchy with identification categories per USP <1663>.

	Identification hierarchy			Title in USP
Level	Information content	Information type	Description per USP <1663>	<1663>
1	Insufficient information to propose a structure or identity (unknown)	Chromatographic and/or spectroscopic data that provides little or no insight into the possible structure or identity	Not used in USP <1663> but inferred to be a compound whose identity cannot even be proposed based on the available data	Unknown
2	Sufficient information to propose a structure but insufficient information to propose a specific identity (Partial)	Chromatographic and/or spectroscopic data that is sufficient to infer a general structure but not sufficient to propose a specific identity	Data has been collected which is consistent with a class of molecule only. Examples of such data include mass spectrometric fragmentation behavior/expert mass spectrum interpretation and mass spectrum matches automated library or literature spectrum	Tentative
3	Sufficient information to propose a specific identity (Tentative)	Chromatographic and/or spectroscopic data that is sufficient to infer a specific identity; however, inference is not supported by additional data	Not used in USP	NA
4	Sufficient information to support a proposed specific identity (Confident)	Chromatographic and/or spectroscopic data that is sufficient to infer a specific identity; inference is supported by additional data	A tentative identification that has been bolstered by additional and sufficient confirmatory information to preclude all but the most closely related structures. Such confirmatory information could include confirmation of molecular weight, confirmation of elemental composition (via accurate mass), and supporting spectral information from an orthogonal method (e.g. NMR)	Confident
5	Sufficient information to confirm a specific identity (Confirmed)	Chromatographic and/or spectroscopic data that is sufficient to infer a specific identity; inference is supported either by comparison to an authentic reference standard of the inferred substance or via a preponderance of collaborating data	A preponderance of evidence confirms the entity in question can only be the identification provided. Although it is possible that a highly confident identification may meet the standard implied by the preponderance of evidence, the only means of providing a confirmed identification is via mass spectral and retention time match with an authentic reference compound	Confirmed

be the basis of the assessment. This is the case as the "margin of error" associated with the identity being tentative is covered by the definitive assessment outcome. However, if the outcome of the suitability for use assessment for a tentatively or speculatively identified compound is equivocal or borderline, then the identification should be elevated to either confident or confirmed by collecting the required additional supporting data.

3.2.6 The Identification Process

The previous section established the various identification categories and described the information necessary to

secure an identification within each category. This section discusses the means and process of securing an identity and for "elevating" an identity as new information is obtained. This process of landing in an identification class, or of moving between classes, is illustrated in Figure 3.17.

When a chromatographic response is obtained and the identification process is initiated, all compounds are classified as an unknown, as it is only through the processing of the response information that an identity is obtained. Depending on the amount and nature of the information obtained from the direct analysis of the sample, the compound can be directly identified to the extent possible with the available information. That is, the available information is sufficient to secure an identification within one of the four identification classes of partial, tentative, confident, or confirmed. For example, if the mass spectrum obtained for a compound cannot be matched to a mass spectral database and provides only minimal structural information (for example, there is mass spectral evidence of specific elements or structural units), then one can use the information to move directly to a one-dimensional partial identification. Note that while in certain circumstances the use of accurate mass MS to secure a compound's chemical formula may be sufficient to secure a partial identification (for example, the chemical formula contains a certain element), the chemical formula itself is insufficient information to secure a tentative identification. On the other hand, if the mass spectral information is sufficiently robust and detailed that either a high-quality match can be made to an external spectral database or if the mass spectrum can be fully elucidated, then one can use the information to move directly to a one-dimensional tentative identification. Alternatively, if the mass spectral information is sufficiently robust and detailed that a high-quality match can be made to an external spectral database or if the mass spectrum can be fully elucidated, then one can use the information to move directly to a two-dimensional confident identification. Lastly, if the mass spectral information and retention time can be matched to the same properties of an authentic reference compound, then one can use this information to move directly to a three-dimensional confirmed identification.

In addition to these direct paths to identification, numerous indirect paths can be envisioned, where the indirect paths can be considered to reflect a multiple-step identification process. For example, consider the case where a one-dimensional tentative identification has been secured via mass spectral matching. To "elevate" the tentative identification to a confident identification, a second, corroborating, dimension of data must be obtained. Figure 3.17 illustrates what types of corroborating information could be used to "elevate" an identification by one class; that is, a partial identification to a tentative identification, a tentative identification to a confident identification, and a confident identification to a confirmed identification. To illustrate this, take the case of a tentative identification secured by mass spectral matching. If the analytical team secured a corroborating chemical formula by reanalyzing the same using high-resolution MS detection, then the tentative identity can be "elevated" to a confident identity based on having two-dimensional data. Once the confident identification level has been achieved, then a confirmed identification can be pursued by considering possible additional, third-dimensional, corroborating information. For example, consider the case where an organic extractable

has been confidently identified as an antioxidant degradation product. If the analytical team secures information that establishes that the test article in fact contained the antioxidant as an intentional additive, then this corroborating information would be sufficient to "elevate" the confident identification to a confirmed identification.

As one moves from tentative to confident and especially confident to confirmed, one understands that there are numerous combinations of possible data that would support either a two-dimensional confident identification or a three-dimensional confident identification. While some of these combinations are shown in Figure 3.17, many other combinations can be envisioned, where the number and type of combinations are limited only by the analytical chemist's ability to secure identification-enabling information.

The process of securing identities for organic extractables and leachables is considered in greater detail in the Appendix.

3.2.7 "Quantitation"

Whereas a compound's identity is used to establish its "no-effect" level, a compound's concentration is used to establish the patient's exposure level to the compound. That is, the identity is used to establish the permissible exposure while the concentration establishes the actual exposure. It is often the case that the compatibility assessment simplifies to a comparison of permissible exposure to actual exposure.

As was the case with identification, an analytical method does not produce an analyte's concentration directly; rather, the method's response to the analyte is "calibrated" by some means to infer or estimate the analyte's concentration. As was also the case with identification, there is the ideal situation and practical reality associated with quantitation. It goes without saying that the ideal situation of obtaining the most accurate estimate of an analyte's concentration in a sample is achieved when the test method's response to a specified analyte has been calibrated via the generation of a calibration curve obtained by the analysis of multiple standards specific to the analyte and prepared to contain the analyte at known concentrations. However, given the large and diverse population of potential extractables and leachables and the circumstance that most extractable or leachables profiles consist of numerous and largely unpredictable substances, generation of a calibration curve for each individual extractable (or leachable) is impractical, if not impossible, and hence exact or accurate quantitation is rarely performed during screening. Rather, the analyte's concentration is estimated by making certain assumptions about its response curve, typically in regard

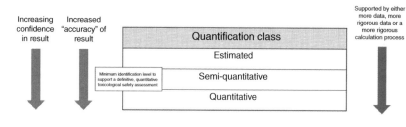

Figure 3.18 The quantitation hierarchy, which establishes quantitation categories. As the rigor of the means by which the concentration is obtained increases, the likelihood that the obtained concentration reflects the true concentration increases and the confidence in the reported concentration increases.

to the reference analyte surrogate standard. Recognizing that accuracy of (or inaccuracy in) a concentration estimate depended on the validity of the response assumptions made, the ISO 10993:18 standard for the chemical assessment of medical devices [1] established categories of quantitation as follows (see Figure 3.18) [18]:

Estimated quantitative analysis. An analytical approach that estimates an analyte's concentration by using the response from a surrogate substance chosen without specifically addressing or considering the relative responses of the analyte and the surrogate.

Semi-quantitative analysis. An analytical approach that provides an analyte's concentration by using the response from a surrogate substance (or substances), specifically accounting for the relative responses of the analyte and the surrogate.

Quantitative analysis. An analytical approach that establishes the most accurate estimate of an analyte's concentration by using a response function (calibration curve) generated specifically for the analyte via the use of a reference standard.

It should be clear from these definitions that estimated quantitative analysis is generally less accurate than semi-quantitative analysis, which is generally less accurate than quantitative analysis.

From the previous discussion, one notes that a truly quantitative analysis is rare in chromatographic screening for organic extractables and leachables while quantitative analysis is the method of choice for a target analysis.

The means by which quantitative analysis is accomplished in extractables and leachables targeting and screening is clear enough. A calibration curve is generated for each relevant substance via the analysis of standards containing known amounts of the substances, where the calibration curve is the best-fit mathematical relationship between the analyte's concentration and its instrumental response (which, for a chromatographic method, is either peak height or peak area). The analyte's concentration in a sample (extract or drug product) is obtained by inputting the analyte's response into the calibration curve and solving. So long as the analyte's concentration in the sample falls within the calibration range, the concentration thus obtained is highly accurate.

Both an estimated and a semi-quantitative concentration are obtained in a similar way. In both cases, a surrogate standard compound is added to the sample (extract or drug product) at a known concentration prior to the testing of the sample. The sample is analyzed and the responses for the surrogate standard and the analyte of interest is noted. These responses are used to establish the analyte's concentration as follows:

$$C_{\text{analyte}} = (R_{\text{analyte}}/\text{RRF}_{\text{analyte}}) \times (C_{\text{SS}}/R_{\text{SS}}) \tag{3.5}$$

where

SS = surrogate standard
R_{SS} = response of the SS
C_{SS} = concentration of the SS
R_{analyte} = response of the analyte
C_{analyte} = concentration of the analyte
$\text{RRF}_{\text{analyte/SS}}$ = relative response factor, analyte versus surrogate standard

The difference between an estimated concentration and the semi-quantitative concentration is the value that is used for $\text{RRF}_{\text{analyte/SS}}$. In the case of an estimated concentration, it is assumed that the response to all potential analytes is the same and thus $\text{RRF}_{\text{analyte/SS}}$ is arbitrarily assigned a value of 1 for all analytes and all surrogate standards. On the other hand, a semi-quantitative concentration is obtained using Equation (3.5) with an experimentally derived $\text{RRF}_{\text{analyte/SS}}$ obtained by analysis of a standard prepared to contain a known quantity of both the analyte and the surrogate standard.

To illustrate the various quantitation approached and categories, consider the following scenario. Chromatographic screening of an extract using two surrogate standards (SS#1 and SS#2) has produced two identified analyte peaks (AN#1 and AN#2). The objective is to quantify AN#1 and AN#2. It should be noted that the RF and RRF calculations are typically performed using the peak area, not the peak height. However, for the sake of illustration, quantitation will be conducted using peak height, as the logic of calculation remains the same.

Step 1: Obtain response factors by injecting a standard containing 10 ppm of each of the four compounds. This produces the chromatogram shown in Figure 3.19.

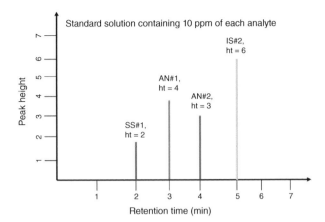

Figure 3.19 Example standard chromatogram illustrating the quantitation of extractables AN1 and AN2 versus the surrogate standards SS#1 and SS#2. Each compound is present in the sample at a concentration of 10 ppm.

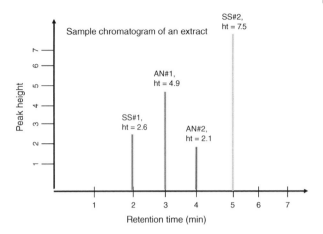

Figure 3.20 Example extract chromatogram illustrating the quantitation of extractables AN1 and AN2 versus the surrogate standards SS#1 and SS#2. Both surrogate standards are present in the sample at a concentration of 10 ppm.

The response factors (RFs) are calculated as the ratio of peak height (R) and the concentration (C):

$$RF_{SS} = R_{SS}/C_{SS} \text{ and } RF_{Compound}$$
$$= R_{Compound}/C_{Compound} \quad (3.6)$$

$$RF_{SS\#1} = 2/10 \text{ ppm} = 0.2$$

$$RF_{AN\#1} = 4/10 \text{ ppm} = 0.4$$

$$RF_{AN\#2} = 3/10 \text{ ppm} = 0.3$$

$$RF_{SS\#2} = 6/10 \text{ ppm} = 0.6$$

The RRFs are calculated as follows:

$$RFF_{Compound} = RF_{Compound}/RF_{SS} \quad (3.7)$$

versus SS#1

$$RRF_{AN\#1/SS\#1} = 0.4/0.2 = 2$$

$$RRF_{AN\#2/SS\#1} = 0.3/0.2 = 1.5$$

$$RRF_{SS\#2/SS\#1} = 0.6/0.2 = 3$$

versus SS#2

$$RRF_{SS\#1/SS\#2} = 0.2/0.6 = 0.333$$

$$RRF_{AN\#1/SS\#2} = 0.4/0.6 = 0.6667$$

$$RRF_{AN\#2/SS\#2} = 0.3/0.6 = 0.5$$

Step 2: Obtain the responses from the sample chromatogram in Figure 3.20:

$$R_{AN\#1} = 4.9$$

$$R_{AN\#2} = 2.1$$

$$R_{SS\#1} = 4.9$$

$$R_{SS\#2} = 7.5$$

After step 2, the calculation processes diverge, depending on the quantitation approach employed.

Approach A: Assume RRF versus SS = 1 for all analytes ("simple quantifications").

Step 3: Calculate the Analyte Concentrations using Equation (3.5):

Process A.1: Concentration using SS#1, assuming RRF versus SS#1 = 1 for all analytes – "simple quantification" or "estimated concentration"

For AN#1:

$$R_{AN\#1} = 4.9; RRF_{AN\#1/SS\#1} = 1 \text{ (assumed)};$$
$$C_{SS\#1} = 10 \text{ ppm}; R_{SS\#1} = 2.6 \text{ (peak height); and}$$
$$C_{AN\#1} = 18.8 \text{ ppm}$$

For AN#2:

$$R_{AN\#2} = 2.1; RRF_{AN\#2/SS\#1} = 1 \text{ (assumed)};$$
$$C_{SS\#1} = 10 \text{ ppm}; R_{SS\#1} = 2.6 \text{ (peak height); and}$$
$$C_{AN\#1} = 8.08 \text{ ppm}$$

For SS#2:

$$R_{SS\#2} = 7.5; RRF_{SS\#2/SS\#1} = 1 \text{ (assumed)};$$
$$C_{SS\#1} = 10 \text{ ppm}; R_{SS\#1} = 2.6 \text{ (peak height); and}$$
$$C_{SS\#2} = 28.8 \text{ ppm}$$

Process A.2: Concentration by SS#2, assuming RRF versus SS#2 = 1 for all analytes – "simple quantification" or "estimated concentration"

For AN#1:

$$R_{AN\#1} = 4.9; RRF_{AN\#1/SS\#2} = 1 \text{ (assumed)};$$
$$C_{SS\#2} = 10 \text{ ppm}; R_{SS\#2} = 7.5 \text{ (peak height); and}$$
$$C_{AN\#1} = 6.53 \text{ ppm}$$

For AN#2:

$$R_{AN\#2} = 2.1; RRF_{AN\#2/SS\#2} = 1 \text{ (assumed)};$$
$$C_{SS\#2} = 10 \text{ ppm}; R_{SS\#2} = 7.5 \text{ (peak height); and}$$
$$C_{AN\#1} = 2.8 \text{ ppm}$$

For SS#1:

$$R_{SS\#1} = 7.5; RRF_{SS\#1/SS\#2} = 1 \text{ (assumed)};$$
$$C_{SS\#2} = 10 \text{ ppm}; R_{SS\#2} = 7.5 \text{ (peak height); and}$$
$$C_{SS\#1} = 3.47 \text{ ppm}$$

Approach B: Use actual RRF versus SS for all analytes, with individually recorded RRF for each analyte.

Step 3: Calculate the analyte concentrations via Equation (3.5) and using Equation (3.7) to calculate RRF:

Process B1: Concentration by SS#1, with the actual; RRF for each analyte – "semi-quantitative"

For AN#1:

$$R_{AN\#1} = 4.9; RRF_{AN\#1/SS\#1} = 2; C_{SS\#1} = 10 \text{ ppm};$$
$$R_{SS\#1} = 2.6 \text{ (peak height); and}$$
$$C_{AN\#1} = 9.42 \text{ ppm}$$

For AN#2:

$$R_{AN\#2} = 2.1; RRF_{AN\#2/SS\#1} = 1.5; C_{SS\#1} = 10 \text{ ppm};$$
$$R_{SS\#1} = 2.6 \text{ (peak height); and}$$
$$C_{AN\#2} = 5.38 \text{ ppm}$$

For SS#2:

$$R_{SS\#2} = 7.5; RRF_{SS\#2/SS\#1} = 3; C_{SS\#1} = 10 \text{ ppm};$$
$$R_{SS\#1} = 2.6 \text{ (peak height); and}$$
$$C_{SS\#2} = 9.61 \text{ ppm}$$

Process B2: Concentration by SS#2, with the actual; RRF for each analyte – "semi-quantitative"

For AN#1:

$$R_{AN\#1} = 4.9; RRF_{AN\#1/SS\#2} = 0.6667; C_{SS\#2} = 10 \text{ ppm};$$
$$R_{SS\#2} = 7.5 \text{ (peak height); and}$$
$$C_{AN\#1} = 9.80 \text{ ppm}$$

For AN#2:

$$R_{AN\#2} = 2.1; RRF_{AN\#2/SS\#2} = 0.5; C_{SS\#2} = 10 \text{ ppm};$$
$$R_{SS\#2} = 7.5 \text{ (peak height); and}$$
$$C_{AN\#2} = 5.60 \text{ ppm}$$

For SS#1:

$$R_{SS\#1} = 2.6; RRF_{SS\#1/SS\#2} = 0.333; C_{SS\#2} = 10 \text{ ppm};$$
$$R_{SS\#2} = 7.5 \text{ (peak height); and}$$
$$C_{SS\#2} = 10.4 \text{ ppm}$$

Approach C: Concentration using each analyte's own RF value :

$$C_{Compound} = R_{Sample}/RF_{Compound} \tag{3.8}$$

$$C_{AN\#1} = \text{peak height}_{AN\#1}/RF_{AN\#1} = 4.9/0.4 = 12.3 \text{ ppm}$$
$$C_{AN\#2} = \text{peak height}_{AN\#2}/RF_{AN\#2} = 2.1/0.3 = 7.00 \text{ ppm}$$
$$C_{SS\#1} = \text{peak height}_{SS\#1}/RF_{SS\#1} = 2.6/5 = 13.0 \text{ ppm}$$
$$C_{SS\#2} = \text{peak height}_{SS\#2}/RF_{SS\#2} = 7.5/1.667 = 12.5 \text{ ppm}$$

As the previous calculations have established, once one has queried the standard chromatogram, there are five ways of calculating the analyte's actual concentration in an extract (three separate ways and two of the ways use the two individual surrogate standards):

1. Using the assumption that RRF = 1 for all surrogate standards and analytes:
   a) Using SS#1
   b) Using SS#2
2. Using the actual calculated RRF values:
   a) Using SS#1
   b) Using SS#2
3. Using the analyte's own RF calculated from the standard chromatogram.

It is noted in considering option number 3 that injection-to-injection variation, which is typically accounted for by a surrogate standard, is not accounted for in this approach.

The results of these quantitation activities are shown in Table 3.9. As expected, the calculated concentrations for the two analytes and the other surrogate standard are higher using the low response surrogate standard SS#1 and lower using the higher response surrogate standard SS#2. The analytical error in either calculation is as much as a factor of 3 (that is, SS#1 overestimates the analyte's true concentration by nearly a factor of 3 while SS#2 underestimates the analyte's true concentration by nearly a factor of 3). It is clear that the actual analyte concentration falls somewhere between the two values obtained using the two surrogate standards.

The concentrations calculated using the actual RRF values are relatively consistent, SS#1 versus SS#2, with the difference between them being injection-to-injection variations that are not fully accounted for by the individual surrogate standards. These calculated concentrations are very close to the expected concentrations, with a calculation error of less than 10%.

The concentrations obtained with the analyte's absolute response factor are close to the expected result, but are not as close as the concentrations obtained using the individual RRF values, as these concentrations do not take into account analytical variations between injections. In this

Table 3.9 Results of quantifying analytes based on different approaches.

Sample	Concentration, ppm					
	RRF = 1 versus SS#1	RRF = 1 versus SS#2	Actual RRF versus SS#1	Actual RRF versus SS#2	Versus its own RF	Actual
AN#1	18.8	6.53	9.42	9.80	12.3	≈9.6
AN#2	8.08	2.80	5.38	5.60	7.00	≈5.5
SS#1	—	3.47	—	10.4	13.0	10
SS#2	28.8	—	9.61	—	12.5	10

example, the relative error involved with the concentration calculation is approximately 30%.

In conclusion, the intrinsic inaccuracy in quantitation assuming all analytes and surrogate standards have the same RRF of 1 has been illustrated. Accuracy is significantly improved via the use of experimentally determined RRF, regardless of the surrogate standard used. It is for this reason that "quantitation" based solely on a surrogate standard's absolute response has been termed an "estimated concentration" in ISO 10993:18 and that quantitation based on RRFs has been termed "semi-quantitative" analysis in the same document. The term "quantitative" has been reserved for quantitation accomplished with a unique standard curve generated for each individual analyte.

The three quantitation categories are all based on certain simplifying assumptions that may or may not be valid. For example, an estimated concentration is obtained based on the assumptions that (i) the response factor for all analytes is the same as that of the chosen surrogate standard (ii) the response factors are unaffected by the absolute and relative concentrations of the analytes and the surrogate standard. Given the known and potentially large variation in response factors for either GC/MS or LC/MS screening methods (for example, references [11, 12, 20, 21]), it is clear that the first assumption is more wishful thinking than reality and hence the use of the term "estimated" to describe this quantitation category. Although this type of quantitation may have been referred to (incorrectly) as "semi-quantitative" by practitioners of the art, the term "semi-quantitative" implies a level of rigor and a degree of confidence that is not justified considering the means by which the estimated concentration is obtained.

The second category, semi-quantitative, takes care of the first concern as semi-quantitation is based on an analyte-specific RRF, determined by the analysis of a standard solution containing the compound of interest and the surrogate standard at known (and typically equal) concentrations. However, the accuracy of semi-quantitative concentrations could be adversely affected by concentration differences (mismatches) between the analyte of

interest and the surrogate standard and the analyte in the sample versus analyte in the standard.

The third category, quantitation, takes care of both concerns with respect to response variations across analytes and across concentrations, as the reported concentration is based on each analyte's unique response function (calibration curve). It is noted that quantitation is a requirement for, and is most easily achieved in, target analysis.

As was the case with the identification hierarchy, it is relevant to consider what level of quantitation is necessary to support rigorous impact assessment. On the surface, it is reasonable that one would conclude that an estimated concentration is never appropriate as the basis of a rigorous impact assessment. However, the validity of this statement depends somewhat on the outcome of the assessment. As was done previously for identification, consider safety assessment and the margin of safety (MoS) as an example. In circumstances where the interpretation of the MoS is definitive (for example, either MoS > 10 and it is definitely concluded that the patient safety risk is negligible or MoS < 0.1 and it is definitely concluded that the patient safety risk is possibly considerable), uncertainty in the estimated concentration may be irrelevant to the assessment outcome. For example, consider the case where the error in the estimated concentration is sufficiently large that its estimated concentration is a factor of 5 less than its true concentration. In the case of the MoS > 10, an MoS adjusted for the error would still be MoS > 2, which would still support the conclusion of negligible safety risk. In the case of the MoS < 0.1, an MoS adjusted for the error would still be MoS < 0.1, which would still support the conclusion of considerable possible safety risk. Thus, in these cases, an estimated concentration could be an adequate basis of the toxicological safety risk assessment.

However, when the MoS based on an estimated concentration is between the values of 0.3 and 3 (for example), then the error in the estimated concentration is sufficiently high that there is little or no "margin for error" in the assessment. In such cases, the estimated concentration is not the proper basis of the safety assessment. However, in these cases an estimated concentration becomes the basis of a screening decision to secure a more accurate quantitation to serve as the basis of the assessment.

Semi-quantitative and quantitative concentrations, given their higher certainty and greater confidence, are both suitable starting points for impact assessment. It is unclear to this author whether the increased certainty and confidence in quantitative concentrations is necessary, especially considering the practical difficulties associated with securing such information. Nevertheless, if quantitative concentrations can be obtained, then they should be the basis of the impact assessment.

The associated relative inaccuracies of these "quantitation" strategies are discussed in greater detail is Section 3.2.8 following.

3.2.8 Uncertainty Adjustments to Estimated Quantitative Data

As was noted previously, the issue with respect to estimated quantitative data is the well-known compound-to-compound variation in response factors for surrogate standard (SS) candidates and extractables/leachables. Application of the SS response factor to a compound that responds better than the SS will mean that the compound's concentration will be overestimated while application of the response factor to a compound that responds more poorly than the SS will mean that the compound's concentration will be underestimated. Underestimation is particularly unfortunate in impact assessment (e.g. toxicological safety assessment) as an assessment based on an underestimated concentration will underestimate the impact. Considering safety, assessing an under-estimated extractable or leachable could lead to the conclusion that the substance has no meaningful adverse impact on safety, when in fact the opposite is possible.

One method that can be adopted to reduce the occurrence and risk of underestimation is to "adjust" estimated concentrations, obtained via an SS response factor, upward via the same UF that is used to adjust the AET. In considering the merits of such a proposal, we go back to Figure 3.5, where the distribution of response factors is normal and the SS is a compound whose response factor is the mean (and median) of the distribution. In this case, the 50% of the compounds that have a response factor greater than that of the SS (the right side of the curve) will not be underestimated; in fact, the higher the compound's response factor, the more its concentration will be overestimated. Thus, no SS adjustment is needed for these compounds and if they are adjusted, their overestimation will only be exacerbated. While this may not present a safety assessment issue, as overestimation is protective of patient safety, it ignores the downside that overestimation magnified by response factor corrections potentially means that perfectly safe products either do not get to the market (because their risk is overassessed) or getting them to the market is delayed and more costly as unnecessary testing may be required to undo an erroneous assessment.

Furthermore, consider the impact of a one standard deviation adjustment on those compounds that will be underestimated (the left side of the curve). A one standard deviation adjustment will mean that those compounds that fall in the part of the distribution between the mean and the mean − 1 standard deviation will have an adjusted concentration that is greater than or equal to their true concentration. Thus, 34.1% of the entire population and $34.1/50 = 68.2\%$ of the compounds that would benefit from adjustment will be adjusted sufficiently. For the remainder of the compounds that would benefit from adjustment, even with the adjustment, their adjusted estimated concentrations will be below their true concentrations (which is 15.8% of the entire population and 31.8% of the population that would benefit from adjustment).

So, let us recap what an adjustment via the UF will accomplish:

1. For the 50% of the compounds that are OK (that is, whose reported concentrations are already equal to or greater than their true concentrations), adjustment increases their extent of overestimation.
2. Of the 50% of the compounds that would benefit from adjustment, 68.2% of them (34.1% of the entire population) would be sufficiently adjusted that their adjusted estimated concentration would be greater than or equal to their true concentration.
3. Of the 50% of the compounds that would benefit from adjustment, 31.8% of them (15.8% of the entire population) would be insufficiently adjusted and their adjusted estimated concentration would still be less than their true concentration.

This situation suggests that the one standard deviation concentration adjustment does not accomplish its objective.

This is especially true when there exists a better way to correct for compound-to-compound response variations, which is to use the compound's own response factor to quantify it. After all, the response factors are in the database so why don't we use them? In essence this means that each individual compound is individually adjusted for its difference in response versus the SS via its RRF and that the adjusted concentration will equal the true concentration, within experimental error.

While the UF adjustment could be considered to be a generic or average adjustment, the RRF adjustment is a customized adjustment that further reduces overprotection.

A question that remains using either approach is how to quantify unidentified compounds that obviously will have unspecified response factors. In the UF approach, the concentrations of the unidentified compounds are estimated using the SS response factor and then adjusted up with the UF. In the RRF approach, the concentration could be estimated using the mean RRF and the same UF adjustment. This would produce the same outcome, as the response factor of the SS is the mean of the response factors for the entire population. In those more common circumstances where the response factor of the SS is not the population mean, then using the mean response factor could be more or less protective, depending on the relationship of the mean and SS response factors. If the mean response factor were greater than the response factor of the SS, then use of the SS response factor and the UF correction would be more protective. If the mean response factor were less than the response factor of the SS, then use of the mean response factor and the UF would be less protective. Thus, the proper answer to the question "what should I do about unknowns?" depends on the relative relationship between the response of the SS and the population's mean response.

If one is compelled (or tempted) to try to quantify unidentified compounds, then the proper approach depends on the relative relationship between the response factor of the SS and the population's mean response factor:

- If the SS response factor is less than the population mean response factor, then the unidentified compounds could have their concentrations estimated using the SS response factor adjusted with the UF, as this would be more protective.
- If the SS response factor is greater than the population mean response factor, then the unidentified compounds could have their concentrations estimated using the mean response factor adjusted with the UF, as this would be more protective.

Lastly, it is noted that this assessment considers GC/MS data, whose response factors are generally recognized as being relatively constant, compound to compound. It is anticipated that the value of the UF, and therefore the raising of the reported concentration resulting from application of the UF, would be more significant for an analytical method such as LC/MS, where it is generally accepted that the response variation between compounds is much greater.

3.2.9 Reducing Variation via Process Optimization

The previous discussion on response factor variation establishes a means for mitigating this issue but does not necessary address the issue. It is possible to address the issue in some fashion by remembering that screening is accomplished not by a single analytical method (for example, GC/MS) but rather by the combination of multiple complementary and orthogonal analytical methods (for example, GC/MS and LC/MS). Moreover, each of the multiple analytical methods could use multiple detection methods (for example, GC/MS and GC/FID or LC/UV, LC/MS + ion and LC/MS − ion). Thus, the screening process is likely to result in multiple responses for the same analyte. This possibility was recognized by Christiaens et al. when they proposed that the magnitude of the response factor could be a means for deciding what compound concentration to report when a compound is quantified by multiple screening methods [22]. Specifically, these authors suggested that the proper concentration to report, that is, the concentration that is most likely to be accurate, is one derived by the method whose RRF for the analyte is closest to 1.

This concept can be expanded in a more general sense as a means of reducing response factor variation via a process that I will refer to as reporting optimization. The concept of reporting optimization is that for every possible extractable (or leachable) there is an optimal analytical approach that will provide the most reliable identification and the most accurate and precise quantitation. A database of relative responses factors, collected over multiple analytical approaches, allows one to identify which analytical approach is the optimal approach for quantifying each specific compound in the database. As noted previously, this optimum approach is the approach that produces a relative response closest to the value of 1. Once the optimum approach has been identified for each compound in the database, the testing laboratory adopts a reporting strategy that says, in essence, "I will only report the concentration for each analyte obtained by its optimal approach." In so doing, the testing laboratory increases analytical accuracy and decreases response factor variation.

This outcome is illustrated in the following example, shown in Table 3.10. In this example there is a database of 14 compounds, each of which was prepared at a concentration of 10 mg/l and had a 15th compound as an SS, also at 10 mg/l. The ratio of the responses was taken as the response factor. The compounds were injected into two analytical systems, one system with detectors 1 and 2 and the other system with detectors 3, 4, and 5. Thus we have five analytical approaches.

Within this approach, the variation in RRFs is established via the appropriate statistical analysis. As shown in Table 3.10, the RSD values obtained for the individual methods are fairly typical of what we have come to expect, with the methods for system 1 (Methods 1 and 2) being

Table 3.10 An example response factor database to illustrate reporting optimization.

Analyte	Measured response factor versus a surrogate standard					
	Method 1	**Method 2**	**Method 3**	**Method 4**	**Method 5**	**Optimum**
A	**1.02**	0.45	0.24	0.54	*ND*	1.02
B	0.83	0.57	2.07	**1.16**	0.76	1.16
C	1.21	**1.09**	3.54	0.23	2.27	1.09
D	1.93	**0.69**	0.09	*ND*	0.46	0.69
E	0.45	**1.19**	0.75	2.47	0.57	1.19
F	*ND*	*ND*	**0.63**	3.58	1.69	0.63
G	0.75	0.82	**0.89**	0.56	0.78	0.89
H	0.56	0.38	*ND*	**0.99**	0.46	0.99
I	1.35	**0.99**	0.35	1.37	1.67	0.99
J	2.09	**1.28**	1.97	1.42	*ND*	1.28
K	0.31	*ND*	**0.47**	*ND*	0.25	0.47
L	**0.89**	0.21	0.79	0.41	*ND*	0.89
M	**1.23**	0.48	*ND*	2.86	3	1.23
Mean	1.0517	0.7409	1.0718	1.4173	1.1910	0.9631
STD	0.5491	0.3576	1.0414	1.0976	0.9195	0.2453
RSD	0.5221	0.4827	0.9716	0.7744	0.7721	0.2547
% RSD	52.2120	48.2709	97.1630	77.4431	77.2059	25.4707

more typical for GC and the methods for system 2 (Methods 3, 4, and 5) being more typical for LC. However, what is important in Table 3.10 is the column labeled Optimum, which reflects, for each analyte, the response factor for any method that is closest to a value of 1. For example, the optimum response factor value for Analyte A is 1.02, obtained by Method 1.

As was noted previously, the reported concentration for analyte A that is most likely to be accurate is that obtained with Method 1. Thus, if the result obtained from the identified optimum method for each analyte is the result that is reported, then all reported results are likely to be the most accurate. Moreover, one notes that the variation in the optimum response factors is reduced versus the variation in the responses for the individual methods. Thus, reporting optimization has the dual benefit of reporting the most accurate concentration for all analytes and reducing the analytical response variation for all analytes. Reducing the analytical variation in response factors could have ramifications, for example, in the UF applied to an AET adjustment.

This concept has been realized in research published by Jordi et al. [20]. In this study, the authors have collected RRF for 217 organic extractables collected over six analytical approaches; GC/FID, GC/MS, LC/UV, LC/CAD, LC/MS (EI, + ion and − ion). The essential argument put forth by these authors was that their "multi-detector" approach could reduce the value of the UF that is required for the AET to be appropriately protective.

To understand the basis of this argument, the use of large UFs in the adjustment of the AET can be referred to as the AET Challenge. That is, as the UFs become larger, the analytical challenge is to be able to have sufficient analytical sensitivity to be able to achieve the low adjusted AET. In a certain way, the AET Challenge can be viewed as a one-dimensional problem; that is, the Challenge exists in a situation where extracts are screened for organic extractables and drug products are screened for organic leachables using a single analytical method. As it is generally recognized and required good practice that extractable and leachables screening is accomplished using a combination of multiple orthogonal methods, it is appropriate to re-visit the AET Challenge and consider whether a combination approach to screening provides some respite from the Challenge. For the sake of convenience, the use of multiple orthogonal approaches to accomplish E&L screening will be referred to as the "multi-combination" approach.

To gain an understanding of the essence of the multi-combination approach, let us first consider a process with which we may be familiar, liquid–liquid extraction. A significant issue associated with the use of liquid–liquid extraction for sample preparation is poor partitioning of polar and semi-polar analytes between the organic and aqueous phases. If a polar analyte preferentially remains in the aqueous phase during the liquid–liquid extraction, the objective of the extraction is not achieved. Although there are several means of "stacking the deck" so that the analyte's partitioning is enhanced (e.g. salting out), a more brute force approach is to perform multiple sequential extractions. For example, consider an analyte whose organic solvent/water partition coefficient is 1 (the analyte partitions equally between the organic and aqueous phases). If the volume of organic and aqueous phases used in the extraction is equal, at the end of the extraction 50% of the analyte will have transferred from the aqueous phase to the organic phase (50% recovery). If the resulting aqueous phase is contacted with a second equal volume of the organic phase, the extraction will again be 50% efficient and the second portion of the organic phase will contain 25% of the initial amount of the analyte in the aqueous sample. If the portions of the organic phase from the first and second extraction steps are combined, the cumulative transfer of the analyte to the combined organic phases is 75%. As additional extraction steps are performed and the resulting organic phases are combined, the cumulative transfer (recovery) of the analyte to the organic phase exceeds 90+% (see Table 3.11).

The purpose of this discussion of liquid–liquid extraction is to illustrate the concept of multi-combination analytical testing and its use to "solve" the AET challenge. Consider, for example, a single chromatographic method reflecting the combination of a separation process and a detection process (for example GC coupled with mass spectrometric detection, GC/MS). Consistent with the liquid–liquid extraction example, let us assume that this method is able to detect 50% of all possible extractables and leachables. To improve the analytical coverage of the universe of extractables, a second, orthogonal, analytical method is added to the analytical process. By orthogonal, one means that the fundamental mechanisms of operation of one of the analytical processes (separation or detection) are different, for example liquid chromatography versus GC or mass spectrometric detection versus UV absorption detection. In the circumstance where this second method is absolutely orthogonal (meaning that there is no overlap between the two methods), the combination of the two methods will, to

a first approximation, detect 75% of all possible extractables and leachables, consistent with the liquid–liquid extraction example. Add a third orthogonal method and the detection efficiency approaches 90%.

With this detectability example, we are close to addressing the AET challenge but are not quite there. So instead of assuming that the first analytical method is able to detect 50% of the extractables and leachables, let us instead assume this method is such that the response of 50% of all extractables and leachables is greater than the response for a chosen SS. This means that 50% of all extractables and leachables would be correctly captured by an AET defined by the SS if they were present in a sample at the AET level. Combining this method with a second, truly orthogonal method, which also has a 50% AET efficiency, would mean that the combination of the methods would be 75% effective in terms of meeting the AET Challenge without the use of an adjustment. As more orthogonal methods are combined, the coverage increases.

Although this discussion illustrates the concept of the multi-combination approach, establishing the practical utility of the approach in terms of considering methods that are, or could be, used in organic E&L screening requires laboratory experimentation. This brings us back to the work performed by the Jordi Labs investigators. In this study, Jordi scientists secured authentic reference standards for 217 semi-volatile and non-volatile organic substances that they had encountered in extraction and leaching studies they had performed. The selected substances included compounds of varying physicochemical properties, which these researchers believed adequately represented the entire universe of possible organic extractables and leachables. Having secured the reference standards, the Jordi researchers prepared standards containing the reference compounds at multiple levels, injected the standards into multiple analytical combinations of chromatographic separation and detection methods, and created calibration curves for each compound in each analytical combination, taking the slope of the calibration curve as the response factor of the specific compound in the specific analytical combination. The six orthogonal analytical combinations used by the Jordi researchers, all of which are either in common use in E&L screening or which have been reported as being applicable for this purpose, included:

1. GC with mass spectrometric detection (GC/MS)
2. GC with flame ionization detection (GC/FID)
3. Liquid chromatography with mass spectrometric detection, atmosphere pressure chemical ionization, + ion mode (LC/MS – APCI+)
4. Liquid chromatography with mass spectrometric detection, atmosphere pressure chemical ionization, – ion mode (LC/MS – APCI–)

Table 3.11 Use of multiple sequential extractions to improve the recovery of a liquid–liquid extraction.

Sequential extraction step	Efficiency of the extraction step	% of analyte recovered in the extraction step	Cumulative % of analyte recovered
1	50%	50%	50%
2	50%	25%	75%
3	50%	12.5%	87.5%
4	50%	6.25%	93.75%
5	50%	3.125%	96.875%

5. Liquid chromatography with UV absorption detection (LC/UV)
6. Liquid chromatography with charged aerosol detection (LC/CAD)

The nature of the study was such that the six combinations reflected two chromatographic separations, each of which was linked to several detectors, mentioned above. Thus, these were not two separate GC separations, each with their own detection method, but rather a single GC separation where the column effluent was split between an MS and an FID detector. Similarly, these were not four different LC separations, each linked to a separate detection method, but rather a single LC separation combined, either sequentially or via column effluent splitting, with all four detection methods.

The Jordi database, consisting of response factors for 217 compounds and 6 analytical methods, provides one with the opportunity to experimentally establish the actual effect of multiple analytical combinations on the utility of the AET.

Table 3.12 provides a means of reviewing the Jordi data for this purpose. In this table, the data is presented as the percentage of compounds in the database that would properly be reported as being at or above the AET based on a comparison of the analyte's response to that of an SS. Thus, for example, one notes that use of the combination of GC/MS with LC/MS without an AET adjustment (UF = 1) results in the situation where 54% of the 217 compounds in the database would correctly be flagged as being at or above the AET (the blue box in Table 3.12). If one were to add FID detection to the GC/MS, the AET effectiveness of the resulting three procedure combinations (meaning the combination approach's ability to correctly flag compounds at or above the AET) is increased to 63% (the green box in Table 3.12). Continuing the process by adding detectors to the LC method ultimately produces the situation where the combination of all six approaches has a 75% "efficiency" in terms of properly flagging compounds above the AET (red box).

Although the previous discussion effectively illustrates the utility of Table 3.12, the highest AET efficiency obtainable with a UF = 1, 75% for all six combinations, is generally recognized as being not sufficiently protective. Thus, another way of interpreting Table 3.12 is to specify an acceptable degree of protection and then establish what conditions are able to meet the acceptance criterion. For example, if an analytical approach is deemed to be acceptably protective if its AET efficiency is greater than 90%, then one observes that this criterion can be achieved for many groupings of UF value and combinations of analytical approaches (gray boxes in Table 3.12). For example,

- The combination of GC/MS and LC/MS meets this criterion with a UF = 3.
- This criterion can be met with a UF = 2 and a minimum of five combinations (either GC/MS-FID + LC/MS-UV or GC/MS + LC/MS/UV/CAD, noting that LC/MS is actually the combination of APCI+ and APCI−).

What has been established in this assessment is to demonstrate that the application of the multi-combination approach can meet the AET challenge without resorting to excessive UF values and their associated issues.

To aid in understanding the impact of the multi-combination approach, consider the following analogy. Before the advent of one coat coverage paints, it was often necessary to apply two coats of paint to cover large spots on a surface or to cover a dark color with a lighter color. This had the unfortunate outcome of making painting more laborious than one might have hoped it would be. Now that we have "one coat covers all" paints, painting is a breeze (or at least the commercials would have you believe that it is).

In a similar vein it is possible to look at the multi-combination solution to the AET Challenge and conclude that "this is an awful lot of work with a somewhat limited benefit." However, in so-doing one fails to recognize one of the more significant features of this approach, which loosely can be described as "one coat covers all." What is meant by this is the observation that in order for the multi-combination approach to be effective in addressing

Table 3.12 Percentage of compounds correctly reported as at or above the AET.

Detector	UF = 1		UF = 2		UF = 3		UF = 4	
	GCMS	GCMS-FID	GCMS	GCMS-FID	GCMS	GCMS-FID	GCMS	GCMS-FID
LCMS	54%	63%	85%	86%	91%	92%	93%	93%
LCMS-UV	63%	70%	90%	91%	95%	95%	95%	96%
LCMS-CAD	62%	71%	90%	91%	95%	95%	96%	96%
LCMS-UV-CAD	69%	75%	93%	94%	97%	97%	97%	97%

the AET Challenge, only one of the multiple combinations needs to flag the peak in question as being above the AET.

This circumstance is illustrated in Figure 3.21, which is the chromatographic response to Bisphenol S via the six multi-combination methods used in the Jordi research. One notes that in only one combination method, LC/UV, is the Bisphenol S response actually greater than the AET line. At first glance this outcome could be perceived as a problem, as it appears to be "non-reproducible." However, this one result, and this one result alone, is all that is required to establish that the level of Bisphenol S is above the AET and thus that Bisphenol S must be reported for toxicological safety risk assessment. What the various chromatograms establish is that this analyte is a poorly responding analyte, compared to an SS, in five of the six combinations. However, this rather unfortunate circumstance is rendered irrelevant by the one case where the analyte's response is at least as strong, if not stronger, than that of the SS.

To put this in perspective, what student does not like a test where you need get only one out of six questions right to get a passing grade?

Although collaborating evidence in the form of multiple combinations producing a response that is greater than the AET response is satisfying, only a single result above the AET is necessary to properly "pull the trigger" and report the compound for assessment.

Lastly, the use of the multi-combination approach to reduce analytical uncertainty is illustrated in Table 3.13, which captures the Jordi Labs data in terms of the results obtained first for each analytical method and lastly for the dataset of optimal result (best RRF for each analyte by any method, where the best RRF is established to be that RRF whose value is closest to 1.0). Rather large % RSD values for the individual analytical methods (with LC/MS providing the largest % RSD values) are significantly lessened when the optimal response obtained for any one detector is considered.

3.2.10 Screening Errors and Their Mitigation – The Extractables/Leachables Database

As noted previously, because chromatographic methods are uniquely suited for analyzing test samples for organic constituents, it has become standard industry practice to accomplish extractables and leachables screening via chromatography, as the chromatographic process is a necessary and effective means of separating individual analytes from one another. This separation is an essential aspect in terms of securing identities and concentration estimates. Moreover, the detection methodologies typically used with chromatographic separations fulfill the requirements for sensitivity, universal response, and information-rich response. Lastly, the various available

types of chromatography are broad in scope in the sense that they are complementary, overlapping, and orthogonal.

In essence, the chromatographic separation is a means of preparing an extract (or drug product) for detection. It is the detection that specifically accomplishes discovery, identification, and quantitation. The need to perform these functions for a large and diverse population of potential analytes establishes the requirements for appropriate detection mechanisms, where the ideal detection mechanism produces:

1. A response for the widest possible set of potential analytes (discovery)
2. A response that contains information with which or from which an analyte's identity can be inferred (identification)
3. A response whose magnitude can be correlated to the analyte's concentration in the tested sample (quantitation)

In the ideal or perfect situation shown in Figure 3.22, leachables in a drug product and extractables in an extract are:

*   Completely and fully accounted for (meaning that all leachables present in the drug product, or all extractables present in the extract, above an established and justified analytical threshold have been revealed)
*   Fully and exactly identified
*   Accurately and precisely quantified as such information is essential to assessing potential incompatibilities

In the real and imperfect world, the possibility exists that leachables or extractables screening is subject to errors, including (Figure 3.23) [22]:

*   The Error of Omission (discovery), where analytes in the sample go undetected,
*   The Error of In-exact Identification, where analytes in the sample are either not identified, mis-identified, or identified with a reduced level of confidence,
*   The Error of Inaccurate and Imprecise Quantitation, where analytes in the sample are inaccurately quantified.

These errors, shown in Figure 3.24, can hinder, complicate, bias or even thwart, individually or cumulatively, an impact assessment, producing an incomplete or erroneous assessment that could lead to either the commercialization of compromised drug products or the rejection of acceptable, necessary, and effective drug products.

3.2.10.1 Error of Omission (Discovery) [22]

Extractables and leachables profiles obtained by analytically screening either extracts or drug products should ideally consist of every organic extractable present in an

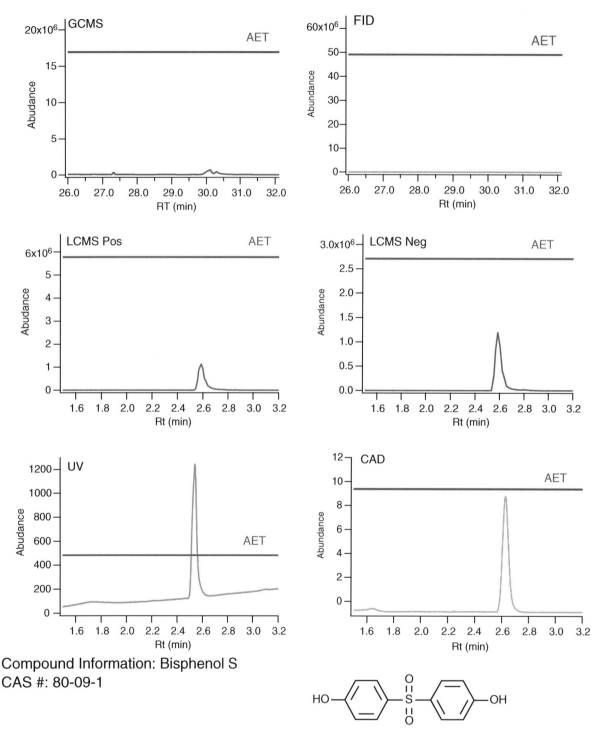

Compound Information: Bisphenol S
CAS #: 80-09-1

Figure 3.21 Multi-detector AET determination for Bisphenol S. In this example, the concentration of Bisphenol S in the sample is equal to the AET. However, due to the nature of the Bisphenol S response versus the internal standard, the peak for Bisphenol S is only above the AET for the one method (LC/UV) whereas the response factor for Bisphenol S is greater than the response factor of the surrogate standard (SS). Although this result for this method is an outlier with respect to the other methods, it is nevertheless sufficient to yield the right conclusion that Bisphenol S is above the AET and thus must be reported for assessment. Note that this conclusion is reached in this example without the use of the AET adjustment.

Table 3.13 Normalized relative response factor distribution statistics.

Item	Result for a given method						
	LC/MS +	LC/MS −	LC/UV[a]	LC/CAD	GC/MS	GC/FID	Optimal
Mean[b]	1.0	1.0	1.0	1.0	1.0	1.0	1.0
% RSD	108%	119%	60%	65%	52%	54%	≈20%
Minimum	0.0013	0.0065	0.0364	0.0336	0.0318	0.0098	Not reported
Maximum	6.1140	7.9816	2.8493	2.6467	2.8401	3.1218	Not reported
% Detected	72%	34%	52%	56%	59%	59%	100%
% of compounds with RF between 0.6 and 1.4	33%	22%	44%	41%	61%	55%	91%

a)   At 230 nm.
b)   The authors "normalized" the response factors by dividing them by the population mean. Thus, the reported mean is 1.0.

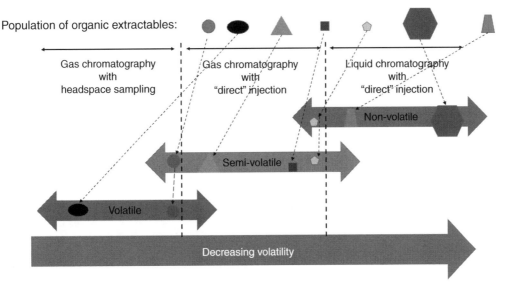

Figure 3.22 Ideal situation for screening extracts for organic extractables. In this ideal situation, all the individual extractables in the extractables profile are accounted for by one or more of the screening methods. Additionally, the screening methods do not "obscure" the extractable's identity (indicated by the shape) and/or do not distort the extractable's concentration (indicated by the size of the shape). (*See insert for colour representation of the figure.*)

extract (or organic leachable in a drug product) above a defined and justified evaluation or reporting threshold (for example, the AET). An error of omission occurs when the analytical screening process fails to account for all such extractables and leachables. In chromatographic methods used to screen for organic extractables/leachables, an error of omission is the absence of a recognizable chromatographic peak attributable to an analyte of interest.

Commission of an error of omission is a fatal error as the assessment of the extractables or leachables profile is irreversibly compromised by committing the error. An extractable or leachable that is not accounted for by the analytical process is an extractable or leachable that cannot and will not be assessed. Thus, an error of omission causes the assessment of the effect of the extractable/leachable to be incomplete.

There are two major "flavors" of omission errors, each of which are considered in greater detail as follows:

1. Squeezing through the cracks
2. Failing to see the tree in the forest

Substances Squeezing Through the Cracks   The error of omission that is termed "squeezing through the cracks" is undoubtedly the omission error that comes most readily to people's minds, which is the circumstance where the analytical method(s) do not produce a response to an analyte that has thus "squeezed through the cracks" of the method(s). Although most modern analytical screening strategies for extractables/leachables employ multiple overlapping but orthogonal analytical methods to close as many cracks (gaps) as possible, it is well-recognized that there are significant and meaningful gaps in both

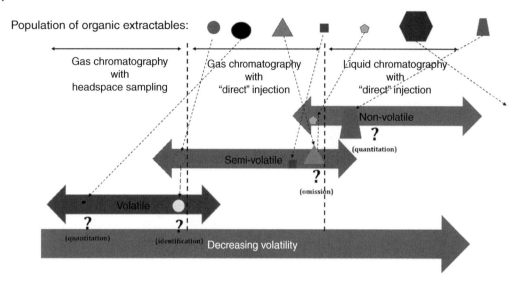

Figure 3.23 General actual situation for screening extracts for organic extractables. In this actual situation, some of the individual extractables in the extractables profile are not accounted for by any of the screening methods (for example the red rhombus) Additionally, the screening methods "obscure" the extractable's identity (indicated by the color) and/or distort the extractable's concentration (indicated by the size of the shape). For example, the concentration of the extractable represented by the black oval is underestimated by the volatiles method while the concentration of the extractable represented by the green rhombus is overestimated by the non-volatiles method. Additionally, one notes that although the extractable represented by the brown circle is "picked up" by both the volatiles and non-volatiles, it is identified as different compounds by both methods and neither of the identities is the correct one. Lastly, the analyte represented by the red hexagon is completely missed by all three of the screening methods. (*See insert for colour representation of the figure.*)

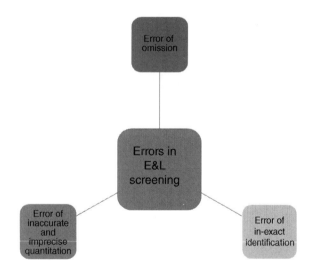

Figure 3.24 Errors in extractables and leachables screening.

the individual analytical methods and in the combined, multi-method analytical strategy.

There are three aspects to consider in addressing this type of error of omission: (i) how to determine that such an error has occurred, (ii) how to correct the error once it has surfaced, and (iii) how to minimize the possibility that such an error occurs. Considering aspect (i), addressing the means of establishing whether extractables or leachables have been missed by the analytical screening process, concepts such as Total Organic Carbon (TOC) reconciliation have been proposed as one means of establishing that one or more substances has "fallen through the cracks," as noted previously in Section 3.2. Moreover, a highly evolved internal database of extractables and leachables data could provide alerts to potential omissions if the database is associative. For example, a number of degradation products of Irganox-related antioxidants have been reported as extractables [23, 24]. In an associative database, noting that a test article contains an Irganox-type antioxidant could produce a list of probable additional Irganox-related extractables. If extractables screening of an Irganox-containing test article revealed only some of these known related extractables, it would be reasonable for the analyst to ask "I wonder where the other related substances are?" and to answer the question by more closely examining the chromatographic data, possibly "finding" a nearly omitted substance.

However, once it testing has been established that an error of omission has occurred, the only means of correcting the error is to expand or augment the analytical screening process by either optimizing the existing screening methods to close the identified gaps or by including additional screening methods that are capable of responding to compounds that fall outside the capabilities of the initial screening approach.

Failing to See the Trees in the Forest In an ideal world, individual extractables or leachables produce individual and unique responses in screening methods that are readily distinguishable from all other analytical responses. In the real world, it is often the case that analyses for extractables/leachables produce a multitude of responses, which, for one reason or another, might hide or obscure other responses.

Consider extractables screening as an example. It is generally the case that one produces an extract that is well-suited for analysis, meaning that it is more or less readily amenable to analysis and it produces minimal background response. Thus, extractables can generally be readily differentiated from the background response associated with the extraction solvent.

Nevertheless, certain extractable profiles themselves can be sufficiently complex that it is difficult to differentiate individual extracted substances from one another. Although a response is produced for all analytes, responses for individual analytes might be either unresolved, unrecognizable, or otherwise not useful.

In certain cases, leachables screening can be even more complicated, as the drug products that are screened for leachables may be much more analytically "busy" than extraction blanks, meaning that the drug product matrix itself may produce a multitude of analytical responses that interfere with, mask, or obscure the analytical responses associated with leachables.

For example, consider the paired chromatograms shown in Figure 3.25, reflecting a drug product blank (drug product stored in an inert vessel) and a drug product sample (drug product stored in its container closure system). As analytical responses produced by the drug product itself obscure entire regions in the chromatogram, it is possible that responses to leachables that elute in the obscured regions could go unrecognized, resulting in an error of omission.

Thus, an error of omission occurs when one cannot differentiate one tree (the response produced by a single extractable or leachable) from the forest of trees (all responses produced by either the test sample itself or other extractables/leachables).

3.2.10.2 Error of Inexact Identification [25]
Although a screening assay produces a response that contains information that can be used to infer an identity, the response itself is not an identity. It is only with further processing and/or interpretation that the response's information can lead to an identification. Thus, screening methods do not identify substances; rather, the screening assay produces data that is further interpreted to provide an identity.

Rigorously speaking, an error of inexact identification occurs when (i) the response contains no identifying information, (ii) an identity cannot be inferred from the response data, or (iii) the inferred identity is not the correct identity. Understanding and addressing errors of inexact identification is facilitated if one understands the data interpretation process that is most commonly used for compound identification. The process is based on the observation that an identification derived from information represents a guess at the identity, where the confidence one has in the guess depends on the amount and the nature of the information that suggests, supports, and ultimately confirms the identification. The more information that is available, and the more rigorous the available information, the greater confidence one can have that the inferred identity is the correct identity. Thus, one can "grade" an identification based on the level of confidence one has that the identification is correct. For example, such grading of identifications was discussed previously in Section 3.2.5.

Given the "grading" scale, it is clear that an unidentified analyte reflects one type of error of inexact identification (that is, the inability to secure an identity). Analytes that have been tentatively identified can be subject to the other error of inexact identification, which is mis-identification. Given the additional confirmatory information that is required to secure a confident or confirmed identification, these identifications are not generally prone to inexact identification errors.

Commission of an error of inexact identification is a fatal error because such an error precludes a proper assessment. If the error of inexact identification is that an identity cannot be secured, then clearly the substance's impact on the drug product's suitability cannot be assessed as the link between the extractable and its relevant related information cannot be established. If the error of exact identification is that the wrong identity is secured (see Figure 3.26 for an example of this type of identification error), then clearly the substance's impact on the drug product's suitability cannot be correctly assessed. This is the case as the assessment is based on information relevant to what could be a completely unrelated compound.

3.2.10.3 Error of Inaccurate and Imprecise Quantitation [25]
Once an extractable has been discovered in an extract (or a leachable in a drug product) above a reporting threshold, it becomes a candidate for impact assessment; for example, establishing its potential adverse effect on patient safety. The impact assessment considers two factors, the leachable's intrinsic ability to produce an effect (for example,

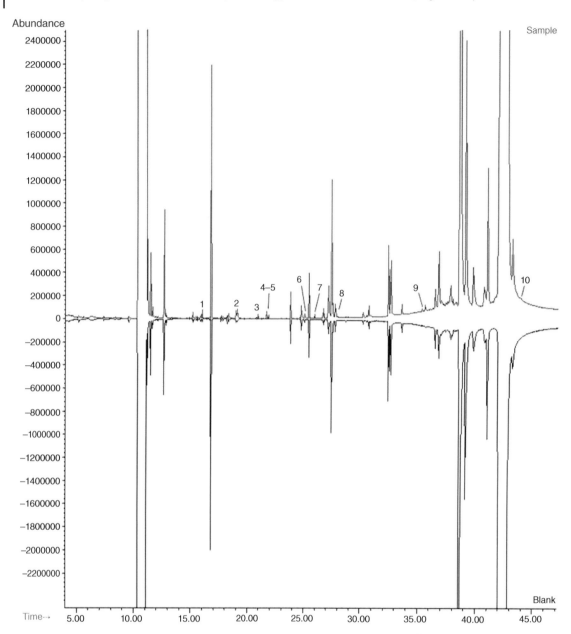

Figure 3.25 An example of the "Failing to See the Tree in the Forest" omission error. The bottom chromatogram is the response to the injection of a drug product (blank solution) while the top chromatogram results from a drug product spiked to contain 10 potential leachables. One notes that the blank chromatogram contains chromatographic regions where the matrix interference would be so great that leachable responses would be obscured. In fact, it is difficult to distinguish the responses for several of the 10 intentionally added substances, even knowing that the drug product had been spiked with them.

the leachable's safety hazard) and the amount of the leachable available to cause the effect (for example, a patient's exposure to the leachable). Specifically, it is the substance's identity that links it to its relevant effect-indicating information (for example, toxicity data), thus facilitating its proper impact evaluation. However, the impact assessment cannot be completed without quantifying the patient's or drug product's exposure to the substance, which is established only by accurately determining the concentration of the substance in either the extract or the

drug product. This is the case for safety assessment as a patient's exposure to a leachable (in terms, for example, of mass per day) is determined as the product of a leachable's concentration in a drug product and the daily dose volume of the drug product.

As a consequence, the most relevant and meaningful impact assessment is based on a sufficiently accurate and precise quantitation of the extractable in the extract (or the leachable in the drug product). Screening methods, however, do not always achieve these desired quantitation

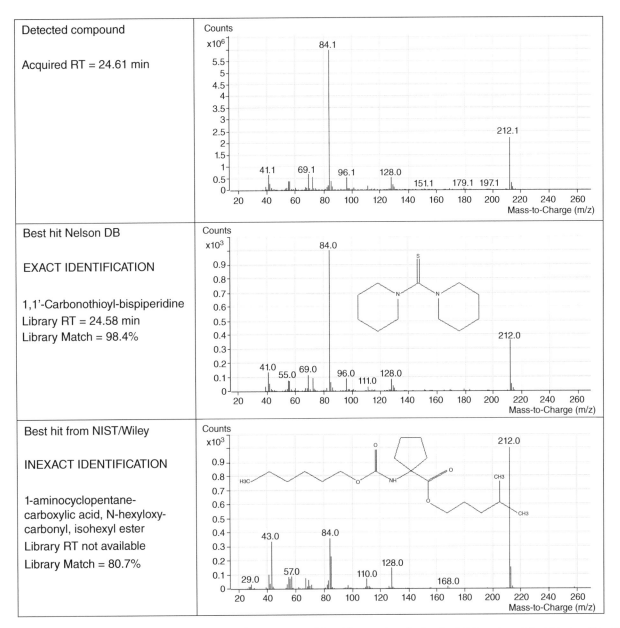

Detected compound Acquired RT = 24.61 min	
Best hit Nelson DB EXACT IDENTIFICATION 1,1'-Carbonothioyl-bispiperidine Library RT = 24.58 min Library Match = 98.4%	
Best hit from NIST/Wiley INEXACT IDENTIFICATION 1-aminocyclopentane- carboxylic acid, N-hexyloxy- carbonyl, isohexyl ester Library RT not available Library Match = 80.7%	

Figure 3.26 Example of an exact identification via mass spectral fit with an authentic standard (1,1'-carbonothioyl-bispiperidine) in an internal database (with retention time confirmation) versus an incorrect identification, relying only on the mass spectral fit with the external NIST/WILEY mass.

goals as the most commonly employed approaches to quantitation are prone to errors of inaccurate and imprecise quantitation. For example, Section 3.2.7 discussed in detail the types of errors in quantitation which can occur in screening when specific quantitation strategies are applied.

In addition to errors in the means of performing quantitation, there are also errors of extrapolation. Generally, the most efficient and generally applied quantitation strategy is to use response factor values obtained at a single concentration of both extractables and the SS. Concentration estimates obtained using such response factor values

may have associated extrapolation errors when they are applied to situations where there is a considerable difference in concentration between the analyte of interest and the SS in a sample (extract or drug product). Such extrapolation errors may be caused either by exceeding the dynamic range of the response curve or by different response curves between a compound and its SS.

Going Beyond the Dynamic Range This first type of extrapolation error has its roots in the assumption that RRF values remain constant as a function of the absolute and relative concentrations of the SS and the compound of interest. Over a certain range of concentration, the response functions for the SS and the analyte of interest will be well-defined and the RRF will be more or less constant. However, at some concentration either one or both of these compounds will exceed its dynamic range. Beyond this point, the assumed correlation between the responses of the SS and the target compound may become invalid. The accuracy of the estimated RRF-corrected concentration consequently deteriorates outside the dynamic range of either compound.

This issue could occur when RRFs are applied to numerous extractables of varying concentrations. Optimally, the SS concentration is selected to be within what is hopefully a narrow concentration range exhibited by all extractables in the extractables profile. The facts that (i) this range is not necessarily known during the design or acquisition of an internal database and (ii) the concentration range exhibited by real extractables in a real extract is likely to be large, can lead to a disconnect between the concentrations of the SS and the compounds that are present and need to be quantified in a sample. Such a disconnect could therefore cause extrapolation errors, especially for compounds present in a sample at a concentration that is substantially different from the concentration at which the RRF was determined.

Generally, modern analytical detectors applied in chromatographic screening exhibit a dynamic range that spans several orders of magnitude. The selection of the SS and its applied concentration should take this range into account. It is generally considered good practice that the SS results in a response that is linear within at least one order of magnitude centered on its applied concentration. This range is then considered to result in concentration estimates with an acceptable accuracy. For example, for an SS at 1 mg/l in a sample and with a dynamic range of 25, estimated analyte concentrations within the 0.2–5.0 mg/l range are consequently considered acceptably accurate whereas values outside this range can be compromised due to dynamic range issues. To extend the quantitation range, two SSs can be added at different concentrations, e.g. one compound at 0.1 mg/l and the other at 1 mg/l. This practice will increase the original quantitation range from a factor of 25 (0.2–5.0 mg/l) to a factor of 250 (0.02–5.0 mg/l). It should be noted, however, that response differences between the two SSs must be properly accounted for.

Response Functions The second type of extrapolation error is rooted in the assumption that all compounds detected in screening mode exhibit the same concentration–response relationship as the SS. Response factors that are determined at a specific SS concentration do not account for differences in response functions. Such dissimilar relationships are usually caused by varying physicochemical properties between a compound and its corresponding SS. This is especially relevant for techniques with inherently complex response functions (e.g. LC/MS). In such situations, the error of extrapolation can be considerable since it can be anticipated that the concentration-response functions are non-uniform for all detected compounds.

Depending on a compound's concentration–response function, different situations can occur, as illustrated in Figure 3.27. In **Case 1**, the concentration–response curves of the compound and the SS are identical. Consequently, the RRF is 1 within the entire applied range and no extrapolation errors are encountered.

In **Case 2**, the response functions for the analyte and the SS are both linear within the applied range but with different slopes. The average RRF can be calculated by dividing the slope of the response curve of the extractable by that of the SS. The RRF accounts for the different slopes between the curves and prevents extrapolation errors within the applied range. This RRF value is applicable to the entire applied range and no extrapolation errors are encountered when the RRF is used.

From the third case on, differences in response functions for the analyte and the SS result in extrapolation errors within the applied range. This is the case when response functions have similar slopes but variable intercepts (**Case 3**), when both intercepts and slopes between curves differ (**Case 4**) and certainly when the response functions are dissimilar and nonlinear (**Case 5**).

It should be noted that the error of extrapolation is small within the limited concentration region where the analyte and the SS are present in a sample at concentrations close to the respective concentrations at which their RRF was estimated. Beyond this point, however, the magnitude of the extrapolation error will increase as the concentration difference between the analyte and the internal standard increases and as the analyte's and SS's response functions diverge.

3.2.10.4 Reducing, Mitigating, and Eliminating Errors in E&L Screening

At one point in time, the state of the art in E&L screening was such that the screening errors of discovery, identification, and quantitation were taken as "part of the game" and generally recognized as unavoidable and uncorrectable. However, screening methods and data processing and collection practices have evolved to the point where it is practical to "change the rules of the game" so as to manage these issues. Thus, Section 3.6 discusses means, specifically associated with an extractables/leachables database, for reducing, mitigating, and eliminating errors in E&L screening.

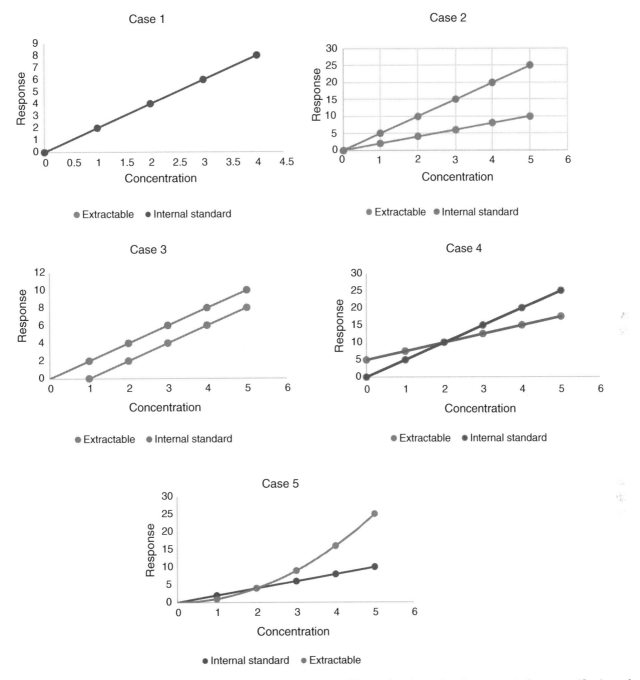

Figure 3.27 Concentration–response function examples to illustrate different situations related to extrapolation errors. (*See insert for colour representation of the figure.*)

3.2.11 Method Qualification

An analytical method is qualified to establish that it is suited for its intended purpose. In extractables or leachables studies, analytical methods serve one of two purposes; screening samples for unspecified analytes and testing samples for specified (targeted) analytes. As these purposes are quite different, and the analytical methods used for each purpose might be different, it is reasonable to suspect that their qualification would differ. In fact,

qualification of targeted methods is typically referred to as validation, which is addressed in greater detail in Section 3.3.4.

The qualification of an analytical method is documented in a Qualification Protocol, which establishes:

- Relevant qualification parameters;
- Experimental means by which the qualification parameters will be assessed;
- Performance expectations for each parameter.

Qualification parameters that are specifically relevant to screening methods include:

- Sensitivity, as it is expected that the method's LOQ be less than or equal to the reporting threshold (such as the AET); refer back to Figure 3.15. However, in cases of very low reporting thresholds, it may not be possible to achieve an LOQ that is less than or equal to the reporting threshold. In such cases, the lowest reasonably attainable LOQ should be used, and all analytes above this LOQ should be reported with concentrations. Analytes between the LOQ and the AET should also be reported, but obviously no concentrations can be provided as they are below the LOQ. When the LOQ is higher than the AET, this circumstance should be explained and justified, as extractables whose concentration is between the AET and the LOQ are potentially unsafe but cannot be fully addressed (as they cannot be quantified). Should an even more extreme case occur where the AET is above both the LOQ and the LOD, then potentially unsafe substances cannot even be detected, let alone be identified and quantified.
- Specificity, which is the ability to assess unequivocally the analyte in the presence of other constituents that may be expected in the sample.
- Accuracy, taken as the ability to produce a response that is comparable to the true value (for example, a measured concentration in a spiked extract that is comparable to the spiked amount). Accuracy for screening tests is typically accomplished using surrogate substances that are either extractables themselves or representative of extractables. For screening methods an accuracy of 50–200% (± a factor of 2) is generally regarded as acceptable.
- Precision, taken as the variation in replicate analyses of either the same extract or a standard solution containing extractables or leachables and considering both repeatability (intra-lab or intra-day) and reproducibility (inter-lab or inter-day). Intra-day and inter-day precisions of 15% RSD and 25% RSD are generally regarded as acceptable.
- Dynamic range, taken as the concentration range over which the response and the analyte concentration producing that response can be related by a simple mathematical function. Dynamic range can be established by analysis of surrogate or standard solutions at various concentrations. A dynamic range of an order of magnitude, centered around a concentration threshold, is generally the minimum acceptable outcome, although many methods can provide larger dynamic ranges.

Qualifying that a method is rugged is relevant for both screening and targeting methods.

Additional performance parameters can be included in a qualification at the discretion of the method's user and with appropriate justification. These additional parameters could include robustness, efficiency (for a chromatographic separation this might include resolution), matrix effects, and sample and standard stability.

Given their different purpose and function, the processes of qualifying a screening or targeting method will be different, even if the qualifying criteria are generally the same. For example, while both screening and targeted methods are qualified for accuracy, the nature of the qualification activity is different. While accuracy is established in a targeting method specifically for the analyte(s) of interest, in a screening method accuracy is established more generally by considering a group of surrogate analytes. Additionally, as screening methods provide concentration estimates, the acceptance criterion for accuracy is less rigorous than the acceptance criterion for accuracy in targeted analysis, where the calculated concentration is expected to be highly quantitative.

The same concept is applicable to precision, as it is generally accepted that the precision expectations for a targeting method are more rigorous than are the precision expectations for a screening method.

Specificity is important in a screening method as the identification of individual extractables is facilitated if the chromatographic peak associated with an extractable is produced by only that extractable. In a targeting method, specificity for the targeted compound, meaning that the target compound's chromatographic peak is pure, is necessary to provide the required degree of accuracy and precision. Because a targeted substance is established in advance of implementing the method, specificity can be established up-front. However, since it is not possible to establish up-front what analytes might be discovered in screening, specificity in screening methods is typically established at the time of use. Thus, specificity could be measured and judged quite differently in screening versus targeting.

A method is considered to be qualified (that is, suited for its intended use) when:

1. It has been established that the method is able to routinely meet the performance expectations contained in the Qualification Protocol and
2. Appropriate system suitability testing has been established.

In addition to having documented performance capabilities, qualified analytical methods should have additional controls that may include, but are not limited to:

- Documentation of the method in the form of a standard operating procedure (SOP) that is controlled in a document change system.

- An approved and specified Scope, captured in the method's SOP.
- A detailed scientific description and justification of the method, establishing its suitability for the intended use.
- A requirement that the qualified method is implemented by an appropriately qualified and trained staff.
- A requirement that the qualified method is implemented on calibrated/qualified instrumentation.

Considering system suitability specifically, establishing system suitability is a time of use assessment and qualification that addresses three performance aspects of the method:

1. That the method has been set up and implemented properly,
2. That the method as set up is capable of performing at the same level it performed at during its initial qualification, and
3. That the method has performed acceptably throughout its use.

System suitability assessment [26] should focus on that minimum number of performance characteristics that individually and in aggregate demonstrate that these three performance criteria were achieved. The system suitability parameters to be assessed and their associated acceptance criteria should be rigorous enough to ensure that the method produces data of acceptable quality but not so rigorous that potentially acceptable analytical runs are rejected on a frequent basis. Properly collected and statistically evaluated system suitability data can provide diagnostic evidence of imminent method failure.

3.3 Organic Substance Analysis: Targeting

3.3.1 Intent and Purpose

As we have seen in screening, the contents of the sample being screened are largely unknown and it is the analyst's primary job to discover all substances present in the sample at a concentration above a defined threshold. Once the compounds have been discovered, then it becomes the analyst's task to identify and quantify the discovered compounds.

Although screening of extracts for extractables and drug products for leachables is a common analytical task in extractables and leachables assessment, there are circumstances where the analytical challenge can be more focused, addressing a smaller population of compounds whose members can be specified upfront. That is, specific extractables or leachables can be established as targets for the analysis. As the targets are known, the primary purpose of the target analysis is to quantify the targets in either the extracts or the drug product. As quantitation is the primary purpose and the number of targeted substances is generally small, targeted methods are optimized in terms of attributes relevant to quantitation, such as accuracy, precision, and sensitivity.

The objective of analyzing a drug product over shelf-life for targeted leachables is to establish trends in the leachables' accumulation over shelf-life, thus allowing the impact assessor to determine the leachables' maximum concentrations over shelf-life. This is important as the maximum concentration generally establishes the maximum effect (for example, the maximum patient exposure to the leachables). Knowing the maximum effect enables the risk assessment; for example, in the toxicological safety risk assessment of the leachables, where the maximum patient exposure is compared to a toxicologically meaningful quantity such as the PDE.

It is deceptively easy to forget that there is a secondary purpose in target analysis. While securing a compound's identity was a primary purpose of screening, confirming identity (that is, confirming that the compound believed to be the target in the sample is in fact the targeted compound) is a secondary purpose of targeting, performed to ensure that the compound that has been identified as the target compound in the test article is in fact the target compound and not an imposter. Thus, in addition to the quantitation attributes mentioned previously, a target method must be selective so that false positive results are not obtained and reported. In fact, it is the aspect of specificity that is sometimes forgotten in the choice of target methods. It is not unheard of, for example, for a screening method for leachables in a drug product to be validated with freshly-prepared samples of the drug product, only to find out that drug product degradation products that form slowly over the shelf-life interfere with the target analysis at later testing timepoints.

3.3.2 Selection and Justification of Targets

The single most important factor for selecting a substance to target is that the selected compound has a potential impact. For example, consider the case of targeting extractables as leachables in a drug product for the purpose of generating leachables data over the shelf-life for toxicological safety assessment. Logically, the extractables that are selected as targets should be those extractables that have the potential to adversely affect patient safety. If it has already been established that the extractable, as a leachable, would have a negligible adverse effect on patient safety (allowing for the uncertainty in the data upon which the assessment is based), then there is little benefit gained by targeting this extractable, except to use the leachables data to confirm the extractables data (for example, to confirm that the highest leachable levels projected from the extractables study are in fact the highest leachables levels reached over the shelf-life).

Although the potential to have an impact may be the primary factor in selecting targets, it is not the sole factor as targeting may serve purposes other than an impact assessment (for example, as noted previously, for confirming the results of simulation studies). Moreover, although it could be possible to review extractables data and conclude that there are no extractables that present credible patient safety risks (and thus that no extractables need be targeted as leachables), it is typical regulatory practice to expect targeted leachables data as part of a product registration file. Thus, other factors to consider in selecting targets include:

1. Projected accumulation level. Generally speaking, one targets those extractables that accumulate to the highest level.
2. Analytical expediency. The nature of the leachable and the product solution may be such that an analytical method appropriate for the leachable's quantitation is not practically viable. Less frequently, an analytical method with the required level of performance cannot be developed. Generally speaking, then, a target leachable should be one for which an appropriate analytical method can be developed, validated, and routinely implemented.
3. Ability to represent a particular compound class or material source. It may be the case that a product's leachables profile is such that (a) the profile includes functionally similar compounds or (b) materials of construction of the product produce more than one leachable. In such cases, it is possible that one can assess the impact of a related family of compounds by studying the accumulation of a smaller subset of family members.
4. Regulatory or historical precedence. There are certain substances that are historically or regulatorily linked to certain types of products. An example is di-(2-ethylhexyl) phthalate associated with plasticized polyvinyl chloride (PVC). The examples of N-nitrosamines, polynuclear aromatic hydrocarbons, and 2-mercaptobenzothiazole for OINDPs have been previously discussed. Such compounds would become targets based on these precedents, as it is likely that their exclusion from the migration study would result in an inquiry to a product registration package.
5. Present in most materials (for comparative purposes). It is possible, albeit not common, that a migration study at the product development stage would include an assessment of more than one material or product. An example of this would be the evaluation or two or more container materials for a certain solution product. In such cases, a leachable that is associated with all the materials under investigation might be chosen as a target as the relative accumulation of the target might be a criterion for choosing one material versus the others.

3.3.3 Establishing Reporting Practices

A key facet of the analytical approach taken to analyze samples for targets is the calibrated concentration range of the analytical method(s) employed. In general, two options are possible:

1. The calibration range can encompass concentrations that are near to the impact threshold (for example, the PDE). In so doing, the leachables' approach to the impact threshold can be monitored and an "early warning" can be given if a leachable's level begins to approach the threshold. This approach is referred to as the impact-relevant calibration range.
2. The calibration range can encompass concentrations near the projected or expected accumulation level of the leachables. In so doing, trending of the leachables accumulation over time is facilitated. This approach is referred to as use of the leachables accumulation-relevant calibration range.

From an analytical perspective, the best-case situation occurs when the calibration ranges for impact assessment and trending overlap, as a single calibration curve facilitates achieving both objectives. From an impact perspective, however, the best-case situation occurs when the levels of the targets are well below any impact threshold, specifically option 2.

This discussion of calibration range is relevant when the impact-relevant range and the accumulation-relevant range are quite different as analytical methods do not have an infinitely large linear response range. When these two ranges are similar and fall within the linear response range, then the desired actions of trending and impact assessment can be accomplished together. When the ranges are sufficiently different that they are broader than the linear response range, then the analyst is faced with a choice in terms of the primary objective of the testing (that is, the relative importance of trending versus impact assessment).

To illustrate this point, consider the following example. Based on a simulation study, it is anticipated that an extractable's concentration as a leachable will not exceed

5 mg/l. Based on a toxicological safety assessment of this extractable/leachable, the concentration associated with an unacceptable safety risk is 100 mg/l. Two analytical methods are available for quantitating this leachable; method A has a linear calibration range of 1–10 mg/l while method B has a linear calibration range of 25–250 mg/l. Although use of method B will produce data that establishes whether the leachable's concentration exceeds the concentration associated with the unacceptable safety risk (and provides some "early warning"), it is not likely that use of this method will provide trending data, as the most likely outcome of the test is a result that is reported as "less than the low calibration standard." Although use or Method A will likely provide trending data, if the anticipated behavior is not exhibited and the leachable's level exceeds the extractable's level, then it is possible that the leachable's level would increase beyond the method's linear range, rendering the method useless for establishing impact.

3.3.4 Method Validation

All analytical methods, screening and targeted, that generate product-qualifying information must be established to be suitable for this purpose. Previously (Section 3.2.11), the process to establish the suitability of screening methods was termed method qualification, as opposed to the more rigorous term method validation, based on the premise that "it is exceedingly difficult to validate a method for an unspecified number of a large population of potential analytes that are present in the sample at unknown and potentially variable concentrations." However, in targeted analysis, the analytes of interest have been chosen and the linear calibration of range of interest has been established. Thus, the more rigorous process of validation is fully applicable to target analysis.

Validation parameters that are specifically relevant to targeting methods include:

- Sensitivity, relevant in the circumstance that the method's range includes, or is near to, the LOQ.
- Specificity, as described for screening methods.
- Accuracy, as described for screening methods. However, as opposed to screening tests, accuracy in targeting is evaluated with the actual substances being targeted. Additionally, validation may require that accuracy be established at multiple concentrations and the acceptance criterion for validation (70–130% expressed as % recovery) is more rigorous than the criterion for screening (50–200%).
- Precision, as described for screening methods. However, as opposed to screening tests, precision in targeting is accomplished with the actual substances being targeted.

Additionally, validation may require that precision be established at multiple concentrations and the acceptance criteria for validation (10% RSD for intraday precision and 20% RSD for interday precision) are more rigorous than the criteria for screening.

- Dynamic range, taken as the concentration range over which the response and the analyte concentration producing that response are relatable by a simple mathematical function. Dynamic range can be established by analysis of standard solutions at various concentrations. An acceptable dynamic range for a targeted method depends on the purpose of the method.
- Goodness of fit, taken as the degree to which a simple mathematical function can express the relationship between an analyte's concentration in a standard and the method response obtained when the standard is analyzed. Although the desired mathematical function is generally a linear function, simple, nonlinear functions can be used if they are able to meet the acceptance criteria for goodness of fit.

Qualifying that a method is rugged is relevant for both screening and targeting methods.

Additional performance parameters can be included in a validation at the discretion of the method's user and with appropriate justification. These additional parameters could include: robustness, efficiency (for a chromatographic separation this might include resolution), matrix effects, and sample and standard stability.

Given their different purpose and function, the processes of qualifying a screening or validating a targeting method will be different, even if the qualifying criteria are generally the same. For example, while both screening and targeted methods are qualified for accuracy, the nature of the qualification activity is different. While accuracy is established in a targeting method specifically for the analyte(s) of interest, in a screening method accuracy is established more generally by considering a group of surrogate analytes. Additionally, as screening methods provide concentration estimates, the acceptance criterion for accuracy is less rigorous than the acceptance criterion for accuracy in targeted analysis, where the calculated concentration is expected to be highly quantitative.

The same concept is applicable to precision, as it is generally accepted that the precision expectations for a targeting method are more rigorous than are the precision expectations for a screening method.

Specificity is important in a screening method as the identification of individual extractables is facilitated if the chromatographic peak associated with an extractable is produced by only that extractable. In a targeting method, specificity for the targeted compound, meaning that the target compound's chromatographic peak is pure, is necessary to provide the required degree of accuracy and precision. Because a targeted substance is established in advance of implementing the method, specificity can be established up-front. However, since it is not possible to establish up-front what analytes might be discovered in screening, specificity in screening methods is typically established at the time of use. Thus, specificity could be measured and judged quite differently in screening versus targeting.

A method is considered to be validated (that is, suited for its intended use) when:

1. It has been established that the method is able to routinely meet the performance expectations contained in the Validation Protocol and
2. Appropriate system suitability testing has been established.

In addition to having documented performance capabilities, validated analytical methods must have additional controls that may include, but are not limited to:

- Documentation of the method in the form of an SOP, which is controlled in a document change system;
- An approved and specified Scope, captured in the method's SOP;
- A detailed scientific description and justification of the method, establishing its suitability for the intended use;
- A requirement that the qualified method is implemented by an appropriately qualified and trained staff; and
- A requirement that the qualified method is implemented on calibrated/qualified instrumentation.

Considering system suitability specifically, establishing system suitability is a time of use assessment that addresses three performance aspects of the method:

1. That the method has been set up and implemented properly,
2. That the method as set up is capable of performing at the same level it performed at during its qualification, and
3. That the method has performed acceptably throughout its use.

System suitability assessment should focus on that minimum number of performance characteristics that individually and in aggregate demonstrate that these three performance criteria were achieved. The system suitability parameters to be assessed and their associated acceptance criteria should be rigorous enough to ensure that the method produces data of acceptable quality but not so rigorous that potentially acceptable analytical runs are rejected on a frequent basis. Properly collected and statistically evaluated system suitability data can provide diagnostic evidence of imminent method failure. It is also important to note that while validation focused primarily on the quantitative aspects of the method, system suitability should consider performance characteristics of the method. For example, for chromatographic methods, system suitability testing should include measures of chromatographic performance such as resolution, peak shape, chromatographic efficiency, etc.

3.4 Organic Substance Analysis: Targeted Screening

Previously, the term screening was defined as the analytical process of discovering, identifying, and estimating the concentration of all relevant, but unspecified, analytes in a test sample above an established reporting threshold. This is contrasted to targeting, which was defined as the analytical process of quantifying specified analytes in a specified test sample over a specified concentration range.

The previous discussion of these two processes implied that they were mutually exclusive but that is not necessarily the case as it is possible to reconcile the concepts of screening and targeting. That is, it is possible to screen a test article for specified analytes via a process called targeted screening. In targeted screening, one preserves the essential aspect of screening, which is establishing what analytes are present in the sample above an established threshold, but refines the universe of potential analytes to a smaller population that has been defined (targeted) up-front. In so doing, one combines the virtues of screening (broad scope, wide applicability) and targeting (more well-defined focus, more accurate quantitation approaches).

Extracted elements testing is an example of targeted screening. Although the sample is analytically screened for extracted elements, the universe of extracted elements is relatively small and well-defined and thus while the extract is screened for extracted elements, each of the elements that are included in the screen are specified (targeted).

Targeted screening is differentiated from target analysis in three major ways:

1. The number of analytes in targeted screening is much larger than the number of analytes in targeting. Whereas the number of analytes in a targeted analysis is typically no more than 10 (and is frequently as small as 1), the number of analytes covered by a targeted screening method is limited only by the number of compounds that have been established to produce a response in the targeted screening method.
2. The method used in targeted screening is a screening method. This means that the trade-offs of screening versus targeting (for example, broad scope versus more accurate quantitation) come into play.

3. The method used in targeted screening will have been qualified as a screening method and not by the more rigorous process of validation of a targeted method.

The key issue in terms of applying targeted screening to organic extractables and leachables involves establishing the targets and then optimizing the screening method(s) so that it (or they) properly responds to the targets in the tested sample. That is, in order for one to claim that the targeted screening method is applicable for a particular extractable, the nature of the method's response to that extractable (e.g. retention time, RRF) must be established (typically by analyzing the extractable via the target screening method and recording the method's response).

In should be clear, then, that a method for targeted screening is supported by a database of compounds that have been analyzed by the method and been shown to produce an adequate response. Without the supporting database, target screening is no more quantitative than screening, where concentrations are estimated. It should also be noted that it is "natural" for a screening method to evolve into a targeted screening method over time. For example, when the identity of an extractable is confirmed by analysis of an authentic reference standard, the data required for identification is sufficient to "qualify" the screening method as a targeted screening method for the identified extractable. As more and more identifications are confirmed in the screening method, the screening method becomes a targeted screening method for an expanding population (database) of compounds.

Concluding, when screening is replaced, at least partially with targeted screening facilitated by a database of extractables/leachables, the following three benefits are derived:

1. The extractables identities secured by the targeted screening are confirmed identities.
2. The concentration estimates secured by the targeted screening are more accurate than those secured by screening (use of RRF as opposed to an SS response factor).
3. The data processing process is more efficient.

3.5 Extractables/Leachables Correlations

3.5.1 Linking Extractables and Leachables

It is logical and appropriate to believe that there should be a cause-and-effect relationship between extractables and leachables; that is, leachables should generally be linked to extractables in the same way that products are linked to reactants. In fact, existing or proposed regulatory Guidance strongly encourages that such a linkage be established. For example, the following language appears in FDA Guidance for Nasal Spray and related products [27] and Metered Dose Inhaler and Dry Powder Inhaler drug products [28]:

> "Identity and concentration profiles of the leachables in the drug product or placebo formulation … should be determined through the end of the drug product's shelf life and correlated, if possible, with the extractables profile(s) of the container and closure components."

In fact, the concept that one "profiles extractables as probable leachables" is valid only when extractables and leachables are correlated. Furthermore, one could suggest that if leachables cannot be correlated to extractables then extractables profiling has no useful or meaningful purpose.

The clarity and strength of the link between an extractable and a leachable varies greatly from circumstance to circumstance. Sometimes the link between an extractable and leachable is clear and obvious. An example of this is the case where the leachable and extractable are one and the same and the extractable is a known, intentionally added component of the container/closure. Thus, for example, it would be relatively easy to establish a link between di-(ethylhexyl) phthalate (DEHP) that is measured in a drug product as a leachable and the DEHP that is the major plasticizer in a PVC container/closure. At other times, however, the link may be more difficult to establish. For example, Ulsaker and Teien documented the presence of trace amounts of caprolactam in parenteral solutions stored in PVC bags [29]. These authors document a fairly extensive effort to confirm their identification of this leachable, since it is reasonable to observe that the caprolactam cannot be linked to an intentional additive in the primary packaging (the PVC containers themselves and their associated closure system). Ultimately, this extracted substance was correlated to an adhesive used in laminated secondary packaging (plastic overwrap). Thus, although the correlation was not obvious, a correlation could be established once all sources of the leachables were properly accounted for.

Another perplexing correlation involved protein aggregation, which was reportedly occurring in biopharmaceutical drug products packaged in glass pre-filled syringes. After considerable detective work, this phenomenon was linked to tungsten that had leached into the drug product over shelf-life [30, 31]. But what was the source of this unexpected leachable; what extractables could this leachable be linked to? Tungsten is not an element that is associated with glass as either an intentional or unintentional additive. Extraction studies performed on the source glass confirmed that the glass itself was not the source of the tungsten. Further investigation revealed that the tungsten was present in the formed syringe as a manufacturing residual [32, 33]. During syringe manufacturing, a cavity is made for positioning the syringe's needle. This cavity is formed by inserting a tungsten pin into the syringe as it is being shaped at high temperature. In this process, tungsten is deposited in the cavity area and is oxidized to form soluble tungsten anions, which can leach from the syringe to the contained drug product. Again, after much investigation, a definitive extractable–leachable correlation could be established.

Regardless of the difficulty in establishing an extractable–leachable correlation, such a correlation "validates" the completeness and accuracy of both the extractables and leachables profiling activities. A general rule of thumb is that the link between an extractable and leachable is simpler to establish if the presence of the extractable in the relevant test item and the presence of the leachable in the drug product are readily ascertained.

3.5.2 A Hierarchy for Linkages Between Extractables and Leachables

The rigor with which linkages between substances needs to be established depends on how such a linkage is used. With this premise in mind, a hierarchy is proposed that categorizes and ranks various types of linkages based on the rigor of the supporting information (Table 3.14) [34]. The hierarchy is proposed based on the observation that, while the most rigorous linkage is a mathematically quantified one, such a rigorous linkage may be difficult to obtain and in fact may be beyond the needs of the situation for which the linkage will be used.

The simplest linkage in the hierarchy is termed a connection. A connection can be established between two substances if the chemical similarity between the

Table 3.14 A hierarchy of linkages between extractables and leachables.

Linkage level	Linkage description
Connection	A chemical similarity between the compounds can be readily established and universally recognized.
Association	A viable chemical mechanism defining the relationship between the compounds can be proposed.
Affiliation	The chemical mechanism defining the relationship between the compounds can be completely enumerated and demonstrated.
Correlation	The relationship between the compounds can be mathematically quantified (for example, a mass balance).

substances can readily be illustrated and universally recognized by professionals reasonably competent in the pharmaceutical and related sciences. As an example of such a linkage, consider the connection that can be established between DEHP as an extractable and MEHP as a leachable (see Figure 3.28). It is reasonable to expect that competent pharmaceutical professionals would look at the structures of these compounds, note the obvious similarities, and conclude, without additional information, that they are chemically linked, extractable to leachable.

A somewhat more rigorous linkage between two substances could be termed an association. An association is a chemical linkage that is supported by a proposed mechanism that explains or defines the association. An association is typically established between an extractable and a leachable whose linkage is not obvious from a cursory inspection of their structures. Considering the connection between DEHP and MEHP further, the previously discussed connection could be "upgraded" to an association if one proposed that MEHP arises via the hydrolysis of one of DEHP's ester side chains.

The most rigorous qualitative linkage between extractables and leachables is termed an affiliation and is a linkage for which the chemical mechanism that explains or defines the association can be completely enumerated and demonstrated. Considering the association between DEHP and MEHP further, one notes that the hydrolysis of DEHP to MEHP should also produce 2-ethyl-1-hexanol. Additionally, it is reasonable to expect that both ester side chains would be prone to hydrolysis and thus that 1,2-benzenedicarboxylic acid would also be formed. Thus, the identification of all three of these leachables in drug product solutions packaged in a DEHP-plasticized PVC container/closure system would be the kind of information that could be used to justify an affiliation between the leachables (MEHP, others) and the extractable (DEHP).

A qualitative relationship between two entities, no matter how well understood and enumerated, can always be made more rigorous if the relationship is quantified. Thus, the most rigorous linkage between extractable and leached substances is termed a correlation and reflects the situation where the relationship between the substances can be mathematically quantified. Such a quantification may be based on first principles (e.g. mass balance considerations) or may be empirically obtained. As an example of an empirically derived correlation, consider the following. Ten lots of a DEHP-plasticized PVC container/closure system are generated and extracted with a pharmaceutical product. In this situation the container/closure system is characterized with respect to its DEHP level while the pharmaceutical product is tested for leached MEHP. If a mathematical relationship can be established between the DEHP levels in the

Extractable: 1,2-Benzenedicarboxylic acid, bis(2-ethylhexyl) ester (DEHP), $C_{24}H_{38}O_4$, formula weight = 390.6, CAS RN 117-81-7

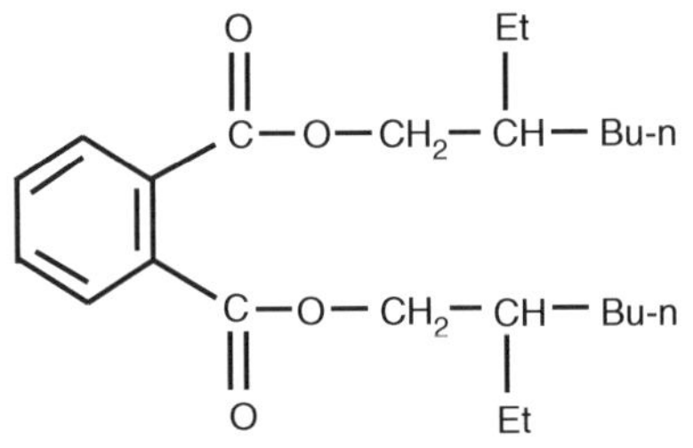

Associated Leachables:

1-Hexanol, 2-ethyl-, $C_8H_{18}O$, formula weight = 130.23, CAS RN 104-76-7

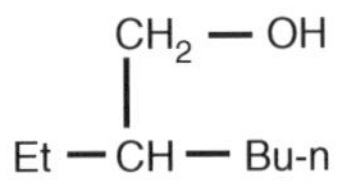

1,2-Benzenedicarboxylic acid, $C_8H_6O_4$, Formula weight = 166.1, CAS RN 88-99-3

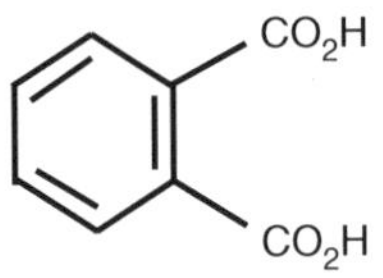

1,2-Benzenedicarboxylic acid, mono(2-ethylhexyl) ester (MEHP), $C_{16}H_{22}O_4$, formula weight = 278.3, CAS RN 4376-20-9

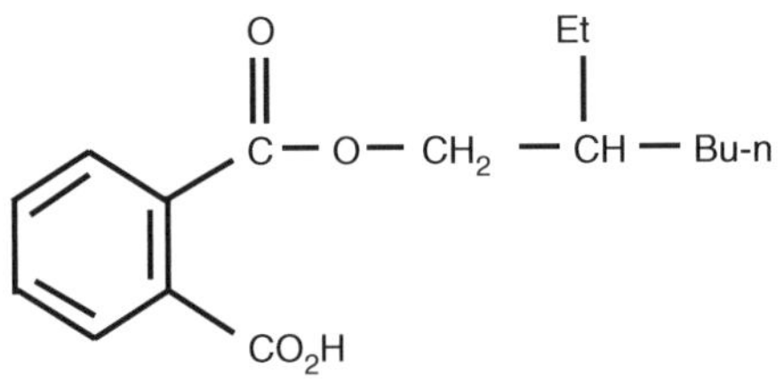

Figure 3.28 DEHP and its associated leachables. DEHP is a plasticizer used in polyvinyl chloride materials.

container/closure and the MEHP levels in the specific pharmaceutical product, then in this situation the extractable DEHP and leachable MEHP have been correlated.

3.5.3 Decisions Concerning the Required Rigor for Linkages

As noted previously, while there are many instances where it is useful to establish a link between extractables and leachables, it is not always necessary, or even appropriate, for all linkages in all situations to be correlations. The degree of rigor associated with a particular application depends on (i) the nature of the application, (ii) the nature

of the extracted substance, and (iii) the product impact associated with the leachable's accumulation.

To more effectively define the nature of the application, let us define a product's lifecycle as consisting of four stages: development, internal justification, registration, and distribution. It is clear that a leachable presents the greatest risk to users of a product when the product is being distributed to end users versus when the product is in its initial stage of development. As a rule of thumb, it is appropriate to use linkages to manage risk; that is, the greater the risk, the more rigorous the linkage. Thus, it may be adequate to possess no more than a general understanding

Table 3.15 Decision tree for the type of linkages required between extractables and leachables.

Location of application in product lifecycle	Minimum level of linkage
Development	Connection
Justification	Association
Registration	Affiliation
Commercialization (including manufacturing)	Correlation

of the container/closure system and its interaction with the extractables and leachables during early product development. During later product development, when a company is formulating its internal product justification, strategies for demonstrating product/container compatibility are being developed. For example, the choice of target leachables to be monitored during stability testing must be made and justified. Such a justification may be supported by a more rigorous linkage, such as an association. During product registration, the demonstration of a complete chemical understanding of the container/closure system may require an even more rigorous linkage such as an affiliation. Finally, product commercialization may require the demonstration of some type of manufacturing control on the accumulation of extracted substances. If such control includes the testing of incoming raw materials, the linkage between the extractable in the raw material and the leachable in the final product must be a quantitative correlation.

This "decision tree" for assigning the correct level of rigor to a linkage is further enumerated in Table 3.15. It is noted that this decision tree represents minimum requirements. Thus, for example, while Table 3.15 identifies an association as the minimum level of rigor for a linkage between reactive leachables and extractables during product justification, it may be the case that in some situations a more rigorous linkage may be necessary.

3.5.4 Circumstances Requiring the Linking of Extractables and Leachables

As noted previously, there can be a considerable benefit derived from establishing a link between an extractable and a leachable. It is also true that a considerable benefit can be derived from establishing linkages among leachables and/or linkages among extractables. For example, consider the following situations. The first situation involves the design of a formal leachables assessment of a container/closure system (that is, monitoring the leachables' levels in the finished product "on stability"). Typically, a leachables assessment does not involve a complete delineation of the fate of all leachables but rather focuses on a subset of leachables that are specifically targeted for an in-depth evaluation. While several factors may be utilized to segregate the smaller group of targeted leachables from the entire leachables profile, one important factor is the ability of a target to represent, mimic, or model the behavior of several leachables. The argument goes that if one studies the fate of the target, the fate of the other compounds can be inferred. The link between a target and its modeled compounds can be established from empirical data (for example, the level of the target is always twice the level of the modeled compound) or on a chemical basis (that is, a structural or compositional link can be established between the target and the modeled compound). In either circumstance, the use of the target can be justified by the fact that a link can be, or has been, established between it and those compounds it models.

The second circumstance addresses the control of the levels of leachables in finished products. It is clear that it is most beneficial, from a productivity perspective, to exert control over leachables as early in a system's manufacturing process as possible. This is true because such a strategy surfaces issues before significant value has been added to the construct (or the final drug product) as a result of its manufacture. If, for example, control over leachables is achieved by final product testing, a test result outside the acceptable limit could result in destruction of the entire product batch. Alternatively, if control over the leachables is achieved by incoming raw material testing, unacceptable test results would only cause the rejection of the incoming material. Although this situation may present its own practical issues (e.g. stock out of the raw material), the impact of the failure is minimized as no value has been added to the raw material via processing. While the advantages of raw material testing for establishing control of leachables are clear, such a strategy is only justifiable if a link between the leachable and the extractable that is monitored in the raw material is established.

The third circumstance can be described as the need to establish scientific credibility. For example, when one registers a pharmaceutical product and its associated packaging, one can reasonably anticipate that the regulatory agency reviewing the registration documentation will expect that the registrant would possess (and demonstrate) a complete and detailed chemical understanding of how the product and container/closure system interact. It is realistic to expect that the ability to link extractables and leachables would be viewed by a registering body as being a desirable and necessary part of that chemical understanding.

3.5.5 Examples of Linkages Between Extractables and Leachables

Examples of related extractables and leachables are illustrated in Figures 3.29–3.33. The linkages illustrated in these figures represent information obtained from the literature (e.g. references 35–39) and gleaned from the author's experiences with leachables and extractables in a wide diversity of applications. Figures 3.29 and 3.30 reflect more or less typical linkages, where a highly hydrophobic extractable (which does not accumulate in an aqueous solution) is linked to one or more related leachables. Thus, in Figure 3.29, for example, the essentially water-insoluble extractable Irgafos 168 is linked to its more soluble reactive leachables.

Extractable: Irgafos 168, Phenol, 2,4-bis(1,1-dimethylethyl)-, phosphite (3:1), $C_{42}H_{63}O_3P$, formula weight 646.9, CAS RN 31570-04-4.

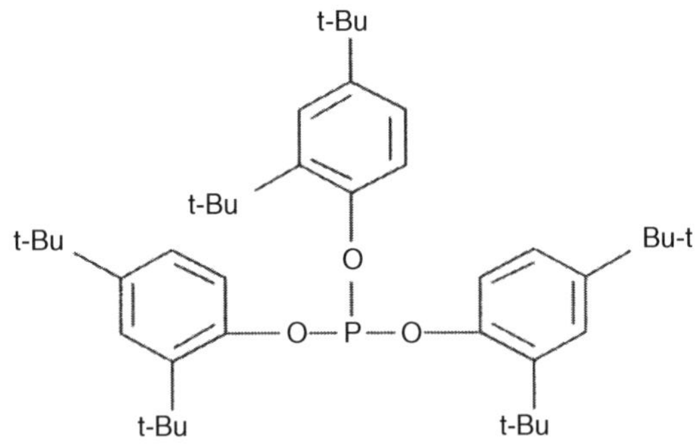

Associated extracted substances (Leachables):

2,4-di-t-butyl phenol, $C_{14}H_{22}O$, formula weight = 206.3, CAS RN 96-76-4

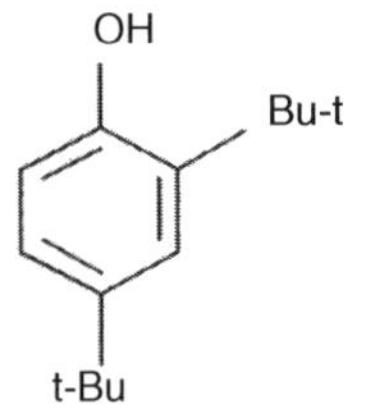

Phenol, 2,4-bis(1,1-dimethylethyl)-, phosphite (2:1), $C_{28}H_{44}O_4P$, formula weight = 475.3

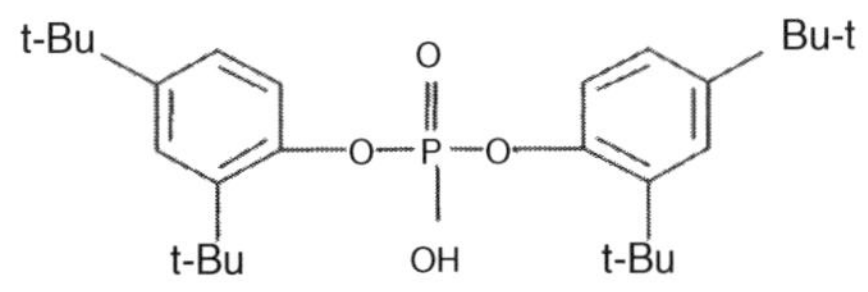

1,3-Bis(1,1-dimethylethyl)benzene, $C_{14}H_{22}$, formula weight, 190.3, CAS RN 1014-60-4

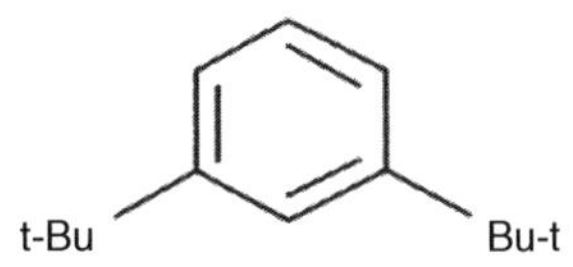

Figure 3.29 Irgafos 168 and its associated extracted substances (leachables). Irgafos 168 is a common antioxidant used in polypropylene and polyethylene resins. The link between Irgafos 168 and 2,4-di-t-butyl phenol is discussed in reference 33.

Extractable: Irganox 1076, Benzenepropanoic acid, 3,5-bis(1,1-dimethylethyl)-4-hydroxy-,octadecyl ester, $C_{35}H_{62}O_3$, formula weight = 530.9, CAS RN 2082-79-3.

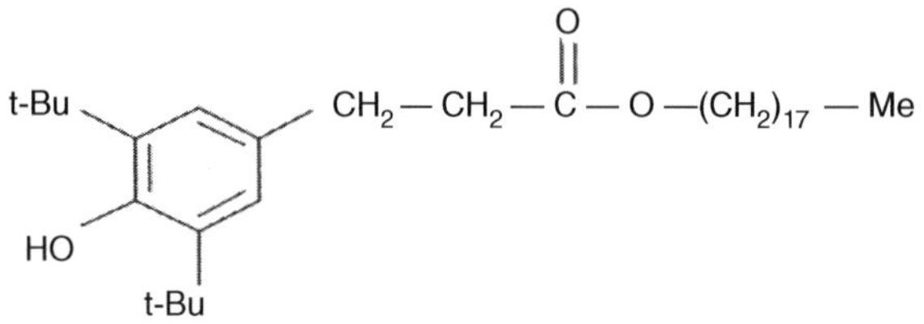

Associated extracted substances (Leachables):

3,5-bis(1,1-dimethylethyl)-4-hydroxy-benzenepropanoic acid, $C_{17}H_{26}O_3$, formula weight = 278.4, CAS RN 20170-32-5.

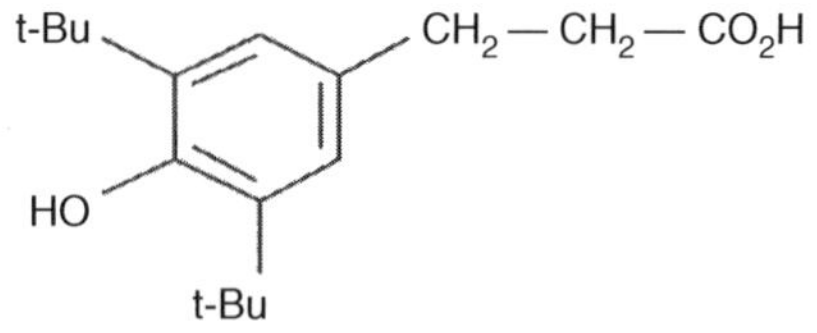

3,5-bis(1,1-dimethylethyl)-1-hydroxy-4-oxo-2,5-cyclohexadiene-1-propanoic acid), $C_{17}H_{26}O_4$, formula weight = 294.4, CAS RN 83237-15-4.

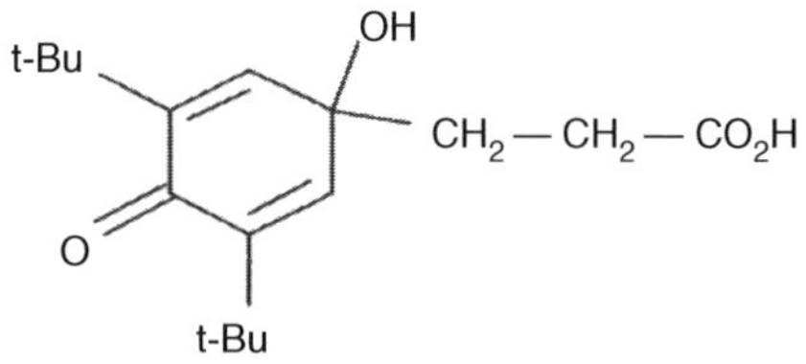

7,9-di-t-butyl-1-oxaspiro[4,5]deca-6,9-diene-2-8-dione, $C_{17}H_{24}O_3$, formula weight =276.4, CAS RN 8230466-3

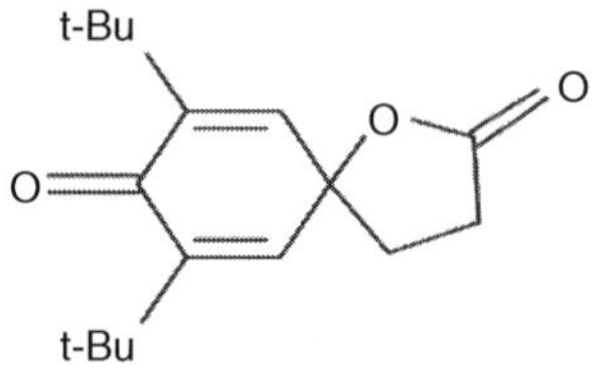

1-Octadecanol, $C_{18}H_{38}O$, formula weight = 270.5, CAS RN 112-92-5

$$HO - (CH_2)_{17} - Me$$

Figure 3.30 Irganox 1076 and its associated extracted substances (leachables). Irganox 1076 is a common antioxidant used in polypropylene and polyethylene resins. The link between Irganox-type antioxidants and their related decomposition products is discussed in references 33–36.

Figure 3.31 illustrates a situation that is not altogether uncommon. The leachables illustrated in Figure 3.31 all demonstrate similar functional characteristics and can be readily traced back to a known additive in the plastic formulation (in this case the initiator Lupersol 221). However, as the Lupersol 221 is essentially completely used up in the polymerization of ethylene and vinyl acetate to create the ethylene vinyl acetate co-polymer, it is not, rigorously speaking, an extractable since it is not present in the material at readily detectable levels. Thus, the carboxylic acid diester extractables and leachables are essentially related to and derived from a compound that cannot be analytically identified in an extract of the raw material because it is not present in the raw material. The only way to determine that this "ghost" compound exists (and thus link it to the observed extractables and leachables) is by knowing the polymerization process (and its associated participating reagents) because this compound is not present in (and thus cannot be extracted from) the finished polymer.

Extractable: Lupersol 221, Peroxydicarbonic acid, dipropyl ester, $C_8H_{14}O_6$, formula weight = 206.2, CAS RN 16066-38-9.

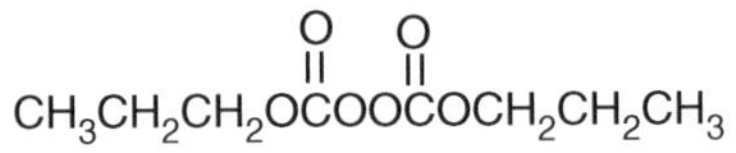

$$
\underset{\substack{|| \\ O}}{CH_3CH_2CH_2OCOOCOCH_2CH_2CH_3}
$$

Associated extracted substances (Leachables):

Carbonic acid, n-prophyl ester, diester with 1,2-ethanediol, $C_{10}H_{18}O_6$, formula weight = 234.2.

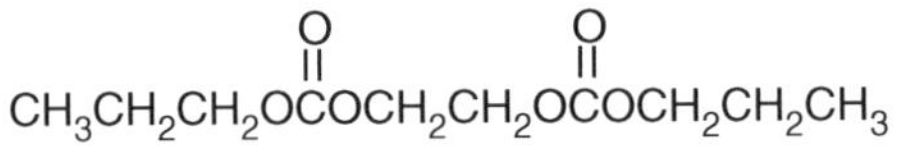

$$
CH_3CH_2CH_2OCOCH_2CH_2OCOCH_2CH_2CH_3
$$

Carbonic acid, n-prophyl ester, diester with 1,4-butanediol, $C_{12}H_{22}O_6$, formula weight = 262.3, CAS RN 96620-38-1.

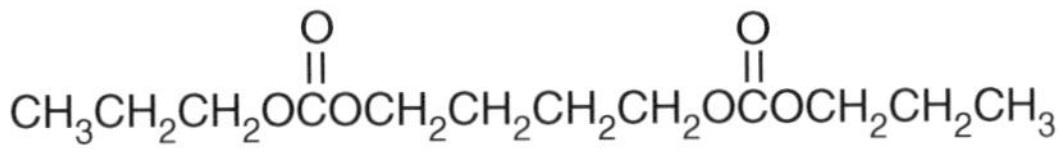

$$
CH_3CH_2CH_2OCOCH_2CH_2CH_2CH_2OCOCH_2CH_2CH_3
$$

Carbonic acid, n-prophyl ester, diester with 1-acetoxy-1,2-ethanediol, $C_{12}H_{20}O_8$, formula weight = 292.4.

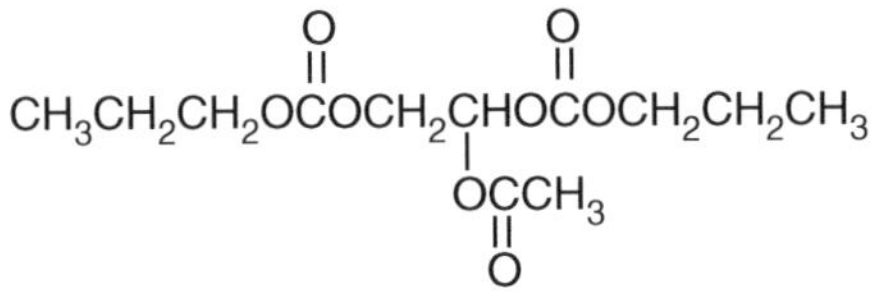

$$
CH_3CH_2CH_2OCOCH_2CHOCOCH_2CH_2CH_3 \\
\quad\quad\quad\quad OCCH_3 \\
\quad\quad\quad\quad O
$$

Carbonic acid, n-prophyl ester, diester with 1-acetoxy-1,4-butanediol, $C_{14}H_{24}O_8$, formula weight = 320.3.

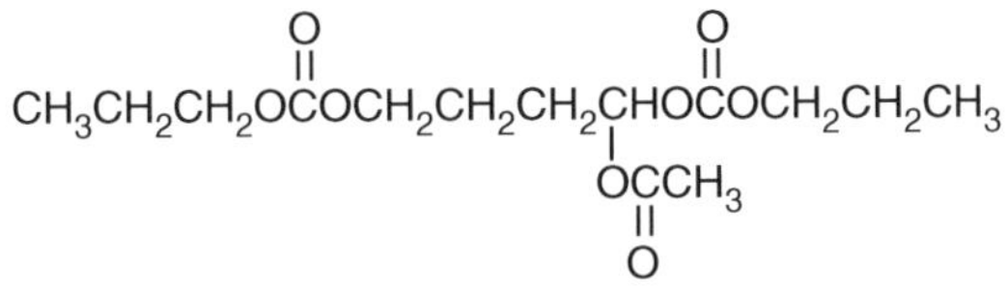

$$
CH_3CH_2CH_2OCOCH_2CH_2CH_2CHOCOCH_2CH_2CH_3 \\
\quad\quad\quad\quad\quad\quad OCCH_3 \\
\quad\quad\quad\quad\quad\quad O
$$

Figure 3.31 Lupersol 221 and its associated extracted substances (leachables). Lupersol is a common initiator used to produce ethylene-vinyl acetate (EVA) resins.

Figures 3.32 and 3.33 illustrate classes of extractables and leachables that are linked to one another, not in terms of cause and effect but rather because they share a similar chemical genesis and functionality. Figure 3.32 presents caprolactam and a series of related oligomers. These substances, which reflect incomplete polymerization residuals present in nylon-type polymers, are both extractables, as they are present in the plastic material, and leachables, as they have sufficient solubility to accumulate to measurable levels in aqueous drug products. Similarly, Figure 3.33 presents a second class of related compounds, referred to as cyclic esters, that have been associated with polyurethane adhesives used to bind the layers of the laminated materials together. It is hypothesized that the cyclic esters are formed during the first step of the manufacture of the polyester prepolymer, which is the formation of the linear polyester. Depolymerization of the linear polyester via ester exchange could result in cyclic ester formation.

The nomenclature that can be used to describe the various cyclic esters refers to their molecular weight. Thus, for example, CE 216 is a cyclic ester with a molecular weight of approximately 216. Where isomers exist, they are denoted trivially via a Roman numerical (e.g. CE228 (I)), where the increasing Roman numbers refer to the order in which the isomers were identified.

While cyclic esters with higher molecular weights than those illustrated in Figure 3.33 have been discovered, their solubilities, even in very lipophilic solutions, are low and they accumulate at only very low concentrations in most drug products.

3.5.6 Practical Realities Associated with Linking Extractables and Leachables

In a perfect world, one would characterize a container/closure system for its complement of extractables

Associated extractable/extracted substances (leachables):

Caprolactam, 2H-Azepin-2-one, hexahydro-, $C_6H_{11}NO$, formula weight = 113.15, CAS RN 105-60-2.

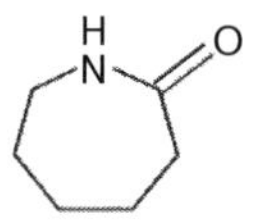

Caprolactam cyclic dimer, 1,8-diazacyclotetradecane-2,9-dione, $C_{12}H_{22}N_2O_2$, formula weight = 226.3 CAS RN 56403-09-9.

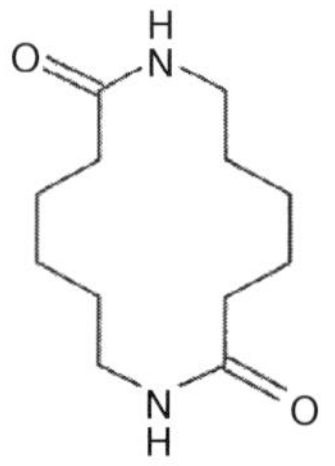

Caprolactam cyclic trimer, 1,8,15-triazacycloheneicosane-2,9,16-trione, $C_{18}H_{33}N_3O_3$, formula weight 339.47, CAS RN 56403-08-8.

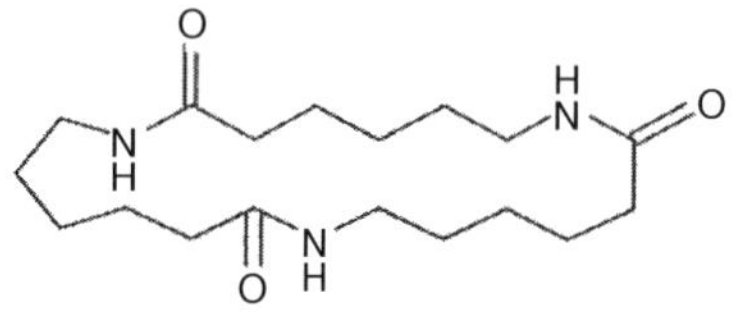

Caprolactam cyclic tetramer, 1,8,15,22-tetraazacyclooctacosane-2,9,16,23-tetrone, $C_{24}H_{44}N_4O_4$, form weight = 452.63, CAS RN 5834-63-9.

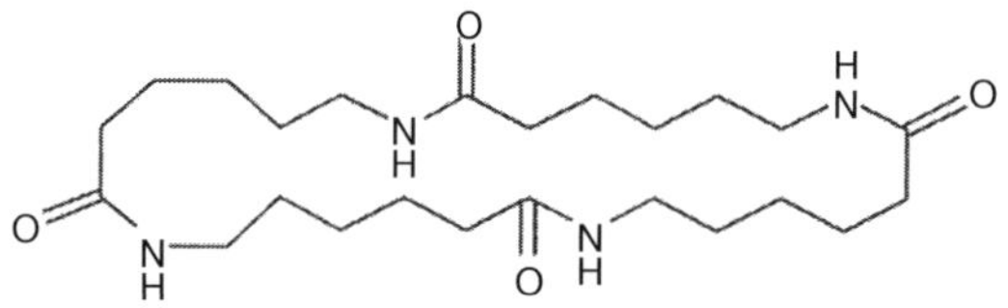

Hydrolyzed caprolactam dimer, Hexanoic acid, 6-[(6-amino-1-oxohexyl)amino]-, $C_{12}H_{24}N_2O_3$, formul weight = 244.33, CAS RN 2014-58-6.

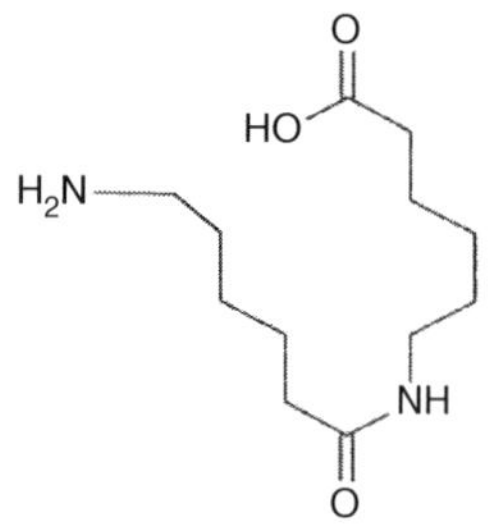

Figure 3.32 Caprolactam and its associated oligomeric extracted substances (leachables). Caprolactam and its related oligomers are typically associated with Nylon 6 and represent incomplete polymerization residuals (reference 37).

and characterize a contacted drug product for its leachables and find (i) that the lists of extractables and leachables were the same and (ii) that the amount of each leachable in the drug product could be mathematically related to the amount of the same extractable in the container/closure. This circumstance would be analogous to obtaining a complete correlation between all extractables and leachables, but it is rarely obtained (or even obtainable) in practice. Several reasons for this reality follow.

Reasons why an extractable may not be a leachable (i.e. why can we find it in the container/closure but not in the drug product?):

1. The extractable is practically insoluble in the drug product.
2. The extractable is degraded or chemically altered during drug product contact.
3. The extractable is in the drug product but eludes analytical detection.

Associated extractable/extracted substances (leachables).

CE200, 1,6-dioxacyclododecane-7,12-dione, $C_{10}H_{16}O_4$, formula weight = 200.23, C = 59.99%, CAS RN 777-95-7, adipic acid + tetramethylene glycol

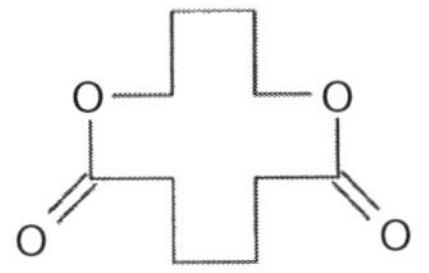

CE216, 1,4,7-trioxacyclotridecane-8,13-dione, $C_{10}H_{16}O_5$, formula weight = 216.22, CAS RN 6607-34-7, adipic acid + diethylene glycol

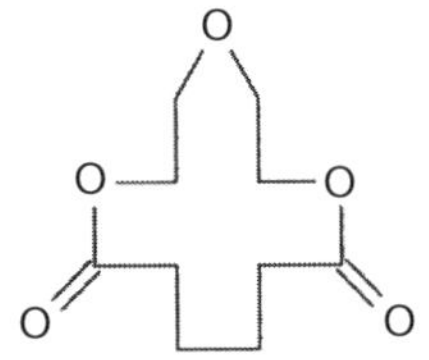

CE228(I), 1,4-dioxacyclotetradecane-5,14-dione, $C_{12}H_{20}O_4$, formula weight = 228.28, CAS RN 5578-82 5, sebacic acid + ethylene glycol

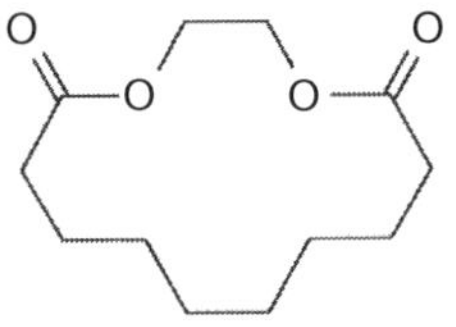

CE228(II), 1,8-dioxacyclotetradecane-2,7-dione, $C_{12}H_{20}O_4$, formula weight = 228.28, C = 63.14%, CAS RN 13926-69-7, adipic acid + hexamethylene glycol

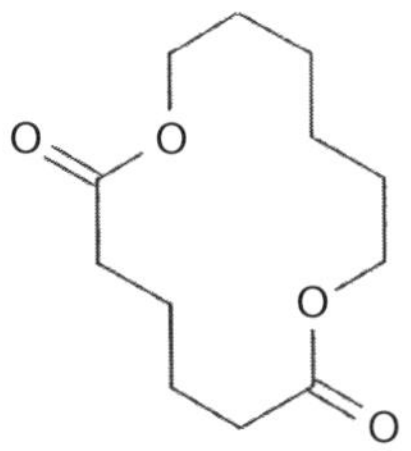

Figure 3.33 Cyclic ester-type extracted substances (leachables). These compounds are associated with polyurethane adhesives. It is hypothesized that the cyclic esters are formed during the first step of the manufacture of the polyester prepolymer, specifically during the formation of the linear polyester. Depolymerization of the linear polyester via ester exchange could result in cyclic ester formation. The cyclic esters typically are derived from glycols and higher molecular weight di-acids. Another type of cyclic ester (lactone) is formed through cyclization of hydroxy acids such as 6-hydroxyhexanoic acid.

Reasons why a leachable may not be an extractable (i.e. why can we find it in the drug product but not in the container/closure?):

1. The leached substance is formed during drug product contact.
2. The container/closure is not tested under fully processed conditions (and the extracted substance is process-related).
3. The extracted substance is in the container/closure but eludes analytical detection.

A reason why an extractable and a leachable cannot be mathematically correlated:

The leaching of substances from a container/closure into a drug product is a complex physiochemical process that is affected by a whole host of product-, container/closure-, and application-specific factors. Many of these circumstances are application-specific and are not amenable to generalization. However, one can examine the circumstance of solubility in some detail, as the factors influencing solubility are well understood and have been effectively modeled. For example, Yalkowsky and Valvani proposed that the aqueous solubility (S_w) of a non-electrolyte (specifically rigid and short-chain molecules) could be estimated as a function of the molecule's octanol/water partition coefficient ($P_{o/w}$) and melting point (MP) via the expression [40]:

$$\log S_w = -\log P_{o/w} - (0.1 \times MP) + 1.05 \qquad (3.9)$$

Table 3.16 summarizes $P_{o/w}$ values and calculated estimates of S_w for many of the substances illustrated in Figures 3.28–3.33. This information provides partial explanations for some of the discrepancies that can arise between extractable and extracted (leached) substances. Consider DEHP, Irganox 1076, and Irgafos 168 as examples. While such substances may be present in a container/closure at readily discernible levels, their aqueous solubility is so small that most analytical test methods would be insufficiently sensitive to find them in a drug product. Thus, one can understand why these compounds are extractables but not leachables. Alternatively, consider a product application where the ratio of the weight of the container/closure to the volume of the contacted solution (e.g. a pre-filled syringe) is exaggerated versus the same ratio used in the extraction study. In such a circumstance, a highly soluble but trace-level substance in the container/closure (e.g. cyclic esters or caprolactam and related oligomers) could

"concentrate" in the drug product. One could envisage circumstances where the substance would escape detection in the extract of container/closure but be readily detectable in the drug product, reflecting the circumstance of a leachable that has no apparent linked extractable.

3.6 Intra-laboratory Variation in Extractables/Leachables Profiles

Strategically and at a high level, the process of leachables and extractables profiling is well understood and standardized, as all practitioners of extractables and leachables profiling use the same general approach. Specifically, extractables and leachables profiling involves two steps: generating the test sample and screening the test sample. For leachables, the test sample is the drug product, properly aged consistent with its conditions of storage and distribution. For extractables, the test sample is an extract, generated by

Table 3.16 Estimated aqueous solubility of extractables and leachables.

Extractable	Associated compounds	Molecular weight	Log $P_{o/w}$[a]	Solubility, mg/l
DEHP	DEHP	390.56	8.695	0.005
	MEHP	278.34	4.665	37.1
	1,3-benzenecarboxylic acid	166.1	1.021	1580
	2-ethyl-1-hexanol	130.2	2.817	2230
Irganox 1076	Irganox 1076	568.1	13.93	<0.001
	7,9-di-t-butyl-1-oxaspiro [4, 5] deca-6,9-diene-2-8-dione	276.4	3.272	66.0
	3,5-bis(1,1-dimethylethyl)-4-hydroxy-benzenepropanoic acid	278.4	4.479	2.02
	3,5-bis(1,1-dimethylethyl)-1-hydroxy-4-oxo-2,5-cyclohexadiene-1-propanoic acid)	294.4	3.055	89.9
Irgafos 168	Irgafos 168	646.9	17.56	<0.001
	1,3-bis(1,1-dimethylethyl)benzene	190.3	5.595	4.31
	2,4-di-t-butylphenol	206.3	4.859	8.62
Lupersol 221	Lupersol 221	206.2	2.997	2330
	Carbonic acid, n-propyl ester, diester with 1,4-butanediol	262.3	3.030	2750
Caprolactam	Caprolactam	113.2	0.324	>10 000
	Caprolactam dimer	226.3	−0.649	>10 000
	Caprolactam trimer	339.5	−0.973	>10 000
	Caprolactam tetramer	452.6	−1.297	>10 000
	Hydrolyzed Dimer	244.2	−0.323	>10 000
Cyclic Esters	CE200	200.2	−0.061	>10 000
	CE216	216.2	−1.000	>10 000
	CE228 (I)	228.3	1.203	>10 000
	CE228 (II)	228.3	0.898	>10 000
	CE270	270.4	2.624	7210
	CE384	384.4	5.120	32.7

a) Octanol/water partition coefficients were obtained from the CAS REGISTRY Database, available through CAS SciFinder. Copyright 2005, American Chemical Society (http://www.cas.org/SCIFINDER).

extracting the test article with a solvent under defined conditions of contact. In either case, the test sample is screened by appropriate analytical methods to discover, identify, and quantify the substances of interest.

Considering analytical screening for organic extractables/leachables, there is a certain level of standardization in the tactics used. It is generally accepted that a proper screening approach for organics includes the following orthogonal techniques:

- GC with headspace sampling (headspace GC) for volatiles,
- "Direct injection" GC with flame ionization and mass spectrometric detection (GC/FID/MS) for semi-volatiles, and
- Direct injection liquid chromatography with ultraviolet (UV) absorption and mass spectrometric detection (LC/UV/MS) for non-volatiles.

Furthermore, there is some level of standardization in the specific methods used for each one of these techniques. For example, most LC screening methods are based on a reversed-phase separation accomplished with a C18 stationary phase and a mobile phase gradient consisting of an aqueous buffer and one or more organic solvents such as methanol and acetonitrile. Similarly, most GC methods employed in screening use capillary columns packed with a common stationary phase (USP designation G27, 5% phenyl, 95% dimethylpolysiloxane high temperature) and similar oven temperature profiles.

It is at the next level of tactical detail, operational conditions, where standardization is lost, as the specific conditions used by individual practitioners begin to diverge rather significantly. For example, one practitioner might use an LC column of certain dimensions packed with a C18 stationary phase from a certain vendor, while another practitioner might use a column with different dimensions and a different stationary phase from a different vendor but that still fits the general C18 classification. Similarly, the multiple critical operating parameters for the MS detector used may vary from one laboratory to the next.

If two practitioners use generally similar methods with different operating conditions to generate a chromatogram (which is the basis of an extractables or leachables profile), their chromatograms might differ suggesting that their extractables (or leachables) profile elucidated from the chromatogram could differ.

This possibility, different profiles generated by different testing laboratories, has led to much debate in the E&L community of practice and to much consternation on the part of users of E&L data specifically toxicological safety risk assessors. If it is the case that an extractables or leachables profile might differ depending on the testing

laboratory preforming the analytical profiling, then it can be (and has been) argued that extractables or leachables data might not be sufficiently reliable to be the basis of a safety risk assessment.

Considering variation in extractables or leachables profiles, there is some evidence, largely anecdotal, that E&L screening results are inadequately accurate and reproducible, lab to lab, to support rigorous impact assessment. At face value this seems improbable, given the pride the E&L community expresses in terms of its devotion to "good science." However, the assertion merits closer, more logical, examination.

If one seeks to minimize variation in any process, the approach to do so is through standardization. Standard methods, resulting from rigorous development and optimization and properly implemented, produce reproducible outcomes. Different methods that are not individually optimized to suppress variation and that are implemented without adequate control can only lead to highly variable results. However, routinely performed E&L operations, such as extraction and extract analysis, are rarely standardized across and between E&L laboratories and are generally not optimized and routinely assessed (via system suitability) to reduce or manage variation. At a high level, all practitioners employ GC/MS to address volatile and semi-volatile organic substances. Furthermore, the methods are similar as (i) it is logical that different laboratories would come to a similar solution to the same problem and (ii) GC/MS instrumentation is highly standardized. However, the method operating details may not be the same lab-to-lab and it is reasonable to expect that this will lead to increased lab-to-lab variations. The choice of IS and/or SS is not the same from lab to lab and, as we have seen previously, this surely will result in lab-to-lab variation in quantitative results. Although the screening methods are optimized for certain parameters, it is rare that reduction of variation is an optimization parameter that is intentionally addressed. Moreover, although most labs use system suitability testing to ensure methods are properly implemented at the time of use, it is rare that the system suitability tests include a consideration of response variation in general (although injection-to-injection response variation is typically addressed).

Lastly, this discussion has yet to consider other sources of variation that strictly speaking are not analytical in nature, such as variation in the generation and treatment of extracts and variation in laboratory operations (such as how and when the IS and/or SS is added to the extract).

The available information that addresses lab-to-lab variation in reported extractables profiles is limited and largely circumstantial. Nevertheless, results from four studies will be examined.

3.6.1 PQRI Controlled Extraction Study [41]

To establish thresholds and best demonstrated practices for performing controlled extraction studies specifically relevant for polymers and elastomers used in container closure systems for parenteral and ophthalmic dosage products (PODPs), a PQRI PODP Leachables and Extractables Working Group initiated a study that considered the processes by which extracts are generated and analyzed and the means by which test results are evaluated and interpreted. The purpose of this study was to generate and interpret data from controlled extraction studies performed on multiple polymeric and elastomeric materials of construction commonly encountered in PODP packaging systems. These materials were subjected to different extraction conditions and the resulting samples (i.e. extracts) were then characterized for extracted substances to establish how the different experimental parameters affected the resulting extractables profiles. Thus, this study was designed and executed to be exploratory in nature and differed significantly, in terms of design, execution, and interpretation, from studies whose objective is to establish the suitability for use of marketed packaging.

Details and comments about the study design include (see Figure 3.34):

- Five materials commonly encountered in PODP packaging, PVC, brominated isobutylene-isoprene rubber, low-density polyethylene (LDPE), polycarbonate (PC), and a cyclic olefin copolymer (COC), were solvent extracted under specified conditions using extraction solvents of varying "leaching power" (polarity, pH) and various extraction processes (e.g. Soxhlet (where appropriate), reflux (where appropriate), sonication, closed vessel).
- The resulting extracts were screened for organic extractables using specified chromatographic methods including GC/FID/MS and LC/UV/MS.
- The aqueous extracts were screened for extracted elements using a specified ICP/MS method.
- The test materials were thermally extracted and the evolved organic gases were analyzed using specified analytical conditions.
- It was intended that the resulting extractables data would be used to establish best demonstrated practice

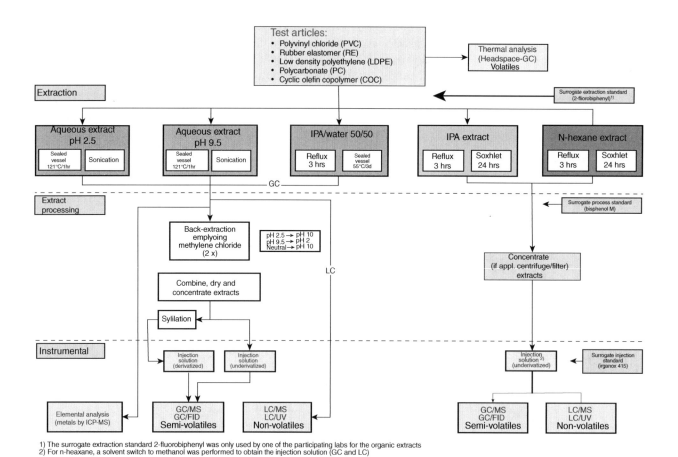

1) The surrogate extraction standard 2-fluorobiphenyl was only used by one of the participating labs for the organic extracts
2) For n-heaxane, a solvent switch to methanol was performed to obtain the injection solution (GC and LC)

Figure 3.34 Study design, PQRI controlled extraction study on five materials. (*See insert for colour representation of the figure.*)

recommendations for performing extractions and testing extracts.

- The study was supported by volunteer efforts of established E&L CROs.
- Not all laboratories were given all samples to do all the extractions and test all the extracts by all the methods, as it was not the intended purpose of the study to investigate lab-to-lab variations in processes and outcomes. Thus, any ability of the study to address lab-to-lab variations is coincidental and not intentional.

Observations about the study design and execution are as follows:

- Although extracts were generated and tested according to protocol, minor deviations in particular method details and extraction procedures were occasionally made by individual participating laboratories to accommodate different laboratory facilities and instrumentation.
- Individual laboratories used the instrumentation available to them, resulting in different (but generally equivalent) equipment being used across laboratories. Thus, minor variations in operating parameters for chromatographic methods, lab-to-lab, were sometimes necessary.
- This present study ... "did not definitively establish the identities of all observed extractables (and) did not extensively investigate perceived anomalies in the data ...," meaning that no attempt was made (and likely would have been difficult to make) to reconcile inconsistencies.
- This study was designed to identify extracted substances at a concentration of 10 µg/g or higher (meaning the study participants intentionally did not "play in the grass").
- Samples that contained particulate matter included the reflux and Soxhlet extracts of PVC and the pH 9.5 sealed vessel extracts of the PVC and rubber (poor thinking on the part of the study designers for asking the labs to generate and test compromised extracts).

Practical lessons learned from this study include:

- A study of this size requires a significant amount of effort (both anticipated and unanticipated), which in and of itself is a contributor to variation.
- It is a significant burden to ask CROs to use analytical methods and processes that differ greatly from their own standard practices. When push comes to shove, the labs will revert to their own standard methods.
- Practical issues will invariably pop up and unavoidably cause deviations to the study design.
- The study design will invariably put the participating laboratories in difficult positions in which there are no good paths forward.
- The concept of time and deadlines lose all meaning.

More or less typical results obtained in this study are summarized in Table 3.17. The most important feature of this table with respect to analytical variation is the reporting convention for concentrations. After reviewing the data provided by the participating CROs, it became clear to the study's director that the degree of lab-to-lab agreement in the reported concentrations was no better than an order of magnitude; that is, the values reported by the various CROs agreed to within an order of magnitude, regardless of whether the data was obtained by GC or by LC.

Conclusions that can be drawn for this study are as follows:

1. Unless a study is specifically designed to address interlaboratory variation, interpreting the results in terms of interlaboratory variation should be undertaken with great caution.
2. As more variables are added to the study (e.g. multiple test articles, multiple solvents, multiple extraction methods), the study becomes so cumbersome to execute that the study design itself becomes a major and unmanageable source of variation.

Table 3.17 Organic extractables profile of the cyclic olefin copolymer; identified compounds reproducibly extracted at levels of approximately 1 µg/g or greater.

Compound Information				Extracted concentration, µg/g[a]	
Name	CAS RN	Formula	Mol. wt	IPA	Hexane
Cis-decahydronapthalene	493-01-6	$C_{10}H_{18}$	138.25	1–10	10–100
Di-(2-ethylhexyl) phthalate related[b]	117-81-7	$C_{24}H_{38}O_4$	390.56	10–100	1–10
Oleamide	301-02-0	$C_{18}H_{35}NO$	281.48	1–10	10–100
Dimethylterephthalate	120-61-6	$C_{10}H_{10}O_4$	194.18	1–10	1–10
Hexadecoanic (palmitic) acid	57-10-3	$C_{16}H_{32}O_2$	256.42	—[c]	—

a) Reflux extraction.
b) Sum of both DEHP and MEHP.
c) Not detected at a level above 1 µg/g.

3. Study designs should be carefully considered and pressure-tested upfront to be sure that study participants are not presented with "no good choice" decisions (e.g. how to deal with compromised extracts).
4. Generally, testing laboratories consistently discovered the major extractables and generally provided the same identities for those major extractables.
5. The concentrations reported by the different testing laboratories for the major extractables generally agreed to within a factor of 10.
6. The reporting of minor extractables was highly variable between testing laboratories in terms of the compounds discovered and their reported identities and concentrations.

In the end, the study provided a useful insight into lab-to-lab variation in reporting identities and concentrations of extractables within the limitations of the study and its design. As the study design was too ambitious and poorly conceived for the purpose of performing a lab-to-lab comparison, practical issues encountered in the execution of the study limited the participating lab's ability to "be on their best behavior." In all fairness, the outcome of the study was the classical "glass either half empty or half full." For the major extractables, identities were generally consistent, reported concentrations agreed to within an order of magnitude, and it was rare that a major peak was missed by one or more labs. For the minor extractables, however, conflicting identities and broadly disagreeing concentrations were more common and instances of orphan peaks (peaks not reported by all labs) were more frequent.

3.6.2 Investigation of Lab-to-Lab Variation Between Four Testing Laboratories

The information about this study was gleaned from a presentation made to the Society of Toxicology in 2018 (M. Posgia, R. Przygoda, Y. Zhou, and D. Malek, Variation in Chemical Characterization Results from Four Test Laboratories, Presented to the Society of Toxicology, San Antonio, TX, 14 March 2018). As the results of this study have not been published, the ability to dig deeply into the details of the study and its outcome is somewhat limited. In essence, however, the purpose of the study was to investigate interlaboratory variations related to compound identification and quantitation and evaluate whether the lab selection impacts test results.

Details and comments about the study design include (see Figure 3.35):

- The study involved the solvent extraction (IPA and purified water) of three test items (polyurethane and polyisoprene materials and an unspecified medical device composed of many different (unspecified) materials) under specified conditions ($50\,^\circ$C for 72 hours, $3\,cm^2/ml$).
- The extracts were tested by NVR/FTIR.
- The extracts were screened for organic extractables using the labs' own chromatographic methods (with extract preparation as appropriate) including GC/MS and LC/MS.

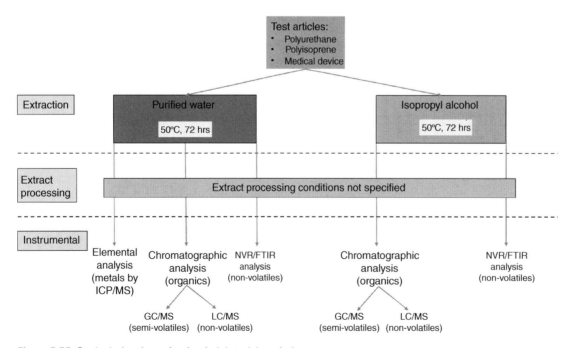

Figure 3.35 Study design, investigation in lab-to-lab variation.

Figure 3.36 Test results, NVR, lab-to-lab comparison. MD = medical device; PU = polyurethane; PI = polyiosprene. (*See insert for colour representation of the figure.*)

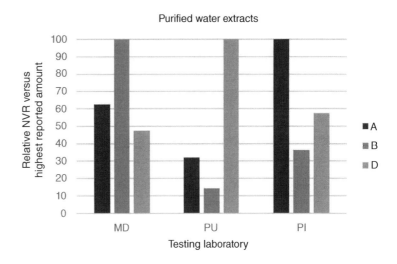

Purified water extracts

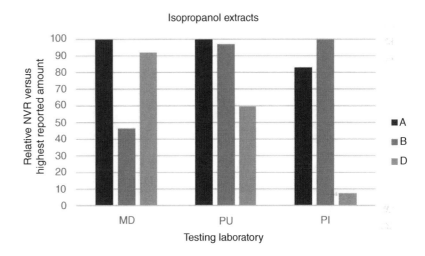

Isopropanol extracts

Note: MD = medical device; PU = polyurethane; PI = polyiosprene

- The purified water extracts were screened for extracted elements using the labs' own ICP/MS methods (with sample preparations as appropriate).
- The study was submitted to established E&L CROs as actual projects.

Test results were as follows:

For NVR, see Figure 3.36. Note that testing lab C had analytical issues and could not report a result. The agreement between laboratories was not outstanding.

For non-volatile organic extractables (measured by LC/MS), see Figure 3.37. For semi-volatile organic extractables (measured by GC/MS) see Figure 3.38. The metric of number of peaks reported is a very poor representation of analytical variation. Much better metrices would include the identities and concentrations reported for the detected peaks. Considering identification, the authors noted "The detail level (of the provided

identities) was insufficient for conducting a comparative risk assessment." Considering the number of reported peaks, there was considerable lab-to-lab variation. Considering quantitation, certain of the reported GC/MS results are contained in Table 3.18. Of the three organic extractables reported, only two of the four test labs reported each individual extractable.

For extracted elements, see Table 3.19. Laboratory C was unable to report results. All three of the other labs reported extracted levels of nickel and zinc; however, the reporting of other extracted elements was more sporadic between laboratories.

Considering the results of their study, the authors noted that significant differences were observed among the test laboratories in compound identification, confidence in the identification and quantity reported, and therefore that the selection of a test laboratory could have a significant impact on a risk assessment.

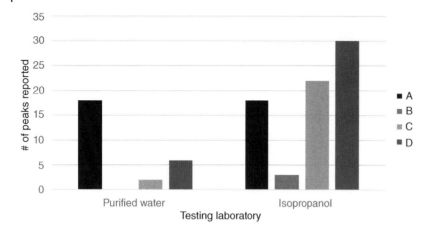

Figure 3.37 Test results, non-volatile organic extractables by LC/MS, lab-to-lab comparison. (*See insert for colour representation of the figure.*)

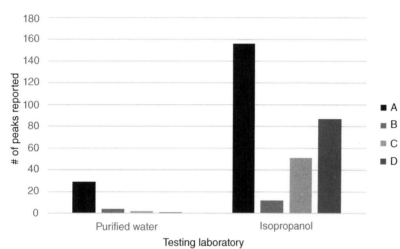

Figure 3.38 Test results, semi-volatile organic extractables by GC/MS, lab-to-lab comparison. (*See insert for colour representation of the figure.*)

Table 3.18 Semi-volatile organic extractables reported in the IPA extract of the polyisoprene test article.

Extractable		Reported level (μg/test article)			
Name	CAS RN	Lab A	Lab B	Lab C	Lab D
Aniline	62-53-3	380	690	NR[a]	NR
Benzothiazole	95-16-9	590	840	NR	NR
2-Mercapto-benzothiazole	149-30-4	NR	2700	NR	170

[a]NR = not reported.

3.6.3 Investigation of Extraction Conditions for Medical Devices

ISO 10993:12 contains instructions on how to extract medical devices for the purpose of biocompatibility assessment [42]. Considering the choice of the specified extraction conditions, the document notes "The extraction conditions described above, which have been used to provide a measure of the hazard potential for risk estimation of the device or material, are based on historical precedent."

Table 3.19 Extracted elements reported in the purified water extract of the medical device.

	Reported level (μg/test article)			
Element	Lab A	Lab B	Lab C	Lab D
Nickel (Ni)	5.4	1.8	N/A[b]	570
Zinc (Zn)	260	150	N/A	240
Calcium (Ca)	NR[a]	85	N/A	NR
Potassium (K)	NR	7.1	N/A	NR
Iron (Fe)	4.0	6.1	N/A	NR
Magnesium (Mg)	0.75	3.5	N/A	NR
Tin (Sn)	0.83	NR	N/A	NR
Barium (Ba)	NR	0.77	N/A	NR

[a]NR = not reported.
[b]N/A = results not available.

Unfortunately, the historical precedent is so ancient that the original rationale for choosing the specified conditions is lost in antiquity.

To shed some light on the effect of extraction conditions (including an extraction solvent), an ISO working group commissioned a study that examined these effects.

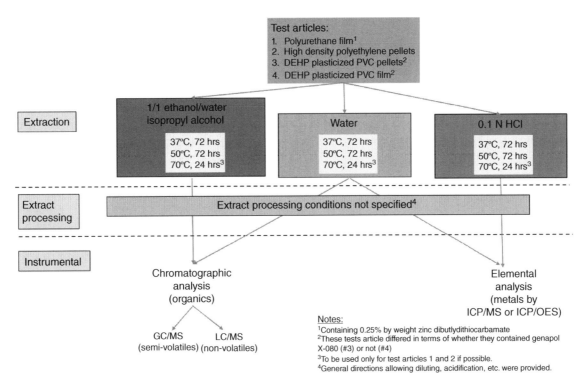

Figure 3.39 Study design, investigation of ISO 10093:12 extraction conditions. (*See insert for colour representation of the figure.*)

The information about this study was gleaned from a presentation made at the Smithers Rapra Extractables & Leachables USA 2019 Conference (A.Y. Sawyer, The Results of the Interlaboratory Investigation of ISO 10993 Extraction Conditions, Arlington, VA, 8 May 2019). As the results of this study have not been published, the ability to dig deeply into the details of the study and its outcome is somewhat limited. In essence, however, the purpose of the study was to answer the question "Are extracts prepared according to ISO 10993:12 for ISO 10993 biological test systems adequate for the identification of potential hazards/toxic chemicals in medical devices?"

Details and comments about the study design include (see Figure 3.39):

- Four materials that had previously demonstrated biological reactivity were solvent extracted under explicitly specified conditions using extraction solvents of varying polarities and various extraction conditions of temperature, duration, and surface area/volume ratio (specified in ISO 10993:12).
- The extracts were screened for organic extractables using specified chromatographic methods (with extract preparation as appropriate) including GC/MS and LC/MS.
- The aqueous extracts were screened for extracted elements using a specified ICP/MS or ICP/OES method (with sample preparation as appropriate).

- The intent was to use the resulting extractables data to establish the effect that the extraction variables have on the extractables profiles.
- The study was supported via volunteer efforts of six experienced E&L testing laboratories.

The following observations were made about the actual conduct of the study:

- It was not the intended purpose of the study to investigate lab-to-lab variation in processes and outcomes. This is important as data relevant to the issue of lab-to-lab variation was incidental to the purpose of the study.
- The PVC extracts were quite cloudy. Furthermore, the organic extracts of the PVC contained so much DEHP that they needed to be substantially diluted prior to chromatographic analysis. For these reasons, it is possible that the PVC data is unreliable and unsuitable for interpretation.
- The HDPE material was established as containing few extractables at relatively low levels. This suggests that the detailed interpretation of the HDPE data may be somewhat unreliable.
- A study of this size requires a significant amount of effort (both anticipated and unanticipated) which in and of itself is a contributor to variation.
- Due to large demands on the lab's equipment, personnel changes, and a few instrument issues, two laboratories dropped out and deadlines were extended.

- Since only two labs completed the 70 °C study, the 70 °C data was excluded from the comparative analysis.
- The testing laboratories used different chromatography methods and instrument operating parameters.
- The testing laboratories used different spectral databases and strategies for the identification of organic extractables.
- The testing laboratories used different reference standards or response factors for concentration estimation.

These observations, along with similar observations made for the previously discussed studies, suggest that the studies were almost pre-ordained to produce difficult-to-interpret results, as the complex experimental designs could not be readily implemented and completed and poorly chosen test articles and test conditions led to the production of compromised extracts that were challenging to properly analyze. The steps the participating laboratories were forced to adopt to deal with these circumstances were themselves major contributors to lab-to-lab variations.

As was the case with the previously discussed study, this study used the imperfect metric of number of reported peaks to establish outcomes. Nevertheless, results for organic extractables, illustrated in Figures 3.40 and 3.41, clearly show major differences between the results reported by the participating laboratories. Looking at the data somewhat more closely, Table 3.20 summarizes the five semi-volatile organic extractables reported by each laboratory at the highest extracted concentrations. Although the results for three of the labs are moderately consistent, at least in terms of the identified extractables, the results reported for the fourth laboratory leave one wondering "did this lab even test the same sample?"

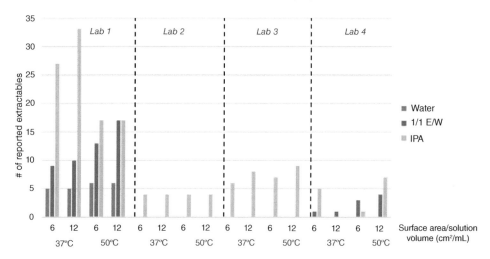

Figure 3.40 Test results, semi-volatile organic extractables by GC/MS, polyurethane test article. (*See insert for colour representation of the figure.*)

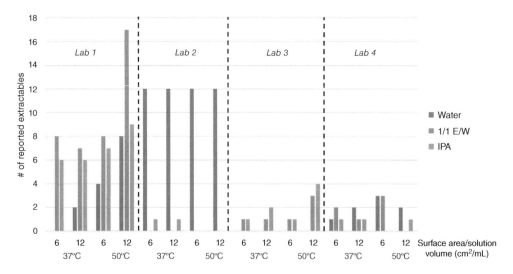

Figure 3.41 Test results, non-volatile organic extractables by LC/MS, polyurethane test article. (*See insert for colour representation of the figure.*)

Table 3.20 Summary of results, polyurethane test article, semi-volatile organic extractables (GC/MS Data), and IPA extracts.

Extractable's identity (top 5 peaks)	Quantities (µg/cm²)			
	Lab 1	Lab 2	Lab 3	Lab 4
Methylene diphenyl diisocyanate	3.4	7.4	9.4	—
Hydroxyl containing aliphatic compound (possible polymer fragment)	1.6	—	—	—
Zinc dibutyldithiocarbamate	1.0	10.4	25	—
Hydroxyl containing aliphatic compound (possible polymer fragment)	0.86	—	—	—
Hydroxyl containing aliphatic compound (possible polymer fragment)	0.70	—	—	—
4,4,5,6-Tetramethyltetra-hydro-1,3-oxazin-2-thione	—	—	2.4	—
1,4-Butanediol	—	—	2.1	—
2,2-Dimethyl-3-octanone	—	—	—	3.9
2,5-Hexanedione	—	—	—	3.7
2-Pentanone, 3-ethyl	—	—	—	3.1
Hydroperoxide, 1-methylpentyl	—	—	—	61
Hydroperoxide, 1-ethylbutyl	—	—	—	44
Unknown	—	3.1	—	—

This study also included testing of the extracts for extracted elements. These results, summarized in Figure 3.42, are much more consistent across laboratories, reflecting the more consistent analytical technique employed (ICP/MS).

Based on the accumulated results, these investigators concluded that:

1. Lab-to-lab variability in the number of compounds identified was noticeable.

2. The lack of agreement between chemistry test laboratories calls into question the reliability of chemical extract data to support toxicological risk assessments used in the biological evaluation of medical devices. Generally, testing laboratories consistently discovered the major extractables and generally provided the same identities for those major extractables.

3. Limited trends were observed for variable temperature, sample amount, or solvent selection (addressing the actual purpose of the study).

3.6.4 If You Play This Game Long Enough, You Are Bound to Repeat Yourself

I recently came across an unintentional situation that involved the organic extractables profiling of the same test article by two different testing laboratories. In this case, the test article, a laminated plastic/foil pouch commonly used as barrier packaging of medical devices, was extracted by filling pre-irradiated pouches with two extraction solvents (water and 50% ethanol in water) and storing the filled pouches for seven days at 55 °C, conditions expected to achieve asymptotic (equilibrium) extraction. As the individual studies were performed months apart, they did not share the exact same test articles (the same type of pouch was used but the batch number was different and each set of pouches had been irradiated separately). Each laboratory produced the extracts independently and then screened the extracts for organic extractables using their own methods for the three typically employed chromatographic screening techniques (headspace GC/MS, "direct injection" GC/MS, and LC/MS).

In comparing the organic extractables profiles produced by the two laboratories, I took the position that it mattered less what the individual test methods' results were and that it mattered more what the actual reported extractables

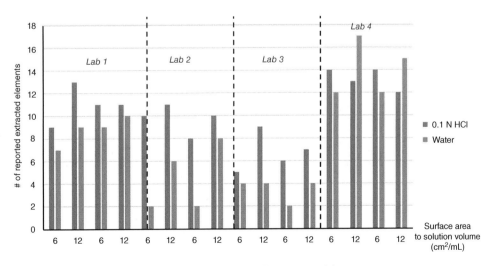

Figure 3.42 Test results, extracted elements, polyurethane test article.

profiles were. This position is reasonable in that it is the reported extractables profile that is assessed to establish the extractables' safety and/or product quality impact. Although the results obtained with the individual test methods are important and relevant, in the end it is the extractables profile, which represents a composite of the individual test results, that is actionable.

The comparison of the extractables profiles reported by the two testing laboratories is shown in Table 3.21. Unfortunately, the names of individual organic extractables cannot be disclosed due to confidentiality considerations; nevertheless, differences in the profiles reported by the two testing laboratories are still readily apparent. The first point to make about the comparative data is the consistencies in the profiles reported by the individual laboratories. Several extractables were reported by both laboratories and the reported concentrations were well within an order of magnitude of one another, even better agreement than was reported in the PQRI study. The second point about the comparative data is the inconsistencies between the individual laboratories, where there are several extractables reported by one laboratory that were not reported by the second laboratory.

Considering the differences in the reported extractables profile, Table 3.21 represents the worst case in terms of comparing the two laboratories, as it implies that there were extractables that were reported by one lab that were not even detected by the second lab. While this may be the case, it is possible that some of these "difference" extractables were detected by both labs but identifiable by only one lab (that is, the second lab reported the extractable as an unknown). Although both laboratories reported several unknown extractables, it was not possible to reconcile known (reported with identities) and unknown extractables across the two labs.

It is my opinion that these results are typical of the circumstance where a study is properly designed, albeit unintentionally, for lab-to-lab comparisons. With only one test article, two extraction solvents, and one extraction process, executing the study was manageable and did not require heroics, thus avoiding these complicating factors that plagued the previously noted studies. Additionally, the test article and the extraction conditions were such that the extracts were not compromised (e.g. lack of particulate), thus eliminating the need to design and implement variation-producing sample preparation processes to render the extracts analyzable. Lastly, the test article and extraction conditions were such that a robust extractables profile was generated, meaning that there were sufficient extractables at sufficiently high levels that an appropriate comparison could be made.

3.6.5 Addressing Variation in Environmental Testing [43]

The field of E&L is hardly unique in the sense that it involves characterizing (screening) samples for unspecified constituents at trace levels. Another field that shares this common task is environmental monitoring. Therefore, it should come as no surprise that environmental scientists have considered and addressed the aspect of lab-to-lab variation in reported results.

Preliminary results have recently been reported for EPA's Non-Targeted Analysis Collaborative Trial (ENTACT). ENTACT is a round-robin or ring-trial style project involving laboratories from academia, government, and the private sector. Nearly 30 laboratories have enrolled in ENTACT and used a variety of chromatography, mass spectrometry, and data processing approaches to characterize 10 synthetic chemical mixtures, containing 95–365 chemical substances each, three standardized media (human serum, house dust, and silicone band) extracts, and thousands of individual substances. These mixtures were distributed to participating labs to be analyzed using a variety of non-targeted analysis (NTA) methods. Participating laboratories were instructed to analyze study samples (including blanks) in accordance with their existing SOPs and/or methods. Dilution, concentration, and/or solvent exchange was performed at the discretion of each lab. The study was divided into two phases, with Phase 1 involving the blinded testing of the study samples while phase 2 involved a review of the blinded test data by the participating laboratories after the contents of the test samples had been revealed to them.

At the time the referenced article was written, 16 groups had submitted data for Phase I (blinded analysis) of parts 1 (ToxCast mixtures) and, to a lesser degree, part 2 (media extracts) of ENTACT. Seven groups had submitted data for Phase 2. Considering the analytical methodologies employed, two laboratories used GC methods, with both one- and two-dimensional chromatography used, and both low- and high-resolution MS. Thirteen laboratories used LC methods and the vast majority used a C18 column. Seven of the LC labs chose methanol and five chose acetonitrile as their organic phase, with the most common mobile phase modifiers being formic acid and ammonium formate. Eight labs used time-of-flight (TOF) style instruments, six used Orbitrap style instruments, one used a triple quadrupole, and one used an ultra-high resolution Fourier transform-ion cyclotron resonance (FT-ICR) style instrument. Two laboratories also included ion mobility as an orthogonal technique. Both laboratories performing GC analyses used electron ionization (EI) with one adding CI. While all laboratories performing LC used electrospray

Table 3.21 Organic extractables profile of a laminated foil pouch as reported by two E&L testing laboratories.

Extractable	Reported extracted concentration, µg/l (ppb)	
	Lab 1	Lab 2
Identified organic extractable #1	13 000	21 000
Identified organic extractable #2	5700	4900
Identified organic extractable #3	2500	—
Identified organic extractable #4	360	1200
Identified organic extractable #5	—	740
Identified organic extractable #6	720	—
Identified organic extractable #7	—	660
Identified organic extractable #8	480	390
Identified organic extractable #9	30	460
Identified organic extractable #10	350	194
Identified organic extractable #11	310	—
Identified organic extractable #12	15	235
Identified organic extractable #13	220	560[a]
Identified organic extractable #14	180	—
Identified organic extractable #15	160	—
Identified organic extractable #16	—	185
Identified organic extractable #17	120	150
Identified organic extractable #18	—	125
Identified organic extractable #19	—	125
Identified organic extractable #20	96	115
Identified organic extractable #21	33	89
Identified organic extractable #22	—	85
Identified organic extractable #23	85	—
Identified organic extractable #24	77	—
Identified organic extractable #25	12	77
Identified organic extractable #26	—	75
Identified organic extractable #27	25	74
Identified organic extractable #28	73	—
Identified organic extractable #29	70	—
Identified organic extractable #30	25	52
Identified organic extractable #31	16	51
Identified organic extractable #32	—	59
Identified organic extractable #33	50	—
Identified organic extractable #34	—	42
Identified organic extractable #35	40	—
Identified organic extractable #36	26	—

Compounds shaded are extractables reported by both laboratories.
a) Lab 2 tested the extracts by GC/FID/MS after derivatization and reported this compound.

ionization (ESI), three added atmospheric chemical ionization (APCI); nearly all labs used both positive and negative ionization modes. Nearly all labs included MS/MS experiments, with most using data-dependent acquisition (DDA).

Initial results show that nearly all participants detected and reported more compounds in the mixtures than were intentionally added. The authors suggest possible causes of this outcome, including double counting of reported peaks (e.g. across ionization modes, multiple isomers) and additional compounds present in the samples that were not intentionally added. The additional compounds could originate from impurities in the neat standards used to prepare the mixtures, reaction or breakdown products from chemicals added to the mixtures, or laboratory contaminants added during the handling of the samples, with large interlab variability in the number of reported compounds. Furthermore, the authors noted that overall, the information returned was highly variable, likely due to the differences in instrumentation, approaches, and workflows across labs.

Considering the results in somewhat greater detail, of the 1269 unique substances added to one or more of the mixtures, 1074 substances were detected and correctly identified by at least one of the methods and 195 substances (15.4%) were not identified by any method. The authors hypothesized that it is quite possible that other methods and laboratories might be able to detect and identify more of the 195 not detected compounds. Furthermore, the authors note that a majority of identified substances were detected by only one of the three methods (575 compounds, 45.3%), a slightly smaller number were identified with two methods (462 compounds, 36.4%), and the smallest number were identified by all three methods (37 compounds, 2.9%). Including replicate detections across chromatography and ionization methods, 809 substances were detected by GC and 801 by LC (539 by ESI+ and 262 by ESI−).

Lastly, considering the practical aspects of participating in the study, the authors noted that the requested timeline for the study (180 days for Phase 1 and 270 days for Phase 2) proved overly ambitious, as the samples and analyses were complex, and resource (personnel, instrumentation) shortages were common.

3.6.6 Concluding Thoughts

Given the noted studies and the outcomes, what can one conclude on the topic of lab-to-lab variation in reported extractables profiles? The first conclusion is that most of the studies purported to provide insight into the issue although they are poorly suited to the task. For example:

1. Most study designs were too ambitious and beyond reasonable expectations for acceptable execution. Thus, the execution was flawed.
2. Several of the studies were designed to investigate a phenomenon other than lab-to-lab variation. Thus, the data produced was sub-optimally related to the phenomenon for which it was interpreted.
3. Many studies contained design flaws that complicated, rather than simplified, data interpretation (for example, cloudy extracts in several studies). Thus, the entire set of generated data could not be used for interpretation.
4. The metrics used in several studies, such as the number of peaks reported, were not effective indicators of the phenomenon being investigated. Thus, the credibility of the conclusions drawn is open to question.
5. Most studies unsuccessfully controlled critical variables. Thus, the effects of critical variables could not be differentiated from the effects of non-critical variables.

When one:

- Does not control the essential variables of a multi-variable experiment,
- Starts an experiment with insufficient resources to finish the experiment,
- Relies on resources that are outside of one's control, and
- Uses the simplest and likely least relevant metrics,

how can one expect to be able to draw and defend meaningful conclusions from the results of the experiment?

The second conclusion merely re-states a point made earlier. Variation and standardization go hand-in-hand. When standardized approaches are used to produce an outcome, the outcome is more consistent than in the case where approaches vary widely among practitioners. Although the analytical approaches used in extractables screening are standardized in terms of technique (e.g. GC/MS, LC/MS, etc.), the actual methods used to accomplish screening (operating parameters and operational properties such as column type) can vary significantly from laboratory to laboratory. Furthermore, data processing methods, such as those used for the purposes of identification and quantitation, vary widely among laboratories. Given these circumstances, it is hardly unexpected that there is some level of lab-to-lab variation in reported extractables profiles.

However, I tend to see the glass as being half full on this topic. It is my firm belief that starting with comparably generated extracts, the top tier laboratories engaged in extractables and leachables screening will produce comparable extractables profiles for the major extractables (those extractables present in the extracts in quantities well above the analytical methods' limits of sensitivity). This means that the top tier laboratories will produce the same identities for the same detected extractables and will report concentration estimates that agree to within a factor of 10 (or better). However, as the laboratories begin to "play in the weeds" and deal with extractables whose levels approach the sensitivity capabilities of their methods, then there will be greater lab-to-lab variation in the entities that are detected and the identities and concentrations reported for the detected entities.

References

1 ISO 10993:18 (2020). *Biological Evaluation of Medical Devices – Part 18: Chemical Characterization of Medical Device Materials within a Risk Management Process*. Arlington, VA: Association for the Advancement of Medical Instrumentation (AAMI).

2 Safety thresholds and best practices for extractables and leachables in orally inhaled and nasal drug products. PQRI Leachables and Extractables Working Group, 8 September 2006.

3 Jenke, D. (2018). Identification, analysis and safety assessment of leachables and extractables. *TrAC* 101: 56–65.

4 Valcarcel, M. (2000). *Principles of Analytical Chemistry – A Textbook*. Berlin: Springer-Verlag.

5 Norwood, D.L., Jenke, D., Manoleseu, C. et al. (2009). HPLC and LC/MS analysis of pharmaceutical container closure system leachables and extractables. *J. Liq. Chromatogr. Relat. Technol.* 32: 1768–1827.

6 Legrand, P., Desdion, A., Boccadifuoco, G. et al. (2018). Development of an HPLC/UV method for the evaluation of extractables and leachables in plastic: Application to a plastic packaged calcium gluconate glucoheptonate solution. *J. Pharm. Biomed. Anal.* 155: 298–305.

7 Scherer, N., Marcsekova, K., Posset, T., and Winter, G. (2019). New studies on leachables in commercial scale protein drug filling lines using stir sorptive extraction coupled with TD-GC-MS and UPLC/QTOF-MS/MS. *Int. J. Pharm.* 555: 404–419.

8 ISO 10993:17(2002) (2002). *Biological Evaluation of Medical Devices – Part 17: Establishment of Allowable Limits for Leachable Substances*. Arlington, VA: Association for the Advancement of Medical Instrumentation (AAMI).

9 ICH Q3D (R1) (2020). Elemental impurities: Guidance for industry. US Department of Health and Human Services: Food and Drug Administration (FDA), Center for Drug Evaluation and Research (CDER), Center for Biologics Evaluation and Research (CBER).

10 ICH M7(R1) (2018). Assessment and control of DNA reactive (mutagenic) impurities in pharmaceuticals to limit potential carcinogenic risk: Guidance for industry. Food and Drug Administration (FDA), Center for Drug Evaluation and Research (CDER), Center for Biologics Evaluation and Research (CBER).

11 Mullis, J.O., Granger, A., Quin, C., and Norwood, D.L. (2008). The analytical evaluation threshold (AET) concept, sensitivity and analytical uncertainty. In *Conference Proceedings, Leachables and Extractables*, Smithers Rapra, Dublin, Ireland.

12 Jenke, D. and Odufu, A. (2012). Utilization of internal standard response factors to estimate the concentration of organic compounds leached from pharmaceutical packaging systems and application of such estimated concentrations to safety assessment. *J. Chromatogr. Sci.* 50: 206–212.

13 Jenke, D. (2020). Correcting the analytical evaluation threshold (AET) and reported extractable's concentrations for analytical response factor uncertainty associated with chromatographic screening for extractables/leachables. *PDA J. Pharm. Sci. Technol.* 74 (3): 348–358.

14 Jenke, D. and Heise, T. (2021). The implications of chromatographically screening medical products for organic leachables down to the AET adjusted for response factor variation. *PDA J. Pharm. Sci. Technol.* 75(3): 273–288.

15 Oktem, B. (2020). CDRH scientific perspective on chemical analysis for medical devices. *Presented at the PharmaEd Extractables & Leachables Virtual Summit 2020*.

16 Mellon, D. (2019). Nonclinical review of extractable leachable studies: Practical advice from an FDA reviewer. *Presented at the Smithers Rapra Extractables & Leachables USA Conference*, Arlington, VA.

17 Jenke, D. (2009). *Compatibility of Pharmaceutical Products and Contact Materials: Safety Considerations Associated with Extractables and Leachables*, 50. Hoboken, NJ: Wiley.

18 Jenke, D. (2020). Identification and quantitation classifications for extractables and leachables. *PDA J. Pharm. Sci. Technol.* 74 (2): 275–285.

19 USP <1663> Assessment of extractables associated with pharmaceutical packaging/delivery systems. USP 43/NF 38, pp. 8442. Effective 1 Dec 2020.

20 Jordi, M.A., Rowland, K., Liu, W. et al. (2020). Reducing relative response factor variation using a multidetector system for extractables and leachables (E&L) analysis to mitigate the need for uncertainty factors. *J. Pharm. Biomed. Anal.* 186 (15): 113334.

21 Jenke, D. and Liu, N. (2016). Chromatographic considerations in the standardization of liquid chromatographic methods used for extractables screening. *J. Liq. Chromatogr. Relat. Technol.* 39 (13): 613–619.

22 Christiaens, P., Beusen, J.M., Verlinde, P. et al. (2020). Identifying and mitigating errors in screening for organic extractables and leachables: Part I: Introduction to errors in chromatographic screening for organic extractables & leachables and discussion of the error of omission. *PDA J. Pharm. Sci. Technol.* 74 (1): 90–107.

23 Jenke, D. and Ruberto, M. (2014). Using the correlation between material composition and extractables and leachables to forecast extractables and/or leachables profiles. Part 2: Polyethylene and Bromobutyl rubber case studies. *Pharm. Outsourcing* 15 (2): 22, 24, 26–31.

24 Jenke, D. Chapter 10: Product Maintenance. In: *Compatibility of Pharmaceutical Products and Contact Materials; Safety Considerations Associated with Extractables and Leachables.* Wiley: Hoboken, NJ, 2009, 302–303.

25 Christiaens, P., Beusen, J.M., Verlinde, P. et al. (2020). Identifying and mitigating errors in screening for organic extractables and leachables: Part 2: Errors of inexact identification and inaccurate quantitation. *PDA J. Pharm. Sci. Technol.* 74 (1): 108–133.

26 Jenke, D. (2020). Identifying and mitigating errors in screening for organic extractables and leachables: part 3; considering errors of implementation and the use of a database to judge and promote good science and efficient practices. *PDA J. Pharm. Sci. Technol.* 74 (1): 134–146.

27 Food and Drug Administration (2002). *Center for Drug Evaluation and Research. Section III.G. Guidance for Industry: Nasal Spray and Inhalation Solution, Suspension, and Nasal Spray Drug Products—Chemistry, Manufacturing and Controls Documentation.* Rockville, MD: FDA.

28 Food and Drug Administration (2018). *Center for Drug Evaluation and Research. Section III.G. Draft Guidance for Industry: Metered Dose Inhaler (MDI) and Dry Powder (DPI) Drug Products—Chemistry, Manufacturing and Controls Documentation.* Rockville, MD, Revision 1: FDA.

29 Ulsaker, G.A. and Teien, G. (1992). Identification of caprolactam as a potential contaminant in parenteral solutions stored in overwrapped PVC bags. *J. Pharm. Biomed. Anal.* 10: 77–80.

30 Bee, J.S., Nelson, S.A., Freund, E. et al. (2009). Precipitation of a monoclonal antibody by soluble tungsten. *J. Pharm. Sci.* 98: 3290–3301.

31 Jiang, Y., Nashed-Samuel, Y., Li, C. et al. (2009). Tungsten-induced protein aggregation: solution behavior. *J. Pharm. Sci.* 98: 4695–4710.

32 Nashed-Samuel, Y., Liu, D., Fujimori, K. et al. (2011). Extractable and leachable implications on biological products in prefilled syringes. *Am. Pharm. Rev.* 14 (1): 74–80.

33 Van Hoecke, K., Catry, C., and Vanhaecke, F. (2013). Determination of elemental impurities in leachate solutions form syringes using sector field ICP-mass spectrometry. *J. Pharm. Biomed. Anal.* 77: 139–144.

34 Jenke, D.R. (2005). Linking extractables and leachables in container/closure applications. *PDA J. Pharm. Sci. Technol.* 59 (4): 265–281.

35 Haider, N. and Karlsson, S. (2002). Loss and transformation products of the aromatic antioxidants in MDPE film under long-term exposure to biotic and abiotic conditions. *J. Appl. Polym. Sci.* 85: 974–988.

36 Yu, K., Block, E., and Balogh, M. (2000). LC-MS analysis of polymer additives by electron and atmospheric pressure ionization: identification and quantification. *LC/GC North Am.* 18 (2): 162, 164, 166, 168, 170, 172, 174, 176, 178.

37 Chen, J.T., Mossoba, M.M., Varner, S.L. et al. (1989). GC/matrix isolation/FTIR studies of decomposition products of Irganox 1010 in an aqueous ethanol system. *Proc. SPIE – Int. Soc. Optical Eng.* 1145: 600–601.

38 Singh, G., Gollapalli, R., Binder, A., and Patel, M. (2018). Identification of leachable impurities in an ophthalmic drug product originating from a polymer additive Irganox 1010 using mass spectroscopy. *J. Pharm. Biomed. Anal.* 152: 197–203.

39 Jenke, D.R., Poss, M., Sadain, S. et al. (2005). Identification of caprolactam oligomers and related compounds in aqueous extracts of Nylon 6. *J. Appl. Polym. Sci.* 95: 1262–1274.

40 Yalkowsky, S.H. and Valvani, S.C. (1980). Solubility and partitioning I: Solubility of nonelectroyltes in water. *J. Pharm. Sci.* 69: 912–922.

41 Jenke, D., Castner, J., Egert, T. et al. (2013). Extractables characterization of five materials of construction representative of packaging systems used for parenteral and ophthalmic drug products. *PDA J. Pharm. Sci. Technol.* 67 (5): 448–511.

42 ISO 10993:12(2012) (2012). *Biological Evaluation of Medical Devices - Part 12: Sample Preparation and Reference Materials.* Arlington, VA: Association for the Advancement of Medical Instrumentation (AAMI).

43 Ulrich, E.M., Sobus, J.R., Grulke, C.M. et al. (2019). EPA's non-targeted analysis collaborative trial (ENTACT): Genesis, design, and initial findings. *Anal. Bioanal.Chem.* 411: 853–866.

4

General Principles of Chemical Assessment: Analytical Testing for Extractables and Leachables; Elements, Anions, and General Testing

4.1 Elemental Analysis

4.1.1 The Mandate for Extractable Element Testing

ICH Q3D [1] and the aligned USP <232> [2] monograph establish elemental impurities requirements for finished drug products, including a list of elements that, if present in drug products at levels above established permissible daily exposures (PDEs), could represent a patient safety issue. Furthermore, both documents clearly establish container closure systems as sources of elemental impurities, noting that "elemental impurities (that) have the potential to be leached out into the drug substance and drug product from the container closure system." Moreover, ICH Q3D notes that:

> "The identification of potential elemental impurities that may be introduced from the container closure system should be based on a scientific understanding of the likely interactions between a particular drug product type and its packaging. When a review of the materials of construction demonstrates that the container closure system does not contain elemental impurities, no additional risk assessment needs to be performed. … Studies to understand potential leachables from the container closure system … should be performed. The source of elemental impurities will typically be addressed during evaluation of the container closure system for the drug product."

Although there may be questions related to the nature, extent, and form of the elemental testing required for packaging systems, manufacturing components, and medical devices and their materials of construction, there can be no doubt that Q3D sets the stage for system, component, and materials of construction testing. The above text is clear on this subject and the intent is unmistakable. Moreover, Q3D definitively promotes the concept of risk assessment and requires its use to address elemental impurities. Since the

proper foundation of a valid and rigorous risk assessment is data and information, a risk assessment performed on a container closure system must be based on data secured for that system and its components and/or materials of construction.

However, even if ICH Q3D was more equivocal on this subject, other factors necessitate the testing of systems and materials of construction as potential sources of elemental impurities. For example, consider quality by design, which, simply stated, is the concept that quality is best served by building it into the product as opposed to establishing that it has been achieved by testing the product. Surely QbD supports the concept that "selecting materials of construction based on their ability to contribute elemental impurities is an important step in ensuring that the drug product will not contain excessive levels of packaging-related elemental impurities." Moreover, it is logical that the selection of materials should be based on a consideration of relevant information and data concerning the materials' ability to contribute elemental impurities.

Lastly, extractable elements data for systems and their components or materials of construction may be the basis of an efficient and effective change control process for container closure systems. It would be most unfortunate and inefficient indeed if change control of container closure systems could only be exercised by elemental impurity profiling of packaged drug products.

Once the mandate for testing container closure systems to determine their potential to contribute elemental impurities to packaged drug products has been established, the practical question becomes how to accomplish this task. Keys aspects to consider include:

1. What test articles should be testing and for what purpose?
2. How should the articles be tested?
3. What elements should be targeted in the testing?
4. How should the results of the testing be reported and interpreted?

Extractables and Leachables: Characterization of Drug Products, Packaging, Manufacturing and Delivery Systems, and Medical Devices, First Edition. Dennis Jenke.
© 2022 John Wiley & Sons, Inc. Published 2022 by John Wiley & Sons, Inc.

Each of these questions will be addressed as follows.

4.1.2 Identifying Proper Articles for Extracted Elements Testing

Figure 3.25 illustrates the high-level process for developing a packaged drug product from the perspective of its container closure system. In its simplified and generalized form, the flow diagram captures the essence, if not detail, of actual processes used by packaging or drug product development organizations.

The process starts with the selection of the container closure system's materials of construction. In fact, the actual first step is the generation of a Requirements Definition listing the requirements that the packaging system and its materials of construction must meet. For the sake of this discussion, assume that a Requirements Definition has been generated and contains a requirement that "the container closure system does not cause the packaged drug product to contain elemental impurities at levels that adversely affect the drug product's suitability for its intended use" (note that in making this statement, we have already moved beyond Q3D, which addresses only one aspect of suitability for intended use – patient safety).

As noted in Figure 4.1, actions performed at the materials of construction stage differ from actions performed at the other stages. While selection and qualification are performed for both components and container closure systems, only selection is specified for materials of construction. To understand this difference, one must first understand the difference between selection and qualification. Selection is a decision to use (or not use) a specific article in a particular application, presumably based on the article's ability to meet defined requirements. Selection is a not a guarantee that the specific material will produce an acceptable item; selection is a risk-based decision that the material is likely to produce an acceptable item. In contrast, qualification is used to confirm a material's ability to meet defined requirements (acceptance criteria), where such requirements ensure that the produced item will be acceptable. Thus, qualification can be considered as a "guarantee" that the produced item is acceptable.

While one understands the desire to (and the value in) qualify materials of construction, the propriety of doing so is questioned based on the simple observation that materials are selected so early in the product development project. This is important for two reasons. First, materials are typically highly processed by the time the final container closure system is constructed. Thus, it is questionable whether tests performed on unprocessed materials could be definitively linked to performance requirements for fully processed, packaged drug products. Moreover, the means by which materials of construction are processed to produce packaged drug products vary so widely, drug product to drug product, that it would be challenging to devise test methods for unprocessed materials that would be applicable in a majority of packaging applications.

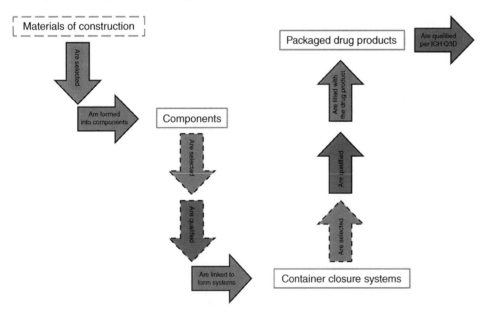

Figure 4.1 Process flow diagram, development and testing of a packaged drug product from the perspective of the container closure system. Green arrows reflect selection steps, red arrows reflect qualification steps, and blue arrows refer to construction steps. An arrow whose outline is solid denotes processes that must occur in development and testing while arrows with dashed outlines are processes that could occur in the development and testing at the discretion of the process owner. The dashed box surrounding materials of construction indicates that the process of Material Selection is typically completed by a component's vendor and not necessary by the drug product development team. (*See insert for colour representation of the figure.*)

The second reason addresses the question of "whose responsibility is to perform materials of construction testing?" Note that this is a different question than "whose responsibility is it to possess materials of construction data?" whose answer surely is "users of the materials of construction." Since "who should generate the data?" is a different question than "who should possess the data?" it should not be a surprise that the answers are different. To provide the answer, consider the simple mathematics of economics. For example, consider a vendor of a material for which there are 100 potential users. Is it not more economical for the vendor to do the testing once and provide the test data to all 100 users than for the users to independently perform the same testing 100 times?

The conclusion that vendor testing of materials is appropriate is another reason why materials are tested for selection and not for qualification. In the example of the one vendor and 100 users, if each potential use of the material is profoundly and fundamentally different and if the vendor accepts the responsibility for qualification, then the vendor would have to perform 100 individual tests to match each user's conditions of use. While it may be a reasonable request that a vendor implement a well-defined, efficient, and effective test protocol once, it is surely less reasonable to demand that the vendor perform 100 customized assessments.

Most certainly there are other dimensions of the "who should do the test" question that need to be considered. However, the point is not to make a definitive statement as to who should do the test but rather to note that because there is a question of who should do the test, testing at the materials stage is geared toward selection and not qualification.

Considering components, one can envision container closure development and/or manufacturing processes where the term component has little meaning, as materials of construction are directly converted to packaging systems. Even when components are a relevant consideration, selection may be irrelevant (as the optimal component may already have been specified) or qualification may not be appropriate (as organizations may decide to perform qualification solely at the system level). Nevertheless, component selection may be a critical part of product development and could rely on extracted elements data as a selection criterion.

Considering the container closure system stage, the required action is most likely qualification as it is a global regulatory requirement that a container closure system be qualified as being suited for its intended use. It is less frequent that container closure systems are tested for the purpose of selection, but one cannot completely discount this possibility.

This is a very long discussion indeed to establish that the answer to the question "what test articles should be tested?" is "materials of construction, components, and container closure systems." However, because selection and qualification serve two different purposes, selection and qualification are best accomplished by two different testing and reporting strategies. Furthermore, because materials are typically selected and systems are typically qualified, material testing and reporting and system testing and reporting will be different. Lastly, because components may (or may not) fall in the middle, components can either be selected or qualified (or both) based on methods that are suited for these individual purposes.

In summary:

- Materials of construction are tested for the purpose of selection.
- Components are tested for selection and/or qualification.
- Systems are tested for qualification.
- Testing for selection and testing for qualification are different by necessity and purpose.
- While users are responsible for possessing test data, vendors may be the more economic choice for generating test data, specifically if the data is used to support selection.

4.1.3 Establishing Proper Test Parameters

Generally, testing articles to establish their ability to contribute elemental impurities to packaged drug products consists of two independent processes, preparation of the article for testing and testing of the prepared article. Each is considered separately as follows.

4.1.3.1 Generating the Extract

This question of how to prepare a container closure system to test its ability to contribute elemental impurities to packaged drug products is easily answered, as the answer has already been provided in regulatory guidelines and is well understood in the technical community. Specifically, the ability of the container closure system to contribute leachables to a packaged drug product is best addressed by profiling the packaged drug product for leachables over their shelf-life, as leachables information is used to support the qualification of the container closure system. While this is a straightforward answer for organic leachables, there is a catch when it is applied to elemental impurities. Specifically, when a packaged drug product is screened for elemental impurities, the results obtained reflect all contributions to elemental impurities, including not only the container closure system's contribution but also the combined contributions of the drug substance, excipients, diluent (for example, water), and the manufacturing system. While the total levels of elemental impurities in a packaged drug product is necessary and relevant information, it does not delineate the contribution of the container closure system.

As measuring elemental impurities in a packaged drug product does not establish the container closure's contribution to the elemental impurities, this information must be obtained by other means, such as a controlled extraction study of the container closure system performed using extraction conditions that reflect, mirror, or simulate the drug product-filled container's clinical conditions of use. Because the conditions for performing such a controlled extraction (simulation) study closely match the clinical conditions of use of the packaged drug product, it is anticipated that the resulting extractable elements profile will be similar to the leached elements profile, both in terms of which elements are measurable and their measured levels.

Because drug products differ so greatly in terms of their inherent "leaching power" for elemental impurities, because the clinical conditions of use of drug products can vary widely, and because the scope of the study can vary (e.g. performing a study that considers one packaged drug versus many packaged drugs), the extraction conditions used in these controlled extraction (simulation) studies vary from circumstance to circumstance and thus are customized as opposed to standardized.

The question of how to prepare a material of construction for testing is less clear as there are two potential options, digestion and extraction. In digestion, the material is completely solubilized, typically by the action of strong acids and elevated temperatures and pressures. Thus, the prepared sample (the digest) contains all the potential elemental impurities that were present in the material at a level reflecting the total amount of the potential elemental impurities present in the material. As such, a digestion establishes what is present in the material, reflecting its greatest potential to contribute elemental impurities (i.e. the situation where "it all comes out") and providing the absolute worst-case situation.

Alternatively, the material can be prepared for testing by extraction, as was the case for the container closure system. An extraction, as opposed to a digestion, establishes "what can come out," which more directly addresses the issue at hand (that is, substances in a drug product have to be leached from the test article before they can be leachables). It is for this reason that extractions are the appropriate means of preparing a material of construction for testing. This is a compelling reason, as it is well established that generally only small portions of an item's total pool of elements are leached into drug products under conditions of use [3, 4].

Although both materials and systems are prepared for testing by extraction, the nature of the extraction is likely to be different, material versus system, as the purpose of the material testing is selection, as opposed to qualification. If

the selection is made between multiple candidate materials, there is considerable value if the same conditions of extraction are employed for all candidates, as this ensures that materials can be compared on an "apples to apples" basis. This is the exact opposite case with qualification, which is a process that is customized to match each situation and which places a high premium on testing items with the conditions under which they are used.

If the goal for material testing is standardization, then it is relevant to consider the extent to which extraction methods can be standardized across material classes. In general, materials used in pharmaceutical packaging fall into four broad classes: plastics, elastomers, glasses, and metallics. Each of these materials is sufficiently different that it is unlikely that a single extraction solvent and process can be established that is equally relevant to and appropriate for each class. Thus, different extraction solvents and different extraction processes may be needed for these different classes of materials.

For any material, aggressive extraction using an appropriate extraction medium (an inorganic acid, for example), may be justifiable as being a worst-case exaggeration of pharmaceutical contact conditions. Use of an aggressive extraction will likely lead to two possible conclusions:

1. If I judge the test article to be suitable for use under aggressive experimental conditions, it is likely to be suitable for use under actual use conditions.
2. If I judge a test article to be unsuitable for use under aggressive experimental conditions, then additional testing will be needed to establish whether the test article might still be suitable for use under actual use conditions.

A similar discussion and conclusion are relevant for components. For the purpose of selection, components should be tested via an approach that is similar to the approach for materials. To the extent that it is practical, components should be tested for selection in a manner that is consistent with the way in which the component contacts the drug product during the product's storage in its packaging. Furthermore, components should be tested for selection in their fully processed form. When tested for qualification, the word "should" in the last two sentences must be replaced with the term "must."

4.1.3.2 Testing the Extract

In theory, elemental impurities could encompass the entire periodic table, although practically speaking many elements are unlikely to be encountered in pharmaceutical packaging systems [3, 4]. Thus, the method for testing the extracts must be comprehensive, addressing a large population of elements. As the adverse effect of some elements may occur at low levels, the test method should be highly sensitive. Alternatively, as the levels of extracted elements could vary substantially, the test method should have a wide dynamic range. Given the chemical complexity of test articles and extraction solvents, the test method should be highly selective and robust.

Considering the current state of the art in elemental testing, atomic spectroscopic methods, such as inductively coupled plasma atomic emission spectroscopy (ICP-AES) and inductively coupled plasma mass spectrometry (ICP-MS) are recommended for elemental impurity testing (see, for example, USP <233> [5]).

It is difficult, if not impossible, to talk about analytical testing without addressing the aspect of validation. In the pharmaceutical world, it is clear that test methods that directly address and establish a test article's suitability for use (qualification) must be validated using appropriate procedures and specifications. USP <233> is an appropriate reference to consider when designing validation studies and setting acceptance criteria. While it is also clear that test methods for selection must be suited for this intended use, it is less clear that they need to be rigorously validated, as selection at an earlier stage is always followed by qualification at a later stage. For this and other reasons, methods for selection are typically qualified for use via a process that is less rigorous than formal validation.

In either case, selection or qualification, methods must be established as effective at the time of use via an appropriate system suitability assessment.

In summary,

- Materials are extracted under aggressive, standardized conditions, which may be optimized for different material types, regardless of the material's possible use.
- Components are extracted using conditions that simulate their intended use.
- Systems are extracted using conditions that simulate their actual use.
- Extracts are tested for targeted elements using atomic spectroscopic methods as indicated in USP <233>.
- Test methods for selection must be qualified to establish their suitability for use.
- Test methods for qualification must be more formally and rigorously validated to establish their suitability for use.
- Test methods for either selection or qualification must include appropriate system suitability assessment at the time of use.

4.1.4 Selecting Target Elements

On the one hand, the answer to this question is simple enough; elements that are present in the test article and which present a potential hazard should be targeted. On the other hand, the individual aspects of "what elements are present" and "what elements are potential hazards" are sufficiently complex that this answer is difficult to execute.

From a practical perspective, the analytical methods referenced previously, ICP-AES and ICP-MS, have the capability to provide data for many more elements than could be present in materials used in pharmaceutical packaging and/or which are pharmaceutical hazards. Thus, one possible answer is to "target them all." If it were as simple as that, then I might agree with that approach. However, for one thing, data tables that have a long list of targeted analytes and concentration entries such as "ND = not detected" are inefficient. More importantly, since the test methods must be either qualified or validated, each unnecessary element increases the possibility of a test failure, raising the specter of endless analytical runs being rejected because of unnecessary elements. A similar situation could also arise during system suitability testing.

Thus, testing should not include superfluous or non-essential elements.

Now someone will point out that given advances in drug product development and material science one never knows what element will be the next "bad actor." Though I acknowledge the truth in that statement, I point out that the concept here is risk management and not risk avoidance. Should such "bad actors" be surfaced, then of course the list can be adjusted to include them.

I would also expect that the reverse would be true; that is, once sufficient evidence has established that certain elements are not associated with certain materials then those elements should be removed from the target list for those materials.

Considering the potential impact of elemental impurities on patient safety, ICH Q3D provides a list of 24 "elements to be considered in the risk assessment if intentionally added (all routes)." As such, this list is the foundation of our generic answer to the question, as generally speaking it is difficult to always know (a) whether a particular element is intentionally added or not and (b) what dosage form a particular item will be used with. However, this list is only a foundation as it is understood that the Q3D Table 5.1 list is not a comprehensive list of potential hazards. If we consider hazard in the strict context of safety, Q3D itself acknowledges that elements that have "differences in regional regulations are not addressed in this guidance." This is why aluminum, manganese, and zinc, for example, are not assigned a PDE in Q3D, even though their presence in certain drug products poses a potential patient safety risk. If we consider hazard in the broader sense of being a potential threat to a product's suitability for use, Q3D also notes that it does not address product quality issues, pointing to tungsten's documented ability to adversely affect therapeutic proteins as an example of a potential element of interest. Thus, in additional to providing Table 5.1, Q3D notes that "some of the (additional) elements (to be) considered include aluminum, boron, calcium, iron, potassium, magnesium, manganese, sodium, tungsten and zinc."

Thus, the answer to "what elements should be targeted in testing?" is "the minimum number of targeted elements includes the 24 elements listed in Q3D Table 5.1 as well as the 10 elements specified in the text as 'considered elements'."

In summary:

- To address safety, the 24 elements in Table 5.1 of ICH Q3D are targeted.
- To address safety and/or quality, 10 additional elements specified in ICH Q3D are targeted.
- These targeted elements establish a baseline that can augmented with additional elements as appropriate for a specific test article material.
- With experience, elements may be deleted from the list if it is established that they are not relevant to a particular test article.

4.1.5 Reporting and Interpreting Extracted Elements Results

The question specifically addressed here is "should there be acceptance criteria for extracted elements?" To support the answer of "no, there should not be specifications," consider the following. Materials and components can be significantly altered as they are used to construct packaging systems. Moreover, as was stated previously, there is some question as to whether "extractable elements" measured under standardized and aggressive extraction conditions correlate well (and reproducibly) to leached elemental impurities. Thus, it is questionable whether extractable elements results for materials or components can easily and rigorously be linked to elemental impurity limits in drug products. Therefore, elemental impurity specifications for finished drug products cannot be easily associated with extractable element specifications for materials. This is not only a question of whether the link can be made mathematically but whether the link can be justified scientifically. If specifications for materials cannot (or should not) be linked to elemental impurity specifications for drug products, then it is not clear what else would serve as the appropriate basis for material specifications. If there is no appropriate basis for material specifications, then there should be no material specifications.

Moreover, selection processes, and therefore selection criteria, may vary across organizations and situations. Given the great diversity in organizations and situations, it is unlikely that useful, relevant, and generally applicable specifications can be established to aid material selection. It is the responsibility of a given organization to justify their material and component choices and it is unreasonable to expect that an organization with standards-setting authority would publish universally applicable specifications for selection.

Thus, extracted element results for materials are not correlated with general specifications, unless specific specifications have been developed by the material's vendor and/or user.

On the other hand, complete container closure systems are tested under conditions that simulate their intended use and it may be acceptable and relevant under such circumstances to interpret extractable element results in the context of Q3D PDEs, provided that the clinical use of the packaged drug product is well understood and can be linked to the extraction conditions. However, it is noted that PDEs as acceptance criteria do not address product risks other than safety.

It is clear that in order for extractable elements to be interpreted in either manner, the results of extracted element testing must be reported in terms of concentration units and not in general terms such as "pass" or "fail."

Once it has been established that extractable element test results are reported as absolute concentrations, a reasonable follow-up question is "at what concentration level should they be reported?" Somehow the obvious answer that they should be reported at levels at which they are actionable and relevant seems imprecise. Considering selection, extractable element information is actionable and relevant at levels consistent with the capabilities of appropriately state of the art methods, expertly applied. Thus, for example, if it is the consensus of qualified analytical experts that elements extracted into a certain extraction solvent can be reliably and routinely measured above a level of X ng/ml, then X ng/ml becomes the reporting threshold, where values above the threshold are reported and values below the threshold are not.

Considering qualification, the action taken is the assessment of whether the extracted element will adversely affect the quality of the drug product into which it is leached. Thus, the extracted elements should be reported at levels well below the level at which quality is adversely affected. Considering patient safety as a critical product quality attribute, it is noted that ICH Q3D has established 30% of an element's PDE as the control threshold above which control measures should be implemented to assure that the level of the element in the drug product does not exceed the PDE. Since this is a control threshold for elements in the drug product, it is not directly applicable to extracted metals because it does not consider other contributors of elements to the drug product. In this context, then, extractable elements are reported at values somewhat less than 30% of published PDEs (for example, 5% of the PDE if such levels are analytically achievable), so that an element's approach to such control thresholds can be anticipated.

In summary,

- Results to inform selection should be reported as concentrations above a reporting threshold based on analytical capability. It is the prerogative of the user to interpret the results in terms of their own selection process and selection criteria.
- Results to support qualification should be reported as concentrations to enable interpretation with respect to safety by considering drug product PDEs and the clinical use of the packaged drug product. Results should be interpreted in the context of other product quality aspects by means consistent with the circumstances and the individual organization's risk management processes.

4.1.6 Other Considerations

4.1.6.1 Extracted Elements and the Product Lifecycle

When the process of drug product development is started from the beginning, then a staged testing strategy of material selection, component selection/qualification, and system qualification makes sense from a QbD perspective. However, if one already has a fully developed packaged drug product that is compliant with Q3D, it makes little sense to go back and perform the testing required to qualify the container closure system, select or qualify the components, and select the materials, as obviously the system, components, and materials are suited for their intended use if the packaged drug product is suited for its intended use. Similarly, if one has already qualified a container closure system, it makes little sense to go back and perform the testing required to select or qualify the components and select the materials. Thus, it is always the case that qualification at a later stage of development satisfies the need, if such a need exists, to test, for qualification or selection, at an earlier stage of development. This makes intuitive sense as testing performed to qualify a system is inherently more rigorous than the testing performed to select a system's materials of construction and the acceptance criteria for qualification are likely to be more restrictive than the criteria for material selection.

4.1.6.2 Global Standards for Extractable Elements Testing

The reality of the global marketplace is that globally marketed drug products must be tested via local procedures and meet local requirements. It is unfortunate that globalization of tests and specifications have not kept pace with the globalization of the marketplace and the frustrating and inefficient reality is that the same material, component, and system tested and used in one market region may need to be tested by varying and differing means to be used in a different market region.

When regional differences are driven by science or overwhelming practical circumstances, then they may be justifiable, if inconvenient. When regional differences arise because of concepts such as "not invented here" or "that's the way we've always done it (even if we can't remember why)," then they are unjustifiable and unacceptable. Surely it is desirable to drive toward the global alignment of extractable element testing of materials, components, and systems, even if the mechanism for doing so is unclear.

At the time this book is being submitted for print, a monograph in the European Pharmacopeia is being circulated for public comment. This monograph, 2.4.35, is titled: Extractable elements in plastic materials for pharmaceutical use [6]. The monograph is based on:

- Autoclave extraction (121 °C for 1 hour) of 40 g of plastic in 100 ml of 0.1 M hydrochloric acid (HCl).
- Analysis of the resulting extracts by ICP/MS.

Targeted elements and their limits are shown in Table 4.1. Additionally, the Pharm Eur monograph contains method validation procedures and requirements.

Should this monograph be adopted, and if it is based on sound scientific principles, then perhaps it can serve as the global standard for testing materials of construction for extractables elements for the purpose of material selection.

4.1.6.3 Use and Value of Vendor-Supplied Information

The role of the vendor of a material or component in the selection and qualification processes has been previously considered, specifically addressing who could perform

Table 4.1 Target elements and their limits per Pharm Eur 2.4.35.

Target element	Reporting limit, mg/l in the extract
Antimony (Sb)	0.10
Arsenic (As)	0.01
Cadmium (Cd)	0.01
Cobalt (Co)	0.01
Copper (Cu)	0.30
Lead (Pb)	0.01
Lithium (Li)	0.30
Mercury (Hg)	0.01
Nickel (Ni)	0.01
Vanadium (V)	0.01

testing. However, vendors can supply relevant information without actually performing testing. Thus, for example, it is not untypical for vendors to supply potential users of their articles with documents that address elemental impurities by specifying what elements the vendor did not intentionally add to the article. For example, if a vendor certified that "none of the ICH Q3D elements are intentionally added," this could be valuable information for the article's potential users.

However, one must understand the limitations of such certifications, as they do not specify:

1. What elements the vendor has intentionally added,
2. What elements might have been added by others involved further upstream in the manufacturing process,
3. What elements might have been unintentionally added.

Given these limitations, it must be recognized that while "not intentionally added" certificates have value in terms of assessing an item's risk of contributing elemental impurities, at best their contribution provides only an incomplete assessment of risk.

In other circumstances, vendors may supply extractable elements data from controlled extraction studies they have performed or commissioned. As was the case with certificates, this information is valuable and can be the basis for thoughtful material or component selection. However, vendor to vendor comparisons can be difficult if the conditions of extraction are not consistent across vendors. Of course, consistency among vendors would be facilitated by a standardized test procedure.

In summary:

- While sequentially testing materials, components, and systems is logical and appropriate during the packaging system or drug product development process, retroactively testing materials or components in a system that has been tested and found to be satisfactory provides little additional value and should not be required.
- When scientifically feasible and practically appropriate, test methods for extracted elements should be aligned across geographic and market-defined regions.
- While vendor information related to extractable elements is always useful, particularly in regard to selection, the decision to use vendor data should be based on a careful consideration of the conditions under which, and the means by which, the data was generated.

4.1.7 Existing Knowledge Concerning Extracted Elements in Materials and Systems

Global guidelines that address the levels of elemental impurities in finished drug products suggest that packaging and manufacturing systems for such products can be sources of these elemental impurities; however, the guidelines do not provide specifications regarding elemental entities in such systems and/or their materials of construction. Knowledge about elemental entities present in pharmaceutical packaging and manufacturing systems and the leaching characteristics of these entities may facilitate setting meaningful, actionable, and appropriate specifications on the composition and properties of such materials and systems. Two published reviews on this subject contain the information compiled from the available body of literature and consider how such information might be used to set relevant specifications for elemental entities [4, 5]. While such a compiled body of information could be used to address a number of issues and topics, the following questions were the primary focus of this work:

1. What elemental entities are present in the polymers and materials that are typically used in packaging and at what levels are they present?
2. To what extent are elemental entities present in polymers and materials leached from these items under conditions relevant to the manufacturing and storage/distribution of solution drug products?

Considering the topic of elemental entities in general and these two questions specifically, the following observations are made and the following conclusions are drawn:

1. Pharmaceutical packaging and manufacturing systems and their materials of construction contain elemental entities as such entities are either:
   a. Known and intentionally used as materials of construction,
   b. Known and intentionally included as additives in the materials of construction,
   c. Known or logically inferred impurities in the materials of construction and/or their additives;
   d. Known or logically inferred residuals arising from the manufacturing of the systems and/or their materials of construction. The inferred presence of elemental entities in systems and/or their materials of construction have been confirmed by laboratory testing of said materials and systems.
2. The sources of these elemental entities can be grouped into four classes:
   a. Elemental entities that are actual materials of construction (for example, metal oxides in glass, stainless steel),
   b. Elemental entities that are intentionally present in the materials of construction because they are important and necessary components in the materials (e.g. metal stearates as acid scavengers),

c. Elemental entities that may be unintentionally present in the materials of construction because they are used in the generation of the materials and may not be completely removed by the generation process (e.g. residual organometallic catalysts),

d. Elemental entities that are unintentionally, coincidentally, and unpredictably present in the materials of construction because they are ubiquitous contaminants that cannot be avoided.

3. Because elemental entities in classes (a) through (c) above are either intentionally present or expected to be present, their ability to be leached from the materials must be established, especially for products that are capable of leaching (such as liquid formulations). However, because these elemental entities are known, such leachables testing involves quantifying known and targeted entities and not screening for unanticipated entities.

4. Because elemental entities in class (d) above cannot be anticipated, they can only be evaluated by screening. Nevertheless, available data suggests that such unforeseen and unpredictable elemental entities are rarely encountered and when they are encountered their levels in materials and systems are low. Furthermore, when such elemental entities are present in materials and systems, only a small fraction of the total available amount of the entity can be leached under conditions that are relevant to packaged drug products. Thus, while sources of certain elemental impurities may be ubiquitous in the natural environment, they are not ubiquitous in materials used in pharmaceutical packaging and manufacturing systems and when they are present, they are not extensively leached under relevant conditions (see, for example, Table 4.2). Therefore, there are a number of elemental impurities that may be relevant for packaged drug products that are much less relevant as elemental entities in packaging and manufacturing systems. The value of including such elemental entities in tests that are performed on packaging and manufacturing systems (and their materials of construction) for the purpose of establishing or controlling the levels of elemental impurities in drug products is low because the risk that such elemental entities would accumulate as elemental impurities at levels sufficiently high to produce an adverse effect is low. As I have suggested previously [7], this would be the basis for removing irrelevant elements from the list of elements addressed in routine screening of pharmaceutical packaging and manufacturing materials, components, and systems.

5. The available information establishes that certain elemental entities could be relevant targets for assessing the impact of packaging and manufacturing systems on

Table 4.2 Percentage of elements extracted in a controlled extraction study.

Material	Metal	% Extracted
Polyethylene terephthalate	Sb	< 0.01%
Polyethylene terephthalate	Mg	0.05%
	Al	0.001%
	Co	0.003%
	Ge	0.007%
	Sb	0.004%
Polyethylene terephthalate	Cr	0.0009%
	Ni	0.0004%
	Cd	< 0.5%
	Pb	< 5%
	Sb	<1%
Polyethylene	Sb	0.04%
Polyethylene (Low-density)	Mg	0.009%
	Al	0.49%
	Co	0.01%
	Cr	0.18%
	Sb	0.008%
	Zn	1.0%
Polyethylene (High-density)	Al	10.1%
	Ca	87.5%
	Fe	<20%
	Ti	0.026%
	Zn	25.6%
Polyethylene (High-density)	Al	2.7%
	Cr	0.054%
	Zr	0.002%
Poly (vinyl chloride)	Zn	1.8–10.8%
	Ba	< 0.07%
	Cd	< 1.4%
	Sn	< 0.93%
	Pb	< 0.49%
Poly (vinyl chloride)	Al	4.8%
	Zn	3.8%
Poly (vinyl chloride)	Zn	0.25%
	Ca	0.47%
Ethylene vinyl acetate	Zn	6.7%
	Ba	< 0.26%
Polypropylene	Zn	< 2.1%
	Ba	<0.39%
Polypropylene	Al	0.004%
	Cr	0.077%
Glass	Al_2O_3	<0.005%
	SiO_2	<0.005%
	B_2O_3	<0.005%
	N_2O	<0.005%

(Continued)

Table 4.2 (Continued)

Material	Metal	% Extracted
Acrylonitrile butadiene styrene (ABS)	Mg	0.002%
	Zn	0.067%
Polycarbonate	Ca	41.7%
	Zn	15.4%
Poly(methyl methacrylate) (PMMA)	Mg	0.19%
	Zn	0.40%
Polystyrene	Mg[a]	0.004%
	Mg[a]	22.6%
	Al	0.17%
	Co	0.17%
	Cu	0.18%
	Zn[a]	1.2
	Zn[a]	21.7%

a) These values were obtained using two different extraction solvents.

the safety of the packaged drug product. Furthermore, this same information is sufficient to establish that many elemental entities are not relevant targets for this purpose. Lastly, this information could be used to support the development of acceptance criteria for elemental entities in systems and their materials of construction.

These two broad and comprehensive surveys of the literature establish that in most cases the levels of elemental entities in materials and components used in packaging and manufacturing systems is low and that the fraction of these elemental entities that leach into either the packaged drug product or the manufacturing process stream and become actual elemental impurities is small. Thus, it is unlikely that manufacturing and packaging systems are major sources of elemental impurities in drug products. While this statement is generally true for pharmaceutical packaging, it is almost always true for single-use components used in biopharmaceutical manufacturing. The validity of these statements, supported by the information obtained in the cited review articles, suggests that screening of packaging and especially manufacturing systems for all possible extractable elements generally serves no useful purpose, as packaging and manufacturing systems are not generally an important source of elemental impurities in drug products. Nevertheless, a comprehensive risk management process for drug product safety and quality should consider the potential for elemental impurities from drug product packaging and manufacturing systems.

4.2 Ionic Extractables

It comes as no surprise to material scientists that polymeric materials used in pharmaceutical applications are sources of extracted inorganic and low molecular weight organic acids. For example, the chloride in poly(vinylchloride), PVC, is a clear signal that this material is a source of extractable chloride, especially when exposed to sterilizing radiation. The formation of acetic and formic acids from polyolefin materials after irradiation is well documented [8–10]. Brominated and chlorinated butyl rubbers are obviously potential sources of extractable bromide and chloride. Polymeric materials stabilized with phosphorous-containing antioxidants (e.g. Irganox 168), are likely sources of extractable phosphate, again especially after irradiation. Lastly, it is reasonable to expect the presence of extractable sulfate from pharmaceutical packaging items containing adhesives, such as multi-layered films and labels.

Ion chromatography (IC) has been applied to the screening of aqueous extracts for common inorganic anions (chloride, bromide, nitrate, sulfate, phosphate) and low molecular weight organic acids (e.g. acetic and formic acids) [11–13]. In many ways, screening pharmaceutical items for extractable anions fulfills the analytical chemist's fondest dreams. As the anions are highly water soluble (they are, after all, ionic), water is the ideal extraction solvent, providing the worst-case extracted ions profile. Since water is an ideal and analytically expedient sample matrix for IC analyses, extracted anions can be screened in water extracts with high precision, accuracy, and sensitivity.

Screening drug products for extracted anions is a bit more analytically challenging, depending on the drug product's composition. Clearly non-aqueous or aqueous drug products containing significant quantities of organic additives present a challenge, primarily due to matrix interferences and incompatibility with IC columns and instrumentation. Dilutions required to reduce the interferences or incompatibilities to manageable levels render the IC sensitivities inadequate for trace analysis. Even aqueous drug products can be challenging if they intrinsically contain the anions as intentional ingredients. For example, many drug products are formulated with saline (sodium chloride) diluents, have active ingredients that are hydrochloride salts (to improve aqueous solubility), or are pH adjusted with hydrochloric acid. Not only do these situations complicate the determinations of extractable chloride, as the levels of extracted chloride are much lower than the drug product's intrinsic chloride level, but the large chloride peak in the chromatogram may interfere with the determination of closely eluting anions. A similar problem can be encountered with acetate and phosphate, as salts of these ions are used in drug products as pH buffers.

Analytical issues notwithstanding, IC screening is not universally practiced in extractables and leachables screening for the simple reason that the anions that are addressed are generally regarded as safe. For example, both acetic and formic acid are classified as relatively safe residual solvents in ICH Q3C, with PDEs of 50 mg/day [14]. In general, the quantity of these anions that could leach are sufficiently low that this safety threshold would be nearly impossible to reach. Chloride does not pose a safety issue, since, as noted above, drug products can contain chloride in quantities well in excess of 1000 mg/l and phosphate in lesser amounts, quantities that are rarely, if ever, achieved by these extracted anions. Thus, information on extracted or leached anions is rarely significant from a safety assessment perspective.

However, knowledge of extractable anions may forecast drug product stability issues. As the extracted anions are acids, their leaching into a poorly buffered drug product could change the pH of the drug product sufficiently that its stability and compositional compliance is adversely affected. Phosphate is a precipitating anion and if the drug product contains alkaline earth metals (particularly calcium), phosphate salts may form particulate matter.

Based on these considerations, the routine use of IC testing for screening of extracts and drug products for anions is not recommended. However, if the pharmaceutical packaging system consists of components that are sterilized by irradiation or that are likely sources of specific anions, then this may be a sufficient cause to merit anion screening.

4.3 Scouting via General Chemical Analysis

4.3.1 General Discussion

Extracts and extraction blanks are screened and/or profiled by relevant and appropriate analytical methods to establish the identity and concentration of all extracted substances at levels higher than an established threshold. This is the case as extractables' identities and concentrations are essential information required to assess the effect of extracted substances (should they become leachables) on certain critical drug product quality attributes, including safety, efficacy, stability, and compliance.

One of the challenges faced by an analytical chemist performing extractables screening is assessing the completeness of the screening process. Although extractables screening strategies based on multiple, orthogonal, and complementary analytical methods are broad in their scope and may be appropriately sensitive and information-rich, there is no guarantee that even the most well-conceived and the most effectively implemented analytical strategy can respond to, and provide reliable information for, all possible extractables. If the "guarantee of completeness and correctness" cannot be supplied by the analytical strategy itself and the analytical methods themselves, then the "guarantee" must be obtained by other means, as the possibility that key extractables have been missed or have been improperly identified or quantified leads to ineffective, incomplete, or flawed extractables impact assessments.

Additionally, it can be an analytical challenge to identify an "unknown" extractable if nothing is known about the general chemical nature of the extractable. "Simple" compositional knowledge, such as "the unknown is an acid," may facilitate the identification process by suggesting relevant analytical approaches and by providing a "differentiator" between candidate structures.

Thus, extracts and extraction blanks may be tested for certain chemical properties or characteristics that are not related to single extractables but which may be associated with a certain group of extractables. This process is referred to as scouting, where the intent of scouting is to establish the general characteristics of extractables or leachables as a group. These general characteristics may relate to the total amount of extractables or the general chemical characteristics of the extractables. Such tests are termed General Chemistry tests and typically include pH (or some other measure of acidity or alkalinity), UV absorbance, and Total Organic Carbon (TOC). Historically, conductivity and gravimetric measurements, such as non-volatile residue (NVR), have also been applied as General Chemistry tests, although they were not necessarily described using this terminology.

Although the general chemical properties or characteristics may not be as directly relatable to product quality attributes as are extractables' identities and concentrations, they may be useful in extractables profiling as they are indicative of the nature and/or amount of certain extractables or groups of extractables. As such, the General Chemistry data may provide insights that guide extractables profiling in terms of the testing required to "find" the type and to measure the amount of organic extractables that should be contained in a complete profile of organic extractables. Furthermore, the General Chemistry data can be used to reconcile, either qualitatively or quantitatively, the extractables profile, where the term "reconcile" is defined as the process of comparing the General Chemistry data with the extractables profile and establishing whether they are consistent.

The value derived from performing these tests is as follows:

1. pH: Differences in pH, extraction blank versus an extract, may indicate that either acidic or basic substances have been extracted. Titrimetric measurements such as acidity and alkalinity can provide an estimate of the total amount (concentration) of acidic or basic extractables.
2. UV absorbance: Differences between the UV absorbance of an extract and an extraction blank are attributable to organic extractables. The nature of the UV absorption spectrum and the magnitude of the absorbance at specific wavelengths may be indicative of the nature (e.g. aromatic versus aliphatic) and concentration of certain organic extractables.
3. TOC: The magnitude of the TOC difference, extract minus extraction blank, reflects the total amount of measurable extracted organic carbon. Depending on the means by which the TOC measurement is performed and the way the results are calculated, the measured and reported TOC may reflect some or all of volatile, semi-volatile, and non-volatile organic extractables.

The following guidelines help ensure that the General Chemistry data is reported, discussed, and interpreted in a consistent and appropriate manner.

4.3.2   pH

The concept that the difference in pH between an extract and an extraction blank is indicative of the extraction of acidic or basic extractables is straightforward. If the pH of the extract is less than the pH of the extraction blank, then acids have been extracted, and if the pH of the extract is greater than the pH of the extraction blank, then bases have been extracted.

However, the interpretation of pH differences, extract versus extraction blank, becomes more complicated in practice for the following reasons:

1. The extraction of both acids and bases would tend to "cancel each other out" and would complicate the interpretation of the pH difference.
2. The logarithmic nature of the pH scale complicates the interpretation of "delta pH," where delta pH is the magnitude of the difference extract pH – extraction blank pH. For example, consider the following two cases: Case 1, extraction blank pH = 5, extract pH = 4; Case 2, extraction blank = pH 3, extract pH = 2. While the delta pH for these two scenarios are both −1, the amount of acid extracted in case 2 is 100 times greater than the amount of acid extracted in case 1.

3. Certain media used as extraction solvents are such that they tend to mask the presence of acidic or basic extractables. For example, some extraction solvents are buffered and thus the buffer would "absorb" either hydrogen or hydroxide ions without changing pH substantially. In essence, the buffer masks the acidic or basic extractables. Additionally, some extraction solvents may have such a low or high pH that only large amounts of acidic or basic extractables would cause a measurable pH difference.
4. The relationship between the delta pH and the concentration of acidic or basic extractables is simple if the extracted substances are strong acids or strong bases. However, many of the organic extractables typically encountered in pharmaceutical applications are weak acids and bases and thus the relationship between delta pH and extractable concentration is more complicated, requiring a consideration of the extractable's acid or base dissociation constant.
5. Factors other than the extraction of acids or bases may affect an extract's pH. For example, the pH of a basic solution open to the atmosphere will decrease over time as the solution absorbs atmospheric carbon dioxide. Thus, if the atmospheric exposure of the extract and the extract blank differ, then a difference in pH can arise even in the absence extracted acids or bases.
6. Although pH measurement within an extract is highly reproducible, pH measurements may vary somewhat across replicate extracts. This variation should be taken into account in terms of reporting pH differences, specifically addressing the question "how large must the delta pH be for it to be real?"

For these reasons:

1. The delta pH or pH of an extract is never reported alone and is always accompanied by the pH of the extracts and the extraction blank. For example,

	Measured pH		Delta
Extraction medium	In extracts	In extraction blank	pH[a]
Solvent A	3.00	4.00	−1.00

a) Delta pH = pH in extracts – pH in extraction blank.

The delta pH is rarely used as a quantitative means of reconciling an extractables profile. Rather, delta pH or pH difference is properly used as a qualitative means of guiding the extractables profiling process (by suggesting that acids or basic extractables may be present in certain approximate amounts) or as a qualitative means of evaluating the completeness of the extractables profile (e.g. "the pH difference suggested the presence of small levels of acidic acids and the extractables profile included small levels of acetic acid").

2. The minimum reportable delta pH value is 0.1 unit. Delta pH values less than 0.1 are within the imprecision of the measurement and cannot be confirmed to be a meaningful difference, extract versus extraction blank.

4.3.3 UV Absorbance

An increase in UV absorbance, extract versus extraction blank, is an indication that organic extractables, which themselves absorb UV radiation, have been extracted. To a limited extent, the wavelengths at which the increase occurs may be indicative of the structure of the organic extractables, although in general a UV absorption spectrum lacks the specificity and diversity to be a particularly powerful tool for securing an extractable's identity. Although the magnitude of the increase is related to the amounts of the extractables, the proportionality between UV absorbance and the extractable's concentration varies widely across extractables. Thus, the following points are relevant:

1. Net absorbance, an extract–extraction blank, is always reported to be consistent with the following points. There is little value, in terms of extractables profiling, in reporting the actual absorbances of the extract and the extraction blank.
2. Negative net results, where the absorbance of the extraction blank is higher than the absorbance of the extracts, are not reported. Rather, this result is reported as "ND, No discernible increase in absorbance due to extraction." Slightly negative net results can reasonably be anticipated (due to analytical variation) when there is really no difference in UV absorbance, extract versus extraction blank. However, large negative net results strongly suggest an analytical issue, which could include contamination of the extraction blank.
3. Absorbance values should be reported to three decimal places, e.g. 0.002. The analytical uncertainty and imprecision in the UV absorbance is sufficiently large that a fourth decimal place is not significant.
4. The minimum reportable net absorbance value is 0.005 units. Net absorbance values less than 0.005 are within the imprecision of the measurement and cannot be confirmed to be a meaningful difference, an extract versus extraction blank. If the net absorbance value is less than 0.005 units, then the net absorbance is reported as "less than 0.005 absorbance units (< 0.005)." If the net absorbance is 0.005 units or greater, then the net result is reported. The following is an example of the reporting of net absorbances:

	Net absorbances[a]			
Extraction medium	**At 210 nm**	**At 220 nm**	**At 240 nm**	**At 280 nm**
Solvent A	0.010	< 0.005	N/A[b]	ND[c]

a) Net absorbance = absorbance extract – absorbance extraction blank = absorbance due to extractables.
b) N/A = not applicable, neither the extract nor the extraction blank had a discernible absorbance at this wavelength.
c) ND = no discernible increase in absorbance due to extraction.

5. The net absorbance is only cautiously used as a quantitative means of reconciling an extractables profile. Rather, net absorbance is properly used as a qualitative means of guiding the extractables profiling process (by suggesting that UV active extractables may be present in certain approximate amounts) or as a qualitative means of evaluating the completeness of the extractables profile (e.g. "the net absorbance suggested the presence of small levels of aromatic extractables and the extractables profile included small levels of DEHP"). In general, the simpler the profile (i.e. the fewer the number and the lesser the chemical diversity of the extractables), the more effective a UV-based reconciliation can be.

4.3.4 TOC

4.3.4.1 Use of TOC in Screening

An increase in levels of TOC, extract versus extraction blank, is an indication that organic extractables have been extracted. TOC differs from the other general chemistry parameters (pH and UV absorbance) in two important regards:

1. Unlike pH and UV absorbance, which may provide an indication of the chemical nature of the extractable, TOC provides no such indication.
2. While pH and UV absorbance are largely qualitative measures of the concentration of extractables, TOC provides a more quantitative measure of the total amount of organic extractables.

Thus, the following points are relevant:

1. Net TOC, extract–extraction blank, is always reported consistent with the following points. There is little value, in terms of extractables profiling, in reporting the actual TOC values for the extract and the extraction blank.
2. Negative net results, where the TOC of the extraction blank is higher than the TOC of the extract, are not reported. Rather, such a result is reported as "ND, no discernible increase in TOC due to extraction." It is noted that slightly negative net results can reasonably be anticipated (due to analytical variation) when there is really no difference in TOC levels, extract versus

extraction blank. However, large negative net results strongly suggest an analytical issue, which could include contamination of the extraction blank.

3. TOC values should be reported to no more than two decimal places, e.g. 5.42 mg/l, and no more than three significant figures, e.g. 5.42 mg/l, 10.2 mg/l, and 109 mg/l. The analytical uncertainty and imprecision in the TOC measurement are sufficiently large that a third decimal place or a fourth significant figure is not justified.

4. The minimum reportable net TOC level is 0.2 mg/l. Net TOC levels less than 0.2 mg/l are within the imprecision of the measurement and cannot be confirmed to be a meaningful difference, extract versus extraction blank. If the net TOC value is less than 0.2 mg/l, then the net TOC is reported as "less than 0.2 mg/l (<0.2 mg/l)." If the net TOC level is 0.2 mg/l or greater, then the net result is reported. The following is an example of the reporting of net TOC values:

Extraction medium	Net TOC, mg/l[a]
Solvent A	2.3
Solvent B	<0.2 mg/l
Solvent C	NA[b]
Solvent D	ND[c]

a) Net TOC = TOC in the extract – TOC in the extraction blank = TOC due to extractables.
b) NA = not applicable, neither the extract nor the extraction blank had a measurable level of TOC.
c) ND = no discernible increase in TOC due to extraction.

5. The net TOC can be used to qualitatively reconcile an extractables profile. Thus, net TOC can be used as a qualitative means of guiding the extractables profiling process (by suggesting that organic extractables may be present in certain amounts) or as a qualitative means of evaluating the completeness of the extractables profile (e.g. "the net TOC was 10 mg/l, consistent with the approximately 15 mg/l of total measured individual organic extractables").

6. In certain situations, net TOC can be used as a quantitative means of reconciling an extractables profile. For example, net TOC can be used as a quantitative means of guiding the extractables profiling process (by suggesting that organic extractables may be present in greater or lesser amounts than were detected by screening methods). Additionally, a quantitative TOC reconciliation can be used to establish the level of completeness of the extractables profiling. Such a process is discussed in greater detail later in this section. Finally, a quantitative TOC reconciliation can be used to estimate the concentration of unknown (unidentifiable) extractables.

4.3.4.2 Performing a TOC Reconciliation

The TOC reconciliation is calculated as follows:

1. The TOC concentration of an individual extractable (TOC_e) is calculated as the product of the extractable's measured concentration in the extract (C_m) and the extractable's carbon fraction (as derived from the extractable's empirical formula, C_f):

$$TOC_e = C_m \times C_f \qquad (4.1)$$

2. The calculated TOC for the entire extractables profile (TOC_{calc}) is obtained by adding the individual TOC concentrations (TOC_e) for all extractables:

$$TOC_{calc} = \sum TOC_e \qquad (4.2)$$

3. The TOC reconciliation (TOC_{rec}), as a percent, is calculated as the ratio of the calculated TOC for the entire profile (TOC_{calc}) and the net measured TOC (TOC_{meas}):

$$TOC_{rec} = (TOC_{calc}/TOC_{meas}) \times 100\% \qquad (4.3)$$

4.3.4.3 Using TOC Reconciliation to Judge the Completeness of Extractables Profiling

Once the TOC reconciliation has been calculated, the inevitable question follows: "What range in TOC reconciliation values represents a complete extractables profile?" Generally, TOC reconciliation is insufficiently precise to lend itself to generalization in terms of "what range of TOC reconciliations means that the profiling activity is done?" Rather, the TOC reconciliation is a semi-quantitative tool that is most effectively used to establish those cases when the extractables profiling clearly is not complete or is inaccurate. In this context, it is suggested that [15–17]:

1. If the TOC reconciliation is less than 50%, this is a clear indication that the extractables profile has not been fully delineated and that there are additional organic extractables needing to be found.

2. If the TOC reconciliation is between 75 and 125%, this is a clear indication that the extractables profile has likely been effectively delineated and that all major organic extractables have been found and accounted for.

3. If the TOC reconciliation is greater than 150%, this is a clear indication that the extractables profile has not been effectively delineated. Such a result typically would suggest that the individual extractables' concentrations have been overestimated.

4.3.4.4 Using Unreconciled TOC to Estimate the Levels of Unknown Extractables

The difference between TOC_{meas} and TOC_{calc} represents the TOC that is due to substances that were unaccounted for in the reconciliation, which are typically those substances identified as being "unknowns" in the extractables profiling. The concentration of unknown extractables is this difference multiplied by the unknown's carbon fraction:

$$\text{Concentration of unknowns} = (\text{TOC}_{\text{meas}} - \text{TOC}_{\text{calc}}) \div C_{\text{f}}$$

(4.4)

Of course, the carbon fraction of the unknowns is itself unknown as it depends on the unknown's chemical formula. In the absence of any information about the unknown's structure, a carbon fraction of 0.60 can be used as a fairly representative value for substances known to be extractables.

For example, consider the case where the measured TOC is 5 mg/l and the total calculated TOC of all measured organic extractables is 3 mg/l. In this case the concentration of the undiscovered or unknown extractables becomes:

$$\text{Concentration of unknowns} = (\text{TOC}_{\text{meas}} - \text{TOC}_{\text{calc}}) \div C_{\text{f}}$$
$$\text{Concentration of unknowns} = (5\,\text{mg/l} - 3\,\text{mg/l}) \div 0.6$$
$$\text{Concentration of unknowns} = 3.3\,\text{mg/l}$$

When reporting a concentration of unknowns calculated in this manner, one must be aware of the accumulated uncertainty in the calculation. What this means is that the TOC calculation carries with it all the accumulated inaccuracies of all the individual TOC measurements for the individual extractables. The greater the number of extractables that are reconciled, the larger is the uncertainty in the TOC reconciliation.

References

1 ICH Q3D (R1) (2020). Elemental impurities: Guidance for industry. US Department of Health and Human Services: Food and Drug Administration; Center for Drug Evaluation and Research (CDER), Center for Biologics Evaluation and Research (CBER).

2 USP <232> (2017). Elemental impurities – limits. First Supplement to USP 40–NF 35, pp. 8065–8069.

3 Jenke, D.R., Stults, C.L.M., Paskiet, D.M. et al. (2015). Materials in manufacturing and packaging systems as sources of elemental impurities in packaged drug products: A literature review. *PDA J. Pharm. Sci. Technol.* 69 (1): 1–48.

4 Jenke, D. (2020). Materials in manufacturing and packaging systems as sources of elemental impurities in packaged drug products: An updated literature review. *PDA J. Pharm. Sci. Technol.* 74 (3): 324–347.

5 USP <233> (2017). Elemental Impurities – Procedures. First Supplement to USP 40–NF 35, pp. 8069–8072

6 Ph Eur. 2.4.35. Extractable elements in plastic materials for pharmaceutical use. *Pharmeuropa*, 32 (2). PA/PH/Exp. 16/T (19) 22 ANP.

7 Jenke, D. (2019). How one might experimentally determine if container closure systems and their components and materials of construction contribute elemental impurities to packaged pharmaceutical drug products. *PDA J. Pharm. Sci. Technol.* 73 (2): 181–190.

8 Jenke, D., Zietlow, D., and Sadain, S. (2004). Leaching of organic acids from irradiated EVA plastic as a function of solution pH and polarity. *PDA J. Pharm. Sci. Technol.* 58 (1): 24–32.

9 Azuma, K., Tanaka, Y., Tsunoda, H. et al. (1983). Identification of the volatiles from low density polyethylene film irradiated with an electron beam. *Agric. Biol. Chem.* 47: 855.

10 Yao, S. (1984). Detection of odorous compounds in plastic packaging materials induced by the electron beam irradiation and the preventative measures. *Shipin Yu Fajiao Gongye.* 2: 8.

11 Tsui, V., Somma, M.S., and Zitzner, L.A. (2009). Leachables evaluation for bulk drug substance. *PDA J. Pharm. Sci. Technol.* 63 (2): 168–183.

12 Ding, W. and Nash, R. (2009). Extractables from integrated single-use systems in biopharmaceutical manufacturing. Part 1: Study on components (Pall Kleenpak™ connector and Kleenpack filter capsule). *PDA J. Pharm. Sci. Technol.* 63 (4): 322–338.

13 Jenke, D.R., Zietlow, D., Garber, M.J. et al. (2007). Accumulation of organic compounds leached from plastic materials used in biopharmaceutical process containers. *PDA J. Pharm. Sci. Technol.* 61 (4): 286–302.

14 Q3C: Tables and list; Guidance for industry (2018). US Department of Health and Human Services: Food and Drug Administration, Center for Drug Evaluation and Research (CDER), Center for Biologics Evaluation and Research (CBER). ICH, Revision 4.

15 Jenke, D., Couch, T.R., Robinson, S.J. et al. (2014). The use of TOC reconciliation as a means of establishing the degree to which chromatographic screening of plastic material extracts for organic extractables is complete. *PDA J. Pharm. Sci. Technol.* 68 (3): 256–270.

16 Tribon, V. and Creasey, J. (2013). Measuring potential leachables in single-use assemblies using total organic carbon (TOC) analysis. *Am. Pharm. Rev.* 16 (3): 18–26.

17 Jenke, D. (2009). Reconciliation as a survey tool. In: *Compatibility of Pharmaceutical Products and Contact Materials; Safety Considerations Associated with Extractables and Leachables*, 140–142. Hoboken, NJ: Wiley.

5

Extractables, Leachables, and the Product Lifecycle Through Product Approval and Launch

5.1 Introduction

A pharmaceutical product or medical device has a well-defined lifecycle, starting with concept generation, moving though development, securing registration, achieving launch, realizing a fixed period of sustainable sales, and culminating with its ultimate retirement (discontinuation in the market). It is reasonable and correct to anticipate that the issues associated with the product's contact with items (and both the means of addressing those issues and the requirements that the product must meet) are different at the various stages of the product's lifecycle. It is also clear that certain activities related to compatibility assessment are most effectively performed at certain stages in a product's lifecycle. Indeed, as a product proceeds from concept (as a potential product candidate) through prototyping, registration and approval, commericalization, and ultimately retirement, it is not uncommon or inappropriate for that product's compatibility issues to have been addressed in several different ways, in several different contexts, and with several degrees of rigor. For example, at the product concept stage, the focus of a compatibility assessment may take the form of a screening evaluation (i.e. provide answers to the question of whether the material has any general chemical characteristic, including composition, that would suggest that it would be unsuited for use in the intended application). It is clear from a cost and time perspective that an in-depth compatibility assessment (leachable's identification and accumulation levels and toxicological assessment thereof) may not be appropriate at the concept stage as moving the product from concept to reality may include the consideration of many materials, some of which ultimately will be rejected as viable candidates for use with the product for any number of reasons. Rather, the concept phase should be that time in the compatibility assessment process where inappropriate materials are quickly identified and eliminated from consideration. It is also equally clear that at the time of a product's registration, a general screening compatibility approach, which might have been appropriate at the concept stage, will not be sufficient to secure regulatory approval.

Thus, a product and its associated contact items may undergo several levels or phases of chemical testing as the product moves through its lifecycle.

The processes for performing a compatibility assessment are very clearly linked to where the product is in its lifecycle. Thus, the product lifecycle serves, in fact, as a Master Flow Diagram for chemical compatibility assessments. This relationship is illustrated in Figure 5.1.

It is the intent of this chapter to describe and discuss a chemical compatibility assessment process that links assessment activities to that point in the product lifecycle where the activities are most efficiently and effectively performed. While this will be illustrated by taking a specific pharmaceutical product through its lifecycle, the process and its discussion is relevant to most pharmaceutical products and medical devices.

5.2 Situation

The situation that our discussion will consider is an organization (the manufacturer) that markets liquid pharmaceutical products that are packaged in container-closure systems that consist, in whole or in part, of plastic or polymeric materials. While the liquid pharmaceutical product and its packaging are in contact, the product and the packaging will interact. The interaction of specific concern to this discussion is the leaching of substances present in the packaging system into the liquid pharmaceutical product, as the presence of such packaging-related foreign impurities (leachables) in the product could adversely affect its critical quality attributes, such as safety, efficacy, and/or stability, thereby making the product and its packaging incompatible.

Establishing the potential impact of leachables on the quality attributes of a packaged liquid pharmaceutical product can be accomplished by two means. On the one hand, the impact of leachables on a quality attribute can be

Extractables and Leachables: Characterization of Drug Products, Packaging, Manufacturing and Delivery Systems, and Medical Devices, First Edition. Dennis Jenke.
© 2022 John Wiley & Sons, Inc. Published 2022 by John Wiley & Sons, Inc.

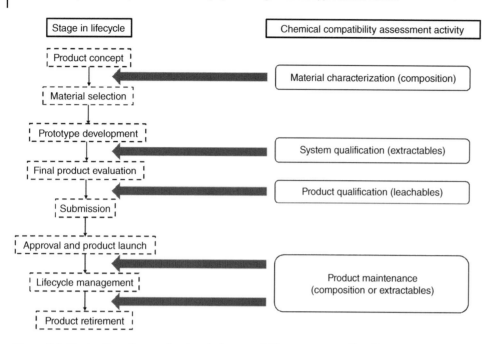

Figure 5.1 Master flow diagram for chemical compatibility assessment. The diagram linking stages in a pharmaceutical product's lifecycle with actions and activities performed as part of the chemical compatibility assessment.

established by directly measuring the quality attribute over the period of time that the liquid pharmaceutical product and the packaging system are in contact (referred to as the product's shelf-life). Thus, for example, the effect of leachables on the stability of a liquid pharmaceutical product can be assessed by measuring the stability-indicating attributes of the liquid pharmaceutical product over its shelf-life and establishing that those attributes meet their associated specifications over the entire shelf-life. Alternatively, the impact of leachables on a quality attribute can be inferred if all the leachables are individually identified and quantified and the effect of each individual leachable is established. Thus, for example, the safety impact of leachables can be established by identifying and quantifying all leachables present in the packaged liquid pharmaceutical product above a defined "no effect" threshold and then determining the safety effect of the individual leachables via a toxicological safety risk evaluation. As it is reasonable to expect that the levels of leachables may change over the course of the shelf-life of a packaged liquid pharmaceutical product, leachables identification and quantitation will likely be performed multiple times over the course of the product's shelf-life.

The process of developing a packaged liquid pharmaceutical product, from concept to registration, is typically long and multi-faceted. Furthermore, it is envisioned that once a packaged liquid pharmaceutical product has been successfully registered it will be marketed for an extended period of time. Clearly, the packaged liquid pharmaceutical

product must be qualified as being suitable for its intended clinical use by the end of its development and the qualification must be maintained throughout the product's market lifetime. Considering safety qualifications specifically, one can envision a one-time process that involves establishing the leachables profile of a packaged liquid pharmaceutical product over its shelf-life during registration stability studies and then toxicologically assessing each leachable by considering the highest level it reaches over its shelf-life. Although such a process alone may constitute an effective and efficient compatibility assessment in certain circumstances, it is typically the case that most circumstances require that such a process be augmented by other activities. For example, consider the aspect of risk management. Registration stability studies are typically performed late in the product development process. If the only means employed to address the potential safety risk of packaging were to perform leachables testing "on stability," then the first opportunity to surface a potential safety issue is after product development is essentially complete. Should it be the case that one or more leachables are surfaced that have a potential safety issue, finding this out after development is complete is a poor means of risk management, as the cost and time delay associated with redevelopment to manage a potentially troublesome leachable would be most unfortunate. Thus, if one were to consider safety qualification to be a process consisting of multiple events instead of a single event, the proper process would include actions prior to leachables profiling that mitigate the risk of impactful leachables being surfaced during registration stability testing.

Additionally, it is noted that it is not uncommon for packaging systems to undergo changes during the time that a packaged liquid pharmaceutical product is "on the market." Supply disruptions, ongoing product improvement, vendor-related initiatives and other circumstances can all lead to the modification of the marketed product's packaging system and/or its materials of construction. When such changes occur and are implemented in a marketed product, it is reasonable to expect that adoption of the changes should be predicated on establishing that the changes do not alter the product's leachables profile, as an alternation of the leachables profile might adversely impact the product's compatibility attributes of stability, efficacy, and safety.

Furthermore, a strategy based on "leachables testing only" presumes that the activities required in leachables testing can be readily accomplished. Specifically, the "leachables testing only" strategy presumes that the leachables can be discovered in the liquid pharmaceutical product and that the discovered leachables can be identified and quantified in the liquid pharmaceutical product. Since leachables typically are present in liquid pharmaceutical products at low levels, as the chemical complexity of the liquid pharmaceutical product increases (i.e. as the product contains more chemically complex ingredients in higher amounts), the activities of discovery, identification, and quantitation become more and more challenging, as the process of leachables profiling becomes akin to "finding an unknown number of needles of varying (and unspecified) sizes and shapes in a haystack." Clearly, when leachables profiling becomes practically prohibitive, it is reasonable to consider and adopt alternate and enabling approaches to safety qualification.

Lastly, multiple liquid pharmaceutical products may be packaged as a single packaging system. Alternatively, packaged liquid pharmaceutical products may be marketed in various configurations (e.g. container sizes). In either circumstance it may be more efficient to adopt a safety qualification process that is linked to the packaging system and not to the packaged pharmaceutical product, as there is only a single packaging system but multiple packaged products.

To address this situation, a strategic process for performing a chemically compatibility assessment for packaged liquid drug products and their packaging systems over the entire product lifecycle, starting from ideation and ending with retirement (i.e. from cradle to grave), is discussed, focusing specifically on the compatibility parameter of patent safety.

5.3 Definitions

A **material** is an entity, consisting of a single substance, that is used in a packaging system. While a material may perform a simple function, it must typically be fashioned into a component to perform a more complex function. In the situation considered herein, the drug product manufacturer does not produce materials and thus those materials are supplied to the manufacturer by an external vendor.

Considering this definition, it is clear that a material includes many disparate items. For example, a vendor may supply the manufacturer with a resin (typically of a single substance) that the manufacturer then processes by some means to produce a film or part used in a packaging system. An example of this type of material would be a polypropylene resin that the manufacturer extrudes to produce the sheeting that is used as the container body of a packaging system. Alternatively, a vendor may supply the manufacturer with a processed part that the manufacturer uses to construct a packaging system. An example of this type of material is an elastomeric septum, which is used as an injection site in a packaging system.

A **component** is a part, used in a packaging system, that consists of one or more materials. A component may perform a simple function but typically must be assembled into a system to perform a more complex function. Components can either be supplied to the manufacturer by an external vendor or produced by the manufacturer from materials it has obtained from external vendors.

For example, if a vendor supplied the manufacturer with a multi-layered film that the manufacturer fashioned into a packaging system's container body, then the multi-layered film would be a component. Similarly, if the manufacturer itself fashioned the multi-layered film from individual resins it received from vendors, then the multi-layered film would be a component.

A **system** is an entity, consisting of multiple materials or components, that performs a specific overarching function. Systems can be supplied to the manufacturer by an external vendor or produced by the manufacturer from materials and components that it either purchased from an external vendor or produced itself.

A **packaging system** is the sum of components and materials that perform the overarching function of containing and protecting the drug product that is in the packaging system. For example, a simple packaging system might consist of a minimal number of components such as a vial and its closure (stopper). More complex packaging systems, such as a pre-filled syringe, might consist of a larger number of components.

It is important to note that while a material and a component may perform a specific function within a system, the material or component cannot perform the full function of the system. It is this point that differentiates a component from a system.

5.4 The Chemical Compatibility Assessment Process

In developing the chemical qualification process (strategy), one leverages the key stages in a packaged liquid pharmaceutical product's lifecycle, as were previously noted in Figure 5.1. While this figure links these stages to specific chemical assessment actions and activities in general, more detail concerning the chemical compatibility assessment process is illustrated in Figure 5.2. The process is based on the concept that the strategy should require a specific and appropriate action at each one of the key stages in a product's lifecycle, as performing the right activity at the right stage is the most efficient and effective means of risk assessment and risk management. Rather than addressing and retiring all the safety risk in a single comprehensive and complex step, the process seeks to address and remove risk in stages by means that are optimized by matching their complexity with their value.

The key activities in the process are considered, in general terms, in the following sections.

5.4.1 Material Selection

From the perspective of the chemical compatibility assessment of a packaging system, the first significant event in early-stage product development is the selection of the materials that could be used to construct the packaging system and/or its components. In the same way that the proper materials are selected based on their functional characteristics, proper materials should also be selected based on their potential safety impact as so doing reflects the simple observation that the best means of ensuring a safe packaging system is to use safe materials of construction. Thus, the operational question for the chemical compatibility assessment is not whether materials of construction should be tested for safety-indicating chemical properties but rather which chemical properties are appropriate and necessary as safety indicators. Given that this activity occurs early in product development, it is unlikely that the individual materials will be subjected to comprehensive extractables testing followed by a toxicological safety assessment. Rather, a material's composition is reviewed, typically to identify potential "bad actors."

5.4.2 System Qualification

The next significant milestone in the development of a packaging system occurs when development is essentially complete, prototyping has stopped, and the design, composition, and construction process of the packaging system is "frozen." It is logical and appropriate that the extensive process of generating the system's extractables profile is initiated once the freeze has occurred. Extractables profiling prior to such a design freeze is extractables profiling that will either have to be repeated or augmented as the design changes during development.

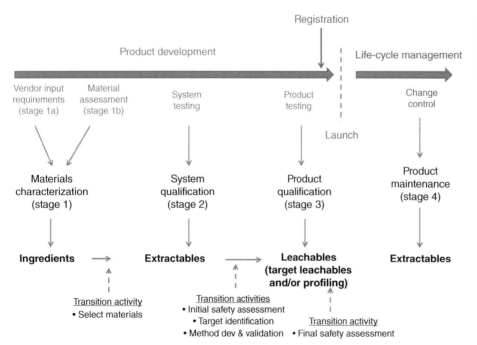

Figure 5.2 The chemical compatibility assessment process applied to a family of liquid drug products and their packaging system.

Extractables profiling of the packaging system is that stage in the chemical compatibility assessment process where the safety risk of the packaging system is most directly and rigorously addressed. This is true as system testing can provide an accurate indication of the safety of the packaged liquid pharmaceutical product and is the regulatory "point of entry" for safety assessment (meaning that regulatory submission for packaged liquid pharmaceutical products requires information related to the packaging system's extractables profile).

5.4.3 Product Qualification

Although the packaging system development activity is complete after the design freeze, the qualification of the packaged liquid pharmaceutical product is not complete until the product has been placed into its packaging system and established to be compatible with that system, as registration of the product with the proper regulatory agency requires compatibility data. Thus, that point in time where the compatibility of the packaging system and the product is demonstrated from a product stability perspective (i.e. stability studies) is the proper point in time to demonstrate compatibility from a product safety perspective. The proper means of establishing compatibility from a safety perspective is to test the packaged liquid pharmaceutical product for leachables and to establish the potential safety impact of the leachables via a toxicological safety risk evaluation of each leachable.

5.4.4 Product Maintenance and Change Control

The ideal situation where the packaging system for a liquid pharmaceutical product remains unchanged during the product's entire commercial lifecycle is rarely achieved and thus change control is a necessary post-launch exercise. Specifically, a change to the packaging system may become necessary after the registered product has been commercially launched. Change control is that process by which changes are evaluated and ultimately approved or rejected. The means by which change control is exercised from a chemical compatibility assessment perspective can be established in a Comparability Protocol, which in essence is an agreement between the manufacturer and the relevant regulatory authority that defines how the manufacturer will perform the change control activity and how the change control assessment will be reviewed by the agency.

5.4.5 Component Testing

One reason that the definitions provided earlier are important is that the chemical compatibility assessment process does not specifically address components. The materials, the system components, and the packaged drug product are tested but there is no place where component testing is specified. Components are considered in the process by different means, depending on whether they are components that the manufacturer makes or components that the manufacturer buys. In the case that the manufacturer makes the component itself, such component testing would be in addition to the material selection and system qualification activities that the manufacturer would perform. Generally speaking, if the manufacturer has characterized a component's materials of construction and will qualify the system that uses a component, testing of that component provides little value added and is neither recommended nor required. However, there may be reasons that component testing, albeit not required, is desired. In such circumstances the manufacturer must establish the means and the methods for component qualification. It may be the case that simple components (for example, components made from a single material) could be qualified by a process that resembles material characterization and selection. Generally, however, more complex components (for example, components made from multiple materials) would be qualified by a process that is more similar to system qualification (e.g. extractables profiling).

When a manufacturer procures a component from an external vendor, the situation is both similar to and different from the case above. One difference is that the manufacturer will not have characterized or selected the component's materials of construction. The second difference is that the manufacturer may want data to support the decision to use the components [1]. The similarity is that the manufacturer will qualify the packaging system that is derived from the externally purchased component. In this situation the manufacturer is faced with a choice: does the manufacturer want to take the risk that the component is safe and forgo any testing until system qualification or does the manufacturer want to reduce the risk by characterizing the component? An important input to this question is the information that can be provided by the component's vendor. In the best possible scenario, the vendor will have selected the component's material by characterizing those materials via the manufacturer's Stage 1 approach. In this circumstance the manufacturer would choose to qualify the component in the same way it would qualify the component if it had made the component itself (including the possibility of no qualification).

If the component's vendor has not selected the component's materials based on the manufacturer's Stage 1 approach, then the manufacturer has essentially four choices:

1. Characterize the component's materials as per Stage 1,
2. Qualify the component based on whatever material or component information the vendor can supply,

3. Qualify the component via component testing (by a process the manufacturer deems to be appropriate), or
4. Perform no testing and require no vendor information, accepting the risk at the component stage and deferring safety qualification to the system stage.

Regardless of whether a component is made by the manufacturer or procured by the manufacturer, if a component is qualified by testing then the component should be tested in its finished form, that is, the form in which it exists when it is assembled with the other components into a system. Thus, for example, if a component is sterilized by irradiation prior to attaching the component to other components to produce a system, then the component should be qualified after sterilization and not before.

5.5 Elements of the Chemical Assessment Strategy

Previously, the four stages of the chemical compatibility assessment process were identified and briefly described. The following discussion provides a more detailed consideration of each stage and will establish the type of information generation and interpretation activities that are performed in each stage.

5.5.1 Stage 1: Material Selection

5.5.1.1 General Discussion
One answer to the question "what testing is required at the material selection stage?" is "full testing," where full testing establishes the material's extractables profile. While full testing is the most rigorous approach, it is also the most arduous approach as (i) extractables profiling of individual materials is time intensive and costly and (ii) a packaging system can consist of numerous materials, meaning that many extractables profiles will be generated. While time and cost cannot be the overriding factors in designing a safety assessment process, they are necessary aspects to consider and one recognizes that the most effective safety assessment process is neither the most nor the least costly or time intensive but rather is the most cost and time effective. Thus, the manufacturer will not require full extractables profiling of all materials, as selecting appropriate materials can be effectively accomplished with data that is more readily secured than are individual extractables profiles. Three considerations go into this observation. First, converting materials into packaging systems typically means that the materials undergo processing that involves "stressful" conditions such as elevated temperatures, mechanical stress, and others. It is reasonable to suggest that a material's extractables profile might be altered due to such processing. Furthermore, the concept that "the extractables profile of a packaging system is the sum of the extractables profiles of its individual materials of construction" is most likely inaccurate. Therefore, profiling materials for extractables produces information that is unnecessarily rigorous as the rigor of an individual material's extractables profile is essentially "diluted" by the less rigorous process of correlating materials profiles to system profiles.

Second, safety assessment at the material selection stage does not need to be as rigorous and comprehensive as the assessment performed for system qualification as the staged safety assessment strategy is designed so that no single stage need address and eliminate the full risk. Thus, the function of the individual stages is to reduce the risk and therefore each stage can be designed for risk reduction and not risk elimination. Considering materials of construction specifically, any safety assessment made at the materials stage will be revisited, re-enforced, and replaced by more substantial and rigorous assessments performed at the System and Product stages.

Lastly, it is noted that materials are not the point of regulation for packaging systems. While regulatory submissions are facilitated by a discussion of material selection, such a discussion and justification are not required parts of such submissions and such a discussion and justification is not likely to weigh heavily in a regulatory approval decision. Thus, if the manufacturer were to generate information on materials, it would be for the manufacturer's internal purposes and not to directly fulfill a specific regulatory requirement.

Therefore, material characterization is not extractables profiling; rather, as noted in Figure 5.2, it is a process of data collection and interpretation. Specifically, the manufacturer receives critical information about candidate materials from their vendors, augments that information to fill gaps in the vendor information, performs a review of the historical behavior of similar materials, and then assesses that information in the context of a potential safety impact to establish whether candidate materials are suited for use in the application being considered. Materials that are established to be suited for use continue as candidates in the packaging system development while materials established as unsuited for use are eliminated from further consideration (see Figure 5.3).

5.5.1.2 Vendor Information
Since the goal of Stage 1 is material selection, the key to performing a Stage 1 assessment is establishing that set of information that is necessary to determine the suitability of a material. As noted previously, it is the manufacturer's position that full extractables profiling is not necessary to

Figure 5.3 Process flow diagram, Stage 1, material characterization. The manufacturer secures information from the vendors of candidate materials, performs a historical review of the behavior of similar materials, and establishes whether the candidates are suited for use or not.

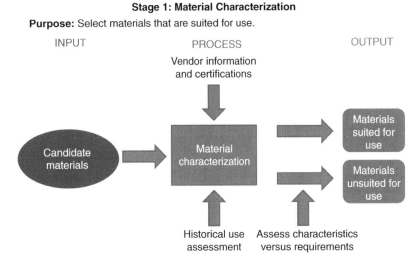

Stage 1: Material Characterization

Purpose: Select materials that are suited for use.

enable material selection. Rather, it is the manufacturer's position that a material can be established as being suited or unsuited for a specific use if (see Figure 5.4):

1. The material is fully described,
2. The material's composition is known,
3. The material's manufacturing process is understood (focusing specifically on process-related impurities),
4. The material's Material Safety Data Sheet (MSDS) is provided,
5. The material has been certified as compiling with the relevant compendial requirements,
6. The material has been certified as compiling with relevant Indirect Food Additive regulations, and
7. The material has been certified as compiling with regulatory (and/or the manufacturer's) requirements concerning banned or regulated substances.

It is reasonable to expect that the material's vendor has information relevant to these aspects and that the vendor agrees to provide that information to the manufacturer. In the unlikely and undesirable situation that the vendor is unable to provide such information, then the manufacturer must perform whatever testing is necessary to procure the missing information. This is the case as the selection decision cannot be made on the basis of partial data. This requirement for full data to enable material selection is a cornerstone of the manufacturer's chemical compatibility assessment process.

Certain of the above points require some clarification. For example, considering point 1, it is appropriate to ask "what is the difference between a full description and an incomplete description?" The answer is that a full description enables material selection while a partial description does not. To clarify this somewhat, consider the case where

Figure 5.4 Vendor information required as input into the material characterization process.

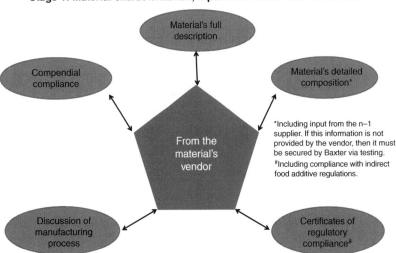

Stage 1: Material characterization; input information from the vendor

a material is described as a "polyisoprene." While such a description differentiates the material from other material types (for example, polypropylene), it is unlikely that such a generic description would facilitate the manufacturer's review of its historical information related to materials.

Considering point 2, a material's composition drives its extractables profile, as extractables are typically the material's ingredients themselves or impurities in or decomposition products of the ingredients. In fact, one of the reasons that experimentally establishing a materials extractables profile is not required for material selection is that the material's extractables profile can be inferred based on its composition (and considering its manufacturing process).

Considering point 2 further, it is noted that the composition of a material may not be established solely by the vendor who supplies the material to the manufacturer. For example, consider the case of a polyisoprene septum supplied to the manufacturer from vendor A. In this case, the value added provided by vendor A is that they take a polyisoprene material they procure from a supplier (suppler B) and mold it into a specific form. In so-doing, vendor A adds a certain antioxidant to stabilize the material during the molding process. In this case the compositional information that the manufacturer's vendor might supply (polyisoprene raw material + antioxidant) might not reflect the materials' complete composition, as it is likely that supplier B to the manufacturer's vendor A formulated the polyisoprene in such a manner that it was suitable for use by the manufacturer's vendor A. In this case, the most useful material composition information is the combination of information supplied by the manufacturer's vendor A and vendor A's supplier B. It is for this reason that the manufacturer would require that its vendor A engage their own suppliers when specifying a material's composition for the manufacturer.

Considering point 3, it is clear that the means by which a material is manufactured is germane to establishing its suitability for use. For example, exposure of the material to a high temperature or harsh or undesirable chemicals during manufacturing may significantly impact its extractables profile and therefore its suitability for use. The emphasis here is to understand how the material's manufacturing process can influence its extractables profile, for example, by either process-induced degradation of the material's additives or contamination of the material with process residuals.

Point 4 is self-evident as clearly a material's MSDS can contain meaningful information about its potential use in packaging systems.

Considering point 5, it is a universal regulatory requirement that materials used in packaging systems comply with the relevant compendial specifications. In a certain manner, point 5 is a "check the box" aspect. However, even if point 5 were not mandatory it would have value in terms of material selection. This is the case as compendial testing can provide information relevant to material selection. Consider, for example, a plastic material of construction whose use in packaging of products registered in the US is governed by USP <661.1> [2]. Part of the testing suggested in USP <661.1> is Extractable Metals and it is certainly the case that a review of Extractable Metals information is relevant to the material selection process.

Considering point 6, many of the plastics used in pharmaceutical packaging are also used in food packaging. Regulation of plastics for food packaging allows such plastics to be approved for certain applications based on their compliance with a set of regulations that can be loosely termed Indirect Food Additive Compliance. These regulations are described in this way as the exact regulations and requirements are not globally universal and vary somewhat from region to region. While approval of a plastic's use in food packaging may not be the sole criterion for approval of the plastic's use in certain pharmaceutical applications, it is generally recognized that compliance with food additives regulations is relevant to assessing the material's suitability for use in pharmaceutical applications. In fact, there was a time when compliance with indirect food additive regulations was the only safety data required for a packaging system used with certain "low-risk" dosage forms (see Chapter 7, Section 7.3 for a further discussion on this).

Considering point 7, it is noted that in certain circumstances regulatory bodies and/or standard-setting organizations have restricted the use of certain substances in plastics and polymers used for pharmaceutical applications. Thus, for example, it is common practice that vendors of plastics used in pharmaceutical applications certify that their materials do not intentionally contain certain entities or that the entities that the materials do contain are not derived from certain sources. Securing all necessary and appropriate certifications from vendors is a critical and necessary part of the material qualification process, as the absence of relevant certificates could be the justification for rejection of candidate materials. While it is reasonable and appropriate for the manufacturer to expect material vendors to supply the appropriate certifications, it is the manufacturer's responsibility to establish what the appropriate certifications are.

Additionally, the manufacturer itself may have internal restrictions or guidelines concerning the use or presence of certain substances in plastics and polymers. As was the case with regulatory restrictions, while it is reasonable and appropriate for the manufacturer to expect material vendors to supply the appropriate certifications, it is the manufacturer's responsibility to establish what the appropriate certifications are.

Items 1 through 7 reflect a minimum set of information that can be provided by the material's vendor. It may be the case that certain vendors will have additional information that would be relevant to material selection. Examples of this additional information include:

1. Impurity profiling of the material,
2. Extractables profiling of the material,
3. Biological testing of the material beyond the compendial requirements (for example, USP class VI testing or Ames testing of extracts).

Although such information might not be available from all vendors, the manufacturer should routinely seek to establish whether such information is available and to secure that information when it is available.

Lastly, it is observed that the process of material characterization produces a body of information that could be significant in situations other than material selection. For example, the same information used to select a material may be useful in terms of managing changes associated with that material. Furthermore, once the material characterization information has been collected, it can be used to support situations where the characterized material may be used in applications other than the initial application that triggered the material characterization process. For these reasons it is suggested that the material characterization information should be captured in a format and maintained in a location that facilitates the storage, retrieval, and use of that information. This suggestion is captured in the *Material Characterization and Selection Report*, discussed in Section 5.5.1.4.

5.5.1.3 Material Assessment

Ultimately, the material selection process requires that the material's assessor consider the available material information and forecast the material's likely behavior in a specific application. Based on the forecasted behavior, the material is either selected or rejected. One means of accomplishing this objective is by considering the material characterization data provided by the material's vendor. Thus, for example, information about a material's composition can facilitate the forecasting of the material's likely extractables profile. Another means of accomplishing the objective is to answer the question "how do similar materials behave in the packaging situation I am considering?" For example, if a particular material is being considered as a candidate in a packaging system, one means to forecast the behavior of the candidate in that system is to review the behavior of a similar material used in a similar packaging system.

Thus, the second aspect of the Material Characterization stage is the Material Assessment (Figure 5.5). Using the vendor information, the material's assessor searches any internal and external sources (such as the manufacturer's Materials Database) to identify previously characterized materials that are close compositional matches to the candidate material. The closely matched, previously characterized material becomes a surrogate (or predicate) for the candidate material. In essence, the assessor searches information resources for qualified packaging systems that use the surrogate and then establishes if any of those packaging systems are comparable to the system for which the candidate material is being considered. If the candidate material and its anticipated packaging system can be matched to a predicate material used in a similar packaging system, then information about the predicate material and its similar packaging system (for example, the system's

Figure 5.5 The material assessment performed to support material qualification.

Stage 1: Material characterization; material assessment

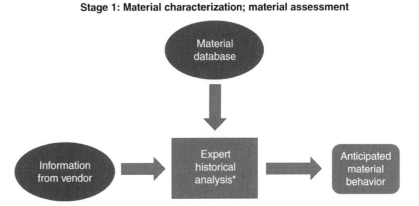

*Based on vendor information, an expert assess or examines the material database, identifying comparable materials (candidates versus historical). The historical information is used to forecast the behavior of candidate materials.

extractables profile) can be used to forecast the properties of the candidate material and the packaging system that the candidate material is being considered for.

To illustrate this concept, consider the following example. A polypropylene (PP) candidate material is being considered for use as a layer in a multilayered film that is being developed as the container body for terminally sterilized aqueous drug products under development. After searching the Materials Database, the assessor finds an appropriate predicate PP that itself is used in a layer of a multilayered film that serves as the container body for currently marketed terminally aqueous sterilized drug products. The fact that these are marketed drug products means that the packaging system and the predicate material have been established to be compatible with aqueous drug products (for example, they have extractables that have a minimal adverse effect on user health). Because the marketed system and the predicate material are appropriate models for the candidate material and its anticipated system, the behavior of the predicate (specifically that it is safe) can be forecasted as the behavior of the candidate material and its anticipated system, at least in terms of material selection.

5.5.1.4 The Material Characterization and Selection Report

Once the process of Material Characterization has been completed and the Material Selection decision has been made, it is necessary and appropriate to document and justify the decision. This is accomplished in a Materials Characterization and Selection Special Report, which has the following format:

1. Introduction. This section identifies the candidate material and its anticipated application and establishes the purpose of the report as "documenting the basis of, the justification for and selection decision associated with this candidate material in this specified application." The identification of the candidate in the Introduction is not the complete material description provided by the material's vendor but rather general information (material class, lot #) sufficient to distinguish the candidate material from other materials.
2. Material Characterization. This section contains all the material characterization information provided by the vendor or otherwise procured by the manufacturer. The individual sub-sections in this section mirror the inputs illustrated in Figure 5.4 and include:
   a) Material Description
   b) Material's Complete Composition
   c) Discussion of Material's Production Process
   d) Material Safety Data Sheet

e) Compendial Compliance (supported with data tables as appropriate). The actual reports provided by the vendor would be captured in one or more Appendices.
   f) Certificates for Indirect Food Additive Compliance. The actual certificates themselves will be captured in one or more Appendices.
   g) Certificates of Regulatory Compliance (typically a list or table of such certificates). The actual certificates themselves will be included in the report as one or more Appendices.
   h) Other documents and information sources.
3. The Material Assessment.
4. Selection Decision and Justification.
5. Conclusion. This section identifies the candidate material and the anticipated application (drawn from the Introduction) and states the Selection decision (for example, "Material x is (or is not) deemed to be suitable for use, from the perspective of user safety, in the solution contact layer of the manufacturer's XYZ packaging system, which will be used to package ....")
6. References.
7. Appendices (capturing Vendor communications, letters, reports, etc., containing information related to the items in Report Section 2 – Material Characterization).

5.5.2 Supporting Product Development Between Stages 1 and 2

In general, that period of product development that occurs between material selection and the generation of a viable system prototype is characterized by extensive product development activities but few chemical safety qualification activities (see the following discussion of Stage 2). However, certain circumstances that occur during development may require some chemical safety qualification efforts, including revision control during product development and the request for supplemental testing.

5.5.2.1 Revision Control During Product Development

The ideal and highly desirous situation that the originally selected materials of construction perform up to requirements and that the prototype system consists solely of these materials is infrequently encountered. It is more frequently the case that one or more of the originally selected materials of construction do not perform up to requirements and thus must be replaced, triggering a revision. Additionally, the originally selected materials of construction may themselves be altered during product development. In either circumstance the development team is faced with the circumstance of exercising revision control.

The strategy for revision control during product development is different from the strategy for change control after product commercialization (Stage 4). On the one hand, there is no external reporting requirement. Additionally, revision control in development need not be as rigorous as change control after commercialization because revision control in development will be followed by full extractables profiling (Stage 2). Thus, the risk that the revised material may have an adverse impact on safety will be more rigorously assessed by the profiling activity and for this reason the revision control activity itself can be less rigorous. Alternatively, change control after commercialization is an activity unto itself and thus the change control activity must necessarily be more rigorous.

The revision control strategy for revisions made during product development is captured in a re-statement of the testing requirements for material selection. As established previously, all materials of construction used in a packaging system for a marketed product must be intentionally selected based on a Stage 1 assessment. This requirement applies equally to materials that were originally selected at the start of product development and to materials that are selected during product development. The revision control activity during product development is that the revised material must be assessed per Stage 1 and that the revised material must be selected for use based on that assessment.

5.5.2.2 Supplemental Testing During Product Development

As noted in the Stage 2 discussion, performing excessive chemical safety testing during that period of time when product design and composition is in flux is potentially inefficient as one will be testing materials that may not be used in the finished systems and testing components and prototypes that may bear little resemblance to the final marketable product. In these cases, the generated chemical data, which will be secured at some cost and over some period of time (thus affecting project budgets and timelines), will have little or no regulatory significance and will not facilitate the process of securing regulatory approval. However, one can envisage certain circumstances where some level of chemical assessment would facilitate, guide, or drive the product development effort, specifically around managing safety risks.

In these circumstances, chemical safety studies must be designed and implemented to meet a clear and appropriate objective and such studies would require the same level of testing necessary to fully qualify a test article for use. In any event, such testing is not common and thus is not a well-defined aspect of the chemical compatibility assessment process.

5.5.3 Stage 2: System Qualification

5.5.3.1 Discussion

Once Material Selection has been completed, one can anticipate that there will be a considerable period of packaging system development during which time the packaging system evolves from a prototype to something that very closely resembles its actual commercial configuration. While new candidate materials may have been needed to address packaging system development challenges and these new materials had to be characterized to support their selection, there is little or no chemical assessment activity that is performed to address the evolving packaging system's potential safety impact. This is the case as the next significant stage in the chemical safety assessment strategy is the qualification of the packaging system, where qualification is defined as the process by which the packaging system is established to be compatible with the packaged drug product and therefore suited for its intended use.

Although there may be a great temptation to perform system qualification while packaging development is still ongoing, such an activity is generally counterproductive. This is the case as (i) the process of system qualification can be time- and cost-intensive and (ii) if the packaging system changes after it was qualified, then the qualification needs to be adjusted, revised, or redone to be valid for the final commercial packaging system. Thus, it is logical and appropriate that the extensive process of qualifying the system is initiated only after the design, composition, and construction process of the packaging system is "frozen."

This analysis notwithstanding, a considerable challenge to a strategy that "delays" system qualification until the design freeze is that performing a robust qualification that addresses the end of shelf-life adds a significant length of time to the product development process. Thus, an efficient qualification process must include accelerated shelf-life testing to reduce the qualification's negative impact on product development time.

5.5.3.2 The System Qualification Process

The purpose of System Qualification (Figure 5.6) is to establish that the packaging system and the packaged drug product are compatible. Incompatibilities between the drug product and its packaging system occur when the drug product and its packaging system are in contact during product manufacturing, distribution, and storage and clinical use. During contact, the drug product and its packaging system can interact, potentially producing incompatibilities. Specifically, a packaging system can interact with the packaged drug product in two ways: (i) entities in the packaging system can leach out into the drug product where they could have an adverse effect on the drug product's quality attributes or (ii) entities in the drug

Stage 2: System qualification
Purpose: Qualify container-closure system as suitable for use.

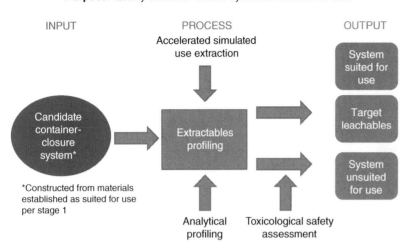

Figure 5.6 Process flow diagram, Stage 2, system qualification. A packaging system, constructed using well-characterized materials, is assessed for suitability by the process of extractables profiling.

product can be sorbed by the packaging system, potentially affecting drug product potency and stability.

Sorptive interactions between a drug product and its packaging systems are typically addressed as a specific part of the drug product development process. Generally speaking, sorptive interactions are specifically addressed in stability studies, which establish that the drug product, in its packaging system, is capable of meeting established compositional specifications over its specified shelf-life. Because sorptive interactions are addressed in stability studies (and may have also been assessed during product development), System Qualification as defined in this document excludes sorptive interactions and focuses solely on leaching interactions.

It is clear and well-recognized that testing a product for packaging-derived substances that leach into it over its shelf-life is the most direct and appropriate means of qualifying both the product and the packaging. However, it is also well recognized that the process of profiling products for leachables may be challenging, most notably as leachables tend to be trace level impurities that are present in a chemically complicated drug product. Thus, extractables profiling of packaging systems is a means of addressing leachables in a drug product. In this circumstance, extractables profiling is an analytically-expedient surrogate for leachables profiling.

Risk management offers another reason for performing system qualification via extractables profiling. In theory and in practice, a risk assessment performed on leachables can only be performed at the end of stability studies, as leachables are typically measured during stability studies and leachables levels increase with time. It would be an unfortunate circumstance indeed to discover at the end of shelf-life testing that a leachable accumulated in the

drug product to a level at which it had a significant and adverse effect on drug product quality. A much better risk management approach would be the pre-stability characterization of the packaging system for extractables. In this way, potential leachables issues are surfaced early enough in the development process that potential adverse effects can be properly and efficiently managed.

It is clear from this discussion that system qualification via extractables profiling is effective only if the packaging system's extractables profile mirrors the drug product's leachables profile, at least in terms of all leachables being represented by extractables and the maximum levels of extractables matching (or exceeding to a limited extent) the maximum levels of leachables. Accomplishing this objective dictates the proper design of the extractables profiling study.

5.5.3.3 Extractables Profiling
The process of extractables profiling, more extensively discussed in Section 2.2 in Chapter 2, includes two steps: generating the extract and characterizing (testing) the extract. Each is considered in greater detail as follows.

Extract Generation
Extraction is generally a process of treating a material with a solvent to remove soluble substances. Extraction is a complex process influenced by variables that include time, temperature, surface area to solution volume ratio (i.e. stoichiometry), extraction medium, and the "extractability" of the material. The design of an extraction study, and thus the setting of these variables, is dictated by the purpose of the extractables assessment. In the case of System Qualification, the purpose of the extractables assessment is to produce an extractables profile for the packaging system that mimics the

packaged drug product's leachables profile. Given this purpose, these variables must have the following values:

1) The extraction medium (also called the extraction solvent) used in the extraction study must have the same "leaching power" as the drug product.
2) The surface area to solution volume ratio used in the extraction study must be at least equal to the surface area to solution volume ratio of the packaging system filled with the drug product.
3) The time and temperature of the extraction study must match the manufacturing, storage, and distribution conditions specified for the packaged drug product.

Considering the aspect of extraction time and temperature, the extraction study seeks to establish patient exposure to leachables. Furthermore, as noted previously, it is highly desirable that the overall process of extractables profiling be more time efficient than the process of leachables profiling over the course of a drug product's shelf-life (so that the extractables study can be completed in less time than a full shelf-life). Since the process of generating the extract is a significant contributor to the amount of time required for extractables profiling, any means of shortening the extraction time versus product shelf-life makes the process more time efficient. For example, consider the case of a packaged drug product with a labeled shelf-life of two years at ambient temperature. An extraction study that exactly mirrored this shelf-life would require two years to merely generate the extract, let alone perform the necessary extract testing. However, using the well-accepted convention that storage at 40 °C for six months mimics two years of storage at ambient temperature reduces the duration of the extraction study by a factor of 4.

Considering extraction acceleration further, it is noted that leachables testing seeks to establish the greatest exposure a user of the drug product has to leachables, which occurs at that point in a product's shelf-life when the leachables levels in the drug product are at their greatest. As the time in a product's shelf-life that this greatest leachables level occurs is generally unknown, leachables studies typically include multiple time points for testing, as the results obtained at the multiple time points establish when the leachables' levels are at their maximum. Alternatively, if the point at which maximum leachables levels were achieved could be established without testing (for example, by mathematical migration modeling), then extractables studies could be designed with fewer test intervals.

Thus, mathematical migration modeling has the potential of reducing the duration of extractables studies as well as reducing the number of test intervals included in such studies.

Guidelines for establishing the composition of the extraction medium, for establishing the proper surface area to solution volume ratio and for accelerating extraction via conventional means have been established previously in Section 2.2.5 in Chapter 2. Considering the acceleration of the extraction process specifically, one notes that other mechanisms may exist to define and justify acceleration factors greater than those provided by conventional means. This is significant as the greater the acceleration factor, the shorter is the duration of extract generation and the more time efficient the extraction process becomes. For example, use of mathematical migration modeling to identify and justify greatly accelerated extraction conditions could greatly reduce the time required for the extractables profile, thereby reducing the impact of extraction study duration on the total product development timeline.

Extract Testing

Once the extract is generated, the extracted is tested to establish what substances are present in the extract. That is, the extract is screened by appropriate analytical methods to discover, identify, and quantify the extractables. In this screening process, extractables are "discovered" when they produce a response in one of the appropriate analytical methods, are identified based on the characteristics of the response, and are quantified based on the magnitude of their responses.

As extractables testing is currently practiced, an extractable's identity is established by comparing the extractable's response to a database of known responses produced by known substances, termed a reference response. When the extractable's response has the same or nearly the same characteristics as the reference response, then the extractable is identified as the compound that produced the reference response. An identity procured in this manner is termed a tentative identity as it is based on a single dimension or attribute. Alternatively, identities may be elucidated via an expert interpretation of the available analytical information. Such an identity is termed a speculative identity as it is based on interpretation as opposed to matching.

Additionally, an extractable's concentration in the extract is estimated, often by comparing the response obtained for the extractable to the response obtained for a "surrogate standard," which is a substance that is either intentionally added in a known quantity to the extract during sample processing or which is present in a sample, at a known quantity, which is tested contemporaneously with the extract. An extractable's concentration obtained in this manner is an estimate as the concentration-dependent magnitude of the response for the extractable may not be the same as the concentration-dependent magnitude of the response for the surrogate standard.

The generated tentative identities and estimated concentrations in an extractables assessment can be appropriate and adequate to establish the suitability of the packaging system in terms of the potential leachables. However, one must understand that a compatibility assessment based on tentative (or speculative) identities and estimated concentrations is an assessment that is largely uncertain, imprecise, and prone to error. To provide greater certainty and gain greater confidence in the assessment, one can envision a means by which identifications can become more confident and concentration estimates can become more accurate. Such a process starts by analyzing extractables standard reference materials by the analytical screening methods. Multi-dimensional characteristics of the responses obtained for the reference standards are captured in a database. The database is then linked to the screening methods. Responses obtained when an extract is analyzed by a screening method are compared to the database. When a multi-dimensional match is made between a response in the extract and a reference standard in the database, the match is the basis for identification and quantitation of the extractable. Because such an identification is based on a multi-dimensional match and because the match data was generated by analysis of a reference standard, the match is a confirmed identification as opposed to a tentative identification. Because the concentration estimate is obtained with a compound-specific response factor, the estimate is more accurate than the estimate obtained using a generic response factor. Furthermore, the confirmed identity and more accurate concentration estimate are obtained automatically by the analytical system. In this case a trained analyst confirms the assessment made by the analytical system. In the current situation, the trained analyst actually makes the identification and performs the concentration calculation. Since it requires less time to confirm an assessment than it does to perform an assessment, the database-driven process is more resource efficient than the current process.

Over time, the investment made in generating the database is recouped by the higher quality of the generated data (greater certainty in the impact assessment) and the reduced use of trained analyst time.

5.5.3.4 Impact Assessment of the Extractables Profile

The process of profiling a packaging system with respect to its extractables profile produces data, specifically a list of identified extractables and their concentrations in the extract. While this data has value as it is the input to impact assessment, the data is not the impact assessment itself and thus the extractables profile has to be interpreted to establish its effect on the quality attributes of the packaged drug product.

An important aspect of product quality that can be affected by leachables is the product's impact on the health of patients who are administered the drug product in a therapeutic setting. The process of establishing the potential patient safety impact of leachables is termed toxicological safety risk assessment. Although the process of toxicological safety risk assessment is considered in greater detail in Chapter 10, it can be generally described as follows. Given the identity of the individual extractables (as potential leachables), the toxicological safety assessor obtains the relevant toxicological safety data for that compound and if the data is sufficiently relevant and robust uses the data to establish a tolerable daily intake (TDI) for the extractable. Additionally, given the concentration of the extractable in the extract and the dosing regimen for the drug product, the safety assessor can calculate a patient's maximum daily exposure (MDE) for the extractable. The comparison of the TDI with the MDE establishes the potential patient safety impact of the individual extractables. If the MDE is less than the TDI, it is generally concluded that the extractables present a negligible risk of producing an adverse health effect in patients.

Additional drug product quality attributes that could be affected by leachables include product stability and potency. It has previously been noted that the effect of leachables on these product quality attributes is established in stability studies. In such studies the effect of individual leachables is not established; rather, the effect of the entire leachables profile is addressed along with all other stability- and potency-limiting drug product characteristics.

Generally speaking, it is not possible to infer the effect of an individual leachable on product stability or potency in the same way that the effect of a leachable on patient safety was inferred by toxicological assessment because the amount of published literature documenting stability or potency effects of leachables is many orders of magnitude smaller than the amount of published toxicology data.

5.5.3.5 The Analytical Evaluation Threshold (AET)

It is a practical reality that all analytical methods used in extractables screening or profiling have limits in terms of their ability to detect, identify, and quantify individual extractables. The analyte concentration at which an analytical method loses these abilities is termed the analytical action threshold (AAT).

Furthermore, it has been established and is generally well accepted that leachables, regardless of their identity, can have an adverse effect on a patient using drug products only if the patient's daily exposure to the leachables is above a certain threshold value, which has been termed the Safety Concern Threshold (SCT). Practically speaking, leachables whose dose is less that the SCT are deemed to be safe regardless of their identity. Such leachables do not have to be detected, identified, or quantified to establish that a packaging system is safe.

If the dosing regimen of the drug product is known, the SCT can be converted to the concentration of the leachable in the drug product. The concentration analog to the SCT is termed the Analytical Evaluation Threshold (AET). Thus, the AET establishes the concentration of a leachable in the drug product above which the leachable must be detected, identified, quantified, and reported for toxicological safety assessment. In cases where extractables = leachables, the AET is relevant for extractables as well.

It is clear that the relationship between the AAT and AET is a critical aspect in determining whether analytical methods used in extractables scouting are suited for that purpose or not. When the AAT is less than or equal to the AET (AAT ≤ AET), then the method is fully capable of performing the functions of detection, identification, and quantitation with the required sensitivity. If the ATT is greater than the AET (AAT > AET) then there is the possibility that the method will not detect, identify, and quantify potentially impactful extractables. The greater the difference between the AAT and the AET, the greater the risk that the analytical method will miss potentially impactful extractables.

Because the AET is applied to all extractables, it is clear that the AET is based on a very conservative SCT. In most cases, a compound-specific extractables threshold (CET), such as a TDI, would have a higher value than the AET, as most compounds would be safer than the worst-case situation captured by the AET (that is, a compound's TDI is larger than the SCT). This situation highlights another advantage of extractables profiling against a database of targeted compounds versus extractables screening for unknowns. In the case of extractables screening for unknowns, the performance expectation is driven by the AET. In the case of extractables profiling for targets, the performance expectation is driven by each target's specific CET, which is expected to be higher than the AET. Furthermore, target analysis against a database that contained CET data would, in essence, be a toxicological assessment. Specifically, if an extractable's concentration in the extract is less than the CET, that extractable is deemed to be safe without further toxicological assessment.

It is noted that because the AET is specifically tied to the product quality attribute of safety, it is directly applicable only to safety. That is to say that, rigorously speaking, the AET cannot be interpreted to be a relevant threshold for considering other product quality attributes, such as stability and potency. Nevertheless, given the generally low value of the SCT, it may be the case that the AET is somewhat applicable to these other product quality attributes.

The derivation and use of the AET was discussed in greater detail in Section 3.2.3 in Chapter 3.

5.5.4 Stage 3: Product Qualification

5.5.4.1 Discussion

It is an accepted and expected practice in the pharmaceutical marketplace that drug products that are ready for commercialization are subjected to a formal investigation, which confirms that the drug product can meet its established quality specifications through the product's labeled self-life. Such an investigation is termed a stability study. Information collected in the stability study is a key component in the drug product's regulatory submission.

Performing a stability study as the last key activity of the product development process affords the opportunity to establish the drug product's leachables profile. The process of testing drug products for leachables has been termed a migration study. In deciding whether a migration study should be performed, it is important to consider what information is available at this stage in the product development process to establish that the packaging system is suited for its intended use. This information includes:

1. The material characterization data that was used to choose and justify a packaging system's materials of construction (Stage 1),
2. The extractables profile of the packaging system, toxicologically assessed and deemed to be suitable from a patient safety perspective (Stage 2),
3. Product development studies addressing and confirming packaging–product compatibility.

Because the extractables profile was obtained via a simulated extraction study, the extractables profile should establish the complete and total population of leachables and the highest possible accumulation of those leachables over shelf-life.

Given the above, one can propose that leachables testing of a drug product on stability is unnecessary to establish the safety of the packaging system, as this was already accomplished during packaging system qualification. Nevertheless, one can envisage performing leachables testing on stability for the following purposes:

1. To confirm the close similarity of the leachables and extractables profiles (establish an extractable/leachables correlation),
2. To clarify or "fine tune" the extractables profile,
3. To satisfy regulatory expectations.

Considering these, point 1 is self-evident. Considering point 2, it is understood that even the most well-designed simulation extraction study may not be an exact match for the product under its clinical use conditions. Furthermore, the implementation of the system qualification extractables study is such that the concentrations obtained for the extractables are somewhat imprecise. Thus, there may be some value in quantifying individual leachables during stability studies to establish the exact leachables level in the actual drug product that has experienced its clinical use conditions. This could be especially true for extractables that were present in the extracts at levels close to the relevant toxicological safety threshold (TDI).

Considering point 3, the relevant guidelines for packaging systems in many global jurisdictions support the concept that leachables studies are not required if their exclusion can be justified. For example, Section 5.1 of the European Medicines Agency's Guideline on Plastic Immediate Packaging Materials [2] states that:

> "Migration studies may only be omitted if, based on the outcome of extraction studies, the calculated maximum amount of individual leachable substance that may be present in the active substance/medicinal product leads to levels demonstrated to be toxicologically safe. When a migration study is not considered necessary and thus is not conducted, a justification should be provided."

A similar concept is contained in the FDA Container Closure Guidance [3]. Although Section II.B of the Guidance states that "the ultimate proof of the suitability of the container closure system and the packaging process is established by full shelf-life stability studies," Section II.B.c. establishes that:

> "For a drug product such as an injection, inhalation ophthalmic, or transdermal, a comprehensive study is appropriate. This involves two parts: first an extraction study on the packaging component to determine which chemical species may migrate into the dosage form (and at what concentration); and, second, a toxicological evaluation of those substances which are extracted to determine the safe level of exposure via the label specified route of administration."

The Guidance does not specify that such an extraction study needs to be followed by or augmented with a migration study.

The specific regulatory language notwithstanding, it is general practice that the manufacturer includes some type of migration study in its global regulatory submissions. Figure 5.7 illustrates the process for the product qualification process (leachables profiling, migration study).

5.5.4.2 The Concept of Target Leachables

Given the previous discussion, there is little or no reason that a migration study performed on a packaged drug product during stability testing would routinely or necessarily involve screening the drug product for all leachables. Rather, the three purposes that were presented previously for performing a migration study can be achieved if the migration study is performed on a set of extractables that are specifically targeted as leachables. Such target leachables would be chosen from the entire population of extractables as follows:

1. Primary target leachables would be chosen based on their potential to impact product quality. For example, extractables that were identified as having a potential adverse impact on product quality would be targeted as leachables. Thus, an extractable whose MDE was close

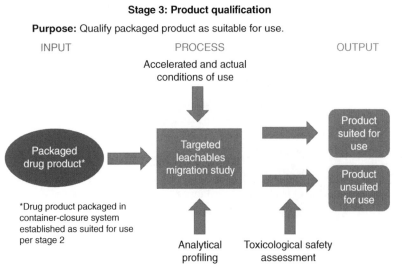

Stage 3: Product qualification

Purpose: Qualify packaged product as suitable for use.

INPUT   PROCESS   OUTPUT

Accelerated and actual conditions of use

Packaged drug product*

Targeted leachables migration study

Product suited for use

Product unsuited for use

*Drug product packaged in container-closure system established as suited for use per stage 2

Analytical profiling   Toxicological safety assessment

Figure 5.7 Process flow diagram, Stage 3, product qualification. The packaged product, including a qualified container-closure system, is assessed for suitability for use by performing a targeted migration study.

to its TDI could be targeted as a leachable to establish its exact accumulation behavior.

2. Secondary target leachables would be chosen based on their ability to confirm the extractables profile. While such secondary targets would not be anticipated to impact product quality, they would be considered to be diagnostic or characteristic of the packaging system. For example, if the extractable at the highest concentration in the extracts was not targeted due to its potential product quality impact, it might be targeted simply due to its high concentration relative to the other extractables.

5.5.4.3 Product Qualification via a Targeted Migration Study Performed on Stability

Consistent with the previous discussion, the migration study performed as part of the stability study for a packaged drug product will be an exercise in quantitating targeted leachables. While this approach establishes the nature of the migration study, there are two aspects of the migration study that require further consideration: quantitation levels and frequency of testing.

Quantitation Levels

When considering the quantitation of targeted leachables during a migration study, a significant question is "at what level should the method be operating?" There are three possible answers to this question:

1) At the lowest achievable level to allow one to establish the full accumulation profile over time,

2) At the approximate level at which the target accumulated as an extractable in the extraction study, as this is the most likely level the leachable will achieve in the migration study, or

3) At or near the level at which the leachable begins to have a potential product quality impact, as this is the level of impact and the whole purpose of the migration study is to establish a potential impact.

If these three levels are similar, then the answer to the question must be "operate the method at the lowest achievable level, as this is the only way to assess impact and/or confirm the extractables results." However, if the three levels are different, then the following answers are appropriate:

1) For primary leachables, chosen for their potential impact, the method should be operated at or near the level the leachable would have to achieve to have an impact. In this case the upper end of the method's operating range would be somewhat above the impact level (for example, impact level plus 50%), as there is little point in quantifying accumulation at higher levels after the impact level has been breached. Additionally, the lower end of the method's operating range should be well below the impact level (for

example, 10% of the impact level) in order to allow for both "early warning" that the target's concentration is approaching the limit and trending as the target's concentration approaches the impact level.

2) For secondary leachables, chosen for their ability to confirm that leachables behave similarly to extractables, the method should be operated at or near the maximum level that the extractable achieved in the extraction study. In this case, the lower end of the method's operating range should be well below the extractable's maximum accumulation level (for example, 10% of the maximum level) in order to allow for trending as the target's concentration approaches the extraction study level. The high end of the operating range should exceed the maximum accumulation level by a significant amount (for example, two times the maximum) to account for a reasonable level of imprecision in the estimated maximum amount.

This process for establishing the method's operating range is illustrated as follows. Extractable substance A accumulated in the extraction solution at a level of 10 mg/l, which is only a factor of two lower than the extractable's CET of 20 mg/l. Thus, this extractable was targeted as a primary leachable. A second extractable B was present in the extracts at the highest level of all extractables, 30 mg/l, but this level was 10 times lower than B's CET. Thus, extractable B was targeted as a secondary leachable. Lastly, the method developed for this application has an AAL of 0.1 mg/l when it is optimized.

In this example, the operating ranges of the methods for the two target leachables are set as follows:

1) The method is not operated near the AAL for either A or B as the AAL is much lower that the anticipated accumulation levels for both A and B.

2) The operating range for primary target leachable A starts at 2 mg/l, which is 10% of its CET (20 mg/l). The high end of the operating range is established as 30 mg/l, which is 150% of its CET.

3) The operating range for secondary target leachable B starts at 3 mg/l, which is 10% of the extractable's maximum level in the extraction study. The high end of the operating range is set at 60 mg/l, which is two times the target's maximum accumulation in the extract.

It is noted that the entire operating ranges noted above may not be achievable by the analytical method. In such a case the operating range must be truncated to match the method's capabilities with the purpose of the study (assessment versus confirmation).

Frequency of Testing

Since the targeted migration study occurs as part of the stability study, the testing intervals for leachables quantitation are generally limited to the testing intervals established for the other quality attributes. However, quantifying leachables at every time point in the stability study might be unnecessary in the context of achieving the objectives of the migration study. At a minimum, the migration study should include the following three test intervals:

1) Initial time point (time zero),
2) End of shelf-life time point, and
3) Accelerated end of shelf-life time point.

In many cases, testing at the beginning and end of shelf-life will establish the maximum accumulation level for a targeted leachable. Both the actual end of shelf-life and accelerated end of shelf-life intervals are included as it may be the case that a drug product's registration dossier is submitted with accelerated stability testing and then augmented with end of shelf-life data post-approval.

It is clear that these minimum testing timepoints do not allow for leachables trending. If trending the target leachable's accumulation is desired or required, it is clear that additional test intervals will need to be added to this testing schedule.

5.5.4.4 Impact Assessment of Target Leachables

As was the case with extractables, data obtained in the targeted migration study must be interpreted to establish the effect of the targeted leachables on product quality. This interpretation is facilitated by the interpretations that were made for extractables in the system qualification stage. Considering patient safety, a Compound-specific Evaluation Threshold (CET) was established for each extractable (and thus any target leachable) when the extractables profile was safety assessed. Thus, the toxicological safety assessment performed on target leachables is simply a comparison of the target leachable's concentration measured in the migration study and the CET. When the function of the target leachable is to confirm the results of the extractables profiling, the assessment involves comparing the levels of targeted substance measured in the extraction study with the levels of the targeted substance measured in the Migration Study. If the levels of the targeted substance as an extractable are greater than or equal to the levels of the targeted substance as a leachable, then the migration study corroborates the results of the extraction study and any safety conclusions drawn on the basis of extractables data are valid for leachables in the drug product. If the levels of the targeted substance as a leachable are greater than or equal to the levels of the targeted substance as an extractable, then the toxicological safety risk assessment performed for the extractable must be recalculated using the higher leachable's concentration.

5.5.4.5 Leachables Screening

If the sole activity in leachables profiling is target analysis, then eventually the following question will be raised: "how are you sure that there are no new leachables, that is, leachables that are not also extractables?" This is one of those questions to which there is no satisfactory answer because the fact of the matter is that there is no 100% guarantee one can give. If the extractables profiling was based on a properly designed and executed simulation study, then the possibility that there would be leachables that were not picked up as extractables would be low (but not zero). Thus, there will always be the nagging doubt of "what if I missed something?"

If a drug product can be screened for unspecified leachables down to levels approaching the AET, then performing this screening as part of the product qualification stage of the chemical compatibility assessment process may be worthwhile if for no other reason than to avoid questions for which there is no good answer. However, even when the drug product can be screened to AET levels, the decision to implement leachables screening as part of the product qualification process has significant practical ramifications. In theory, the screening process, which requires multiple assays, would have to be performed with every product lot that is used in the stability study and at every timepoint in that stability study. Perhaps the appropriate approach in such a circumstance is to employ statistical design of experiment principles to establish the appropriate leachables screening schedule to implement, leveraging the experimental design to preclude screening every lot at every interval.

Of course, there is always the possibility that it is not analytically viable to screen the drug product down to the AET. In this case, the value of screening would be lessened as the screening test methods would have gaps in them that could potentially include new leachables of toxicological safety concern.

5.6 Advanced Topics in Chemical Compatibility Assessment

5.6.1 Application of Design Space Concepts to Chemical Compatibility Assessment

The intent of having a chemical compatibility assessment process is that it can be applied in each case where a chemical safety qualification is required. However, it is reasonable to expect that the process would not be perpetual in the sense that it is applied in perpetuity without review and potential modification. Over an extended period of time the process will be applied to many cases and it may be that continued application of the process reveals certain trends in outcomes that might suggest that further application of the process in commonly encountered cases provides little or no value in terms of safety risk management. For example,

- After a certain number of materials of a certain type (for example, polypropylene) have been characterized, it is noticed that these materials are always deemed to be suitable for use in films for containers for aqueous drug products. If the composition of these materials can be used to define a design space, then can additional materials that fit in this design space be approved for use without testing?
- After a certain number of extractables studies and toxicological assessments, it is established that a specific extractable's MDE is always less than its TDI. In this case, can the extractable be established as generally recognized as safe (GRAS) and thus removed from future extractables studies and safety assessments?
- After a certain number of targeted migration studies have been performed for a certain number of drug products in a specific container system, it has been established that the targeted leachables always meet their specifications. Rather than continuing to perform targeted migration studied on additional similar drug products in the same container system, if the compositions of the tested drug products can be used to define a design space, then can migration studies be discontinued for additional drug products whose composition falls within the design space? It is noted that data supporting such an approach (applied to aqueous parenteral drug products packaged in flexible containers) has been published [1, 4].
- A specific type of process improvement is being universally applied to all materials of a certain plastic type and thus the drug product manufacturer is receiving the same change control notification from multiple vendors of multiple materials for that type of plastic. An example of this is the replacement of phthalate-based catalysts used in polypropylene production. After a certain number of materials have been assessed under the Change Control process and the change has been approved for these materials, is it possible to conclude that the change in general is acceptable and therefore suspend change control testing for the remaining effected materials?

The point of this discussion is that when a process is applied in multiple similar situations and when a consistent compliant output is obtained, it may be appropriate to conclude that the process is under control. In such a situation, a design space can be created, where the design space captures those input properties that affect the output and establishes values for those properties that produce an acceptable output. Armed with such a design space, one can examine new inputs in terms of the critical properties. If an input's critical properties fall within the design space, it can be concluded that the process output will be complaint without testing the output. The implication of this thought process to chemical safety assessment is that experience gained in applying this process may suggest opportunities to reduce or eliminate certain testing required by the process.

5.6.2 Migration Modeling to Reduce Study Duration

One of the practical challenges in performing extractables and leachables studies is the facet of study duration. This facet is challenging given that:

1. The shelf-life of a typical packaged product is many months (typically the longest possible shelf-life is desired, which can be as long as three years) and
2. The extractables or leachables study must reflect both the beginning and the end of shelf life.

Rigorously speaking, an extractables or leachables study that directly reflects the end of shelf-life will have a duration as long as the shelf-life.

This circumstance is mitigated somewhat by using well-accepted conventions for accelerated aging. For example, it is generally accepted that storage for six months at 40 °C is an appropriate acceleration of a two-year, ambient temperature (~25 °C) shelf-life. However, even in this case, an extractables study duration of six months aging plus the required analysis time (three to four months) can be excessive in terms of the impact on the product development timeline or the ability to rapidly address changes.

Given this circumstance, there is significant interest and value in reducing the duration of accelerated aging. However, such a reduction would ultimately require scientific justification as a prerequisite for regulatory acceptance. As the leaching of substances from packaging into a packaged product is largely controlled by the rate at which the substances migrate (or diffuse) through the packaging itself, migration models based on diffusion would provide the required justification. Such migration modeling would impact an extraction study duration in two ways:

- By establishing that an equilibrium steady state is established at durations much shorter than the actual shelf-life, meaning that the time that must be accelerated is shorter (and thus, that the accelerated duration is shorter), or
- By justifying the use of atypically high extraction temperatures to reduce duration (as higher extraction temperatures lead to shorter extraction durations).

The aspect of migration modeling and specifically the means of designing and justifying accelerated studies was considered in greater detail in Section 2.2.5.1 in Chapter 2 and in Section 11.6 in Chapter 11.

5.6.3 Extractables and Leachables Database to Facilitate the Transition from Scouting to Profiling

When a test sample is analyzed to establish all the compounds that are present in that sample, the process of discovering, identifying, and quantifying the compounds is termed scouting. For organic extractables, the predominant method for scouting an extract for extractables is chromatography. An extractable is discovered by chromatographic analysis when a peak in the chromatogram is observed. The identity of the compound that produced the peak is tentatively inferred by characteristics of the peak, such as retention time, mass spectrum, etc. The concentration of the substance is estimated by comparing the magnitude of the substance's peak with the magnitude of a peak related to a second compound, which was intentionally added to the extract at a known level (termed an internal standard). Confirming that the tentative identity is correct and procuring a more accurate concentration requires additional analytical effort.

Profiling is an alternate analytical approach to scouting. In profiling, the chromatographic peak obtained via analysis of a test sample is compared to a database of chromatographic peaks compiled via the analysis of reference standards containing known compounds at known concentrations. If the sample peak is matched to a peak in the database, then the substance that is related to the

sample peak has been confirmed to have the identity of the reference standard and the accuracy of the estimate of the substance's concentration in the sample has been improved by taking into account the exact response characteristics of the reference substance itself. Securing such a confirmed identity and a more accurate concentration estimate requires no additional work. Thus, in profiling, confirmation of a tentative identification and generation of a more accurate concentration estimate comes with a reduced analytical effort versus scouting.

5.6.4 Regulatory Status

A chemical compatibility assessment process involves three distinct actions and stages during product development and prior to product registration is consistent with regulatory expectations and best demonstrated practice recommendations (Table 5.1). For example, the FDA 1999 Container Closure Guidance [3] requires that the chemical composition of all materials of construction (plastics, elastomers, adhesive, etc.) be revealed. Extraction studies are specifically called out in numerous places in the Guidance. Product assessment (leachables) testing is captured in the FDA Guidance under the classification of other studies as appropriate. The three-stage nature of safety assessment is more clearly reflected in the EMEA 2005 Guidelines for Plastic Immediate Packaging Materials [2], which are based on General Information (complete qualitative composition of the plastic material), Extraction Studies, and Migration Studies (corresponding to the material characterization, simulation study, and product assessment portions of the Chemical Safety Qualification process). The concepts of compositional information, (controlled)

Table 5.1 Examples of three phase approaches to chemical safety assessment.

Stage	FDA [3]	EMEA [2]	PQRI OINDP [5]	BPSA [6]	This author [7, 8]	USP [9]	ISO10993:18 [10]
Material Screening and Selection	Chemical Composition	General Information and Specifications	Compositional Information, Qualitative Controlled Extraction Study	Compositional information	Material Characterization	USP <661.1> Plastic Materials of Construction	Compositional profiling
Simulation Study	Extractables	Extraction Studies	Quantitative Controlled Extraction Study	Extractables study	System Qualification	USP <661.2> Plastic Packaging Systems for Pharmaceutical Use, Extractables	Extractables profiling
Product Assessment	Other studies as appropriate	Interaction Studies (Migration Studies)	Leachables Study	Migrants study	System Validation	USP <661.2> Plastic Packaging Systems for Pharmaceutical Use, Leachables	Leachables profiling

extraction studies, and leachables (migrants) studies are central to the best demonstrated practice recommendations proposed for Orally Inhaled and Nasal Drug Products (OINDP) [5] and for single-use manufacturing systems. Additionally, the three stages are consistent with the Material Characterization, System Qualification, and System Validation activities that have been previously proposed by this author [8]. Lastly, a three-stage approach to chemical characterization of medical devices (compositional profiling, extractables profiling, and leachables profiling) is the cornerstone of the general chemical characterization process for medical devices (Figure 1 in ISO 10993:18 [10]).

References

1 Jenke, D., Gill, M., Liu, J. et al. (2014). Application of quality by design (QbD) principles to extractables/leachables assessment for terminally sterilized aqueous parenteral solutions stored in a plastic packaging system. *CNPPA Magazine: Pharmacy & Packaging* 36 (6): 4–10. and 37(1): 5–10 (2015).

2 Guideline on Plastic Immediate Packaging Materials. European Medicines Agency. CPMP/QWWP/4359/03. EMEA/CVMP/205/04. 5/19/05.

3 Guidance for Industry (1999). Container Closure Systems for Packaging Human Drugs and Biologics. US Department of Health and Human Services, Food and Drug Administration, Rockville, MD.

4 Jenke, D. (2010). Application of quality by design (QbD) principles to extractables/leachables assessment. Establishing a design space for terminally sterilized aqueous drug products stored in a plastic packaging system. *PDA J. Pharm. Sci. Technol.* 64 (6): 527–535.

5 Safety Thresholds and Best Practices for Extractables and Leachables in Orally Inhaled and Nasal Drug Products. PQRI Leachables and Extractables Working Group. (2006). http://www.pqri.org/pdfs/LE-Recommendations-to-FDA-09-29-06.pdf.

6 Martin, J., Fitzgerald, R., Pothier, N., and Ding, W. (2010). *Recommendations for Testing and Evaluation of Extractables from Single-Use Process Equipment*. Bio-Process Systems Alliance.

7 Jenke, D. (2009). *Compatibility of Pharmaceutical Solutions and Contact Materials; Safety Considerations Associated with Extractables and Leachables*. New York: Wiley.

8 Jenke, D. (2012). A general strategy for the chemical aspects of the safety assessment of extractables and leachables in pharmaceutical drug products; The chemical assessment triad. *PDA J. Pharm. Sci. Technol.* 66 (2): 168–183.

9 USP <1661>. Evaluation of Plastic Packaging Systems for Pharmaceutical Use and Their Materials of Construction. USP43-NF38 – 8434. Official as of 1 November 2020.

10 ISO 10993:18 (2019). *Biological Evaluation of Medical Devices – Part 18: Evaluation and Testing within a Risk Management Process*. Arlington, VA: Association for the Advancement of Medical Instrumentation (AAMI).

6

Extractables, Leachables, and the Product Development Lifecycle: Change Control

6.1 Stage 4: Lifecycle Management and Change Control via a Comparability Protocol

As was noted previously, it is not uncommon for pharmaceutical items including packaging systems, manufacturing components, and medical devices to undergo changes during the market lifetime of a pharmaceutical or medical product. Change control is that process by which changes to an approved and registered pharmaceutical item are evaluated and ultimately accepted or rejected. Generally speaking, this change control process manages the risk that a safe, effective, stable, and compliant pharmaceutical or medical product is rendered unsafe, ineffective, unstable, or non-compliant by the change. The means by which change control is exercised from a chemical safety assessment perspective is established in a Comparability Protocol (CP), which in essence is a guideline, procedure, or specification that defines how the change control activity will be performed by the manufacturer. The concept of a CP was established by the FDA to address changes in marketed drug products and is defined by the FDA as "a comprehensive, prospectively written, plan for assessing the effect of a proposed CMC post-approval change(s) on the identity, strength, quality, purity and potency of a drug product or a biological product as these factors may relate to the safety or effectiveness of the product (i.e. product quality)" [1]. As explained by the FDA, "in many cases, using a CP will facilitate the subsequent implementation and reporting of CMC changes." Further, FDA notes that "you can use a CP for changes to the container closure system" and that "the CP can either apply to components or processes of the packaging system."

As noted in Figure 6.1, the CP consists of a series of tiered evaluations where change assessment moves through the tiers based on the magnitude of the product risk associated with the change and the ability of the evaluations performed in the tiers to manage that risk. As the CP establishes test methods and contains acceptance criteria,

it reduces the change control activity to standard work; that is, the change control activity becomes a process of performing the same actions for all changes that fall within the scope of the CP. One envisions a situation where separate CPs exist for each of the manufacturer's commercial pharmaceutical products. It is reasonable to expect that the CPs for the various pharmaceutical products would be similar in strategy but could be different in tactics and acceptance criteria, as tactics and criteria may need to be customized for each product.

Ultimately, completing the change assessment (finishing the CP) can have only two outcomes, acceptance of the change or rejection of the change.

Two critical aspects of a CP include the concepts of a Tiered Approach and the Two Levels of Change. These concepts are considered further as follows.

6.1.1 A Tiered Approach to Change Control

Practically speaking, one notes that there is a great diversity of changes that can occur to an approved and registered pharmaceutical item, both in terms of magnitude and potential impact. Furthermore, it is logical to suggest that there be a correlation between the magnitude and potential impact of a change and the amount of testing required to evaluate that change. Too much testing to support more simple changes is inefficient while too little testing to support more complex changes is insufficient.

For these reasons, a tiered approach (Figure 6.2) can be adopted for change control. The tiered approach represents a structured, systematic, and step-wise analysis of the potential impact the change may have on the key quality attributes of the product. In essence, each tier is defined by a certain evaluation process. The tiers are such that the evaluation becomes more rigorous, extensive, and robust as one moves from the lowest tier to the highest tier. Between tiers there is a decision point, where the decision is "does the evaluation performed in the tier allow the change to be approved (accepted)?" If the answer to this question is "yes," then clearly the evaluation is completed, no further

Extractables and Leachables: Characterization of Drug Products, Packaging, Manufacturing and Delivery Systems, and Medical Devices, First Edition. Dennis Jenke.
© 2022 John Wiley & Sons, Inc. Published 2022 by John Wiley & Sons, Inc.

Stage 4: Lifecycle management

Purpose: Qualify changes to approved and marketed packaged drug products

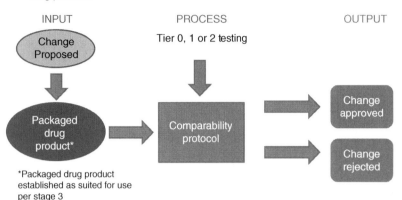

*Packaged drug product established as suited for use per stage 3

Figure 6.1 Process Flow Diagram, Stage 4, Lifecycle Management (Change Control). The process flow is illustrated for the case of a change to the packaging system of a packaged drug product. Changes to registered and marketed packaging systems are assessed via a Comparability Protocol to establish whether the change is acceptable or not.

The tiered approach process; moving through tiers

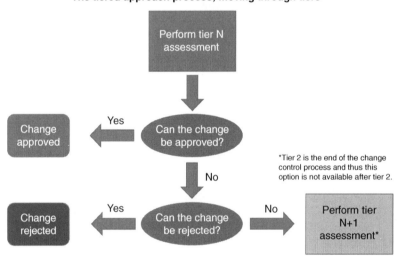

*Tier 2 is the end of the change control process and thus this option is not available after tier 2.

Figure 6.2 Moving through the Tiers of the Change Control Process. Evaluations performed in each Tier establish whether the change can be accepted or not. If changes can neither be rejected or accepted in a Tier, the evaluation moves to the next Tier.

evaluation is necessary, and the change is approved. If the answer to this question is "no," then the "no" answer is more closely examined. One possible reason for the "no" answer could be that the change is so impactful that the evaluation concludes that the change is not approvable regardless of how much additional testing is performed. In this case the outcome of the assessment is to reject the change with no further assessment. A second possible reason for a "no" answer would be that the evaluation was insufficient to support either the acceptance or rejection of the change. In this case, the only path forward is to continue the evaluation to the next tier.

In its essence, the tiered approach to change assessment is comparable to the product development process. Thus, for example, Tier 0 is typically an information gathering process, akin to Material Characterization. Tier 1 change control can be considered to be the link to system qualification (extractables) and Tier 2 change control can be linked to product qualification (leachables).

Generally, speaking, Tier 0, the first step of the process, involves two actions. The first action is collecting and evaluating all background information related to the change. For example, if the change to an item was "we are replacing one material with another," then appropriate background information could include certificates of compendial compliance for the replacement material. In essence, the Tier 0 evaluation is guided by a checklist that establishes what information is necessary to evaluate the change. A significant portion of the Tier 0 activity is obtaining this necessary information ("checking the boxes"). Since the information was deemed to be necessary, one would envisage that Tier 0 is concluded only when all the available information was collected ("all boxes checked").

The second action involves the collection of relevant information defining the exact nature of the change and reconciles this nature of the change with the patient exposure to the change. The purpose of this action is to establish the severity of the risk, as the severity of the risk establishes the level of evaluation that will be required to retire the risk. Specifically, the second Tier 0 action provides the answer to the question "Is this collected information sufficient to approve or reject the change?" As noticed previously, answering this question requires that the value, or information content, of the collected information be reconciled with the risk associated with the change. If the value or content of the collected information matches the risk, then a decision to either reject or accept the change can be made and the Change Control activity is completed. If the value or information content of the collected information is not sufficient to match the risk, then the only means of either approving or rejecting the change is to continue the Change Control activity by moving to Tier 1 testing.

Tier 1 differs from Tier 0 in that the primary activity in Tier 1 is the generation of data, as opposed to the collection of available information in Tier 0. Generally speaking, Tier 1 also differs from Tier 0 in terms of data assessment. While the decision on how to proceed out of Tier 0 may be somewhat subjective, the decision on how to proceed out of Tier 1 should be the objective process of comparing test results with acceptance criteria. Presumably, by completing Tier 1 and meeting all of the established acceptance criteria, one has demonstrated that the changed item is appropriate for use and thus that the change should be accepted. Failure to meet one or more of the acceptance criteria triggers either the rejection of the change (if the failures are deemed to be catastrophic) or further consideration of the change with the more extensive and rigorous Tier 2 testing.

Tiers 1 and 2 are closely aligned in terms of approach and process and differ in terms of the stage that they address in the product's shelf-life. Tier 1 testing focuses on the earliest stage in a product's shelf-life, specifically post-production. For most packaging items, for example, post-production includes exposure of the packaged product to elevated temperatures, including pasteurization and sterilization (autoclaving) processes. On the other hand, Tier 2 testing focuses on the latest stage in a product's shelf-life, specifically at expiry. In this regard Tier 2 addresses the circumstance where packaging leachables levels increase over the course of the product's shelf-life.

Because Tier 2 is the "end of the road" action in Change Control, there can be only two Tier 2 outcomes; (i) the risk is sufficiently low that the change is accepted or (ii) the risk is sufficiently high that the change must be rejected.

6.1.2 Tier 0 Assessment for Two Levels of Change

When a pharmaceutical item is changed, it can be changed at two levels, the material level and the system level. The distinction between material level and system level changes is important at the Tier 0 assessment stage. At the Tier 1 stage, material level and system level change control processes converge (see Figure 6.3).

6.1.2.1 Material Level Changes (Tier 0)
A change at the material level is a change that is specifically targeted at a material and involves either changing a material or replacing a material. Thus, for example, a material may be changed by changing its additive package. Alternatively, a system may be changed by replacing a material that is currently in the system with a material that is not currently in the item (system). In either case, we understand that the essential requirement for all materials used in an item is that the material be selected based on its characterization. Thus, an essential requirement for material level changes is that that complete characterization be available for the changed material. The information that reflects a complete characterization was defined previously in the Material Characterization section of this book, Section 5.5.1, Chapter 5. Collecting this information is a Tier 0 activity. The decision on how to proceed

Figure 6.3 The relationship and movement between the Tiers in Change Control. Different Tier 0 evaluations merge into a single process at the Tier 1 stage. The example illustrated here is a change to the packaging system of a packaged drug product.

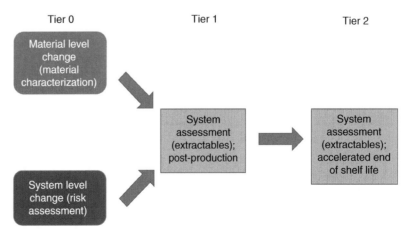

The tiered approach; material level and system level processes when exiting Tier 0 is reflected in the question "does the collected information establish that the material is suited for its intended use?" One notes that this is exactly the same question one asks in considering whether to move a material ahead in the product development process. If the answer to this question is "yes," then clearly the change is accepted. If the answer to this question is "no," the change is either rejected (if the material is clearly unsuited for its intended use) or further investigation is deemed to be necessary (Tier 1 testing).

6.1.2.2 Item Level Changes (Tier 0)

A change at the item (system) level is a change made to the item that may (or may not) involve a material but does not involve changing that material. For example, an item level change that involves a material would be changing the amount or location of that material in the item. An item level change that does not involve a material could be a change in the process used to assemble the items from its various materials and components.

Whereas the Tier 0 question for a material level change was "is the material suited for its intended use?" the Tier 0 question for an item is "is the change to the item so small that I can reasonably approve the change without testing the item?" To illustrate how this question might be evaluated, consider two examples of an item-level change. In the first example, consider the situation in which the assembly location (site) of the item is being changed. In investigating the change, the following points are established:

1. Both locations use the same components, sourced from the same component vendors.
2. Both locations use the same assembly process. There are no steps in the process and no reagents used in the process that are different from the current and proposed locations.
3. All machinery used in the process is operated within the same specifications.
4. The assembly environments are the same in both locations.
5. The storage of the assembled item prior to its final disposition is the same in both locations.
6. Items assembled at both locations meet the same item specifications.

In such a situation it has been established that there are no intentional means by which the item assembled in either location would be exposed to different process impurities. Furthermore, process-induced impurities should also be similar across the locations. Lastly, both processes are operating "under control" as acceptable output is generated in both locations. On this basis and this basis

alone, it may be acceptable to accept the change at the Tier 0 level without further item and product testing in Tiers 1 and 2.

The second example considers a more significant change. In this example, the pre-assembly processing of a port tube on a bag is being changed from ethylene oxide exposure to gamma irradiation. The port tube represents approximately 20% of the total weight of the packaging system (bag plus port) and is directly bonded on to the bag, potentially exposing portions of the port tube to the solution that is stored in the bag. Furthermore, the packaged product is terminally sterilized and has a two-year shelf-life at ambient temperature. Knowing that (i) gamma irradiation is a high-energy process whose ability to change a component's extractables profile is well-known and (ii) there is a clear path and sufficient thermal driving force for port tube extractables to become entrained in the packaged product as leachables, it is most likely that this change would not be approved in Tier 0 and thus that change approval would require Tier 1 assessment.

6.1.3 Tier 1 Assessment

The starting point of the Tier 1 assessment is that the registered item was profiled for extractables as part of the product development process and that the resulting extractables assessment was established to be acceptable in terms of the low risk that the extractables (as leachables) would adversely affect critical product quality attributes (such as patient safety, product stability, etc.). Thus, if the change were to adversely affect the product's critical quality attributes, the change would have to produce a modification in the extractables profile. If the change had no discernible effect on the extractables profile, or the changes to the profile did not have an associated product quality risk, then the item's change could be approved.

Thus, one approach to Tier 1 Change Control is to generate the extractables profile of the changed item and to compare that profile with a profile that was previously generated for the current item. These two profiles, current versus changed, would be examined for potentially significant differences, as similarities or insignificant differences between the current and changed item would be irrelevant.

There are several means for performing such a difference assessment. On the one hand, the extractables profile for the changed item could be compared to a historical extractables profile obtained for the current item. Otherwise, a side-by-side comparison can be made between current and changed items that are extracted as control and test articles in the same study.

Alternatively, the changed item could simply be tested to establish its own extractables profile. Such a profile would be assessed from an impact perspective solely on its own basis.

From a practical perspective, the choice of whether to perform a differential assessment or a stand-alone assessment depends on the number of anticipated difference extractables versus the number of current extractables. When the number of difference extractables is small compared to the total number of extractables, a difference study may have certain practical advantages versus the stand-alone assessment.

An additional consideration is the ability to obtain an appropriate control. For the difference study to be most effective, the only difference between the test and control should be the change being investigated. However, it is frequently the case that obtaining a truly matched test and control article is difficult. When an appropriate control cannot be obtained, stand-alone profiling of the changed item is the proper approach. It is the author's experience that the failure of difference studies to produce actional outcomes is often the result of poorly matched test and control articles.

Tier 1 studies are typically performed on complete items, as (i) items are more likely to have predicate extractables profiles and (ii) the full effect of the change on the product is most appropriately assessed on an item basis. However, Tier 1 assessment can be performed on components or materials of construction. Nevertheless, the following point is relevant when considering whether to perform Tier 1 tests on materials, components, or complete items. A properly designed and executed difference study performed on a complete item definitively establishes how the change impacts the item's suitability for use. On the other hand, properly designed and executed studies performed on materials or components may be equivocal and the true impact of the change on the item's suitability for use might only be ascertainable by follow-up testing at the item level. Thus, all things being equal, performing difference studies on complete items is the quickest route to generating definitive and actionable information.

It is logical that there are two possible differences in extractables profiles, test item versus control item, that can lead to the conclusion that the change is not acceptable as it results in an incompatibility; new extractables, and increased extractables. Both terms are evident; new extractables resulting from a change to an item are extractables associated with the test article (changed item) that were not associated with the control article (current item). Increased extractables are extractables associated with both the test and control articles, but with higher levels in the test article. Clearly, both a new extractable and increased levels of current extractables could produce an incompatibility.

It is possible that an item change can also produce a "better" extractables profile, such as the loss of an extractable from the currents item's extractable profile or a reduction in the level of an extractable. Such changes are trivial to assess and approve, as these changes cannot worsen an item's compatibility with its contact medium.

6.1.4 Tier 2 Assessment

The Tier 2 assessment is triggered by two circumstances. The first circumstance is when the Tier 0 and Tier 1 assessments were not sufficiently definitive to either accept or reject the change. This would be the case if the extractables assessment established that an extractable associated with the changed system could accumulate in the drug product as a leachable at potentially impactful levels. In this circumstance the purpose of the Tier 2 assessment would be to establish the accumulation properties of the extractable, targeted as a leachable.

The second circumstance would be the situation in which the Tier 1 assessment was sufficiently definitive to accept or reject the change but additional confirmation of this outcome is desired. In this circumstance, the purpose of the Tier 2 assessment would be to establish that the maximum level achieved by the target leachable in the product over its shelf-life is less than the maximum level achieved by the extractable in the extraction study, as it is the maximum level that is used to assess product impact.

In either case, a Tier 2 assessment would be performed to determine the levels of extractables at the end of a drug product's shelf-life. The Tier 2 assessment, leveraging accelerated aging to reduce study duration, essentially repeats the extractables profiling activities performed in Tier 1. Since it can be anticipated that many of the extractables encountered in the Tier 2 assessment are the same extractables that were addressed in the Tier 1 assessment, the Tier 2 assessment should be more readily accomplished than the Tier 1 assessment.

6.1.5 Additional Lifecycle Management Considerations

6.1.5.1 Refreshing a System's Extractables Profile

As noted previously, post-launch changes made to the packaging system of a marketed product will be assessed individually within the context of each individual change (Stage 4). This is the case as changes rarely occur all at once and rather are interspersed throughout the product's lifecycle. While such a strategy addresses the effect of each change individually, it might not address the accumulated effect of all changes, which may be meaningful over a long commercial lifetime punctuated by multiple changes. Thus, at some point in a product's commercial lifetime it may be that a packaging system has been so greatly modified as a result of small individual changes that the changed system is readily recognized as being different, in a potentially meaningful way, from the system that was safety qualified and registered.

In such a circumstance, it is necessary and appropriate to "refresh" the packaging system's extractables profile by performing a full Stage 2 assessment on the changed system. Circumstances that would trigger such a refreshing of the extractables profile should be clearly defined in a packaging system's CP.

Although this discussion focuses on changes to a packaging system, it is equally relevant to changes in other pharmaceutical items such as manufacturing components and medical devices.

6.1.5.2 Control of Incoming Materials

It is reasonable to suppose that a marketed product will be manufactured from different and multiple lots (batches) of materials of construction over the course of its commercial lifetime. Although the Stage 2 extractables and Stage 3 leachables safety assessments of a marketed product or its packaging system might address lot-to-lot variations in the composition of materials of construction, and establish the effect of such changes on extractables and leachables profiles to the extent that is adequate for the initial registration, such assessments may not be sufficient to address lot-to-lot variation over the entire commercial lifetime.

Although the use of quality control measures such as the routine testing of incoming lots of materials is considered in the regulatory guidelines (for example, Section III.B.2.b in reference [2]), it is noted in that reference that "currently there is no general policy concerning the monitoring of a packaging system and components with regard to safety. One notable exception involves inhalation drug products for which batch-to-batch monitoring of the extraction profile for the polymeric and elastomeric components is routine." Nevertheless, in certain other higher risk circumstances it may be appropriate and prudent to implement a well-defined process of incoming materials qualification for extractables and/or their sources (ingredients). A critical part of such a process is the means by which such testing would be suspended after process capability has been adequately established. Additionally, such a process would be based on an established and well-defined correlation between the material property that is being measured and the packaging system's extractables profile (or the packaged drug product's leachables profile).

6.2 Stage 4: Life-Cycle Management and Change Control via a Risk-Based Approach

6.2.1 Discussion

In the previous section, a change control strategy based on a CP was discussed. In this section, we consider a second approach to change control, based on risk management principles and concepts. It could be the case that such a risk-based approach could exist inside a CP.

As was done in the previous discussion, this Discussion will use the case of a packaging system for the purpose of illustration. However, the general concepts are equally applicable to the other types of pharmaceutical items.

Practically speaking, one notes that there is a great diversity of changes that can occur to an approved and registered packaging system, in terms of the nature, magnitude, and potential impact of the change. Thus, it is logical to expect that different changes could be addressed in different ways. Furthermore, it is logical to suggest that there is a correlation between the magnitude and potential impact of a change and the nature of the assessment required to evaluate that change, with smaller, less-impactful changes requiring lesser assessment and evaluation and larger, more-impactful changes requiring greater assessment and evaluation. In any case, it is necessary to properly match the risk with the level of assessment/evaluation as too much assessment to support more simple changes is inefficient while too little assessment to support more complex changes is insufficient to provide a valid assessment.

At one extreme, simple, small, or less impactful changes may obviously have a limited associated risk and in such cases the assessment may be limited to an explanation and documentation of the circumstances that establish the low risk. Such changes are termed **trivial changes**. At the other extreme, complex, large, or highly impactful changes may obviously have a sufficiently high associated risk that the risk can only be managed by an extensive assessment that will include laboratory testing. Such changes are termed **essential changes**. Representing a middle ground, certain changes may be such that although they cannot be justified via a limited statement of certain obvious aspects of the change, it is unlikely that justifying the change will require laboratory testing. Such changes are termed **moderate changes**. In the middle circumstance of a moderate change, a detailed discussion of available data and the relevant science may be sufficient to justify the change.

These change categories are illustrated in Figure 6.4.

There is a certain logic that the magnitude of testing required to approve a change is directly proportional to

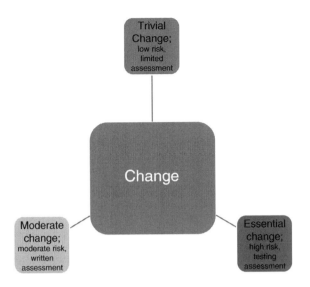

Figure 6.4 The three categories of changes to a packaging system.

the expected impact that change will have. Furthermore, there is ample regulatory precedent in terms of regulatory support for risk-based practices. The key to applying a risk-based approach is performing an initial appraisal to categorize the change and to establish the likely type of evaluation and assessment that will be required to justify the change. Categorization is an essential aspect of change control as it is used to establish the nature and amount of evaluation and assessment that is required to approve or reject the change. Proper and reproducible categorization leads to efficient and effective change control. If a change can be categorized as trivial, then the initial appraisal is both the documentation that the change is trivial and the documentation of the facts that support this change categorization. As such, the initial appraisal is essentially the entire evaluation and leads directly to the assessment that the change is approvable. If the change is established as either moderate or essential, then the initial appraisal documents this categorization and the change is directed to that technical function that is responsible for designing and implementing the evaluation, be it information gathering or testing. Upon completion of the evaluation, the resulting information is reported and technically assessed, via the generation of a Special Report (in the case of a moderate change) or a Testing Protocol and Report (for an essential change). The conclusion drawn from the information is either approval or rejection of the change.

These approaches to change control are illustrated in Figure 6.5.

The purpose of this section of the book is to provide a generalized means by which changes can be categorized as a prerequisite to change evaluation and assessment and as the first step in the change control process illustrated in Figure 6.5. Use of this categorization process will enable a change assessor to establish the type of evaluation and assessment that is required to address the change and to more readily and reproducibly recognize changes that are likely to be either trivial, moderate, or essential. In order to provide sufficient context and detail, this section addresses changes to packaging systems, although the approach may be applicable to other changes and to other pharmaceutical items. Although making a change to an existing and approved packaging system can potentially have many and varied effects on the stability, efficacy, purity, and quality of a packaged drug product, this document considers only the adverse effect that a packaging change could have on patient safety.

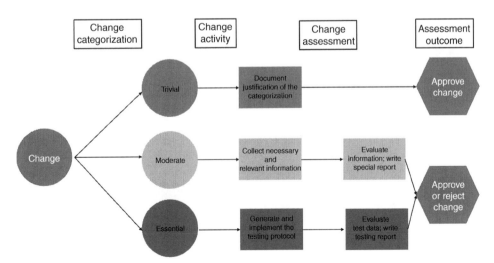

Figure 6.5 Process Flow Diagram, Change Control. The activity of change characterization is discussed in greater detail in Section 6.2. The change activity and change assessment would typically be addressed as part of a formal change control process such as the Comparability Protocol. (*See insert for colour representation of the figure.*)

6.2.2 Definitions

Change Control. A formal process used to ensure that changes to a product or system are introduced in a controlled and coordinated manner and that the change does not adversely impact key product quality and performance attributes.

Principles of Risk Assessment. Risk assessment, in the context of safety and change control, refers to the identification of potential hazards associated with a change and a consideration of the likelihood that the hazard will manifest itself in the changed drug product at a level sufficiently large that patient safety is adversely affected. The identification of potential hazards addresses the *severity* aspect of risk assessment and the assessment of the probable impact on the hazards addresses the *probability* aspect of risk assessment.

Safety Risk. Although in general change control deals with ascertaining the effect of a change in a packaged drug product's key quality attributes, perhaps the most critical quality attribute is patient safety. Thus, the safety risk associated with a change to a packaged drug product's packaging system is that the change will alter the leachables profile of the drug product to such an extent that the packaged drug product will contain one or more leachables at unsafe levels (meaning that the leachables are likely to have an adverse impact on patient health).

Safety Risk Assessment. This is a risk assessment process that considers the ability of a change to a drug product's packaging system to adversely affect patient safety via the biological action of leachables associated with the change.

Categories of Changes. Changes are categorized in terms of their probable effect and the means by which they are addressed and assessed.

Trivial Changes. Changes whose effect on the key quality attributes of the packaging system and the packaged drug product are likely to be minimal and generally unimportant. Such changes are assessed via a proper entry in the change control documentation.

Moderate Changes, Changes that could affect the key quality attributes of the packaging system and the packaged drug product, although generally the anticipated effect is small and unimportant. Such changes are assessed via a review of the circumstances of the change and the available information that addresses possible impacts of the change. Based on the review, the change is assessed via a Special Report and the conclusion of "no significant effect" is documented. Although it is desirable that moderate changes be addressed without the generation of experimental data, limited laboratory studies designed to "fill in the gaps" in essential information may be required.

Essential Changes. Changes that most likely will affect the key quality attributes of the packaging system and the packaged drug product and whose effect could be significant and important. In such cases, it is unlikely that available information will be sufficient to address and assess the change and thus laboratory studies, such as extractables and leachables studies, will be necessary to produce the information that addresses the probable impacts of the change. The design and justification of the design is contained in a testing protocol that specifies test methods and acceptance criteria. A Final Report is written to capture both the testing performed and the assessment of the test data and ultimately contains the conclusion of either "no significant effect" (accept change) or "significant effect" (reject change).

Stages of Categorization. Categorization of changes occurs in two stages.

Stage 1: Stage 1 categorization is based on the packaged drug product's sensitivity to change. If a drug product is sensitive to changes, meaning that small changes can have an important effect on a key quality attribute, then the change moves into Stage 2 for categorization. If a drug product is relatively immune to changes, or if the effects of change are muted by some property of the drug product, then the change is categorized as Trivial and addressed and assessed in the appropriate manner.

Stage 2: Stage 2 categorization is a defined process by which the probable impact of the change on the drug product is established. Depending on the probable impact, the change is categorized as Trivial, Moderate, or Essential.

Types of Changes. Changes can be classified based on the nature of the change as follows:

Elimination: Some aspect of the packaging system or of the packaged drug product is eliminated, either process or composition.

Adjustment: Some aspect of the packaging system or of the packaged drug product is changed by an incremental amount (adjusted), either process or composition.

"Like for Like" Replacement: Some aspect of the packaged drug product, either process or composition, is eliminated and replaced by a process step or component that is comparable (and generally equivalent) to the eliminated entity.

Substitution: A replacement in which the replacing entity is not comparable to the entity it is replacing.

Figure 6.6 Process Flow Diagram, Change Categorization.

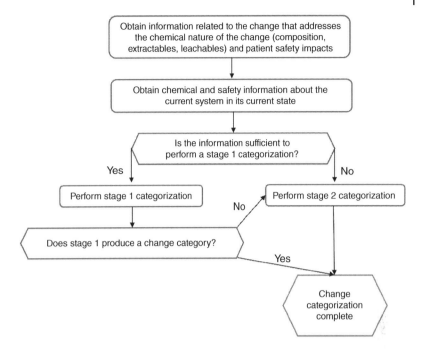

Addition: A new aspect, related to either a process or to composition, is added to the existing packaging system.

Classes of Packaging. Packaging systems are classified in terms of the nature of their contact with the packaged drug product.

Primary Packaging: Components of a packaging system that are in direct contact with the drug product at some time during the manufacturing storage or use of the drug product.

Secondary Packaging: Components of a packaging system that are never in direct contact with the drug product during the manufacturing storage or use of the drug product as they are separated from the drug product by the primary packaging.

Tertiary Packaging: Components of a packaging system that are never in direct contact with the drug product during the manufacturing storage or use of the drug product as they are separated from the drug product by the primary and secondary packaging.

6.2.3 Essential Principles of Change Categorization

A key to categorizing a change is understanding what drives the patient safety risk associated with the change. Thus, the following discussion establishes the effect of the change on the leachables profile as the key driver of the safety risk. The document then goes on to describe a two-stage process for categorizing changes: Stage 1, which is based on knowledge of the safety of the existing packaging system (prior to the change), and Stage 2, which is based on a consideration of the nature of the change itself (see Figure 6.6).

6.2.4 Understanding the Absolute Risk

Patient safety is potentially impacted by drug product packaging via leachables derived from the packaging. These leachables represent foreign impurities in the drug product; as drug product impurities, the patient is exposed to the leachables during the clinical use of the packaged drug product. If the exposure levels are sufficiently high and/or the inherent toxicity of the leachable is sufficiently great, then exposure to leachables could adversely impact a patient's health.

Thus, the patient safety issue associated with a change to a drug product's packaging system is reflected in how the change affects the packaged drug product's leachables profile. Two circumstances are relevant as they could lead to an increased patient exposure to leachables:

1. The change increases the level of an existing leachable to a value sufficiently high that the leachable represents a threat to patient safety and/or
2. The change results in a new, previously unreported, leachable whose level is sufficiently high that it represents a patient safety risk.

6.2.5 Stage 1 Categorization: Use of Prior Knowledge to Drive Categorization

During the development of a packaged drug product, the patient safety impact of the packaging is unknown and can only be established by extractables profiling of the packaging and/or leachables profiling of the drug product. The situation is markedly simpler in the situation of change control. Prior to the change, the starting packaging system has already been established to be safe. The question associated with change control is therefore not "is the changed system safe?" but rather "has the change rendered a safe system unsafe?"

Considering existing leachables, the issue of "is the change large enough that it could produce unsafe leachables?" can be addressed by asking "how safe was the initial packaging system prior to the change?" For example, let us consider two cases. In case 1, the toxicological safety assessment of a packaging system revealed that the lowest margin of safety (MoS) for an individual leachable was 10 (MoS = 10), meaning that the patient's exposure to the leachable was 10 times lower than the leachable's daily tolerable intake. This leachable with the lowest margin of safety is termed the trigger leachable. In this case, the change would have to increase the level of the specific trigger leachable by a factor of 10 before the leachable would become a safety issue.

In case 2, the toxicological safety assessment of a second packaging system revealed the same leachable as in case 1, but in case 2, the leachable's MoS is 1. In this case, almost any increase in the leachable's level resulting from the change would decrease the MoS to a value of less than 1, meaning that the patient's exposure would be greater than the tolerable daily intake, which is an unsafe circumstance.

These cases can be used to illustrate how one can use this information to categorize changes. In either case, the change would have to affect the safety trigger leachable for safety to be affected by the change. If it can be established that the change does not affect the safety trigger leachable, then the safety risk for the existing leachables is low and the change is trivial from that perspective.

If the change does affect the safety trigger leachable, then the safety risk is dictated by the MoS for the trigger leachable. In case 1, the high value of the MoS suggests that the change would have to be large and significant to increase the level of the safety trigger leachable to the extent that it becomes a potential safety hazard. Thus, any change that is neither large nor significant can be concluded to be a trivial change with respect to existing leachables, which is likely to be the most common situation. However, in case 2, the low value of the MoS suggests that even a small change could increase the level of the safety trigger leachable to the extent that it becomes a potential safety hazard. Thus, in case 2, any change, regardless of size, will likely be an essential change. Of course, circumstances of an intermediate MoS are categorized on a case-by-case basis.

Once the change has been categorized with respect to existing leachables, then it must be categorized with respect to new leachables via the Stage 2 process.

This overall Stage 1 categorization process is illustrated in Figure 6.7.

In certain circumstances, it may not be possible or practical to perform a Stage 1 categorization. In such

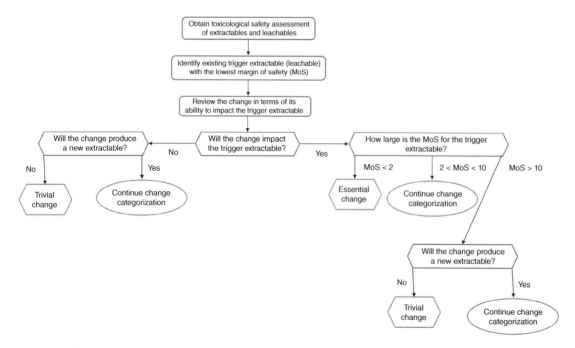

Figure 6.7 Process Flow Diagram, Stage 1 Categorization.

circumstances, the Stage 2 categorization is performed as a first step.

6.2.6 Stage 2 Categorization: General Process

6.2.6.1 Change Dimensions

Categorization of a change requires that the change be evaluated with respect to those critical dimensions of the change, and of the packaging system being changed, which are related to the change's ability to affect patient safety. As noted in Figure 6.8, four critical dimensions have been established:

- Where in the packaging system is the change being made?
- What type of change is being considered?
- How is the packaging system being changed?
- What is the magnitude of the change?

Classification of a change with respect to these four dimensions allows for an estimation of the relative safety risk associated with the change, thereby allowing the change to be categorized.

6.2.6.2 Dimension 1, Where in the Packaging System Is the Change Being Made?

As noted previously, a change can adversely impact the safety of a packaging system if it alters the system's leachables profile. The system's leachables profile is changed if (i) the change increases the level of extractables in the packaging system or introduces a new extractable to the packaging system and (ii) the extractable(s) leach into the drug product.

Considering factor (ii), it is noted that the likelihood that an extractable will become a leachable depends on the nature of the contact between the packaging system component that is the source of the extractable and the packaged drug product. If there is direct contact between the source component and the drug product, then leaching of the extractable is more likely to occur versus if there is no direct contact. This is the case as leaching in the case of indirect contact can only occur if the extractable crosses whatever barrier defines the indirect contact.

The term primary packaging has been applied to packaging system components that have direct contact with the drug product. Thus, for example, the plastic bag in which the drug product is stored is primary packaging. Because there is no barrier between the bag and the drug product, leaching of bag extractables by the drug product is likely. Thus, a change to a primary packaging component that results in new extractables or a higher level of existing extractables is likely to produce a change in the drug product's leachables profile, reflecting a greater safety risk.

Packaging components that are separated from the drug product by the primary packaging are termed secondary

Figure 6.8 Change dimensions considered during Stage 2 Categorization.

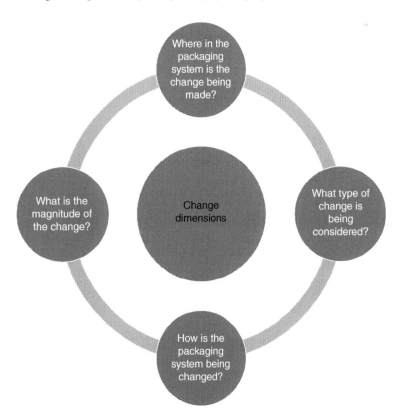

packaging. An example of secondary packaging is a bag's overpouch or dust cover. Other examples of secondary packaging include a port tube assembly (when separated from the drug product by a membrane), printing or a label placed on the outside of the bag, an oxygen scavenger, a drain bag, and associated connectology (e.g. tubing). As the primary packaging (bag) is a barrier to the leaching of extractables from secondary packaging (overpouch), it is less likely (but not impossible) that extractables from the secondary packaging will leach into the drug product, reflecting a lower safety risk.

Lastly, packaging components that are separated from the drug product by both the primary and secondary packaging are termed tertiary packaging. Examples of tertiary packaging include printing or labels affixed to the outside of an overpouch and shipping cartons. As both the primary and secondary packaging are barriers to the leaching of extractables from secondary packaging, it is much less likely that extractables from tertiary packaging leach into the drug product, reflecting an even lower safety risk.

The concepts of primary, secondary, and tertiary packaging and their relationships to safety risk is illustrated in Figure 6.9.

While both primary and secondary packaging can be barriers to leaching, this does not mean that leaching from secondary and tertiary components does not occur. Unless they are specifically designed to be absolute barriers (for example, a foil overpouch), primary and secondary packaging is semi-permeable and thus movement of substances across these barriers will occur given a sufficiently long contact time (for example, a two-year ambient temperature shelf-life) or a significant thermal driving force (e.g. terminal sterilization by heat).

6.2.6.3 Dimension 2, What Type of Change Is Being Considered?

A second dimension to consider when categorizing a change is the type or nature of the change. In general, three types of changes can be envisioned (Figure 6.10):

1. A process change, meaning that something about the process of either generating the packaging system or preparing the packaging system for use, is being changed. Examples of processing changes include:
   - Adding, changing, or eliminating a washing step for a packaging system component,

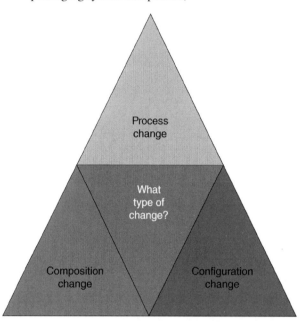

Figure 6.10 Options associated with Change Dimension 2. What type of change is being considered? (*See insert for colour representation of the figure.*)

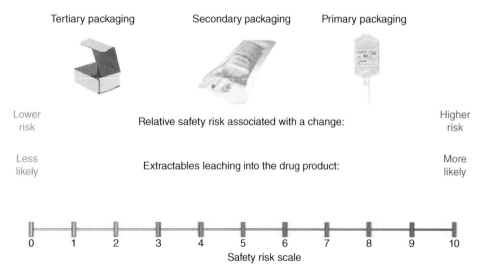

Figure 6.9 Change Dimension 1, where in the packaging system is the change? The concept of primary, secondary, and tertiary packaging and their associated safety risks.

- Changing the terminal sterilization cycle,
- Changing the manufacturing process or location,
- Pre-sterilizing a previously unsterilized packaging component.

2. A configuration change, meaning that although the components of the packaging system are not being changed, the structure or configuration of the components in the packaging system is being changed. Examples of configuration changes include:
   - Increasing the thickness of a layer of a multi-layered film,
   - Changing the structure of a multi-layered film (for example, starting with an A–B–A structure and changing it to a A–A–B structure),
   - Removing or adding a membrane to a port tube,
   - Changing the size of a rubber stopper.

3. A composition change, meaning that an ingredient, material, or component that is currently in the packaging system is being eliminated, adjusted, replaced or that an ingredient, material, or component that is not currently in the packaging system is being added. Examples of composition changes include:
   - An antioxidant in a material used in the system is being replaced by another antioxidant,
   - The catalyst used to produce a polyolefin material that is used in the system is being replaced,
   - A material from Vendor A is being replaced by the same material sourced from Vendor B,
   - A polyethylene material is used as a substitute for a polypropylene material that is currently used in the system,
   - A new component is being added to the packaging system.

Unlike the options in Dimension 1, the options in Dimension 2 cannot be unilaterally ranked in terms of risk, although generally speaking the safety risk associated with a process change is lower than the risk associated with a configuration change, which is lower than the risk associated with a composition change. Rather, the three Dimension 2 options are unique and unrelated as they are fundamentally different from one another and thus must be addressed during categorization in fundamentally different ways.

6.2.6.4 Dimension 3, how Is the Packaging System Being Changed?

The third change dimension to consider when categorizing a change is the means by which the change is being accomplished. For example, a compositional change can be made in a number of different ways:

- The composition of a material in the packaging system can be *adjusted*, meaning that the proportion of its existing ingredients is changed,
- The composition of the packaging system is changed by *eliminating* an existing material,
- The composition of the packaging system is changed by performing a *"like-for-like" replacement*, replacing one material with a similar material (e.g. one polypropylene with another polypropylene),
- The composition of the packaging system is changed by *substituting* a new material for an existing dissimilar material (e.g. using a polyethylene in place of a polypropylene),
- The composition of the packaging system is changed by *adding* a new material.

A similar distinction between the various means of change can be envisioned for process changes. Alternatively, configuration changes are generally considered to be accomplished as adjustments.

Certain generalizations can be made in terms of the safety risk associated with each of these means (Figure 6.11). For example, eliminating a material from a packaging

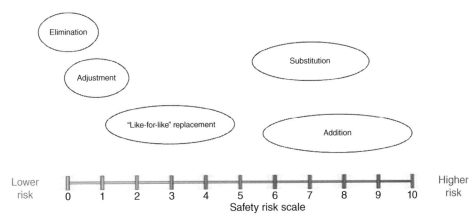

Figure 6.11 The relative safety risk associated with various means of changing a packaging system (Dimension 3).

system is not only considered to be a low risk to patient safety, it could be considered to be a benefit to patient safety, as eliminating a material also eliminates a source of extractables. However, as is the case with many generalizations, one must be vigilant in terms of allowing for the exceptions to the rule.

6.2.6.5 Dimension 4, What Is the Magnitude of the Change?

The fourth dimension considered in change categorization reflects the generalization that "there are big changes and there are little changes and the little changes should be assessed to a lesser extent than the bigger changes." While it is most difficult to provide useful generalizations on how to differentiate a minor change from a major change, it is noted that the concepts of minor and major have two facets: the absolute magnitude of the change and the criticality of the item that is being changed. To illustrate this concept, consider the Magnitude of the Change Assessment Grid shown as Figure 6.12. This two-dimensional grid, which is established by two measures (minor and major) of the dimension "Importance of the Item Being Changed" and two measures (minor and major) of the dimension "Magnitude of the Change," consists of four separate possibilities. One of these possibilities, corresponding to a minor change made to a minor item, has the lowest associated risk, while another possibility, corresponding to a major change being made to a major item, has the highest associated risk. The other two possibilities are assigned a roughly equal, intermediate risk.

To illustrate the application of the Grid, consider the following four potential changes:

1. A manufacturer is qualifying an alternate vendor for the DEHP that is used as a plasticizer in its plasticized PVC film. The film contains 30% by weight DEHP.
2. A specific polypropylene is used in the outer-most layer of a five-layer film that makes up the body of an i.v. bag. The weight percent of the polypropylene in the entire film is 5%. The amount of antioxidant in the polypropylene is being increased from 0.10 to 0.15%.

3. A label is being added to the outside of an overpouch post-sterilization.
4. The solution contact layer of a multi-layered film contains 20% polyethylene. The vendor of the polyethylene has informed the manufacturer that it is changing the manufacturing location of the polyethylene. The vendor has provided the manufacturer with a statement that establishes that the raw materials used in manufacturing and the manufacturing processes are the same in both locations. Additionally, the vendor has provided the manufacturer with statements that document the fact that the polyethylene manufactured from the proposed location meets the relevant international compendial requirements and has been established to comply with international indirect food additive regulations.

These four changes are placed in the Grid as follows:

1. Using a new vendor to source a raw material is generally considered to be a large change, as there is every reason to believe that the impurity profiles of the currently used and proposed raw materials are different. Because the DEHP is 35% by weight of the bag material and because the bag material will be in direct drug content, the item being changed is important. Thus, this change is classified as major item, major change, which is the highest risk.
2. The material being changed is not solution contact. The material being changed makes up only a small portion of the total film by weight. Thus, the material being changed is a minor item. The change itself is small, increasing the amount of the antioxidant by only 0.05% by weight. Thus, this change is classified as minor item, minor change, which is the lowest risk category.
3. The addition of a label is generally a major change to a packaging system. However, this label is being added to the outside of an overpouch, which means it is tertiary packaging. Furthermore, the label is being added post-sterilization, which means the label is not present during the thermal stress of autoclaving. Thus, the item being changed (the overpouch) is considered

		Importance of the item being changed	
		Minor item	Major item
Magnitude of the change	Minor change	Lowest risk	Intermediate risk
	Major change	Intermediate risk	Highest risk

Figure 6.12 The magnitude of the Change Assessment Grid (Dimension 4).

to be minor. Therefore, this change is classified as major change, minor item, which is an intermediate risk.

4. The material being changed is in the product contact layer of the container and makes up a substantial portion of that contact layer. Thus, the change is being made to a major component. Given the level of documentation provided by the materials' vendor, the magnitude or impact of the change is likely low. Therefore, this change is classified as a minor change, major item, which is an intermediate risk.

As there is a certain measure of subjectivity associated with the use of the Grid and the differentiation between minor and major, the risk assessor is advised to be appropriately conservative in the assessment.

6.2.7 Completing Stage 2 Categorization

The cumulative use of the four dimensions to categorize a change is enabled via the use of Figure 6.13, which represents a decision matrix for the Stage 2 change categorization process.

The decision matrix is used as follows:

1. The change is associated with the place in the packaging system where the change occurs (primary, secondary, tertiary packaging) per dimension 1.
2. Within the dimension 1 category, the change is linked with the proper descriptor associated with "What type of change is being considered?" dimension 2 (process, configuration, composition).
3. Within the dimension 1 and 2 categories, the change is linked with the proper descriptor associated with "How is the packaging system being changed?" dimension 3 (elimination, adjustment, replacement, substitution, addition).
4. Within the dimension 1, 2, and 3 categories, the change is linked to the proper descriptor associated with "What is the magnitude of the change," dimension 4 (minor item, minor change; major item, minor change; minor item, major change; major item, major change).
5. Completion of steps 1 through 4 places the change into a specific category (box) on the right-hand side of Figure 6.13. The color of the category box establishes the category of the change (trivial change, moderate change, and essential change).

It is noted in a review of Figure 6.13 that there are four change categories (colors), but to this point the discussion has only established three change categories. What has happened in generating Figure 6.13 is that the highest risk essential change category has been broken into two sub-categories, depending on the amount of testing required. In certain situations, the necessary testing may be

limited to highly specific and targeted controlled extraction studies. In other situations, it may be necessary to perform extensive extractables (and possibly leachables) testing, designed to essentially re-set the baseline extractables profile of the changed packaging system.

To illustrate the use of the decision matrix, consider a change that was discussed previously. A certain polypropylene is used in the outer-most layer of a five-layer film that makes up the body of an i.v. bag. The weight percent of the polypropylene in the entire film is 5%. The amount of antioxidant in the polypropylene is being increased from 0.10 to 0.15%. The categorization of this change proceeds as follows:

1. Dimension 1: the change is to primary packaging.
2. Dimension 2: the change is one of composition.
3. Dimension 3: the change is an adjustment.
4. Dimension 4: as established previously, the change is minor item, minor change.

This process is illustrated in Figure 6.13 by the progression of arrows and ultimately ends with the change being classified as a trivial change. Thus, in this case the change assessor documents the thought process that led to this outcome, documents the outcome and approves the change, completing the change assessment process.

As a second example of use of the decision matrix, consider a change that involves the addition of an adhesive label to a plastic package. In this case, the label is applied to the outer surface of the bag before the bag is placed in a foil (barrier) overpouch and terminally sterilized. The label contains critical product information and covers roughly 20% of the bag's surface. The categorization of this change proceeds as follows:

5. Dimension 1: the change is to secondary packaging.
6. Dimension 2: the change is one of composition.
7. Dimension 3: the change is an addition of a new component.
8. Dimension 4: the change is judged to be major item, major change, given the size of the label, the circumstance that it added prior to sterilization, and the fact that the labeled bag is inside a barrier overpouch (preventing volatile label-related substances from escaping the system).

This process is illustrated in Figure 6.13 by the progression of stars and ultimately ends with the change being classified as a major change carrying the highest risk. Thus, in this case the change assessor concludes that testing is necessary, documents the thought process that led to this outcome, and generates a Testing Protocol, which initiates the actions that must be taken to complete the change assessment process.

Where is the change?	What type of change?	How is the change being made?	How big is the change?
Tertiary packaging (e.g., label on overpouch, shipping carton)	Process change	Adjustment of an existing step	Minor adjustment to a minor step
			Major adjustment to a minor step
			Minor adjustment to a major step
			Major adjustment to a major step
		Elimination of an existing step	Elimination of a minor step
			Elimination of a potentially important step
			Elimination of a major step
		Replacement of an existing step	Replacement of a minor step
			Replacement of a potentially important step
			Replacement of a major step
		Addition of a new step	Addition of a minor step
			Addition of a potentially important step
			Addition of a major step
	Configuration change	Re-ordering of existing configuration	Re-ordering of minor components
			Re-ordering of major components
		Dimensional changes to existing configuration	Minor change to a minor component
			Major change to a minor component
			Minor change to a major component
			Major change to a major component
	Compositional change	Adjustment to an existing component	Minor adjustment to a minor component
			Major adjustment to a minor component
			Minor adjustment to a major component
			Major adjustment to a major component
		Elimination of an existing component	Elimination of a minor component
			Elimination of a potentially important component
			Elimination of a major component
		Like for like replacement of an existing component	Replacement of a minor component
			Replacement of a potentially important component
			Replacement of a major component
		Substitution for an existing component (not like for like)	Substitution for a minor component
			Substitution for a potentially important component
			Substitution for a major component
		Addition of a new component	Addition of a minor component
			Addition of a potentially important component
			Addition of a major component

Figure 6.13 Decision Matrix, Change Stage 2, Change Categorization. (*See insert for colour representation of the figure.*)

Where is the change?	What type of change?	How is the change being made?	How big is the change?
		Adjustment of an existing step	Minor adjustment to a minor step
			Major adjustment to a minor step
			Minor adjustment to a major step
			Major adjustment to a major step
	Process change	Elimination of an existing step	Elimination of a minor step
			Elimination of a potentially important step
			Elimination of a major step
		Replacement of an existing step	Replacement of a minor step
			Replacement of a potentially important step
			Replacement of a major step
		Addition of a new step	Addition of a minor step
			Addition of a potentially important step
			Addition of a major step
	Configuration change	Re-ordering of existing configuration	Re-ordering of minor components
			Re-ordering of major components
		Dimensional changes to existing configuration	Minor change to a minor component
			Major change to a minor component
			Minor change to a major component
			Major change to a major component
Secondary packaging (e.g., overpouch, port assembly separated by a membrane, oxygen indicator in space between primary bag and overpouch) ★	Compositional change ★	Adjustment to an existing component	Minor adjustment to a minor component
			Major adjustment to a minor component
			Minor adjustment to a major component
			Major adjustment to a major component
		Elimination of an existing component	Elimination of a minor component
			Elimination of a potentially important component
			Elimination of a major component
		Like for like replacement of an existing component	Replacement of a minor component
			Replacement of a potentially important component
			Replacement of a major component
		Substitution for an existing component (not like for like)	Substitution for a minor component
			Substitution for a potentially important component
			Substitution for a major component
		Addition of a new component	Addition of a minor component
			Addition of a potentially important component
			Addition of a major component ★

Figure 6.13 (Continued)

Where is the change?	What type of change?	How is the change being made?	How big is the change?
Primary packaging (e.g., solution bag)	Process change	Adjustment of an existing step	Minor adjustment to a minor step
			Major adjustment to a minor step
			Minor adjustment to a major step
			Major adjustment to a major step
		Elimination of an existing step	Elimination of a minor step
			Elimination of a potentially important step
			Elimination of a major step
		Replacement of an existing step	Replacement of a minor step
			Replacement of a potentially important step
			Replacement of a major step
		Addition of a new step	Addition of a minor step
			Addition of a potentially important step
			Addition of a major step
	Configuration change	Re-ordering of existing configuration	Re-ordering of minor components
			Re-ordering of major components
		Dimensional changes to existing configuration	Major change to a minor component
			Major change to a minor component
			Minor change to a minor component
			Minor change to a major component
			Major change to a major component
	Compositional change	Adjustment to an existing component	Minor adjustment to a minor component
			Major adjustment to a minor component
			Minor adjustment to a major component
			Major adjustment to a major component
		Elimination of an existing component	Elimination of a minor component
			Elimination of a potentially important component
			Elimination of a major component
		Like for like replacement of an existing component	Replacement of a minor component
			Replacement of a potentially important component
			Replacement of a major component
		Substitution for an existing component (not like for like)	Substitution for a minor component
			Substitution for a potentially important component
			Substitution for a major component
		Addition of a new component	Addition of a minor component
			Addition of a potentially important component
			Addition of a major component

Key:
- Green shading = trivial change, lower risk; address change with a entry to the history file that records that the risk was assessed as low (including details), requiring no additional assessment or testing to approve change.
- Orange shading = moderate change, moderate risk; address change via data review and generation of a Special Report. If insufficient data is available to generate the special report, limited testing may be necessary.
- Red shading = essential change, higher risk; address change by limited extractables testing.
- Blue shading = essential change, highest risk; address change by full extractables testing.

The arrows placed in the figure represent the assessment of the first example change (change of an antioxidant), as documented in the text. The stars placed in the figure represent the assessment of the second example change (addition of a label), as documented in the text.

Note:
These are generalizations that are applicable to most circumstances. In certain cases, a decreased or increased level of assessment may be necessary. For example, consider the case of a change to a three-layered overpouch (secondary packaging). A change to the overpouch layer furthest from the drug solution is inherently of lower risk than a change to the overpouch layer that is closest to the drug solution. Thus, the same change applied to the closest layer might require a more rigorous assessment than the same change made to the furthermost layer.

As another example, consider the case of an overpouch (secondary packaging) that is a known migration barrier (e.g. constructed with an aluminum foil layer). In this case, any changes to tertiary packaging outside of the barrier overpouch is risk level green, regardless of the nature of the change.

Figure 6.13 (*Continued*)

References

1 Guidance for Industry (Draft Guidance) (2016). *Compara-bility Protocols for Human Drugs and Biologics: Chemistry, Manufacturing, and Controls Information*. Food and Drug Administration, Center for Drug Evaluation and Research (CDER), and Center for Biologics Evaluation and Research (CBER).

2 Guidance for Industry (1999). *Container Closure Systems for Packaging Human Drugs and Biologics*. Rockville, MD: US Department of Health and Human Services, Food and Drug Administration.

7

Applications – Pharmaceutical Drug Products – Packaging

When a patient is dosed with a pharmaceutical drug product, the patient is exposed to foreign impurities contained within that drug product. These foreign impurities can be derived from the drug product's manufacturing and/or its packaging systems. When a patient is treated with a medical device, the patient is exposed to substance(s) present in or on the device if those substances leach from the device under the device's conditions of clinical use.

Thus, a pharmaceutical drug product and a medical device are alike in the sense that a patient could be exposed to non-therapeutic substances as a result of their clinical use. Accordingly, one might suggest that at a very high level the patient safety risk posed by exposure to non-therapeutic substances for drug products would be strategically addressed in a similar manner to medical devices. However, it is clear and obvious that a pharmaceutical drug product and a medical device can differ quite dramatically in terms of the identities of the non-therapeutic substances, their exposure amounts, and the conditions of exposure. Such fundamental differences between a pharmaceutical drug product and a medical device generally dictate that the tactics employed to address and assess the potential patient safety risks associated with non-therapeutic substances will be substantially different.

Complicating things further, pharmaceutical drug products and medical devices are generally regulated by different organizations within the same regional regulatory authority. For example, the US FDA regulates pharmaceutical drug products in its Centers for Drug Evaluation and Research (CDER) and for Biological Evaluation and Research (CBER) while medical devices are regulated in the FDA's Center for Devices and Radiological Health (CDRH). Given that chemical characterization remains a developing science, it is not wholly unexpected that the various organizations would develop somewhat different approaches to safe use evaluation.

Consequently, the chemical characterization of pharmaceutical drug products and medical devices are considered separately in this book. Furthermore, foreign impurities in pharmaceutical drug products will be considered separately based on whether the impurity's potential source is the drug product's packaging, or its manufacturing system.

7.1 Classification of Packaging Systems and Components

The following definitions, obtained from USP <659>, Packaging and Storage Requirements [1], is the foundation of the discussion of pharmaceutical packaging:

1. **Container.** A receptacle that holds an intermediate compound, API, excipient, or dosage form, and is in direct contact with the article (e.g. ampules, vials, bottles, syringes, and pen injectors).
2. **Closure.** A material that seals an otherwise open space of a Container and provides protection for the contents. It also provides access to the contents of the Container (e.g. screw caps and stoppers).
3. **Packaging component.** Any single part of the package or container–closure system, including: the container (e.g. ampules, syringes, vials, and bottles); closures (e.g. screw caps and stoppers); ferrules and overseals; closure liners (e.g. tube cartridge liners); inner seals; administration ports; overwraps; administration accessories; labels; cardboard boxes; and shrink wrap.
4. **Packaging system** (also referred to as a container–closure system). The sum of packaging components and materials that together contain and protect the article. This includes primary packaging components as well as secondary packaging components when such components are required to provide additional protection.
5. **Primary packaging component.** A packaging component that is in direct contact with or may come into direct contact with the article.
6. **Secondary packaging component.** A packaging component that is in direct contact with (*the*

Extractables and Leachables: Characterization of Drug Products, Packaging, Manufacturing and Delivery Systems, and Medical Devices, First Edition. Dennis Jenke.
© 2022 John Wiley & Sons, Inc. Published 2022 by John Wiley & Sons, Inc.

outside of) a primary packaging component and may provide additional protection for the article.

7. **Tertiary packaging component.** A packaging component that is in direct contact with (*the outside of*) a secondary packaging component and may provide additional protection for the article during transportation and/or storage.

8. **Ancillary component.** A component or entity that may come into contact with (*the outside of*) a Tertiary packaging component during the distribution, storage, and/or transportation of the packaged article (e.g. pallets, skids, and shrink wrap).

9. **Associated component.** A packaging component that is typically intended to deliver the drug article to the patient but is not stored in contact with the article for its entire shelf life (e.g. spoons, dosing cups, and dosing syringes).

10. **Materials of construction.** The materials (e.g. glass, plastic, elastomers, and metal) of which a packaging component consists.

Note that the phrase in italics (*the outside of*) is not part of the USP definition but was added here to provide greater clarity.

Perhaps it is worth some time working through these various terms, as the USP definitions are curious and not altogether clear. A packaging system includes all those items (packaging components) that contain and protect a pharmaceutical drug product. Packaging components in turn are made from packaging materials. Packaging components within a packaging system are classified on the basis of how they contact the drug product during storage and distribution (otherwise called the shelf-life). Classifying packaging in this manner is significant as the extent to which a drug product and its associated packaging components interact is controlled by the nature of the contact between the drug product and the packaging component. That is, the various natures of contact are the specific dimensions upon which the classification is based. All other things being equal, the more intimately and directly a drug product contacts a packaging component, the greater will be the interaction between the drug product and the packaging component. Furthermore, such a classification may be useful in terms of establishing how the packaging components are tested and assessed for extractables as potential leachables.

Primary packaging includes those packaging components of the packaging system that are either in continuous direct contact with the drug product during storage and distribution or which do not contact the drug product directly and continuously but rather could be in transient direct contact with the drug product during a portion of the storage and distribution. The common factor between continuous and transient direct contact items is that there is no barrier that physically separates the drug product from the primary packaging. As a result of the direct contact between the drug product and primary packaging, there is a greater likelihood that the drug product and primary packaging will interact, and should an interaction occur, there is a greater likelihood that the interaction will occur to a greater extent (versus packaging components that are not in direct contact with the drug product).

Secondary packaging is packaging that contains both the drug product and the primary packaging that the drug product is contained in. Secondary packaging is typically used to provide an extra layer of physical protection (e.g. abrasion, coring, puncture) or to counterbalance a deficiency in the primary packaging. For example, consider a drug product that is oxygen sensitive. For various reasons, it may be that the primary packaging is not a sufficient barrier to oxygen to preserve the drug product. As opposed to adding an antioxidant to the drug product, the drug product could be stabilized by placing it and its primary packaging inside a barrier secondary package.

It is intuitive and obvious that the primary packaging acts as a physical barrier between the drug product and the secondary packaging. It is less obvious whether the primary packaging acts as a chemical migration barrier between the drug product and the secondary packaging. This distinction is important because it is, and has been, too easy to confuse these two very different barrier properties. It was not too long ago that secondary packaging was dismissed as a potential source of leachables, both by the regulatory community and by the community of practice. And then the "unexpected" leachables began to be reported and both communities realized that they needed to look beyond the obvious and address the underlying science. My own education on this topic occurred during the development of a drug product containing calcium. Knowing that stearic acid and/or its salts is a common additive in the polyolefin materials used in the primary packaging (multi-layered bag), an extensive container development program was initiated to identify stearate-free polyolefin resins to use as materials of construction so as to preclude the possibility of calcium stearate particulates. Appropriate candidates were located and the absence of extractable stearate in the primary packaging was established via extraction studies performed on both the individual resins and the manufactured bag. However, roughly two months into the stability shelf-life study, the drug product began to form particulate matter, which was subsequently identified as calcium stearate. Where was the stearate coming from?

Well, I would like to say that we immediately recognized that the source was the polyolefin overpouch that was used as secondary packaging. But that was not the case. After all, the drug product was not in direct contact with the secondary packaging. Stearic acid is not a volatile substance and thus it was not diffusing through the primary packaging as a gas. How could the overpouch be a source of a largely non-volatile, non-polar leachable?

It was not until I was removing product samples from an autoclave (the product was terminally sterilized) that I understood the reality of the situation. Removing the product from the autoclave, I noticed that in lying flat on the autoclave tray, the primary container was resting directly on top of the overpouch. In essence, the overpouch might as well have been the outer layer of the primary packaging. What was happening here was that under the thermal driving force of the high-temperature sterilization, coupled with the product's multi-year ambient shelf-life, stearic acid was migrating from the overpouch into the primary packaging and eventually through the primary packaging and into the drug product, where it encountered dissolved calcium and precipitated. This hypothesis was readily confirmed when the overpouch's vendor confirmed that calcium stearate was an intentional additive of one of the layers of the overpouch.

Since this discovery, numerous other instances of extractables breaching the barrier of primary packaging have been reported, both verbally and in print. In fact, the "leachables from the overpouch" mystery had been previously reported by Ulsaker and Teien while investigating leached caprolactam originating from the overwrap used with PVC iv bags [2] and by Akapo and Mccrea who reported the accumulation of overwrap-related leachables in drug products stored in LDPE bottles [3]. The lesson has been learned by practitioners and regulators alike: unless the primary packaging is an absolute and impermeable barrier to migration, extractables from secondary packaging can and will breach the physical barrier of the primary packaging and accumulate in the packaged drug products as leachables.

It is worth noting that the term secondary packaging can be a bit deceiving. While it is easy to recognize an overpouch as secondary packaging, it may not be so easy to understand that the printing or a label applied to the outside of the primary packaging is secondary packaging, not so much by the USP definition (as the label or the printing does not afford the drug product with extra protection) but by the definition noted previously that stresses that secondary packaging includes any packaging component that is separated from the drug product by the primary packaging. The accumulation of label-related leachables in packaged drug products is well-documented [4–8]. Other examples of secondary packaging include oxygen scavengers and other sachets that are placed between

the drug product's primary and secondary packaging and injection sites that are isolated from the drug product by a membrane in the port tube (closure) of a bag.

This discussion of drug product leachables arising from secondary packaging has helped us understand that direct contact between the drug product and the packaging component is not necessary for component extractables to become drug product leachables. Once we understand this, we can re-examine our beliefs about direct contact and primary packaging. Another personal experience further illustrates this point. This case involved the use of a packaging system (bag), approved for aqueous drug products, as a container for lipid emulsions (e.g. parenteral nutrition solutions). In this case, the closure consisted of a port tube with an elastomeric injection site at its end. The port tube contained a membrane; thus, by definition the injection site was secondary packaging with no direct contact with the drug product. As was the case with the previous example, aqueous drug products packaged in this system were terminally sterilized and had shelf-lives that included storage at ambient temperature for several years.

The containers themselves had been profiled for extractables and aqueous drug products stored in this packaging systems had been profiled for leachables and thus it was expected that transition of the packaging to lipid emulsion drug products would present few challenges. However, leachables screening of the packaged lipid emulsions revealed the presence of BHT in considerable quantities, specifically after a relatively long period of storage. Where was the BHT coming from and why had the BHT not been seen in the aqueous drug products as a leachable?

Eventually, after all other possible sources of this leachable had been considered, attention was focused on the elastomeric injection site. Extractables profiling of the injection site revealed that it contained BHT, which was confirmed as an intentional additive by the injection site's vendor. So now a source was identified. But again, BHT is not particularly volatile and the injection site is separated from the liquid drug product by the membrane. Furthermore, the port tube is quite narrow and it is physically unlikely that the drug product solution would move into the port tube. Thus, the only way that BHT could leach into the drug product would be for it to transfer from the injection site to the injection port material, migrate from the injection site to the bag surface, and leach from the bag surface into the drug product. To investigate this possibility, the injection port and portions of the bag, at various locations on the bag (close to and further away for the injection port) were extracted and tested for BHT. In fact, the presence of BHT in the injection port was confirmed and a concentration gradient for BHT in the bag was observed, with the largest BHT levels being recorded for those portions of the bag that were closest to the injection site.

At this point, the origin and mechanism were confirmed. But why did the BHT accumulate to such high levels in the lipid emulsion but at undetectable levels in the aqueous drug products? Clearly, it is simply a matter of solubility. Whereas BHT has minimal water solubility, it is highly soluble in the lipid emulsion. Thus, while BHT was readily available in the bag for leaching into any drug product contained in the bag, it could only leach into those drug products in which it was soluble.

A final example of the circumstance of direct versus indirect contact is leachables accumulation in packaged solid drug products, specifically a lyophilized drug powder stored in a stoppered glass vial. At first glance, this situation seems simple enough as there are literally two strikes against significant leaching of substances from the stopper. First, it is clear that the drug product is physically separated from the only possible source of organic leachables (the stopper) and unlike the previous BHT example, glass is not an effective conduit for migration of stopper-related organic extractables. This "no direct" contact situation is even further strengthened by the fact that many rubber stoppers used as closures for vials are coated with a barrier material. Second, the fact that the drug product is a solid means that it will not participate in "active leaching" in the same way that a liquid drug product would. Taking these two factors into account, it has been conventional wisdom, captured in regulatory practice, that solid dosage forms are at very low risk for leachables. These factors notwithstanding, stopper-related extractables accumulate as leachables in the lyo cake [9, 10], as the enclosed headspace of the vial is an excellent transport medium for volatile and semi-volatile extractables from the stopper. Once volatilized, these extractables are trapped in the headspace of the vial and eventually can be sorbed on or in the lyo cake and thereby become leachables, even in the absence of direct contact and "active leaching."

A side note to this discussion of direct and indirect contact involves the practical handling of product samples for leachables testing. For example, it is practical reality that packaged drug product samples have to be labeled in some manner so that they can be identified and recognized throughout storage, testing, and disposal. Simple labeling processes, such as affixing a secondary label to the sample units, writing on the outer surface of the unit and the like can have the unfortunate outcome of introducing "leachables that are not extractables" into the drug product. Similarly, storage of packaged drug product units during aging must be performed with care, less environmental contaminants similarly become "leachables that are not extractables." Even the simplest sample handling actions

can complicate the profiling of packaged drug products for true leachables. For example, Singh and associates have reported the "adulteration" of a packaged drug product with diethylphthalate (DEP) from a non-packaging source [11]. Apparently, these authors used scotch tape to wrap packaged drug product units together while the units were in storage in a stability chamber for shelf-life testing. Screening of the drug product units for leachables over the shelf-life lead to the reporting of DEP as a measurable leachable, despite the fact that DEP had not been reported as an extractable in the drug product's primary (LDPE bottle) and secondary (cardboard carton) packaging. The author's data established that (i) the scotch tape contained DEP and (ii) that the DEP migrated from the scotch tape into the drug product crossing the physical barriers provided by the primary (LDPE container closure system) and secondary packaging (carton and label).

Discussing the classification further, tertiary packaging are those packaging components that contain the drug product in its primary and secondary packaging. Thus, the tertiary packaging is separated from the drug product by two physical barriers, the primary and secondary packaging. For example, consider the case of the drug product stored in a bag that is further packaged in an overwrap. If the outer surface of the overpouch is printed or a label is affixed to the outer surface, then the printing or label becomes tertiary packaging. If the individual drug product units, contained within the primary and secondary packaging, are placed, either individually or collectively, in a shipping carton, for example, the shipping carton becomes tertiary packaging. If items such as package inserts are placed in the cartons, then the package inserts become tertiary packaging.

Considering leachables, substances derived from the tertiary packaging would have to breach the two barriers that are respectively the secondary and primary packaging and be soluble in the drug product in order for them to accumulate in the drug product as leachables. This involved circumstance alone makes it unusual to encounter leachables derived from tertiary packaging, as the conditions that would allow these circumstances to be realized are limited. For example, there is the juxtaposition of the time it would take a compound to breach the barriers and the duration and temperature of the contact. If the contact time is short and/or the contact temperature is low, then either the primary and secondary packaging must be easy to breach (e.g. are thin or are highly permeable) or the compound of interest must have a large diffusion coefficient in the packaging. If either the primary or secondary packaging has reasonably good migration barrier properties or the compound of interest has a low diffusion coefficient, then either the contact temperature has to be high or the contact duration has to be long (or likely both) for significant migration to occur. Even in these circumstances, the migrating compound must be either solubilized in or sorbed by the drug product for the migrating compound to be entrained into the drug product as a leachable.

Published examples of leachables from tertiary packaging for pharmaceutical drugs products are difficult to find, although the possibility of tertiary leachables is raised in numerous publications.

Lastly, USP <659> considers even more remote classes of what it calls packaging components, including ancillary components and associated components. However, when one considers the examples that are provided for these packaging classes, for example, pallets, skids, and shrink wrap for ancillary packaging and spoons, dosing cups, and dosing syringes for associated components, it is clear that these items do not fit the definition of packaging and that substances derived from these items do not fit the definitions of extractables and leachables (especially items such as pallets and skids). In this author's opinion, such sources of drug product impurities are more environmental in nature and less packaging-related and thus are more properly referred to as environmental impurities as opposed to leachables. Thus, a consideration of substances from these items is outside the scope of this book, which specifically addresses extractables and leachables. That is not to say, however, that contamination of drug products by environmental impurities is irrelevant. In fact, the presence of environmental impurities in packaged drug products can have considerable ramifications. For example, in 2009, consumers of Tylenol Arthritis Pain caplets (McNeil Consumer Health, a division of Johnson and Johnson, Fort Washington, Pennsylvania) complained about a "moldy" odor associated with the product, prompting its recall. The odor was eventually traced to the wooden pallets used to ship the plastic bottles in which the medicine was packaged. A fungicide used to treat the pallets, 2,4,6-tribromophenol, had decomposed to produce 2,4,6-tribromoanisole, which had migrated into the bottles and created the odor [12]. Similarly, Ortho-McNeil-Janssen Pharmaceuticals, Inc. recalled 40 000 pill bottles in 2011 because there were two consumer reports that the product had an "uncharacteristic odor" [13]. TBA (2,4,6-tribromoanisole) was again identified as the offending environmental contaminant causing the odor.

From an extractables and leachables perspective, packaging components are classified as primary, secondary, tertiary, and ancillary because the classification establishes the relative likelihood that packaging-related extractables could accumulate in the packaged drug products as leachables and thus suggests the extent to which the components would be screened for extractables. It goes without saying that primary packaging components must, without exception, be appropriately and completely profiled for all classes of extractables given their proximity to the packaged drug product and the high probability that primary packaging extractables will become leachables. Furthermore, secondary packaging must be properly profiled for relevant classes of extractables, which have the potential to become leachables, taking the migration barrier properties of the primary packaging and the conditions of contact between the packaged drug product and the packaging into account. For example, if the primary packaging is a good migration barrier (e.g. glass), there would be little use in profiling the secondary packaging for extractables as potential leachables. However, if the primary packaging was a poor migration barrier (e.g. LDPE), then the secondary packaging should be fully profiled for extractables. Extractables profiling of tertiary packaging should be less common than profiling of primary and secondary packaging components and should address only those substances that have the requisite "mobility" to breach the multiple physical barriers representing the primary and secondary packaging, likely focusing on volatile substances. Routine extractables profiling of ancillary packaging is not recommended unless there is compelling evidence to suggest otherwise (please do not make me follow the packaged drug product through its entire distribution chain and test every item it might encounter for extractables). Extractables profiling of ancillary packaging should only be performed when the primary, secondary, and tertiary packaging, taken as a whole, is a poor barrier to migration and/or when the drug product is in contact with the ancillary packaging for either long periods of time or at elevated temperatures. Should extractables testing of ancillary packaging be deemed appropriate, such testing should focus on the volatile organic substances that have sufficient mobility to breach the multiple barriers presented by the primary, secondary, and tertiary packaging.

While it is useful to classify packaging with respect to its proximity to the drug product, it is also useful to classify packaging with respect to its proximity to the patient who is dosed with the drug product to produce a desired therapeutic outcome, perhaps in a manner similar to the way medical devices are classified for the purpose of establishing necessary and relevant biocompatibility testing per ISO 100993:1(2019) [14]. As was the case with classification of packaging components with respect to drug product contact, classification of packaging with respect to patient contact could be a means of establishing and justifying the appropriate extractables testing procedures and processes.

Considering Table A.1 of ISO 10993:1(2019), it surely is the case that packaging is properly placed in the category of externally communicating, as the packaging indirectly "communicates" with the patient through the medium of the drug product. Although there are numerous routes of administration for packaged drug products (e.g. oral, inhaled, parenteral, ophthalmic), the ultimate objective is for the drug substance to circulate in the body, performing its desired therapeutic function when it arrives at the target site. Thus, it seems reasonable to classify packaging contact as "blood path, indirect." Considering the classification further, the next step is to classify the packaging in terms of its duration of contact with the patient, which is dictated by the therapeutic use of the packaged drug product. Thus, while packaging systems can generally be classified as externally communicating, blood path indirect, they must individually be classified in terms of the duration of contact.

The classification of a packaging system as if it was a medical device is relevant to extractables only to the extent that guidelines for characterizing medical devices for leachables can be extended to cover packaging. To consider this, one turns to ISO 10993:18(2020) [15], which is the ISO standard for the chemical characterization of medical devices. In a certain way, it seems that ISO 10993:18(2020) anticipates the previous classification of packaging systems as the normative text of Section 5.6 includes the following:

> "The nature of use for some medical devices (e.g. indirect contact devices such as saline infusion bags) can obviate the need for extractables testing, as the conditions of use associated with the maximum human exposure to leachables can be replicated and the clinical use solutions can be analyzed in a straightforward manner. In such cases, extractables testing could reasonably be replaced by leachables testing."

Although the text could be confusing when taken out of context, the statement essentially reinforces the suggestion that was communicated previously in Section 3.2 of Chapter 3 in this book, which was "if one can screen the drug product for leachables, then this is what one should do and avoid extractables testing." Should an extractables study be performed, footnote c in Table 2 suggests that "Exaggerated (extraction) conditions can be appropriate for external communicating or non-absorbable surface contact devices, with justification." Further points to consider and recommended practices for performing extractions are provided in Part 18's informative Annex D, some of which could be relevant to packaging systems.

7.2 Pharmacopeial Monographs

Regulatory authorities are charged with the considerable responsibility of certifying that pharmaceutical and medical products are safe and effective. Marketers of pharmaceutical and medical products are charged with

the considerable responsibility of supplying the relevant regulatory authority with the information and data that regulators need to assess and establish the product's safety and efficacy. Both parties, regulators and marketers, benefit when the regulatory expectations and requirements have been clearly and unambiguously established, justified, and communicated. Moreover, both parties benefit when the expectations and requirements are based on best demonstrated and achievable practices grounded in good and achievable science, practically applied. Lastly, both parties benefit when expectations and requirements are standardized enough to be efficient but are not so restrictive and inflexible that they cannot be optimized for the myriad "special cases" and "extenuating circumstances" that are the norm, rather than the exception, for a diverse population of items such as pharmaceutical and medical products.

In the absence of a single authoritative, harmonized, and widely accepted standard for extractables and leachables, it is the nature of the current state of affairs that various organizations have accepted partial responsibility for establishing best demonstrated practices for extractables and leachables characterization of packaging materials, components, and systems and interpreting those practices in the context of establishing regulatory expectations. To a certain extent this circumstance is understandable and even justifiable given the complexity of extractables and leachables testing and reporting, both as a science and as a practical endeavor. Furthermore, it requires the wisdom of Solomon, if not more, to find that common ground where:

- The regulator's desire for complete and compelling information is matched by the marketer's ability to supply such information.
- The regulator's desire for standardized practices and procedures is matched by the marketer's desire to uses practices and procedures that are well-tailored to their specific products and their specific circumstances.

Lastly, as chemical characterization is an evolving, rather than a mature, science, it can be difficult to reach a consensus as to what is good and practically achievable science and, once a consensus is reached, to preserve or adjust that consensus as new science is discovered, communicated, and commercialized.

Guidance on the science and practice of chemical characterization can be found in four general sources, two of which have clear and direct legal authority for setting requirements and standards (pharmacopeia and regulatory guidance documents), one of which is granted legal authority for setting requirements and standards by virtue of recognition and adoption by regulatory bodies (international standards) and the last of which has no legal authority but derives its influence via its adoption by the community of practice (best demonstrated practices documents). Each of these will be considered in greater detail as follows.

7.2.1 The Role of the Pharmacopeia in Setting Standards for Chemical Characterization of Packaging

A pharmacopoeia, pharmacopeia, or pharmacopoea, in its modern technical sense, is a resource containing directions for the identification and compounding of medicines and is published by the authority of a government or a medical or pharmaceutical society. In a broader sense it is a reference source for pharmaceutical drug specifications. All pharmacopeias share the goal of advancing public health by helping to ensure the quality and consistency of medicines, thereby promoting the safe and proper use of medications. The role of a modern pharmacopoeia is to furnish quality specifications for drug substances and general requirements for dosage forms. The existence of such specifications and requirements is necessary for the proper functioning or regulatory control of drugs.

If a purpose of the pharmacopeia is to ensure the quality and safety of drug products and the quality and safety of drug products are affected by the drug product's packaging, it is logical that a pharmacopeia would contain monographs related to packaging. Thus, the major global Pharmacopeias contain monographs related to packaging systems and materials of construction for those packaging systems. It is generally expected by global regulatory authorities that drugs products approved in the authority's jurisdiction are packaged in packaging systems that comply with the relevant compendial monograph and that the packaging system is constructed from materials that comply with their relevant monographs. This expectation is clearly communicated in regulatory guidance such as the FDA container-closure guidance of 1999 [16] and the European Medicines Agency (EMEA) guidelines for plastic immediate packaging [17], both of which establish the importance that chemical compendial testing plays in regulating pharmaceutical packaging.

An example of how the various pharmacopeia accept this regulatory responsibility is contained in USP <659> [1]:

"Packaging materials must not interact physically or chemically with a packaged article in a manner that causes its safety, identity, strength, quality, or purity to fail to conform to established requirements. Any plastic material used to construct a Packaging system must meet the applicable requirements of Plastic Materials of Construction ⟨661.1⟩. (Official 1-Dec-2025) All Packaging systems must meet the applicable requirements specified in Containers—Glass ⟨660⟩, Plastic Packaging Systems and Their Materials of Construction ⟨661⟩, Plastic Packaging Systems for Pharmaceutical Use ⟨661.2⟩, (Official 1-Dec-2025) and Auxiliary Packaging Components ⟨670⟩. All elastomeric closures must meet the applicable requirements in Elastomeric Closures for Injections ⟨381⟩."

7.2.2 Pharmacopeial Guidance via Compendial Monographs

In considering the contribution of compendial monographs to chemical characterization standards, it is important to manage expectations up-front. The compendial monographs related to the chemical testing of packaging components and systems and their associated materials of construction do not address extractables (or leachables) directly and furthermore do not provide cookbook-type instructions on how to perform extractables or leachables studies. In fact, currently official compendial packaging monographs do not prescribe or address mandatory extractables and/or leachables testing and the results of performing the compendial tests are not extractables profiles for packaging materials, components, or systems. Rather, official compendial packaging monographs are currently designed to produce two pieces of information: (i) general chemical properties of largely aqueous extracts of the packaging items and (ii) composition of the packaging items. Although this information may be relevant to chemical characterization as (i) physicochemical properties of water extracts may be "early warning" indicators of potential extractables issues and (ii) composition allows the forecasting of extractables profiles, the information provided by compendial tests for packaging does not provide the quantitative information necessary to establish packaging's potential adverse impact on a packaged drug products quality.

Generally speaking, the compendial monographs are fairly straightforward in content and context and the reader is referred to the individual monographs themselves to understand the testing requirements and specifications. However, the compendial monographs associated with plastic packaging will be given more attention as these monographs either have been recently revised or they are currently in the process of being revised.

7.2.3 The United States Pharmacopeia (USP) Monographs on Packaging Materials and Systems

7.2.3.1 General Discussion
The USP addresses packaging systems and their materials of construction largely on the basis of type of material

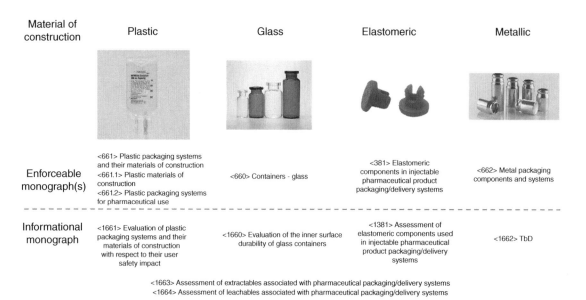

Figure 7.1 The USP monographs concerning to the chemical characterizations of pharmaceutical packaging.

(see Figure 7.1); thus, there are USP chapters that address plastic packaging (<661> [18], <661.1> [19], and <661.2> [20]), glass packaging (<660> [21]), and elastomeric components used as closures (<381> [22]). A USP chapter addressing metal-based packaging, <662>, is under development. Each of these enforceable monographs has, or will have, an associated informational monograph that discusses the purpose and intent of the enforceable monograph, establishes the scientific and practical basis for the monograph's tests and specifications, provides insight into the proper use of the monograph, and generally provides best practice recommendations. Thus, for example, USP <1661> [23] is the informational companion chapter to the plastic packaging monographs <661>, <661.1>, and <661.2>.

Additionally, USP information monographs <1663> [24] and <1664> [25] provide best practice information regarding extractables and leachables respectively.

The numbering of USP monographs is intentional, with monographs numbered below 1000 being "mandatory" (that is, monographs that are intended to be enforced by regulatory authorities with enforcement responsibility) and with monographs numbered above 1000 being "advisory" (that is, monographs that are intended to inform interested parties in terms of understanding the application and genesis of mandatory chapters and/or in terms of communicating best demonstrated practices generally concerning a specific technical discipline relevant to either the Pharmacopeia's mission or to specific mandatory monographs). Thus, the packaging monographs mentioned in the previous paragraph are mandatory (or normative), with the exception of <1661>, meaning, for example, that plastic packaging systems used for USP-labeled drug products must be tested as described in <661.2> and must meet the specifications contained in <661.2> and that the plastic materials of construction must be tested as described in <661.1> and meet the specifications contained in <661.1>. On the other hand, <1661> is an advisory (informative) monograph that contains suggestions, as opposed to requirements, and information, as opposed to specifications.

7.2.3.2 Plastic Packaging Systems and Their Materials of Constructions, <661>, <661.1>, <661.2>, and <1661>

At the time this book was being written, the USP monograph for plastic packaging, <661> Plastic Packaging Systems and Their Materials of Construction, was undergoing a transformation. Effective December 1, 2025, this monograph will be replaced by three monographs: <661> with the same title, which essentially becomes an introductory monograph that contains no testing specifications, <661.1> and <661.2>. The generation of the <661.1> and <661.2> monographs reflects the evolution of compendial testing to embrace the role of extractables and leachables in assessing packaging's impact on drug product quality and the attempt by the USP to align with the European Pharmacopeia (Pharm. Eur.) in terms of tests, test methods, and specifications, at least for materials of construction. Anticipating that these monographs will be in common use once

Figure 7.2 USP testing strategy for plastic pharmaceutical packaging materials and systems. Figure 1 from USP <1661> [22].

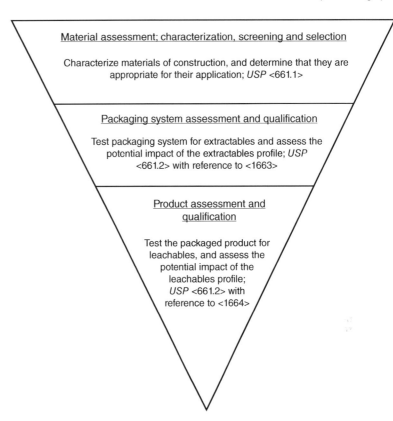

Material assessment; characterization, screening and selection

Characterize materials of construction, and determine that they are appropriate for their application; *USP <661.1>*

Packaging system assessment and qualification

Test packaging system for extractables and assess the potential impact of the extractables profile; *USP <661.2>* with reference to <1663>

Product assessment and qualification

Test the packaged product for leachables, and assess the potential impact of the leachables profile; *USP <661.2>* with reference to <1664>

this book is in use, the following discussion will focus on the replacement monographs.

The objective of USP plastic packaging standards is to establish the tests and acceptance criteria to ensure packaging systems do not materially and adversely impact the effectiveness of the drug product. Given the complex nature of packaging systems and their development and manufacturing processes, multiple testing procedures are needed to establish their suitability with a specific drug product. The logical development and manufacturing process for packaged drug products starts with the packaging system's materials of construction, continues with the packaging system itself, and ends with the packaged drug product. This progression forms the basis of a three-stage approach to packaging systems qualification, as illustrated in Figure 7.2. The process for establishing a packaging system's suitability includes: characterization of its materials of construction (ingredients); testing and assessment of the system itself (extractables); and testing and assessment of the packaged drug product (leachables). The initial step of chemically characterizing candidate materials of construction is performed so that the selection of materials can be rationally made and scientifically justified. The intermediate step of system assessment is useful and necessary because it bridges the risk assessment gap between testing starting materials and testing the finished drug product, while providing a means for optimizing drug product testing. The third stage of the process includes

product assessment, specifically leachables testing of the packaged product and impact assessment, which considers the user's exposure to the leachables.

Considering USP <661.1>, testing and characterizing materials of construction provide a rational basis for material selection and minimizes the risk that a system made from one or more materials will be unsuitable. Additionally, chemical characterization data may also provide the basis for effective and appropriate change control.

The basic tenet of materials assessment is that knowing the general composition and certain general characteristics of a material of construction allows one to:

- Rationally assess the potential suitability of materials with a degree of certainty that is appropriate for early product development and/or manufacturing.
- Forecast with some degree of accuracy the identity of extractables from that material of construction and from systems that use that material of construction.
- Use the assessment and forecast to establish and justify the selection (or rejection) of a particular material for use in a particular packaging system.

To this end, ⟨661.1⟩ defines a well-characterized plastic material of construction as one whose (see Table 7.1 and Figure 7.3):

- Identity has been definitively established.
- General physicochemical properties have been established.

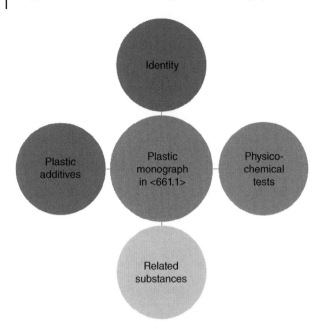

Figure 7.3 Anatomy of USP <661.1> material sections. In order for a material to be judged to be compliant with <661.1>, it must be tested for these various parameters and must meet the specified acceptance criteria.

- Additives, impurities and/or related substances have been quantified.
- Biological reactivity has been established.

Generally, the portions of <661.1> that address individual materials are similar in their respective contents, each describing a specific test and providing acceptance criteria. Thus, each material is characterized to establish its identity, general chemical characteristics, and composition. Identification tests typically rely on thermogravimetric or infrared spectroscopic analysis. General chemical characteristics (physicochemical tests) are assessed by performing a water extraction (reflux for five hours) of the test material and testing the resultant extract for general chemical properties such as UV absorbance, acidity/alkalinity (and pH), and total organic carbon (TOC). A material's composition is established by performing generally chromatographic assays to identify and quantify the material's additives, impurities, and related substances (such as residual solvents).

At the time that this book is published, <661.1> contains tests, methods, and acceptance criteria for the following materials: cyclic olefins, polyamide 6, polycarbonate, polyethylene, polyethylene terephthalate, polyethylene terephthalate G, poly(ethylene-vinyl acetate), polypropylene, polyvinyl chloride, and polyvinyl chloride, plasticized. It is the USP's intent to add monographs for additional materials as appropriate test methods for relevant materials can be established and qualified. Until

such monographs can be developed, <661.1> speaks to "unaddressed materials" and establishes that unaddressed materials must be characterized and meet acceptance criteria established in ways that are comparable to those used for the materials specified in this chapter. Specifically, the unaddressed material of construction must be identified by appropriate methodology and be tested with consideration of the dosage forms for which it is used.

There are three aspects of <661.1> that require further explanation: risk-based testing, extractable elements testing, and biological reactivity testing. Considering the concept of risk-based testing, <1661> notes that while the testing required and acceptance criteria for materials of construction contained within ⟨661.1⟩ are relevant to and applicable for all dosage forms, it is well-established that risk-management tools and principles can be used to define the nature and magnitude of assessment (including testing), where low-risk situations require reduced or alternate assessment (testing) versus high-risk situations. Thus, as noted in Table 7.1, the testing requirements for biological reactivity and chemical tests differ for low-risk dosage forms (such as oral and topical) versus high-risk dosage forms (such as inhalation and injections).

Addressing extractable elements, the authors of <661.1> found themselves in a quandary. It is universally accepted that knowledge of the potential extractable elements present in the materials of construction for plastic packaging components is important in establishing that the material is well-characterized and that knowledge of the elements that are likely to be present and the concentrations at which they may be extracted provides information to determine potential drug product quality. However, materials of construction can vary widely in terms of their intentionally and unintentionally added elements, the levels of the elements, and the chemical form of the elements. Because of this variety in composition, it is challenging to provide universally effective and efficient test methodologies for extracted elements (including extraction conditions), lists of elements to target, and reporting requirements. In fact, consensus among the various stakeholders regarding extracted elements testing could not be reached. Therefore, it is the material user's responsibility to evaluate the need for extractable elements testing and, if such testing is necessary, to establish and justify the means by which testing is accomplished, taking into account extraction conditions, target elements, extract analysis, and reporting requirements.

In general, however, the following points are relevant:

- Extraction conditions should be such that elements that could leach into pharmaceutical drug products are fully extracted from the test material. Thus, the combination of extraction solvent(s), extraction temperature,

Table 7.1 Application of tests, USP <661.1>.[a]

Test parameter	Oral and topical dosage forms[b]	All other dosage forms
Identification	X	X
Chemical tests		
Physicochemical test – UV absorbance	X	X
Physicochemical test – Acidity/Alkalinity	X	X
Physicochemical test – TOC	X	X
Composition – Plastic additives	–[c]	X
Composition – Related substances	–[c]	X
Composition – Impurities	–[c]	X
Biological reactivity		
USP <87> Biological reactivity tests, in vitro	–	X[d]

a) Adopted from Table 1 in <661.1> [19].
b) For aqueous-based oral drug products that contain cosolvents (or if, for any reason, it may be expected to extract greater amounts of substances from plastic packaging components than water), additional extractables information may be needed to determine suitability. If additional information is required, perform Composition tests as directed in this table.
c) Provide appropriate reference to the Indirect Food Additive regulations in 21 CFR 174–186, specifically those addressing the purity criteria and limitations pertaining to use.
d) Biological reactivity testing in support of plastic packaging materials used for final pharmaceutical product packaging/delivery systems (drugs and drug/device combination products) provides baseline information and will often not be sufficient to assess the final suitability for use expectations of regulatory authorities. Thus, it is important to work with the appropriate regulatory authority for guidance regarding a product specific application.

extraction duration, extraction process, and extracted surface area to extraction solution volume should be carefully considered. It is important to remember that the objective is to perform an extraction, not to digest or dissolve the test material. Furthermore, the extraction process should produce an extract that is analytically expedient, meaning that the extract is amenable to analysis by the chosen test methodology with the necessary levels of sensitivity, accuracy, and precision.

- Target elements should minimally reflect those elements that have been intentionally added (e.g. catalyst and processing aids) to the material, those elements that could be unintentionally added to the material (e.g. processing contaminants, ingredient impurities), and those elements of potential toxicological concern, whether or not they are intentionally or unintentionally added.

- Analytical test methods should be compatible with the extract and should have the necessary and specified sensitivity, accuracy, and precision.
- Reporting requirements must reflect the analytical capabilities of the method and the levels at which an extracted element could have an adverse effect on drug product quality. The desired situation is that the test method is capable of producing reliable information at levels near those at which an element could have such an adverse effect. In this case, the reporting requirement is established considering the juxtaposition of analytical capability and potential effect.

Considering biological reactivity testing, such testing provides baseline information and will often not be sufficient in and of itself to fulfill the final suitability for use expectations of regulatory authorities. Thus, <661.1> takes a minimalistic approach to biological reactivity testing, providing a minimum standard of testing for higher risk dosage forms (that is, test per USP <87>). Although not required in <661.1>, biological reactivity testing involving in vivo tests (e.g. USP <88>) may be appropriate in certain circumstances. Ultimately, it is noted that it is important to work with the appropriate regulatory authority for guidance regarding a product-specific application of biological reactivity tests.

As is likely to be the case with every USP monograph, there are "grey areas" in application and applicability that can make use of the monograph challenging. Foremost among the challenges is scope. The authors of <661.1> clearly understood the issues involved with applying a USP monograph in situations and circumstances for which it was not intended, usually because "there is nothing else out there and doing something is better than doing nothing" (talk about a sure path to disaster!). Thus, <1661> clearly establishes that application of <661.1> to materials of construction used for items other than finished drug product packaging is beyond the scope of the monograph, although the concepts and principles could be applied to materials for construction for other systems (e.g. manufacturing systems for pharmaceutical products, packaging systems for drug substances, drug delivery items, and medical devices).

Second, there is the very practical issue of "who is responsible for doing this testing?" Addressing this point, <1661> puts the responsibility for possessing test results squarely on the shoulders of the drug manufacturer and/or the holder of the drug product license, noting that "The holder of the drug product application and drug product manufacturer … bear primary responsibility … for ensuring that the requirements of the chapters are met." However, having the results and actually performing the tests are two different matters and <1661> goes on to state that "the means by which the holder of the drug product application and drug product manufacturer obtain information … is at the discretion of the holder." To this author, the laws of economics suggest that it is more economically efficient if the <661.1> testing is performed by a material's vendor and supplied to each potential user of the material, as opposed to the same testing being repeated by every potential user, the only caveat being the case where the user subjects the material to a meaningful processing step, for example, irradiation. Although it may be reasonable to expect a material's vendor to test a material in the form that it is sold, it is less reasonable to expect a vendor to provide test data on test samples that have been processed consistent with each potential user's processing methods.

Another practical issue with application of <661.1> is what I call the "chicken and the egg dilemma". That is to say, when a packaging system is under development, it makes perfect sense to perform <661.1> testing as the very purpose of <661.1> testing is to enable effective material selection. However, let us imagine the sponsor who has already gone well beyond packaging design and is either in late-stage drug product development or has already launched a commercial product. To suggest to such sponsors that they need to go back and do <661.1> testing of materials of construction could undoubtedly "rub some sponsors the wrong way," as performing such tests could be perceived as adding little value in terms of establishing the safe use of the packaging system. Thus, <1661> notes "There are two means of demonstrating that a plastic material of construction has met the requirements of <661.1>. The second is the use of the material in a packaging system that meets the requirements of <661.2>." Specifically, if a packaging system meets the requirements of <661.2> then every material used in that packaging system is deemed to have met the requirements of <661.1>.

Another aspect of scope is addressing which materials in a drug product packaging system require testing. As was noted previously in Section 7.1, a packaging system can consist of components that directly contact the drug product, indirectly contact the drug product, or are isolated from the drug product. Logically, there is little value in testing materials that are isolated from the drug product and thus cannot contribute leachables to that drug product. Thus <1661> is clear on this point and notes:

"Testing of materials of construction via <661.1> is predicated on the circumstance that the material will most likely interact with the packaged drug product when the material is used in a packaging system. It is not necessary for a material used in a packaging system to be well-characterized if there is little or no chance of the material and the packaged drug product interacting. Under these conditions

the materials of construction would be considered non-interacting and would be exempt from <661.1> and <661.2> testing."

Therefore, <1661> differentiates materials into three types: directly contacting, indirectly contacting, and non-contacting. Directly contacting is clear without further definition and it is obvious that directly contacted materials must be tested per <661.1> and meet the requirements of <661.1>. Indirectly contacting and non-contacting both refer to the circumstance where the drug product and the material are separated from one another by a physical barrier. In the case of the indirectly contacting material, the barrier is physically, but not chemically, impermeable. Because there is a possibility of material extractables becoming drug product leachables, indirectly contacting materials must also be tested per <661.1> and meet the requirements of <661.1>. In the case of a non-contacting material, the barrier between the drug product and the material is both physically and chemically impermeable and there is no possibility of material extractables becoming drug product leachables. Thus, non-contacting materials do not require <661.1> characterization.

Lastly, <661.1> testing is not a guarantee that packaging systems constructed from plastic materials meeting <661.1> requirements will be suitable for their intended use. Characterization of a material using <661.1> establishes the composition or characteristics of the material, which can be used to determine if the material is an appropriate candidate for use in a packaging system. It is highly likely, although not a certainty, that a material that does not meet the requirements of <661.1> will not be suitable for use in a packaging system. On the other hand, it is highly likely, although not a certainty, that a material that meets the requirements of <661.1> will be suitable for use in a packaging system.

Armed with data from <661.1> testing (and presumably information on other material characteristics), appropriate materials of construction have been selected, a packaging system has been designed, and prototypes of the system have established its potential as viable packaging. In this circumstance, one transitions from material testing to enable material selection (<661.1>) to system testing for safe use qualification (<661.2>, packaging system assessment and qualification, as shown in Figure 7.2). As established in <661.2>, a packaging component or system is chemically suited for its intended use if:

- The packaging component's or system's general physico-chemical properties have been established.
- The packaging component's or system's biological reactivity has been appropriately established.

- The packaging component or system has been established to be suitable by means of the appropriate chemical suitability for use assessment.

One notes the similarity between this list of requirements and the list of requirements for materials of construction. As with materials of construction, the packaging component or system is tested for general physicochemical properties and biological reactivity as these tests are relatively easy to perform and serve as an "early warning" of potential safe use issues. As with materials of construction, the packaging component or system is chemically characterized to provide data for assessment. However, unlike materials of construction, chemical testing for packaging components and systems does not seek to establish composition (that is, what is present in the test article) but rather seeks to establish the extractables profile (that is, what can be extracted from the test article as a means of projecting potential patient exposure to leachables).

In the case of physicochemical and biological reactivity testing, one might question the necessity to perform such arguably redundant testing, as materials of construction were tested in this manner. To address this question, first, as was previously noted, there may be cases where <661.1> testing was not performed on the materials of construction. In this case, testing the packaging system is the first opportunity to collect the diagnostic physicochemical and biological reactivity data. Second, it is noted that materials of construction undergo potentially significant processing as they are converted into a packaging component or system. Furthermore, construction aids (for example, adhesives or solvent) may be required to construct the packaging system. In such cases, the results of physicochemical and biological reactivity testing on a packaging system could be quite different from the sum of the test results of the individual materials of construction.

It is worth noting that similar to <661.1> for materials, <661.2> for packaging uses a risk-based approach for biological reactivity testing, requiring only USP <87> for dosage forms including orals and topicals.

The term Chemical Suitability Assessment is used by <661.2> to describe the process of chemically characterizing a packaging system for safe use assessment and notes that "an appropriate and rigorous chemical suitability for use assessment may include extractables testing of the packaging component or system and leachables testing of the packaged drug product."

Although <661.2> establishes the requirement for performing extractables and/or leachables testing, it does not specify the means of performing such testing. In view of the considerable diversity of packaging systems, dosage forms, and packaged drug products, <661.2> notes that it is not possible to provide specific test conditions for performing extractables and leachables studies for all possible situations and products. Nevertheless, it is expected that the design of the extractables and leachables study would be based on sound and justifiable scientific principles, and that the studies themselves would be consistent with (i) the nature of both the packaging system and packaged drug product, (ii) the clinical use of the packaged drug product, and (iii) the perceived safety risk associated with the packaging system and dosage form. To provide users of <661.2> with some practical knowledge on how to properly design and conduct extractables and leachables studies, <661.2> notes that "general essential principles and demonstrated best-practices recommendations for extractable and leachable studies can be found in USP <1663> and <1664>, respectively" and that "these chapters may serve as helpful resources for designing and justifying rigorous and appropriate studies."

Although no dosage form is excluded from a chemical suitability for use assessment, <661.2> stipulates that the nature and degree of testing should be dosage form-dependent and consistent with a risk-based approach. For example, the testing of packaging components or systems for low-risk dosage forms, such as solid and aqueous-based oral and topicals, should be consistent with the low risk associated with these dosage forms

Lastly, <661.2> states that "alternative testing strategies for chemical suitability for use assessment may be appropriate in justified circumstances, subject to agreement by an appropriate regulatory authority," thus affording sponsors at least the possibility of "playing by their own rules".

7.2.3.3 Plastic Packaging: Extractables and Leachables, <1663> and <1664>

Considering that they are directly referenced in enforceable monographs, <1663> for extractables and <1664> for leachables are prominent informative monographs. Each will be considered in greater detail as follows.

Monograph <1663> "presents a framework for the design, justification, and execution of an extractables assessment for pharmaceutical packaging and delivery systems. The chapter establishes critical dimensions of an extractables assessment and discusses practical and technical aspects of each dimension." It further notes that "achieving the objectives of an extractables assessment requires performance of an extraction study in order to create extractables profiles. An extraction study has two critical dimensions: laboratory generation of the extract (extraction) and testing the extract (characterization)." It then provides guidance on how to perform the processes of extract generation and testing.

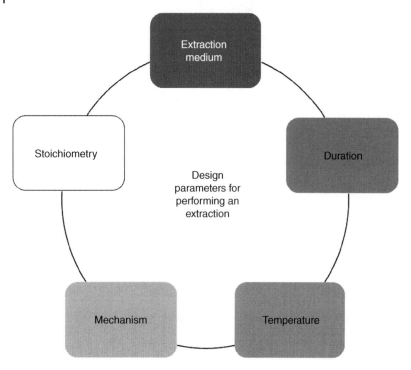

Figure 7.4 Critical design parameters for performing an extraction per USP <1663>.

Considering the generation of an extract, critical design parameters for performing an extraction include (Figure 7.4):

- The chemical nature of the extracting media
- The duration of the extraction process
- The temperature at which the extraction is performed
- The stoichiometry of the extraction process (extracted surface area per unit volume of extracting solution)
- The mechanism or process by which the extraction is accomplished

With respect to the extraction medium, <1663> notes that "the extracting medium is the most critical (design parameter) because it is the extracting medium that accomplishes the extraction, and all other parameters merely facilitate the extraction" and suggests that "the most logical tactic … is to use the formulation itself as the extracting medium and in the absence of complicating factors, such an approach is recommended." However, when the use of the formulation as an extracting medium complicates extract testing to such an extent that it is impractical, the drug product vehicle, or placebo, is suggested as an alternate extracting medium, based on the generalization that the drug substance itself does not typically establish the "leaching power" of a drug product but rather that it is the formulation's ingredients (drug product vehicle) that establish the drug product's ability to leach substances from a contacted item.

These points notwithstanding, there will be circumstances that require the use of surrogate solvent(s) as the extraction medium. For example, even the placebo itself may be challenging to profile for extractables. Alternatively, solvents may be used to represent multiple drug products that may use the same pharmaceutical package. When extraction solvents are used, it is necessary to establish and justify the composition of these solvents, considering all the physicochemical characteristics of a formulation and/or surrogate solvent that influence its "extracting power." When considering surrogate solvents, <1663> offers the following insights:

- As the extracting power of polar aqueous drug products consisting of soluble ingredients (such as an injectable with a drug substance, buffers, and diluent) is driven primarily by the drug product pH, replacing the drug product (or placebo) with an analytically viable buffer system with a similar pH may be appropriate and justifiable.
- Largely non-polar drug products can be replaced with analytically expedient organic solvents; for example, chlorofluorocarbon and hydrofluoroalkane propellants used in metered dose inhalers (MDIs) can be replaced with dichloromethane as an extracting solvent and isopropanol can be used to replace ethanol, a common co-solvent in MDI formulations.
- Semi-polar drug products, such as "aqueous" drug products that contain stabilizers, solubilizing agents, chelating agents and buffers, lipid-containing products, and biotechnology products containing proteins, peptides, and blood-derived products, can be replaced by binary

mixtures of miscible solvents (such as alcohol/water) that have been polarity-matched to the drug product.

- The use of multiple solvents (or extracting media) with different polarities, pH, ionic strength, or extracting powers is recommended for high-risk dosage form packaging components and systems.
- For inorganic extractables, utilization of a surrogate solvent having similar metal-chelating properties as the drug product vehicle may be appropriate and justifiable.

Extraction temperature and duration are frequently addressed as a single aspect of an extraction study as they are closely related. Although the nature of the extraction solvent establishes the magnitude of the extraction (i.e. the amount of substances that can be extracted from a material at equilibrium), the combination of extraction time and temperature establishes the magnitude of the driving force and the degree to which equilibrium is actually achieved. Frequently, an elevated extraction temperature is used to increase the extraction rate, so that a short experimental time may simulate longer leaching times.

When selecting an elevated extraction temperature (and a shorter extraction duration), it is noted that extraction is a diffusion-controlled process wherein the relationship between the diffusion rate and temperature can be expressed empirically by the Arrhenius equation. The mathematics involved in a process that is driven by Arrhenius kinetics have been established in ASTM F1980-16 [26], which may be a useful guide for establishing accelerated contact conditions, recognizing that the proper use of this, or any other, migration model requires an understanding of the model's basis and essential principles, assumptions, and limitations.

Extraction stoichiometry refers to the ratio of the physical mass and/or surface area of the test article relative to the volume of the extracting medium. Extraction stoichiometry is frequently manipulated to presumably produce a more concentrated extract (that is, an extract that contains levels of extractables higher than the levels of leachables in a drug product). Production of a more concentrated extract generally improves the sensitivity of the extractables study, resulting is more extractables that can be discovered, properly identified, and accurately quantified.

Another aspect of extraction stoichiometry that is considered in <1663> is sizing (e.g. cutting, grinding, etc.) of test articles. While noting the potential benefits of sizing, the monograph warns about sizing's potential for alternating an extractables profile and advises that "careful consideration should be given to the effect of physical sizing of test articles on the extractables profile before such sizing methods are employed in extraction studies."

Considering extraction methods (that is, methods by which the extraction is accomplished), <1663> establishes several types of extraction methods including:

1. Maceration (solvent soaking) – in which the test article is allowed to soak for a period of time in an organic or aqueous extracting solvent at temperatures below the solvent's boiling point. Analysts can also fill packaging system units with extracting solvent and store them at relevant temperatures.
2. Reflux – in which the test article is immersed in boiling solvent for a period of time.
3. Soxhlet – in which the test article is placed in the "thimble" of a Soxhlet extraction apparatus that is slowly filled with redistilled solvent from a boiling flask/condenser system and, periodically, the extracting solvent (containing extractables) is siphoned back into the boiling flask and the process begins again (for as many times as required to attain equilibrium).
4. Sealed vessel – in which the test article and extracting solvent are sealed inside a container capable of withstanding elevated temperatures and pressures, placed into a laboratory autoclave, and heated with steam for a period of time.
5. Instrument-based solvent extraction – in which the test article is placed inside a sealed apparatus and extracted in an automated cycle; examples include pressurized fluid extraction, microwave-assisted extraction, and supercritical fluid extraction.
6. Sonication – in which the test article and extracting solvent are placed into a glass container and partly immersed in water inside an ultrasonic bath.

The monograph is not prescriptive in the sense that it advocates the use of certain extraction methods in certain circumstances; rather, the monograph notes practical issues with certain methods (e.g. potential loss of volatile extractables during refluxing and difficulty in using water as the extraction solvent for Soxhlet) and advises practitioners that "it is necessary that the means of performing the extraction match the objectives of the extractables assessment."

Monograph <1663> closes its discussion of extraction by noting that "not all drug product or material-contact situations are solution mediated and not all issues related to leaching of material-derived entities involve a solution phase." Considering volatile organic extractables specifically, the monograph notes that volatile extractables are most effectively addressed by using thermal means of extraction, usually directly coupled to analytical instruments, such as headspace analysis (as headspace gas chromatography, HS/GC), direct thermal desorption (usually coupled to gas chromatography, TD/GC), and thermogravimetric analysis (TGA/GC).

Once an extract has been generated, the next objective is to perform a thorough chemical characterization of the extract, where the purpose of the characterization is to discover, identify, quantify, and report all extractables present in the extract at a level at or above an established reporting threshold. Much of what is contained in <1663> concerning extract testing, including the actions of scouting, discovery, identification, and quantitation, has been previously discussed, for example, Sections 3.2 (Chapter 3), 4.1, and 4.2 (Chapter 4). The concept of classes of extractables identities, based on the nature and amount of corroborating information (discussed previously in greater detail in Chapter 3, Section 3.2.5), was introduced in <1663> with four extraction classes, including unknown, tentative, confident, and confirmed. Although <1661> describes identification classes, it does not specify which class might be necessary for what purpose, noting instead that "the level of identification required for any individual extractable depends on the intended use of that identification" and that "it is up to the organization responsible for the extraction study to determine this after appropriate consideration of applicable regulatory guidances."

Addressing quantitation, <1663> considers both true quantitation, where analytes are quantified using standards prepared from reference materials, and estimation, where analytes are estimated using their responses (or response factors) relative to internal standards, or other surrogate reference compounds. Should estimation be used as a "quantitation" method, <1663> advocates that "diligence must be exercised in terms of establishing and justifying the choice and use of internal standards."

Monograph <1663> closes with recommendations for designing an extractables assessment intended to establish a rigorous leachables–extractables correlation for a high-risk drug product, where a challenging safety threshold might apply (e.g. 0.15 µg/day), as follows:

1. Generation of extracts should be accomplished with:

   a. Multiple solvents or extracting media with varying extracting power based on the known extracting power of the drug product vehicle;

   b. Multiple and complementary extraction techniques, including those with the capability for volatiles analysis;

   c. Extraction conditions where equilibrium (asymptotic extraction) has been achieved.

2. Characterization of extracts should use:

   a. Multiple and complementary analytical techniques;

   b. Careful sample preparation, keeping the analytical technique(s) in mind;

   c. A systematic process for identification and quantitation of extractables.

Turning from extractables to leachables, monograph <1664> presents a framework for the design, justification, and implementation of assessments for drug product leachables derived from pharmaceutical packaging and delivery systems. Management of leachables is important to pharmaceutical and biotechnology/biologic product manufacturers and regulatory authorities because certain leachables above specific concentrations can present safety concerns for patients and/or compatibility issues for drug product formulations.

Perhaps one of the most significant aspects of <1664> was its modification of the Dosage Form Risk Classification Table that had previously appeared in the FDA 1999 Container-Closure Guidance (discussed in greater detail in this Chapter, Section 7.3.1 following). This Table essentially classifies all drug product dosage forms with respect to the risk that the dosage form could contain unsafe levels of leachables based on two dimensions, Likelihood of Packaging Component–Dosage Form Interaction and Degree of Concern Associated with the Route of Administration, and serves as a way of matching a dosage form with the type of information and testing required to secure regulatory approval. The modified Table appearing in <1664> has certain dosage forms being downgraded to having lower potential for interaction with packaging components, based on experiences gained in testing and safety assessing leachables profiles for the packaged dosage forms.

Monograph <1664> contains commentary on several topics that serves as an excellent primer on leachables assessment including:

- General Concepts for Leachables Assessment
- Safety Thresholds, including special case compounds
- Information Sharing
- Leachables Characterization, including analytical thresholds such as the AET and analytical methods and techniques

"Rules" established in <1664> for conducting a leachables assessment are as follows:

- The assessment is performed on the actual drug product and not simulations thereof.
- The assessment is performed with the actual packaging and delivery system in the form it will be commercialized, not with a prototype or on system components.
- The leachables assessments are accomplished on the same lots of packaging components and systems that were used in the extractions studies, thereby fostering good extractables–leachables correlations.
- The assessment is performed on a product that is manufactured under conditions that reflect the actual commercial processes of production of the drug product and the packaging/delivery system, filling of the drug product into the packaging/delivery system, post-filling

treatment of the filled packaging (e.g. terminal steril-ization), distribution, storage, and clinical use of the drug product. Although leachables studies may include accelerated storage conditions, they cannot be limited to accelerated conditions and must include real-time assessment.

Because leachables data is used to establish critical drug product quality attributes, such as safety and efficacy, it is a regulatory and good science expectation that analytical methods used to generate leachables data are suitable for their intended use (qualified or validated), especially considering quantitation. Monograph <1664> notes that the validation of quantitative leachables methods should be accomplished according to industry accepted prac-tices, criteria, and standards and that relevant validation parameters include: accuracy, precision (repeatability, intermediate precision), specificity, limit-of-detection, limit-of-quantitation, linearity and range, and robustness. Although <1664> does not specify a detailed validation protocol and does not include acceptance criteria, it pro-vides the following guidance with respect to how to address validation parameters.

- **Accuracy and precision**. The validation parameters of accuracy and precision (repeatability and intermediate precision) are evaluated using drug product samples spiked with known amounts of target leachables. The drug product spiking matrix should have little-to-no contact with the packaging materials used in the final drug product, and therefore little-to-no measurable lev-els of endogenous leachables. Suitable spiking matrices can include freshly manufactured drug product and simulated drug product vehicles. Spiking levels should be based on results from accelerated stability studies or estimated from the known amounts of potential target leachables determined from extraction studies. Accu-racy and precision testing is typically performed at three spiking levels.
- **Linearity and range**. Since potential leachables are present in packaging components at widely varying lev-els, actual drug product leachables can likewise appear at widely varying levels. The best accuracy and precision are achieved when the validated linear range considers the potential maximum accumulation levels of each target leachable or chemical class of leachables.
- **Limit of detection/Limit of quantitation.** To detect and quantitate unknown leachables, the limit of quan-titation should be at or below the designated analytical threshold (e.g. AET).
- **Specificity.** Evaluation of method specificity can be accomplished by evaluating chromatographic peak purity in spiked and non-spiked drug product samples. For GC-based quantitative methods, this can be accom-plished by GC/MS. For HPLC-based methods, either

LC/MS or LC/DAD (diode array detection) can be used. Specificity can be qualitatively demonstrated if there are no observable method interferences related to the chemical entities present in the drug product.
- **Robustness.** A design-of-experiments statistical approach with consideration of critical analytical method parameters (e.g. HPLC flow rate, HPLC column, mobile phase gradient, etc.) should be used to create robustness evaluation protocols. Other approaches, such as serial change of critical parameters, can also be applied.
- **System suitability**. Chromatographic methods should include appropriate system suitability criteria for rou-tine method evaluation, including tests for method linearity, precision, sensitivity, and specificity as appro-priate. These parameters should be evaluated with an appropriately constituted test mixture(s) each time the quantitative leachables method is used, and should include appropriate system suitability acceptance crite-ria based on the method validation results.

Furthermore, <1664> is that official guidance that introduces the concepts of extractables – leachables cor-relations, addressed previously in Chapter 3, Section 3.5, and simulation studies, discussed previously in Chapter 2, Section 2.2.2. Addressing simulation studies more specifi-cally, <1664> notes that the results of the simulation study (probable leachables identities and concentrations) may be sufficient to establish patient safety and the quality impact of the actual drug product leachables. To the extent that the simulation study mimics the drug product leachables study, the potential safety or quality impact of a compound as an extractable is an estimate of the potential safety or quality impact of the compound as an actual leachable. If it can be established that a compound quantitated as a pack-aging system extractable has an acceptably small impact on drug product safety and quality, then it follows that the same compound as a leachable in the drug product for-mulation will likely have a similarly low impact on safety and quality. However, as is the case with most approaches used to assess drug product quality, the acceptability of a simulation study for any particular drug product needs to be scientifically justified (likely to a regulatory reviewer) by the drug product applicant.

Lastly, <1664> considers the topic of profiling drug products for leached elements, incorrectly referred to as inorganic leachables in the monograph (as extracted ele-ments can include both inorganic and organic substances such as silicic acid versus siloxanes). As leached elements are just one of several types of elemental impurities, <1664> notes that testing drug products over the shelf-life for leached elements is equivalent to addressing elemental impurities over the drug product shelf-life. Therefore, USP

Table 7.2 Modified FDA/CDER/CBER risk-based approach to consideration of leachables per USP <1664>[a].

Degree of concern associated with the route of administration	Examples of packaging concerns for common classes of drug products		
	Likelihood of packaging component – dosage form interaction		
	High	**Medium**	**Low**
Highest	Inhalation Aerosols and Sprays	Injections and Injectable Suspensions[b]; Inhalation Solutions[b]	Sterile Powders and Powders for Injection[b]; Inhalation Powders[b]
High	Transdermal Ointments and Patches	Ophthalmic Solutions and Suspensions[b]; Nasal Aerosols and Sprays[b]	–
Low	Topical Solutions and Suspensions; Topical and Lingual Aerosols; Oral Solutions and Suspensions	–	Oral Tablets and Oral (Hard and Soft Gelatin) Capsules; Topical Powders[b]; Oral Powders[b]

a) While this table provides a convenient overview of the general level of regulatory concern with various dosage forms regarding leachables, it should not be inferred that "low-risk" dosage forms (e.g. oral tablets) carry no risk for leachables issues.

b) These dosage forms were downgraded versus the original table contained in the FDA 1999 Container-Closure guidance.

chapters that deal with elemental impurities, specifically <232> for elements to target and their associated permissible daily exposures [27] and < 233> for analytical methods [28] are directly relevant for profiling a drug product for elemental leachables.

In addition to the general text on leachables, <1664> contains sub-sections that deal with the unique leachables issues associated with various dosage forms. For example, <1664.1> [29] addresses leachables aspects associated with Orally Inhaled and Nasal Drug Products (OINDP) including metered dose inhalers (MDIs), nasal sprays, inhalation solutions, suspensions, and sprays, and dry powder inhalers (DPIs). Because OINDPs are generally categorized as high-risk dosage forms due to safety considerations related to the route of administration and high probability of packaging component interaction with the formulation (see Table 7.2), OINDPs typically require:

- A leachables stability study for drug product registration that supports intended storage and use conditions throughout the proposed shelf-life ideally on primary drug product stability batches manufactured with the same lots of packaging components used in extraction studies (in order to facilitate a leachables–extractables correlation).
- Sensitive, selective, and fully validated leachables analytical methods.
- Leachables assessments based on safety thresholds (Safety Concern Threshold (SCT): 0.15 μg/day, and

Qualification Threshold (QT): 5 μg/day total daily intake (TDI) for an individual organic leachable; however, note *Special Case Compounds* as exceptions).

- Complete qualitative and quantitative leachables–extractables correlations (which require that extractables assessments be accomplished on all critical packaging components).
- Leachables specifications including acceptance criteria (which assumes a complete extractables assessment for each critical packaging component). (Note that in many cases routine extractables testing for the release of critical components can be used to control drug product leachables in lieu of routine drug product leachables testing, providing that a comprehensive leachables–extractables correlation is established.)

Monograph <1664.1> contains a description of the various OINDP product types and sample calculations for the AET associated with these types; the reader is referred to the actual monograph for details. Additionally, the monograph addresses the concept of special case compounds, which are "individual (or classes of) compounds that have special safety or historical concerns as drug product leachables in OINDP, and therefore must be evaluated and controlled as leachables (and extractables) by specific analytical techniques and technology-defined thresholds." Polycyclic Aromatic Hydrocarbons (PAHs) or Polynuclear Aromatics (PNAs), *N*-nitrosamines, and 2-mercaptobenzothiozole (2-MBT) are considered to be "special case" compounds requiring special characterization studies using specific analytical techniques and methods. Thresholds for characterization of these compounds as extractables or leachables in OINDP are typically based on the limits of these specific analytical techniques and methods. It is noted that all of these special case compounds have been associated with elastomers and thus at the current time there are no OINDP special case compounds that are specifically linked to plastic components.

Although not unique to OINDPs, <1664.1> also addresses the concept of analytical uncertainty, specifically related to the application of the AET for surfacing leachables that must be reported for toxicological safety risk assessment. This monograph advocates that a database of response factors be created for commonly encountered leachables so that the variation in response factors can be used to calculate the AET "uncertainty factor." Specifically, the monograph recommends for OINDP that the AET be lowered by a factor linked to one relative standard deviation in an appropriately constituted response factor database, or a factor of 2, whichever is greater.

7.2.3.4 Containers – Glass, <660> and <1660>

As an inorganic substance, glass itself is typically not a significant source of organic extractables and leachables. Organic extractables and leachables from glass items are not related to the glass itself but rather to coatings that may be applied to a glass surface for practical reasons (for example, silicone oil applied to the inner surface of glass syringe barrels to promote easy movement of the syringe's plunger). Thus, the USP monographs for glass containers, <660> [21] for the enforceable guidance and <1660> [30] for the informational discussion, deal less with chemical aspects of glass suitability for use and more with physical aspects including glass delamination. In fact, the only chemical test contained in <660> that is relevant for extractables is for extractable arsenic, which is likely a historic carry-over from a time when arsenic could be present in the glass raw material (sand). Although arsenic is potentially relevant as it is a Class 1 elemental impurity per ICH Q3D (and therefore possesses a low PDE), the test for Arsenic in <660> does not address patient safety as the threshold ($0.1\,\mu g/g$) is not specifically linked to the ICH Q3D PDE and the extraction is not particularly aggressive (water extraction at $121\,°C$ for one hour).

Glass delamination is described in <1660> as "the appearance of thin flexible flakes of glass (or lamellae) that can range in size from <50 µm to 200 µm in a drug product solution." The appearance of glass lamellae is a visual indicator of an interaction between the drug product and the inner surface of the glass, representing the final stage of a complex glass corrosion reaction. Additionally, mechanical energy from shaking or vial-to-vial contact

during transportation may dislodge the lamellae from the internal surface of a filled vial.

Glass delamination is a serious quality issue and can result in a product recall, but is generally not related to extractables and leachables per say. Although delamination may be a result of leaching, it is unlikely that the leached elements leading to delamination are potential safety hazards. Thus, the reader is referred to <1660> and its cited references for more information on this topic.

7.2.3.5 Elastomeric Components, <381> and <1381>

Certain components used in pharmaceutical packaging/delivery systems must have elastic properties for the system to function properly. Elastomers are a unique family of polymers with such elastic properties, including the ability to recover from being stretched or deformed beyond their original state. This allows components to be flexible, maintain a seal, and be able to reseal after puncturing, thus explaining why elastomeric components are widely used in pharmaceutical packaging.

USP monographs for elastomeric components include <381> [22] for the enforceable guidance and <1381> [31] for the informational discussion. As these monographs are not generally applicable to all elastomeric components used in packaging for all pharmaceutical dosage forms, but rather are specific to injections and drug products, an important part of both monographs is the statement of Scope. Elastomeric components for injectable products that are in scope include, but are not limited to:

- Those used for vials and bottles (stoppers and cap liners),
- Pre-filled syringes (plungers, needle shields, and tip caps),
- Cartridges (plungers and seal liners),
- Flexible bags (injection ports), and
- Blow-fill-seal containers (cap liners).

All elastomeric components in direct or indirect contact with the pharmaceutical product are within scope. Items within scope include elastomeric components of systems or packages that are intended for both long-term storage (shelf-life) and transient product storage and/or product delivery intended for specific pharmaceutical products.

Elastomeric components that are out of scope include:

- Components of containers and closures that do not have direct or indirect contact with the pharmaceutical product or hold intermediate compounds, active pharmaceutical ingredients (APIs), and excipients,
- Elastomeric components used to package dosage forms other than injectables and implanted drug products,
- Elastomeric components of containment and/or transport systems used in product, intermediate compound, API, or excipient manufacturing, and
- Elastomeric components used in products regulated by the Center for Devices and Radiological Health (CDRH).

Similar to the plastic materials of construction monograph <661.1>, <381> contains physicochemical chemical tests to establish the general chemical characteristics of the elastomeric test article and biological reactivity testing requirements. Given the great compositional diversity in elastomers used in pharmaceutical packaging, tests relevant for plastic materials, such as identity and composition, are not included in <381> for elastomeric components.

Considering the physicochemical tests and biological reactivity tests, elastomeric components are tested as finished elastomeric components obtained after the completion of all manufacturing and processing steps (e.g. molding conditions, sterilization, etc.), and surface modifications (such as siliconization, chlorinated surface treatments, fluoropolymer coatings, and films). For the physicochemical tests, the extraction for all elastomeric items is performed as follows:

- Water as the extraction solvent,
- Sealed vessels extraction as the method,
- Autoclaving at 121 °C for 30 minutes as the temperature and duration, and
- 100 cm^2 per 200 ml as the ratio of the extracted surface area to extraction solvent volume.

The tests performed on the extracts include those tests used for plastic materials (acidity/alkalinity, UV absorbance) plus tests specific to elastomers (appearance, color, reducing substances (as an alternate to TOC for plastics), volatile sulfides and ammonium). The purpose of these physicochemical tests is established in <1381> as follows:

- **Appearance** (turbidity/opalescence). This is a non-specific test for all the extractable species in an elastomeric formulation that are not soluble in an aqueous solution. A high turbidity is the indication of a high extractable potential. Species promoting turbidity have numerous origins in an elastomeric formulation, including fatty-acid derivatives, residues of curing systems, and oligomers from the elastomer.
- **Color.** This is a non-specific test indicative of the presence of extractable species in an elastomeric formulation that have the capacity to add color to an aqueous solution. Species that cause color may have several origins in an elastomeric formulation. Aqueous solutions are common in pharmaceutical packaging/delivery systems.
- **Acidity or alkalinity.** This is a non-specific test indicative of the acidic, basic, or buffering power of the aqueous extractables from the elastomeric formulation. High values in the acidity/alkalinity test may need to be evaluated in conjunction with the specifics of a drug solvent vehicle and anticipated acceptance criteria for the pH of the drug product.

- **Absorbance.** This UV spectrum test of an aqueous extract from an elastomeric formulation is indicative of the unsaturated or aromatic character of the chemical species extracted. Unsaturated compounds in the extracts may originate from many raw materials and additives of an elastomeric formulation such as antioxidants and curing or dying agents.
- **Reducing substances.** This is a non-specific test. Extracted species from an elastomeric formulation with potential reducing power may originate from most raw materials of an elastomeric formulation (polymers, curing systems, antioxidants, etc.).
- **Volatile sulfides.** This test is specific for elastomeric formulations containing sulfur. Sulfur and sulfur precursors are often used as components of curing systems for elastomers.
- **Ammonium.** This test is specific for elastomeric formulations with nitrogen-containing raw materials. Ammonium ions can be generated during the curing process. Thiurams and thiazoles are examples of nitrogen-containing curing systems used.

When considering the physicochemical tests, it is important to note that elastomeric components may be classified in two types: Type I elastomeric closures that can meet the strictest physicochemical requirements and are preferred and Type II elastomeric closures that have mechanical properties suitable for special uses (e.g. multiple piercing) but cannot meet the Type I acceptance criteria for Appearance, Absorbance, and Reducing Substances. For these tests, Type II elastomeric closures have alternative acceptance criteria.

As was the case with plastic materials, elastomeric components are tested for biological reactivity per USP <87>. However, whereas plastic materials that do not meet the <87> requirements cannot be used in pharmaceutical packaging, elastomeric components that do not meet <87> requirements can be further tested via USP <88> *in vivo* methods, including the Systemic Injection test and Intracutaneous test. Although not specifically stated in <381>, one presumes that an elastomeric component that does not meet the requirements of <87> but does meets the requirements for the specified <88> tests are deemed to have met the biological reactivity requirement.

In addition to a discussion of the testing of elastomeric components, <1381> contains an excellent discussion of elastomeric components in general, addressing topics such as:

- Material of construction,
- Polymer types and attributes,
- Surface coatings and treatments,
- Manufacturing technology, and
- Sterilization procedures.

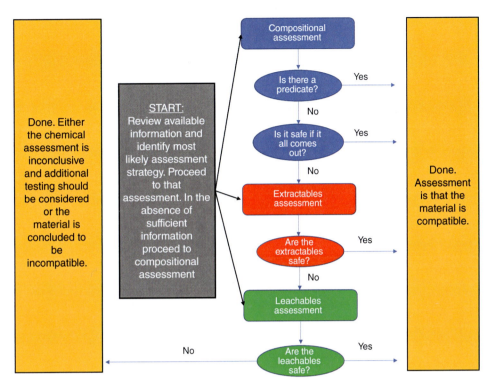

Figure 1.5 Process flow diagram, chemical compatibility assessment process for a material.

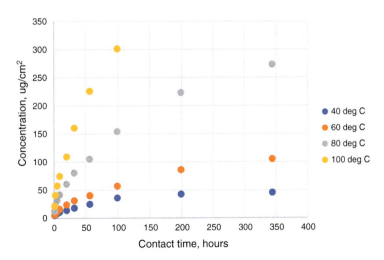

Figure 2.6 Leaching of dipropyl phthalate (DPP) into isopropyl alcohol (IPA) from a bromobutyl rubber (BIIR) disk. The effect of extraction temperature on the equilibrium concentration of this extractable is clearly evident, as the equilibrium concentration increases as the temperature increases.

Extractables and Leachables: Characterization of Drug Products, Packaging, Manufacturing and Delivery Systems, and Medical Devices, First Edition. Dennis Jenke.
© 2022 John Wiley & Sons, Inc. Published 2022 by John Wiley & Sons, Inc.

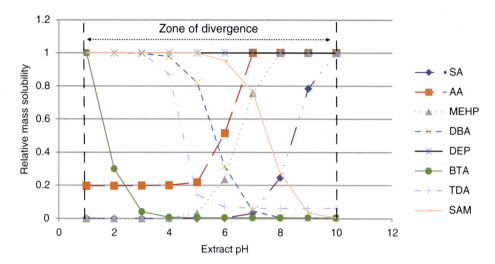

Figure 2.9 The Effect of pH on the reported solubility of extractables. Because DEP is non-ionic, its solubility is unaffected by the pH of the extract. The solubility of the acidic extractables (AA, SA, and MEHP) increases with increasing pH, depending on their specific pK_a values. The solubility of the basic extractables (SAM, DBA, TDA, BTA) increases with decreasing pH, again consistent with their pK_a values. A Zone of Divergence exists between the pH values where the weakest acid (SA) and the weakest base (BTA) achieve their maximum solubilities. If a set of extraction solvents is designed to capture essentially all possible acidic or basic extractables at their likely highest concentration in the extract, then the set must have a pH that spans the values that define the Zone of Divergence. Source: From reference [58] with permission.

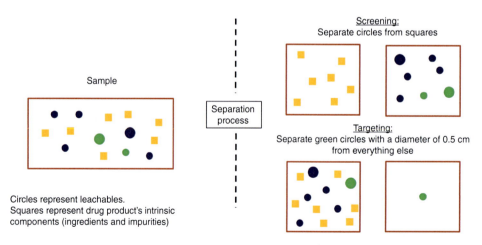

Figure 3.2 Comparison of screening versus targeting. In screening, the separation is at a high level, with the screening method separating the analytes (circles) from the other constituents in the sample (squares). However, in targeting the separation is much more specific, focusing on a specific analyte.

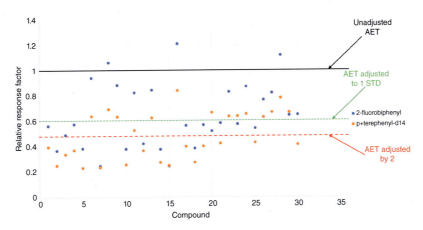

Figure 3.11 Distribution of GC/MS response factors for the database summarized in Table 3.5.

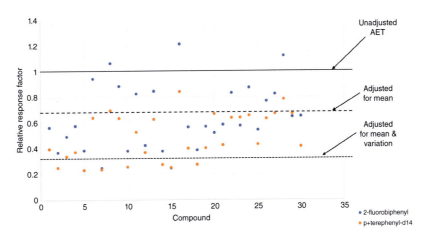

Figure 3.12 Adjustment of the AET for both the magnitude of the mean response factor and the variation in the response factor. The AET adjustment is based on 2-FB as the IS.

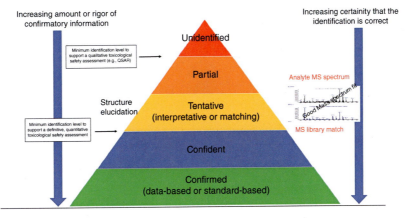

Figure 3.16 The Identification hierarchy, which establishes identification categories. As the amount and rigor of confirming information increase, the information content of the identity increases and the likelihood that the identity is correct increases.

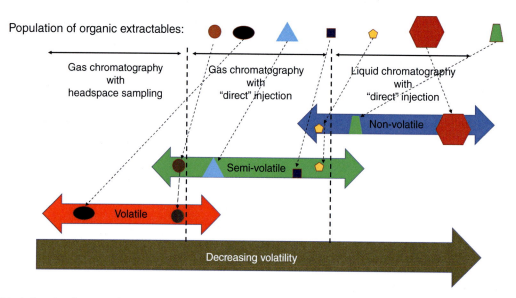

Figure 3.22 Ideal situation for screening extracts for organic extractables. In this ideal situation, all the individual extractables in the extractables profile are accounted for by one or more of the screening methods. Additionally, the screening methods do not "obscure" the extractable's identity (indicated by the shape) and/or do not distort the extractable's concentration (indicated by the size of the shape).

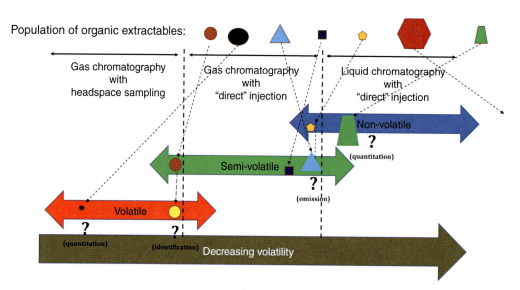

Figure 3.23 General actual situation for screening extracts for organic extractables. In this actual situation, some of the individual extractables in the extractables profile are not accounted for by any of the screening methods (for example the red rhombus) Additionally, the screening methods "obscure" the extractable's identity (indicated by the color) and/or distort the extractable's concentration (indicated by the size of the shape). For example, the concentration of the extractable represented by the black oval is underestimated by the volatiles method while the concentration of the extractable represented by the green rhombus is overestimated by the non-volatiles method. Additionally, one notes that although the extractable represented by the brown circle is "picked up" by both the volatiles and non-volatiles, it is identified as different compounds by both methods and neither of the identities is the correct one. Lastly, the analyte represented by the red hexagon is completely missed by all three of the screening methods.

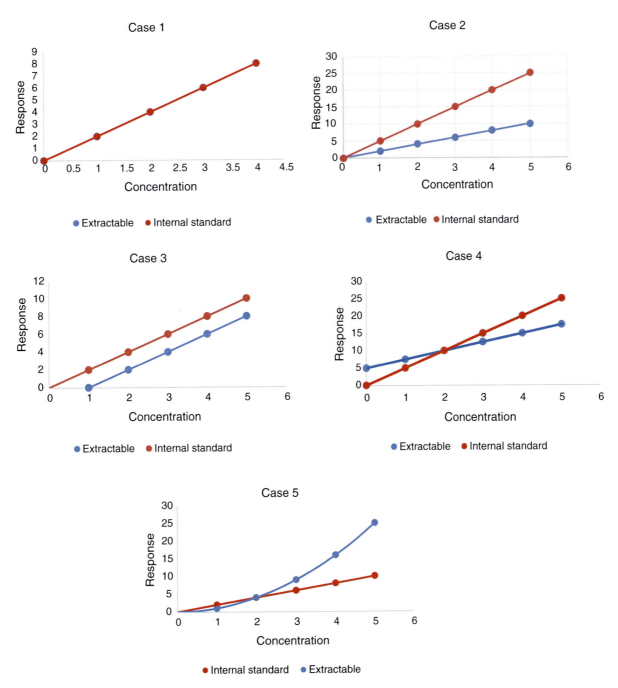

Figure 3.27 Concentration–response function examples to illustrate different situations related to extrapolation errors.

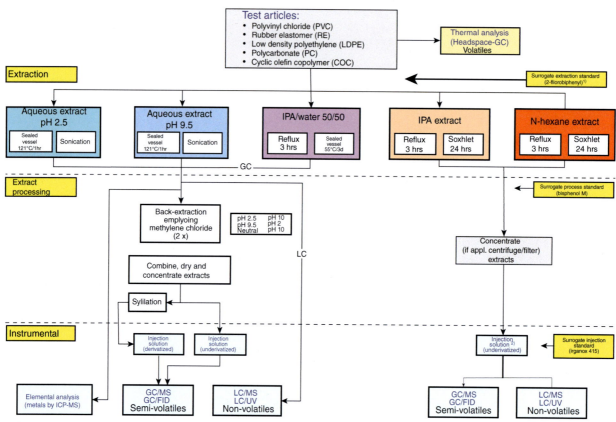

1) The surrogate extraction standard 2-fluorobiphenyl was only used by one of the participating labs for the organic extracts
2) For n-heaxane, a solvent switch to methanol was performed to obtain the injection solution (GC and LC)

Figure 3.34 Study design, PQRI controlled extraction study on five materials.

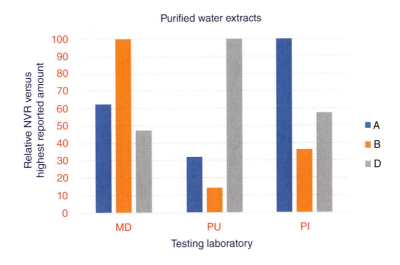

Purified water extracts

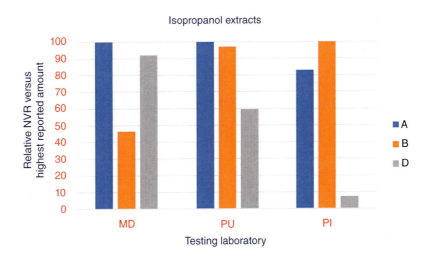

Isopropanol extracts

Note: MD = medical device; PU = polyurethane; PI = polyiosprene

Figure 3.36 Test results, NVR, lab-to-lab comparison. MD = medical device; PU = polyurethane; PI = polyiosprene.

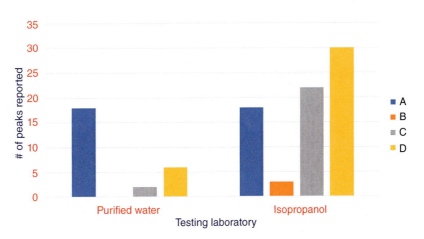

Figure 3.37 Test results, non-volatile organic extractables by LC/MS, lab-to-lab comparison.

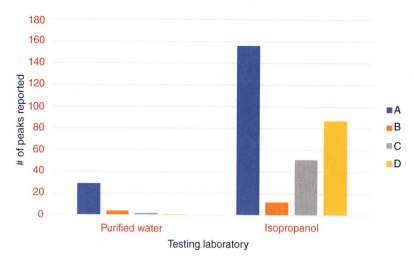

Figure 3.38 Test results, semi-volatile organic extractables by GC/MS, lab-to-lab comparison.

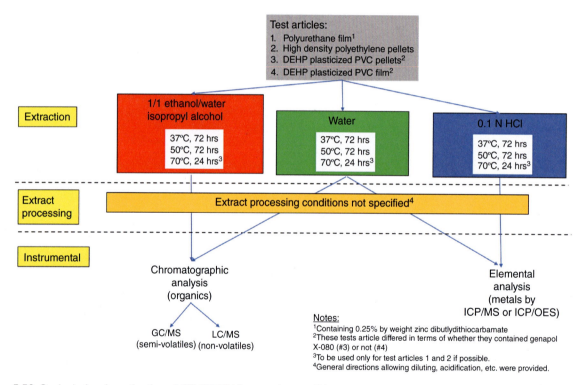

Figure 3.39 Study design, investigation of ISO 10093:12 extraction conditions.

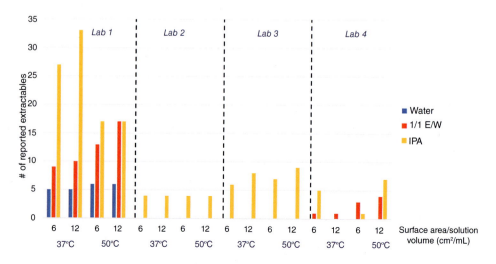

Figure 3.40 Test results, semi-volatile organic extractables by GC/MS, polyurethane test article.

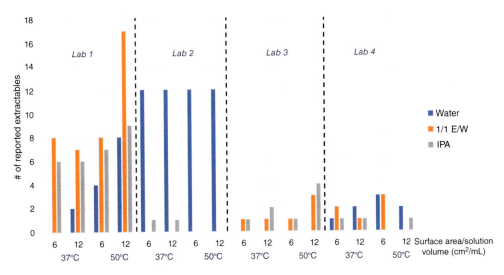

Figure 3.41 Test results, non-volatile organic extractables by LC/MS, polyurethane test article.

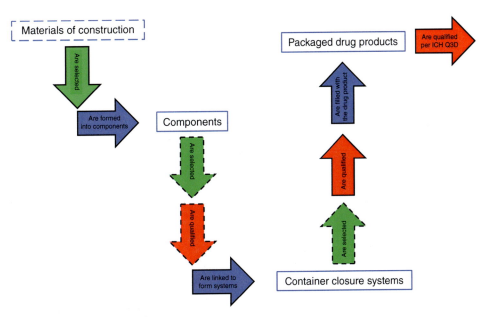

Figure 4.1 Process flow diagram, development and testing of a packaged drug product from the perspective of the container closure system. Green arrows reflect selection steps, red arrows reflect qualification steps, and blue arrows refer to construction steps. An arrow whose outline is solid denotes processes that must occur in development and testing while arrows with dashed outlines are processes that could occur in the development and testing at the discretion of the process owner. The dashed box surrounding materials of construction indicates that the process of Material Selection is typically completed by a component's vendor and not necessary by the drug product development team.

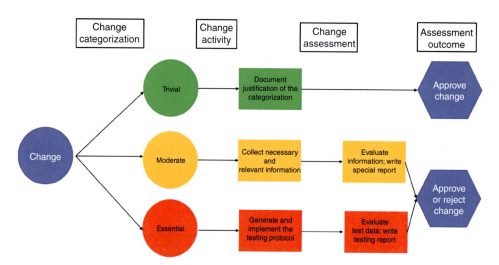

Figure 6.5 Process Flow Diagram, Change Control. The activity of change characterization is discussed in greater detail in Section 6.2. The change activity and change assessment would typically be addressed as part of a formal change control process such as the Comparability Protocol.

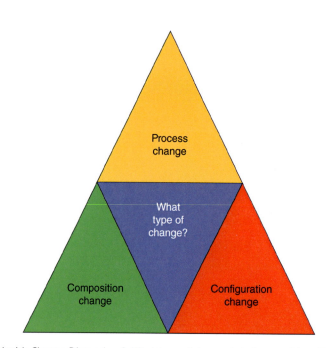

Figure 6.10 Options associated with Change Dimension 2. What type of change is being considered?

Where is the change?	What type of change?	How is the change being made?	How big is the change?
Tertiary packaging (e.g., label on overpouch, shipping carton)	Process change	Adjustment of an existing step	Minor adjustment to a minor step
			Major adjustment to a minor step
			Minor adjustment to a major step
			Major adjustment to a major step
		Elimination of an existing step	Elimination of a minor step
			Elimination of a potentially important step
			Elimination of a major step
		Replacement of an existing step	Replacement of a minor step
			Replacement of a potentially important step
			Replacement of a major step
		Addition of a new step	Addition of a minor step
			Addition of a potentially important step
			Addition of a major step
	Configuration change	Re-ordering of existing configuration	Re-ordering of minor components
			Re-ordering of major components
		Dimensional changes to existing configuration	Minor change to a minor component
			Major change to a minor component
			Minor change to a major component
			Major change to a major component
	Compositional change	Adjustment to an existing component	Minor adjustment to a minor component
			Major adjustment to a minor component
			Minor adjustment to a major component
			Major adjustment to a major component
		Elimination of an existing component	Elimination of a minor component
			Elimination of a potentially important component
			Elimination of a major component
		Like for like replacement of an existing component	Replacement of a minor component
			Replacement of a potentially important component
			Replacement of a major component
		Substitution for an existing component (not like for like)	Substitution for a minor component
			Substitution for a potentially important component
			Substitution for a major component
		Addition of a new component	Addition of a minor component
			Addition of a potentially important component
			Addition of a major component

Figure 6.13 Decision Matrix, Change Stage 2, Change Categorization.

Where is the change?	What type of change?	How is the change being made?	How big is the change?
Secondary packaging (e.g., overpouch, port assembly separated by a membrane, oxygen indicator in space between primary bag and overpouch) ★	Process change	Adjustment of an existing step	Minor adjustment to a minor step
			Major adjustment to a minor step
			Minor adjustment to a major step
			Major adjustment to a major step
		Elimination of an existing step	Elimination of a minor step
			Elimination of a potentially important step
			Elimination of a major step
		Replacement of an existing step	Replacement of a minor step
			Replacement of a potentially important step
			Replacement of a major step
		Addition of a new step	Addition of a minor step
			Addition of a potentially important step
			Addition of a major step
	Configuration change	Re-ordering of existing configuration	Re-ordering of minor components
			Re-ordering of major components
		Dimensional changes to existing configuration	Minor change to a minor component
			Major change to a minor component
			Minor change to a major component
			Major change to a major component
	Compositional change ★	Adjustment to an existing component	Minor adjustment to a minor component
			Major adjustment to a minor component
			Minor adjustment to a major component
			Major adjustment to a major component
		Elimination of an existing component	Elimination of a minor component
			Elimination of a potentially important component
			Elimination of a major component
		Like for like replacement of an existing component	Replacement of a minor component
			Replacement of a potentially important component
			Replacement of a major component
		Substitution for an existing component (not like for like)	Substitution for a minor component
			Substitution for a potentially important component
			Substitution for a major component
		Addition of a new component ★	Addition of a minor component
			Addition of a potentially important component
			Addition of a major component ★

Figure 6.13 (Continued)

Where is the change?	What type of change?	How is the change being made?	How big is the change?
Primary packaging (e.g., solution bag)	Process change	Adjustment of an existing step	Minor adjustment to a minor step
			Major adjustment to a minor step
			Minor adjustment to a major step
			Major adjustment to a major step
		Elimination of an existing step	Elimination of a minor step
			Elimination of a potentially important step
			Elimination of a major step
		Replacement of an existing step	Replacement of a minor step
			Replacement of a potentially important step
			Replacement of a major step
		Addition of a new step	Addition of a minor step
			Addition of a potentially important step
			Addition of a major step
	Configuration change	Re-ordering of existing configuration	Re-ordering of minor components
			Re-ordering of major components
		Dimensional changes to existing configuration	Minor change to a minor component
			Major change to a minor component
			Minor change to a major component
			Major change to a major component
	Compositional change	Adjustment to an existing component	Minor adjustment to a minor component
			Major adjustment to a minor component
			Minor adjustment to a major component
			Major adjustment to a major component
		Elimination of an existing component	Elimination of a minor component
			Elimination of a potentially important component
			Elimination of a major component
		Like for like replacement of an existing component	Replacement of a minor component
			Replacement of a potentially important component
			Replacement of a major component
		Substitution for an existing component (not like for like)	Substitution for a minor component
			Substitution for a potentially important component
			Substitution for a major component
		Addition of a new component	Addition of a minor component
			Addition of a potentially important component
			Addition of a major component

Key:
• Green shading = trivial change, lower risk; address change with a entry to the history file that records that the risk was assessed as low (including details), requiring no additional assessment or testing to approve change.
• Orange shading = moderate change, moderate risk; address change via data review and generation of a Special Report. If insufficient data is available to generate the special report, limited testing may be necessary.
• Red shading = essential change, higher risk; address change by limited extractables testing.
• Blue shading = essential change, highest risk; address change by full extractables testing.

The arrows placed in the figure represent the assessment of the first example change (change of an antioxidant), as documented in the text. The stars placed in the figure represent the assessment of the second example change (addition of a label), as documented in the text.

Note:
These are generalizations that are applicable to most circumstances. In certain cases, a decreased or increased level of assessment may be necessary. For example, consider the case of a change to a three-layered overpouch (secondary packaging). A change to the overpouch layer furthest from the drug solution is inherently of lower risk than a change to the overpouch layer that is closest to the drug solution. Thus, the same change applied to the closest layer might require a more rigorous assessment than the same change made to the furthermost layer.

As another example, consider the case of an overpouch (secondary packaging) that is a known migration barrier (e.g. constructed with an aluminum foil layer). In this case, any changes to tertiary packaging outside of the barrier overpouch is risk level green, regardless of the nature of the change

Figure 6.13 (Continued)

		Chemical risk factor		
		Higher	Moderate	Lower
Toxicological Risk factor	Higher	Risk class 1	Risk class 2	Risk class 3
	Moderate	Risk class 2	Risk class 3	Risk class 4
	Lower	Risk class 3	Risk class 4	Risk class 5

Example 1, 3

Example 4

Example 2

Figure 7.6 Risk classification matrix for packaged drug products. The arrows refer to examples discussed in the text.

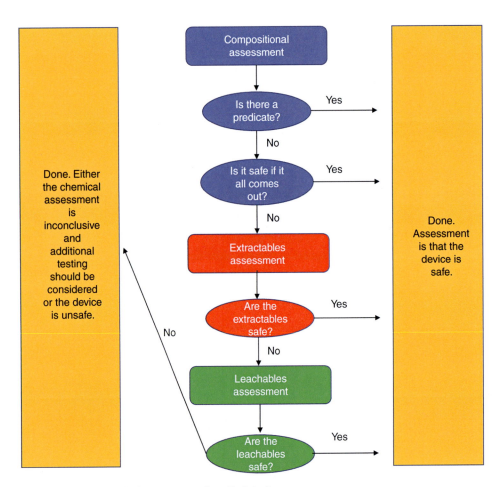

Figure 9.1 General flow diagram, chemical assessment of medical devices.

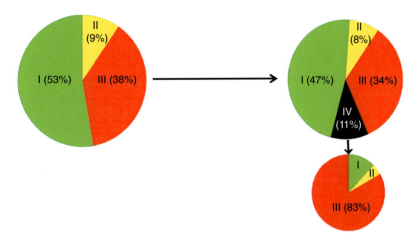

Figure 10.6 Distribution of classifications for 606 organic extractables. The entire database was grouped using the Cramer classification, producing the distribution shown on the left. To address mutagenicity, compounds were also grouped based on DEREK and SARAH, producing the distribution on the right. The "validity" of the DEREK and SARAH Class IV designations was addressed by considering the Cramer classification of the Group IV compounds; 83% compounds with a mutagenicity alert were also classified as Cramer Class III. Source: Data from Paskiet [37].

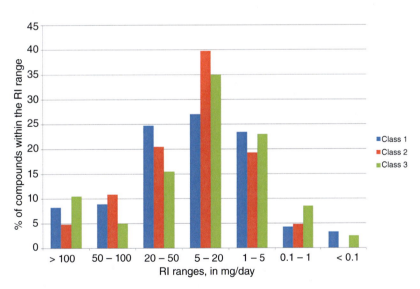

Figure 10.9 Distribution plots, Cramer class versus RI values, all extractables. In general, the distribution of the RI values is similar across the three Cramer classes, although the distribution is shifted slightly toward the lower RI values for the Cramer Class 2 and 3 extractables. Source: From reference [38], with permission.

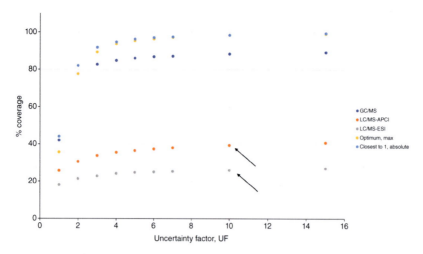

Figure 11.1 Percent coverage of the adjusted AET as a function of the value of the uncertainty factor (UF) applied to the AET; all compounds. The arrows illustrate the point of diminishing return, where further increases in UF do not significantly increase the % coverage. The individual methods do not achieve 100% coverage as there are analytes in the Nelson dataset that do not produce responses by those methods. As the UF increases, the incremental increase in % coverage gets smaller, eventually reaching a point of diminishing returns where large increases in UF produce only small increases in % coverage.

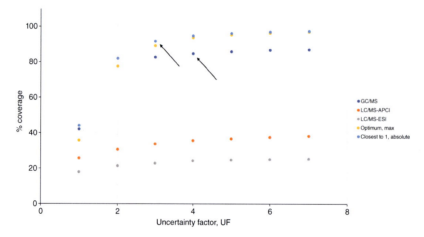

Figure 11.2 Percent coverage of the adjusted AET as a function of the value of the uncertainty factor (UF) applied to the AET; an expanded UF axis versus Figure 11.1 for all compounds. The arrows illustrate the point of diminishing return, where further increases in UF do not significantly increase the % coverage. UF values to be generally used in setting the AET are those that are at the point of diminishing returns.

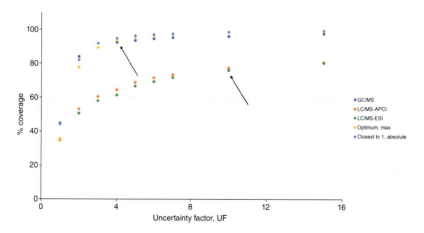

Figure 11.3 Percent coverage of the adjusted AET as a function of the value of the uncertainty factor (UF) applied to the AET; responding compounds only. The arrows illustrate the point of diminishing return, where further increases in UF do not significantly increase the % coverage. The LC/MC methods do not achieve 100% coverage as there are analytes in the Nelson dataset that do not produce small responses by those techniques. As the UF increases, the incremental increase in % coverage gets smaller, eventually reaching a point of diminishing returns where large increases in UF produce only small increases in % coverage.

Table 7.3 Compounds of concern for elastomeric components (Table 4 from USP <1381>).

Compound of concern	Source	Risk
Latex	Associated with compounds containing dry natural rubber or derivatives	Associated with anaphylaxis in individuals allergic to natural rubber latex proteins
Materials of animal origin	Fatty acids and their metal salts used as processing aids or slip agents in polymers	Potential sources of transmissible spongiform encephalopathies (TSEs) in pharmaceutical products before manufacturing
2-Mercapto-benzothiazole (MBT)	Vulcanization accelerator used in the production of elastomers	Potential carcinogen
N-Nitrosamines	Associated with the use of certain secondary amines in the cure system	Potential carcinogens
Phthalates (e.g. orthophthalates such as bis(2-ethylhexyl) phthalate (DEHP))	Used as a plasticizer or softener in polymers	Associated with developmental, reproductive, and endocrine health effects
Polycyclic aromatic hydrocarbons (PAHs)	Associated with carbon black (colorant or reinforcing agent)	Potential carcinogens

Interested readers are directed to <1381> for additional details.

Lastly, <1381> offers a somewhat more expanded list of compounds of concern than were listed in <1664> (see Table 7.3).

7.2.4 The European Pharmacopeia (Pharm Eur) Monographs on Packaging Materials and Systems

Considering the length of the discussion of the USP monographs, the reader might be disappointed with the brevity of the following discussion of the Pharm Eur monographs on pharmaceutical packaging. However, there are certain systemic reasons for the brevity. First, the Pharm Eur does not contain informational monographs as does the USP; rather, the Pharm Eur monographs are all enforceable documents containing tests, test methods, and test specifications with very little discussion. Thus, there are no informational monographs in the Pharm Eur that explain the intent and justify the content of the mandatory monographs. Additionally, there are no general informational Pharm Eur monographs on the topic of extractables and leachables and in general the enforceable Pharm Eur monographs do not address extractables and leachables testing per say. Rather, the Pharm Eur monographs are similar to the USP material monographs in that they contain tests that establish identity, composition, and general chemical nature.

Second, the USP material monographs (<660>, <661.1>, and <381>) have, as a result of recent revisions, become intentionally aligned with the corresponding Pharm Eur monographs. For example, a quick comparison of the USP <661.1> section on polypropylene and the Pharm Eur monograph 3.1.6 for polypropylene for containers and closures for parenteral and ophthalmic preparations reveals a certain degree of agreement between the documents, although there are some frustrating differences, especially in terms of what testing to perform (see Table 7.4). Thus, much of the previous discussion of the USP <661.1> monographs is equally relevant to Pharm Eur Section 3.1 monographs.

Both the USP and the Pharm Eur have individual texts for packaging materials and container themselves; while USP <66.1.> contains individual sections devoted to individual plastic materials, the Pharm Eur has individual monographs for the various plastic materials, grouped together as Section 3, Materials used for the manufacture of containers. The individual Section 3 monographs and the plastic materials they address are listed in Table 7.5.

The Pharm Eur monographs for material of construction strategically differ from the USP <661.1> sections as the individual monographs address both the material and its use. For example, while the USP <661.1> section for polypropylene does not specify the dosage form that would be packaged in the polypropylene-containing container, the Pharm Eur monograph for polypropylene, 3.1.6, specifies that the polypropylene it is applicable to is used for containers and closures for parenteral preparations and ophthalmic preparations. Presumably, 3.1.6 would not be applied to polypropylene used to package other dosage forms.

Like USP monograph <661.2>, the Pharm Eur monographs on actual containers (Sections 3.2 for Containers and 3.3 for Containers used for blood and blood products) include many of the same general chemistry tests that were in the material monographs. Thus, for example, the test

Table 7.4 Comparison of USP <661.1> Section for polypropylene with Pharm Eur 3.1.6 polypropylene of containers and closures for parenteral preparations and ophthalmic preparations.

USP <661.1> Polypropylenes		Pharm Eur 3.1.6. Polypropylene of containers and closures for parenteral preparations and ophthalmic preparations	
Test	**Description**	**Test**	**Description**
Identity	FTIR Spectrometry on thin film	Identity	FTIR Spectrometry on thin film or refluxed toluene extract
	Thermal analysis		–
Physico-chemical Tests, Solution S1	Aqueous extract; 25 g in 500 ml *Purified Water*, reflux for 5 hours	Tests, Solution S1	Aqueous extract; 25 g in 500 ml *water for injections R*, reflux for 5 hours
	–		Appearance of solution; visual inspection
	Acidity or alkalinity; titration		Acidity or alkalinity; titration
	Absorbance; UV spectrometry		Absorbance; UV spectrometry
	–		Reducing Substances, titration
	Total Organic Carbon (TOC), instrumental		–
–	–	Tests, Substances soluble in hexane	Reflux extraction in hexane, gravimetric determination
–	–	Tests, Sulfated ash	Gravimetric residue on ignition
–	–	Tests, Extractable heavy metals	Sample refluxed in 0.1 M HCl for 1 hour. Extracts tested with sulfide precipitation.
–	–	Tests, Extractable aluminum, chromium, titanium, vanadium, zinc	Sample refluxed in 0.1 M HCl for 1 hour. Extracts tested via Inductively Coupled Plasma Atomic Emission Spectroscopy
Plastic Additives – Phenolic antioxidants	Extract in toluene; 2.0 g per 80 ml, reflux for 1.5 hours. Precipitate dissolved plastic with methanol and filter, collecting the filtrate. Evaporate the filtrate to residue and re-dissolve in methylene chloride.	Supplementary Tests, Phenolic antioxidants	Extract in toluene; 2.0 g per 80 ml, reflux for 1.5 hours. Precipitate dissolved plastic with methanol and filter, collecting the filtrate. Evaporate the filtrate to residue and re-dissolve in methylene chloride.
	Test method A: HPLC with UV detection (280 nm). 25 cm x 4.6 mm column; C18 reversed phase stationary phase. 30/70 (v/v) acetonitrile/water mobile phase, flow rate = 2 ml/minute 20 µl injection volume.		Test method A: HPLC with UV detection (280 nm). 25 cm x 4.6 mm column; C18 reversed phase stationary phase. 30/70 (v/v) acetonitrile/water mobile phase, flow rate = 2 ml/minute, 20 µl injection volume.
	Test Method B: same as Test Method A except: 10/30/60 water/tetrahydrofuran/acetonitrile mobile phase, flow rate = 1.5 ml/minute		Test Method B: same as Test Method A except: 10/30/60 water/tetrahydrofuran/acetonitrile mobile phase, flow rate = 1.5 ml/minute
	Test method C: same as Test Method A except: 5/45/50 water/2-propanol/methanol mobile phase, flow rate = 1.5 ml/minute		Test method C: same as Test Method A except: 5/45/50 water/2-propanol/methanol mobile phase, flow rate = 1.5 ml/minute
Plastic Additives – Non-phenolic antioxidants	Similar sample preparation as for phenolic antioxidants. Thin-layer chromatography (TLC). TLC silica gel GF$_{254}$ plate. Mobile phase A = hexane, B = methylene chloride. Detection, spray with Iodine in ethanol detection solution and examine at 254 nm. 20 µl sample volume.	Supplementary Tests, Non-phenolic antioxidants	Similar sample preparation as for phenolic antioxidants. Thin-layer chromatography (TLC). TLC silica gel GF$_{254}$ plate. Mobile phase A = hexane, B = methylene chloride. Detection, spray with Iodine in ethanol detection solution and examine at 254 nm. 20 µl sample volume.
Amides and Stearates	Use sample from Non-phenolic antioxidants. Thin-layer chromatography (TLC). TLC silica gel GF$_{254}$ plate. Mobile phase 2,2,4-trimethylpentane and anhydrous ethanol (75/25). Detection, spray with 2 g/l 2,6-dichlorophenol-indophenol solution in dehydrated alcohol, heat, and examine. 10 µl sample volume.		Use sample from Non-phenolic antioxidants. Thin-layer chromatography (TLC). TLC silica gel GF$_{254}$ plate. Mobile phase 2,2,4-trimethylpentane and anhydrous ethanol (75/25). Detection, spray with 2 g/l 2,6-dichlorophenol-indophenol solution in dehydrated alcohol, heat, and examine. 10 µl sample volume.

Table 7.5 Contents of the European Pharmacopeia, 10.0; Monographs Related to Packaging and Packaging Materials (Volume 1, implementation 01/2020).

Monograph number	Monograph title
	3.1. Materials used for the manufacture of containers
3.1.1	Materials for containers for human blood and blood components (see 3.3.1)
3.1.3	Polyolefins
3.1.4	Polyethylene without additives for containers for parenteral preparations and for ophthalmic preparations
3.1.5	Polyethylene with additives for containers for parenteral preparations and for ophthalmic preparations
3.1.6	Polypropylene for containers and closures for parenteral preparations and ophthalmic preparations
3.1.7	Poly(ethylene - vinyl acetate) for containers and tubing for total parenteral nutrition preparations
3.1.8	Silicone oil used as a lubricant
3.1.9	Silicone elastomer for closures and tubing
3.1.10	Materials based on non-plasticized poly(vinyl chloride)for containers for non-injectable, aqueous solutions
3.1.11	Materials based on non-plasticized poly(vinylchloride) for containers for solid dosage forms for oral administration
3.1.13	Plastic additives
3.1.14	Materials based on plasticized poly(vinyl chloride)for containers for aqueous solutions for intravenous infusion
3.1.15	Polyethylene terephthalate for containers for preparations not for parenteral use.
	3.2. Containers
3.2.1	Glass containers for pharmaceutical use
3.2.2	Plastic containers and closures for pharmaceutical use
3.2.2.1	Plastic containers for aqueous solutions for infusion
3.2.9	Rubber closures for containers for aqueous parenteral preparations, for powders and for freeze-dried powders
	3.3 Containers for human blood and blood components, and materials used in their manufacture; syringes
3.3.1	Materials for containers for human blood and blood components
3.3.2	Materials based on plasticized poly(vinyl chloride) for containers for human blood and blood components
3.3.3	Materials based on plasticized poly(vinyl chloride) for tubing used in sets for the transfusion of blood and blood components
3.3.4	Sterile plastic containers for human blood and blood components
3.3.5	Empty sterile containers of plasticized poly(vinylchloride) for human blood and blood components
3.3.6	Sterile containers of plasticized poly(vinyl chloride) for human blood containing anticoagulant solution
3.3.7	Sets for the transfusion of blood and blood components
3.3.8	Sterile single-use plastic syringes

specified in monograph 3.2.2.1, plastic containers for aqueous solutions for infusion, includes tests for appearance, acidity or alkalinity, absorbance, and reducing substances, all performed on a water extract. Certain other section 3.2 monographs contain tests specific to the container type (for example, 3.2.8, sterile single-use plastic syringes contain a test for ethylene oxide as a sterilization residue) while other monographs include physical testing (for example, 3.3.4, sterile plastic containers for human blood and blood components includes tests for resistance to stretch, leakage, vapor permeability, and others).

Recently, the Pharm Eur has proposed to revise all the existing Section 3.1 general chapters on plastic materials by deleting the heavy metals test and replacing it with an extracted elements test as specified in monograph 2.4.35, Extractable elements in plastic materials for pharmaceutical use. This change to the Pharm Eur is consistent with the application of Quality by Design (QbD) principles to the development and qualification of plastics used in container/closure systems and reflects the evolution of improved analytical testing capabilities.

There are three critical aspects of the 2.4.35 monograph (proposed at the time this book is written):

1. The list of targeted elements,
2. The extraction process,
3. The limits for the targeted elements.

Considering the list of targeted elements, the close alignment between the 2.4.35 monograph and the ICH Q3D elemental impurities guideline is clear and unmistakable, as the elements targeted in 2.4.35 are those elements listed in Q3D Table V.1 as elements to be considered in the risk assessment of a parenteral dosage form, even in the circumstance that the element is not intentionally added to the plastic material under assessment. This alignment is appropriate as (i) the list of elements to be considered for the parenteral dosage form in Q3D includes all elements to be considered in oral dosage forms and (ii) many of the Pharm Eur Section 3.1 monographs are specific to plastics used in containers for parenteral dosage forms.

The above discussion notwithstanding, it is suggested that two additional elements be added to the target list in 2.4.35, aluminum and zinc. It is noted that while these elements do not appear in Table V.1 of Q3D, this omission is not because these two elements are generally recognized as safe. Rather, as noted in Section IV of Q3D, "some elemental impurities for which PDEs have not been established … are not addressed in this guidance … (as) they are addressed by other guidances and/or regional regulations and practices that may be applicable for particular elements." Q3D then specifically points to aluminum (for compromised renal function) and zinc (for patients with compromised hepatic function) as examples of these other elements. Furthermore, it is noted that testing for these elements is included in several 3.1 monographs (3.1.1.1, 3.1.3, 3.1.5, 3.1.6, 3.1.10, 3.1.11, 3.1.14, 3.1.15 for zinc; 3.1.3, 3.1.5, 3.1.6, 3.1.15 for aluminum). Moreover, both aluminum and zinc are known to be present in and extractable from numerous plastic materials.

For these reasons, it is suggested that aluminum and zinc are elements that are relevant to the quality of plastic materials used in container closure systems and thus that they should be added to the target list in 2.4.35.

Considering the extraction process, 2.4.35 includes an extraction in strong acid (0.1 M hydrochloric acid) at elevated temperature (autoclaving at 121 °C for one hour), which is likely to result in a worst-case extraction of the materials versus their clinical conditions of use. These extraction conditions are consistent with extractions already described in 3.1 monographs, replacing refluxing with autoclaving.

Considering the limits that appear in 2.4.35, it is noted that it is difficult to directly correlate the PDEs in ICH Q3D with a limit in 2.4.35 as the correlation could be different for each drug product and its associated container closure system based on the physical dimensions of the container closure system, the container closure system's composition (e.g. weight fraction of a given plastic material), and the clinical usage of the packaged drug product. Nevertheless, there is some relative correlation between the monograph's

limits and the guideline's PDEs. Specifically, a limit of 0.01 mg/l has been assigned to all the Q3D elements with the lowest PDEs, in particular the Q3D Class 1 and 2A elements, while a higher limit of 0.30 mg/l has been assigned to the Q3D Class 3 elements with the highest PDEs (Li and Cu). An intermediate limit of 0.10 mg/l has been assigned to that Class 3 element (Sb) whose PDE is intermediate between the Class 1 and 2A and the Class 3 elements.

Another way to consider the limits contained in 2.4.35 is to calculate the amount of extractable elements that would have to be extracted from the plastic material under consideration in order to exceed the limit. For the case of a Class 1 element, Cd as an example, the amount of Cd that would have to be extractable from the material under consideration to exceed the limit would be:

Extractable Cd (μg/g, ppm)

$$= 0.01 \text{ mg/l} \times (0.1 \text{l}/40 \text{ g}) \times 1000 \, \mu\text{g/mg} \times 2 = 0.05 \, \mu\text{g/g}$$

where 0.01 mg/l is the limit, 0.1 l/40 g is the extraction stoichiometry, 1000 μg/mg is a conversion factor, and 2 takes into account the dilution of the extract required in the test methodology. As 0.05 μg/g is a small portion of the extracted material, it is noted that the test is quite rigorous and that the limit is appropriately challenging. If there is any concern with the limit, it is not that the limit is so loose than potentially unsafe materials could pass the limit (false negative) but rather that safe materials could fail the limit (false positive).

Lastly, monograph 2.4.35 contains validation requirements that are both consistent with ICH Q3D and appropriate for trace level elemental analysis.

7.2.5 Other Pharmacopeia

In addition to the USP and Pharm Eur, the Japanese (JP) and Chinese Pharmacopeia contain monographs related to packaging and packaging materials. Considering the JP XVII, the relevant monographs in Section 7, Tests for Containers and Packaging Materials [32] are focused on the three major materials of construction for containers: glass (7.01 Test for Glass Containers for Injections), plastics (7.02 Test Methods for Plastic Containers), and elastomers (7.03 Test for Rubber Closure for Aqueous Infusions) (see Table 7.6). The tests, test methods, and acceptance criteria specified in the various monographs and sub-sections thereof are similar to tests, test methods, and acceptance criteria in comparable monographs and sections in the USP and Pharm Eur, with some exceptions, especially in the case of glass containers and elastomeric closures. Strategically, there are three major differences between the JP and its USP and Pharm Eur counterparts. First, the JP sections for plastics and elastomers are limited to aqueous infusion drug products, whereas the USP and Phar Eur monographs are more broadly applicable to other dosage forms. Second,

Table 7.6 Packaging monographs in the Japanese Pharmacopeia, JP 17. Chemical and Biological Tests.

7. Tests for containers and packaging materials			
Section	**Subsection 1**	**Subsection 2**	**Chemical test**
7.01 Test for Glass Containers for Injections	–	–	Soluble alkali test
			Soluble iron test for light-resistant containers
7.02 Test Methods for Plastic Containers	1. Test Methods	1.1 Combustion Tests	1.1.1 Residue on Ignition
			1.1.2 Heavy Metals
			1.1.3 Lead
			1.1.4 Cadmium
			1.1.5 Tin
		1.2 Extractable Substances[a]	Foaming test
			pH
			Potassium permanganate reducing substances
			UV spectrum
			Residue on evaporation
		1.7 Cytotoxicity Test	–
	2. Requirements for plastic containers for aqueous injections	2.1. Polyethylene or polypropylene containers for aqueous injections	4 Heavy metals
			5 Lead
			6 Cadmium
			7 Residue on ignition
			8 Extractable substances; foaming test, pH, reducing substances, UV spectrum, residue on evaporation2
			9 Cytotoxicity
		2.2. Polyvinyl chloride containers for aqueous injections	7 Heavy metals
			8 Lead
			9 Cadmium
			10 Tin
			11 Vinyl chloride
			13 Residue on ignition
			14 Extractable substances; foaming test, pH, reducing substances, UV spectrum, residue on evaporation[b]
			15 Cytotoxicity
		2.3. Plastic containers for aqueous injections being not described above	Heavy metals
			Residue on ignition
			Extractable substances; foaming test, pH, reducing substances, UV spectrum, residue on evaporation[b]
			Cytotoxicity
7.03 Test for Rubber Closure for Aqueous Infusions	–	–	1 Cadmium
			2 Lead
			3 Extractables substances (Description, pH, zinc, reducing substances, residue on evaporation, UV spectrum)[b]
			4 Cytotoxicity test
			5 Acute systemic toxicity

a) Water (immersion) extraction of cut containers unless containers are made of composite plastics, in which case they are filled. Extraction at 121 °C for 1 hour is the default extraction conditions. Lower temperatures and longer durations may be used if the container "deforms" at a higher temperature/shorter duration.
b) Water (immersion) extraction of sized items; extraction at 121 °C for 1 hour.

the JP plastics monographs focus on containers exclusively and there are no monographs for plastic materials. Lastly, the JP monographs for plastics do not contain tests for plastic additives and related substances, with the exception being vinyl chloride for PVC-based containers.

As was the case with the Pharm Eur, the JP monographs do not address extractables or leachables per say and the JP does not contain informational chapters.

7.3 Regulatory Guidance

It is an interesting way to start a discussion on regulatory guidance by noting that there is no relevant and meaningful regulatory guidance available for assessing packaging for extractables and drug products for leachables. This is the case as the guidance documents that are available are sufficiently out of date that at best the concepts and requirements are stale and at worst are at odds with essential scientific principles and sound and logical practices. In fact, it is the undeniable truth that if a sponsor were to present a regulatory authority with an extractables and leachables dossier that fully and completely complied with the guidance documents, the sponsor's file would surely be rejected as being deficient.

While the existing regulatory guidance documents may be out of date in terms of context and contents, they nevertheless establish a philosophy for managing extractables and leachables that is as relevant today as it was when the guidance documents were hot off the press. This philosophy is one that is based or risk; that is, the greater the risk that a packaged drug product could contain leachables in potentially impactful quantities, the greater the amount and rigor of extractables and leachables testing and assessment that would be required to establish that the packaging system was suited for its intended use.

Thus, the purpose of the following reviews of the regulatory guidance documents is to highlight this philosophy and illustrate its strategic and tactical implications.

7.3.1 The FDA 1999 Container Closure Guidance

In my opinion, the most meaningful text in the FDA's 1999 Container Closure Guidance [16] that is still relevant today can be found in Section III.A (Qualification and Quality Control of Packaging Components, Introduction) which states the essential regulatory expectation:

> "CDER and CBER approve a container closure system to be used in the packaging of a human drug or biologic as part of the application (NDA, ANDA, or BLA) for the drug or biologic. Each application should contain enough information to show that each proposed container closure system and its components are suitable for its intended use."

Simply put, the regulatory expectation is that the drug product application contains sufficient information to establish that the packaging system is suited for its intended use.

The first question that comes to mind after reading this text is "what is a packaging system's intended use?" as it is necessary to understand intended use before one can establish whether it is suited to that use or not. This question is answered in Section III.B.1, where it is established that a packaging system should:

- Protect the dosage form,
- Be compatible with the dosage form,
- Be safe for use with the dosage form, and
- Function properly.

Of these four characteristics, it is compatibility and safety that are of primary concern from a chemical characterization perspective. Expanding on these characteristics somewhat, the guidance notes that "packaging components that are compatible with a dosage form will not interact sufficiently to cause unacceptable changes in the quality of either the dosage form or the packaging component" and points out the following potential incompatibilities:

- Loss of potency due to absorption or adsorption of the active drug substance,
- Loss of potency due to degradation of the active drug substance induced by a chemical entity leached from a packaging component,
- Reduction in the concentration of an excipient due to absorption, adsorption, or leachable-induced degradation,
- Precipitation,
- Changes in the drug product pH,
- Discoloration of either the dosage form or the packaging component or increase in brittleness of the packaging component.

Thus, the concept of compatibility addresses the effect of the packaging system on key quality attributes of the drug product (e.g. efficacy, stability, or physical/chemical characteristics) due to either the direct action of the packaging system (adsorption or absorption) or the indirect action of the packaging system mediated by leached substances.

Considering safety, the guidance is sufficiently clear when it notes that "packaging components should be constructed of materials that will not leach harmful or undesirable amounts of substances to which a patient will be exposed when being treated with the drug product."

To this point the guidance is clear, appropriate, and relevant in establishing the strategic objectives of a chemical assessment, which is to establish that a packaging system is compatible with the packaged drug product and safe to use. However, knowledge of the purpose of chemical assessment leads immediately to the "how" question; that is, how does one establish compatibility and safety?

Again, the guidance is clear, appropriate, and relevant, as stated in Section III.A: "*The type and extent of information that should be provided in an application will depend on the dosage form and the route of administration*." I have chosen to emphasize this text because it is this text that establishes the fundamental regulatory philosophy that chemical assessment is a risk-based approach. Enumerating this concept of a risk-based approach is the single most enduring part of the container closure guidance. Unfortunately, it is the specifics contained within this sentence that is the single most important reason why the guidance is irrelevant as a current guide to fulfilling regulatory expectations.

I want to clear what I mean by these last two sentences. The concept of a risk-based approach to chemical assessment was profound when it was first proposed and has been supported by all the scientific and practical experience that has been gained in this field since its proposal. All packaging systems do not have the same propensity to be leached, all drug products do not have the same propensity to be "leaching agents," all contact conditions between the drug product and its packaging are not the same, the therapeutic use of all drug products is not the same, and all drug products are not administered to the patient in the same manner. Thus, it defies logic to suggest that there is a "one size fits all" approach to chemical characterization. Furthermore, if risk is not the most appropriate driver for differentiating approaches, then what is?

No, it is not the concept of a risk-based approach that is the issue with the container closure guidance. The problem is the dimensions of risk that the guidance uses to base their risk classification on, dose form, and route of administration. The container closure correctly notes in Section II.B.2.c that "making the determination that a material of construction used in the manufacture of a packaging component is safe for its intended use is not a simple process." The problem with the container closure guidance's approach to establishing risk is that in establishing risk on the basis of only two dimensions (dosage form and route of administration), the guidance over-simplified the many factors that contribute to risk and likely focused on two dimensions that are important but not overly critical. Thus, in defining essentially a two-dimensional risk matrix and then placing dosage forms into the risk matrix, the guidance created placements that defied practical experience and scientific principles and which created clear and obvious mis-matches.

Thus, the container closure guidance is an example of "good idea, faulty execution." To illustrate the "good idea" aspect, the contents of the guidance will be considered in greater detail as follows.

Table 1 of the container closure guidance, shown in revised form as Table 7.2 previously, is the manifestation of the guidance's two-dimensional risk evaluation matrix. The two dimensions of risk, Degree of Concern Associated with the Route of Administration and Likelihood of Packaging Component-Dosage Form Interaction, form the axes of the risk evaluation grid. Each dimension is divided into three "buckets," high risk, moderate risk, and low risk, creating a matrix with nine boxes. The matrix is populated by grading each dosage form on the basis of the two dimensions and placing the graded dosage forms in the grid.

Based on the knowledge and experience available in 1999, the various dosage forms were placed in the original Table 1. Since then, additional experience and knowledge has been gained, suggesting that some of the placements overestimated the likelihood of packaging component-dosage form interaction. Thus, USP monograph <1664> contained a revision of Table 1, essentially "moving several dosage forms to the right," meaning they were placed in lower risk categories. It is curious to this author that some dosage forms (e.g. certain solid dosage forms) were not "moved to the left" and placed in higher risk categories. When establishing the likelihood of component-drug product interactions, the authors of Table 1 were influenced by the thinking that "More detailed information usually should be provided for a liquid-based dosage form than for a powder or a solid, since a liquid based dosage form will more likely interact with the packaging components." A body of scientific evidence has established that, although solid dosage forms might not be doing "actual leaching," leachables can accumulate in solid dosage forms by indirect mechanisms such as volatilization in a closed package and sorption of the volatilized leachables by the solid dosage form.

Again, the point of this discussion of the container closure guidance is to address the strength of the concept and not dwell too long on the shortcomings of the content. Thus, one notes the value in having categorized the various dosage forms into risk classes. However, the next question then becomes, "now that I have populated the risk classes, what do I do with them?" or, more specifically, "how do I use the risk classes to establish and justify my approach to chemical characterization?"

It amazes me that most discussions of the container closure guidance begin and end with a discussion of Table 1. Yes, Table 1 is the proper start but it is not an end unto itself. In order for the container closure guidance to be useful, it must contain directions on how the risk classification enables chemical assessment. It is not enough to place the dosage forms in the individual risk boxes; one then needs guidance as to what to do with the dosage forms in each box.

Table 7.7 Typical suitability considerations for common classes of drug products.

Route of administration/Dosage form	Suitability		Risk Category
	Compatibility	Safety	
Inhalation Aerosols and Solutions, Nasal Sprays	Case 1c	Case 1s	Highest
Injections, Injectable Suspensions	Case 1c	Case 2s	Highest
Inhalation Powders	Case 3c	Case 5s	High
Sterile Powders and Powders for Injection	Case 2c	Case 2s	High
Ophthalmic Solutions and Suspensions	Case 1c	Case 2s	High
Topical Delivery Systems	Case 1c	Case 3s	Medium
Topical Solutions and Suspensions, and Topical and Lingual Aerosols	Case 1c	Case 3s	Medium
Oral Solutions and Suspensions	Case 1c	Case 3s	Medium
Oral Powders	Case 2c	Case 3s	Medium
Topical Powders	Case 3c	Case 4s	Low
Oral Tablets and Oral (Hard and Soft Gelatin) Capsules	Case 3c	Case 4s	Low

1c = Liquid-bases dosage form that conceivably could interact with its container closure system components.
2c = Solid dosage form until reconstituted; greatest chance for interacting with its container closure components comes after it is reconstituted.
3c = Solid dosage form with low likelihood of interacting with its container closure system.
1s = Typically provided are USP Biological Reactivity Test data, extraction/toxicological evaluation, limits on extractables, and batch-to-batch monitoring of extractables.
2s = Typically provided are USP Biological Reactivity Test data and possibly extraction/toxicological evaluation.
3s = Typically, an appropriate reference to the indirect food additive regulations is sufficient for drug products with aqueous-based solvents. Drug products with non-aqueous based systems or aqueous based systems containing co-solvents generally require additional suitability information.
4s = Typically, an appropriate reference to the indirect food additive regulations is sufficient.
5s = Typically, an appropriate reference to the indirect food additive regulations for all components except the mouthpiece for which USP Biological Reactivity Test data is provided.
Source: Abstracted from Table 2 of the FDA 1999 Container Closure Guidance.

This guidance was provided by Table 2, reproduced in its essence here as Table 7.7. On a certain level, the correlation between Tables 1 and 2 in the container closure guidance is obscure, as there is no direct link between the classification in Table 1 and the directions provided in Table 2. However, a subtle correlation can be drawn as follows. As a dosage form, inhalation aerosols and solutions, nasal sprays were classified as the highest risk in Table 1. In Table 2, this dosage form is placed in safety category case 1S, which has the most extensive requirements in terms of safe use information. Alternatively, the dosage form of oral tablets and oral capsules was classified as the lowest risk in Table 1 and in Table 2 is assigned the least information-intensive safety category, case 4s.

While Table 2 is useful in establishing the regulatory expectations, it is not particularly specific in terms of what exact information would meet the regulatory expectations. More specific details in terms of the information that should typically be submitted for certain dosage forms are provided in other tables contained in the container closure guidance and reproduced here as Table 7.8. Again, it is noted that this information is provided not to imply that it would satisfy current regulatory requirements but rather to illustrate a risk-based approach to testing.

Reading through the container closure guidance provides other insights into the regulatory thinking of the time, which again may or may not be relevant in the current regulatory environment. Examples of such insights are as follows (with my own comments added in italics as appropriate:

- A packaging system found acceptable for one drug product is not automatically assumed to be appropriate for another (*although, not automatic, technical rationales for extension of "acceptability" to multiple drug products based on similar composition and similar contact condition to the packaging should be allowed*).
- Approaches that differ from those described in this guidance may be followed, but the applicant is encouraged to discuss significant variations in advance (*good advice that can be difficult to follow*).
- Information intended to establish suitability may be generated by the applicant, by the supplier of the material of construction or component, or by a laboratory under contract to either the applicant or the firm.
- Some interactions between a packaging component and dosage form will be detected during qualification studies on the container closure system and its components. Others may not show up except in the stability studies.

Table 7.8 Information that typically should be submitted for various dosage forms per the FDA 1999 Container Closure Guidance.

Dosage form	Information for safety (for each material of construction, as appropriate)	Information for compatibility (for each material of constructions, as appropriate)
Injectable or Ophthalmic (taken from Table 4)	• Chemical composition of all plastics, elastomers, adhesives, etc.[a] • For elastomeric closures: USP Elastomeric Closures for Injections testing • For glass components: USP Containers: Chemical Resistance – Glass Containers • For plastic components and coatings for metal tubes: USP Biological Reactivity Tests • If the extraction properties of the drug product vehicle may reasonably be expected to differ from that of water (e.g. due to high or low pH or due to a solubilizing excipient), then drug product should be used as the extracting medium. • If the total weight of extracts significantly exceeds the amount obtained from water extraction, then an extraction profile should be obtained. • For plastic or elastomeric components undergoing heat sterilization, it is current practice to request that the extraction profile be obtained at 121 °C/one hour using an appropriate solvent.	• For coatings on metal tubes: Coating integrity testing • For elastomeric components: Evaluation of swelling effects • For plastic components (including tube coatings): USP Containers Physicochemical Tests - Plastics testing • For ophthalmics: Particulate matter and eye irritants • Stability studies also support compatibility
Liquid-based Oral and Topical (taken from Table 6)	• Chemical composition of all plastics, elastomers, adhesives, etc.[a] • For most liquid-based oral drug products: appropriate reference to the indirect food additive regulations • For liquid-based oral drug products with chronic dosing regimens that contain alcohol or a cosolvent: information to establish that exposure to extractables will be no greater than that expected to result from the use of similar packaging components when used with foods,[b] or that the exposure is acceptable based on toxicological data. • For topical drug products (plastic coatings for metal tubes), and plastic drug delivery system components: USP Containers testing • For topical delivery systems: appropriate reference to indirect food additive regulations	• For LDPE and glass components, USP Containers testing[c] • For coatings for metal tubes: coating integrity testing
Solid Oral and Powders (taken from Table 7)	• Chemical composition of all plastics, elastomers, adhesives, etc.[a] • For tablets, capsules, and powders, appropriate reference to the indirect food additive regulation may be submitted, but may not be appropriate for Powders for Reconstitution. • For rayon and cotton fillers, data from USP monographs. For non-USP materials, data and acceptance criteria should be provided. • For dessicants and other absorbent materials: the size and shape should differ from that of the dosage form.	• For glass and plastic containers, data from USP Containers[c] testing.

a) Including any additives used in the manufacture of a packaging component.
b) The materials of construction should be acceptable for contact with foods that have extraction characteristics similar to those of the drug product (e.g. aqueous, acidic, alcoholic, or fatty).
c) Plastics testing should be performed on the packaging component, not on the unformed resin.

Therefore, any change noted during a stability study that may be attributable to interaction between the dosage form and a packaging component should be investigated and appropriate action taken.

- This consideration (safety) is especially important for those packaging components that may be in direct contact with the dosage form, but it is also applicable to any component from which substances may migrate into the dosage form (e.g. an ink or adhesive).

- For a drug product such as an injection, inhalation, ophthalmic, or transdermal, a comprehensive study is appropriate. This involves two parts: first, an extraction study on the packaging component to determine which chemical species may migrate into the dosage form (and at what concentration) and, second, a toxicological evaluation of those substances that are extracted to determine the safe level of exposure via the label specified route of administration. (*At one time, the FDA was satisfied with extraction studies (and not leachables studies) as an adequate demonstration of safety. Contrast that to today, where leachables studies are the gold standard and lack of rigorous and comprehensive leachables data is a sure way to have a file rejected.*)

- For drug products that undergo clinical trials, the absence of adverse reactions traceable to the packaging components is considered <u>supporting</u> evidence of material safety (*underline added by the author*).

- In addition to providing data to show that a proposed container closure system is suitable for its intended use, an application should also describe the quality control measures that will be used to ensure consistency in the packaging components.

- Currently there is no general policy concerning the monitoring of a packaging system and components with regard to safety. One exception involves inhalation drug products for which batch-to-batch monitoring of the extraction profile for the polymeric and elastomeric components is routine.

- Because secondary packaging components are not intended to make contact with the dosage form, there is usually less concern regarding the materials from which they are constructed. However, if the packaging system is relatively permeable, the possibility increases that the dosage form could be contaminated by the migration of an ink or adhesive component, or from a volatile substance present in the secondary packaging component. (For example, a solution packaged in an LDPE container was found to be contaminated by a volatile constituent of the secondary packaging components that enclosed it.) In such a case, the secondary packaging component should be considered a potential source of contamination and the safety of its materials of construction should be taken into consideration.

- Tests described in the USP are typically considered sufficient standards for establishing specified properties and characteristics of specified materials of construction or packaging components.

- For non-USP tests, an applicant should provide justification for the use of the test, a complete and detailed description of how the test was performed, and an explanation of what the test is intended to establish. If a related USP test is available, comparative data should be provided using both methods. Supporting data should include a demonstration of the suitability of the test for its intended use and its validation.

Undoubtedly, this is an incomplete list of all the historically interesting information contained within the container closure document and the reader is referred to Guidance for additional details.

7.3.2 The FDA Guidance for OIDNP

In its container closure guidance, the FDA identifies Inhalation products as having the highest degree of concern with respect to potential container product interactions due to the route of administration and the likelihood that an interaction will occur. Accordingly, the FDA has provided Guidance for these types of products. Inhalers are addressed in the FDA Guidance for Industry, Metered Dose Inhaler (MDI) and Dry Powder (DPI) Drug Products, Chemistry, Manufacturing, and Controls Documentation [33] and nasal spray and inhalation suspensions are covered in the Guidance for Industry, Nasal Spray and Inhalation Solution, Suspension and Spray Drug Products – Chemistry, Manufacturing and Controls Documentation [34]. Considering the level of concern associated with such products, the FDA advises that "For safety considerations, (container closure) materials should be chosen that minimize leachables without compromising the integrity or performance of the drug product."

Central to both guidance documents are the concepts of independent but related extraction and leaching studies. Quantitative controlled extraction studies are performed to characterize the container closure system by establishing its extractables profile. Such controlled extraction studies are to include specified and justified extraction and analyses processes (multiple appropriate solvents and methods recommended). The resulting information forms the basis of a toxicological assessment of the extractable's product impact and serves to support acceptance criteria for container closure components.

Recognizing that the OINDP dosage forms have the highest risk associated with leachables, routine extraction studies, including discriminatory and validated test methods and appropriate and justified acceptance criteria, are used for the routine testing of incoming components of the container closure system. Based on the analytical and toxicological evaluation of the extractables from the controlled extraction studies, the applicant should establish discriminatory test procedures and set appropriate acceptance criteria for the extractable profiles for routine testing for each critical component of the container closure system. An extraction test should be performed on every incoming component batch using water and other suitable solvents selected from the control extraction studies in order to determine the individual and total extractables. For nasal spray drug products, if the level of extractables for each component is relatively low, it may be appropriate to establish a limit only for the total weight of extractables from each individual critical component. It is noted that a reduced acceptance testing schedule may be considered once the applicant establishes the reliability of the supplier's test results and has tested multiple incoming batches.

Considering validation of the analytical methods used in the routine extraction studies, the specificity, linearity, range, accuracy, precision, detection limit, quantitation limit, and robustness of the proposed validated test procedures, including system suitability testing, should be established.

In addition to a characterization of the container closure system itself, both guidance documents require the generation of leachables data (identity and accumulation) representative of the finished drug product through the end of its shelf-life. Such a leachables assessment includes not only the generation of leachables data using validated analytical methods but also the interpretation of the data in terms of established and justified acceptance criteria. Correlating the leachables and extractables data is recommended as such a correlation may obviate the need to do leachables testing in future routine stability studies.

7.3.3 The EMEA Guideline on Plastic Immediate Packaging

This Guideline addresses the information that must be submitted in marketing authorization applications for plastic materials being used as immediate packaging for active substances and medicinal products. As outlined in its Scope, the Guideline is limited to plastic immediate packaging materials that are intended to be in direct contact with the active substance or medicinal product, including the container, the closure or seal, and other parts of the system. The Guideline specifically excludes elastomers and natural and synthetic rubbers from its Scope [17].

Considering extractables/leachables, the directives of the EMEA Guideline fall into four broad categories:

1. Section 3.2, Specifications.
2. Section 4, Extraction Studies.
3. Section 5, Interaction Studies and specifically 5.1, Migration Studies.
4. Section 6, Toxicological Information/Documentation.

An analysis of each section is as follows.

Section 3.2. Specifications

The requirements of this section are straightforward. Quoting directly from the Guideline,
"… (for plastic packaging materials) reference should be made to the appropriate monographs of the European Pharmacopoeia or the monograph of the pharmacopoeia of a Member State. When referring to a monograph, compliance should be demonstrated.

If the plastic material is not described in the European Pharmacopoeia or in the pharmacopoeia of a Member State, an in-house monograph has to be established according to the list below, taking into account the general methods of the pharmacopoeia …."

The issue of compliance to the pharmacopoeia is central to the EMEA strategy for qualifying immediate packaging materials and such compliance represents significantly more than just a matter of fulfilling the specification requirement. Considering Extraction Studies, for example, the Guideline notes that Extraction Studies are considered to be necessary "… if the material is neither described in the European Pharmacopoeia nor in the pharmacopoeia of a Member State, nor has been approved for food stuff packaging." The implication here is that if the materials are covered by and demonstrated to comply with a relevant compendial monograph and have been approved for foodstuff packaging, then extraction studies are not necessary.

Similarly, the Guideline states in Section 6, Toxicological Information/Documentation, that "if the plastic material or additives used are described in the European Pharmacopoeia, the pharmacopoeia of a Member State or have been approved for use in food packaging, toxicological data may not be required." It is thus noted that compliance with a relevant Pharmacopoeia monograph is a prime consideration in the development of an extractables/leachables strategy and that such compliance can greatly focus and simplify such a strategy. However, it is also noted that compliance with a relevant Pharmacopoeia monograph is not, of itself, an acceptable extractables/leachables strategy. This juxtaposition was considered in greater detail in Section 7.2.3 of this Chapter.

Section 4. Extraction Studies

The aim of the Extraction Study is to determine those additives in the material that might be extracted by the preparation or the active substance in contact with the material.

The studies typically involve "exposing … the material to an appropriate solvent system under stress conditions to increase the rate of extraction." While the preferred extraction solvent for a medicinal product is the product itself or a placebo vehicle, this is not a requirement and the specific direction of the Guideline is that the extraction solvent used should have the same propensity to extract substances as the active substance/dosage as appropriate.

Section 5. Interaction Studies

"In order to evaluate the suitability of the selected plastic packaging material for the intended use, the compatibility of the material with the active substance or medicinal product should be demonstrated." Interaction Studies, whose purpose is to demonstrate the compatibility of the material with the active substance or medicinal product, include Migration Studies and Sorption Studies. Migration Studies should demonstrate that substances do not migrate from the material under conditions representative for the intended use in such quantities as to alter the efficacy and stability of the active substance/medicinal product or to present a toxicological risk. Sorption Studies address the possibility that product quality might be altered by sorption of an active substance or excipient by the packaging material.

Section 5.1. Migration Studies

The Guideline is very clear that Migration Studies are a required component of the data to be provided in product registration files. Migration Studies must be performed regardless of whether the materials comply with an existing pharmacopeial Monograph or not. The only two circumstances that preclude Migration Studies are based on the outcome of Extractions Studies and include:

1. The circumstance in which no extracted substances are identified in the Extraction Study.
2. The circumstance in which the calculated maximum amount of individual extracted substances that may be present in the active substance/medicinal product can be demonstrated to be toxicologically safe.

In either circumstance, the decision not to perform migration studies needs to be justified.

Migration Studies are very clearly identified as an activity to take place during the development stage of the medical product. The implication here is that monitoring of leachables during stability studies is not a requirement of this Guideline. However, if Migration Studies are not performed during product development, then the Guideline clearly states that leachables should be monitored during formal stability studies, conducted under normal and accelerated storage conditions.

The purpose of the Migration Study is to demonstrate that substances will not migrate in such quantities as to alter the efficacy and stability of the active substance/medicinal product or to present a toxicological risk. The Guideline is very specific in terms of many aspects of Migration Studies, including the following:

1. The studies are necessary if Extraction Studies have resulted in one or several extractables.
2. The studies should be performed under conditions representative for the intended use.
3. The studies shall be performed on at least one batch of the active substance/medicinal product.
4. The studies must be performed with the active substance/medicinal product, and studies performed with other test media will only be considered to provide preliminary information.
5. Analytical methods used need to be either compendial methods or validated methods.
6. If the plastic material is composed of layers of different plastic materials, the possibility of migration of components from external layers to the medicinal product must be evaluated.
7. It must be demonstrated that no components of agents applied to the outer surface of the container/closure system (e.g. adhesives, inks) will migrate into the medicinal product.

Despite this clarity on the strategic levels, the EMEA Guideline does not provide tactical information such as specific test procedures, methods, process, or techniques.

Section 6. Toxicological Information/Documentation

The Guideline requires that toxicological data be provided for extractables and leachables, depending on their level and chemical structure. Exceptions for materials complying with pharmacopeial Monographs have been previously noted. For non-compendial materials, toxicological information is required, even if the material under consideration has been approved for use in food packaging. The Guideline provides no specific guidance in terms of how such a toxicological assessment is to be performed.

Decision Tree

A Decision Tree that defines the extent of testing required is supplied as Appendix II in the Guideline. Portions of that Decision Tree that are relevant for common non-solid dosage forms are reproduced here as Figure 7.5.

The Decision Tree clearly establishes that the EMEA Immediate Packaging Guideline is similar to the FDA Container Closure Guidance in the sense that both adopt

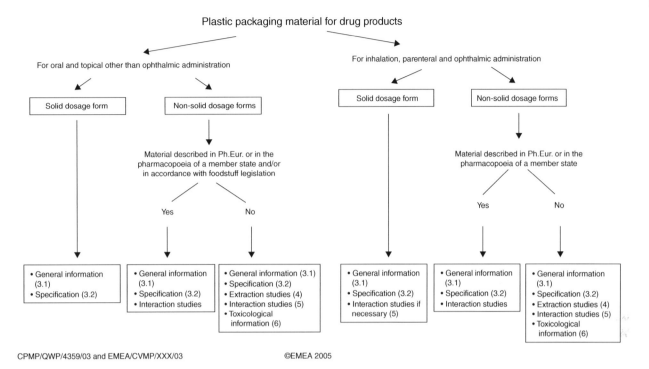

Figure 7.5 Decision tree on the presentation of the documentation of plastic packaging materials. Source: From the EMEA Guideline [17].

a risk-based approach to chemical assessment, based on both the route of administrations and the physical state of the dosage form. Thus, for example, the documentation required for an oral and topical dosage form differs whether that dosage form is a solid or non-solid, with the non-solid dosage form requiring an interaction (leachables) study which the solid dosage does not, presumably because of its lower leachables risk. So too, the documentation requirements for a solid oral or topical dosage form are less than the requirements for a solid inhalation, parenteral or ophthalmic dosage form, with the difference being the need to perform an interaction study.

Lastly, it is clear in the Decision Tree that compliance with an applicable Pharm Eur material monograph (or a comparable monograph from an EU Member State) and/or foodstuff regulation is an important part of the EMEA immediate packaging qualification process, as such compliance eliminates the need to perform extraction studies and to possess toxicological information for solid dosage forms. Such an approach makes some sense with respect to foodstuff regulations, as compliance with such regulations typically involves some type of extractables assessment, either experimental or mathematical (migration modeling). However, using Pharm Eur plastic materials monograph compliance in the place of extraction studies is more of a stretch, given that the Pharm Eur materials monographs produce more general chemical test data and less compound-specific extraction data.

7.4 Best Practice Recommendations

7.4.1 PQRI Recommendations for Orally Inhaled or Nasal Drug Products (OINDP)

Although one expects official sources such as regulatory authorities and pharmacopeia to provide guidance on the subject of chemical characterization, another source that can be tapped for establishing "this is the way things should be done" is the practical experience of the multitude of practitioners who perform chemical characterization studies as a matter of employment (I do not know many people who do E&L for the fun of it). Although on a limited basis this experience and the associated scientific knowledge may be summarized and communicated via publications, presentations, educational seminars, webinars, meetings, and conferences, there is an important place in the literature for documents that "bring it all together" and establish best demonstrated practices and that translate best practice recommendations into strategies and tactics. Even though such documents do not and cannot rise to the level of enforceable guidance, guidelines, and standards, they can be the enablers of such standards and they can be an effective means of providing the practical insights for interpreting and managing the standards in day-to-day laboratory operations.

With respect to packaging, the most historically relevant example of a best demonstrated practice document is the

Safety Thresholds and Best Practices for Extractables and Leachables in Orally Inhaled and Nasal Drug Products, created and published by the Leachables and Extractables Working Group of Product Quality Research Institute in 2006 [35]. Recognizing that "leachables and extractables issues represent some of the most significant challenges facing a pharmaceutical development team responsible for the registration and manufacture of Orally Inhaled or Nasal Drug Products" and noting that although "… guidance documents … from the … US FDA have significantly clarified the pharmaceutical development process for OINDP, … significant uncertainties remain," this L&E team, made up of "highly experienced scientists including toxicologists, analytical chemists and others from industry, government and academia," created a document that proposed numerous new to the world best practices in areas such as:

1. Exposure threshold for extractables and leachables
2. OINDP component selection
3. Controlled Extraction Studies
4. Leachables studies
5. Routine Quality Control Methods

Arguably the most important contribution of the PQRI Leachables and Extractables Working Group was their identification and justification of various safety thresholds, such as the AET, for leachables related to orally inhaled and nasal drug products. While this is a significant contribution in and of itself, an equally significant contribution of this Group was their delineation of best demonstrated practices for the conduct of extractables and leachables studies for these dosage forms. While the recommendations and supporting data generated by this Group are most appropriately directed toward OINDP dosage forms in particular, in general the best demonstrated practices are likely to be applicable, to some extent, to many dosage forms and product applications.

Before the L&E Working Group discussed their best demonstrated practices, the PQRI provided cornerstone recommendations for the process for conducting extractables and leachables studies. This process is as follows:

1. The sponsor (i.e. that company developing the drug product that is contacted by a container closure system) should first select appropriate components based on functionality, availability, physicochemical makeup, and other appropriate factors. The sponsor should collect relevant chemical information from component suppliers to facilitate the selection process. Such information should be reviewed by toxicologists as part of the selection process.

2. The sponsor should understand the drug product configuration, e.g. number of doses per day, total number of doses in a drug product unit.

3. The sponsor should conduct Controlled Extraction Studies and consider reporting individual identified extractables for risk assessment.

4. Extraction and analytical methods for Routine Extractables Testing should be established based on the methods developed in the Controlled Extraction Study and validated according to established parameters. Routine Extractables Testing is conducted for the purpose of ongoing incoming raw material quality control.

5. After Controlled Extraction Studies have been completed, the sponsor should conduct Leachables Studies on the drug product. Such studies would produce the data upon which product risk assessments are based. In addition to monitoring the accumulation of leachables over its shelf-life, this activity should include the generation of a correlation between extractables and leachables.

6. Risk assessments on leachables should be performed.

7. After product launch, Controlled Extraction Studies are performed on incoming raw materials to ensure that the Risk Assessment is still valid in terms of the levels of extractables and leachables.

Once the process for performing a risk assessment is in place, the PQRI team tackled the issues of best-demonstrated practices, which are those tactics by which the risk assessment is accomplished. The first best demonstrated practice established by the PQRI Group is data mining, specifically that "the … team should obtain all available information on the composition and manufacturing/fabrication processes for each component to the extent possible." Pertinent information includes:

- The base material(s) of construction.
- The additive composition of all components, including the detailed chemical composition and the reaction/ degradation chemistry of each additive.
- The polymer forming process, including associated process agents.
- The component fabrication process, including agents used to assist the process. This would include any cleaning/washing processes for finished components.
- Any storage or shipping experienced by the finished components.

If such mined information were sufficiently complete and comprehensive, it would, in theory, preclude the need for a controlled extraction study, at least from the perspective of material selection. It is most often the case, however, that such information provides at best a very coarse screen for material selection. Nevertheless, the Group recommended that the mined information be the basis of a preliminary toxicological assessment whose primary focus is material selection.

In addition to material selection, data mining facilitates the design of a controlled extraction study (designing a focused and effective study based on what is known, and not known, about the material in question).

A cornerstone of the PQRI Best Demonstrated Practices is the Controlled Extraction Study. By definition, the controlled extraction study is "a laboratory investigation into the qualitative and quantitative nature of extractables profiles of critical components of a … container/closure system." Clearly, the primary intent of the controlled extraction study is to produce a complete extractables profile, where complete is defined within the context of the various safety thresholds established by the PQRI. More significant from a practical perspective is how such a complete extraction profile would be used. It goes without saying that the complete extraction profile would be used to complete the process of material selection and for the ongoing toxicological assessment of viable material candidates. Furthermore, the complete extraction profile would be the basis for the development and validation of routine quality control methods and acceptance criteria for incoming materials. Obviously, the complete extraction profile facilitates the correlation between extractables and leachables. Finally, the methods used in the controlled extraction study could be the starting point for the development (and validation) of methods used in leachable studies.

Guided by their professional experience and insight, which was augmented by understanding gained in testing four "model" materials (three elastomers and one plastic), the PQRI Group made the following recommendations on how to conduct a controlled extraction study [35]:

1. Controlled Extraction Studies should employ vigorous extraction with multiple solvents of varying polarity. The PQRI team chose methylene chloride, 2-propoanol, and hexane as appropriate solvents because they represent a wide range of polarities (and therefore solubilizing properties), they represent a wide range of boiling points, they are relatively non-reactive, they are easily and safely handled in a typical laboratory setting, and they are readily available in high purity.

2. Controlled Extraction Studies should incorporate multiple extraction techniques. The PQRI Group chose Soxhlet extraction, sonication, and refluxing because these techniques are in common use in industry laboratories, they have a long history of varied, safe, and effective use, and they employ equipment that is routinely available in a typical analytical laboratory.

3. Controlled Extraction Studies should include careful sample preparation. The PQRI Group correctly notes that an important consideration in the selection and proper utilization of extraction and extract analysis techniques is the issue of mutual compatibility. The PQRI team emphasized the point that the extract and its analysis methods must be mutually compatible and that "strange" things can happen in cases of incompatibility, such as "mystery" extractables (for example, "extractables" that are produced in the injector of a GC system) and analytical artifacts.

4. Controlled Extraction Studies should employ multiple analytical techniques. Analytical techniques specified by the PQRI Group included:
   o Techniques to detect, identify, and quantify individual organic extractables:
     ▪ Gas Chromatography/Mass Spectrometry (GC/MS)
     ▪ Liquid Chromatography/Mass Spectrometry (LC/MS)
     ▪ Liquid Chromatography/Diode Array Detection (LC/DAD)
   o Techniques to detect and quantify organic extractables:
     ▪ Gas Chromatography/Flame Ionization Detection (GC/FID)
     ▪ Liquid Chromatography/Ultraviolet Detection (LC/UV)
   o Techniques capable of detecting, identifying, and quantitating inorganic extractables:
     ▪ Inductively Coupled Plasma/Mass Spectrometry (ICP/MS)
     ▪ Inductively Coupled Plasma/Optical Emission Spectroscopy (ICP-AES)
     ▪ Scanning Electron Microscopy/Energy Dispersive X-ray (SEM/EDX).

5. Controlled extraction studies should include a defined and systematic process for the identification of individual extractables. The PQRI Group recognized that while the identification of an extracted substance is often times akin to art, the art must be grounded in and guided by certain comprehensive and systematic concepts. Thus, for example, the PQRI Group suggested that there are five categories of information that could be used to identify an organic compound, including:
   o Mass spectrometric fragmentation behavior
   o Confirmation of molecular weight
   o Confirmation of elemental composition
   o Mass spectrum matches, an automated library, or a literature spectrum
   o Mass spectrum and chromatographic retention index matching an authentic compound

Depending on which among these supporting information sources are available for a specific organic compound, the compound's identity can be described as tentative, confident, or confirmed.

6. A Controlled Extraction Study uses "definitive" extraction techniques/methods that have been optimized. The definitive extraction techniques used in a Controlled Extraction Study should meet several objectives. First, they should produce extractables profiles that represent at least a worst-case scenario for potential leachables. Second, they should have known and documented performance characteristics, specifically related to their achievement of asymptotic levels for the extractables.

   In addition to providing the above advice, the PQRI Group performed laboratory work directed toward optimizing extraction procedures. For example, they optimized a Soxhlet extraction method with methylene chloride for the sulfur-cured elastomeric test article. The optimized extraction conditions were 7 g of cut-up material, contacted with 200 ml methylene chloride, for eight hours (with a drop rate of solvent in the extractor of 20 drops/min). The Group also optimized a reflux extraction with 2-propanol for the polypropylene test articles. The optimized extraction conditions were 1 g of sample (total surface area of 50 cm^3) per 25 ml of solvent, a solvent mixture of 50 : 50 2-propanol/tetrahydrofuran, and an extraction time of three hours.

7. During the Controlled Extraction Study process, sponsors should revisit supplier information describing component formulation. The PQRI Group noted that the team performing the Controlled Extraction Study should revisit and review the supplier information to surface any analytical anomalies or discrepancies between the results of the Controlled Extraction Study and the supplier information. Failure to find an additive revealed by the supplier or, conversely, finding an extractable that cannot be traced to an ingredient indicated by the supplier are situations that require close examination of the results of the Controlled Extraction Study, the supplier's information and/or other available information.

8. Controlled Extraction Studies should be guided by an Analytical Evaluation Threshold (AET) that is based on an accepted safety concern threshold. The proper use of the AET has been discussed elsewhere in this book.

9. Certain compounds are considered to be "special case" compounds, requiring evaluation by specific analytical techniques and technology defined thresholds. The PQRI Group identified the following compounds as those that have an increased toxicological risk and thus which qualify as "special cases" that require specific and sensitive analytical methodologies:

Polycyclic aromatic hydrocarbons (PAH's) or Polynuclear Aromatics (PNA's): naphthalene, acenaphthylene, acetaphthalene, fluorene, phenanthrene, antracene, fluoranthene, pyrene, benzo(a)anthracene, chrysene, benzo(b)fluranthene, benzo(k)fluranthene, benzo(e)pyrene, benzo(a)pyrene, indeno(123-cd)pyrene, dibenzo(ah)anthracene, benzo(ghi)perylene.
N-nitrosamines: N-nitrosodimethylamine, N-nitrosodiethylamine, N-nitroso-di-n-butylamine, N-nitrosomorpholine, N-nitrosopiperdine, N-nitrosopyrrolidine.
2-Mercaptobenzothiozole (MBT).

These compounds should be considered to be specific to OINDP and not necessarily applicable or limiting for other product types and container/closure systems. However, if these compounds are present in extracts of any particular container/closure system, it is pertinent and necessary that their potential safety impact be established.

10. Qualitative and quantitative extractables profiles should be discussed with and reviewed by toxicologists so that any potential safety concerns regarding individual extractables, i.e. potential leachables, are identified early in the product development process. The appropriate, necessary, proper, and close collaboration of analytical chemists and toxicologists has been discussed elsewhere in this book.

A major driving force for the generation of the PQRI recommendations was the observation that regulatory guidance tends to be strategic rather than tactical; that is, the regulations indicate that a specific task must be performed without necessarily providing details as to how the task is to be performed. While the PQRI recommendations provide further clarity in terms of the various aspects one must consider in designing, performing, and interpreting a controlled extraction study, it is somewhat unfortunate that the recommendations suffer the same practical limitation of being strategic versus tactical. For example, while the recommendations clearly establish the use of multiple solvents, no general insight or guidance is provided in terms of what solvents should be for pharmaceutical dosage forms other than OINDP. This is not necessarily a shortcoming of the PQRI effort, as OINDP was the clear focus of their effort. Nevertheless, the pharmaceutical industry is in great need of the development of established, recognized, and accepted "simulating solvents" and "standardized extraction conditions," such as the simulating solvents and "standardized conditions" that are used with great efficiency and effectiveness in the food industry when qualifying food packaging.

In order to obtain a greater understanding of the PQRI best demonstrated practices and to more directly benefit from the insights, experiences, and findings of this team, the reader is referred directly to the PQRI document [37].

7.4.2 PQRI Recommendations for Parenteral Drug Products (PDP)

The success of the PQRI recommendations for OINDP virtually insured that additional effort would be undertaken to address other dosage forms, particularly those that were also classified as higher risk per Table 1 of the FDA container Closure guidance. A logical next step for the PQRI would be to address parenteral and ophthalmic drug products, collectively referred to as PODP. Given the similar risk classification of OINDP and PODP dosage forms, a PQRI E&L Working Team proposed that best practices suitable for OINDP could be applied to PODP; that is, many of the concepts developed in the OINDP Recommendations could be adapted to and customized for PODP. Specifically, this team developed the following working hypothesis and used that hypothesis to develop best practice recommendations for these dosage forms:

- Threshold concepts that have been developed for the safety qualification of leachables in OINDP can be extrapolated to the evaluation and safety qualification of leachables in PODP, with consideration of factors and parameters that are unique to various types of PODPs.
- The best demonstrated practices that were established for the OINDP pharmaceutical development process can be extrapolated to PODP container closure systems.
- Threshold and best practices concepts provide comprehensive process for characterizing container closure systems for leachable substances and establishing the leachables' impact on PODP safety.

During the development of their recommendations, the PQRI Working Team concluded that parenteral drug products (PDP) and ophthalmic drug products (ODP) are sufficiently different that they could not be readily treated in the same manner. Significant differences noted by this team included:

- Route of administration
- Dosing (frequency and amount)
- Relevant toxicological endpoints, with sensitization and irritation being more important for ophthalmic drug products (ODP) than for parenteral drug products (PDP)

Thus, the Working Team separated these two dosage forms and developed recommendations relevant to both individually. Recommendations relevant to PDP are addressed in this section while ODPs are addressed in Section 7.4.3.

In order to understand the application of the PDP recommendations, one must understand the pharmaceutical products that were grouped as PDPs. Such products include:

- Drug products packaged in pre-filled syringes (PFS) and small and large volume parenterals (SVP and LVP) contained in vials or flexible bags.
- Drug products that are administered by intraocular, intrathecal, intracerebroventricular, intraarticular, epidural, and perineural routes of administration.

The PQRI PDP Recommendations have just been recently published [36]. This circumstance necessarily limits my ability to discuss these Recommendations in great detail. Nevertheless, the PDP Recommendations can be summarized as follows:

1. An SCT approach can be applied to leachables and extractables qualification in parenteral drug products. Based on most Parenteral Drug Product (PDP) formulations, an SCT of 1.5 µg/day for an individual organic leachable can be used to calculate an AET. An SCT that is lower than 1.5 µg per day may be warranted for certain classes of compounds, such as those within the cohort of concern (e.g. aflatoxin-like, N-nitroso-, and alkyl-azoxy compounds).

2. The QT developed for OINDP was evaluated and it was determined that when no concern for genotoxic or carcinogenic potential is identified, a QT of 5 µg/day is appropriate in the absence of supporting general toxicology data and an identified potential for irritation or sensitization in PDP. Above the QT additional toxicology, evaluation is necessary to qualify an individual organic leachable.

3. Extractables assessments and extraction studies for PDP may be considered as appropriate for specific application to materials of construction, finished components, or complete packaging systems (i.e. container closure systems).

4. Extractables assessments and extraction studies for PDP packaging systems should include aqueous-based extraction solvents with appropriate consideration of extraction pH, organic solvent content, and other appropriate extraction conditions (e.g. extraction time, extraction temperature, extraction technique, and sample-to-solvent ratio). Extractable studies for CCS used with complex drug products should consider appropriate solvent propensity to establish the

extractable profile to guide optimization of non-targeted screening methods for placebo or leachables. Note that examples of complex products include complex API (e.g. polymeric compounds, peptides), complex formulations (e.g. liposomes, emulsions, suspensions), complex routes of delivery (e.g. topical), complex dosage forms (e.g. long-acting injectables), and complex drug-device combinations (e.g. pre-filled syringes, autoinjectors).

Where appropriate, extractables assessments, extraction studies, and leachables assessments for parenteral drug products and their packaging systems should consider the possibility of migration across packaging barriers (i.e. drug product labels, adhesives, inks, etc.).

In situations of analytically challenging AETs for certain PDP (e.g. large volume parenterals), a simulation study may supplement and guide subsequent drug product leachables studies. These studies can establish an extractables profile to inform a probable leachables profile of the packaged drug product that the study simulates. Use of a simulation study would need to be appropriately justified.

Biological products have unique considerations compared to chemically synthesized drug products. Comprehensive risk assessments should consider biological activity, efficacy, and safety and may include the following:

1. Leachable interactions affecting product quality attributes, i.e. degradation, oxidation, chemical modification, aggregation, and immune adjuvant activity
2. Material compatibility, surface characteristics, and organic/inorganic alert compounds
3. Individual components and system interfaces, performance, and functionality
4. Leachables assessment performed on the product under accelerated storage/stress conditions and during stability storage

Three aspects of these Recommendations are particularly noteworthy, including the values assigned to the SCT and QT, the concept of a simulation study, and the discussion of the unique risks associated with biological PDPs. As has been noted previously in this book, the SCT and QT derive their importance from the circumstance that they are the points of departure (PoD) for the AET. Practically speaking, higher values for the SCT and QT enable higher values for the AET, where higher values for the AET typically mean that few extractables and leachables require reporting for toxicological safety risk assessment (fewer compounds that need to be discovered, identified, and quantified). Early communications from the PDP Working team suggested that higher values for these quantities, and especially the QT, might be scientifically justifiable [37]. For example,

the toxicology working team evaluated a database of over 600 potential leachables using current toxicological qualification approaches in order to generate data and used this information and other information to support SCT and QT recommendations. More formal toxicological safety risk assessments were performed for a subset of these potential leachables to establish the leachables' PDEs and the resulting PDEs were statistically evaluated to support setting of SCT and QT values. This information notwithstanding, the toxicology community of practice ultimately reached a consensus agreement to support the SCT and QT, as listed above.

The second noteworthy aspect of the Recommendations was the concept of the simulation study as a type of controlled extraction study. Citing the analytical challenge of trying to achieve the low AETs that are common for certain classes of PDPs (for example, large volume parenterals, LVPs), the PQRI Working Team proposed simulations studies as a means of addressing this "AET Challenge." By definition, a simulation study is a controlled extraction study whose purpose is to produce an extractables profile for a packaging system that mirrors, as closely as possible, the leachables profile that a drug product has stored in the packaging system. The intrinsic value of the simulation study rests in its use of simulating solvents that are more analytically expedient than the drug product, thus facilitating the process of detecting, identifying, and quantitating extractables. That is, the simulation study's extractables profile reveals probable leachables at their likely highest accumulation levels. The success of the simulation study depends on the extent to which the "extracting power" of the simulation solvent is consistent with the "leaching power" of the drug product. Thus, extraction solvent(s) must be justified and the justification is essential for establishing the validity and applicability of a simulation study.

As many PDPs are aqueous based and noting the importance of solution pH on the leaching of substances from PDP packaging, the PDP recommendations contain the following points about the pH of a simulating solvent:

1. The proper pH range for extracting solvents used in extraction studies designed to forecast aqueous drug product leachables spans the pH range of the relevant drug products.
2. Extraction solvents with a neutral pH between 6 and 8 should not be used to establish an organic worst-case extractables profile (i.e. the greatest number of extractables at their highest levels).
3. Simulation studies require the use of extraction solvents with pH values closely matching the extreme pH values of the drug products.

4. Expanding the pH range of the extracting solvents beyond the pH range of the drug products produces an extractables profile that will exaggerate the drug products' leachables profile.

The third noteworthy aspect of the Recommendations was the Working Team's recognition that biological PDPs have an increased risk of adverse leachable's safety effects versus their non-biological counterparts. While leachables in biologics can adversely affect safety based on their inherent toxicity (as was the case with non-biologic PDPs), the ability of the leachables to chemically interact with the biological drug substance may result in secondary toxicity effects, such as the formation of particulates (e.g. protein aggregation) and/or the chemical alternation of the biological drug substance to produce a reaction product that itself has an adverse safety effect (e.g. immunogenic effects). Moreover, process equipment-related leachables can adversely affect the efficiency of the manufacturing process for biologics, for example, by reducing cell culture yields. Such effects are not typically encountered in the manufacturing of non-biological drug substances.

Although the PQRI PDP recommendations do not contain specific actions for managing and assessing leachables in biologics, the recommendations do suggest that due to the increasing complexity and diversity of pharmaceutical and biological products, manufacturing systems, drug delivery devices, and CCS, justifications for the AET, extraction conditions, extraction solvents, and analysis methods, should be discussed early in product development with the appropriate regulatory authority.

Two additional items addressed in the PDP Recommendations strike this author as being particularly impactful. The first item is that the PDP Recommendations include a flow chart of the toxicological qualification of leachables in PDP, the essence of which is captured in Table 7.9.

Table 7.9 PQRI PDP qualification process for leachables.

Question	Answer	Action
1. Is the leachable a compound of concern?[a]	Yes	Leachable's threshold may be lower than 1.5 µg/day. Consult with regulatory authority to establish an acceptable exposure.
	No	Proceed to Question 2.
2. Is the leachable's calculated daily exposure < 1.5 µg/day?	Yes	Leachable's adverse effect on patient safety is negligible, document result. Qualification Complete.
	No	Proceed to Question 3.
3. Is the leachable a potential mutagen?[b]	Yes	Follow ICH M7 principles for control of mutagenic impurities.
	No	Proceed to Question 4.
4. Is the leachable a possible sensitizer or irritant?	Yes	Proceed to Question 5.
	No	Proceed to Question 7.
5. Is the leachable's calculated daily exposure < 5 µg/day?	Yes	Proceed to Question 7.
	No	Proceed to Question 6.
6. Can the leachable's calculated daily exposure be reduced to < 5 µg/day?	Yes	Proceed to Question 7.
	No	Perform a risk assessment based on appropriate exposure limits (e.g. PDE). Proceed to Question 8.
7. Does the leachable have any other toxic effects?	Yes	Perform a risk assessment based on appropriate exposure limits (e.g. PDE). Proceed to Question 8.
	No	Leachable's adverse effect on patient safety is negligible, document result. Qualification Complete.
8. Is the leachable risk assessed as safe?	Yes	Leachable's adverse effect on patient safety is negligible, document result. Qualification Complete.
	No	Take corrective action to either reduce the leachable to a safe level or replace the packaging item associated with that leachable.

a)   The PDP document provides generalities around compounds of concern but does not provide a definitive list of such compounds.
b)   Suspicions of possible mutagenicity should be checked with either *in silico* analysis or published literature.

The second item is that the PDP recommendations address the topic of analytical uncertainty, specifically noting four analytical uncertainties:

1. Uncertainty in the proposed structure and elemental composition of leachables
2. Uncertainty in response with regard to detection and quantitation
3. Uncertainty related to sample matrix effects and interference
4. Uncertainty attributable to the quantification approach employed

Although the PDP recommendations do not provide specific directions on how to assess or address analytical uncertainty, they do recommend that the estimated AET values be adjusted for analytical uncertainty when applied to unknown leachables and that the adjustment should be achieved through a rational, scientifically justifiable approach at the discretion of the PDP drug product development team.

7.4.3 PQRI Recommendations for Ophthalmic Drug Products (ODP)

As the actual ODP Recommendations have just recently been published [38], it is challenging to speak to them in great detail. Nevertheless, it is noted that the Recommendations address the unique aspect of ODP packaging (that it is generally permeable and thus that ODPs are at risk of leachables from secondary packaging and environmental contaminants) and establish the principles of toxicological safety risk assessment for this dosage form.

By their very nature, ODP drug products are dosed topically in small aliquots directly to the eye. Thus, one expects the toxicological endpoints of sensitization and irritation to be of primary importance with respect to ODP leachables (and correspondingly of lessor importance to PDP dosage forms). Considering this, the ODP Working Team noted that there currently is not a sufficient database developed on all the relevant toxicity endpoints to allow the Working Team to recommend specific safety thresholds (i.e. sensitization, ocular irritation) for ODP. Thus, the hypothesis that threshold principles could be extrapolated from OINDP to ophthalmic solutions and suspensions lacked sufficient scientific support to develop a recommendation. Stated another way, the concepts of the SCT, QT, and AET cannot be applied to ODPs.

Nevertheless, certain conventions have been adopted for the reporting and assessment of leachables in OPD. These conventions are concentration-based, as opposed to dose-based (as toxicological endpoints such as sensitization and irritation may not have dose-response relationships) and are summarized in Table 7.10.

Table 7.10 Reporting and qualification threshold for ODP leachables (see reference [39]).

Concentration of the leachable in the DP	Required action
<1 mg/l (ppm)	No action required
1–10 mg/l	Report that one or more leachable(s) have been detected
10–20 mg/l	Report the identity of the leachable(s)
>20 mg/l	Qualify (safety assess) the leachable(s)

7.5 Risk Classification of Packaging Systems

Earlier, in Section 7.3, Table 1 of the FDA container closure guidance (and updated in USP <1664>) was criticized as being too simplistic to truly represent the complex nature of chemical characterization, resulting in a classification of dosage forms that provided little practical guidance in terms of establishing reasonable expectations for the strategies and tactics of chemical characterization studies and assessments. In the following text, an alternate risk-based classification tool is developed and discussed. It is noted that Sections 7.5 and 7.6 are further discussed in reference [40].

As a starting point to the risk classification tool, it is useful to establish (i) what risk is being classified and (ii) how the classifications are to be used. Considering (i), the risk that is being classified is the risk that a drug product would contain sufficiently high levels of leachables that the leachables could have an adverse effect on the health of a patient being treated with the drug product in question. Considering (ii), it is the intent that the classification would inform chemical characterization; that is, by classifying drug products one would be able to specify the information that would be necessary to establish the safety of a packaged drug product and to establish those actions (testing and assessment) that would be required to produce the necessary information.

To discuss this risk classification tool, the tool will first be presented (Figure 7.6) and then its genesis and justification will be provided. In this way it may be easier to understand the thought process behind the tool and its contents.

Perhaps the most significant difference between the Risk Classification contained herein and FDA's Table 1 is that the classification contained herein seeks to classify individual drug products whereas Table 1 classified drug product families (dosage forms). It is this author's opinion that one of the major issues with the FDA's Table 1 is

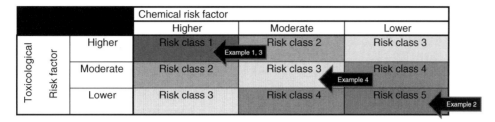

			Chemical risk factor		
			Higher	Moderate	Lower
Toxicological Risk factor		Higher	Risk class 1	Risk class 2	Risk class 3
			Example 1, 3		
		Moderate	Risk class 2	Risk class 3	Risk class 4
				Example 4	
		Lower	Risk class 3	Risk class 4	Risk class 5
					Example 2

Figure 7.6 Risk classification matrix for packaged drug products. The arrows refer to examples discussed in the text. (*See insert for colour representation of the figure.*)

that in grouping drug products by dosage form, incongruous groupings were produced. For example, consider the dosage form grouping of Injections. In this grouping are parenteral products such as small volume parenterals (SVPs) and large volume parenterals (LVPs). Although these types of parenteral products have certain similarities and differences, perhaps the most obvious difference is the volume contained in a single packaging unit (hence the names small and large volume). As patient safety impact is affected by patient daily exposure (which for an aqueous drug product is reflected in the daily dose volume), then surely the patient safety risk of leachables is greater for LVPs (because of their greater daily dose volumes) than for SVPs, and yet both SVPs and LVPs fall into the same risk class in the FDA's Table 1.

The proposed risk classification tool is based on the hypothesis that the safety risk being assessed has two dimensions, a dimension that addresses toxicological factors that relate to a patient's exposure to leachables and a dimension that addresses chemical factors that drive the extent of leaching. As was the case with the FDA's Table 1, each of these dimensions will be divided into three levels of risk, higher, moderate, and lower. Considering the toxicological dimension, the risk that is being managed is that a leached substance would have an adverse effect on patient safety. Thus, lower risk means that it is unlikely a leached substance would adversely affect patient safety while higher risk means there is a good chance that patient safety would be adversely affected by the leached substance. Considering the chemical dimension, the risk that is being managed is that the compounds would be leached for the packaging in significant quantitates. Thus, lower risk means that fewer compounds will be leached in lesser quantities while higher risk means that a greater number of compounds would be leached in higher quantities.

It is clear that the proper use of the Risk Classification Matrix shown in Figure 7.6 is to establish the means by which a particular drug product is placed in the Matrix. That is to say, how does one establish the drug product's toxicological and chemical risk category (lower, moderate, or higher) so that it can be placed in the Matrix?

In answering this question, it is noted that the intent of the Matrix is not to produce a scoring (or ranking) algorithm that quantitatively takes every factor relevant to either the Toxicological Risk or Chemical Risk dimensions into account, as such a quantitative algorithm would be impractical to use and impossible to justify, as (a) the basic and underlying science has not been completely delineated and (b) the algorithm would likely have to consider complex interdependencies between the algorithm's input variables. Rather, it is intended that the Matrix would largely be consistent with the underlying science and be practical to use. In so doing, one is unavoidably forced to rely on certain generalizations and simplifications about the underlying science, a situation that I have referred to as "voodoo" or "intuitive" science.

To this end and with this caveat, the dimension of the Toxicological Risk Factor is broken up into three contributing elements, Route of Administration, Daily Patient Exposure Amount, and Duration of Clinical Therapy. The Route of Administration is chosen as an element consistent with the FDA Table 1 and its revised form in USP <1664>. The Daily Patient Exposure Amount reflects the amount of a drug product that a patient is exposed to in a day and is chosen as an element as it is generally the case that the greater the amount of exposure, the greater the adverse safety impact. For liquid dosage forms, daily patient exposure is more readily recognizable as the maximum daily dose volume; however, to reflect both solid and liquid dosage forms a more compatible measure is the maximum daily dose amount measured as a weight. The Duration of Clinical Therapy is chosen as an element based on the observations that the likelihood of an adverse health effect increases the longer an individual is exposed to a potentially unsafe chemical.

Once the elements have been established, then the individual criteria for scoring within an element must be set. For ease of use and to be consistent with other proposed Risk Evaluation Matrices (for example, the leachables Risk Evaluation Matrices proposed for pharmaceutical manufacturing components [41–43]), each element will encompass a scoring scale of 1–10, with one reflecting

Table 7.11 Ranking (scoring matrix) for the elements of the toxicological risk factor dimension.

	Numerical risk score	Route of administration	Daily patient exposure (grams)	Duration of therapy
	10	Inhalation aerosols and sprays	>100	>10 years to lifetime
		Injections and injections suspensions		
		Inhalation solutions		
		Sterile powders and powders for injection		
	6	Inhalation powders	10–100	1–10 years
		Transdermal ointments and patches		
		Ophthalmic solutions and suspensions		
	3	Nasal aerosols and sprays	1–10	1–12 months
		Topical solutions and suspensions		
		Topical and lingual aerosols		
	1	Oral solutions and suspensions	<1	<1 month
		Oral tablets and capsules		
		Topical powders		
		Oral powders		

Increasing risk (vertical axis with upward arrow)

the lowest risk of the element contributing to an adverse patient safety effect and 10 being the highest risk. Within the scale of 1–10 there will be four risk score gradations reflecting relative risk: 1, 3, 6, and 10. Lastly, the items that fall within each element must be linked to the appropriate risk score. Thus, for example, all dosage forms listed in the FDA's Table 1 must be scored as either 1, 3, 6, or 10.

The outcome of this process is shown in Table 7.11. Scoring of the individual dosage forms under the Route of Administration element reflects the relative placement of these dosage forms in FDA's Table 1 with those dosage forms identified in FDA Table 1 as having the highest degree of concern being assigned a risk score of 10, those dosage forms identified in the FDA Table 1 as having a high degree of concern being assigned a score of 6, and those dosage forms having a low degree of concern being assigned a risk score of either 1 (solid dosage forms) or 3 (solutions, suspension, aerosols).

Scoring of the Daily Patient Exposure element was based on daily dose breakpoints of an order of magnitude, specifically 1g, 10g, and 100 g, corresponding to daily dose volumes of roughly 1, 10, and 100 ml.

Scoring of the Duration of Therapy element was based on the duration categories contained in the staged TTC approach of ICH M7, with the longest M7 duration (> 10 years) being assigned the highest score of 10 and the shortest M7 duration (< 1 months) being assigned a risk score of 1.

Once the scoring rubrics for the individual elements have been established, it is necessary to define the means by which the individual element's scores are combined to produce the overall score for the Toxicological Risk Factor and then to link the overall score to the risk categories of higher, moderate, and lower risk that are used in Figure 7.6. If it is not the case that empirical science has been employed up to this point in designing this risk evaluation process, then it is surely the case that pure science gives way to practicality at this point. On the one hand, the practical use of the risk tools to classify drug products is facilitated if the mathematical means of combining the risk scores from the individual elements is simple (for example, add them all up, multiply them together, or something equally straightforward). From the perspective of rigorous science, it is likely that capturing the impact of the individual elements and the interplay between the various elements requires a sophisticated (and complex) mathematical expression. Erroring on the side of practicality, an additive model is proposed; that is, the overall risk score will be the sum of the individual element's risk scores, with a twist. The twist is rooted in the proposition that a patient's health response to a leachable is affected more by patient exposure (daily amount and duration) than it is by route of administration. Thus, the elements of exposure amount and duration are weighted more heavily than is the element of route of administration by multiplying the score for these elements by a factor of 1.5 before performing the addition.

Therefore, the equation for calculating the overall risk score from the individual element's risk score becomes:

Overall risk score, Toxicological Risk Factor

$$= RS_{route} + (1.5 \times RS_{amount}) + (1.5 \times RS_{duration}) \quad (7.1)$$

where:

- RS_{route} = risk score for the Route of Administration
- RS_{amount} = risk score for Daily Patient Exposure
- $RS_{duration}$ = risk score for Duration of Therapy (Exposure)

This twist noted above, weighing patient exposure more heavily than route of administration, surely requires some level of discussion. The first point is to reaffirm that the route of administration is an important factor and weighting it less heavily does not detract from that. The second point is emphasized when Equation (7.1) is rearranged to produce the form:

Overall risk score, Toxicological Risk Factor

$$= RS_{route} + 1.5 \times (RS_{PDE} + RS_{DoT}) \quad (7.2)$$

From Equation (7.2) it is noted that the "twist" noted previously is not that the factors of Daily Patient Exposure and Duration of Therapy are individually weighted higher than is Route of Administration but rather that it is the combination effect of Daily Patient Exposure and Duration of Therapy that is weighted higher than the effect of Route of Administration.

Daily Patient Exposure establishes the hazard; as Daily Patient Exposure increases the hazard increases. When considering the route of administration, it either mitigates or amplifies the hazard that is established by the Daily Patient Exposure. That is to say, a leachable whose Patient Daily Dose is a certain value may be a more significant hazard if that dose is inhaled rather than ingested. The circumstance that the Patient Daily Exposure is the mitigated factor and route of administration is the mitigating factor suggests that Patient Daily Exposure is a primary factor and route of administration is the secondary factor.

Using this approach produces a range of possible overall risk scores for the Toxicological Risk Factor of 4 to 40. This range now needs to be divided to differentiate between lower, moderate, and higher Toxicological risk, requiring more compromise and generalization. However, perhaps a little bit of introspection can help us out here. For example, consider that the ICH M7 TTC for a chronically dosed drug product is 1.5 μg/day. If the daily dose amount of a drug product were to be 15 g per day, then the level of the leachable in the drug product corresponding to the TTC is 1.5 μg/15 g = 0.1 μg/g = 0.1 ppm. Thus, if one were to propose that a drug product, regardless of the route of administration, that contains 0.1 ppm of a single leachable,

dosed at 15 g daily for a lifetime is a moderate risk, then an overall risk score of 19 would have to fall within the moderate risk category. Similarly, if one were to propose that an inhalation aerosol, dosed daily at a volume of less than 1 g (approximately 1 ml) for a lifetime is high risk, then an overall risk score of 26.5 would have to fall within the higher risk category. Similarly, addressing the lower risk, if one were to propose that an ophthalmic solution, dosed in a daily amount of less than 1 g (approximately 1 ml) for a duration of less than a month is low risk, then an overall risk score of 9 would have to fall in the lower risk category.

Based an assessment of all possible scenarios, the following rubric is posed for the Toxicological Risk Factor:

Overall Risk Score 4.0–13.0 = Lower Risk
Overall Risk Score 13.5 to 25.0 = Moderate Risk
Overall Risk Score 25.5 or higher = Higher Risk

See Table 7.12 for a complete listing of the 64 possible scenarios for the Toxicological Risk Factor. A roughly symmetrical distribution of Toxicological Risk Factors was obtained, with 15, 32, and 17 of the possible scenarios being scored Lower, Moderate, and Higher risk respectively.

A similar process is performed for the dimension of Chemical Risk Factor, which is broken up into four logical contributing elements, Duration of Storage, Temperature of Storage (of the drug product in its packaging over shelf-life), Physical/Chemical Nature of the Drug Product, and Inherent Extractables Risk for Material of Construction. Duration and Temperature of Storage are chosen as they are the storage conditions that drive the kinetics of leaching. The physical/chemical nature of the drug product reflects the effect that the physical state and chemical composition of the drug product has on its "leaching power." Inherent extractables risk for materials of construction is chosen as an element as it is the composition of the material and its propensity to be leached that establish the absolute maximum amount of leaching that can occur.

The same scoring approach used for the Toxicological Risk dimension is used for the Chemical Risk dimension; each element encompasses a scoring scale of 1–10 with gradations reflecting relative risks of 1, 3, 6, and 10.

The outcome of this process is shown in Table 7.13. Scoring of the Storage Duration element reflects reasonable storage conditions from short-term to multi-year shelf-lives. Scoring of the Storage Temperature element reflects the three most commonly encountered storage temperatures while additionally taking into account the significant kinetic driving force provided by high-temperature steam sterilization (autoclaving) of some packaged dosage forms. Scoring of the Physical/Chemical

Table 7.12 Toxicological risk scores and risk factors for all 64 possible situations.

Route of administration	Risk element			
	Daily patient exposure	Duration of therapy	Toxicological risk score	Toxicological risk factor
Oral tablets and capsules	< 1 g	< 1 month	4	Low
Topical powders		1–12 months	7	Low
Oral powders		1–10 years	11.5	Low
		> 10 years	17.5	Moderate
	1–10 g	< 1 month	7	Low
		1–12 months	10	Low
		1–10 years	14.5	Moderate
		> 10 years	20.5	Moderate
	10–100 g	< 1 month	11.5	Low
		1–12 months	14.5	Moderate
		1–10 years	19	Moderate
		> 10 years	25	Moderate
	> 100 g	< 1 month	17.5	Low
		1–12 months	20.5	Moderate
		1–10 years	25	Moderate
		> 10 years	31	Moderate
Topical solutions and suspensions	< 1 g	< 1 month	6	Low
Topical and lingual aerosols		1–12 months	9	Low
		1–10 years	13.5	Moderate
Oral solutions and suspensions		> 10 years	19.5	Moderate
	1–10 g	< 1 month	9	Low
		1–12 months	12	Low
		1–10 years	16.5	Moderate
		> 10 years	22.5	Moderate
	10–100 g	< 1 month	13.5	Moderate
		1–12 months	16.5	Moderate
		1–10 years	21	Moderate
		> 10 years	27	High
	> 100 g	< 1 month	19.5	Moderate
		1–12 months	22.5	Moderate
		1–10 years	27	High
		> 10 years	33	High
Transdermal ointments and patches	< 1 g	< 1 month	9	Low
		1–12 months	12	Low
Ophthalmic solutions and suspensions		1–10 years	16.5	Moderate
		> 10 years	22.5	Moderate
Nasal aerosols and sprays	1–10 g	< 1 month	12	Low
		1–12 months	15	Moderate
		1–10 years	19.5	Moderate
		> 10 years	25.5	High
	10–100 g	< 1 month	16.5	Moderate
		1–12 months	19.5	Moderate
		1–10 years	24	Moderate
		> 10 years	30	High
	> 100 g	< 1 month	22.5	Moderate
		1–12 months	25.5	High
		1–10 years	30	High
		> 10 years	36	High
Inhalation aerosols and sprays, injections and injection suspensions	< 1 g	< 1 month	13	Low
		1–12 months	15	Moderate
		1–10 years	19.5	Moderate
Sterile powders and powders for injection		> 10 years	26.5	High
	1–10 g	< 1 month	16	Moderate
Inhalation powders		1–12 months	19	Moderate
		1–10 years	23.5	Moderate
		> 10 years	29.5	High
	10–100 g	< 1 month	20.5	Moderate
		1–12 months	23.5	Moderate
		1–10 years	28	High
		> 10 years	34	High
	> 100 g	< 1 month	26.5	High
		1–12 months	29.5	High
		1–10 years	34	High
		> 10 years	40	High

Table 7.13 Ranking (scoring matrix) for the elements of the chemical risk factor dimension.

Numerical risk score	Duration of storage	Temperature of storage	Physical/chemical nature of drug product	Inherent extractables risk for materials of construction
10	> 12 months	Autoclave sterilization	Non-polar	Ethylene propylene diene monomer rubber, EPDM
				Chlorobutyl and bromobutyl rubber[a]
				Plasticized Poly (vinyl chloride)
6	1–12 months	Room temperature	Aqueous, pH < 3 or pH > 10	Silicone
			Semi-polar	Pt-cured Silicone
				Polyolefin, PO
				Nylon
				Ethylene vinyl acetate (EVA)
				Ethylene vinyl alcohol (EVOH)
				Polyurethane, PU
3	1 week – 1 month	Refrigerated	Aqueous, 3 < pH < 10	Polyethylene, PE (low and high density)
				Polypropylene, PP
				Polystyrene, PS
				Polyethersulfone, PES
				Polycarbonate, PC
				Polysulfone, PSU
				Perfluoroalkoxy,PFA
				Cyclic olefin copolymer, COC
1	< 1 week	Frozen	Solid	Poly-(tertrafluoroethylene), PTFE
				Type I and Type II Glass
				Poly(vinylidene difluoride), PVDF

Increasing risk ↑

a) Rubber items such as stopper and plungers can be coated with a barrier layer such as Teflon. In this case, the risk score for such components is reduced to a value of 3.

nature of the drug product element addresses the common determinants of "leaching power," including the physical state (e.g. solid versus liquid), pH, and polarity. Scoring of the Inherent Material risk element is based on published information and practical experience related to the composition and "leachability" of certain materials widely used in packaging. For example, glass, given its composition (no organic additives) and resistance to leaching, is scored lower risk while plasticized PVC, known to contain plasticizers and other additives at relatively high levels, is scored higher risk.

The relative scoring of a solid dosage form as lower risk versus a solution dosage form in the element of Physical/Chemical Nature of the Drug Products is justified as follows. First, it is noted that this is consistent with the risk ranking of solids versus solutions in both the FDA [16] and EMEA [17] guidance and guidelines, based on the logic that solids cannot actively leach a packaging system in the same manner as a solution can (e.g. leaching is enhanced

by solvent uptake by the packaging and potential swelling of the polymer). However, it has been established that solid dosage forms can sorb volatile organic leachables when packaged in closed systems (for example, the headspace of a stoppered vial) and thus the risk of leachables being present in a solid dosage form is not zero. Furthermore, it is possible in certain situations that volatile leachables may be more readily sorbed by a solid dosage form than they can be dissolved in a solution drug product, leading to the situation where such leachables are present in a solid dosage form in higher quantities than in a solution dosage form. Nevertheless, this sorption phenomenon does not occur for less volatile organic leachables (as there typically is no avenue of transfer between the source of the leachable and the solid dosage form). For example, the reason that volatile extractables from a vial's stopper can become leachables in a solid drug product is that the gas in a vial's headspace serves as the medium of transfer between the stopper and the solid drug, which typically are not in direct physical contact. Because the stopper and the solid drug product are not in direct physical contact, there is no medium of transfer for non-volatile extractables in the stopper to become leachables in the solid drug product.

In the end, defending a generalization that leachables are always present in solid dosage forms in lesser or greater amounts than liquid dosage forms is a fool's errand. However, it is this author's experience that the circumstances where leachables would be higher in a solid dosage form versus a solution dosage form are rarer than the opposite and thus solid dosage forms are assessed as lower risk than solution dosage forms.

The different classes of polarity that are used to segregate the element of Physical/Chemical Nature of Drug Product in Table 7.13 also requires some discussion, as the different classes are described generally and rather imprecisely as polar, semi-polar, and non-polar. Rather than provide a comprehensive listing of the polarity of all possible dosage forms, which surely is a next-to-impossible task, examples of dosage forms that fit into the three polarity classes are provided. Considering polar as a class, many pharmaceutical dosage forms are formulated in aqueous vehicles; thus, it is appropriate that the lowest risk score for a polar vehicle is linked to aqueous solutions of varying pH in Table 7.13. Such polar vehicles would include aqueous vehicles containing non-solubilizing excipients such as salts, buffers, sugars, etc. Parenteral and ophthalmic dosage forms include many aqueous, polar drug products.

At the other extreme, non-polar, it is first noted that no known pharmaceutical dosage form is formulated in hexane. Thus, although hexane is certainly a non-polar solvent, it is not the non-polar solvent that establishes the high-risk end of the element of Physical/Chemical Nature of Drug Product. Arguably, the most non-polar vehicles are encountered in inhaled dosage forms, where the propellants consist of chlorofluorocarbons, fluorocarbons, hydrocarbons (propane, butane, isobutane), hydrochlorofluorocarbons, and hydrofluorocarbons, and co-solvents such as ethanol are often used as a vehicle or co-solvent to help increase the solubility of the drug or other excipients. Other examples of non-polar dosage forms include topical ointments and injectable suspensions.

Examples of semi-polar dosage forms would include lipid emulsions, blood and blood products, injectables containing solubilizing agents and/or co-solvents, and topical creams and lotions.

Lastly, the classification of the element Materials of Construction is quite challenging. This author recognizes well the potential issues and shortcomings of implying, for example, that all "polypropylenes are inherently of low to moderate risk of leaching." Thus, users of the Risk Evaluation process contained in this manuscript are reminded that following the process blindly is no substitute for thinking one's decisions through. For example, a potential user of the Risk Evaluation process might note that although their material of interest is ranked lower risk, the fact that the material is gamma irradiated before it is constructed into the packaging systems might increase the risk. As another example, the fact that a rubber stopper has been coated with Teflon should be considered as having lowered the assessed risk level of the rubber elastomer.

Once the scoring rubric for the individual elements have been established, it is necessary to define the means by which the individual element's scores are combined to produce the overall score of the Chemical Risk Factor and then the link of the overall score to the risk categories of higher, moderate, and lower risk that are used in Figure 7.6. As was the case for the Toxicological dimension, an additive model is used; that is, the overall risk score will be the sum of the individual element's risk scores, with a twist. Additionally, as was the case for the Toxicological dimension, not all risk elements are weighted equally. Thus, the elements of Physical/Chemical Properties of the Drug Product and Inherent Extractables Risk for Materials are weighted more heavily than are the elements of Duration and Temperature of Storage by multiplying the score for these elements by a factor of 1.5 before performing the addition. This is the case as the elements of Physical/Chemical Properties and Inherent Extractables Risk dictate the amount of leaching that can occur, while the factors Temperature and Duration of Contact dictate the speed with which leaching occurs. Thus, the equation for calculating the overall risk score from the individual element's risk score becomes:

$$\text{Overall risk score, Chemical Risk Factor}$$
$$= RS_{duration} + RS_{temp} + (1.5 \times RS_{nature})$$
$$+ (1.5 \times RS_{risk}) \qquad (7.3)$$

where:

$RS_{duration}$ = risk score for the Duration of Storage

RS_{temp} = risk score for the Temperature of Storage

RS_{nature} = risk score for the Physical/Chemical Nature of the Drug Product

RS_{risk} = risk score for the Inherent Extractables Risk for Materials

Using this approach produces a range of possible overall risk scores for the Chemical Risk Factor of 5 to 50. This range now needs to be divided to differentiate between lower, moderate, and higher chemical risk, requiring more compromise and generalization. To set the inflection points separating the classes, a few examples are considered, similar to what was done for the toxicological risk dimension. For example, it seems reasonable that an aqueous drug product at neutral pH, autoclaved and stored in a polyolefin bag at room temperature with a shelf-life of two years would have a moderate to higher chemical risk. For this to be true, the risk score for this case, 33.5, would have to be close to the boundary between the moderate and higher risk categories. Similarly, addressing lower risk, if one were to propose that a vaccine, sterile filled into a pre-filled syringe with a glass barrel and a Teflon-coated bromobutyl rubber plunger and stored frozen for up to 2 years is low risk, then an overall risk score of 17 would have to fall in the lower risk category.

Based on an assessment of all possible scenarios, the following rubric is posed for the Chemical Risk Factor:

Overall Risk Score 5.0–19.0 = Lower Risk
Overall Risk Score 19.5 to 31.0 = Moderate Risk
Overall Risk Score 31.5 or higher = Higher Risk

See Table 7.14 for a complete listing of the 256 possible scenarios for the Chemical Risk Factor. A roughly symmetrical distribution of Chemical Risk Factors was obtained, with 69, 127, and 60 of the possible scenarios being scored Lower, Moderate, and Higher risk respectively.

Four examples will be provided to demonstrate the use of the risk assessment matrix.

Example 7.1 *Large Volume Parenteral, Nutrition Solution (10% Lipid Emulsion) in Polyolefin Bag (1000 ml)*

Toxicological Risk Assessment per Table 7.11:

Route of Administration: Injections, Risk score = 10
Daily Patient Exposure: One bag per day, > 100 g, Risk score = 10
Duration of Therapy: Nutrition replacement therapy is lifetime, > 10 years, Risk score = 10
Overall Risk Score, Toxicological Risk per Equation (7.1):
    $10 + (1.5 \times 10) + (1.5 \times 10) = 40$
Toxicological Risk Category = Higher risk

Chemical Risk Assessment per Table 7.13:

Duration of Storage: Shelf-life is two years at ambient temperature, Risk score = 10
Temperature of Storage: Product is terminally sterilized in its packaging, Risk score = 10.
Physical/Chemical Nature of the Drug Product: 10% lipid emulsion is classified as semi-polar, Risk score = 6

Inherent Extractables Risk for Materials: Bag is a five-layered polyolefin laminate, Risk score = 6
Overall Risk Score, Chemical Risk per Equation (7.2):
    $10 + 10 + (1.5 \times 6) = (1.5 \times 6) = 38$
Chemical Risk Category = Higher risk
Overall Assessment: see Figure 7.6, Risk Class 1 assessment required. Given the specifics of this example, this outcome is logical and to be expected.

Example 7.2 *Vaccine Packaged in a Pre-filled Syringe (1 ml, dose = 0.5 ml)*
Toxicological Risk Assessment per Table 7.11:

Route of Administration: Injections, Risk score = 10
Daily Patient Exposure: One injection per day, 0.5 ml, < 1 g, Risk score = 1
Duration of Therapy: Two injections two-weeks apart, < 1 month, Risk score = 1
Overall Risk score, Toxicological Risk per Equation (7.1):
    $10 + (1.5 \times 1) + (1.5 \times 1) = 13$
Toxicological Risk Category = Lower risk

Chemical Risk Assessment per Table 7.13:

Duration of Storage: Shelf-life is one year, refrigerated, Risk score = 6
Temperature of Storage: Product is sterile filled, Risk score = 3
Physical/Chemical Nature of the Drug Product: largely aqueous formulation, near neutral pH, Risk score = 3
Inherent Extractables Risk for Materials: pre-filled syringe has a glass barrel, Teflon-coated rubber plunger, Risk score = 3
Overall Risk Score, Chemical Risk per Equation (7.2):
    $6 + 3 + (1.5 \times 3) = (1.5 \times 3) = 18$
Chemical Risk Category = Lower risk
Overall Assessment: see Figure 7.6, Risk Class 5 assessment required. Given the specifics of this example, this outcome is logical and to be expected.

Example 7.3 *MDI Product for Chronic Therapy (Asthma)*
Toxicological Risk Assessment per Table 7.11:

Route of Administration: Inhalation powder, Risk score = 10
Daily Patient Exposure: 100 µg/actuation, 12 actuations/day, < 1 g, Risk score = 1
Duration of Therapy: Lifetime, Risk score = 10
Overall Risk Score, Toxicological Risk per Equation (7.1):
    $10 + (1.5 \times 1) + (1.5 \times 10) = 26.5$
Toxicological Risk Category = Higher risk

Table 7.14 Chemical risk scores and risk factors for all 256 possible situations.

Storage duration	Storage temperature	Nature of DP	Material extractables risk	Chemical risk score	Chemical risk factor
			Risk element		
< 1 week	Frozen	Solid	Group 1[a]	5	Low
			Group 2[b]	8	Low
			Group 3[c]	12.5	Low
			Group 4[d]	18.5	Low
		Aqueous, 3 < pH < 10	Group 1[a]	8	Low
			Group 2[b]	11	Low
			Group 3[c]	15.5	Low
			Group 4[d]	21.5	Moderate
		Aqueous, pH < 3 or pH > 10	Group 1[a]	12.5	Low
		Semi-polar	Group 2[b]	15.5	Low
			Group 3[c]	20	Moderate
			Group 4[d]	26	Moderate
		Non-polar	Group 1[a]	18.5	Low
			Group 2[b]	21.5	Moderate
			Group 3[c]	26	Moderate
			Group 4[d]	32	High
	Refrigerated	Solid	Group 1[a]	7	Low
			Group 2[b]	10	Low
			Group 3[c]	14.5	Low
			Group 4[d]	20.5	Moderate
		Aqueous, 3 < pH < 10	Group 1[a]	10	Low
			Group 2[b]	13	Low
			Group 3[c]	17.5	Low
			Group 4[d]	24.5	Moderate
		Aqueous, pH < 3 or pH > 10	Group 1[a]	14.5	Low
		Semi-polar	Group 2[b]	17.5	Low
			Group 3[c]	22	Moderate
			Group 4[d]	28	Moderate
		Non-polar	Group 1[a]	20.5	Moderate
			Group 2[b]	23.5	Moderate
			Group 3[c]	28	Moderate
			Group 4[d]	34	High
	Room Temperature	Solid	Group 1[a]	10	Low
			Group 2[b]	13	Low
			Group 3[c]	17.5	Low
			Group 4[d]	23.5	Moderate
		Aqueous, 3 < pH < 10	Group 1[a]	13	Low
			Group 2[b]	16	Low
			Group 3[c]	20.5	Moderate
			Group 4[d]	27.5	Moderate
		Aqueous, pH < 3 or pH > 10	Group 1[a]	17.5	Low
		Semi-polar	Group 2[b]	20.5	Moderate
			Group 3[c]	25	Moderate
			Group 4[d]	31	Moderate
		Non-polar	Group 1[a]	23.5	Moderate
			Group 2[b]	26.5	Moderate
			Group 3[c]	31	Moderate
			Group 4[d]	37	High
	Autoclave Sterilization	Solid	Group 1[a]	14	Low
			Group 2[b]	17	Low
			Group 3[c]	21.5	Moderate
			Group 4[d]	27.5	Moderate
		Aqueous, 3 < pH < 10	Group 1[a]	17	Low
			Group 2[b]	20	Moderate
			Group 3[c]	24.5	Moderate
			Group 4[d]	31.5	High
		Aqueous, pH < 3 or pH > 10	Group 1[a]	21.5	Moderate
		Semi-polar	Group 2[b]	24.5	Moderate
			Group 3[c]	29	Moderate
			Group 4[d]	35	High
		Non-polar	Group 1[a]	27.5	Moderate
			Group 2[b]	30.5	Moderate
			Group 3[c]	35	High
			Group 4[d]	41	High

Table 7.14 (Continued)

Storage duration	Storage temperature	Nature of DP	Material extractables risk	Chemical risk score	Chemical risk factor
			Risk element		
1 week – 1 month	Frozen	Solid	Group 1[a]	7	Low
			Group 2[b]	10	Low
			Group 3[c]	14.5	Low
			Group 4[d]	20.5	Moderate
		Aqueous, 3 < pH < 10	Group 1[a]	10	Low
			Group 2[b]	13	Low
			Group 3[c]	17.5	Low
			Group 4[d]	23.5	Moderate
		Aqueous, pH < 3 or pH > 10 Semi-polar	Group 1[a]	14.5	Low
			Group 2[b]	17.5	Low
			Group 3[c]	22	Moderate
			Group 4[d]	28	Moderate
		Non-polar	Group 1[a]	20.5	Moderate
			Group 2[b]	23.5	Moderate
			Group 3[c]	28	Moderate
			Group 4[d]	34	High
	Refrigerated	Solid	Group 1[a]	9	Low
			Group 2[b]	12	Low
			Group 3[c]	16.5	Low
			Group 4[d]	22.5	Moderate
		Aqueous, 3 < pH < 10	Group 1[a]	12	Low
			Group 2[b]	15	Low
			Group 3[c]	19.5	Moderate
			Group 4[d]	26.5	Moderate
		Aqueous, pH < 3 or pH > 10 Semi-polar	Group 1[a]	16.5	Low
			Group 2[b]	19.5	Moderate
			Group 3[c]	24	Moderate
			Group 4[d]	30	Moderate
		Non-polar	Group 1[a]	22.5	Moderate
			Group 2[b]	25.5	Moderate
			Group 3[c]	30	Moderate
			Group 4[d]	36	High
	Room temperature	Solid	Group 1[a]	12	Low
			Group 2[b]	15	Low
			Group 3[c]	19.5	Moderate
			Group 4[d]	25.5	Moderate
		Aqueous, 3 < pH < 10	Group 1[a]	15	Low
			Group 2[b]	18	Low
			Group 3[c]	22.5	Moderate
			Group 4[d]	29.5	Moderate
		Aqueous, pH < 3 or pH > 10 Semi-polar	Group 1[a]	19.5	Moderate
			Group 2[b]	22.5	Moderate
			Group 3[c]	27	Moderate
			Group 4[d]	33	High
		Non-polar	Group 1[a]	26.5	Moderate
			Group 2[b]	28.5	Moderate
			Group 3[c]	33	High
			Group 4[d]	39	High
	Autoclave sterilization	Solid	Group 1[a]	16	Low
			Group 2[b]	19	Low
			Group 3[c]	23.5	Moderate
			Group 4[d]	29.5	Moderate
		Aqueous, 3 < pH < 10	Group 1[a]	19	Low
			Group 2[b]	22	Moderate
			Group 3[c]	26.5	Moderate
			Group 4[d]	33.5	High
		Aqueous, pH < 3 or pH > 10 Semi-polar	Group 1[a]	23.5	Moderate
			Group 2[b]	26.5	Moderate
			Group 3[c]	31	Moderate
			Group 4[d]	37	High
		Non-polar	Group 1[a]	29.5	Moderate
			Group 2[b]	32.5	High
			Group 3[c]	37	High
			Group 4[d]	43	High

(Continued)

Table 7.14 (Continued)

Storage duration	Storage temperature	Nature of DP	Material extractables risk	Chemical risk score	Chemical risk factor
			Risk element		
1–12 months	Frozen	Solid	Group 1[a]	10	Low
			Group 2[b]	13	Low
			Group 3[c]	17.5	Low
			Group 4[d]	23.5	Moderate
		Aqueous, 3 < pH < 10	Group 1[a]	13	Low
			Group 2[b]	16	Low
			Group 3[c]	20.5	Moderate
			Group 4[d]	26.5	Moderate
		Aqueous, pH <3 or pH > 10 Semi-polar	Group 1[a]	17.5	Low
			Group 2[b]	20.5	Moderate
			Group 3[c]	25	Moderate
			Group 4[d]	31	Moderate
		Non-polar	Group 1[a]	23.5	Moderate
			Group 2[b]	26.5	Moderate
			Group 3[c]	31	Moderate
			Group 4[d]	37	High
	Refrigerated	Solid	Group 1[a]	12	Low
			Group 2[b]	15	Low
			Group 3[c]	19.5	Moderate
			Group 4[d]	25.5	Moderate
		Aqueous, 3 < pH < 10	Group 1[a]	15	Low
			Group 2[b]	18	Low
			Group 3[c]	22.5	Moderate
			Group 4[d]	29.5	Moderate
		Aqueous, pH <3 or pH > 10 Semi-polar	Group 1[a]	19.5	Moderate
			Group 2[b]	22.5	Moderate
			Group 3[c]	27	Moderate
			Group 4[d]	33	High
		Non-polar	Group 1[a]	25.5	Moderate
			Group 2[b]	28.5	Moderate
			Group 3[c]	33	High
			Group 4[d]	39	High
	Room temperature	Solid	Group 1[a]	15	Low
			Group 2[b]	18	Low
			Group 3[c]	22.5	Moderate
			Group 4[d]	28.5	Moderate
		Aqueous, 3 < pH < 10	Group 1[a]	18	Low
			Group 2[b]	21	Moderate
			Group 3[c]	25.5	Moderate
			Group 4[d]	32.5	High
		Aqueous, pH <3 or pH > 10 Semi-polar	Group 1[a]	22.5	Moderate
			Group 2[b]	25.5	Moderate
			Group 3[c]	30	Moderate
			Group 4[d]	36	High
		Non-polar	Group 1[a]	29.5	Moderate
			Group 2[b]	31.5	High
			Group 3[c]	36	High
			Group 4[d]	42	High
	Autoclave Sterilization	Solid	Group 1[a]	19	Low
			Group 2[b]	22	Moderate
			Group 3[c]	26.5	Moderate
			Group 4[d]	32.5	High
		Aqueous, 3 < pH < 10	Group 1[a]	22	Moderate
			Group 2[b]	25	Moderate
			Group 3[c]	29.5	Moderate
			Group 4[d]	36.5	High
		Aqueous, pH <3 or pH > 10 Semi-polar	Group 1[a]	26.5	Moderate
			Group 2[b]	29.5	Moderate
			Group 3[c]	34	High
			Group 4[d]	40	High
		Non-polar	Group 1[a]	32.5	High
			Group 2[b]	35.5	High
			Group 3[c]	40	High
			Group 4[d]	46	High

Table 7.14 (Continued)

Storage duration	Storage temperature	Nature of DP	Material extractables risk	Chemical risk score	Chemical risk factor
> 12 months	Frozen	Solid	Group 1[a]	14	Low
			Group 2[b]	17	Low
			Group 3[c]	21.5	Moderate
			Group 4[d]	27.5	Moderate
		Aqueous, 3 < pH < 10	Group 1[a]	17	Low
			Group 2[b]	20	Moderate
			Group 3[c]	24.5	Moderate
			Group 4[d]	30.5	Moderate
		Aqueous, pH <3 or pH > 10 Semi-polar	Group 1[a]	21.5	Moderate
			Group 2[b]	24.5	Moderate
			Group 3[c]	29	Moderate
			Group 4[d]	35	High
		Non-polar	Group 1[a]	27.5	Moderate
			Group 2[b]	30.5	Moderate
			Group 3[c]	35	High
			Group 4[d]	41	High
	Refrigerated	Solid	Group 1[a]	16	Low
			Group 2[b]	19	Low
			Group 3[c]	23.5	Moderate
			Group 4[d]	29.5	Moderate
		Aqueous, 3 < pH < 10	Group 1[a]	19	Low
			Group 2[b]	22	Moderate
			Group 3[c]	26.5	Moderate
			Group 4[d]	33.5	High
		Aqueous, pH <3 or pH > 10 Semi-polar	Group 1[a]	23.5	Moderate
			Group 2[b]	26.5	Moderate
			Group 3[c]	31	Moderate
			Group 4[d]	37	High
		Non-polar	Group 1[a]	29.5	Moderate
			Group 2[b]	32.5	High
			Group 3[c]	37	High
			Group 4[d]	43	High
	Room temperature	Solid	Group 1[a]	19	Low
			Group 2[b]	22	Moderate
			Group 3[c]	26.5	Moderate
			Group 4[d]	32.5	High
		Aqueous, 3 < pH < 10	Group 1[a]	22	Moderate
			Group 2[b]	25	Moderate
			Group 3[c]	29.5	Moderate
			Group 4[d]	36.5	High
		Aqueous, pH <3 or pH > 10 Semi-polar	Group 1[a]	26.5	High
			Group 2[b]	29.5	Moderate
			Group 3[c]	34	High
			Group 4[d]	40	High
		Non-polar	Group 1[a]	33.5	High
			Group 2[b]	35.5	High
			Group 3[c]	40	High
			Group 4[d]	46	High
	Autoclave sterilization	Solid	Group 1[a]	23	Moderate
			Group 2[b]	26	Moderate
			Group 3[c]	30.5	Moderate
			Group 4[d]	36.5	High
		Aqueous, 3 < pH < 10	Group 1[a]	26	Moderate
			Group 2[b]	29	Moderate
			Group 3[c]	33.5	High
			Group 4[d]	40.5	High
		Aqueous, pH <3 or pH > 10 Semi-polar	Group 1[a]	30.5	Moderate
			Group 2[b]	33.5	High
			Group 3[c]	38	High
			Group 4[d]	44	High

(Continued)

Table 7.14 (Continued)

		Risk element			
Storage duration	Storage temperature	Nature of DP	Material extractables risk	Chemical risk score	Chemical risk factor
		Non-polar	Group 1[a]	36.5	High
			Group 2[b]	39.5	High
			Group 3[c]	44	High
			Group 4[d]	50	High

a) Group 1 (lower risk) materials of construction include: Poly(tertrafluoroethylene), PTFE; Type I and Type II Glass; Poly(vinylidene difluoride), PVDF.

b) Group 2 (moderate risk) materials of construction include: Polyethylene, PE (low and high density); Polypropylene, PP; Polystyrene, PS; Polyethersulfone, PES; Polycarbonate, PC; Polysulfone, PSU; Perfluoroalkoxy, PFA; Cyclic olefin copolymer (COC); Teflon-coated chlorobutyl and bromobutyl rubber; polyurethane (PU).

c) Group 3 (higher risk) materials of construction include: Silicone; Pt-cured Silicone; Polyolefin, PO; Nylon; ethylene vinyl acetate (EVA); ethylene vinyl alcohol (EVOH).

d) Group 4 (higher risk) materials include: Ethylene propylene diene monomer rubber, EPDM; Chlorobutyl and Bromobutyl rubber (uncoated); Plasticized Poly (vinyl chloride).

Chemical Risk Assessment per Table 7.12:

Duration of Storage: Shelf-life is one years, ambient, Risk score = 6

Temperature of Storage: ambient temperature storage, Risk score = 6.

Physical/Chemical Nature of the Drug Product: organic non-polar propellant, Risk score = 10

Inherent Extractables Risk for Materials: uncoated rubber contact components, Risk score = 10

Overall Risk score, Chemical Risk per Eq. (7.2): $6 + 6 + (1.5 \times 10) = (1.5 \times 10) = 42$

Chemical Risk Category = Higher risk

Overall Assessment: see Figure 7.6, Risk Class 1 assessment required. Given the specifics of this example, this outcome is logical and to be expected.

Example 7.4 *Ophthalmic Eye Drops for Chronic Therapy*

Toxicological Risk Assessment per Table 7.11:

Route of Administration: Ophthalmic solution, Risk score = 6

Daily Patient Exposure: 2 drops per eye per day, each drop = 0.1 ml, total = 0.4 ml, < 1 g, Risk score = 1

Duration of Therapy: Lifetime, Risk score = 10

Overall Risk score, Toxicological Risk per Equation (7.1): $6 + (1.5 \times 1) + (1.5 \times 10) = 22.5$

Toxicological Risk Category = Moderate risk

Chemical Risk Assessment per Table 7.13:

Duration of Storage: Shelf-life is two years, ambient, Risk score = 10

Temperature of Storage: Ambient temperature storage, no terminal sterilization, Risk score = 6.

Physical/Chemical Nature of the Drug Product: Aqueous, neutral pH, Risk score = 3

Inherent Extractables Risk for Materials: HDPE squeeze bottle, Risk score = 3

Overall Risk score, Chemical Risk per Equation (7.2): $10 + 6 + (1.5 \times 3) = (1.5 \times 3) = 25$

Chemical Risk Category = Moderate risk

Overall Assessment: see Figure 7.6, Risk Class 3 assessment required. Given the specifics of this example, this outcome is logical and to be expected.

7.6 Assessments of Drug Products, Packaging Systems, and Materials of Construction Based on Risk Classification

Section 7.5 ended with the ability to classify packaged drug products in terms of the risk that they would contain unsafe levels of packaging-related leachables. Although the ability to perform such a classification is useful, the classification must have some utility, presumably as a guide to establishing the actions necessary to qualify the packaged drug product and its packaging system as being safe. That is, the risk assessment establishes the possibility that leachables may be unsafe; it is the qualification of the packaged drug product and its packaging system that establishes that the leachables are safe.

As has been proposed previously and consistent with regulatory policy and perspective, the actions taken to qualify a packaged drug product and its packaging system should reflect the risk that packaging-related leachables could be unsafe, with higher risk requiring greater and more rigorous action. This was the purpose of Table 2 in the FDA Container Closure Guidance and it is therefore appropriate that an action strategy be established that is the companion to the risk classification illustrated in Figure 7.6.

To do this, let us consider the extreme classes first. As shown in Figure 7.6, the lowest risk class is Risk Class 5. If one were to compare this to the lowest risk class in the FDA's Table 2, this would be their Case 4s, where "typically an appropriate reference to the food additive regulations is sufficient." It is difficult to imagine that in today's regulatory environment an appropriate reference to food additive regulations would be considered to be sufficient proof that a drug product packaging system is safe from the perspective of no leachables at unsafe levels in the packaged drug product. However, perhaps the concept behind this requirement is worth consideration. That is, perhaps the packaging system for a Risk Class 5 drug product can be established to be safe based solely on information collected about the packaging system's materials of construction with no actual qualification of the packaging system itself. If one accepts this concept, then the qualification issue becomes "what information about the materials of construction is sufficient?" The answer to this question is as follows:

- Proper reference to applicable food additive regulations
- Compliance with the relevant compendial monographs (for example, USP <661.1> or the Pharm Eur Section 3.1 monograph)
- Biological reactivity testing, USP <87> *in vitro* for the packaging's materials of construction (typically available from the material's vendor)
- Compositional compliance statements (e.g. REACH)
- Composition (intentional additives, known processing aids and residuals, identity, and quantity)
- Safe-use justification based on the risk classification and toxicological review of the above information

That is to say that a preponderance of evidence argument, based the information noted above, is adequate and sufficient to qualify a packaging system for a Risk Class 5 packaged drug product as safe for its intended use.

Addressing the highest risk class, Class 1, one can start defining the requirements for high-risk Class 1s from FDA Table 2. These requirements include:

- Full extractables profiling of the commercial packaging system (with qualified methods and adjusted AET)
- Toxicological safety risk assessment of extractables as potential leachables to establish and justify target substances for drug product leachables testing
- Full leachables profiling (targeting and screening) of the commercial packaged drug product over-shelf (three lots with appropriately validated (targeted) or qualified methods (screening) and adjusted AET for screening)

- Leachables screening to include extractables (target screening) and "leachables that are not extractables" (new requirement not originally in the FDA Table 2)
- Qualitative and quantitative extractables–leachables correlation
- Toxicological safety risk assessment of leachables at their highest reported concentrations
- Development of extractables specifications for incoming packaging items (materials, components, or systems)
- Incoming batch testing of packaging materials, packaging components, or packaging systems for relevant extractables (that is, extractables that must be controlled to ensure the safety of the packaged drug product)
- USP Biological Reactivity Test data, both *in vitro* <87> and *in vivo* <88>

From a chemical perspective, this list of requirements is comprehensive, robust, and appropriately rigorous. Furthermore, inclusion of the Biological Reactivity testing addresses toxicological safety endpoints that cannot be adequately addressed by review of chemical data. Thus, this list is adopted, without adjustment or addition, as the qualification requirements for a packaging system for a Risk Class 1 packaged drug product.

Presumably establishing the qualification requirements for the remaining Risk Classes would involve incremental adjustments of the requirements for the extreme Risk Classes. Qualification requirements would be lessened from those of Risk Class 1 as the leachables risk becomes lower and would be increased from those of Risk Class 5 as the leachables risk becomes greater.

In considering the qualifications for the intermediate Risk Classes, the following transitions serve as useful focal points:

- Specifications for extractables and extractables testing of incoming packaging items (as in when they should or should not be required)
- Quantitative versus qualitative extractables – leachables correlations
- Targeted leachables testing (as necessary) versus screening leachables testing
- Transitioning from extractables testing to leachables testing
- Transitioning from material assessment to component or system assessment
- Transitions in Biological reactivity testing

In moving to higher risk from Risk Class 5 to Risk Class 4, the transition from material assessment to component or system assessment is relevant. Furthermore, it is reasonable to propose that the transition from Risk Class 5 to Risk Class 4 would also be the transition from general data to

Table 7.15 Summary of testing requirements for the five packaged drug product risk classes.

	Requirement for each drug risk class				
	Higher risk →				Lower risk
Test	Risk Class 1	Risk Class 2	Risk Class 3	Risk Class 4	Risk Class 5
			SYSTEM TESTING		
Extractables	Full extractables profiling of the commercial packaging system (with qualified methods and adjusted AET) Toxicological safety risk assessment of extractables as potential leachables to establish and justify target substances for drug product leachables testing	Full extractables profiling of the commercial packaging system (with qualified methods and adjusted AET) Toxicological safety risk assessment of extractables as potential leachables to establish and justify target substances for drug product leachables testing	Full extractables profiling of the commercial packaging system (with qualified methods and adjusted AET) Toxicological safety risk assessment of extractables as potential leachables to establish and justify target substances for drug product leachables testing	Full extractables profiling of the commercial packaging system (with qualified methods and adjusted AET) Toxicological safety risk assessment of extractables as potential leachables to establish and justify target substances for drug product leachables testing	–
Leachables	Full leachables profiling (targeting and screening) of the commercial packaged drug product over-shelf (3 lots with appropriately validated (targeted) or qualified methods (screening) and adjusted AET for screening) Leachables screening to include extractables (target screening) and "leachables that are not extractables" Toxicological safety risk assessment of leachables at highest reported concentrations	Full leachables profiling (targeting and screening) of the commercial packaged drug product over-shelf (3 lots with appropriately validated (targeted) or qualified methods (screening) and adjusted AET for screening) Leachables screening for extractables (target screening) Toxicological safety risk assessment of leachables at highest reported concentrations	Full leachables profiling (target extractable only) of the commercial packaged drug product over-shelf (3 lots with appropriately validated methods) – Toxicological safety risk assessment of leachables at highest reported concentrations	Leachables profiling (of the commercial packaged drug product at various points over shelf-life **only** if there are safety-related extractables targets) – Toxicological safety risk assessment of leachables at highest reported concentrations	–
Extractables – Leachables correlations	Qualitative and quantitative extractables-leachables correlation	Quantitative correlation for targets; qualitative correlation for screening	Qualitative extractables-leachables correlations for targets	Qualitative extractables-leachables correlations for targets	–
Extractables specifications	Extractables specifications for incoming packaging items Incoming batch testing of packaging materials, components or systems for extractables which must be controlled to ensure the safety of the packaged drug product	–	–	–	–
Biological reactivity	USP Biological Reactivity, *in vitro* <87> and *in vivo* <88>	USP Biological Reactivity, *in vitro* <87> and *in vivo* <88>	USP Biological Reactivity, *in vitro* <87>	USP Biological Reactivity, *in vitro* <87>	–

MATERIAL TESTING

General Information	–	–	Proper reference to food additive regulations
			Compliance with compendial monographs
			Composition
			Compositional compliance statements
General information assessment	–	–	Safe use justification based on toxicological review of the general information
Biological reactivity	–	–	USP Biological Reactivity, *in vitro* <87>

infer safe use (such as a UV absorption test performed per USP <661.1> to specific data that can be used to assess safe use, such as extractables data). Thus, the requirements for Risk Class 4 packaged drug products become:

- Full extractables profiling of the commercial packaging system (with qualified methods and adjusted AET).
- Toxicological safety risk assessment of extractables as potential leachables to establish and justify target substances for drug product leachables testing, should such testing be necessary.
- Leachables profiling (of the commercial packaged drug product at various points over shelf-life (could be from different lots) *only* if there are targets established in the previous bullet with respect to potential adverse safety effects and only for those targets. If no targets are established, no leachables testing is required.
- Biological reactivity testing, USP <87> *in vitro*, performed on the packaging system.

In moving to lower risk from Risk Class 1 to Risk Class 2, several transitions are made. First, the development of extractables and mandatory testing of incoming lots of packaging items is eliminated, unless extractables/leachables testing and subsequent toxicological safety risk assessment establish that there is need to do so. Furthermore, screening of drug products for "leachables that are not extractables" is eliminated; while drug products will be screened for non-targeted leachables, the screening will be limited to looking for known extractables as leachables. This process is commonly referred to as targeted screening, where the screening methods are used for the testing but the only peaks associated with known extractables are addressed. Lastly, while extractables–leachables correlations still need to be quantitative for targeted leachables, they need be only qualitative for screening leachables. Thus, the requirements for a Risk Class 2 packaged drug product include:

- Full extractables profiling of the commercial packaging system (with qualified methods and adjusted AET)
- Toxicological safety risk assessment of extractables as potential leachables to establish and justify target substances for drug product leachables testing
- Full leachables profiling (targeting and screening) of the commercial packaged drug product over-shelf (three lots with appropriately validated (targeted) or qualified methods (screening) and adjusted AET for screening)
- Leachables screening to include extractables (target screening) only.
- Quantitative extractables–leachables correlations for targeted leachables and qualitative extractables–leachables correlations for screening leachables

- Toxicological safety risk assessment of leachables at their highest reported concentrations
- Biological reactivity testing, USP <87> *in vitro* and <88> *in vivo*

Intermediate Risk Class 3 is a step down (less risk and therefore less testing) from Risk Class 2 but a step up from Risk Class 4 (greater risk and more testing). Considering biological reactivity testing, Risk Class 3 is the transition from <87> to <87> and <88> requirements. Furthermore, Risk Class 3 is the point where some degree of leachables testing becomes mandatory. Moreover, with the advent of leachables testing, qualitative extractables–leachables correlations become necessary. Thus, the requirements for a Risk Class 3 packaged drug product include:

- Full extractables profiling of the commercial packaging system (with qualified methods and adjusted AET)
- Toxicological safety risk assessment of extractables as potential leachables to establish and justify target substances for drug product leachables testing
- Full leachables profiling (target extractable only) of the commercial packaged drug product over-shelf (three lots with appropriately validated methods)
- Qualitative extractables-leachables correlations
- Toxicological safety risk assessment of leachables at their highest reported concentrations
- Biological reactivity testing, USP <87> *in vitro* and < 88> *in vivo* (optional)

A summary of the requirements for all five Risk Classes is shown in Table 7.15.

7.7 Case Studies

An understanding of the complexities and idiosyncrasies of extractables and leachables testing of packaging systems and packaged drug products can be gained from the published chemical literature on this subject. Although it is rare, in fact unprecedented, that the entire contents of the chemical characterization sections of a drug product's regulatory submission would be placed in the public domain, it is generally the case that unusual or interesting situations are published in the chemical literature. Such case studies serve the industry in the sense that shared practices and experiences help to establish the ever-elusive concept of "good science" and to prevent the "re-inventing of the wheel" that can occur in the absence of information and experience exchange.

One must approach published literature on chemical characterization with the right frame of mind. Perhaps the most important point to make here is not to judge the object based on only what one can easily see. Perhaps one remembers the parable of the blind men and the elephant. A group of blind men were asked to describe an elephant, based on their ability to touch only one of the elephant's body parts. One blind man, touching the tail, likened the elephant to a rope. Another, touching the elephant's knee, described it as a tree. The other men, touching other parts, described the elephant in several different ways, none of which was correct.

My point here is this. Case study manuscripts about extractables and leachables are published about interesting cases. For example, while a case study where a potentially unsafe extractable is found makes for interesting reading, publications where extraction studies were meticulously planned and implemented, but which found essentially few compounds at relatively low levels, would likely never be published due to their somewhat uninteresting findings. Thus, one needs to understand published case studies in the proper context, as they tend to be the exception to the rule and not the rule. It is much too easy to fall into the trap of concluding that "all extractables studies must produce unsafe extractables" or "every identification requires the wisdom of Solomon to figure out" or that "every drug product must contain reactive leachables that are not linked to extractables," based on the small number of case studies that are published relative to the large number of extractables and leachables assessments that are performed but remain unpublished. In fact, the opposite is closer to the truth. In today's reality, and fully aware of E&L issues and consequences, suppliers develop and test raw materials (e.g. plastic resins) that are compositionally designed to be safe, packaging engineers design and develop packaging systems using safe materials and safe manufacturing processes and then test those systems to establish their safe use profile, and drug product development teams use packaging systems that are likely to be compatible with the drug product and then perform exacting testing to establish that compatibility was achieved. With the proper due diligence, with clear and rationale assessment, and with the proper amount and rigor of testing, it is the rule, rather than the exception, that packaging systems are fully, completely, and undoubtedly suitable for their intended uses.

However, it is certainly true that the best laid plans can go astray, that there are gaps in the underlying science (and its understanding), that complex issues are oversimplified, that warning signs are either absent or missed, and that low risk does not mean no risk. Taken in the proper context, interesting case studies serve the important purpose of reminding E&L practitioners that every E&L study is potentially a research project and that unanticipated outcomes do occur.

It is beyond the scope of this text to comprehensively account for and detail every recent publication related to

packaging and packaged drug products. A recently published review of the state of the art and current analytical practices with respect to extractables and leachables accomplishes this objective to some extent, especially in its Table 4 [44]. Rather, a few case studies of particular note, addressing a variety of dosage forms, will be considered herein.

7.7.1 Injectable Drug Products

Several research studies have recently been published on plastic bags used as containers for parenteral products. Zhuang and associates studied the accumulation of extractables from a multi-layer coextruded polyethylene bag in an exaggerated aqueous drug product solution consisting of 4% benzyl alcohol (2% is the nominal drug product) in 0.1 M sodium acetate, pH 5 [45]. The bag's three-layer structure included an outer layer of low-density polyethylene, a core layer of an ethylene vinyl alcohol copolymer, and an inner (solution contact) layer of ultralow-density polyethylene. In this study, 500-ml bags were filled with 250 ml of exaggerated drug product solution and stored at ambient temperature. The contents of the bags were screened for organic extractables using UPLC/MS (Q-TOF) after 15, 45, and 90 days of storage. Although the article contains information in terms of the author's strategy for revealing and identifying extracted compounds and includes a consideration of the toxicology of the identified extractables (which were ultimately established to be within their relevant toxicological thresholds), an important aspect of the manuscript is the trending data provided for certain measured extractables (see Figure 7.7).

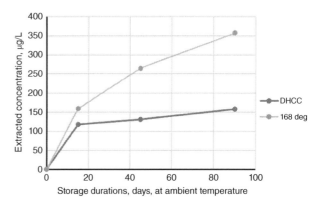

Figure 7.7 Accumulation profile of extractables from a multi-layered polyethylene bag. Although both extractables are approaching asymptotic extraction after 90 days of ambient temperature storage, it is clear that a longer storage time is necessary before the maximum concentration of these extractables is achieved. DHCC = Dimethyl 2-hydroxy-1,3-cyclohexanedicarboxylate. 168 deg = several peaks identified as degradation products of Irgafos 168. Source: Data from reference [45].

Table 7.16 Levels of compounds in extracts of a polypropylene-based infusion bag (extractables) or in a drug product (calcium gluconate glucoheptonate) stored in the infusion bag (leachables).

	Measured concentration, µg/ml			
	Extractables (10 days at 25 °C)			
Compound	100% Acetonitrile	50% Acetonitrile	Water	Leachables, 3 months at 25° or 40 °C
BHT	0.34	< 0.02[a)]	< 0.02[a)]	< 0.2[b)]
Irganox 1330	9.99	1.82	< 0.02[a)]	< 0.2[b)]
Irganox 1076	22.6	< 0.02[a)]	< 0.02[a)]	< 0.2[b)]
Irgafos 168	0.93	< 0.02[a)]	< 0.02[a)]	< 0.2[b)]

a) LoQ.
b) AET.
Source: Data obtained from reference [46].

Additionally, the study illustrated the challenges encountered when reporting extractables at levels lower than the AET. In this study, for example, the AET was established as 360 µg/l, a value well above the method's reported limit of quantitation of 5–15 µg/l (established for three reference compounds). As can be seen in Figure 7.7, two extractables present in the samples near the AET were detected and reported reproducibly. However, three additional extractables were only sporadically reported at various timepoints at levels between 4 and 50 µg/l. Although these levels are above the theoretically established LoQ, the sporadic nature with which these extractables were reported suggests that the method's technical ability to detect and "quantify" a compound is greater than the ability to definitively establish these compounds as true extractables (as opposed to sporadic analytical artifacts). In this study the sporadic nature of these potential extractables is not an issue as their levels are much lower than the AET and therefore they are irrelevant from a safety risk assessment perspective.

Legrand and associates profiled polypropylene- and polyethylene-based plastic bags for extractables and a calcium gluconate glucoheptonate drug product stored in the bags for leachables [46]. Extractions were performed by filling bags with three extraction solvents of varying polarity (water, 50% acetonitrile in water, and 100% acetonitrile) and storing for the filled bags for up to 10 days at 40 °C. The extracts were characterized for extractables by HPLC/UV, targeting potential additives (antioxidants) and noting any non-additive peaks in the chromatograms. Drug product leachables were measured in an aqueous drug product containing calcium gluconate glucoheptanate (calcium supplement, parenteral nutrition), stored in the PP and PE bags for up to three months at either 25 or 40 °C using the same analytical methodology. As one might anticipate and as shown in Table 7.16, the generally non-polar antioxidants were present in measurable quantities only in the most non-polar extraction solvent (100% acetonitrile) and were not present in the aqueous drug product at levels above the calculated AET of 0.2 µg/ml (see Table 7.16). In addition to the known additives, the authors reported several additional organic extractables that were sporadically present in the various extracts at the various test intervals. Moreover, the authors recommended the use of the PE, as opposed to the PP, bag for this drug product, based on fewer extractables at lower levels in the PE versus the PP bags.

Lastly, Sun et al. reported the development of a dispersive liquid–liquid microextraction-based sample preparation for the chromatographic analysis of extracts and applied the method to extracts of a multi-layer coextrusion film infusion bag [47]. Specifically, the infusion bags were filled with one of three extraction solvents (pH 3 acetate buffer, pH 9.2 phosphate buffer, and 0.9% sodium chloride) and aged at 25 °C for nine days, after which the resulting extracts were screened for organic extractables using the developed sample preparation method coupled to UPLC/MS. The resulting accumulation levels of three discovered extractables are listed in Table 7.17.

Table 7.17 Accumulation levels of extractables from a multi-layer coextrusion film infusion bag.

	Concentration in extract, µg/l		
Extractable	Acidic (pH 3) extraction solvent	Basic (pH 9.2) extraction solvent	Salt (0.9% NaCl) extraction solvent
Irganox 235	14	66	24
Irganox 1024	6.8	36	11
Tinuvin 329	ND	5.0	11

Source: Data obtained from reference [47].

Polyethylene terephthalate (PET) is used in pharmaceutical primary packaging such as bottles and several types of medical devices. Carire-Mauriser and associates studied the effect of electron beam sterilization on PET, delineating the effect of irradiation on the formation of non-volatile extractables [48]. In this study, PET sheets were irradiated at doses of 25 kGy and higher, extracted with methanol via refluxing for two hours and the resulting extracts were profiled for extractables using size exclusion chromatography and LC, both with UV detection. These authors report that numerous PET oligomers, from $n = 1$ to 8, were extracted from both irradiated and non-irradiated PET, with the amounts of the oligomers being greater in the irradiated samples. The trimer was the oligomer that was extractable in the greatest quantities. Additional extractables reported by these authors included benzoic acid and terephthalic acid, whose extracted concentrations increased with increasing irradiation dose.

The accumulation of leachables in solid dosage forms has been discussed previously, specifically considering the general misconception that solid dosage forms are low risk with respect to leachables as "solids can't leach." However, Zdravkovic reports that significant accumulation of leachables can occur in lyophilized drug powders stored in stoppered glass vials [10]. To support this observation, the author measured the levels of three stopper-related extractables, BHT and butyl rubber oligomers $C_{13}H_{23}Br$ and $C_{21}H_{29}Br$, in aged drug products including both liquid formulations and lyophilized powders. The author reports significant levels of these substances as leachables in both the liquid and solid dosage forms and concludes that "substances originating from rubber stoppers that comprise the primary packaging system of parenteral drug products have a higher affinity for lyophilized drug formulations as opposed to those stored as a liquid" and that "lyophilized drug products will contain more leachable compounds in higher concentrations compared to liquid drug products when all other variables are equal." Although reporting of leachables accumulating in solid dosage forms is a noteworthy contribution, the author's conclusions are essentially an "apples to oranges" comparison of two completely different phenomenon. What the author has essentially discovered is that the aqueous solubility of these leachables is lower than the solid's ability to sorb the compounds; thus, levels of dissolved leachables are lower than levels of sorbed leachables.

It is sometimes possible to find relevant extractables and leachables data for packaging in indirect ways. For example, Kurmi et al. have established the extractables profiles of disposable plastic syringes used for the manipulation and transfer of liquid samples in analytical laboratories [49]. This study is relevant to pharmaceutical packaging considering that pre-filled syringes are used as packaging and delivery systems. These authors extracted numerous commercially available disposable plastic syringes with various solvents including water, organic solvents (e.g. methanol and acetonitrile), and solvent/water mixtures at ambient temperature over relatively short periods of times (minutes) and characterized the resulting extracts via LC/UV, LC/HRMS, and NMR. Identified extractables included: 3-(2-hydroxyethyl)-2-benzothiazolinone (CAS RN 21344-50-3), 3-(2-(2-hydroxyethoxy)ethyl)-2-benzothiazolinone (no CAS RN), 2(3H)-Benzothiazolone (CAS RN 934-34-9), N,N-dimethyl-2-oxobenzo[d]thiazole-3(2H)-carbothioamide (no CAS RN), 2-(2-hydroxyethylmercapto) benzothiazole (CAS RN 4665-63-8), 2-(2-(2-hydroxyethoxy) ethylmercapto)benzothiazole (no CAS RN), benzo[d]-thiazol-2-yldimethylcarbamodithioate (CAS RN 3432-25-5), ethyl 4-ethoxybenzoate (CAS RN 23676-09-7).

Several publications dealt with elements extracted from pharmaceutical packaging materials and/or components. Hladik and associates examined the levels of compositional elements (e.g. Na, Si, B, and Al) from glass vials of varying glass types and with various fill volumes [50]. The extraction was performed by filling the vials with either water or 15% potassium chloride and aging the filled vials at 40 °C for 24 weeks. The resulting extracts were tested for the targeted elements by ICP/MS. Based on their results, the authors concluded that leaching of the glass elements was generally exacerbated by low fill volumes, but could be significantly reduced by rigorous process control over the relevant leaching-influencing factors of vial production. Similarly, Chillon and Zuccato reported the reduction of extractable levels of compositional elements Al, B, Si, and Na from borosilicate glass vials that had been surface-coated via cross-linked silicone oil [51].

Considering the other component of a vial-based packaging system, Paskiet et al. reported extractable metals results obtained for elastomers used as closures, such as vial stoppers and syringe plungers [52]. Specifically, commercially available chlorobutyl and bromobutyl stoppers were aggressively extracted with a mixed strong acid solvent (2% nitric acid, 0.5% hydrochloric acid) at 120 °C for 30 minutes. The resulting extracts were screened for numerous elements by ICP/MS using a qualified methodology. Elements that were extracted in reportable quantities included Al, B, Ca, Fe, Mg, Se, V. and Si (which was extracted at levels higher than the method's dynamic range).

Lastly, Mattiazzi and associates reported on the extraction/leaching of metals (B, Cd, Pb, Sn, and Zn) from polyvinyl chloride, ethylene vinyl acetate, and polypropylene bags into infusion solutions [53]. Portions of the actual bags were profiled for their content of the targeted metals

Table 7.18 Zn in plastic bags for infusion drug products; composition, extractables and leachables.

Sample type	Solution	Source material		
		PP	PVC	EVA
Bag material, mg/kg	–	0.8	1.4–5.71	1.1
Extracted from Bags[a], µg/l	0.9% NaCl	ND[b]	12	ND
	10% Glucose	ND	64	ND
	5% Tween 80	ND	74	ND
Present in commercial product, µg/l	5% Glucose	NT[c]	83[d]	NT
	10% Glucose	NT	50[d]	NT
	0.9% NaCl	NT	16[a]	NT

a) Filled bags were heated at 100 °C for 3 hours to simulate autoclaving and stored at room temperature for 8 months.
b) ND = not detected.
c) NT = product samples in PP or EVA bags were not tested.
d) Mean of 5 product units from 2 different manufacturing lots.
Source: Data obtained from reference [53].

by ashing at elevated temperature followed by dissolution of the ash with a mixture of nitric acid and hydrogen peroxide. Bags were also extracted by filling them with aqueous solutions of 10% glucose, 0.9% sodium chloride, and 5% Tween 80, autoclaving them and then storing the autoclaved bags for up to eight months at room temperature. Lastly, commercially available infusion solutions (e.g. 5% glucose, 10% glucose, and 0.9% saline) were tested directly for the targeted elements. The test method employed was graphite furnace atomic absorption spectroscopy (GFAAS), validated for this purpose.

Of the five targeted metals, all were present in all of the bag materials in measurable quantities, although the levels were generally higher in the PVC bags. However, zinc was the only element extracted in measurable quantities and then only from the PVC bags. The data reported for zinc is presented in Table 7.18. Based on the experimental data, these authors concluded that "although (the metals were) present in the assessed polymeric materials, the levels of leaching were low and unlikely to significantly contribute to the elemental impurity profile of infusion therapy solutions."

7.7.2 Other Dosage Forms

Two recent studies address specific aspects of packaging-related leachables in ophthalmic drug solutions. Singh et al. reported the presence of two well-known degradation products of Irganox 1010, a commonly antioxidant present in polypropylene-based packaging, in an unspecified aqueous ophthalmic drug product [54]. This manuscript speaks to the process of establishing the identities of these leachables, 7,9-di-tert-butyl-1-oxaspiro [4.5] deca-6,9-diene-2,8-dione

and 3-(3,5-bis-tert-butyl)-1-hydroxy-4-oxocyclohexa-2,5-di enylpropanoic acid, but provides no data on their levels in the drug product. A research team from the same organization also report finding a drug product impurity that is produced via the reaction of a leachable and the drug substance present in an ophthalmic drug product [55]. Specifically, ethylene glycol monoformate, leached from the low-density polyethylene bottle used to package a moxifloxacin-containing ophthalmic formulation, reacted with this drug substance to produce a drug product impurity, tentatively identified as 1-cyclopropyl-6-fluoro-7-(1-(2-formyloxy)ethyl)octahydro-6H-pyrrolo[3,4-b]pyridin-6-yl)-8-methoxy-4-oxo-1,4-dihy-droquinoline-3-carboxylic acid ($C_{24}H_{28}FN_3O_6$, MW = 473.19621) during the drug product's shelf-life storage. This impurity was observed at a level of approximately 0.3% in the Moxifloxacin stability samples.

Considering inhaled medicines, Otte published a general review of extractables and leachables issues associated with this dosage form, including some experimental data [56]. Extractables associated with organic solvent extracts of PE materials used in nasal spray devices and dry powder inhalers included Irgafos 168, BHT, Irganox 1010 (and its associated degradation product 7,9-di-tert-butyl-1-oxaspiro [4.5] deca-6,9-diene-2,8-dione), Irganox 245, and a homologous series of PE-oligomers. This author also reports the presence of organic extractables in water extracts of a polyether material, observing different extraction rates as a function of the size (molecular weight) of the extracted compound (see Figure 7.8). Lastly, this author reported the presence of extracted mercaptobenzothiazole in an organic extract of the type of rubber gaskets that is used with pressurized metered dose inhalers.

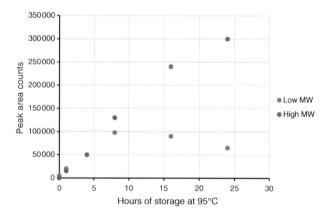

Figure 7.8 Extraction kinetics of organic extractables from a polyether material, water extract. The rate of extraction is clearly influenced by the size (molecular weight) of the extractable, with smaller extractables accumulating at a faster rate than larger extractables. Source: Data obtained from reference [56].

References

1 USP ⟨659⟩ Packaging and Storage Requirements. USP 43/NF 38, 2S. Official 1 April 2021. Current DocID: GUID-48C0134E-8117-4B3B-8DC4-EA70C93EEE8F_6_en-US.

2 Ulsaker, G.A. and Teien, G. (1992). Identification of caprolactam as a potential contaminant in parenteral solutions stored in overwrapped PVC bags. *J. Pharm. Biomed. Anal.* 10 (1): 77–80.

3 Akapo, S.O. and Mccrea, C.M. (2008). SPME-GC determination of potential volatile organic leachables in aqueous-based pharmaceutical formulations packaged in overwrapped LDPE vials. *J. Pharm. Biomed. Anal.* 47 (3): 526–534.

4 Fang, X., Cherico, N., Barbacci, D. et al. (2006). Leachable study on solid dosage form. *Am. Pharm Rev.* 9 (6): 58–63.

5 Jenke, D., Egert, T., Hendricker, A. et al. (2017). Simulated leaching (migration) study for a model container-closure system applicable to parenteral and ophthalmic drug products (PODPs). *PDA J. Pharm. Sci. Technol.* 71 (2): 68–87.

6 Pan, C., Harmon, F., Toscano, K. et al. (2008). Strategy for identification of leachables in packaged pharmaceutical liquid formulations. *J. Pharm. Biomed. Anal.* 46: 520–527.

7 Nerin, C., Canellas, E., Aznar, M., and Silcock, P. (2009). Analytical methods for the screening of potential volatile migrants from acrylic-base adhesives used in food-contact materials. *Food Addit. Contam.* 26 (12): 1592–1601.

8 Tiller, P.R., El Fallah, Z., Wilson, V. et al. (1997). Qualitative assessment of leachables using data-dependent liquid chromatography/mass spectrometry and liquid chromatography/tandem mass spectrometry. *Rapid Commun. Mass Spectrom.* 11: 1570–1574.

9 Zdravkovic, S.A. (2017). Comparison of a lyophilized drug product to other solid and liquid media for the extraction of elastomeric oligomers from a butyl rubber stopper. *PDA J. Pharm. Sci. Technol.* 71 (6): 488–501.

10 Zdravkovic, S.A. (2020). Comparison of the levels of rubber stopper-related organic leachables in commercially available vialed liquid and lyophilized drug products. *Pharm. Res.* 37 (4): https://doi.org/10.1007/s11095-020-02805-z.

11 Singh, G., Gollapalli, R., Blinder, A. et al. (2019). A case study demonstrating the migration of diethyl phthalate from an ancillary component to the drug product. *J. Pharm. Biomed. Anal.* 164: 574–580.

12 https://www.arthritis.webmd.com/news/20091229/tylenol-recall-expands. Accessed 22 November 2020.

13 https://www.prnewswire.com/news-releases/ortho-mcneil-janssen-pharmaceuticals-inc-voluntarily-recalls-one-lot-of-risperdal-tablets-and-one-lot-of-risperidone-tablets-124080244.html. Accessed 22 November 2020.

14 Use of International Standard ISO 10993:1, Biological evaluation of medical devices – Part 1: Evaluation and testing within a risk management process. Guidance for Industry and Food and Drug Administration Staff, US Department of Health and Human Services: Food and Drug Administration; Center for Devices and Radiological Health, Center for Biologics Evaluation and Research. Document issued on 4 September 2020.

15 ISO 10993:18(2020), Biological evaluation of medical devices – Part 18: Chemical characterization of medical device materials within a risk management process.

16 Guidance for Industry. Container Closure Systems for Packaging Human Drugs and Biologics: Chemistry, Manufacturing, and Controls Documentation. US Department of Health and Human Services: Food and Drug Administration; Center for Drug Evaluation and Research (CDER), Center for Biologics Evaluation and Research (CBER). May 1999.

17 Guideline on Plastic Immediate Packaging Materials. European Medicines Agency Inspections; Committee for Medicinal Products for Human Use (CHMP), Committee for Medicinal Products for Veterinary Use (CVMP), CPMP/QWP/4359/03, EMEA/CVMP/205/04. London, 19 May 2005.

18 USP <661>. Plastic Packaging Systems and Their Materials of Construction. USP 43/NF 38, p. 6887. Effective 1 November 2020.

19 USP <661.1>. Plastic Materials of Construction. USP 43/NF 38, p. 6893. Effective 1 December 2025.

20 USP <661.2>. Plastic Packaging Systems for Pharmaceutical Use. USP 43/NF 38, p. 6912. Effective 1 December 2025.

21 USP <660>. Containers – Glass. USP 43/NF 38, p. 6881. Effective 1 May 2015.

22 USP <381>. Elastomeric Components in Injectable Pharmaceutical Product Packaging/Delivery Systems. USP 44/NF 38, p. 6669. Effective 1 December 2020.

23 USP <1661>. Evaluation of Plastic Packaging Systems and Their Materials of Construction with Respect to Their User Safety Impact. USP 43/NF 38, p. 8434. Effective 1 November 2020.

24 USP <1663> Assessment of Extractables Associated with Pharmaceutical Packaging/Delivery Systems. USP 43/NF 38, pp. 8422. Effective 1 Dec 2020.

25 USP <1664> Assessment of Leachables Associated with Pharmaceutical Packaging/Delivery Systems. USP 43/NF 38, pp. 8455. Effective 1 Dec 2020.

26 Standard Guide for Accelerated Aging of Sterile Barrier Systems for Medical Devices. ASTM International:West Conshohocken, PA, 2016.

27 USP <232> Elemental Impurities – Limits. USP43/NF38, p. 6641. Effective 1 December 2020.

28 USP <233> Elemental Impurities – Procedures. USP43/NF38, p. 6645. Effective 1 May 2018.

29 USP <1664.1> Orally Inhaled and Nasal Drug Products. USP43/NF38, p. 8467. Effective 1 August 2015.

30 USP <660> Containers – Glass. USP43/NF38, p. 6881. Effective 1 May 2015.

31 USP <1381> Assessment of Elastomeric Components in Injectable Pharmaceutical Product Packaging/Delivery Systems. USP43/NF38,2S on-line. Current DocID: GUID-59EA5D30-B7DE-436D-A516-2569DABF2B91_2_en-US. doi: https://doi.org/10.31003/USPNF_M11357_02_01. Effective 1 December 2020.

32 JP XVII. The Japanese Pharmacopoeia, 17th edition. The Ministry of Health, Labour and Welfare. Official from April 1, 2016; pp. 166–174.

33 Guidance for Industry (April, 2018). *Metered Dose Inhaler (MDI) and Dry Powder Inhaler (DPI) Drug Products – Quality Considerations.* Rockville, MD: US Department of Health and Human Services; Food and Drug Administration, Center for Drug Evaluation and research (CDER).

34 Guidance for Industry (July, 2002). *Nasal Spray and Inhalation Solution, Suspension, and Spray Drug Products – Chemistry, Manufacturing and Controls Documentation.* Rockville, MD: US Department of Health and Human Services, Food and Drug Administration.

35 Product Quality Research Institute (PQRI), Leachables and Extractables Working Group (2006). *Safety Thresholds and Best Practices for Extractables and Leachables in Orally Inhaled and Nasal Drug Products.* Arlington, VA: Product Quality Research Institute.

36 Product Quality Research Institute (PQRI). Safety Thresholds and Best Demonstrated Practices for Extractables and Leachables in Parenteral Drug Products (Intravenous, Subcutaneous, and Intramuscular). Published by the Parenteral Drug Association. ISBN: 978-1-945584-30-5. 28 October 2021.

37 Paskiet, D., Jenke, D., Ball, D. et al. (2013). The Product Quality Research Institute (PQRI) leachables and extractables working group initiatives for Parenteral and Ophthalmic Drug Product (PODP). *PDA J. Pharm. Sci. Technol.* 67 (5): 430–447.

38 Houston, C., Rodrigues, A.D., Smith, B.B., Wang, T., Richardson, M. Principles for management of extractables and leachables in ophthalmic drug products. *PDA J. Pharm. Sci. Technol.* February 2022, pdajpst.2022.012744; DOI: https://doi.org/10.5731/pdajpst.2022.012744.

39 Ng, L. (2011). FDA Office of New Drug Quality Assessment. Current Regulatory Recommendations for Leachables in Ophthalmic Products. Presented at *PQRI Workshop on Thresholds and Best Practices for Parenteral and Ophthalmic Drug Products*, Bethesda, MD, 22–23.

40 Jenke, D. (2021) A Safety risk-based extractables and/or leachables qualification strategy for packaged drug products. *PDA J. of Pharm. Sci. Technol.* 75(6): 506–535.

41 BPOG Best Practices Guide for Evaluating Leachables Risk in Biopharmaceutical Single-Use Systems. BioPhorum Operations Group. 2018. https://www.biophorum.com/wp-content/uploads/bp_downloads/Leachables-report-12-April-2017-v3-1.pdf. Accessed 5 December 2021.

42 Jenke, D. (2015). Development and justification of a risk evaluation matrix to guide chemical testing necessary to select and qualify plastic components used in production systems for pharmaceutical products. *PDA J. Pharm. Sci. Technol.* 69 (6): 677–712.

43 USP ⟨1665⟩. Characterization and Qualification of Plastic Components and Systems Used to Manufacture Pharmaceutical Drug Products and Biopharmaceutical Drug Substances and Products. USPNF 2022 ISSUE 1 - online Current DocID: GUID-B2665FAF-027F-46AA-8093-2292E0666F03_2_en-US. https://doi.org/10.31003/USPNF_M11136_02_01. To be official 1 May 2022.

44 Cuadros-Rodrígueza, L., Lazúen-Murosa, M., Ruiz-Samblása, C., and Navas-Iglesiasa, N. (2020). Leachables from plastic materials in contact with drugs. State of the art and review of current analytical approaches. *Int. J. Pharm.* 583: 119332. https://doi.org/10.1016/j.ijpharm.2020.119332.

45 Zhuang, J., Xing, X., Wang, D. et al. (2018). Toxicity assessment of the extractables form multi-layer coextrusion polyethylene bags exposed to pH 5 solution containing 4% benzyl alcohol and 0.1 M sodium acetate. *Regul. Toxicol. Pharmacol.* 94: 47–56.

46 Legrand, P., Desdion, A., Boccadifuoco, G. et al. (2018). Development of an HPLC/UV method for the evaluation of extractables and leachables in plastic: Application to a plastic-packaged calcium gluconate glucoheptonate solution. *J. Pharm. Biomed. Anal.* 155: 298–305.

47 Sun, X., Xing, X., and Du, Z. (2020). Application of dispersive liquid–liquid microextraction based on solidification of floating organic drop for the determination of extractables from pharmaceutical packaging materials. *Talanta* 209: https://doi.org/10.1016/j.talanta.2019.120540.

48 Caire-Maurisier, F., Aymes-Chodur, C., Jandard, V. et al. (2019). Effects of electron beam sterilization on polyethylene terephthalate: Physico-chemical modifications and formation of non-volatile organic extractables. *Ann. Pharm. Fr.* 77 (4): 276–285.

49 Kurmia, M., Prasada, S., Pandurangaa, N.S. et al. Identification of extractables by liquid chromatography-highresolution mass spectrometry: A case study to understand the extraction profile of different disposable syringes. *J. Pharm. Biomed. Anal.* 191: https://doi.org/10.1016/j.jpba.2020.113602.

50 Hladik, B., Buscke, F., Frost, R., and Rothhaar, U. (2019). Comparative leachable study for glass vials to demonstrate the impact of low fill volume. *PDA J. Pharm. Sci. Technol.* 73 (4): 345–355.

51 Chillon, A. and Zuccato, D. (2019). Introducing the Alba® primary packaging platform. Part 2: Inorganic extractables evaluation. *PDA J. Pharm. Sci. Technol.* 73 (3): 235–246.

52 Paskiet, D., Kraft, C., Tullo, E. et al. (2018). Assessment of extractable elements form elastomers. *PDA J. Pharm. Sci. Technol.* 69 (1): 49–58.

53 Mattiazzi, P., Bohrer, D., Viana, C. et al. (2019). Extraction/leaching of metal containing additives from polyvinyl chloride, ethyl vinyl acetate and polypropylene bags and infusion sets into infusion solutions. *PDA J. Pharm. Sci. Technol.* 73 (1): 60–69.

54 Singh, G., Gollapalli, R., Blinder, A., and Patel, M. (2018). Identification of leachable impurities in an ophthalmic drug product originating from a polymer additive Irganox 1010 using mass spectrometry. *J. Pharm. Biomed. Anal.* 152: 197–203.

55 Gollapalli, R., Singh, G., Blinder, A. et al. (2019). Identification of an adduct impurity of an active pharmaceutical ingredient and a leachable in an ophthalmic drug product using LC-QTOF. *J. Pharm. Sci.* 108 (10): 3187–3193.

56 Otte, T. (2017). Extractables and leachables testing for inhaled medicines. *Pharm. Technol. Int.* 41 (6): 36, 38, 40, 42, 44, 45.

8

Applications – Pharmaceutical Drug Products – Manufacturing Components and Systems

8.1 Introduction

At one point in time, pharmaceutical manufacturing was a relatively straightforward process. Oversimplifying the complexity of manufacturing processes and equipment somewhat, manufacturing generally proceeded as follows:

1. Drug product ingredients and diluents were mixed in large mixing tanks, as stipulated in Manufacturing Specifications,
2. Samples of the formulated drug product were taken and tested to confirm compliance to compositional requirements,
3. Conforming product was released from the mixing tank and transferred, with little further processing (other than perhaps clarifying filtration), to filling machines,
4. Filling machines dispensed the drug product into its packaging (ending the manufacturing process).

Although there were existing regulatory expectations that "Equipment shall be constructed so that surfaces that contact components, in-process materials or drug products shall not be reactive, additive, or absorptive so as to alter the safety, identity, strength, quality, or purity of the drug product beyond the official or other established requirements" [1], it was generally understood that manufacturing equipment, largely consisting of stainless-steel components, was inert, particularly from the perspective of potential organic contaminants. Thus, other than being a potential source of elemental impurities, manufacturing equipment was generally discounted as a source of foreign impurities in drug products.

The advent and rise of biopharmaceuticals changed all that. Economically, the significant capital cost of constructing "traditional" manufacturing suites became prohibitive. Moreover, such systems tend to be relatively inflexible in terms of the unit operations that can be performed. Furthermore, operating and maintenance costs, including process steps such as cleaning validation, were substantial and became more difficult to justify. Hence, there was a rise of so-called flexible manufacturing based on single-use

(disposable) components, consisting largely of plastic and elastomeric materials.

A predictable outcome of the adoption of single-use technology (SUT) was renewed interest in manufacturing-related foreign impurities in packaged drug products, specifically in regard to their potential adverse effect on patient health. Additional concerns focused on the effect that extracted substances might have on the efficiency and effectiveness of downstream manufacturing processes. In fact, "extractables and leachables" was viewed as a major issue of concern for organizations who were exploring the use of disposable single-use systems [2–5]. In a certain way this view is understandable, as a visual inspection of a manufacturing suite composed of single-use components generally elicits the comment "that sure is a lot of plastic." This visual perception, coupled with a lack of understanding of the confounding scientific and practical aspects of the "extractables and leachables issue," led to a proliferation of opinions on how real the threat of extractables and leachables was and to approaches to address this "extractables and leachables challenge." Although much of the early literature addressing extractables and leachables and plastic manufacturing components is interesting reading from a historical perspective, it is largely irrelevant in terms of both the current perception of the seriousness of issues posed by extractables and leachables and current practices, recommended or mandated, for addressing extractables and leachables and qualifying manufacturing equipment with respect to extractables and leachables.

To distinguish manufacturing system-related leachables from leachables from other sources (for example, packaging), the term process equipment-related leachable (PERL) has been established. A PERL is a substance that has leached into the process stream and is addressed by testing the process stream. Thus, a PERL is differentiated from an extractable derived from a manufacturing component in the same way that a packaging-related extractable is differentiated from a packaging-related leachable; that is, a PERL is measured in the actual process stream under actual manufacturing conditions while a manufacturing-related

Extractables and Leachables: Characterization of Drug Products, Packaging, Manufacturing and Delivery Systems, and Medical Devices, First Edition. Dennis Jenke.
© 2022 John Wiley & Sons, Inc. Published 2022 by John Wiley & Sons, Inc.

extractable is measured in an extract generated under laboratory conditions. Should a PERL persist in the process stream through to final fill, it would become a leachable in the finished drug product.

Experimentally, there is a straightforward way of differentiating a PERL from a packaging-related leachable. Because contact between the drug product and the manufacturing system ends at the point of a final fill, a PERL cannot increase in concentration over a drug product's shelf-life. Thus, a leachable whose concentration increases over shelf-life cannot be a PERL. Although it might be tempting to suggest that a leachable whose concentration stays the same or decreases over its shelf-life is a PERL, such a conclusion is false as certain packaging-related leachables can exhibit these same concentration trends over time.

Considering the case of extractables from manufacturing equipment accumulating in the drug product as foreign impurities (leachables) more closely, in order for manufacturing-related extractables to accumulate in the packaged commercial drug product as a manufacturing-related leachable, two events must occur. First, a manufacturing component must be leached by the manufacturing process stream, resulting in the release of an extractable to the process stream. Second, the released extractable must persist through the entire manufacturing process so that it can accumulate in the drug product. Moreover, even if a manufacturing-related extractable is incorporated in the drug product as a leachable, this circumstance is largely irrelevant unless the leachable accumulates to a level sufficiently high that it adversely affects either the quality or safety of the drug product. Thus, for manufacturing system-related leachables to be a significant issue with respect to packaged drug product quality and safety:

1. Substances must be extracted from the manufacturing components by the process stream,
2. The extracted substances must persist in the process stream all the way to the end of the manufacturing process (final filling of the packaging system with the drug product),
3. The persistent extracted substances must accumulate in the packaged drug product in high enough quantities to have an adverse effect (on, for example, safety).

Although none of these processes are impossible on their own, it is clear that taken together they represent an unlikely circumstance. Consider, for example, the extraction of the manufacturing component by the process stream. In many cases, extraction of significant quantities of substances from the manufacturing component by the process stream is unlikely. For example, many single-use manufacturing components are intentionally constructed from materials that contain only essential additives at an optimized (minimal) level, thereby limiting the pool of extractables to begin with. This is the case as vendors of single-use components understand the potential significance of extractables and have optimized component composition to minimize potential extractables. Many disposable manufacturing components are rinsed or conditioned prior to use and thus are, in essence, pre-extracted prior to use, further reducing extractables levels. The contact conditions between a process stream and a manufacturing component are generally less "harsh" (meaning lower temperature and particularly shorter duration) than the contact conditions between a drug product and its packaging. Many process streams are not as aggressive, in terms of their "leaching power," as are drug products, based on their chemical composition, although some process streams are compositionally pre-disposed to be highly aggressive (e.g. extremes in pH, high content of organic solvents).

It is not the intent of the previous text to suggest that extractable issues are unknown in single-use applications. Rather, the previous text establishes that (i) manufacturing processes are such that the driving force for extraction is generally low and (ii) that the vendor community has taken aggressive actions to minimize the risk of extractables by optimizing the composition of single-use components.

Assuming that a substance is extracted from a manufacturing component by a process stream, that extractable must remain in the process stream through the final manufacturing step for it to become a leachable in the final drug product. When one considers the number of steps in a typical biopharmaceutical's manufacturing process, it is clear that an extractable released into a process stream early (upstream) in the process will encounter a number of plastic components as it moves downstream through the manufacturing process, raising the possibility that the extractable could be removed from the process stream via sorption onto (or into) the plastic components it encounters. Furthermore, the purpose of downstream processing is to purify, enrich, and polish a target substance by removing cells and cell debris from the culture broth, and removing various host cell proteins, potential virus or bacterial contamination, endotoxins, growth promoters, and other chemicals from the process stream. Given this purpose, it is interesting to consider the possibility that these processing steps could also "purify" the process stream by removing extractables. Although mechanisms for the clearance of extractables from the process stream were envisioned as being possible from a theoretical perspective, clearance was not accepted as a mitigating factor for the accumulation of manufacturing-related leachables in the final drug product given the absence of experimental data confirming the theory. As numerous research groups have established that extractables can be almost completely scavenged from the process stream during ultrafiltration/diafiltration [6], clarifying and sterilizing filtration [7], finish/fill operations [8], centrifugation, or depth filtration and virus filtration [9], the scavenger effect has been experimentally validated and downstream processing is recognized as a legitimate factor for reducing the risk of manufacturing-related leachables being present in the finished drug product.

A third factor to consider is essentially a dilution effect. For example, consider the case of a bioprocessing bag that is used to store a drug product intermediate (perhaps the drug substance) between processing steps. As the intermediate is stored in the bag, extractables may accumulate in the intermediate. Once storage has ended, the extractables-laden intermediate is further processed. If the extractables are not cleared, they will be present in the finished drug product batch before it is portioned out into individual doses in individual packaging units. Since the finished drug product is homogeneous, the extractables present in the finished drug product will be evenly portioned into each individual product unit. If the clinical use of the drug product does not require that all of the individual product units be used in a day, then the patient's exposure to the extractables (as leachables) will be less than the total amount of extractables in the finished drug product batch. In a sense, then, the extractables present in the finished drug product batch will be diluted with respect to a patient's daily clinical exposure. For example, consider the case where a single batch of finished drug product produces 1000 individual packaged doses of the drug product, that the patient exposure is to one dose of the drug product per day, and that the drug product is used in an acute therapy. The AET for such an acutely dosed drug product could be assigned a value of 120 μg/day based on ICH M7. Because this AET is "distributed" among all 1000 doses produced in a batch, the AET for the batch becomes 120 μg/day x 1 day/1 dose x 1000 doses/batch = 120 000 μg/batch = 120 mg/batch. Since the entire batch was stored in a single bioprocess container, the AET becomes 120 mg/bag. If the bag weighed 100 g, then the AET becomes 1.2 mg/g = 0.0012 g/g = 0.12 weight percent, which is the approximate level at which additives are present in the materials typically used in bioprocess bags.

In this example then, essentially the entire amount of a typical additive would have to be extracted from the bag by the stored intermediate, and not be cleared by downstream processing, for the levels of the extractable in the drug product to exceed the AET. Since it is by intent that the AET (and thus the TTC upon which it was based) is a conservative threshold, it is likely the case that leachables present in the drug product at the AET would be toxicologically assessed as likely to be safe (that is, most PDEs for leachables would be higher than the ICH M7 TTC). Therefore, even if the bag's ingredient was completely extracted into the process intermediate and was not cleared by downstream processing steps, reflecting an absolute worst-case, the most probable outcome of the toxicological risk assessment of the leachable would be that the leachable is deemed to be safe.

A fourth factor to consider does not involve the actual extraction and accumulation phenomenon per say, but rather addresses how extractables data is often used to establish patient exposure to manufacturing-related leachables. For example, consider a manufacturing process that includes four individual clarifying filtration steps interspersed throughout the process. Consider further that the clarifying filter has effectively been characterized for extractables. One method of calculating patient exposure to filter-related leachables is to take the filter's extractables profile and multiply the extracted quantity by a factor of four to account for all four filters, essentially using an additive model to establish patient exposure. While it may be true that this is the mathematical worst case, it may not be a scientific worst case for the simple reason of solubility. That is to say, once a process stream has been saturated with an extractable (that is, asymptotic extraction has been accomplished) by contact with one filter, then the concentration of the extractable in the process stream cannot be increased by contact with additional filters. In the case where saturation is achieved after the process steam has contacted two filters, then the worst-case patient exposure to leachables is the sum of the extractables from two filters and not four filters. Thus, additive models used to assess the worst-case patient exposure to process-related leachables most likely overestimate patient exposure to a significant extent.

The cumulative result of these four factors is a significant reduction in patient safety-related concerns associated with manufacturing-derived leachables. Whereas extractables and leachables were once considered to be critical factors retarding the adoption of single-use systems, experience, introspection, and experimentation has established that the chances of PERLs accumulating in finished drug products are negligibly low, especially for PERLs derived from downstream manufacturing components.

Even if the patient safety risk related to PERLs is small, a second adverse response to these substances must be considered, specifically, the potential effect of PERLs on the effectiveness and efficiency of the manufacturing process. Although most pharmaceutical and biopharmaceutical process steps are unlikely to be adversely affected by PERLS, the most essential steps of biopharmaceutical manufacturing, including early cell isolation and cultivation, cell banking, and culture expansion of the cells, are at risk from PERLs as the optimal culture environment within a bioreactor is critical to cell performance and can be compromised by leachable compounds [10]. For example, numerous studies have established that the compound bis-(2,4-di-tert-butylphenyl) phosphate, derived from polyolefin-based bioprocessing bags that contain the antioxidant Irgafos 168, is capable of inhibiting cell growth in a variety of CHO cell lines at very low, sub-ppm concentrations [11–14]. Another research team has established that 3,5-dinitro-bisphenol A, leached from single-use Erlenmeyer polycarbonate shaker flasks used for routine cell culture, inhibits the growth of CHO-S cells at the G_1/G_0 phase [15]. Although these are isolated and infrequent situations, the fact that they were unexpected but consequential establishes the fact that vigilance is necessary to ensure that similar outcomes are not repeated in the future.

The importance of assessing manufacturing systems for PERLs has been established; the current state of affairs with respect to available guidance and recommendations for the testing and qualification of manufacturing systems from the perspective of PERLs is discussed as follows.

8.2 Industry Perspectives

8.2.1 Vendor Perspective, Bio-Process Systems Alliance (BPSA)

The Bio-Process Systems Alliance (BPSA) is an industry association whose mission is "to facilitate, globally, the development and manufacturing of biopharmaceuticals and cell therapies through the implementation of robust, safe, precise and sustainable polymeric Single-Use Technologies." BPSA was formed in 2005 as an industry-led international association dedicated to encouraging and accelerating the adoption of single-use manufacturing technologies used in the production of biopharmaceuticals and vaccines. Corporate members include plastic equipment suppliers, service providers, and users in the biopharmaceutical Industry.

Early in its inception, the BPSA formed an Extractables and Leachables Subcommittee to develop a guide to extractables testing of single-use process components and systems for suppliers and users whose purpose was to produce appropriate extractables data in support of equipment selection, process validation, and regulatory filings [16]. To this end, the Subcommittee published two manuscripts (in 2007 and 2008) containing their recommendations for extractables and leachables testing [17, 18]. These manuscripts provided definitions for extractables

and leachables in the context of manufacturing systems, noted the lack of definitive regulatory guidance on the topic, established the concept of product contact with manufacturing components, advocated for a risk-based approach to extractables testing, delineated several key contact factors to establish risk, discussed appropriate analytical technologies, answered frequently asked questions and communicated the expectation that "because leachables testing uses less extreme conditions than does extractables testing does, leachables detected are usually fewer in number and lower in concentration than extractables would be" [18].

These initial publications were followed (in 2010) by a more comprehensive and detailed document entitled "Recommendations for Testing and Evaluation of Extractables from Single-Use Process Equipment" [16]. In this document, the BPSA recognized a gap in application of the terms extractables and leachables when applied to manufacturing equipment. Recognizing that extractables involved testing manufacturing items in the laboratory and that leachables involved testing the finished product for impurities, they asked the question "what does one call an impurity that is identified by testing an actual process stream?" Clearly, such a substance could not properly be called either an extractable (because it is actual process testing and not laboratory testing) or a leachable (because it is not surfaced by testing of the finished product). The BPSA called substances that were present in process solutions equipment migrants because they had been extracted from manufacturing. Although this term never gained much traction, migrants was the precursor of the USP <665> term PERL.

Perhaps more importantly than issues of terminology, the BPSA introduced its Extractables and Leachables Evaluation Decision Tree, which essentially navigated the process of generating and evaluating extractables and leachables data by addressing five essential questions:

Question 8.1. Does the item have product contact? The purpose of this question is obvious. If there are components in a manufacturing system that serve a useful purpose but that do not directly contact the process stream, then logically they cannot contribute PERLs and therefore they do not require further assessment. In this case, the assessment is complete and the circumstance of no contact is documented.

If, however, the item is in direct contact with the process stream, one proceeds to perform a Risk Evaluation, specifically considering the possibility that PERLs from the component could accumulate in the final drug product as leachables. Such an Evaluation considers the following factors:

1. Location of item in the manufacturing process
2. Nature of the product

3. Compatibility of the material
4. Contact temperature and time
5. Contact surface area
6. Pretreatment steps.

The Risk Evaluation prepares the assessor to consider Question 8.2.

Question 8.2. Is an Extractables Evaluation warranted? The nature of the question suggests that the outcome of the Risk Evaluation could be that the risk is so small that extractables testing in deemed unnecessary. Unfortunately, the BPSA Recommendations do not shed light on what combination of factors noted above would produce such an outcome. Nevertheless, should the outcome of the Risk Evaluation be that no extractables testing is necessary, then the Risk Evaluation and its conclusion is documented and the assessment is complete.

If it is concluded that an Extractables Evaluation is warranted, then the assessor moves to Question 8.3.

Question 8.3. Does the item's vendor have suitable extractables data? A mandatory step in the BPSA Extractables and Leachables Evaluation is performing a Toxicity and Quality Risk Assessment on extractables data. An important question is "where does the extractables information come from?" As BPSA is primarily a vendor organization, it is not surprising that its members would suggest that the utility of vendor extractables data be considered, assuming of course that the vendor has, and is willing to share, extractables data. BPSA notes that although vendor information may be generated under extraction conditions that are exaggerated versus the process conditions, the information may an adequate basis for performing a Toxicity and Quality Risk Assessment.

If it is established that the vendor has adequate and applicable data, then that is the data that is used in the Toxicity and Quality Risk Assessment. If not, then adequate and applicable extractables data is generated, presumably by the user.

In any event, adequate and applicable extractables data allows the assessor to move to Question 8.4.

Question 8.4. Does the toxicity of worst-case extractables present a safety or quality risk? Here in Question 8.4, one gets the important answer to a question that may have arisen while considering Question 8.3, which is "what is adequate and applicable extractables data?" The answer, provided in the context of Question 8.4, is that adequate and applicable extractables data is extractables data generated under worst-case conditions, meaning that it is extractables data that is derived from an extraction study where the extraction conditions are exaggerated versus the actual process conditions.

Should the Toxicity and Quality Risk Assessment conclude that the extractables do not present a risk,

then the extractables study, the extractables data, and the Assessment are documented and the Evaluation has been completed. If the Assessment establishes that a possible risk exists, then one is directed to perform a migrants (PERLs) study, which involves generating the process stream under the worst-case production conditions and testing the process stream for migrants (PERLs).

Up until this point, the process established by the BPSA is logical and generally consistent with accepted practice. However, the BPSA recommendations begin to deviate from what is considered to be current good practice when it comes to Question 8.5.

Question 8.5. Are migrants detected? While this question seems simple enough, it is based on a premise that migrants are somehow different from extractables. If extractables and migrants were being tested for in the similar manner, then Question 8.4 for extractables and Question 8.5 for migrants would be the same question. The questions are not the same, reflecting the premise that extractables are likely to be encountered while migrants are less likely to be encountered. This premise further emphasizes the fact that the BPSA approach to an extraction study is such that the extraction study is more highly exaggerated versus the actual process conditions that are used in a migrants study. In any event, according to the BPSA Evaluation Decision Tree, there are only two outcomes of a migrants study. The first outcome is that migrants are not detected, which actually means that migrants are not detected above a reporting threshold such as the AET. With this outcome, the sponsor is directed to document the extractables and migrants data and the Evaluation is completed. This is an interesting instruction as the only migrants data to be reported is that no migrants were detected.

In the event that migrants were detected, the sponsor is directed to identify and quantify the migrants and to perform a Toxicity and Quality Risk Assessment. One possible outcome of the assessment would be that the migrants are deemed to have no consequential effect on safety and quality, in which case the sponsor should document the results and the Evaluation would be completed. In fact, this is the only outcome that the BPSA has in their Decision Tree. However, there is a second possibility, which is that one or more migrants could have an adverse effect on safety and/or quality. In this case, one assumes that the next logical step would be to test the finished drug product for the migrants that have been deemed to potentially be safety-impacting and establish whether the levels of the migrants as leachables is acceptable or not. However, this step in missing from the BPSA Decision Tree, implying that the BPSA expects that all migrants will be assessed as safe and non-quality impacting.

In addition to the Decision Tree, the BPSA Recommendations address the topics of performing an extraction and then testing an extract. Considering performing the extraction itself, the Recommendations include:

- Test articles:
  - The test articles used for extractions should be the same items as used in manufacturing, perhaps downsized for practical reasons (being careful to preserve the surface area to solution volume ratio).
  - Test articles should be processed (e.g. sterilized, flushed) prior to extraction in the same way they are used during manufacturing.
  - Test articles should be tested intact.
- Extraction solvents:
  - Water plus a low-molecular weight alcohol (ethanol, IPA) are required.
  - High and low pH aqueous buffers should be considered as appropriate.
  - N-heptane is optional if the process stream is oil- or hydrocarbon-based.
- Temperature:
  - Use an elevated temperature.
  - Do not exceed 60 °C unless the process step is performed at such a high temperature.
- Time:
  - Biocontainers: 30 days with additional timepoints up to 1 year if appropriate.
  - Bioreactors: 30 days with an additional 90-day timepoint if appropriate.
  - Other components: 4 hours for ethanol, 24 hours for aqueous.
- Agitation:
  - Agitation should be used for all "fill and sit" extractions.
  - Recirculation or "flow through to reservoir" should be used for items that the process stream flows through in actual use.
- Surface area to volume:
  - Two competing parameters: exaggeration versus saturation. Do not increase the ratio of surface area to volume too much.
  - Biocontainers and bioreactors: 0.5 cm^2/ml.
  - Filters: Extract with 1.5 l.
  - Tubing: Fill the tubing with extraction solvent.
- Extraction conditions:
  - Exaggerated extraction is recommended.
  - Soxhlet or microwave-accelerated extraction are not recommended.

Considering the testing of the extracts, the BPSA Recommendations provide high-level information about the gravimetric (NVR), potentiometric (pH), spectroscopic (UV absorbance), chromatographic (GC, LC, IC), and mass spectrometric methods used to screen for extractables. The reader is referred to the Recommendations document to review this information. BPSA notes the use of certain tests (NVR, TOC) that provide information on total extractables and suggests that, if these methods have adequate sensitivity and indicate extractables are not present at levels that pose a safety risk, then this information alone is sufficient for safety risk assessment and selective tests that establish individual extractables may not be necessary.

More recently, BPSA has published an extractables/leachables Guide specifically for cell and gene therapy (CGT) drug products [19]. The Guide notes that certain leachable compounds can have adverse safety implications when delivered directly to patients, especially patients with compromised immune systems and organ functions. In addition, leachables can have compatibility issues with CGT products through chemical reactivity, cell toxicity, and potential initiation of cancerous changes in certain cells through genotoxic and mutagenic effects.

Unlike the BPSA Recommendations for single-use process equipment, the Guide does not provide actionable guidance on how to perform extractables or leachables studies for CGT products. Rather, the Guide provides strategic commentary and cites relevant references for testing. Key points of this discussion are provided as follows: the reader is referred to the actual Guide for more detailed and complete information.

Although classical biopharmaceutical manufacturing and CGT both involve cultivation of living cells, several significant differences are obvious. In classical biopharmaceutical manufacturing, the cells are utilized in the production of the drug substance (DS, e.g. a protein, an antibody, or an enzyme). The cells, cell debris, and host cell-related and process-related impurities are separated and removed from the DS in a downstream purification process. The downstream processing and formulation yield a drug product (DP) with a high purity with very low levels of impurities, including PERLs. In CGT, cells are the product and there is little opportunity to separate impurities from the product. This difference completely changes how extractables and leachables need to be approached.

While manufacturing of CGT products is highly reliant upon single-use technology (SUT), the conditions experienced during CGT (e.g. length of contact material exposure time, solvent/solutions) are typically less invasive than the conditions for manufacturing other biopharmaceuticals. Common SUT contact materials for CGTs may include blood collection bags, processing containers, cell expansion bags, tubing sets, filters, connectors, syringes, and final container vials or bags. SUTs used in monoclonal antibody (mAb) bioprocessing are typically made of the same polymers used in CGT, with a PVC (a plastic more commonly found in medical devices and primary containers) being a notable exception. Special attention should be given to PVC-related extractables and leachables, considering their safety effects, their potential interactions with the cell-based product, and their potential to affect downstream process in CGT production.

Given the extensive use of SUT for CGT manufacturing, the cell-based product and the variety of possible contact surfaces and contact time, it is reasonable to believe that CGT products could be impacted by leachable compounds. An impurity or contaminant (leachable) entering a CGT production stream could affect cell growth and/or expansion in process or ultimately impact a specific critical quality attribute of the final cell-based product. Because there are only very limited procedures to purify the cells, process-related leachables can remain in the product due to absorption and/or adsorption to the cells and thus they can have a negative impact on product quality or patient safety.

Extractables and leachables studies, as they are currently conducted for SUT, may not provide all the relevant data required to qualify CGT processes. Extractables studies using simulating solvents may not capture the ability of cells to sequester leachables and thus would underestimate leachables levels. Leachables studies require that the leachables be measured in the analytically challenging CGT process streams. Thus, it may be difficult to fully quantify patient exposure to manufacturing-related leachables. Moreover, chemical data, either extractables or leachables, might not be interpretable in the context of establishing the leachables' influence on cell viability. Additionally, the question as to whether leachables can cause cell stimulus and/or cell degeneration cannot be answered with analytical tools alone.

8.2.2 User Perspective, BioPhorum Operations Group (BPOG)

8.2.2.1 General Discussion

Since starting as an end-user group in 2008, BioPhorum has grown into an organization consisting of 90 manufacturers and suppliers. BioPhorum's mission is to create environments where the global biopharmaceutical industry can collaborate and accelerate its rate of SUT adoption, for the benefit of all. The organization directs seven focused Phorums, developing and adopting best practices in drug substance, fill finish, process development, and manufacturing IT.

In 2014, BioPhorum (at that time BioPhorum Operations Group (BPOG)) published a standardized extractables protocol [20]. This protocol became widely referred to in the industry as the "BPOG protocol." The BioPhorum's

vision was to accomplish two objectives; (i) to standardize extractables testing to allow for the generation of comparable data and (ii) to review the data generated by its Protocol to understand what testing was necessary and sufficient. In 2019, major changes were made to the 2014 protocol, following an extensive review of relevant extractables data. The review, performed by the BioPhorum extractables workstream, focused on assessing the extraction capability of the recommended solvents and the value of the recommended time points and resulted in the publication of a Best Practices Guide [21], which in this book will be referred to as the "BPOG protocol (2)." A summary of the data that supported these changes to the original BPOG Protocol is contained in reference [22].

The main changes made to the 2014 Protocol and reflected in the Best Practices Guide are:

- Removal of 5M sodium chloride and 1% polysorbate 80 as extraction solvents since these two solvents were shown to have a low unique extraction capability (that is, they produce extractables profiles that are not materially different from the profiles produced by other extraction solvents).
- Elimination of the time zero testing interval as it was shown that compounds observed at this time point were present at higher concentrations at later time points.
- Elimination of the elemental analysis of 50% ethanol extracts.

These updated recommendations reflect the experience of end-users at biopharmaceutical organizations that produce a diversity of biologic products in a variety of regulatory environments. The updated protocol gives guidance on the suggested methods for extractables studies, including sample preparation, extraction conditions, recording test article sampling conditions, and reporting data from the analysis of extracts. Flexibility is deliberately included. The BPOG Protocol (2) is viewed more as a guide and less as a standard in the sense that many of the study parameters can be altered on a case-by-case basis to take into account restrictions based on the conditions of use, physical form and size factors, chemical compatibilities, etc., if valid justifications are provided. It is noted that this flexibility is in stark contrast to a regulatory Standard, which by its very definition is highly specific and offers only limited opportunities for deviation.

As the BioPhorum consists largely of users and BPSA consist largely of vendors, it is not surprising that there are some differences in their respective approaches to extractables and leachables. Moreover, as neither group has regulatory responsibility or authority, it is not surprising that the Recommendations from both organizations do not read, in content and context, as if they were designed for the purpose of establishing necessary regulatory practice. In fact, the BioPhorum document specifically notes that "The extractables testing information package to be provided by a SUS supplier should not be passed directly to a regulatory agency, except where it is essential to include it in the filing to justify leachable targets and test plans. Rather, the purpose of the information package is to allow the SUS end-user to rigorously estimate the types and amounts of leachables that could be generated by the SUS component during its intended use."

An important aspect of the BPOG Protocol (2) is communicated in the statement "For an assembly, the preferred approach is to provide extractables data for each component." Thus, the BPOG Protocol (2) is a protocol for testing components, not materials of construction or complete SUS assemblies. Moreover, this statement provides clarity in terms of responsibilities. The BPOG Protocol (2) is based on the expectation that component suppliers perform the testing specified in the Protocol and provide the test data to component users in a standardized format. It becomes the responsibility of the component user to assemble the data available for all components of a manufacturing system to establish the potential leachables impact of the entire manufacturing system.

Certainly, the concept of standardized testing and reporting is a great advantage for users, as the state of affairs was quite chaotic prior to the BPOG Protocol. When vendors provided extractables data at all, the study designs were often dramatically different, the analytical approaches were profoundly dissimilar, and reporting conventions and practices were literally all over the map. These inconsistencies in approach left the user in the very difficult situation of trying to make apples to oranges comparisons between the various vendor's information, making the challenge of establishing total system extractables or leachables profiles difficult, if not impossible. Standardization, on the other hand, not only makes the users task easier but it results in more rigorous, complete, and robust total system assessment.

The proper place to start a Best Practices document (or a regulatory standard for that matter) is with a statement of Scope. Considering the scope of the BPOG Protocol (2), it is noted that the Protocol applies to, but is not limited to, the following SUS components that come into contact with product or process fluids:

- Films used in bags for storage, mixing, or as bioreactors
- Tubing
- Tubing connectors and disconnectors
- Aseptic connectors and disconnectors
- Platinum-cured molded tube connectors
- Sterilizing-grade and process filters
- Tangential flow filtration cassettes

- Sensors
- Valves
- Elastomeric parts (e.g. gaskets, O-rings, diaphragms, and septa)

BPOG protocol (2) does not cover final container closure systems for drug products. Also, non-fluid contact SUS components, assorted polymeric auxiliary production aid items used extemporaneously for material dispensing or transfer of ingredients and multi-use polymeric components, which are subjected to cleaning validation, are not in scope. Items out of scope include (but are not limited to):

- Vent filters
- Filters using non-polymeric matrices/media (e.g. diatomaceous earth)
- Plugs and end caps
- Sample syringes
- Sampling accessories (e.g. syringes/needles)
- Pipette tips
- Vent valves
- Scoops
- Graduated cylinders
- Beakers
- Weighing dishes
- Chromatography resins
- Any non-fluid contact SUS component

8.2.2.2 Performing Extractions

An extraction study consists of three major activities, performing the extraction, testing the extracts, and interpreting and reporting the test data. Considering the extraction, the heart of the BPOG Protocol (2) is Table 2 (shown here as Table 8.1), which establishes the extraction solvents and extraction durations (at 40 °C) for the major SUS components. In devising this table, the BioPhorum work group used user survey data concerning manufacturing operations and conditions to establish an extraction strategy that was designed to be all-encompassing; that is, the extraction conditions were chosen to encompass both commonly encountered and extreme manufacturing conditions.

Further instructions for performing extractions are contained in the Protocol's Table 1, Testing setup for various SUS components. Additional key recommendations for performing an extraction include:

- During the extraction, a portion of the extraction solvent may evaporate. For this reason, the initial and final volume of the extraction solvent should be recorded. Solvent loss should be handled appropriately:
  o ≤20% loss: correction can be performed but is not necessary
  o >20–50% loss: corrections must be performed
  o >50% loss: justification of the solvent loss and its handling must be provided

Table 8.1 Extraction conditions per the BPOG protocol (2) [21] and USP <665> [23].

Component type	Extraction solvents[a]				Extraction duration (days) at 40 °C			
	WFI	0.1 M H_3PO_4 [f]	0.5 N NaOH[f]	50% Ethanol	1	7	21	70[b]
Bag film, bottles and carboys intended for long term storage	X	X	X	X	X	–	X	X
Tubing intended for storage bags	X	X	X	X	X	–	X	X
Bag ports for storage bags	X	X	X	X	X	–	X	X
Molded stoppers[c]	X	X	X	X	X	–	X	X
Bag film, bottles and carboys	X	X	X	X	X	–	X	–
Bag ports	X	X	X	X	X	–	X	–
Impellers (e.g. in bioreactor mixers)	X	X	X	X	X	–	X	–
TFF cassettes intended for perfusion/continuous processing	X	X	X	X	X	–	X	–
Tubing	X	X	X	X	X[d]	–	X[e]	–
Tubing Connectors and disconnectors, fillings, overmolded junctions	X	X	X	X	X	–	X	–
TFF cassettes	X	X	X	X	X	–	–	–
Aseptic connectors and disconnectors	X	X	X	X	X	X	–	–
Sterilizing-grade filters/process filters	X	X	X	X	X	X	–	–
Filling needles	X	X	X	X	X	–	–	–
Chromatography column housing	X	X	X	X	X	–	–	–
Small parts (e.g. sensors, O-rings, gaskets, check valves, diaphragms, septa)	–	–	–	X	X	–	–	–

a) USP <665> Standard Extraction Protocol conditions are highlighted by shading.
b) The 70-day data point specified for film, tubing, and ports is necessary to support the long-term storage of materials in storage bags.
c) Classified in USP <665> as closures (molded stoppers) for storage containers.
d) Per USP <665>, tubing attached to containers not intended for storage.
e) Per USP <665>, tubing for fluid transport.
f) USP <665> uses a pH 3 extraction solution instead of 0.1M H_3PO_4 (although both are considered to be interchangeable) and a pH 10 buffer instead of 0.5 N NaOH (although 0.5 N NaOH is suitable in certain circumstances).

- For most components, the ratio of a sample's surface area to the volume (cm^2/ml) of extraction solvent to which it is exposed during testing should aim for 6 : 1 or higher. For filters, the ratio of effective filtration area to extraction solvent volume (cm^2/ml) should be minimally maintained at 1 : 1 or better.
- If an item is pre-treated before use in a process, the item should be pre-treated in the same way, e.g. flushing and sterilization before the start of the extraction.
- If the SUS component is intended for use after gamma irradiation, then a gamma-irradiated test article should be used for the extraction study. Due to the degassing of volatile organic compounds from the gamma irradiated components, the time between gamma irradiation and the extraction test should be less than eight weeks to represent the typical worst-case scenario in which the equipment may be used for production.
- If the component is intended for use after autoclaving, then an autoclaved test article should be used for the extraction study. The time between the autoclaving and the extraction test should be less than 24 hours or as soon as practical.

- If the component can be either gamma irradiated or autoclaved, then separate studies for each condition should be performed.
- Negative controls to calculate background levels should be included for all tests, using the same test setup minus the test article. Polytetrafluoroethylene (PTFE) bottles are recommended for inorganic elemental analysis, while clean glass bottles, as well as PTFE bottles, are suitable for organic analysis.

Lastly, the BioPhorum team was concerned with variations in the test results, noting that such variations could arise from:

- Raw material (resin) variability
- Variability due to manufacturing
- Variability during extraction
- Variability during analysis

The following Recommendations were made to combat excessive variations:

- The ideal study would look at components manufactured from two resin lots with separate extractions.
- However, extractables studies should not be delayed unnecessarily given the practical difficulties associated with obtaining test articles from two resin lots. One option would be to use different components from the same family to achieve the testing of two resin lots.
- Alternatively, the following strategies are offered in order of decreasing desirability:
  - Perform two separate extractions from components in the same family manufactured from two different resin lots.
  - Perform two separate extractions from two components in the same family manufactured from the same resin lot in two different manufacturing events.
  - Perform two extractions from one lot of components.

8.2.2.3 Analysis of Extracts

There is nothing unique in the BPOG Protocol (2) about the list of analytical methods that should be employed to screen extracts for organic extractables and extracted elements as it is the same list that has been mentioned numerous times previously. Nevertheless, the BPOG Protocol (2) is unique in certain aspects associated with the listed methods. For example, each method recommendation is annotated with general operating parameters (for example, the GC/MS method's documentation includes a recommended stationary phase (DB-5MS) and a recommended scan range (30–800 m/z)), a method qualification approach and a system suitability test (SST)). Using the GC/MS method as an example, the method's approach for qualification specifies butylated hydroxytoluene (BHT)

and n-octane and/or eicosane as the reference compounds and the use of phenanthrene-d10 as the internal standard. This recommendation for an internal standard is that it is to be used in every instance is critical in attaining lab-to-lab consistency in reported concentrations. The qualification is accomplished by repetitive analysis of a standard containing 1 mg/l of the reference compounds spiked into the various extraction solvents. A method is established to be qualified if:

1. The imprecision of the response obtained from the six replicate injections of the standard is less than or equal to 20% (%RSD ≤ 20),
2. The accuracy (spike recovery) is between 80 and 120%,
3. The limit of detection (LoD) is established and reported (no acceptance criterion is provided for LoD).

System suitability testing is, in certain respects, method qualification at the time of use. However, rather than establishing that a method is qualified for its purpose, system suitability establishes that a qualified method has been properly implemented at the time of use. It is not surprising, then, that method qualification and system suitability testing are both similar and different. Thus, method qualification and system suitability use the same reference compounds, which, in the case of GC/MS is BHT, at the same level (1 mg/l). However, the system suitability standard is prepared directly in the injection solvent as there is no sample preparation involved. As was the case with qualification, precision is established via six sequential injections of the suitability standard, with the acceptance criterion being the same as it was for qualification (% RSD ≤ 20). How LoD is established is also similar to qualification, but the acceptance criterion is different. In system suitability, the concentration of the suitability standard becomes the minimally acceptable LoD. To check the chromatographic performance, the retention time of the reference compounds are compared to historical data. Lastly, if bracketing is employed, then an injection of the standard should occur at least once for every 10 sample injections, with the requirement that the percent difference in response between injections at the beginning and end of each bracket be ≤25%.

Similar qualification and suitability processes and acceptance criteria have been established for the other major screening methods (headspace-GC/MS, LC/MS, and ICP/MS).

The BPOG Protocol (2) also provides guidelines in terms of reporting thresholds. For chromatographic methods used for organic compounds, an individual compound detected at a concentration of 0.1 µg/ml or greater should be reported and, when possible, quantified and identified by using an authentic compound. For extracted elements, the reporting threshold for elements is targeted to be 20 µg/l (ppb). The LoD may be lower or higher than 20 µg/l, depending on the element being detected, the sample matrix, and instrument parameters used.

Lastly, the Protocol (2) addresses extract preparation prior to instrumental analysis, generally recommending that extracts be tested without processing, except for dilution. The only exception to this is sample preparation for GC/MS, where the extraction solvents cannot be injected directly onto the GC columns and where sensitivity may not be adequate to allow for the extracts to simply be diluted before injection. In the case of GC/MS, the recommended sample preparation is solvent exchange with dichloromethane (DCM) and evaporative sample concentration. Specifically, the protocol directs the analyst to:

- Adjust the pH as needed.
- Extract the aqueous samples in 1 : 1 (v/v) ratio with DCM including the internal standard; repeat the extraction three times on each aqueous sample aliquot.
- Combine DCM fractions and evaporate to an appropriate and known volume to concentrate the sample and allow quantitation.
- Redo preparation if the sample is evaporated to dryness.

While the BPOG Protocol (2) speaks about reporting the test data in general terms, the BioPhorum team has developed a standardized report format to facilitate the reporting and communication of test results, which will be discussed in greater detail as follows.

8.2.2.4 Reporting Extractables Test Results

The Extractables Test Report developed by BioPhorum provides comprehensive information on the SUS component tested, including details of the testing setup, testing conditions, and analytical methods applied, and the identity and quantity of extracted compounds. By standardizing the Report format, the BioPhorum team has ensured that all data from all vendors will be summarized and transmitted in a consistent format, facilitating data review and assessment. The BioPhorum team has created a document that describes in detail the structure and contents of its standardized Report [24].

The Extractables Test Report should include, but is not limited to, the following information for each extractables study:

1. *Title page*. The title page should include:
   a. Report title, study identity, report date, report revision
   b. Name and location of lab performing the testing and name and location of sponsor, if applicable
   c. Signatures

2. *Study summary*. The study summary for SUS components tested should consist of:
   a. Short description of background of testing
   b. Short description of the testing setup and experimental part
   c. Short summary of results and conclusions
3. *Study design*. Outline of the study design information must follow the BioPhorum Extractables Data Summary (BEDS) spreadsheet template. This includes information on:
   a. Test article traceability
   b. Pre-treatment(s) of the test article
   c. Extraction conditions, solvents, and time points
   d. Analytical information
   e. Supporting information for the test item
4. *Summary tables*. One summary table for organic compounds and one summary table for elements should be included in the report. The formats of the summary tables must follow the BEDS spreadsheet template. In addition to including the summary tables in the Extractables Test Report, the summary tables need to be made available in BEDS spreadsheet format. It is optional to report structures of identified compounds.
5. *Results from analyses*. Results from each individual analysis technique should be reported separately. It is highly recommended, but not mandatory, to follow the format provided in the BEDS spreadsheet template.
6. *Analytical methods*. Information on each individual analysis technique should be reported separately. It is highly recommended, but not mandatory, to follow the format provided in the BEDS spreadsheet template. Information on analytical methods must include:
   a. Method traceability (e.g. SOP documentation)
   b. Instrument settings
   c. Method qualification
   d. System suitability test (SST)
   e. Sample preparation
   f. Approaches for quantification and identification
7. *Deviations*. Any deviations observed during the study need to be documented.
8. *Terminology*. Acronyms used in the Extractables Test Report need to be explained.
9. *Revision history*. Revision history of the report needs to be included, describing changes made to the Extractables Test Report after initial release.

Additional requirements or recommendations concerning the Extractables Report are as follows:

- It is not required to include chromatograms or spectra in the report, but these shall be made available upon request.

- Chromatographic data should be presented using the total ion current (TIC).
- The final reporting should be within the oversight of the company's quality management system.
- Concerning extracted elements, the mandatory elements are listed in alphabetical order within ICH Q3D class, from class 1 to class 3. In addition to the elements listed in ICH Q3D, it is requested by end-users to analyze Al, Fe, Mg, Zn as these elements may impact drug product quality. It is allowed to add additional elements as needed.
- Providing structures for organic extractables is optional.

The BioPhorum's Guide to the Extractables Test Report [24] contains detailed instructions and provides templates for each individual section of the Extractables Report; the reader is referred to this document for this detailed discussion.

8.2.2.5 Leachables

The BioPhorum E&L team defines leachables as follows: chemical entities that come from SUS components during normal use. This definition is equivalent to the definition provided for PERLs by the USP but differs from the conventional definition of a leachable in that it does not specify that the leachable is present in the finished drug product. In fact, a leachable as defined by BioPhorum is found in the process stream. Thus, per BioPhorum, leachables testing should be carried out using actual process fluids and the actual manufacturing equipment operated under typical production conditions, unless surrogate fluids are used to mitigate analytical interferences.

This distinction is important because it facilitates an understanding of the BioPhorum's Recommendations document for Leachables testing [25]. An important part of this Leachables document is its Risk Evaluation Matrix, which is designed to rank certain manufacturing components and operations involving those components with respect the risk of producing leachables in potentially meaningful quantities. This aspect of the document will be considered further in Section 8.5.3.

The second important aspect of the document is its Leachables Study Design. Review of the Leachables Study Design reveals that it is, in essence, the same study design for extractables except (i) actual process fluids and conditions are used for leachables as opposed to surrogate solvents and exaggerated or accelerated conditions that are used for extractables and (ii) leachables testing includes the possibility for targeted, as well as screening testing for leachables, whereas extractables testing is almost exclusively screening. The concept of targeted analysis is particularly important and is based on the circumstance that the leachables study was preceded by an extractables study. If this is the case, the results of the extractables study

can be used to establish target leachables; that is, leachables that will be targeted in the analysis method employed in the leachables study. By targeting an extractable as a leachable, one is essentially saying "I am interested in this specific substance and thus I want to have high confidence in its reported concentration as a leachable." Because of this expectation for high confidence, analytical methods for targeted leachables are more rigorously developed and validated, specifically for their quantitative aspects. Speaking to method validation specifically, the BioPhorum document advises that:

> "Quantitative methods for leachables should be validated or qualified, as appropriate, when monitoring a known leachable. Depending on the intended use, validation/qualification of quantitative methods may include the characteristics of accuracy, precision, specificity, robustness, detection limit, quantitation limit, linearity and range. Validation/qualification of limit test methods may include specificity and detection limit."

Because the BioPhorum's instructions on how to perform leachables study so closely mirrors the instruction for performing an extraction study (which was discussed previously), the reader is referred to the BioPhorum document for further details. It is also noted that the leachables document includes a chapter on "Key Lessons and Common Pitfalls" which provides "guidance to the risk assessment team in the form of lists reflecting issues that commonly arise during the execution of the leachables assessment process for a SUS." Once again, the reader is referred directly to the BioPhorum leachables document to review these valuable insights.

8.3 Compendial Guidance, USP <665> and <1665>

8.3.1 Perspectives on the Development of a USP Monograph for the Chemical Characterization of Pharmaceutical Plastics

If there ever was a poster child for the trials and tribulations of trying to provide actionable and detailed guidance concerning extractables, leachables, and chemical assessment, it is the USP monographs on plastic components and systems used in pharmaceutical manufacturing systems, USP <665> [23] and <1665> [26]. For individuals who have never participated in the USP process for establishing a monograph, the following is a brief synopsis.

There is a formalized process within the USP by which potential topics of monographs are surfaced, vetted, and adopted for action. For the sake of brevity, this process is described only in its most general sense, as follows. Once a topic has been slated for development into a monograph, the monograph is "adopted" by the specific USP Expert Committee that is most closely aligned with the Scope and objective of the monograph. The responsible USP committee establishes an Expert Panel that has the responsibility for generating the draft version of the monograph. Thus, the first level of review of the monograph and its contents rests with the scientific, regulatory, academic, and industry experts in the Panel, who obviously have to reach a consensus as to the monograph's contents and context. Assuming that the Expert Panel can produce a consensus draft Monograph, the draft then goes to the Expert Committee for their review and approval to proceed. This represents the second level of review and is still internal to the USP.

Should the draft Monograph be approved by the Expert Committee, the draft Monograph is published in USP's *Pharmacopeial Forum*. This action gives the public, including the scientific, industrial, and regulatory communities, their opportunity to review the draft Monograph and opens up a 90-day review period during which interested parties can submit editorial, philosophical, and scientific comments on the draft Monograph to the USP. At the end of the review period, which is generally managed with a fair degree of latitude, the comments are collated and submitted to relevant USP staff members and the Expert Panel for their consideration. Each and every comment received is formally addressed and the USP's response to each comment is recorded internally, regardless of whether the response includes a revision of the draft Monograph or not. Once each and every comment has been addressed, the Expert Panel reviews the revised draft, as it is possible that a "one to one" approach to revision has led to inconsistencies, in context, content, or both, in the text. Once the Expert Panel accepts the revised draft Monograph, the revised draft again goes to the Expert Committee for approval. At this point the Expert Committee has essentially three options: (1) reject the revision, in which case the revision is either sent back to the Expert Panel or the monograph writing process is suspended, (2) approve the revised draft Monograph and recommend its publication in the Pharmacopeia, or (3) approve the revised draft Monograph but require that it undergo a second public review/comment cycle. The deciding factor between approval options 2 and 3 is the extent of the revision; if the extent of the revision is small and largely "cosmetic", then it is typically the case that a second round of public review/comment would serve no useful purpose and thus is not initiated. However, if the revisions are substantial and directly impact the use of the Monograph, then the second round of public review/comment is appropriate and is initiated.

One can certainly envisage that a "new to the world" monograph, addressing an evolving topic such as extractables and leachables in pharmaceutical manufacturing, would require a little bit of discussion to "get it right." However, through four rounds of public comment, the number of comments received in each round did not materially decrease, both in terms of sheer number and changes requested.

The lesson from this story is that there is no consensus that can be achieved on this topic, at least in the near term. There is considerable disagreement as to whether leachables from manufacturing systems are even a cause for concern, especially with respect to the patient safety impact of the finished drug product. When there is agreement that chemical characterization is necessary, there is considerable debate in terms of whether the guidance should be general (that is, these are the metrics that must be met and one can meet them any way one wants, as long as a justification can be provided) or specific (one must do these tests and get the specified results to pass). When there is agreement that specifics are necessary, almost each individual specific is contested and debated, either based on practical considerations, issues of "good science," or "that does not exactly fit my situation."

This reality notwithstanding, the following represents that status of USP chapters <665> and <1665> at the time this book was written (January, 2022). The approved and final versions of these chapters are available on the USP website:

- for <665>, https://online.uspnf.com/uspnf/document/ 1_GUID-F06B057E-8C4D-4F3F-8A5C-9C62012C7947_ 2_en-US;
- for <1665>, https://online.uspnf.com/uspnf/document/ 1_GUID-B2665FAF-027F-46AA-8093-2292E0666F03_ 2_en-US.

On 1 May 2022, these chapters will be published in USP–NF 2022, Issue 1, at which point they will be official. However, a USP general chapter numbered below 1000 becomes compendially applicable and thus is considered a required standard only when:

1. The chapter is referenced in a monograph;
2. The chapter is referenced in another applicable General Chapter numbered below 1000; or
3. The chapter is referenced in General Notices.

When they become official, neither <665> nor <1665> will be referenced in a monograph, another appliable General Chapter numbered below 1000, or General Notices, and therefore they will not be applicable or required standards. Rather, the USP objective is to socialize these chapters to give the E&L community of practice, and specifically its industrial and regulatory members, the opportunity to review the chapters' final versions and to use that review to revise, as necessary and appropriate, registration and/or regulatory practices for manufactured drug products and their manufacturing systems. Unless it is deemed essential to do so, USP will not take steps to revise these chapters, or make them enforceable, until the start of the 2025–2030 revision cycle, at which point the relevant USP Expert Committee, with stakeholder input, will decide how to proceed with the revision of these chapters (if necessary) and with making <665> applicable.

8.3.2 Objectives and Scope

Perhaps the most significant aspects of any Standard or monograph are its intended purpose and its Scope. Considering intended purpose, <665> and <1665> together seek to establish the minimum requirements for the qualification of pharmaceutical and biopharmaceutical manufacturing components and systems as being suitable for their intended use, specifically addressing:

1. The component's and system's ability to add impurities to manufacturing process streams, relevant manufactured drug substances, and finished drug products.
2. The probable effect of the added impurities, specifically focusing on the impact that such impurities could have on patients whose therapies include the use of the manufactured drug products.

To accomplish this objective, <665> establishes the testing requirements for the qualification of components while <1665> communicates the key concepts behind the requirements and provides additional information and guidance regarding the applicability and the application of <665>.

Considering Scope, it was originally envisioned that these USP monographs would be all inclusive in the sense that they would be applied to any plastic manufacturing item (materials, components, and systems) and any manufacturing process encountered in the pharmaceutical industry. This brings us to the first question frequently asked about Scope, which is "what about the other materials (e.g. glass, metal and elastomers) that are used in manufacturing items?" The answer for glass and metals is relatively straightforward. Although the two manufacturing monographs originally were focused on both organic compounds and elements, the practical difficulties in establishing requirements around elements has reduced the text in the Monographs to the suggestion that considering elements is potentially important but that the means of addressing elements has been left to the individual sponsor. Once elements are "off the table" and recognizing that metal and glass are not meaningful sources of organic extractables, glass and metal became out of scope.

The explanation for elastomers, specifically rubber-based elastomers, traditionally being out of scope for the two manufacturing monographs is "territorial," which is to say that since the USP has a monograph for elastomeric components used in pharmaceutical packaging (<381> as discussed previously), the "right" place for a monograph for elastomeric components used in manufacturing would be in <381> as an extension or in its own monograph, which could adopt those tests and specifications in <381> that are relevant to both manufacturing and packaging and then possibly augment those tests with additional tests relevant to manufacturing.

This point notwithstanding, it was ultimately established that a manufacturing component is evaluated and tested in <665> in the same fashion regardless of its materials of construction. For example, a manufacturing component constructed from PP would be tested in the same way if it were constructed from PC instead of PP. Following this logic, the same manufacturing component would be tested in the same manner per <665> as if it was made from a rubber elastomer. Based on this line of thinking, elastomeric components were established to be "in scope" with respect to both <665> and <1665>.

The next aspect of Scope that was frequently questioned in the early drafts of the two manufacturing monographs was whether they should include materials of construction, in much the same way that USP <661> contains <661.1> for materials of construction of packaging components and systems and <661.2> for the components and systems. The first aspect of the question addresses the use of material information. It remains the USP position that "the most effective means of ensuring that a manufacturing system is suitable for its intended use is to use well-characterized and intentionally selected components." However, it is the general opinion of the regulatory and industrial communities that chemically qualifying a manufacturing component's materials of construction is an unproductive process that provides little insight into the ultimate suitability for use of a manufacturing component. Given this opinion of important stakeholders, it would be counterproductive for the two manufacturing monographs to contain tests and specifications for material qualification. Furthermore, since proper material selection does not ensure that a component will be suited for its intended use, the USP felt it was improper to provide tests and specifications for material selection.

Given these two circumstances, <665> does not address materials of construction and <1665> addresses materials of construction only from the perspective of noting that the information on materials of construction enables good component selection, which in turn facilitates the development of suitable components and systems.

This discussion of selection versus qualification can be extended to components and systems; that is, is the purpose of the two manufacturing monographs to address component selection, component qualification, or both? Considering selection, the USP ultimately decided that selection was most appropriately addressed by the individual sponsors themselves, in the context of their own quality processes and systems, primarily because selection is partially based on factors outside of chemical composition or properties. Furthermore, after considerable dialogue with the various stakeholders, the USP became convinced that consensus minimum standards for the qualification of components and systems could be established and thus the purpose of the two manufacturing monographs became the qualification of plastic components used in pharmaceutical manufacturing.

At this point in the discussion of Scope, all plastic manufacturing components are in scope; however, it is quickly understood that some exceptions should apply. Since the purpose of the monographs is to assess interactions between a process stream and a component, it is obvious that components that do not contact a process stream are out of scope. For example, tubing that directs solutions to waste (e.g. rinse solutions) do not contact the process stream and are out of scope and considered to be safe without qualification. Furthermore, one envisions that there are plastic items used in manufacturing where the conditions of contact between the process stream and the item are so minimal, either in terms of contact area, contact temperature, contact duration, or some combination of the three, that the likelihood of significant leaching is so small as to be negligible. These plastic items, referred to as auxiliary items, include scoops, funnels, pipettes, graduated cylinders, weighing dishes, and beakers and poses such a small risk in terms of the transfer of extractables to the process stream that they are outside the scope of the two manufacturing monographs and are considered to be suitable for use without testing.

Additionally, the discussion of Scope to this point has included all process streams. However, the physical state of the process steam is relevant in terms of the process stream's interaction with a component. Oversimplifying somewhat, it is generally accepted that a liquid process stream has a greater "leaching power" than either a solid or gaseous process stream. Although transfer of substances between two solids (e.g. a component and a solid process stream such as a powder) can occur, the transfer is slow and generally inconsequential given the generally short duration of contact between the process stream

and manufacturing components in most manufacturing unit operations. Furthermore, the transfer of substances between a solid component and a gas process stream is limited to volatile substances. Because of the limited propensity for leaching, components in contact with solid and gaseous process streams are out of scope and considered to be suitable for use without qualification. Physical abrasion of process components by a solid process stream is not chemical leaching and thus is outside the Scope of the two manufacturing monographs.

Furthermore, consideration is given to the nature of the product that is manufactured. Pharmaceutical manufacturing processes generally produce one of two types of outputs: a drug substance (DS), which is further processed, or a drug product (DP). At first glance, it might appear that a component in contact with any DS or DP must be qualified per <665>. However, a history of safe use argument can be used to exclude certain DS types from the "qualification by testing" requirement. The term active pharmaceutical ingredient (API) has historically been used for a DS that is a precursor to a non-biologic and non-biopharmaceutical drug product and which is typically manufactured via chemical, as opposed to biological, processes. APIs, also referred to as "traditional" or "small molecule" drug substances, are well-characterized substances that are the result of manufacturing that involves multiple, highly effective purification processes. Such APIs have sufficiently low levels of PERLs that they do not require characterization. Thus, components for manufacturing APIs are out of scope and such components are deemed suitable for intended use without testing per <665>.

Lastly, as matter of practice, <665> is not a retroactive monograph and it should not be applied to manufacturing components and systems that are used to manufacture a commercialized (and therefore regulatorily approved) DS or DP. This is the case as a commercialized DS or DP has been established to be safe and effective by the relevant regulatory authority, who presumably concluded that the manufacturing system was also suited for its intended use. If a manufacturing component or system has been established as suited for use to the satisfaction of the applicable regulatory authority based on a certain set of criteria, then further qualifying the component or system via <665> testing serves no useful purpose and thus is a non-productive utilization of limited resources. This point notwithstanding, modification of the manufacturing process, its operating conditions, or the manufactured DS or DP may trigger the need for <665> qualification. Furthermore, it is a regulator's prerogative to apply <665> retroactively.

These various exclusions, and their suitability of use status, are summarized in Table 8.2.

Table 8.2 Exclusions from the scope of USP <665> and their qualification status.

Out of scope but qualified as suitable for use without <665> testing	Out of scope and requiring qualification outside of <665>	Out of scope because qualification per <665> is inappropriate
• Components that do not contact process streams directly • Components in contact with solid or gaseous process streams • Auxiliary components • Metal components • Glass components • Ceramic components • Components used to manufacture APIs • Components used to manufacture approved commercial DS or DP	• None	• Materials of Construction

8.3.3 The Concept of a Comparator: Initial Assessment

Let us be honest, practically nobody supports unnecessary testing. Organizations that develop and commercialize pharmaceutical drug products have more impactful ways to use their limited resources than performing, reporting, and justifying unnecessary testing. The same is true for regulatory authorities, whose limited resources are better spent reviewing critical and impactful information contained in registration dossiers. Scientific purists object to unnecessary testing as a matter of principle. The general public has little tolerance for unnecessary testing as it can lead to higher healthcare costs and delayed commercialization of life-enhancing and/or life-extending therapies. Even organizations or individuals who make their living designing studies, performing testing, or interpreting test results object to unnecessary testing, as the simple fact is that there is "plenty of necessary testing to go around." One would think, then, that standards would contain appropriate means to eliminate unnecessary testing and that all stakeholders would embrace and support such efforts.

The concept of a comparator is one manifestation of the "let's reduce unnecessary testing" paradigm. The concept is based on the pre-supposition that there is a comparator; that is, a predicate item that has either directly or indirectly been established to be suited for intended use in a defined application under defined circumstances. When one seeks to establish that a second item is suited for its intended use, the concept of the comparator states that if

• The first item can be established to be equivalent to the comparator,

• If the first item's application is equivalent to that of comparator, and
• If the first item's circumstances of use are equivalent to those of the comparator,

then the item can is judged to be suitable for use based solely on it being equivalent to a qualified and "approved" comparator.

Considering a manufacturing component specifically, regulatory approval of a commercially marketed pharmaceutical or biopharmaceutical product implies that the process and system used to manufacture that product has been established to be suitable for their intended purpose. Once a manufacturing system has been established to be suitable for manufacturing a biopharmaceutical DS or pharmaceutical or biopharmaceutical DP by the relevant regulatory authority, then every component used in that manufacturing system has been established to be acceptable. These manufacturing components or systems may then be used as comparators and linking other components or systems to the comparators may be the appropriate and sole means of qualifying the component or system under consideration. In such a case, establishing equivalence between the items in question and the comparator completes the item's qualification.

Equivalence can be established if both the component under consideration and the comparator are:

1. Constructed from the same materials of construction
2. Equivalent in design
3. Equivalent in function
4. Equivalent in the vendor's processing of the item prior to providing it to the component's user
5. Equivalent in their conditions of use in the manufacturing process
6. Prepared for use (e.g. flushed) by the user in an equivalent manner
7. Used to produce the same type of output (biopharmaceutical DS or pharmaceutical or biopharmaceutical DP),
8. The process outputs are administered in the same clinical manner (route of administration and dose).

Reviewing this list and recognizing the challenge in establishing exact equivalence for each of these eight dimensions, one can be excused if they conclude that the concept of a comparator is great in theory but mediocre in practice as the bar has been set too high, perhaps unachievably so. Although it is highly desirable that the equivalence in all eight aspects is exact, such a situation is likely to be rare. Rather, essential equivalence can be established based on strong similarities between the component under consideration and the comparator. The concept of strong similarities implies that the differences in any of the above-listed eight items of comparison are minor, where "minor" means the differences are sufficiently small that the PERLs derived from the component in question:

- Will be the same in identity and amount to the PERLs from the comparator or
- Will have the same process or patient effect as the PERLs derived from the comparator

Minor differences between the component under consideration and a largely representative comparator component are addressed by risk assessing the minor differences. The purpose of this risk assessment is to ensure that a minor difference will have a minor impact on the suitability of a component for its intended use. If it is established that the minor difference will likely have a minor and generally insignificant impact on a component's suitability, then the comparator approach is an appropriate means of qualifying the component. However, if it is possible that the minor difference could have a significant effect on the component's suitability, then the comparator approach is not an appropriate means of qualifying the component and the component must be qualified by testing, as established by risk assessment.

Addressing the issues of Scope and considering the possibility of a comparator are parts of the Initial Assessment that begins the <665> component qualification process (Figure 8.1).

8.3.4 Risk Assessment and Risk Classification

The underlying tenet of the testing specified in <665> is that the amount and rigor of necessary testing is directly linked to the risk that PERLs will accumulate in the final drug product in unsafe quantities. The greater the risk of impactful PERLs, the more extensive and intensive is the testing required for qualification. As was noted previously, two dimensions define this risk (Figure 8.2):

1. The likelihood that PERLs are leached from a manufacturing component by the process stream.
2. The likelihood that the PERL will persist in the process steam to the end of the manufacturing process and thereby become incorporated into the manufactured output (DS or DP).

In essence, a particular manufacturing circumstance is rated (scored) in terms of both the individual risk dimensions and the combination of the individual risk scores is used to establish the circumstance's total risk.

It is almost universally the case that the overall risk is scored using a Risk Evaluation Matrix, which is a mathematical contrivance that attempts to quantify the risk in each of several risk dimensions. This Risk Evaluation Matrix is constructed so that it:

- Establishes the appropriate contributors to, or dimensions of, risk
- Provides a means of quantifying the risk in each of its dimensions

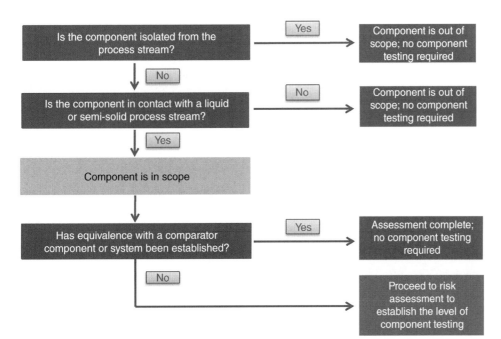

Figure 8.1 The USP <665> Initial Assessment. The Initial Assessment establishes whether a component is within the Scope of <665> and whether the component can be qualified by comparison to a comparator.

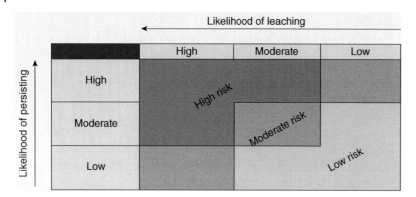

Figure 8.2 Risks associated with process system-related leachables.

- Links the quantified total risk to appropriate characterization strategies

Various Risk Evaluation Matrices have been devised by individual drug companies, trade associations, and standards-setting organizations; representative Matrices are discussed in greater detail in Section 8.5. At this point in the text, it is adequate to know that although there is considerable latitude in terms of the mathematical specifics of the various Risk Evaluation Matrices, <665> requires that the output of all Risk Evaluation Matrices must be that the components and their circumstances of use are classified as either Higher Risk, Moderate Risk, or Lower Risk.

8.3.5 Risk-Based Testing of Plastic Components and Systems

Once the Risk Evaluation process is complete and a particular component is classified as higher, moderate, or lower risk in a particular manufacturing operation, the risk classification is used to establish the testing that is required to qualify that component for the specified application. Two concepts were critical to the development of the <665> risk-based testing strategy:

1. The greater the risk, the more extensive and the more rigorous the required testing.
2. The progression of testing from lower risk to higher risk should be practically progressive, reducing unnecessary testing.

With these concepts in mind, <665> specifies extraction and testing as noted in Table 8.3.

When discussing the rationale behind Table 8.3, it is necessary to rely on generalizations as Table 8.3 is intended to be widely applied to very different manufacturing components used in very different manufacturing operations. While it is certainly the case that there are exceptions to every justification that can be provided to support Table 8.3, the reader is reminded that USP <665> was written to address the typical and is imperfect by design when applied to the extraordinary.

Table 8.3 Guidelines for application of chemical component tests as established by risk from USP <665>.

Risk level	Extraction solutions for chemical testing	Chemical testing of extracts
Lower (A)	• Solution C1 (50% Ethanol/water)	• NVR • UV Absorbance
Moderate (B)	• Solution C1 (50% Ethanol/water)	• Organic Extractables Profiling
Higher (C)	• Solution C1 (50% Ethanol/water) • Solution C2 (pH 3 salt solution) • Solution C3 (pH 10 buffer)	• Organic Extractables Profiling

The first question to address in designing Table 8.3 is the issue of extraction solvents. The range of extraction solvents should be chosen based on the range of compositions of process solutions and should reflect the solution composition factors that affect a solution's "leaching power," which includes polarity and, for aqueous solutions, pH. Considering polarity, it is extremely rare that process solutions are non-polar organic solvents; thus, there is little need for the extracting solutions to include a non-polar organic solvent. Although some organizations have advocated the use of pure ethanol as a standard extraction solvent based on theoretical considerations [27], it is rare that process solutions are "pure" alcohols. Consistent with the philosophy of "aiming for the middle," a 1/1 (v/v) mixture of ethanol and water (semi-polar) and water (polar) were chosen as the extraction solvents to represent the polarity of process streams. The ethanol/water extraction solvent can address process streams that contain organic solvents, organic solubilizing or stabilizing agents, lipids (up to 15% of the process stream's composition by weight), proteins, and blood or blood-derived components.

Once the obvious decision is made to include at least one aqueous extraction solvent in USP <665>, then the influence of pH on "leaching power" must be taken into account. The affect that pH has on an aqueous solution's "leaching power" was discussed previously in Chapter 2, Section 2.2.5.3; for the purpose of the current discussion, it is adequate to merely note that an extraction solution's pH will profoundly affect the extraction of acidic and basic extractables. Thus, the decision to use water as an extraction solvent is actually a decision to use two aqueous extraction solvents, one at low pH and one at high pH. Establishing the exact values of the pH is simply a question of "what are reasonable pH extremes exhibited by process solutions?" This was answered by the USP as pH 3 and pH 10 respectively.

Thus, the USP <665> extraction solvents consist of a 50% ethanol in water solution, a salt solution adjusted to pH 3 and a buffer whose pH is adjusted to a value of 10.

For practical purposes, it is appropriate that the same extraction solvent be used at each risk level and that the extraction solvent represents a worst case. As it is a general rule of thumb that organic extractables profiles are more complex (a greater number of organic extractables at higher individual levels) in ethanol/water mixtures versus aqueous solutions at low and high pH, 50% ethanol in water is the baseline extraction solution that is common to all risk levels. As risk level increases to higher risk, the higher risk is reflected in the use of additional extraction solutions. Thus, for higher risk components, extractions using all three extraction solutions must be employed.

Once the extraction solutions have been established, the tests to be performed on the extracts must be specified. It is simple enough to specify the test for the higher risk situations as the expectation is clear; that is, full organic extractables profiles must be generated in each extraction solution.

The challenge facing USP <665> is to establish the required testing for the lower and moderate risk situations. At this point, it has already been established that 50% ethanol in water will be the extraction solution that is common to all three risk classifications. Thus, the transition from higher risk to moderate risk is accomplished by the transition from three extraction solutions (low pH, high pH, and ethanol water) to one extraction solution (ethanol/water). The testing requirement for moderate risk situations for this single extract solution is full extractables profiling.

The transition from moderate risk to lower risk cannot be managed by the extraction solution, as both moderate and lower risk use the same extraction solution (ethanol/water mixture). Therefore, the transition is accomplished by managing the testing required, specifically that compound-specific extractables profiling for moderate risk is replaced by general chemical property tests for lower risk. The concept behind the use of general chemical property tests is as follows. The classification of a given manufacturing circumstance (manufacturing

component contacted by a process stream) as lower risk is based on a numerical and "theoretical" assessment of the component, the process stream, and the conditions of contact. If no testing were performed for a lower risk situation, there would be no experimental verification that this numerical and "theoretical" outcome was correct. Thus, extracts are generated and tested for lower risk situations specifically and solely for the purpose of confirming the risk scoring outcome, not for the purpose of safety risk assessment. It is already concluded in <665> that lower risk situations most likely result in an acceptable safety risk and therefore that testing for lower risk situations does not need to produce information that can be safety risk assessed. Rather, the sole purpose of testing for lower risk situations is to provide a check that the risk assessment outcome of "lower risk" is, in fact, the right outcome.

With this purpose in mind, it is clear that the objective of lower risk testing is to produce some type of indication that "maybe the risk is higher than we numerically determined." General chemical properties of an extract that are affected by the nature and quantity of organic extractables are appropriate candidates to fulfill this purpose. However, in specifying that the extraction solution as 50% ethanol in water, <665> was limited in the number and type of general chemical tests that could be applied. Commonly employed diagnostic tests such as pH and total organic carbon (TOC) are not suited for testing alcohol/water mixtures, essentially leaving UV absorbance (aromatic extractables) and non-volatile residue (NVR) as the remaining possibilities.

Once the tests have been specified, the question of acceptance criteria becomes relevant. The acceptance criterion for tests involving the generation of an extractables profile is obvious. If the purpose of <665> testing is to qualify components as likely being safe, then it is clear that the organic extractables profiles generated for moderate and higher risk situations must be toxicologically safety risk assessed and that the outcome of the safety risk assessment must be the conclusion that the extractables have a negligible ability to adversely affect patient health and safety were they to become leachables.

However, it is less obvious what the acceptance criteria should be for the lower risk situation's tests of UV absorbance and NVR. While it is obvious that, by intent, the acceptance criterion must be "a value sufficiently high that it indicates that the risk associated with the situation is greater than numerically established," it is not so obvious what the threshold between "low enough" and "inappropriately high" is. Thus, for example, a high NVR value would indicate that although the numerical risk assessment concluded that the likelihood of leaching was low, in fact, for whatever reason, the likelihood of leaching appears to be high. Although such a supposition is logical enough, it fails to establish the threshold between an acceptable value and a value that would essentially elevate the situation being investigated from the lower risk category to the moderate risk category.

The possible situations that could be encountered in manufacturing operations are so great and so diverse that the USP could not specify a single threshold or thresholds for NVR or UV absorbance that it believed would be acceptable to the E&L community. Thus, it is the responsibility of the entity performing the manufacturing system qualification to interpret the NVR and UV results in the context of their specific manufacturing situation and to decide what values for NVR and UV are acceptable.

8.3.6 The Standard Extraction Protocol per USP <665>

The Standard Extraction Protocol (SEP) essentially describes the testing that is performed to qualify a higher risk situation (see Tables 8.1 and 8.2). A discussion and justification of the SEP follows.

When designing a risk-based strategy, one can take one of two approaches. Approach number one is to eliminate all risk to the greatest extent possible while the second approach is to focus on addressing the risks that are most commonly encountered. In terms of the chemical characterization of manufacturing components, the first approach sounds something like this: obtain process information from a large cross-section of pharmaceutical manufacturers and then design a chemical characterization process that encompasses all manufacturing conditions, even the most extreme. The ramifications of this approach are that extractions are performed using solvents at extreme pH or organic content, at the highest possible contact temperature and over a range of contact conditions, from the shortest duration to the longest duration. This is, in essence, the approach that was reflected in the first BPOG extractables protocol.

Adopting the second approach to chemical characterization starts off with the same activity as the first approach, collecting process intelligence. In this case, however, the collected process information is used to establish those conditions that encompass a vast majority of typical manufacturing conditions and then to match the chemical characterization process to these typical conditions. For extractions, this means extraction solvents that are less extreme in terms of pH and organic content and extractions of intermediate durations. This is the approach the USP took in writing its <665> monograph. This difference in philosophy is illustrated in Figure 8.3.

It is clear that both approaches have pros and cons. In the first approach, all operations are covered by an

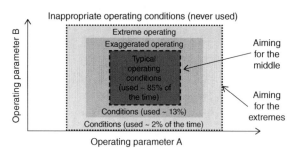

Figure 8.3 Philosophy of risk management in the chemical characterization of manufacturing components. In this figure, a process is established to have two operating parameters, (a) and (b). Pharmaceutical manufacturers who use this process are surveyed to establish the settings they use during process operations. The resulting information is used to establish two design spaces. The first design space encompasses the entire operating space, including extremes in the operating conditions. The second design space "aims for the middle" by only taking into account those process conditions that are most commonly employed as "typical" operations.

experimental design that encompasses the extremes. If the experimental design produces data that meets the acceptance criteria, then operators have the freedom to operate under any and all possible conditions.

However, there are three issues with this approach. The first issue is that the approach is based on the simplifying assumption that extractables profiles will be worst (most compounds at the highest levels) when the extraction conditions are worst (e.g. harshest solvent, highest temperature, longest duration, largest surface area to solution volume ratio). There is no basis in science for this simplifying assumption and examples abound where this is not the case. The second issue is that a majority of operators will be performing tests that are grossly exaggerated versus their conditions of operation. As noted above, if the test results meet the acceptance criteria, then all is good and everyone is happy with the grossly exaggerated testing. Of course, if the test results do not meet the acceptance criteria, then all of a sudden people are unhappy because they are being asked to test with unreasonable conditions and to unreasonable specifications. The third issue considers inefficiency; that is, a testing paradigm based on the worst case for every testing parameter may include test conditions that do not produce worst case outcomes. An example of this was the use of 5 M NaCl as an extraction solvent in the original BPOG protocol. This extraction solvent was included in the original BPOG protocol because it reflects an extreme salt content (ionic strength) for process solutions. However, scientifically, it is reasonable to expect that the levels of organic leachables will be lower in 5 M NaCl than for other aqueous extraction solvents (for example, water) due to a salting-in effect (the high ionic strength shifts the plastic/aqueous solvent partition coefficient to favor the plastic and thus retards leaching). Thus, if the SEP requires both dilute aqueous extraction solutions and 5 N NaCl, the 5 M NaCl extraction solution becomes redundant as it will not produce worst-case data.

These above-mentioned issues with the extreme approach are the strengths of the middle of the road approach. For a majority of operations, the test conditions defined by the "middle of the road" will be reasonable exaggerations of their own operating parameters and thus the test results obtained will more closely resemble the test results they would get by running their operations under these parameters. This is why USP <665> adopted a "middle of the road" approach.

However, what about the operators who work at the extremes? They will potentially be getting an easy ride because the test conditions will be "gentler" than their process conditions and the test results will be less extreme than their operation results. The answer to managing this "free ride" is to acknowledge the boundaries of the "middle of the road" approach and to require that "if you are operating outside the defined boundaries, then you will need to either augment the standard testing procedure with your own extreme conditions or you will need to substitute your extreme conditions into the standard testing procedure and independently establish that your process is qualified to produce acceptable output."

The SEP essentially describes the testing that is performed to qualify a higher risk situation (see Tables 8.1 and 8.2). The extraction solutions specified in the SEP were justified in the previous section; however, there is as much discussion in the E&L community about what solutions are not included in the SEP as there is about what solutions are specified by the SEP. The justification for not including a high ionic strength salt solution as a required extraction solution was addressed previously. The justification for not using solutions containing surfactants such as Polysorbate 80 is based on two points: (i) that the "leaching power" of an extraction solution containing reasonable quantities of surfactants can be mimicked by the 50% ethanol in water solution that is already specified (and thus that use of a surfactant-containing extraction solution is redundant) and (ii) that solutions containing surfactants are analytically challenging when it comes to screening them for organic PERLs at low levels. The justification for not requiring water as an extraction solution is also based on redundancy. As has been noted previously, the extraction of acidic extractables will be enhanced by using a high pH extraction solution, the extraction of basic extractables will be enhanced by using a low pH extraction solution and the extraction of neutral extractables will be largely unaffected by pH. Although a profile of organic extractables may be more complex in a water extraction solution versus either

a low pH or high pH extraction solvent individually, the organic extractables profile in water will not be and cannot be more complex than the **combination** of the extractables profiles for the low pH and high pH extraction solutions. Since the SEP already includes both a low pH and high pH extraction solution, use of a water extraction solution is redundant and not required in USP <665>.

Although the selection of low pH and high pH aqueous extraction solutions has been justified, the exact pH values of these extraction solutions have not been considered. Generally speaking, the values of pH 3 for the low pH extraction solution and pH 10 for the high pH extraction solution were chosen based on two considerations. Consistent that the general strategy upon which <665> is based ("aim for the middle"), it is observed that most process streams routinely used in pharmaceutical and biopharmaceutical manufacturing have pH values between 3 and 10. Scientifically, it is observed that as most basic extractables have acid dissociation constant (pK_a) values greater than 3, reducing the pH of an extraction solvent to a value less than 3 will not increase the solubility of basic extractables to any great extent. This outcome has been confirmed experimentally in extraction studies that have been performed with extraction solutions at pH 3 and using the BioPhorum low pH extraction solvent (0.1 M phosphoric acid, pH ≈ 2.2) and have concluded that the resulting organic extractables profiles are comparable. Considering acidic extractables, it is observed that as most acidic extractables have acid dissociation constant (pK_a) values less than 7, increasing the pH of an extraction solvent to a value greater than 10 will not increase the solubility of acidic extractables to any great extent. Such an outcome has only been partially confirmed experimentally in extraction studies that have been performed with extraction solutions at pH 10 and with the BioPhorum high pH extraction solvent (0.5 N NaOH, pH ≈ 13.6) and which have concluded that there can be differences in the resulting extractables profiles. However, it is likely that the differences in extractables profiles, pH 10 versus 13.6, are not related to just increased solubility of acidic extractables at the higher pH but rather are also related to the decomposition of the plastic and/or extractables that occur at the extreme pH.

The other experimental design parameters of the SEP include the other factors that affect the extent of extraction, notably extraction temperature, extraction duration, and extraction stoichiometry, typically reflected in the ratio of the extracted item's surface area versus the extraction solution's volume.

The combination of extraction temperature and duration influences the extent of an extraction, i.e. the levels to which extractables accumulate in an extract. In some cases, extraction temperature and duration are chosen so that asymptotic extraction occurs; that is, either the total amount of an extractable in the extracted item is extracted or the extraction solution becomes saturated with respect to the extractable. Although asymptotic extraction could be achieved in some manufacturing operations (for example, long-term non-frozen storage of intermediates), for many manufacturing operations the contact temperature is too low and the contact duration is too short to achieve equilibrium. Thus, the extraction conditions of the SEP are not designed to achieve asymptotic extraction, although asymptotic extraction may occur in some cases. Rather, it is the intent that the extraction conditions accelerate and slightly exaggerate the more typically used manufacturing operating conditions. Thus, the SEP extractions are performed at 40 °C, which is elevated versus the temperatures most commonly encountered in manufacturing operations and therefore provides some reasonable degree of acceleration, and at a duration that is consistent with the duration of the manufacturing operation.

Considering the extraction stoichiometry, the USP <665> target SA/V of 6 cm^2/1 ml was adopted from relevant regulatory guidelines and standards.

Many manufacturing operations involve flowing or agitated process streams. Additionally, extractions may be facilitated by movement and mixing of extraction solutions by either agitation or recirculation. Thus, USP <665> specifies that dynamic extractions are to be performed on components via either agitation or recirculation.

8.4 Flexibility in the Implementation of USP <665>

As was mentioned earlier, a significant factor that has hampered USP <665> from becoming official is the juxtaposition between the structure that a monograph and/or regulatory standard should provide and the flexibility that is desired by the monograph's users. Thus, <665> attempts to balance the unreconcilable gap between these two positions by allowing some flexibility in its application.

For example, as has been noted previously, while USP <665> requires the classification of risk, it does not mandate a specific process for performing the risk classification. Rather, <665> establishes the dimensions and factors that a risk classification process must include.

USP <665> also offers considerable flexibility in terms of performing the extraction, the clearest of which concerns incompatibilities between the component being extracted and the extraction process. As it is highly unlikely that manufacturing components deform, degrade, or dissolve under typical operating conditions, it is inappropriate that an extraction of a manufacturing component would

cause the component to deform, degrade, or dissolve. If an extraction solution or process specified in the USP <665> is incompatible with the component being tested (i.e. the component is rendered non-functional by the extraction), then an alternate extraction solution or process should be established and used to replace the conditions specified in <665>. Furthermore, if the alternate extraction process involves an alternate extraction solution, the extraction solution must be analytically compatible with the analytical techniques specified in <665>.

Considering aqueous extraction solutions more specifically, discrepancies between the low and high pH solutions exist between <665> and the BioPhorum Protocol (2). At low pH the discrepancy is easily managed as it has been demonstrated that the extractables profiles produced in both solutions (pH 3 for USP <665> and pH 2.2 for the BioPhorum Protocol (2)) are sufficiently similar that the solutions can be considered to be interchangeable. Thus, an extraction performed with 0.1 M phosphoric acid as the extraction solution meets the extraction requirements in ⟨665⟩ for Solution C2. Furthermore, extraction solutions that achieve pH 3 by various means (e.g. the use of different acids to achieve the pH) are considered to be appropriate surrogates for the pH 3 solution specified in the SEP, so long as the surrogate solution is analytically viable.

The discrepancy at higher pH is more challenging to handle as it has been demonstrated that the extractables profiles at the USP <665> pH of 10 and the BioPhorum pH of 13.5 can be different. USP <665> deals with this situation by stipulating that its pH 10 extraction solvent is to be used unless the following criteria are met:

1. The pH of the contacting process solution is >10.
2. There is a reasonable expectation that extractables resulting from the action of the high pH contacting solution will remain in the process stream.
3. Assessment of the risk associated with this contact situation produces a high-risk characterization (as other risk characterizations do not require a high pH extraction).
4. The contact solution itself, or a proposed simulating solution, are analytically expedient.
5. If these criteria are met, then the <665> pH 10 extraction solution can be replaced with an extraction solution with a higher pH, including the BioPhorum's 0.5 M NaOH solution.

The flexibility contained in <665> with respect to the choice of the extraction solvents is illustrated in Figure 8.4.

Considering the required surface area to solution volume ratio of 6 cm^2/1 ml, specifically when such an extraction ratio cannot be readily achieved, an alternate extraction should be designed so that it is based on the highest possible surface area to solution volume ratio and a justification

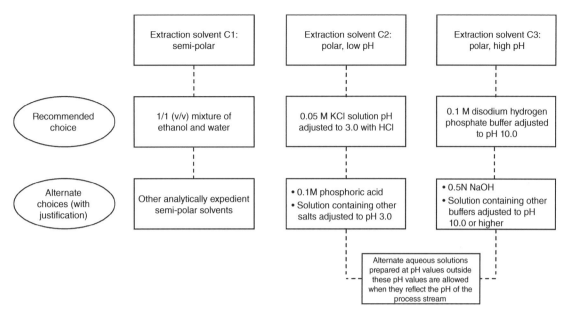

Figure 8.4 Extraction solvents specified in USP <665>.

for the alternate extraction ratio must be provided. Meeting the required surface area to solution volume ratio can be particularly challenging for filters and tubing.

Since the USP <665> extractions were designed to be applicable in a majority of the most commonly encountered manufacturing circumstances, one can encounter extreme manufacturing circumstances where the contact conditions between the process stream and a manufacturing component are more aggressive, in terms of extraction potential, than the extraction conditions specified in <665>. When this is the case, then the more aggressive operating conditions must replace the extraction conditions established in <665>.

Now it is also possible that the extraction conditions specified in USP <665> are more aggressive than the manufacturing conditions of contact. The substitution of a less aggressive extraction for the extraction specified in the SEP is not appropriate, as it is the intent of the SEP to produce worst-case data.

Perhaps the ultimate flexibility of <665> is captured in Section 6.3, Alternate Approaches for Qualification, in <1665>, which reads as follows:

"Chapter <665> establishes qualification procedures and qualification acceptance criteria applicable to manufacturing components that are within its scope. Alternative chemical qualification procedures and acceptance criteria may be appropriate in justified circumstances, subject to agreement by an appropriate regulatory authority. When considering such alternate approaches, the sponsor is advised to consult with the regulatory authority during the

development of such alternate procedures so that regulatory concurrence with the outcome of the alternate qualification is more likely."

8.5 Risk Classification of Manufacturing Components

The need to classify manufacturing components considering their conditions of use is a necessary part of qualifying the components as being suitable. However, the means by which the components have been classified are many and varied. In general, however risk dimensions or factors are established, a component, the contact process solution, and the contact conditions are scored for the individual risk dimensions, the scores for individual dimensions are tallied to produce an overall risk score, and the overall risk score is linked to a risk category based on a "grading scale." The workings of the risk evaluation process, for example, are often captured in a tabular format – hence, the name Risk Evaluation Matrix.

Requirements for and manifestations of various Risk Evaluation Matrices are discussed as follows.

8.5.1 General Requirements per USP <665>

At one point in its development, <665> contained a mandatory Risk Evaluation Matrix. However, the public comments provided, particularly those from industry, did not support the concept of a mandated Matrix, noting essentially that "we have developed our own Matrix and have received regulatory approvals with it so we are not going to change." Bowing to the inevitable, <665> now notes that "It is the sponsor's responsibility to establish and justify the means by which the risk assessment is accomplished. In its justification, the sponsor must consider the multitude of factors that impact the likelihood of leaching and the likelihood of leachables to persist." USP <665> then specifies what considerations the risk evaluation process must address:

1. The chemical and physical nature of the contacted component, establishing the component's "propensity to be leached"
2. The chemical nature of the contacting process stream, establishing the process stream's "leaching power"
3. The conditions of contact, addressing the "driving force" for leaching
4. The ability of downstream process operations (such as filtration) to either eliminate, remove, or clear the PERL from the process stream or dilute the PERL to the extent that an adverse effect is unlikely
5. The inherent risk associated with the manufactured DP, considering such factors as the route of administration of the manufactured dosage form (e.g. inhalation solution (higher risk) versus aqueous-based oral and topical dosage form (lower risk)), the clinical dosing of the DP (i.e. daily dose volume), and the duration of the clinical therapy (i.e. acute versus chronic)

Thus, a Risk Evaluation Matrix complies with the requirements of <665> if the Matrix includes a consideration of the five listed factors, regardless of how those factors are individually or collectively addressed in the Matrix.

8.5.2 A Scientifically Rigorous Risk Evaluation Matrix

A personal objection that I have to Risk Evaluation Matrices in general is that in order to make the Matrices simple, the Matrices are based on intuitive scientific generalities, which may or may not be true. For example, consider the temperature at which a manufacturing operation is performed. Almost everyone will agree that an operation's temperature will affect the degree of leaching of a manufacturing component by a process stream and that the higher the temperature, the greater the extent of leaching. Thus, it is not usual for a Risk Evaluation Matrix to have identified contact temperature as an important dimension to be included in the Matrix. To allow for the assessment of temperature, a grading scale of 1 to 10 (or other values) is set up, where 1 is a low risk of leaching at the lower specified temperature and 10 is a high risk of leaching at a higher specified temperature. Since operating temperatures can be generalized as Frozen, Refrigerated, Ambient,

and Elevated, the operating temperature can be "scaled" as follows:

- Frozen is assigned a risk score of 1,
- Refrigerated is assigned a risk score of 3,
- Ambient is assigned a risk score of 6, and
- Elevated is assigned a risk score of 10.

All this seems fine from a qualitative and intuitive perspective but it does not reflect scientific reality. For example, consider the added risk of leaching that occurs when the temperature of contact is raised from a refrigerated temperature, taken here as 5 °C, and an ambient temperature, taken here as 25 °C. Per the kinetic concepts discussed previously in Chapter 2, Section 2.2.5.1, the effect of a temperature difference of 20 °C on the rate of diffusion (and thus the rate of leaching) can be approximated by the "factor 10 rule," which states that the rate of diffusion will increase by an order of magnitude with each 20 °C increase in temperature. Scientifically, the risk of PERLs accumulating in a process stream is therefore 10 times greater at ambient temperature than it is at refrigerated temperature. However, according to the risk scoring, the risk is only doubled (a score of 6 at ambient versus a score of 3 at refrigerated). Thus, the Risk Evaluation Matrix does not reflect the actual magnitude of the increased risk associated with the higher temperature, which leads to the possibility that the certain manufacturing circumstances are mis-classified into inappropriately low risk classes.

Of course, the challenges in establishing a more scientifically rigorous Risk Assessment Matrix includes the complexity of the Matrix, the difficulty of securing input information for the Matrix, and the circumstance that not all factors contributing to the risk of extractables can be scientifically quantified, meaning that even the most rigorously developed Risk Assessment Matrix is likely to contain certain simplifying assumptions.

Nevertheless, asserting that "unless the specifics of risk evaluation matrices can be scientifically justified, the matrices are, at best, a qualitative tool that provides only a simplistic, general or approximate, and potentially erroneous answer to the complex question of what type and amount of testing is necessary to obtain a rigorous safety assessment of production systems," this author developed and published a "scientifically developed and justified" risk evaluation matrix for manufacturing components [28].

As a starting point, the Risk Evaluation Matrix was based on the five Risk Factors, each of which consisted of multiple contributing Aspects (see Table 8.4).

Each aspect within a Risk Factor was scientifically evaluated to establish and justify a range in risk scores within that aspect and then a means of combining the aspect scores for all aspects within a Risk Factor was developed

Table 8.4 Risk factors used in the scientifically developed risk evaluation matrix.

Risk factor	Aspect	Description
Risk Factor I: Kinetic Factor; Kinetic Driving Force for Leaching	IA	Temperature of Contact
	IB	Duration of Contact
	IC	Dimensional Factor (such as surface area to solution volume ratio)
Risk Factor II: Thermodynamic Factor; The "Leaching Power" of the Contact Solution	IIA	Effect of pH
	II B	Effect of Polarity
Risk Factor III: Material Factors; the Propensity of the Component to be Leached	IIIA	Total Pool of Additives as Potential Extractables
	IIIB	Relative Diffusion Rates
	IIIC	Material Processing Effects (e.g. Irradiation)
	IIID	Ingredients with Safety Alerts
Risk Factor IV: Process Factors; Proximity of the Component to the Final Fill Operation	IVA	Concentrating Manufacturing Operations
	IVB	Diluting Steps between Contact and Final Fill
	IVC	Clearance Steps between Contact and Final Fill
Risk Factor V: Final Product Factors	VA	Final Product's Route of Administration
	VB	Drug Product's Duration of Clinical Use
	VC	Drug Product's Daily Dose Volume

to produce a Factor score for that Factor. Considering the aspect of Temperature of Contact, which appears within the Risk Factor or Kinetic Factor, the following risk scoring was proposed, reflecting the factor 10 rule noted previously:

- If the temperature is <5 °C (frozen), the risk score is 1.
- If the temperature is 5 °C (refrigerated), the risk score is 10.
- If the temperature is 20 °C (ambient), the risk score is 100.
- If the temperature is 40 °C (elevated), the risk score is 1000.
- If the temperature if 60 °C (highly elevated), the risk score is 10 000.

After Risk Scores had been established for each Risk Factor, the individual Risk Scores were combined to produce the Total Risk Score via a combined model using both additive and multiplicative operations. That is, Risk Factors I, II, and III were established to be factors that represent the extent of leaching and which together contribute to

the levels of PERLs. Alternatively, Risk Factors IV and V were established to be mitigating or exacerbating factors; that is, once PERLs had extracted into a process stream Risk Factors IV and V were factors or processes that would reduce or mitigate the effect that the PERLs would have on patient safety.

Obtaining the Total Risk Score was a two-step process. In the first step, the individual factor Risk Scores for Factors I through III were added together. In the second step, the sum of Risk Scores for Factors I through III were multiplied by the Risk Scores for the so-called mitigating factors IV and V:

$$\text{Total Risk Score} = [(\text{RS}_I + \text{RS}_{II} + \text{RS}_{III}) \times \text{RS}_{IV}] \times \text{RF}_V \tag{8.1}$$

As is the case with all Risk Evaluation processes, once the Total Risk Score was calculated it was compared to a "grading scale" so that the Score could be translated into Lower, Moderate, or Higher Risk.

As noted in the Introduction to this Risk Evaluation Matrix, its scientific rigor was counterbalanced by the practical reality that it is difficult to use. For this reason, this Matrix was never embraced by the E&L community and was therefore not implemented in routine use.

8.5.3 BioPhorum Leachables Risk Evaluation Matrix

As part of their Best Practice Recommendations for single-use manufacturing components, the BioPhorum developed their own Risk Evaluation matrix. The model was developed using the collective experience of a group of subject matter experts across the BPOG Extractables and Leachables workstream. The developed model accepts the inherently subjective nature of any risk assessment and does not attempt to eliminate subjectivity completely but ensures that all the relevant scientific aspects are considered in each risk assessment. The risk assessment model considers three dimensions, the composition of the polymeric components of SUS, the processing conditions to which the component is exposed, and the distance along the production stream, defines the factors that are relevant to each dimension, and then scores each factor using an established scale (Table 8.5).

The Leachables Risk Rating, equivalent in concept to a total risk score, is calculated as the weighted average of the individual Risk Dimension risk scores:

$$\text{Risk Rating} = (0.40 \times \text{DAS}) + (0.15 \times \text{ET})$$
$$+ (0.15 \times \text{ED}) + (0.15 \times \text{PFI})$$
$$+ (0.15 \times \text{DR}) \tag{8.2}$$

Table 8.5 The BioPhorum leachables risk evaluation matrix.

Risk dimension	Risk factor	Rating (Score)
Distance along the production stream (DAS)	Upstream	1
	Purification	3
	Bulk drug substance	5
	Final formulation, fill/finish	9
Exposure temperature (ET)	$<0\,°C$	1
	$0–8\,°C$	3
	$>8°$ to $30\,°C$	5
	$>30\,°C$	9
Exposure duration (ED)	Transient (≤ 60 minutes)	1
	Short (≤ 24 hours)	3
	Medium (≤ 7 days)	5
	Long (> 1 week)	9
Process fluid interaction (PFI)	Limited penetration into polymeric component (i.e. water)	1
	Low solvation power or low penetration of polymer component (e.g. neutral pH with organics, surfactants, etc.)	5
	Medium solvation power or medium penetration of polymeric component (e.g. surfactant, low-concentration organic solvents, high/low pH without organics/detergents)	7
	High solvation power or high penetration of polymeric component	9
Dilution ratio (DR)	$<1\times10^{-3}\ \mathrm{m^2/l}$	1
	$1\times10^{-2}\ \mathrm{m^2/l}$ to $1\times10^{-3}\ \mathrm{m^2/l}$	3
	$1\times10^{-1}\ \mathrm{m^2/l}$ to $1\times10^{-2}\ \mathrm{m^2/l}$	5
	$>1\times10^{-1}\ \mathrm{m^2/l}$	9

The BioPhorum Matrix was developed to act as "a road map allowing biopharmaceutical developers/manufacturers to standardize their organization's approach, while still allowing individual end users to apply their perception of the risk to a product associated with a specific process." Thus, "the definitions and numerical scale can be adjusted for an individual organization's needs." Specifically, the numerical ratings (scores) contained in Table 8.5 "represent examples only" and "individual companies should develop their specific (values) according to their internal policies/standard operating procedures." Additionally, the "weighting levels used in (Equation (8.2)) represent examples only. Individual companies may use an equal weighting distribution or may assign weighting levels according to their internal policies."

Table 8.6 Use of the BioPhorum leachables risk rating to establish the requirements for components.

Range of leachables risk rating	Risk level	Component requirements
1.0–3.6	Low	Meets compendial requirements (e.g. USP Class VI, Pharm Eur, etc.)
3.7–6.2	Medium	Low risk requirements plus extractables data evaluation that brackets the intended use (per BPOG protocol)
6.3–8.0	High	Medium risk plus satisfactory extractables data evaluation that brackets the intended use and relevant extractable or leachables profile

To properly use the Leachables Risk Rating to enable appropriate testing, BioPhorum has established a "grading scale" relating the numerical Leachables Risk Rating to the three mandatory levels of risk, low, medium, and high, and then have linked those risk levels to its recommendations for component testing (Table 8.6).

8.5.4 USP Risk Evaluation Matrix per <1665>

In developing a Risk Evaluation Matrix, the USP wanted to move away from the false impression that a number-based algorithm somehow added a level of scientific validity to a process that was based on simplifying generalizations that themselves were over-simplifications, if not actually falsehoods. Rather than using a more or less arbitrary scoring process for each dimension and then employing an unsubstantiated means of mathematically combining the dimension scores to produce a single total score, the USP based their matrix on the simple concept that the higher the risk in each individual dimension, the higher the total risk. Thus, if each dimension was ranked lower, moderate, or higher risk, the more dimensions that were ranked higher risk, the higher the total risk.

Moreover, the USP recognized that there were dimensions that established the risk and then there were dimensions that mitigate the risk. Thus, the USP developed a two-step process where the total or overall risk was established and then the risk was managed by considering mitigating factors.

The USP Risk Evaluation Matrix was based on four factors that have a definitive effect on the leaching of manufacturing components by process streams (see Table 8.7). These four factors are the same factors that were contained in other risk evaluation matrices, including the duration of contact, the temperature of contract, the "leaching power"

Table 8.7 The USP <1665> risk evaluation matrix.

Risk level descriptor	Risk dimension			
	Duration of contact (Driving force for extent of leaching)	**Temperature of contact (Driving force for leaching)**	**Chemical nature of the process solution ("Leaching Power")**	**Chemical nature of the component (Propensity to be leached)**
Level 1 (Lower)	<24 hours	Frozen (<−10 °C)	Aqueous[a]	Low Risk[b]
Level 2 (Moderate)	one to seven days	Refrigerated (2°–8 °C) or Ambient (15°–25 °C)	Somewhat Organic[c]	Moderate Risk[d]
Level 3 (Higher)	>7 days	Elevated (>30 °C)	Highly Organic[e] or Extreme pH[f]	High Risk[g]

a) Aqueous means ≤5% organic v/v; pH ≥3 and pH ≤ 9.
b) Corresponding to a total level of plastic additives in the component that is ≤0.1% by weight.
c) Somewhat organic means a level of organic constituents >5% but ≤40% v/v.
d) Corresponding to a total level of plastic additives in the component that is >0.1% and ≤ 1% by weight.
e) Highly organic means a level of organic constituents >40%.
f) The pH is considered to be extreme if it is < 3 or > 9.
g) Corresponding to a total level of plastic additives in the component that is >1% by weight.

of the process stream, and the propensity of the component to be leached. The dimensions of duration and temperature were numerically defined based on gradations that are consistent with other risk evaluation matrices; for example, the USP gradations of frozen, refrigerated, ambient, and elevated (see Table 8.7) roughly correspond with the BioPhorum gradations for temperature in Table 8.5.

The USP Risk Evaluation Matrix differs from other Matrices in three important ways. First, the dimension of surface area to solution volume ratio is not used in the USP Matrix. There are two reasons for this. First, and most important, is that the effect of surface area to solution volume ratio is complex and goes well beyond the oversimplification that the level of an extractable is directly proportional to the surface area to solution volume ratio. Thus, most Matrices that utilized the surface area to solution volume ratio in their matrix did so in error. Second, the surface area to solution volume ratio for the most commonly used manufacturing components (bags, tubing, and filters) is generally high and promotes leaching. Thus, most manufacturing circumstances would likely be ranked equally, which defeats in part the purpose of having a ranking scale.

The second difference between the USP Matrix and other Matrices is that the individual risk dimensions are broken up into only three categories (lower, moderate, and higher), whereas the dimensions in most numerically-based matrices are divided into four or more categories. The three-category approach was adopted by the USP based on its belief that the categorizations in numerical risk Matrices were based more on appearance

and less on science (that is, a Matrix appears to be ordered if a risk dimension is broken up into evenly spaced categories, even if there is no scientific basis for the spacing).

The third difference between the USP Matrix and the other Matrices is that the gradations in each matrix are defined descriptively (lower, moderate, and higher) as opposed to numerically. This is the case as the USP does not intend to mathematically combine the risk results for each dimension to produce a numerical total risk score. As was the case with the other differences, the USP adopted this approach to move away from the mis-conceptions that because the other Matrices were based on numbers and equations this made them "scientific."

Establishing the gradations in the four dimensions was simple in concept but difficult in practice. The division of the dimensions of temperature and duration of contact was not difficult as manufacturing operations easily fall into categories that are almost universally recognizable. On the other hand, the categories of "leaching power" for the contact solution and "propensity to be leached" for the manufacturing components were more difficult to address. In both cases, it is clear that the property being categorized is impacted by the composition of the item being considered; thus, for example, the composition of a process solution surely affects its "leaching power." Considering this dimension further, it has been previously noted that a solution's "leaching power" is affected by both its pH and polarity. Differentiating "leaching power" on the basis of pH has been previously discussed and naturally leads to the gradations in pH shown in Table 8.7. Differentiating the "leaching power" on the basis of polarity is a little bit more complex and difficult to accomplish and thus it becomes a challenge to capture the aspect of polarity in the matrix in a simple manner. The approach taken by the USP is based on the simplifying assumption that most process streams are aqueous solutions and that the polarity of aqueous solutions is driven by the amount of organic additives (such as surfactants and co-solvents) present in the aqueous solution. Thus, the USP differentiated between aqueous, somewhat organic, and high organic solutions on the basis of the portion of the solution that was organic. The following specific compositional details were used to achieve this differentiation:

- Process streams containing organic solvents:
  - If the level of organic solvents in the process stream is ≤5% by volume, then the process stream is "aqueous."
  - If the level of organic solvents in the process stream is >5% and ≤ 40% by volume, then the process stream is "somewhat organic."
  - If the level of organic solvents in the process stream is >40% by volume, then the process stream is "highly organic."
- Process streams containing surfactants:
  - If the level of surfactants (e.g. polysorbate 80) in the process stream is ≤0.1% by weight, then the process stream is "aqueous."
  - If the level of surfactants in the process stream is >0.1% and ≤ 0.5% by weight, then the process stream is "somewhat organic."
  - If the level of surfactants in the process stream is >0.5% by weight, then the process stream is "highly organic."
- Process streams containing blood or containing blood-derived substances:
  - If the process stream contains blood-derived substances (e.g. albumin) at levels <1% by weight, then the process stream is "aqueous."
  - If the process stream contains blood-derived substances (e.g. albumin) at levels 1–25% by weight, then the process stream is "somewhat organic."
  - If the process stream is blood or if it contains blood-derived substances (e.g. albumin) at levels of ≥25% by weight, then the process stream is "highly organic."
- Process streams containing lipids and proteins:
  - If the level of lipids or proteins in the process stream is <1% by weight, then the process stream is "aqueous."
  - If the level of lipids or proteins in the process stream is 1–5% by weight, then the process stream is "somewhat organic."
  - If the level of lipids or proteins in the process stream is ≥5% by weight, then the process stream is "highly organic."

- Process streams containing other agents:
  - The previous text considers solubilizers in the context of surfactants and the like. Solubilizers could also include agents such as cyclodextrins and chelators. As the effect of such agents on leaching has not been extensively studied and quantified, distinctions such as those applied to surfactants cannot be proposed for these additional agents and it is therefore the risk assessor's responsibility to properly account for these agents in the risk assessment.

The same type of argument was developed for the risk dimension of "propensity to be leached," which is applied to a component and is intended to reflect the component's resistance to leaching. In this case, the USP developed the gradations in this dimension based on the observation that the extent of leaching of a material is strongly affected by its composition and that extractables are related to a material's ingredients. That is, components consisting of polymers that do not have either a large number of ingredients or large quantities of individual ingredients will have fewer extractables at lower levels than will components whose materials of construction contain either a large number of ingredients or individual ingredients at higher levels. Thus, the "propensity to be leached" dimension of the USP Matrix is divided up based on the weight percentage of all additives in a component as follows:

- If the total level of plastic additives in the component is ≤0.1% by weight, then the component is considered to be "low risk."
- If the total level of plastic additives in the component is >0.1% and ≤ 1% by weight, then the component is considered to be "intermediate risk.".
- If the total level of plastic additives in the component is >1% by weight, then the component is considered to be "high risk."

The dimension that addresses "propensity to be leached" is a challenging one to risk assess as it is difficult to enumerate and reconcile all the material properties that can influence a material's "propensity to be leached." One way to address this challenge is to focus on the material itself and not on the material's properties. That is to say, if the leaching characteristics of materials are known, then the materials can be ranked in terms of their individual extractables profiles. More inert materials with modest leachables profiles (e.g. glass, PTFE) would be classified as lower risk. Less inert materials with complex extractables profiles (e.g. elastomeric rubbers, plasticized PVC) would be classified as higher risk.

The challenge in devising such a risk classification for materials is twofold: making sure that all materials used in manufacturing components are on the list and then

Table 8.8 Risk score matrix, material ranked in terms of their "Propensity to be Leached" based on documented leachables profiles.

Risk category	Risk score	Material
Low	1	Poly-(tertrafluoroethylene), PTFE
		Type I and Type II Glass
		Poly(vinylidene difluoride), PVDF
Slight	4	Polyethylene, PE (low and high density)
		Polypropylene, PP
		Polystyrene, PS
		Polyethersulfone, PES
		Polycarbonate, PC
		Polysulfone, PSU
		Polyolefin, COC
		Perfluoroalkoxy,PFA
		Cellulose
		Cellulose acetate
Moderate	7	Silicone
		Reinforced Silicone
		Pt-cured Silicone
		Poly (vinylidene difluoride) – hydrophilic (surface modified)
		Polyolefin, PO
High	10	Ethylene propylene diene monomer rubber, EPDM
		Plasticized poly (vinyl chloride), PVC

properly placing the materials within the list. Ensuring that the list is comprehensive is relatively straight-forward; establishing the materials' proper place on the list is a bit more challenging and undoubtedly involves a certain amount of subjectivity.

Table 8.8 provides an incomplete example of such a material-based classification.

Furthermore, it was recognized that components are often processed before they are used in manufacturing and that such processing could affect the component's "propensity to be leached." Thus, the risk level of a component could be adjusted based on the following types of processing:

- If assembling of the component requires the use of bonding solvents, adhesives, or other chemical means of connecting a component's materials or subassemblies, then the component must be considered either "intermediate risk" or "high risk," depending on the amount of bonding

agents used and documented evidence of their loss from the component prior to its use.

- If the component is flushed or rinsed during use and the flush/rinse solution is removed from the process stream, then the flushing and rinsing has the potential to reduce the number or amount of extractables. If the ability of flushing/rinsing to eliminate or reduce extractables has been established, then such flushing/rinsing can be used to reduce the material reactivity terms by one level, for example, a rinsed "high risk" component becomes an "intermediate risk" component.

The Risk classification of a component used in a particular manufacturing process is accomplished by first evaluating each dimension in Table 8.7 and assigning it a level (1, 2, or 3 corresponding to lower, moderate, and higher risk). This produces a label or risk code consisting of four digits, each digit corresponding to a dimension's level. For example, the four digits 3333 becomes a risk code that corresponds to a situation where the risk level for each dimension was a higher risk (level 3). Conversely, the four digits 1111 would correspond to a situation where the risk level for each dimension was a lower risk (level 1). This combined code reflects the risk that PERLs can be leached from components by the process stream.

The next step in Risk classification is to link the risk code with an overall level of total risk, as noted in Table 8.9. As has been noted previously, Table 8.9 is based on the logic that the more dimensions that are individually rated at a higher risk, the higher the total risk will be. Furthermore, the total risk classifications produced by use of Table 8.9 were all examined to see if they "made sense" in terms of what total risk levels were produced for which combinations of process stream, component, and process conditions. For example, if a particular combination of process stream composition, component composition, and process conditions produced an outcome of Level C, this outcome was reviewed to establish whether it seemed appropriate for the conditions from which it was generated.

It was during this "validation" of the risk classification process that the USP established that not all four of its risk dimensions contributed to the risk of leaching equally. That is, the dimensions of temperature, duration, and "propensity to leach" likely have a greater effect on the likelihood of leaching than did the dimension of "propensity to be leached." This observation is important, especially at the boundaries between the risk levels; for example, when the risk could either be classified as higher versus moderate or moderate versus lower. Thus, footnotes were added to Table 8.9 to help assessors "manage the boundaries" and assign the proper Characterization Levels to specific circumstances.

Table 8.9 Linking the USP <1665> numerical risk sequence to the risk level.

If ...	And ...	Then the characterization level is ...
All four of the dimension scores are level 3 (e.g. 3333)	There is no additional qualifier	Level C (Higher Risk)
Three of the dimension scores are level 3	The other dimension score is level 2 (e.g. 3332) or level 1 (e.g. 3331)	Level C
Two of the dimension scores are level 3	The other two dimension scores are both level 2 (e.g. 3322)	Level C
	One of the dimension scores is level 2 (e.g. 3321)	Level B (Moderate Risk) or C (Low Risk)[a]
	The other two dimension scores are level 1 (e.g. 3311)	Level A or B[b]
One of the dimension scores is level 3	All of the other dimension scores are level 2 (e.g. 3222)	Level B
	One of the other dimension scores is level 1 (e.g. 3221)	Level B
	Two of the other dimension scores are level 1 (e.g. 3211)	Level A or B[c]
	All of the other dimension scores are level 1 (e.g. 3111)	Level A
None of the dimension scores is level 3	All of the dimension scores are level 2 (e.g. 2222)	Level B
	At least one of the dimension scores is level 1 (e.g. 2221, 2211, 2111, 1111)	Level A

In these cases, the temperature, solvent, or duration dimensions have a greater influence on risk than do material considerations.
a) If the level 2 score is in the temperature, solvent, or duration dimensions, then level C; otherwise, level B.
b) If one of the level 1 scores is in the material considerations dimension, then level A; otherwise, level B.
c) If one of the level 1 scores is in the material considerations dimension, then level A; otherwise, level B.

The assessment to this point has produced a Characterization Level of A (higher risk), B (moderate risk), and C (lower risk), accounting for the likelihood that component PERLs will be present in the process stream. The assessment continues by considering mitigating factors that might impact either the PERL's ability to persist in the process stream as the stream moved through downstream process steps and become entrained in the finished drug product or the effect that the PERL could have on patient health if it was entrained in the finished drug product. Thus, the first mitigating factor considers the ability of the downstream manufacturing process to remove (clear) the PERL from the process stream, as it is obvious that if the PERL is removed from the process stream it cannot accumulate in the finished drug process. If it can be established that downstream processing of the process stream removes PERLs, then the risk of PERLs accumulating in the finished drug product is mitigated and the Characterization Level for a particular situation is downgraded by one class. For example, if the assessment established that a given situation had a Characterization Level of A (higher risk) and it was further established that downstream processing removed PERLs from the process stream, then the Characterization Level would be reduced to B (moderate risk).

A second mitigation factor was applied to reflect the potential impact that a PERL could have on patient safety should it persist through downstream manufacturing and become entrained in the finished drug product. Speaking at a high level, the effect that a leachable could have on patient health depends on a drug product's route of administration (with some routes being higher risk than others), duration of clinical use (with shorter durations of use having lower risk), and daily dose amount (with smaller dose amounts having lower risk). Although these factors do not influence the magnitude to which a component will be leached or the extent to which a PERL will be cleared, they do influence the effect the PERL will have on patient safety and therefore should be taken into account when establishing the testing that is required to qualify a manufacturing component.

The second mitigating factor is used as follows:

- If the dosage form is a solid oral, liquid oral (not containing surfactants, co-solvents such as alcohols, or other solubilizing agents), or topical, then use the mitigating factor.
- If the duration of clinical use is <7 days, then use the mitigating factor.

- If the daily dose volume is <10 ml, then use the mitigating factor.
- Otherwise, do not use the factor.

For example, if the assessment established that a given situation had a Characterization Level of C (higher risk) and it was further established that the finished drug product was clinically used only once (for example, a vaccine), then the Characterization Level would be reduced to B (moderate risk).

It is noted that both mitigating factors could be applied to reduce a Characterization Level. Thus, if the circumstance being evaluated had both mitigating factors noted in the previous examples (downstream clearance of PERLS and acute finished drug product use), then the Characterization Level would be reduced by two levels, essentially meaning that in circumstances where both mitigating factors can be applied the only possible Classification Level is A (lower risk).

To establish a component's required qualification testing, the resulting mitigated Characterization Level is linked to test procedures in Table 8.3.

8.6 Case Studies

As extractables and leachables from manufacturing components is one of the current hot topics in this discipline, there are several recent publications on this topic. Recent publications that provide information on the identity and concentrations of extractables from SUS components are discussed as follows.

Researchers from the National Institute for Bioprocessing Research and Training (NIBRT) and ThermoFisher have published the results obtained from testing bioprocess bags from four unspecified vendors for extractables using accelerated solvent extraction (ACE) coupled with high-resolution accurate mass (HRMS) LC/MS analysis [29]. These authors report that 101 compounds in total were extracted from the plastic films, representing diverse functional groups spanning a wide range of polarities and including hindered phenolic antioxidants, phthalates as plasticizers, siloxanes commonly used as lubricants and coupling agents, polymer degradation products, monomers, production intermediates, processing aids, adhesives, surface active agents, and UV absorbers, among others. Antioxidants and their degradation products formed the bulk of the compounds identified and were mainly composed of sterically hindered phenols and tert-butyl groups, as well as phosphite antioxidants. These authors noted that many of the compounds (40.6%) were common across the four investigated plastic films, but that some of the compounds were found in one only type of plastic film (35.6%), while 11.9% of compounds were found in either two or three plastic films. The highest number of compounds (75) were identified in a film whose inner layer is made of EVA.

In an investigation of polyolefin-based films used in bioprocessing bags, researchers from Sartorius Stedim Biotech identified Irgafos 168-related degradation products that inhibited cell growth, including bis(2,4-di-tert-butylphenyl)phosphate (bDtBPP), 3,3',5,5'-tetra-tert-butyl-2,2' dihydroxybiphenylphosphate (TtBBP) and bis(pnonylphenyl) phosphate (bNPP) [30]. Although all three compounds exhibited cytotoxicity, as established by means of cell culture spiking experiments and flow cytometry analysis, bDtBPP produced the most pronounced effect.

PERLs from protein drug filling lines were investigated by a team of researchers from Roche Diagnostics GmbH and Ludwig Maximilian University using an analytical approach that coupled stir bar sorptive extraction as a sample preparation technique with thermal desorption GC/MS and LC/MS/MS analysis [31]. The filling lines consisted of stainless-steel mixing and storage vessels, silicone connecting tubing, disposable drug substance, and buffer storage bags, sterilizing filters, and connectors. These researchers preformed extraction studies on the individual components, focusing on the silicone tubing, buffer bags, and filters, and also secured actual samples of the manufacturing process solutions at various points along the filling line. Considering extractables, the extractions were performed at 70 °C for 8–72 hours using a 50/50 ethanol/water extraction solvent. Reported results were as follows:

- *Silicone tubing*. The extractables were predominately a series of homologous series of polydimethyl-siloxanes with a general structural formula of $[(CH_3)_2SiO]_n$. Compounds reported in lesser quantities included phthalate-related plasticizers and their associated related substances and degradation products. Tubing process residuals, including lubricants, intermediates, solvents, and preservatives, were also reported.
- *Buffer bags*. The buffer bag was a multi-layered structure including polyethylene and ethylene vinyl alcohol layers. The bag had a port consisting of a thermoplastic elastomer and silicone tubing. The bag's extractables profile consisted of:
  - A homologous series of polydimethyl siloxanes, attributable to the tubing,
  - Alkanes, antioxidants and anti-oxidant degradation products attributable to the bag's PE layers,
  - "Sulfurous" compounds, such as di-tert-dodecyl disulfide, the butyl ester of 6-tertadecanesulfonic acid, and octadecyl pentyl ester sulfurous acid, attributable to the thermoplastic port.

- *Filter.* The filter consisted of a polyvinylidene fluoride (PVDF) membrane and a polypropylene housing. Reported extractables included alkanes and antioxidants (including BHT) attributable to the PP housing and phthalates (e.g. diethyl phthalate) and adipates (e.g. di-(2-ethylhexyl) adipate and di-(2E)-2-buten-1-yl adipate) generalized as plasticizers but not attributed to a specific source.
- *Polysulfone connectors.* The reported extractables included tetramethylbutyl phenol, trimethylolpropane trimethacrylate, 1,1′-(1-methylethylidene)bis[4-methoxy]-benzene, and 1-methoxy-4-[(E)-2-phenylethenyl] benzene, the latter two of which are building blocks of the polysulfone material.

Despite the relatively larger number and high amount of extractables, these researchers report that only minimal quantities of PERLs (referred to as leachables) were present in the actual process solutions. A majority of those leachables that were present in the process solutions were present in the API that was added at the beginning of the process (presumably from upstream manufacturing). Rather than measuring an increase in leachables' levels over the entire filling process, the researchers reported that leachables were cleared from the process stream, presumably due to dilution of the API with the drug product's diluent and by absorption of the leachables attributed to the filters and silicone tubing.

A research team from Jordi Labs and Agilent Technologies reported extractables from various SUS components including thermoplastic elastomer tubing, a nylon hydrophilic disk filter, a polypropylene filter housing, and a bioprocessing bag consisting of LDPE, EVOH, and nylon layers [32]. These items were extracted in water, 50% ethanol/water, and saline solutions at 50 °C for 72 hours and the extracts were screened for organic extractables by GC/MS (Q-TOF) and LC/MS (Q-TOF). The authors listed the top 10 extractables present in the extracts at levels above an AET of 37.5 µg/test article for each component and each extraction solvent; extractables reported in the ethanol extracts are summarized in Table 8.10.

As part of an investigation into the theoretical and practical considerations of selecting solvents for use in SUS extraction studies [27], a research team from Sartorius Stedim extracted two films used in bioprocessing bags (one film consisting of PE, EVOH, PA, and PET layers and the other consisting of EVA and EVOH), TPE and silicone tubing, and a cartridge filter with multiple extraction solvents and characterized the extracts for organic extractables using the relevant chromatographic test methods and extracted elements by ICP/MS. In addition to reporting certain of the extracted entities, these researchers drew

Table 8.10 Organic compounds extracted from SUS components with ethanol [32].

SUS component	Reported organic extractables
Bag	• Caprolactam • Bis(2,4-di-tertbutylphenyl) phosphite (2 isomers) • Bis(4-nonylphenyl)hydrogenphosphate • Stearic acid • Erucamide • Irgnox 1076 degradation products • Irgafos 168 degradation products • Trioctyl trimellitate • Tris(nonylphenyl) phosphite degradation products • Ethylene vinyl acetate oligomers
Filter	• Caprolactam • Polyether sulfone oligomers • Hexadecamethylcyclooctasiloxane
Tubing	• Bis(2,4-di-tertbutylphenyl) phosphite (2 isomers) • Stearic acid • Irgafos 168 degradation products • Numerous alkyl siloxanes • Degradation product(s) of Irganox 3114

conclusions about the applicability of the various extraction solvents that were employed as follows:

- Ethanol extracted alkanes, antioxidants and their byproducts, amides, and carboxylic acids at levels high enough to facilitate their identification.
- Ethanol/water mixtures with an ethanol content of less than 80% have extraction properties that are more like those of water than ethanol. While polar extractables such as alcohols, aldehydes, and ketones will migrate into ethanol/water mixtures, more hydrophobic compounds such as antioxidant degradation products dissolve poorly in such mixtures. A 50% ethanol solution can be regarded as suitable for simulation studies.
- Pure water extracts contained a large variety of hydrophilic compounds such as ketones and alcohols at low concentrations.
- 1M hydrochloric acid is more suitable than the other solvents for extracting elements.
- A basic solution, such as 1 M NaOH reveals carboxylic acids and phenols.
- Other investigated solvents, including 1% polysorbate 80, and an aqueous mixture with DSMO, did not result in extractables that were not already observed in other extraction solvents such as water, ethanol, an acid solution, and a basic solution.

Based on their experimental data and other published information [33], these authors advocate the use of a suite of extraction solvents including water, ethanol, strong acid, and strong base to cover a large panel of extractable substances.

An extensive study of single-use bags used for bioprocessing was published by a group of investigators from the National Institute for Bioprocessing Research (NIBRT), ThermoFisher Scientific, Pfizer Ireland BioMarin International Limited, and University College Dublin [34]. In this study, 34 plastic films from single-use bags used for cell cultivation were extracted with five extraction solvents (water, 0.1 M phosphoric acid, 50 mM ammonium acetate buffer at pH 9.5, 1/1 isopropanol/water, and a laboratory-produced cell culture medium) at 50 °C for two days. The resultant extracts were characterized for organic extractables using screening chromatographic methods (GC/MS and LC/MS) and for extracted elements by ICP-MS. The films were also tested for volatile organic extractables using headspace GC/MS.

In addition to compiling information for over 100 organic extractables, these researchers reported the following observations:

- Semi-volatile compounds identified by GC/Ms. included hindered phenolic antioxidants and their degradation products, phthalates, siloxanes, and polymer fragments. Degradation products of the antioxidants formed the bulk of the extractables identified, with di-tert-butylbenzene being the most frequently reported extractable.
- A total of 103 organic extractables were identified via LC/MS, with most of the compounds (71%) being present in the IPA/water extracts. The major extractable compounds reported were antioxidants and their degradation products, plasticizers, polymer-related compounds including oligomers, and residues from the polymerization process such as initiators and catalysts.
- A total of 26 organic solvents were measured as volatile extractables. Of these solvents, 2-propanol, methanol, and n-hexane were thermally extracted in the largest quantities, while acetone and 1,4-dioxane were extracted in smaller quantities. The other 19 solvents were all extracted in trace quantities.
- Targeted extracted elements included the 30 that are listed in ICH Q3D. Elements that were not detected in any of the films included As, Se, Ru, Rh, Pd, Os, Tl, and Pt. Al and Fe were the elements extracted in the largest amounts in most films and Cu and Zn were extracted from most films but in lesser amounts. Fe, Zn, Ni, and Cr were extracted from all the tested bags in measurable quantities; other commonly extracted elements included Pb, Al, and Sb (30 bags) and Hg and Cu (28 bags).
- Multivariate analysis of primarily the GC/MS extractables data grouped the tested bags in terms of commonalities in their extractables profiles. The resulting groups differentiated bags primarily in terms of manufacturing date and polymer type, suggesting to the authors that over time the additive content of the bags had been changed consistent with the reported effect of certain extractables on the viability of cultured cells.

References

1 Equipment Construction. Code of Federal Regulations, Foods and Drugs Title 12, Part 211.65. US Government Printing Office, Washington, DC. Revised 1 April 2006.

2 Auerbach, M. (2010). What lies within. *Pharm Process.* 25 (9): 12, 13.

3 Sandle, T. and Sayhee, M.R. (2011). Some considerations for the implementation of disposable technology and single-use systems in biopharmaceuticals. *J. Commer. Biotechnol.* 17: 319–329.

4 Langer, S. What's holding back single use adoption? https://www.pharmaceuticalprocessingworld.com/whats-holding-back-single-use-adoption/. Accessed 16 December 2020.

5 Rudge, S. Single-use systems for biotechnology products. https://www.europeanpharmaceuticalreview.com/article/50764/single-use-systems-biotechnology-products-2/. Accessed 16 December 2020.

6 Margarian, N., Lee, K., Nagpal, K. et al. (2016). Clearance of extractables and leachables from single-use technologies via ultrafiltration/diafiltration operations. *Biotechnol. Prog.* 32 (3): 718–724.

7 Huak, A., Jurkiewicz, E., Pahl, I. et al. (2018). Filter membranes – Scavengers for leachables? *Eur. J. Pharm. Sci.* 120: 191–198.

8 Scherer, N., Marcseková, K., Posset, T., and Winter, G. (2019). New studies on leachables in commercial scale protein drug filling lines using stir bar sorptive extraction coupled with td-gc–ms and uplc/qtof-ms/ms analytics. *Int. J. Pharm.* 555: 404–419.

9 Paudel, K., Hauk, A., Maier, T.V., and Menzel, R. (2020). Quantitative characterization of leachables sinks in biopharmaceutical downstream processing. *Eur. J. Pharm. Sci.* 143: https://doi.org/10.1016/j.ejps.2019.105069.

10 Kelly, P.S., Dorival, N., Pare, S. et al. (2019). Improvements in single-use bioreactor film material composition leads to robust and reliable Chinese hamster ovary cell performance. *Biotechnol. Prog.* 35 (4): e2824.

11 Hammond, M., Nunn, H., Rogers, G. et al. (2013). Identification of a leachable compound detrimental to cell growth in single-use bioprocess containers. *PDA J. Pharm. Sci. Technol.* 67 (2): 123–134.

12 Hammond, M., Marghitoiu, L., Lee, H. et al. (2014). A cytotoxic leachable compound from single-use bioprocess equipment that causes poor cell growth performance. *Biotechnol. Prog.* 30 (2): 332–337.

13 Kelly, P.S., McSweeney, S., Coleman, O. et al. (2016). Process-relevant concentrations of the leachable bDtBPP impact negatively on CHO cell production characteristics. *Biotechnol. Prog.* 32 (6): 1547–1558.

14 Min, H., Teresa, H.S., Cindy, Z., and Satish, S.K. (2011). Impact of extractables/leachables from filters on stability of protein formulations. *J. Pharm. Sci.* 100 (11): 4617–4630.

15 Peng, J., Zhao, Y., Hong, Y. et al. (2018). Chemical identity and mechanism of action and formation of a cell growth inhibitory compound for polycarbonate flasks. *Anal. Chem.* 90: 4603–4610.

16 Recommendations for Testing and Evaluation of Extractables from Single-Use Process Equipment. Bio-Process Systems Alliance. 2010.

17 Colton, R. (2007). Recommendation of extractables and leachables testing. Part 1: Introduction, regulatory issues, and risk assessment. *BioProcess Int.* 5 (11): 36, 38, 40, 42, 44.

18 Colton, R. (2008). Recommendation of extractables and leachables testing. Part 2: Executing a program. *BioProcess Int.* 6 (1): 44, 48, 50, 52.

19 Extractables/Leachables Considerations for Cell & Gene Therapy Drug Product Development. Bio-Process Systems Alliance. 2020.

20 Ding, W., Madsen, G., Mahajan, E. et al. (2014). Standardized extractables testing protocol for single-use systems in biomanufacturing. *Pharm. Eng.* 34 (6): 1–11.

21 BioPhorum Best Practices Guide for Extractables Testing of Polymeric Single-use Components Used in Biopharmaceutical Manufacturing. BioPhorum Operations Group Ltd. April 2020.

22 A Comprehensive Review of BioPhorum Standardized Extractables Testing Data: A Deep-dive into Similarities, Differences and Trends Across Extraction Solvents and Timepoints. BioPhorum Operations Group Ltd. April 2020.

23 USP ⟨665⟩ Plastic components and systems used to manufacture pharmaceutical drug products and biopharmaceutical drug substances and products. USPNF 2022, Issue 1. Official as of 1-May-2022.

24 Guide to the Extractables Test Report. BioPhorum Operations Group Ltd. April 2020.

25 BPOG Best Practices Guide for Evaluating Leachables Risk in Biopharmaceutical Single-Use Systems. BioPhorum Operations Group Ltd. April 2020.

26 USP <1665> Characterization and qualification of plastic components and systems used to manufacture pharmaceutical drug products and biopharmaceutical drug substances and products. USPNF 2022, Issue 1. Official as of 1-May-2022.

27 Dorey, S., Pahl, I., Uettwiller, I. et al. (2018). Theoretical and practical considerations when selecting solvents for use in extractables studies of polymeric contact materials in single-use system applied in the product of biopharmaceuticals. *Ind. Eng. Chem. Res.* 57: 7077–7089.

28 Jenke, D. (2015). Development and justification of a risk evaluation matrix to guide chemical testing necessary to select and qualify plastic components used in production systems for pharmaceutical products. *PDA J. Pharm. Sci. Technol.* 69 (6): 677–712.

29 Dorival–García, N., Galbiati, F., Kruell, R. et al. (2020). Identification of additives in polymers from single-use bioprocessing bags by accelerated solvent extraction and ultra-high performance liquid chromatography coupled with high-resolution mass spectrometry. *Talanta* https://doi.org/10.1016/j.talanta.2020.121198.

30 Budde, D., Maier, T.V., Jurkiewicz, E. et al. (2020). Identification and evaluation of cell- growth-inhibiting bDtBPP-analogue degradation products from phosphite antioxidants. *Anal. Bioanal. Chem.* 412: 4505–4518.

31 Scherer, N., Marcsekova, K., Posset, T., and Winter, G. (2019). New studies in commercial scale protein drug filling lines using stir bar sorptive extraction coupled with TD-GC-MS and UPLC/QTOF-MS/MS analytics. *Int. J. Pharm.* 555: 404–419.

32 Jordi, M.A., Khera, S., Roland, K. et al. (2018). Qualitative assessment of extractables form single-use components and the impact of reference standard selection. *J. Pharm. Biomed. Anal.* 150: 368–376.

33 Pahl, I., Dorey, S., Uettwiller, I. et al. (2018). Development of a standardized extractables approach for single-use components. *BioProcess Int.* 16 (10): 3–11.

34 Dorival-Garcia, N., Carillo, S., Ta, C. et al. (2018). Large-scale assessment of extractables and leachables in single-use bags for biomanufacturing. *Anal. Chem.* 90 (15): 9006–9015.

9

Applications – Medical Devices

9.1 Introduction

A medical device is any device intended to be used for medical purposes. Medical devices benefit patients by helping health care providers diagnose and treat patients, helping patients overcome sickness or disease, and improving their quality of life. Medical devices vary in both their intended use and indications for use. Examples range from simple, low-risk devices such as tongue depressors, medical thermometers, disposable gloves, and bedpans to complex, high-risk devices that are implanted for the purpose of sustaining life.

A more detailed definition of a medical device is provided by the US FDA:

Per Section 201(h) of the Food, Drug, and Cosmetic Act, a device is an instrument, apparatus, implement, machine, contrivance, implant, *in vitro* reagent, or other similar or related article, including a component part, or accessory which is:

- Recognized in the official National Formulary, or the United States Pharmacopoeia, or any supplement to them,
- Intended for use in the diagnosis of disease or other conditions, or in the cure, mitigation, treatment, or prevention of disease, in man or other animals, or
- Intended to affect the structure or any function of the body of man or other animals, and which does not achieve its primary intended purposes through chemical action within or on the body of man or other animals and which is not dependent upon being metabolized for the achievement of its primary intended purposes.

The EU Medical Device Regulation (MDR, EU2017/745) defines a medical device as: Any instrument, apparatus, appliance, material or other article, whether used alone or in combination, including software necessary for its proper application, intended by the manufacturer to be used for human beings for the purpose of:

- Diagnosis, prevention, monitoring, treatment, or alleviation of disease;
- Diagnosis, monitoring, treatment, alleviation of or compensation for an injury or disability;
- Investigation, replacement, or modification of the anatomy or of a physiological process;
- Control of conception.

and which does not achieve its principal intended action in or on the human body by pharmacological, immunological, or metabolic means, but which may be assisted in its function by such means.

In common with all medical items, medical devices are intended to provide a therapeutic benefit to the person in or on whom the medical device is used (hereafter referred to as the patient). However, it is an unavoidable reality that the use of a medical device exposes the patient to biological hazards. During the use of a medical device, the patient and the medical device come into contact, either directly or indirectly. Direct contact occurs when the device is in physical contact with some part of the human body. Examples of direct contact medical devices include a transdermal patch (skin-contact), urinary catheter (mucosal-contact) and an angioplasty balloon (implant). Indirect contact occurs when there is an intermediary between the medical device and the patient. An example of an indirect contact device (also called an externally communicating device) is an administration set used to deliver an i.v. drug product. Although the device itself does not physically contact the patient, the administered i.v. drug product serves as the medium through which the indirect contact occurs. By virtue of the direct or indirect contact, a device is said to "communicate" with the patient.

As a result of the patient and the medical device "communicating" during its clinical use, patients are exposed to chemicals that are leached from the device. This exposure can potentially impact patient health when the chemicals are transferred (leached) from the device to the patient during use. Chemicals that are transferred from

Extractables and Leachables: Characterization of Drug Products, Packaging, Manufacturing and Delivery Systems, and Medical Devices, First Edition. Dennis Jenke.
© 2022 John Wiley & Sons, Inc. Published 2022 by John Wiley & Sons, Inc.

the medical device to the patient during clinical use are termed leachables.

The biological effect that leachables could have on a patient can be established by direct testing or indirectly by inference. Considering direct testing, one could observe patient responses during clinical use of the medical device and thereby establish any health impacts, the practical and ethical implications of which are considerable and obvious. Thus, the biological effect of leachables on patient health is inferred by leaching the medical device and testing the resultant leachate. Leachates of the medical device can be tested biologically (e.g. subjecting a biological system to the leachate and observing the effect) or chemically (e.g. establishing what chemicals are present in the leachate and then inferring the chemicals' likely impact on health). Biological testing of leachates to establish the biological effect of leachables on patients is outside the scope of this book; the reader is referred to relevant guidelines on biological testing, such as ISO 10993:1 [1, 2] and USP monographs <87> [3] and <88> [4].

To chemically establish the potential patient safety impact of a medical device, leachables transferred from the medical device to the patient must be detected, identified, and quantified, as a leachable's identity (chemical structure) establishes its toxicity (hazard) while a leachable's concentration establishes the extent of patient exposure (severity).

For direct contact devices, it is the contact item (organ, tissue, or body fluid) that does the leaching and is the recipient of the leachables. For indirect contact devices, it is the contact medium that is transferred by or through the device that does the leaching and it is the item (organ, tissue, body fluid) to which the medium is delivered that is the recipient of the leachables.

It is clear that leachables are not measured by testing the medical device directly but rather by testing the contacted item (in the case of direct contact) or contacting medium (in the case of indirect contact). That is, the only way to definitively establish what chemicals and what amounts of chemicals a patient is exposed to by contact with the medical device is to test the contacted item or contacting medium. Considering an indirect contact device, this is generally straightforward, as the contacted medium is typically a liquid delivered through or by the device. If the liquid leaving the device is collected, the collected liquid can be analytically screened for leachables. On the other hand, analytical screening becomes more complicated for certain types of direct contact devices. If the device contacts a body fluid directly, then it is possible that if the fluid could be collected after contact with the device, the fluid could be screened for leachables. However, if the device contacts an organ or tissue directly, it would be the tissue or organ

that would need to be screened for leachables, which might present interesting and possibly insurmountable sampling issues. Thus, while leachables profiling of indirect contact medical devices is technically feasible, leachables profiling of direct contact medical devices is generally impractical, if not impossible.

Corresponding to leachables are extractables, which are substances that are present in laboratory extracts of the medical device and which could be leached (i.e. are potential leachables). In circumstances where leachables cannot be measured (such as direct contact medical devices), extractables information is the only viable surrogate for leachables data and it is the extractables data that stands in the place of leachables data as the basis of a toxicological safety risk assessment.

In circumstances where leachables can be more or less readily measured (such as indirect contact medical devices), extractables information can facilitate leachables testing but generally does not replace leachables testing. Thus, for example, if one has characterized a device for extractables, these extractables can be targeted as leachables when leachables testing is performed. When targets have been specified, the analytical methods used for testing can be optimized and qualified to produce reliable analytical data for the targets.

Surely it is the case that it is always desirable that extractables and leachables be well-correlated, meaning that all leachables are extractables and the levels of leachables are all similar to but less than the levels of extractables. If there is not a good correlation between extractables and leachables and the extractables information is the basis of the safety risk assessment, it is clear that the safety risk assessment based on extractables is likely to be flawed, either overestimating or underestimating the true patient safety risk associated with leachables. Neither underestimation nor overestimation is a desirable outcome. The issue with underestimation is clear; an assessment based on an underestimated leachables profile could result in the commercialization of an unsafe medical device. The issue with overestimation is a little more subtle. A medical device that is established to be unsafe based on an overestimated leachables profile will not be commercialized. It may be, however, that the rejected medical device is safe to use under its prescribed conditions of use. In this case, patients who need the rejected medical device to fulfill a medical need will be deprived of the device and will have their medical needs unmet.

Neither of these two outcomes is acceptable. However, it is a generally accepted philosophy, particularly in the regulatory community, that the greater sin is approving an unsafe medical device. Consistent with this philosophy is the regulatory bias toward studies that produce extractables profiles that overestimate, in some cases substantially, the leachables profile.

The degree to which an extractables profile represents a leachables profile is driven by the design of the extraction study. An extraction study that accurately simulates the conditions of leaching will produce an extractables profile that closely mirrors the leachables profile. Conversely, an extraction study whose extraction conditions are quite different from the conditions of leaching will produce an extractables profile that differs from the leachables profile, oftentimes substantially.

In any event, medical devices are chemically characterized to establish those chemicals to which a patient may be exposed during the device's clinical use. The process of chemical characterization is addressed as follows.

9.2 Regulatory Landscape: ISO 10993; Generating an Extract

The International Organization for Standardization (ISO) has created a series of Standards, collectively designated as ISO 10993:n, where n is the number of the individual Standard, whose collective objective is "the protection of humans from potential biological risks arising from the use of medical devices" [1]. This series of Standards "is intended to be a (set of) guidance document(s) for the biological evaluation of medical devices within a risk management process, as part of the overall evaluation and development of each device."

Different individual Standards within ISO 10993, referred to as "Parts," address different aspects of the biological evaluation of medical devices. It is noted that the term "biological evaluation" is not limited to biological testing; rather, it includes chemical tests whose results are interpreted in the context of the chemicals' biological actions. The Parts of ISO 10993 that specifically address the preparation of extracts for biological evaluation include Part 12, Sample preparation and reference materials [5], and Part 18, Chemical characterization of medical device materials within a risk management process [6].

9.2.1 ISO 10993:12(2012)

ISO 10993:12 specifies requirements and gives guidance on the procedures to be followed in the preparation of samples and the selection of reference materials for medical device testing in biological systems in accordance with one or more of the other Parts of ISO 10993. In addition to providing definitions for key terms involving extractions and extractables, the generation of extracts for biological testing

is addressed in Section 3, Terms and definitions (normative text), Section 10, Preparation of extracts of samples (normative text), and informational Annexes C, Principles of test sample extraction, and D, Exhaustive extraction of polymeric materials for biological evaluation. Important concepts contained within these sections of Part 12 are summarized as follows.

Section 3, Terms and Definitions

Extract. Liquid that results from extraction of the test sample or control

Accelerated extraction. Extraction that provides a measure of the leachable or extractable materials of the device or material, using conditions that shorten the time for leaching of the substances into the extraction vehicle but do not result in a chemical change of the substances being extracted. Means of accomplishing an accelerated extraction include the use of elevated temperature or agitation or changing of the extraction vehicle.

Exaggerated extraction. Extraction that is intended to result in a greater amount of a chemical constituent being released as compared to the amount generated under the simulated conditions of use. Note that it is important to ensure that the exaggerated extraction does not result in a chemical change of the material.

Exhaustive extraction. Extraction conducted until the amount of extractable material in a subsequent extraction is less than 10% by gravimetric analysis of that detected in the initial extraction.

Simulated-use extraction. Extraction conducted to demonstrate compliance with the requirements of this Part of ISO 10993 by evaluating leachable material levels available to the patient or user from devices during the routine use of a device, using an extraction method that simulates product use.

It is noted that the burden of validation on the analytical laboratory is to demonstrate that the simulated-use extraction is carried out under conditions that provide the greatest challenge to the intended use. Product-use simulation is carried out assuming that the device is assigned to the most stringent category possible for the duration of exposure and takes into consideration both the tissue(s) exposed and the temperature of exposure.

Extractables: Substances that can be released from a medical device or material using extraction solvents and/or extraction conditions that are expected to be at least as aggressive as the conditions of clinical use.

Leachables. Substances that can be released from a medical device or material during clinical use.

Section 10, Preparation of extracts of samples

The following aspects of the extraction process are addressed in Section 10.

General Statement (Section 10.3.11). If extracts of the device are required for a test procedure, the extraction vehicles and conditions of extraction used shall be appropriate to the nature and use of the final product and to the purpose of the test, e.g. hazard identification, risk estimation, or risk assessment. The physicochemical properties of the device's constituents, leachable or extractable substances or residues, shall be considered when choosing the extraction conditions.

Extraction Solvents (Section 10.3.5). Extraction using both polar and non-polar extraction vehicles shall be performed. The following are examples of extraction vehicles:

a. Polar extraction vehicle: water, physiological saline, culture media without serum

b. Non-polar extraction vehicle: freshly refined vegetable oil (e.g. cottonseed or sesame oil) of the quality defined in various pharmacopoeias;

c. Additional extraction vehicles: ethanol/water, ethanol/saline, polyethylene glycol 400 (diluted to a physiological osmotic pressure), dimethylsulfoxide and culture media with serum.

It is noted that other extraction vehicles appropriate to the nature and use of the device or to the methods for hazard identification may also be used if their effects on the material and the biological test system are known.

Extraction Temperature and Duration (Section 10.3.1). Extraction conditions are based on common practice and are justified on the basis of providing a standardized approach that is, in many ways, an appropriate exaggeration of product use.

Extraction shall be conducted under one of the following conditions:

a. (37 ± 1) °C for (72 ± 2) h,
b. (50 ± 2) °C for (72 ± 2) h,
c. (70 ± 2) °C for (24 ± 2) h,
d. (121 ± 2) °C for $(1 \pm 0,1)$ h.

These extraction conditions, which have been used to provide a measure of the hazard potential for risk estimation of the device or material, are based on historical precedent. Other conditions that simulate the leaching that occurs during clinical use, or that provide an adequate measure of the hazard potential, may be used but shall be described and justified. The effects of higher temperatures or other conditions on extraction kinetics and the identity of the extraction vehicle(s) should

be considered carefully if accelerated or exaggerated extraction is used.

Extraction Stoichiometry (Section 10.3.3). Requirements for extraction stoichiometry are captured in Table 9.1. Additional guidance provided on this topic includes:

- Other surface-area-to-volume extraction ratios, e.g. those related to evaluation of porous materials, can be used if they simulate the clinical use conditions or result in the appropriate measure of hazard potential.

- Materials shall be cut into small pieces before extraction to enhance submersion in the extraction solution, except when otherwise inappropriate (except for Elastomers, coated materials, composites, laminates, etc., which shall be tested intact whenever possible).

Extraction Process.

- Extractions shall be performed with agitation or circulation (Section 10.3.6).

- If fluids circulate through the device under normal conditions of use, e.g. extra-corporeal devices, extraction via recirculation may be used. When possible, one or more of the extraction conditions shall be exaggerated, e.g. temperature, time, volume, flow rate (Section 10.3.13).

- For hazard identification of polymeric devices, exhaustive extraction conditions shall be considered. The extraction vehicle and conditions of extraction shall be selected on the basis of the physicochemical properties of the material and/or expected extractable substances (Section 10.3.10).

Table 9.1 Standard surface areas and extract liquid volumes for the generation of medical device extracts per ISO 10993:12(2012).

Thickness, mm	Extraction ratio (surface area or mass to volume ratio)	Examples of components with the specified thickness
< 0.5	6 cm^2/ml	Film, sheet, tubing wall
0.5 to 1.0	3 cm^2/ml	Tubing wall, slab, small molded items
> 1.0	3 cm^2/ml	Larger molded items
> 1.0	1.25 cm^2/ml	Elastomeric closures
Irregularly shaped solid devices	0.2 g/ml	Powder, pellets, foam, non-absorbent molded items
Irregularly shaped porous devices (low-density materials)	0.1 g/ml	Membranes, textiles

Annex C, Principles of test sample extraction

Key points made in informative Annex C are as follows:

1. If extracts of the device are prepared, the extraction vehicle and conditions of extraction used should be appropriate to the nature and use of the final product, as well as to the operating principles and parameters (such as test purpose, rationale, sensitivity, etc.) of the test method (Section C.1).
2. Extraction conditions and processes should ideally reflect not only actual conditions of use of the products but also the purpose, principles, and operating parameters of the tests (Section C.1).
3. Extractions differ for various test purposes (Section C.1):
   a. Exaggerated and exhaustive extraction is appropriate for hazard identification.
   b. Simulated-use extraction is applicable for the generation of a safety factor for use in human health risk assessments.
   c. Exhaustive extraction is applicable for the assessment of the safety of polymeric devices for long-term use in order to estimate the highest possible amount of leachables that could be released to the patient.
4. The period of extraction should be sufficient to maximize the amount of substances extracted. In practice, use of the(se) standard conditions of time and temperature for extraction (Section 10.3.1) are recommended in lieu of other unvalidated or non-standard conditions (Section C.3).
5. Extraction should not cause significant degradation of the material, unless the material is intended to dissolve or be resorbed during use (Section C.5).
6. The ratio of the surface area of the device to the volume of extraction vehicle or solvent should be sufficient to (Section C.6):
   a. Attain the maximum amount of extractable substance(s) in an appropriate test volume for biological testing (i.e. dosage volume within physiological limits) or chemical analysis;
   b. Demonstrate the hazard potential of using the device in humans;
   c. Immerse the material in the extraction volume.
7. In practice, the use of a standard area and solvent volume is recommended (described in Section 10.3.3) in lieu of device specific parameters.
8. The solvent(s) selected as the extraction vehicle should (Section C.7):
   a. Be suitable for use in the specific biological test systems;
   b. Simulate the leaching that occurs during clinical use of the device;
   c. Maximize the amount of extractables or leachables.

In practice, the use of standard polar and non-polar solvents (as specified in Section 10.3.5) is recommended.

Annex D, Exhaustive extraction of polymeric materials for biological evaluation

Key points made in informative Annex D concerning the extraction process itself are as follows:

1. For hazard identification, it is recommended that, as much as possible, the total amount of extractable substance be obtained from the polymeric device (Section D.1).
2. Exhaustive extraction using organic solvents is … useful in hazard identification of polymeric devices, especially those for long-term use (Section D.1).
3. Extractions can be done at 37 °C or at elevated temperatures in order to accelerate the experiment. However, analysts are advised that elevated temperatures can cause chemical reactions (or decomposition), producing additional or secondary extractables or leachables (Section D.3.2).
4. When the level of the residue for the nth successive extraction is one tenth (0.1) of the level in the first extraction, the (exhaustive) extraction may be deemed complete (Section D.3.3).
5. The individual residue levels are combined to give the cumulative value and, via the sample/solvent ratio, are converted to sample-related levels and finally back to device-related levels so that patient exposure can be determined (Section D.3.3).
6. The selected solvent should fully extract non-polar low molecular weight extractables, but not dissolve the polymer itself. Care should be taken that the solvent does not chemically alter the material, creating artifacts (Section D.2).

Because ISO 10993:12 is a monograph for the preparation of extracts for biological testing, part 12 contains no guidance on the analytical chemistry testing of extracts.

9.2.2 ISO 10993:18(2020)

This document specifies a framework for the identification and, if necessary, quantification of constituents of a medical device, allowing the identification of biological hazards and the estimation and control of biological risks from material constituents. These actions are accomplished using a generally stepwise approach to the chemical characterization which can include … the estimation (using laboratory extraction conditions) of the potential of the medical device, or its materials of construction, to release chemical substances under clinical use conditions (extractables).

Sections addressing extractions include the following and are summarized as follows:

- Section 3, Terms and definitions (normative text)
- Section 5.6, Estimate the chemical release; perform extraction study (normative text)
- Annex D, Principles of sample extraction (informative text)

Section 3, Terms and definitions (normative)

Key terms related to extractions and the extraction process include:

Accelerated extraction. Extraction whose duration is shorter than the duration of clinical use but whose conditions do not result in a chemical change to the substances being extracted.

Exaggerated extraction. Extraction that is intended to result in a greater number or amount of chemical constituents being released as compared to the amount generated under the clinical conditions of use.

Note: It is important to ensure that the exaggerated extraction does not result in a chemical change of the material or the substances being extracted.

Exhaustive extraction. Multi-step extraction conducted until the amount of material extracted in a subsequent extraction step is less than 10% by gravimetric analysis (or achieved by other means) of that determined in the initial extraction step.

Simulated-use extraction. Extraction using a method that simulates clinical use.

Note: A simulated-use extraction is performed to estimate the type and amount of substances that are expected to be released from a medical device during its clinical use. A simulated-use extraction is designed to produce an extractables profile that represents the worst-case leachables profile, meaning that all leachables are also extractables and the levels of all individual extractables are at least equal to the level of all individual leachables.

Extractable. Substance that is released from a medical device or material of construction when the medical device or material is extracted using laboratory extraction conditions and vehicles.

Leachable. Substance that is released from a medical device or material during its clinical use.

Note: For many medical devices, a leachables study is not practical due to challenges with reproducing actual clinical conditions, so simulated-use extraction studies are often performed instead. See the definition for simulated-use extraction.

Section 5.6, Estimate the chemical release; perform extraction study (normative)

Key points contained within this section include:

1. The primary objective of the extraction is to produce an extractables profile that is at least as comprehensive as

a device's leachables profile, meaning that the extractables profile includes all leachables as extractables and that the concentration of the extractables is at least as great as the concentrations of leachables.

2. An extractables profile that overestimates the leachables profile, specifically by overestimating the extractables concentrations versus leachables concentrations, provides an added margin for uncertainty in the toxicological risk assessment and can be appropriate in many circumstances.

3. Care must be taken to limit the extent of overestimation, as overly aggressive extraction conditions can lead to an altered extractables profiles.

4. The extraction conditions used, exhaustive, exaggerated, or simulated use, shall be documented and justified.

5. The recommended extraction conditions (see Table 9.2) will, in many circumstances, provide such an appropriate overestimation.

6. In certain circumstances, the overestimation provided by the recommended exhaustive extraction conditions will be excessive and thus the recommended extraction circumstances are not appropriate.

7. For all device classifications, alternative credible extraction conditions can be considered and used if deemed appropriate. The use of alternative extraction conditions shall be documented and justified.

Annex D, Principles of sample extraction (informative)

General key points in this informative Annex are as follows:

1. The goal of the extraction is to produce an extractables profile that equals or exceeds the leachables profile generated in clinical use while not causing deleterious effects to the materials (such as degradation) or the extractables profile (such as chemical alteration of the extractables). Doing so provides an extractables profile that is at least as extensive as the medical device's leachables profile, meaning that the extractables profile minimally accounts for all leachables at their highest possible levels.

2. In general, the extraction conditions should not alter the test article, unless justified, as alteration of the test article could change the amount and/or type of extractables released from the test article.

3. Extraction conditions are expected to be at least as aggressive as the conditions of clinical use.

Specific points addressing extraction solvents are as follows:

1. It is generally appropriate for long-term implants that a minimum of two extraction solvents of differing

Table 9.2 Recommended extraction conditions per ISO 10993:18(2020).

Contact category	Recommended extraction conditions	Suitable alternatives
Limited contact devices	Simulated use conditions[a]	Exaggerated conditions
Prolonged contact devices	Exhaustive conditions	Exaggerated conditions[b,c]
Long-term contact devices	Exhaustive conditions	Exaggerated conditions[b–d]

a) Certain regulatory authorities (e.g. FDA) will request exaggerated extraction, unless otherwise justified.
b) Circumstances where exhaustive extraction would not typically be required include:
  – Single-use devices used for less than 24 hours, where repeat use of a new device each day would result in categorization as prolonged or long-term contact;
  – Single-use devices used for several days, where repeat use of new devices would result in categorization as prolonged or long-term contact;
  – Reusable devices, where a patient may be exposed to repeated use of the same device, resulting in categorization as prolonged or long-term contact; when an exaggerated extraction is used for a reusable device, the extraction should properly account for the duration of each individual use.
c) Exaggerated conditions can be appropriate for external communicating or non-absorbable surface contact devices, with justification.
d) An example is a device comprised entirely of non-absorbable metal (e.g. a vascular stent), because migration of constituents from within the material is not possible, and the constituents of interest are related to the surface only and exaggerated extraction can be adequate to generate a complete extractables profile.

polarity be employed; for example, polar and non-polar vehicles consistent with ISO 10993:12. For some regulatory regions, such as the US, three solvents (e.g. polar, non-polar, and semi-polar) are recommended for long-term implants, unless justified.

2. For medical devices with indirect contact, it can be appropriate to use a single extraction solvent that replicates the expected contacting fluid.
3. The choice of extraction vehicle(s) shall be justified.
4. Examples of possible extraction vehicles are presented in Table 9.3. Inclusion of these solvents in Table 9.3 serves only as a starting point for extraction vehicle selection and does not constitute a complete justification for their use.

Specific points concerning an exhaustive extraction are as follows:

1. An exhaustive extraction establishes the maximum amounts of extractables that can be extracted from the medical device or material and thus defines the upper bound for the amount of leachables that could potentially be released by the device or material during clinical use.
2. For long-term implanted medical devices, an exhaustive extraction is recommended. If an exaggerated extraction is used, then its use should be justified.
3. If total extractables from an exhaustive extraction of a long-term implant medical device exceed a permissible daily exposure, the extraction kinetics might need to be evaluated (e.g. to determine maximum daily release) or a leachables study performed, if possible.
4. Exhaustive extraction involves sequential extraction of the test article using relevant extraction conditions and with a relevant extraction vehicle and is achieved when the level of extracted substances by gravimetric

Table 9.3 Solvents commonly used for extraction of polymeric medical devices per ISO 10993:18(2020).

	Solvent[a]	Polarity index
Polar	Water[b]	10.2
	Dimethyl sulfoxide	7.2
	Acetonitrile	5.8
Semi-Polar	Methanol	5.1
	Acetone	5.1
	Ethanol[c]	4.3
	Tetrahydrofuran	4.0
	n-Propyl alcohol	4.0
	i-Propyl alcohol	3.9
	Dichloromethane	3.1
	Toluene	2.4
Non-Polar	Cyclohexane	0.2
	Heptane	0.1
	n-Hexane	0.1

a) These solvents serve only as a starting point for solvent vehicle selection, and their inclusion here does not constitute a complete justification for their use.
b) Physiological saline and aqueous buffer systems such as phosphate buffered saline (PBS) are also considered polar solvents. Although specific values for their polarity index are not listed herein, the presence of relatively small amounts of dissolved salts is not expected to markedly change their extracting power.
c) Aqueous solutions of ethanol will have polarities between those of pure ethanol and water; their polarity indexes can be calculated.

(or other analysis) in a subsequent extraction step is less than 10% of the level of the same extracted substances in the initial extract.

5. Use extraction conditions and extraction vehicles (including those described above) that do not result

in a chemical change of either the test article or the extracted chemical entity.

6. Traditional extraction techniques, including Soxhlet extraction, boiling under reflux, closed vessel extraction, and sonication, are widely used and are more or less easy to implement using basic laboratory apparatus. As the traditional techniques have been used for an extended period of time, their capabilities and performance are well-known and well-documented. Nevertheless, they can have significant practical shortcomings including low extraction efficiencies, long extraction times, and the use of large quantities of environmentally inopportune extraction vehicles.

7. Modern extraction techniques include microwave-assisted extraction, pressurized fluid extraction, and supercritical fluid extraction, which typically employ instrumental means to increase the heat and/or the pressure at which extraction occurs or the power of the extraction vehicle.

8. The use of any technique, "traditional" or "modern," should carefully and fully consider the technique's technical and practical limitations and relevance to the clinical use of medical devices.

9. Sequential extraction is facilitated when the extraction sequence consists of the fewest number of extraction steps while not degrading the materials or its additives and ingredients.

10. Although an exhaustive extractables profile can be relevant to the clinical use of certain medical devices (e.g. long-term implants as noted previously), in many cases the clinical leaching of the medical devices is not exhaustive and thus an alternate extraction process, such as exaggerated or simulated extraction, produces a more appropriate extractables profile for the purpose of toxicological risk assessment.

Specific points concerning an exaggerated extraction include:

1. The purpose of an exaggerated extraction is to produce an extractables profile that is at least as complete and complex as the worst-case leachables profile.

2. The exaggerated extractables minimally include all leachables at levels equal to or greater than the highest levels reached by leachables.

3. The exaggerated extraction is accomplished by using extraction conditions that are, in one or more dimensions, exaggerated versus the conditions of clinical use:

    a. At a temperature that exceeds the clinical use temperature (typically referred to as an accelerated extraction);

    b. With a duration that exceeds the duration of clinical use;

    c. With a vehicle whose extraction power exceeds that of the solution that mediates the clinical contact between the medical device and the patient;

    d. With a surface area/volume ratio that exceeds clinical use exposure;

    e. Via the use of exhaustive (sequential) extraction for limited or prolonged contact medical devices.

4. If an exaggerated extraction cannot be justified or experimentally verified, then its use for producing chemical information that is the basis of a toxicological risk assessment is not recommended.

5. When an exaggerated extraction is employed, it is necessary to account for the exaggeration both in designing the extraction and in interpreting the result of the extraction study.

6. Exaggerated extraction conditions should not alter the extractables profile.

7. As altered extractables profiles can be obtained when greatly exaggerated extraction conditions are employed, it is recommended that exaggerations be kept as small as is necessary, minimizing potential complicating effects such as degradation.

8. The use of exaggeration in toxicological risk assessment should be rigorously justified and documented. While such a justification could be derived from scientific first principles, it is always the case that the most definitive means of justifying an exaggeration is to verify the exaggeration with experimental data.

Specific points made addressing simulated use extractions are as follows:

1. The purpose of a simulated extraction is to produce an extractables profile that closely matches the leachables profile that results when a device is used under typical clinical conditions.

2. A simulated-use extraction establishes the actual amount of leachables that will be released from the medical device or material during clinical use.

3. The simulated extraction is performed in those circumstances where either the clinical conditions of use cannot be achieved in the laboratory or when use of the clinical conditions produces a solution for testing that cannot be analytically profiled for leached substances.

4. The simulated extraction uses extraction conditions (i.e. temperature and duration) that mimic the conditions of clinical use.

5. The simulated extraction can be performed with a vehicle whose extraction power equals that of the solution that mediates the clinical contact between the medical device and the patient.

6. Possible simulating extraction solvents include the following:

a. For blood, a mixture of ethanol in water could be an appropriate simulating vehicle. If an ethanol/water mixture is used, it should be demonstrated to extract comparable levels of the target leachables with respect to blood. Other simulating vehicles can be used if justified.

b. If the medical device communicates with the potentially affected individual via an aqueous solution, then the appropriate simulating vehicle is either physiological saline, adjusted and buffered to a relevant pH, or an appropriate pH adjusted salt solution whose composition is justified. If the clinical application of the medical device involves contact with numerous solutions with varying pH (e.g. solution administration sets), then the pH range of the possible solutions should be properly bracketed by two simulating vehicles, one adjusted to a pH of 2 and the other adjusted and buffered to a pH of 10. If the pH range of solutions encountered in clinical use is smaller than this range, simulating extraction solutions bracketing the smaller range can be used.

c. If the medical device communicates with the patient via a solution with lipophilic properties (e.g. lipid emulsions, drug products containing solubilizing agents such as polysorbate 80), then an appropriate simulating vehicle should be identified and scientifically justified. In many situations, an alcohol/water mixture whose proportion of alcohol to water is justified can serve as a suitable simulating vehicle.

7. The surface area/volume ratio that is used for an extraction should be the same ratio that is experienced during clinical use, where possible.

8. A simulated extraction can be performed under accelerated conditions, at a temperature that exceeds the clinical use temperature and a duration that is shorter than clinical use, provided that the accelerated extraction conditions are fully and rigorously justified.

9. Proper accelerating conditions are those that reduce the extraction duration to a value shorter than the duration of clinical use but that do not result in a chemical modification of the device itself or to the type and amount of extracted substances.

Annex D of Part 18 addresses one of the confounding factors encountered when the extraction solvents recommended in Part 12 are used for the purpose of chemical characterization, which is that an extraction vehicle that is appropriate for biological testing (as recommended in Part 12) might not be amenable to chemical testing because of its composition. In such circumstances,

1. A surrogate extraction vehicle should be found to facilitate the chemical testing or

2. The extract for biological testing should be manipulated to make it analytically viable.

3. If a surrogate extraction vehicle is used, such a surrogate extraction vehicle should have similar extracting properties as the extraction vehicle used for biological testing.

4. If a chemical manipulation (e.g. derivatization) of the extract is used, care should be taken to avoid a chemical change of one or more extractables.

Although Annex D lists specific surrogate extraction vehicles as substitutes for biological vehicles that are challenging from the perspective of chemical testing, these suggestions are not formally recognized by the US FDA.

9.2.3 Reconciling the Extraction Conditions Contained in Parts 12 and 18

At this point it is obvious that there is some degree of inconsistency between the guidance contained in Parts 12 and 18 with respect to generating an extract. Because Part 12 deals specifically with the generation of extracts for biological testing and Part 18 deals specifically with the generation of extracts for chemical testing, it is unavoidable that these two documents would differ at least in terms of the specified extraction solvents. Extracts generated for biological testing must be compatible with the biological test systems; conversely, extracts generated for chemical testing must be compatible with the chemical testing instrumentation. As extract properties that make extracts biologically compatible can also make them analytically incompatible (and vice versa), it is not unexpected that the recommended extraction solutions in Part 12 and Part 18 would differ and that some reconciliation of extraction solutions would be necessary.

Furthermore, there is a difference in the scientific basis of the extractions established in either document. As Part 18 has been recently revised, including a major revision of its requirements and recommendation for performing extractions, and Part 12 has not been meaningfully revised in recent memory, especially with regard to its requirements and recommendation for performing extractions, it is reasonable to expect that the two documents would differ in terms of their extraction recommendations being consistent with "good science" and "good scientific practices."

Moreover, philosophically the two documents approach the subject of extraction from two different directions. Extraction conditions are justified in Part 12 as being "based on common practice and are justified on the basis of providing a standardized approach that is, in many ways, an appropriate exaggeration of product use." No scientific evidence is provided to substantiate the latter statement and, in fact, a scientific review of the extractions required in Part 12 suggests that the statement is largely inaccurate. Despite the assertions in Part 12 that extractions should reflect clinical use, it is clear that the extractions contained in Part 12 do not fulfill this objective. Although Part 12 refers to the use of validated extraction methods, there is no formal validation of the extraction processes contained within Part 12.

These shortcomings in Part 12 notwithstanding, Part 18 was revised in such a way as to be as well aligned as Part 12 as possible. Nevertheless, the extraction processes specified in Part 18 are justifiable in the context of the currently available "good science" and are designed to provide extractables profiles that reasonably reflect and appropriately exaggerate patient exposure to medical device-related leachables.

9.3 Linking Extraction Conditions to Medical Device Categories

It is a true statement that Part 18 is not definitive in terms of specifying exact extraction conditions that are applicable to specific types or classes of medical devices. The following discussion attempts to provide more definitive recommendations linking extraction conditions with the medical device categories that are established in ISO 10993:1.

In developing such recommendations, it is proper to ask "is there a 'One Size Fits All' Approach to Extraction?" That is to say, can a single extraction be designed that can produce a relevant extract for all medical devices and all clinical applications?

To address this question, the following generalization is made: the absolute worst-case patient exposure to medical device leachables occurs when each and every leachable present in the device is released from the device and transferred, in total, to the patient under the device's clinical conditions of use. Thus, an extraction study that reveals all leachables present in a medical device and the total quantity of each leachable in the device is an extraction study that establishes the absolute worst-case patient exposure.

There are two circumstances where such a worst-case extractables study is undoubtedly relevant:

1. When one makes the decision to toxicologically safety assess the absolute worst-case, regardless of the clinical conditions of use, and
2. When the clinical conditions of use are such that it is likely that this worst-case is realized (meaning essentially that "it all comes out during clinical use") and thus that a worst-case assessment in mandatory.

Circumstance 1 is consistent with the hierarchical approach to chemical assessment that is advocated in

Part 18 and is discussed in greater detail in Section 9.5. Oversimplifying somewhat, the chemical characterization approach advocated in Table 1 of Part 18 is "one works down from the worst-possible case to the actual clinical case until the resultant chemical data can be toxicologically safety risk assessed as minimal and acceptable risk. At that point, the chemical assessment is completed and stops." Thus, for example, if the total and complete worst-case leachables profile of a medical device, which is established on the premise that all leachables substances in the device are leached in the total amount that they are present in the device, can be toxicologically safety risk assessed as acceptable, then the medical device need not be tested for extractables or leachables under clinical use conditions. Only in the circumstance that the worst-case leachables profile is likely (or actually) assessed as potentially unsafe is it pertinent and necessary to perform either an extraction study (to establish the clinical worst-case) or a leachables study (to establish the actual clinical case).

Additionally, Part 18 was constructed to be responsive to certain practical realities. Thus, as the converse to the situation considered above, if it is likely that the toxicological safety risk assessment of the absolute worst-case leachables will produce an outcome that concludes that the leachables could pose a significant patent safety risk, then there is no value, in this context, in performing an extraction study that attempts to establish the worst case. Rather, it is more efficient and effective to skip the exhaustive extraction and immediately go to either clinically-relevant extractables or leachables profiling.

These points notwithstanding, establishing a medical device's total leachables profile (all leachables in their total quantity in the medical device) via an exhaustive extraction may be, in fact, the "one size fits all" approach that is being pursued.

Considering this possibility, if the device's clinical conditions of use are likely to result in the worst-case leachables situation of "it all comes out," then it is necessary, without question and recourse, to perform a toxicological safety risk assessment that is based on an absolute worst-case leachables study. The practical point to consider in this context is "do all medical devices fit into this category?" If the answer to this question is "yes," then a worst-case (exhaustive) leachables study becomes the only standard, necessary, expected, mandatory, and compliant approach. However, if only certain medical devices fit into the worst-case category, then worst-case leachables studies are mandatory only for those devices that fit into the category and optional for those devices that do not fit. It is certainly the case that entire classes of medical devices clearly fall into the category of "it all comes out." For example, an implanted medical device that dissolves in situ clearly falls into this category as it is obvious that if the device dissolves under clinical use, the patient will be exposed to all leachables in the total amount that they are present in the device. Another example is an implanted medical device whose clinical conditions of use are established to be permanent. Although it is possible that leachables are not fully leached from permanently implanted devices, in the absence of leaching data it is more reasonable to assume that a device's total pool of leachables will be leached during the permanent use. Thus, it is clear that an absolute worst-case leaching study is mandatory and required for these permanently implanted medical devices.

At the other extreme, it is certainly possible to identify medical devices whose composition, construction, and conditions of use are such that complete leaching is a most unlikely circumstance. For example, externally communicating devices, such as administration sets, and surface contact devices, such as dermal patches, are unlikely to be completely leached during clinical use, especially if the clinical use is relatively short in duration. In such cases, an absolute worst-case exhaustive leachables study would surely overestimate patient exposure to leachables under clinical conditions of use.

The question of whether to perform exhaustive worst-case extraction studies on medical devices is a moot point if the toxicological safety risk assessment of the resulting worst-case leachables profile would always produce the conclusion that the leachables are likely to have a negligible adverse safety risk. If this acceptable outcome was achieved for every medical device and every circumstance of use, then there would be no downside in performing such a worst-case extraction. However, such an outcome is not assured and there will be situations, perhaps a large number of situations, where a toxicological safety risk assessment of a worst-case leachables profile will result in the conclusion of a "potential adverse safety risk," leading to either additional studies or rejection of the device.

On one hand, the impact of such an outcome can seem innocuous, as the worst-case study can be followed by either an extractables study or a leachables study that more closely simulates the conditions of clinical use. However, such a two- (or three)-step process is not without its practical impacts on cost and time-to-market. Lest this "penalty" be trivialized, it is important to observe that increased R&D cost can lead to increased patient cost and that delay to market can lead to patients who needlessly suffer from unaddressed circumstances because the medical device that would alleviate those circumstances is still back in the lab being tested, as opposed to in a clinician's hands being used.

Given the diversity of medical devices and their clinical conditions of use, a "one size fits all" standard approach

to chemical characterization is not viable, scientifically, philosophically, or practically. On the other hand, the opposite approach of "everybody does whatever they think they can justify" turns the qualification of medical devices into chaos and catastrophically complicates the medical device's regulatory submission, review, and approval process.

An intermediate approach to these two extremes can be described as "divide and conquer." If the entire universe of medical devices and their clinical applications is too large to establish a single standard approach, then perhaps the universe can be divided into groups of medical devices that share common characteristics (for example, nature and duration of patient contact) and a standard extraction process can be established for each group.

Considering the grouping aspect of the "divide and conquer" approach, it is noted that this has already been accomplished for medical devices in ISO 10993:1, Table 1. The remaining challenge, therefore, is to establish extraction conditions to apply to each group.

To simplify the extraction options available to address medical devices, the following generalizations are made:

1. The purpose of the extraction is to exaggerate, to a limited degree, clinical use.
2. Three combinations of extraction/temperature/duration will be used; 37 °C for 24 hours to exaggerate limited contact duration devices, 50 °C for seven days to address longer contact durations devices, and 50 °C for three days to address repetitive (exhaustive) extractions.
3. To further differentiate between prolonged and permanent contact and to account for the exaggerating effect of implantation on leaching, multiple step (exhaustive) extractions will be required for permanent contact situations and implanted devices with prolonged contact.
4. Two extraction solvents are adequate to bracket the "leaching power" of the medium of contact between the patient and the medical device.

Justifications for these generalized extraction conditions are as follows:

- Extractions performed at 37 °C for 24 hours are appropriate for limited contact durations as 24 hours is the longest duration in this category and the temperature of contact between a medical device and the medium of body contact is rarely higher than body temperature, taken here as 37 °C.
- Extractions performed at 50 °C for seven days are appropriate for prolonged contact durations as this temperature is higher than the typical temperature of contact between a medical device and the medium of body contact, thereby accelerating leaching, and seven

days is within the duration range for this category. Establishing a single extraction duration for this category is challenging as the category encompasses a wide range of contact durations and it is not possible to establish a single extraction duration that properly addresses the contact at the longest duration of 30 days but which does not overly exaggerate the contact at the shortest duration of 24 hours.

- Extractions performed at 50 °C for three days are appropriate for multi-step (exhaustive) extractions as these conditions can make multi-step extractions more efficient by approaching asymptotic extraction in each step. The extraction duration of each extraction step is shorter versus the extraction for a prolonged device so that multi-step exhaustive extractions do not take excessively long to complete.
- Multi-step (exhaustive) extractions are reserved for those medical device applications where it is likely that the medical device will be fully leached during its clinical use. Clinical conditions that increase the likelihood that a medical device would be fully leached include permanent duration contact and/or implantation.
- The contact medium between the patient and the medical device is rarely so chemically diverse that more than two extraction solvents are required to bracket the medium's "leaching power." Furthermore, it is rarely the case that the contact medium can truly be considered to be non-polar and it is never the case that the contact medium is hexane. Thus, the use of hexane as an extraction solvent is only appropriate in cases of direct and invasive contact, such as those experienced by an implanted medical device.

These generalizations lead to the Medical Device Extraction Matrix shown in Table 9.4.

It is impossible to standardize extractions for medical devices and not run into situations where the Standardized Extraction Matrix flies in the face of the logic of "good science" and "common sense." Some of these situations can be managed by adding a certain level of this "common sense" to the interpretation of the standardized extractions. Thus, for example, the issue with respect to the extraction conditions used for prolonged contact devices (that the duration may be too long, in which case it overexaggerates shorter clinical durations of use, or the duration may be too short, in which case it underestimates the clinical duration of use) is addressed by allowing some flexibility in the extraction duration, when justified. The possibility that there are situations where a semi-polar extraction solvent more closely mimics the "leaching power" of a contact medium than a non-polar extraction solvent or vice versa is addressed by allowing some flexibility in the

choice of the extraction solvent, again when justified. The possibility that a medical device could be used with multiple aqueous contact media with varying pH values is addressed by noting that in some circumstances it may be necessary to replace the single polar extraction solvent (water) with duplicate aqueous extraction solutions at alternate extremes in pH.

However, even with such "common sense" modifications, there will be recommended extractions that are difficult to scientifically justify as being appropriate or adequate representations of a device's clinical conditions of use. Rather than allow investigators complete freedom in terms of modifying the recommended standardized extraction conditions illustrated in Table 9.4, it is suggested that the difference between clinical reality and extraction conditions be addressed in the toxicological safety risk assessment of the extractables data.

A final consideration in performing extractions is to reconsider the extraction that is appropriately applied to permanent implant medical devices. As noted in Table 9.4, it was recommended that each sequential extraction be performed at 50 °C for three days, as it is likely that each extraction step achieves equilibrium (highest possible extracted amount) under these conditions. Achieving equilibrium is important because it makes the sequential extraction process required for an exhaustive extraction more efficient; that is, the exhaustive extraction requires the fewest number of steps when equilibrium is established at each step. However, such an extraction does not facilitate a consideration of extraction kinetics in the toxicological safety risk assessment. This is the case as the duration of extraction (three days) is not the same exposure duration that is the basis of the risk assessment, which is one day (that is, tolerable intakes and PDE for leachables are expressed in units of mass per day and the risk assessments are performed based on daily exposure).

As the assessment is performed on a "per day" basis, the numerator in a release rate is expressed in units "per day." Thus, if an extraction is performed to establish a release rate, the extraction should be completed in one day (or the accelerated equivalent to one day).

For an implanted device, the nominal contact temperature is the body temperature of 37 °C. Thus, an extraction that establishes the amount released in one day is performed at 37 °C for 24 hours. If it desired that the extraction be accelerated, for instance, trying to complete the required sequential extraction in as short a time as possible, then an elevated temperature extraction can be employed. For example, one estimate is that an extraction performed at 37 °C for 24 hours is roughly equivalent to an extraction performed at 55 °C for eight hours.

Table 9.4 Standardized medical device extraction matrix.

Medical device categorization			Extraction conditions for chemical characterization		
Nature of body contact		Contact duration: A – Limited (≤ 24 hours) B – Prolonged (> 24 hours - 30 days) C – Permanent (> 30 days)	Temperature/Duration: A – 37 °C, 24 hours B – 50 °C, 7 days C – 50 °C, 3 days	Exhaustive extraction (multiple sequential steps)	Extraction solvents: A – Polar, Water B – Semi-polar, IPA C – Non-polar, Hexane
Surface device	Skin	A	A	–	A, B
		B	B[a]	–	A, B
		C	C	X	A, B
	Mucosal membrane	A	A	–	A, B
		B	B[a]	–	A, B
		C	C	X	A, B
	Breached or compromised surface	A	A	–	A, B
		B	B[a]	–	A, B
		C	C	X	A, B
External communicating device	Blood path indirect	A	A	–	A[b], B
		B	B[a]	–	A[b], B
		C	B	–	A[b], B
	Tissue/bone/ dentin	A	A	–	A, B
		B	B[a]	–	A, B
		C	C	X	A, B
	Circulating blood	A	A	–	A, B
		B	B[a]	–	A, B
		C	C	X	A, B[c]
Implant device	Tissue/bone	A	A	–	A, C[c]
		B	B[a]	X	A, C
		C	C	X	A, C
	Blood	A	A	–	A, C[c]
		B	B[a]	X	A, C
		C	C	X	A, C

a) Depending on the actual clinical duration of contact, extraction durations either longer or shorter than the 7 days may be used with proper justification.

b) Depending on the chemical nature of the communicating medium (or media), the use of one or more pH-adjusted aqueous extraction solvents may be appropriate. The use of pH-adjusted extraction solvent(s) and the pH of those solvent(s) must be justified.

c) Depending on the chemical nature (polarity) of the medium (or media) of contact, either a semi-polar or non-polar extraction solvent may be appropriate. The choice of which solvent is used must be justified.

If the sequential extraction is performed in this manner, then the information that is produced with each successive sequential extract reflects the release rate per day for successive days of contact.

In cases where either the total pool of an extractable in a medical device is large, or the extractable's rate of diffusion in the device is low, or if the extractable is "buried" inside the device so that there is a significant time lag before it reaches the surface of the device and can actually be leached, the "penalty" for performing an extraction in this manner may be that a large number of extraction steps are required before the extraction approaches exhaustion.

9.4 Chemical Testing of Extracts

9.4.1 Test Methods

Once the extractions are complete and the next step is to analytically characterize the extracts for extractables, then

Table 9.5 Test methodologies for extractables and leachables per ISO 10993:18.

Material type	Characteristic	Example methods[a]	Qualitative (Identification)	Quantitative
All	Organic extractables, Volatiles	HS–GC or GC with FID and/or MS*	X	X
		Total organic carbon (TOC)[b]	–	X
	Organic extractables, Semi-volatiles	HS-GC and GC, with FID and/or MS*	X	X
		HPLC, with UV, CAD, ELSD and/or MS*		
		Total organic carbon (TOC)[b]	–	X
		NMR	X	X
	Organic extractables, Non-volatiles	HPLC, with UV, CAD, ELSD and/or MS*	X	X
		NMR	X	X
		Total organic carbon (TOC)[b]	–	X
		Non-volatile residue*	–	X
	Elemental extractables	ICP-AES, ICP-MS*[b]	X	X
	Anions and cations	Ion chromatography[b]	X	X

a) Not comprehensive or exclusive listing. Methods denoted with a (*) are the most typically and commonly employed for the indicated purpose and are generally considered sufficient. The selection of the appropriate methods should be carried out by qualified personnel, in accordance with the composition of materials of construction and their manufacturing.
b) Generally employed for aqueous extracting solvents (e.g. water, buffers).

an extraction study performed on a medical device is no different from one performed on pharmaceutical packaging or manufacturing components, as the same analytical methods are used to accomplish the same objectives in all three cases. Section 6.5 of Part 18 summarizes relevant analytical methods as shown in Table 9.5

In addition to providing the general guidance contained in Table 9.5, it is possible to provide more specific recommendations concerning the appropriate use of each analytical method in terms of the nature of the extraction solvent. That is, certain analytical methods are best applied to extracts generated in a certain extraction solvent because of the nature of the solvent and the nature of the

analytical methods (see Table 9.6). A good example of this is inductively coupled plasma (ICP)-based methods used for testing samples for elements, including metals. In general, metals are extracted from plastic materials in their ionic form (that is, they generally are not extracted as an organometallic organic substance, for example) and thus they are generally present at their highest levels in aqueous extracts (specifically low pH extracts). Moreover, as an analytical method, ICP does not tolerate organic solvents and samples containing organic solvents must be significantly diluted to allow for their analysis. Thus, testing for extracted elements by ICP is typically performed in aqueous extracts only.

Table 9.6 Medical device chemical characterization; test methods.

Extraction medium	Extractables screening test methods					
	Extracted elements by ICP/OES or ICP/MS[a]	Volatile organic extractables by headspace GC/MS	Semi-volatile organic extractables by GC/MS[b]	Non-volatile organic extractables by LC/MS (APCI ionization)	Non-volatile organic extractables by LC/MS (ESI ionization)	Anions by ion chromatography
Polar (Water)[c]	X	X	X	–	X	X
Semi-polar (IPA)	X[d] (optional)	X[d]	X	X	–	–
Non-polar (Hexane)	–	–	X	X	–	–

Both GC and LC separations can be coupled to additional detectors other than MS; see, for example, Table 9.5.
a) As long as the method has the requisite sensitivity and freedom from analytical interferences, ICP/OES or ICP/MS can be used at the sponsor's discretion.
b) This testing may or may not be facilitated by appropriate extract preparation methods such as liquid/liquid extraction. It is the responsibility of the sponsor to justify and qualify any extract preparation process employed.
c) This includes pH-adjusted extraction solvents that may be used in place of water.
d) Performing this analysis will require that the extract be diluted prior to analysis to manage the adverse effect of the extraction medium.

A similar discussion is also true for measuring extracted anions by ion chromatography. However, when deciding whether to perform IC analysis, one should consider the possible use of this information in safety risk assessment. IC analysis is a cross between screening and targeted analysis, typically referred to as targeted screening. The specific IC method used for anion screening is a targeted method in the sense that it responds to a limited number of commonly encountered anions including acetate, formate, chloride, fluoride, bromide, nitrate, nitrite, phosphate, and sulfate. However, it is also a screening method as it is applied in a semi-quantitative manner. It is the aspect of targeting that brings into question the relevance of anion testing to safety risk assessment as the anions that are targeted are generally recognized as safe. For example, acetate and formate are ICH Q3C [7] Class 3 residual solvents whose PDE of 50 mg/day is rarely exceeded. Chloride and phosphate are major components of solutions and buffers that are recommended as extraction solutions and the remaining analytes have no specifically serious toxicities associated with them.

Considering organic extractables, volatile extractables can be addressed by either thermal extraction of the device itself followed by chromatographic analysis of the evolved gases or by heating the extracts and then testing the evolved gases. When extracts are tested, there are the practical issues of organic extraction solvents being volatilized along with the analytes, complicating the chromatographic testing significantly, which is why volatiles, if they are measured in extracts, are typically measured in aqueous extracts only.

Semi-volatile organic extractables are measured in all three extraction solvents (polar, semi-polar and non-polar) as semi-volatile extractables consist of polar, semi-polar, and non-polar compounds. Moreover, the extracts themselves are generally amenable to analysis by GC, either by direct injection (e.g. non-polar extraction solvent) or after solvent switching (aqueous and semi-polar solvents).

The same points are equally applicable to non-volatile analysis by LC, except that in the case of LC, most extraction solvents can be analyzed by direct injection as they are directly compatible with LC mobile phases. The complicating factor with respect to MS detection for LC is the ionization method itself. As electron impact (EI) MS is more suited for polar analytes, it is recommended in the testing of aqueous and/or semi-polar extracts. As atmospheric pressure chemical ionization (APCI) is more suitable for semi-polar and non-polar analytes, it is the recommended ionization method for semi-polar and especially non-polar extracts. Although arguably both EI and APCI methods could be used in the analysis of all extracts, this has the practical complicating factor that all extracts would have to be tested twice, as EI and APCI are typically not performed simultaneously. It has not yet been definitively established whether the potential analytical benefits of analyzing all extracts by both ionization methods (presumably detecting a greater number of analytes with stronger and more useful responses) outweigh the practicalities associated with performing both analyses individually.

Lastly, Table 9.6 focuses on chromatographic methods with MS detection, as MS is the only detection method that enables the identification of detected analytes. However, additional detection methods can be used with both types of chromatographic separations, as noted in Table 9.5. These additional detection methods can be employed to better detect and/or quantify analytes that do not respond well (if at all) to MS-based detectors.

Part 18 contains several analytically-related definitions in its Terms and Definitions Section 3. For example, Part 18 recognizes the two types of analytical approaches applied to extractables and leachables testing, screening, and targeting.

Analytical screening method. The method whose purpose is to discover, identify, and semi-quantitatively estimate the concentration of all relevant analytes in a test sample above an established reporting threshold (such as the AET).

Analytical targeting method. The method whose purpose is to quantify, with an appropriately high degree of accuracy and precision, specified analytes in a specified test sample over a specified concentration range.

Part 18 also notes the two purposes for which analytical testing is performed, identification and quantitation.

Identification. The process of assigning a molecular structure and chemical name to an organic compound or assigning constituent elements or molecular structures as appropriate, and a chemical name to an inorganic compound.

Quantification. The process of assigning a concentration to an analyte present in a sample.

Considering quantitation, Part 18 establishes three levels or varieties of quantitation:

Estimated quantitative (qualitative) analysis. The analytical approach that estimates an analyte's concentration by using the response from a surrogate substance chosen without specifically addressing or considering the relative responses of the analyte and the surrogate.

Semi-quantitative analysis. The analytical approach that provides an analyte's concentration by using the response from a surrogate substance (or substances), specifically accounting for the relative responses of the analyte and the surrogate.

Quantitative analysis. The analytical approach that establishes the most accurate estimate of an analyte's concentration by using a response function (calibration curve) generated specifically for the analyte via the use of a reference standard.

Establishing the level of a reported concentration is important as it alerts the toxicological safety risk assessor to the accuracy of the reported analytical result. If the accuracy of the reported concentration is low, the assessor may want to take that circumstance into account in the assessment. Estimated quantitative analysis is generally less accurate than semi-quantitative analysis, which is generally less accurate than quantitative analysis.

Section 6.5 is the only Section in normative portion of Part 18 that deals with analytical methods. It speaks to the two types of analyses performed (screening and targeting), as noted previously, and establishes general expectations for the qualification of these types of methods, referring to informative Annex F for more details. Specifically, Part 18 notes that "appropriate analytical methods shall be developed and qualified for these purposes, where qualification is defined as the process by which a method is established to be suited for its intended use." Because of their different purposes and nature, Part 18 establishes that screening and targeted methods will be qualified in a different manner and to a different extent. As it is generally the case that the potential population of analytes, which is addressed by analytical screening methods, is large and diverse, it is not possible that a screening method can produce highly accurate and precise concentration estimates for all potential analytes. Thus, analytical methods used for screening should be qualified using a set of surrogate analytes representative of the entire population of possible analytes. For example, when an analytical method is employed to screen an extract for extractables above the AET, the method shall be qualified using a set of potential extractables as surrogate analytes.

Alternatively, a method used to analyze a test sample to establish the levels of targeted analytes is generally optimized for this purpose. Because the targeting method targets a small and defined list of analytes, the qualification of the method addresses the performance of the method specific to each and every targeted analyte.

9.4.2 Test Method Qualification

Part 18 addresses the qualification of chemical analytical test methods for extractables and leachables in its informative Annex F, Qualification of analytical methods used for extractables/leachables. An analytical method is qualified to establish that it is suited for its intended purpose. Annex F notes that "analytical methods serve one of two

purposes; screening samples for unspecified analytes and testing samples for specified (targeted) analytes. As these purposes are quite different, it is reasonable to suspect that their qualification would differ" and points out that "A method is considered to be qualified (that is, suited for its intended use) when:

- It has been established that the method is able to routinely meet the performance expectations contained in the qualification protocol and
- Appropriate system suitability has been established."

Qualification parameters for screening and targeted methods are summarized in Table 9.7.

In addition to having documented performance capabilities, qualified analytical methods should have additional controls that may include, but are not limited to:

Table 9.7 Performance characteristics addressed in qualifications of analytical methods for extractables and leachables from medical devices.

Performance characteristic	Addressed in screening methods	Addressed in target methods
Sensitivity	Yes[a),b)]	Yes[c),d)]
Specificity[e)]	Yes	Yes
Accuracy[f)]	Yes[b)]	Yes[d)]
Precision[g)]	Yes[b)]	Yes[d)]
Dynamic range[h)]	Optional[b)]	Yes[d)]
Goodness of Fit[i)]	No	Yes[d)]
Ruggedness	Yes	Yes
Chromatographic performance[j)]	Yes	Yes

a) It is expected that the LoQ and/or LoD be lower than the reporting threshold.
b) Established using a set of surrogate standard compounds.
c) It is expected that the LoQ be below the concentration equivalent to each target compound's PDE.
d) Established for each target compound.
e) Defined as the ability to assess unequivocally the analyte in the presence of other constituents that can be expected in the sample.
f) Defined as the ability to produce a response that is comparable to the true value (e.g. a measured concentration in a spiked extract that is comparable to the spiked amount).
g) Defined as the variation in replicate analyses of either the same extract or a standard solution containing extractables or leachables.
h) Defined as the concentration range over which the response and the analyte concentration producing that response are relatable by a simple mathematical function.
i) Defined as the degree to which a simple mathematical function can express the relationship between an analyte's concentration in a standard and the method response obtained when the standard is analyzed.
j) For example, resolution between critical pairs, efficiency, tailing factor, etc.

- Documentation of the method in the form of a standard operating procedure (SOP), which is controlled in a document change system;
- An approved and specified Scope, captured in the method's SOP;
- A detailed scientific description and justification of the method, establishing its suitability for the intended use;
- A requirement that the qualified method is implemented by an appropriately qualified and trained staff;
- A requirement that the qualified method is implemented on calibrated/qualified instrumentation.
- A requirement that the method includes relevant system suitability testing to be performed at the time of use.

Considering the last requirement, system suitability testing can be construed to be a form of qualification at time of use, which establishes that:

a. The method was set up and implemented properly;
b. The method as set up was capable of performing at the same level it performed at during its qualification;
c. That the method performed acceptably throughout its use.

The system suitability parameters to be assessed and their associated acceptance criteria should be rigorous enough to ensure that the method produces data of acceptable quality but not so rigorous that potentially acceptable analytical runs are rejected on a frequent basis. Properly collected and statistically evaluated system suitability data can provide diagnostic evidence of imminent method failure.

Although Part 18 discusses method qualification and system suitability in the context of noting that performance specifications are necessary and must be established, the document contains minimal information or recommendations in terms of what appropriate specification values are, referring the reader to USP <1058>, Analytical instrument qualification [8].

9.4.3 Reporting of Test Data

The value proposition for reporting extractables or leachables data in a standardized format has been stated previously for manufacturing components (Section 8.2.2.4) and is equally valid for medical devices. Although Part 18 did not go to the same extreme as did the BioPhorum in their standard report for SUS extractables data, normative Section 7, Reporting of the chemical characterization data, speaks to this point, requiring that reports:

a. Clearly state the purpose and objectives of the chemical assessment that has been performed on the test article (material or medical device) description and details of sample preparation;

b. Describe and justify analytical methods and extraction conditions (e.g. choice of extraction vehicles, extraction duration and cycles, extraction temperature, extraction/sample ratio, agitation method, and speed during extraction), including sufficient information to establish the appropriateness of the analytical processes employed;

c. Document system suitability testing and its outcome;

d. Discuss the value for, and justification of, the reporting threshold (e.g. AET);

e. Contain all relevant qualitative data generated (e.g. extractable's identities, including a description of the identification procedure);

f. Include all relevant quantitative data generated (e.g. extractable's concentrations, including a description of the quantification procedures and providing the classification of the quantitative data as estimated quantitative analysis, semi-quantitative analysis, or quantitative analysis);

g. Report information in quantities that enable the estimation of a patient's clinical exposure to extractables (e.g. analyte amounts in µg/device).

More detailed recommendations for reporting chemical qualification data are contained in informative Annex G. Reporting details for analytical methods and chemical data are summarized as follows:

- Reporting of analytical data to facilitate toxicological risk assessment:
  - Qualitative data (e.g. extractable's identities).
  - Quantitative data (e.g. extractable's concentrations, including a discussion of the quantification approach and providing the classification of the quantitative data as estimated quantitative analysis, semi-quantitative analysis, or quantitative analysis).
  - A discussion and justification of the reporting threshold and its relevance to toxicological risk assessment (e.g. safety thresholds).
  - List of chemical compounds above the reporting threshold. Such a list can be provided in a tabular format and the table should contain the chemical compounds including their mass, proposed structure, chemical formula, IUPAC chemical name, common chemical name(s) and abbreviation(s), CAS registry number, their identification status (e.g. confirmed, confident, tentative, speculative), and their measured levels in the relevant samples. Additional information, such as chemical structure, may be provided in the document. When multiple candidate identifications are found (e.g. a class of compounds such as is often reported in tentative identifications), all should be reported.

o Appropriate figures, diagrams, etc., that illustrate the analytical data and/or facilitates data review and/or interpretation (e.g. labeled chromatograms, migration curves).

o Approach and rationale addressing cohorts of concern substances.

- Details concerning the extraction of the medical device:
  o Appropriate and complete description of the test article, including relevant processing details (e.g. sterilization, rinsing);
  o Extraction method with justification (e.g. refluxing, sealed vessel);
  o List of extraction vehicles with justification;
  o Extraction vehicle/sample ratio (e.g. extracted surface area to extraction solution volume ratio);
  o Extraction time and temperature;
  o Number of extraction cycles (e.g. single vs. exhaustive);
  o Methods for determining when exhaustive extraction endpoints are reached (as appropriate);
  o Description of changes to the vehicle or test article (e.g. medical device) post-extraction, to include physical state, appearance, color, clarity, or presence of particles;
  o If particles are present in the extract, a description of how they were addressed, including, if performed, the means by which they were separated from the extract prior to analysis and the means by which they were chemically characterized.

- Details concerning the preparation of the extract for analysis:
  o Description of any dilution, concentration, and other significant processing steps (e.g. exchange).
  o Justification for all significant processing steps.
  o Description of any sample filtering/particle separation that was performed.
  o Description of the storage conditions of the extracts between completion of the extraction and preparation for analysis.

- Description of the analytical methods used for testing prepared extracts:
  o Justification of relevance and appropriateness of the analytical method.
  o Relevant operating conditions (e.g. chromatographic mobile phase, methods, flow rates, gradient run time, column temperature).
  o Analytical instrumentation manufacturer, model, principal components.
  o For methods based on mass spectrometric detection include ionization technique (APCI, ESI), polarity mode (positive, negative), mass range (or specific masses analyzed for ICP-MS data), nominal mass resolution.

o For methods using UV detection, include detection wavelength.

o For other detection methods, include key operational parameters.

o For semi-quantitative methods, discuss surrogate standard(s) used, with justification, and resulting response factor to be applied.

o For quantitative methods, discuss the quantification approach used, with justification.

o A description of how confidence in identifications was determined and assigned (e.g. definitions of categorization terms or match scores), with justification.

o Means used to address unknowns (e.g. additional analytical testing to identify or risk mitigation per ISO 10993:1).

o Determination, justification, and application of reporting thresholds (such as the AET).

- System suitability metrics for the analytical methods (all that apply):
  o LoD and LoQ (including how they were established);
  o Linearity (calibration curve(s));
  o Specificity;
  o Recovery (accuracy);
  o Precision;
  o Dynamic range;
  o Other relevant parameters as appropriate.

9.5 The Chemical Characterization Process for Medical Devices per ISO 10993:18(2020)

9.5.1 The Chemical Characterization Flow Chart

As important as the extraction and analytical testing guidelines on Part 18 are, arguably, the most important information contained in Part 18 is found in Section 5, Characterization procedure, which describes the circumstances in which extractions are necessary and establishes how analytical data is used to qualify medical devices as being chemically suited for their intended use with respect to patient safety.

Greatly oversimplified, the purpose of chemical assessment is to collect that chemical information that, when interpreted, supports the conclusion that use of the medical device will not adversely affect patient health because of leached substances (i.e. that the device is safe for its clinical use). With respect to this purpose, the phrase "there is more than one way to skin a cat" is relevant. That is, there are several types of chemical information that can support a "safe for use" assessment, for example:

- If the composition of a medical device under assessment is known and can be established to be equivalent to an approved, marketed, and established safe medical device used in the same clinical manner, then the device under assessment is established to be safe by association.

- If the composition of a medical device under assessment is known, the absolute worst-case patient exposure can be established as "it all comes out," meaning that ultimately the patient is exposed to the entire contents of the medical device during the devise's clinical use (absolute worst-case). If a device's entire contents can be toxicologically assessed as having a negligible adverse effect on patient health, then the device is concluded to be safe by compositional assessment.

- A device can be extracted using appropriate extraction conditions and the resulting extracts analytically profiled to establish those device constituents (extractables) that could be released from the device and to which the patient could be exposed during clinical use of the device. By design, such an extraction study establishes the highest possible patient exposure, both in terms of the number of extractables that the patient could be exposed to and their exposure levels. If the extractables profile can be toxicologically assessed as having a negligible adverse effect on patient health,

then the device is concluded to be safe by extractables assessment.

- A devices' ability to leach chemicals (leachables) to the patient can be established directly by monitoring the chemicals during clinical use or during laboratory simulation of clinical use. Such a study establishes the exact patient exposure to leachables, both in terms of the number of leachables and their levels. If the leachables profile can be toxicologically assessed as having a negligible adverse effect on patient health, then the device is concluded to be safe by leachables assessment.

Part 18 attempts to communicate this approach to assessment via a flow chart, specifically Figure 1, General chemical characterization process. A simplified version of this flow chart is shown here as Figure 9.1. Each of the four possible assessment strategies just discussed is illustrated in Figure 9.1, and the use of assessment to secure the required outcome and exit the chemical characterization process is illustrated.

The problem with illustrating the assessment process as a flow chart is that it implies that there is only one way to navigate the process; that is, one starts at the top and proceeds through the process until they are able to exit from it. That

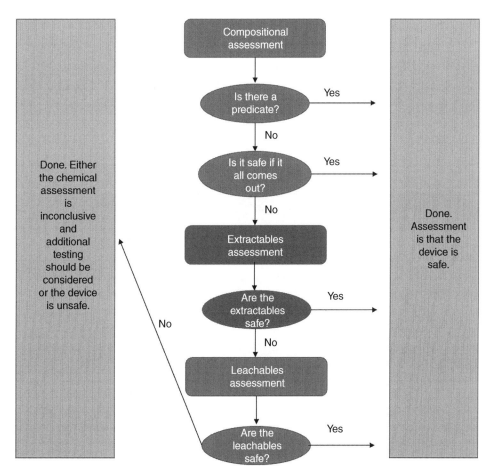

Figure 9.1 General flow diagram, chemical assessment of medical devices. (*See insert for colour representation of the figure.*)

is, Figure 9.1 implies that one needs to perform a compositional assessment first and only proceed to extractables, and then, as necessary, to leachables assessment only after the compositional assessment has failed to secure the desired outcome.

Although there may be circumstances where the "starting at the top and working one's way down" approach may be appropriate, it can be an unnecessarily long (and expensive) process. For example, let us assume that the device under consideration is new to the world and thus that it is likely that a predicate does not exist. Furthermore, let us assume that the device's composition is known (as it should be), meaning that its materials of construction have been specified and the materials' formulation (ingredients) have been established. It may be clear from the device's composition and clinical use that an assessment of "it all comes out" will produce a conclusion that the medical device is potentially unsafe. In this circumstance, a compositional assessment clearly will bear no fruit and the effort to perform such an assessment will largely be wasted. In this case it is not the intent of the Part 18 flow chart to force a sponsor to perform the unnecessary and fruitless work of compositional assessment. Rather, it is the intent of the Part 18 flow chart to say "if you are not going to perform a compositional assessment, then you must perform either an extractables and/or leachables assessment."

This same thought process can be applied to extractables versus leachables. On one hand, if the nature of the device and its clinical use is such that a viable and appropriate leachables study can be performed, then the Part 18 flow chart says "go ahead and perform the leachables study without doing a compositional or extractables assessment." On the other hand, if a leaching study cannot be performed then performing an extractables study is the only viable alternative.

This discussion notwithstanding, it is noted that there is considerable value in following the Part 18 flow chart in its entirety. Philosophically, a sponsor should always know the composition of the device in question; the idea that a medical device could be a "black box" of unknown composition flies in the face of the sponsor's responsibility for establishing that the device is safe and effective. Furthermore, knowledge of the device's composition enables extractables profiling in the sense that extractables are likely to be derived from a device's ingredients. Similarly, knowledge of a device's extractables profile enables leachables profiling as (i) leachables are likely to be either extractables or derived from extractables and (ii) a leachables study should focus on those extractables that are potentially unsafe and need not necessarily address extractables that have been established as likely to be safe.

9.5.2 Compositional Assessment

Should it be decided that the safe use assessment of a medical device will include a compositional assessment, the compositional assessment proceeds as illustrated in Figure 9.2.

It is strategically important that the first step in the compositional assessment is to gather, not create, compositional

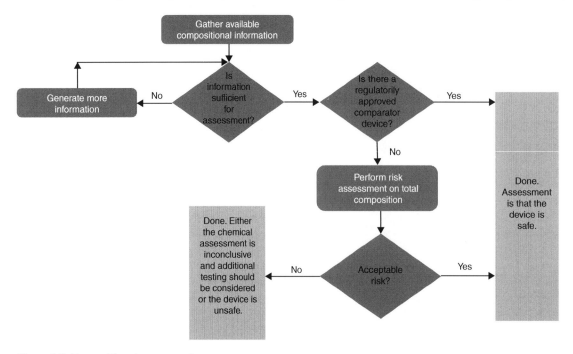

Figure 9.2 Compositional assessment.

information. It is important to recognize that compositional assessment is based on the assessment of information, not strictly the generation of information.

In the previous section I made the point that it is a medical device sponsor's responsibility to know the device's composition, as composition is the single largest contributor to extractables and leachables. In this section I follow up on that point by suggesting that the logical, appropriate, and proper source of the compositional information is the vendor(s) of the device's materials of construction. It is unimaginable, proprietary interests aside, that a material vendor to the medical device industry would intentionally withhold compositional information from a user of that material. It is equally unimaginable that a device's vendor would use a material in the devices whose composition was unknown. As was noted previously, the idea that it is acceptable to sell a material into the medical device industry without divulging its composition to the buyer or that it is acceptable to use a material whose composition is unknown are ideas that have no place in current practice.

Of course, there are certain limitations in terms of what constitutes the vendor's responsibility to provide compositional information and the mechanics for supplying the information. It is necessary that the material of construction be fully described in terms of its intentional additives, both identity and amount. If the material's vendor intentionally uses processing aids, solvents, and/or other manufacturing chemicals in the production of a material or part, then these items should be revealed to the material's users, at least in terms of identity. If the vendor has test information establishing the residual levels of these aids in the material, then this information would be useful and should be supplied.

It can be the case that the genesis of a material is complex, involving several parties (suppliers to the material's vendor) and several processes. Although in an ideal world a material's vendor will have secured compositional information from its suppliers, holding the material's final vendor responsible for providing its own supplier's compositional information to a potential user goes beyond current expectations.

This situation is one of the shortcomings of vendor-supplied compositional information; that is, that the information is limited to how the vendor itself formulated and produced the material. A second shortcoming is that the information provided by the vendor is limited to intentional additives and does not address chemicals that might have been unintentionally added. Lastly, the compositional information provided by the material's vendor will reflect the composition at the time the material was provided to the user. If the user processes the material in the construction of the medical device (e.g. irradiation), then it is clear that the composition that is relevant to safety assessment is the composition after processing and not necessarily the composition before processing.

It is important to note that while the material's vendor has the responsibility for sharing compositional information, the material's vendor is not responsible for the potential implications of that information. That is, it is the user of the material who made the decision to use that material and therefore it is the user of the material who is responsible for the implications and outcomes of using that material.

Furthermore, the material's vendor has the reasonable expectation that the compositional information provided to the user remain confidential information.

The purpose of this rather lengthy exposition is to reiterate that the first step in the compositional assessment is not the generation of new information but the collection of existing information. Should the information that a device's sponsor collects be adequate for supporting an assessment, no new information need be generated. Only if the gathered information is insufficient for supporting an assessment does the new information need to be generated. Presumably the new information would fill gaps and manage any deficiencies in the gathered information.

Part 18 provides general guidance in terms of the nature and amount of information required for compositional assessment, noting that "the extent of physical and/or chemical characterization required depends on what is known about the material formulation, what nonclinical and clinical safety and toxicological data exist, and on the nature and duration of body contact with the medical device." Part 18 suggests that "at a minimum, the characterization shall address the constituent chemicals of the medical device and possible residual process aids or additives used in its manufacture."

As noted in Figure 9.2, compositional information can be assessed in two ways. The first approach is an associative approach where the intent of the assessment is to establish a link between the device being assessed and a predicate device that has secured regulatory approval and might have been commercialized. The logic of this associative assessment is that a predicate device has been, by virtue of its regulatory approval, established to be safe for its intended clinical use. If a device being assessed can be established to be compositionally equivalent to a predicate device, then this equivalence is the foundation for an associative assessment. Of course, the associative assessment also requires that the device being assessed and its predicate be used under the same clinical conditions to address the same medical need.

Thus, the compositional information can be used to determine whether the medical device under consideration is equivalent, in configuration, composition, manufacturing, processing, and intended use, to its clinically established predicate.

When an equivalent clinically established medical device can be identified and justified for the device being assessed, then the chemical characterization process is deemed to have been completed. When a clinically established equivalent medical device cannot be established and justified, additional elements of chemical assessment, including a risk assessment of the chemical characterization information must be pursued.

Undeniably, the greatest potential chemical impact of a medical device would be realized if the device's entire composition were to transfer to the patient during clinical use. This would be accomplished, for example, if an implantable medical device were to dissolve during clinical use or if an externally communicating device were to be completely leached during clinical use. When exposure to a medical device's entire composition can be established as being acceptable (e.g. by comparing the exposure to an established safety threshold established), then the chemical characterization process is deemed to have been completed. When exposure to a medical device's entire composition is established to be potentially unacceptable, then the chemical characterization process can be continued by moving to one of the next process steps (extractables or leachables testing).

9.5.3 Extractables Assessment

If the assessment of the absolute worst case of total release of a device's ingredients and impurities ("it all comes out") concludes that there is a possible adverse effect on patient health, then a more realistic estimation of patient exposure to device-related chemicals must be obtained, either via extractables testing or leachables testing. Extractables testing is performed when it is not practical or possible to perform leachables testing or in the circumstance that a limited number of extractions can represent, or bracket, numerous clinical applications, each one requiring its own leachables testing.

The process for performing an extractables assessment is illustrated in Figure 9.3. The first step in performing the assessment is establishing the AET, as the AET is the reporting threshold for extractables (see Chapter 3, Section 3.2.3). Thus, the AET is the tool that allows the analytical chemist to establish which detected organic extractables must be identified and reported for safety risk assessment.

The primary objective of extractables testing is to produce an extractables profile that is at least as comprehensive as a device's leachables profile, meaning that the extractables profile includes all leachables as extractables and that the concentration of the extractables is at least as great as the concentrations of leachables. An extractables profile that overestimates the leachables profile, specifically by overestimating the extractables concentrations versus

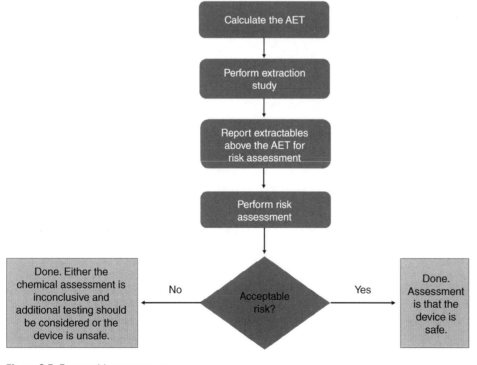

Figure 9.3 Extractables assessment.

leachables concentrations, provides an added margin for uncertainty in the toxicological risk assessment, and can be appropriate in many circumstances, so long as the exaggeration is not excessive.

Considerations in designing and implementing a proper extractables study, including generation and testing of the extract, have been discussed previously in this Chapter.

Once the extractables profile has been generated, the extractables at or above the AET are reported for the purpose of toxicological safety risk assessment and the toxicological safety risk assessment is performed. When exposure to a medical device's extractables profile can be established as being acceptable (e.g. by comparing the exposure to an established safety threshold established), then the chemical characterization process is deemed to have been completed. If the toxicological risk assessment of the extractables data determines that an extractable or extractables could be a risk to the patient using the extractables data, a more clinically relevant experiment (leachables testing) can be performed to estimate more precisely the amount of the chemical or chemicals released from the medical device during clinical use.

9.5.4 Leachables Assessment

The decision to perform a leachables assessment can be made after the compositional assessment (that is, skip extractables and go right to leachables), after the extractables assessment or as the first and only action to take. The nature of the leachables assessment depends upon where in the chemical assessment process the decision to perform a leachables assessment was made. If the decision to perform a leachables assessment was made without having performed a prior extractables assessment, then the leachables testing will involve the screening of the leachate for all possible leachables, in the same way that the extractables testing involved screening the extracts for all possible extractables (see Figure 9.4). In this case the leachables assessment is performed in the same way the extractables assessment was performed:

1. The AET is determined.
2. The leachates are generated.
3. The leachates are tested.
4. All leachables above the AET are identified, quantified, and reported for toxicological safety risk assessment.
5. The toxicological safety risk assessment is performed.

If the decision to perform a leachables assessment was made after having performed a prior extractables assessment, then the existing extractables information can be used to focus the leachables testing on only those extractables that are potential safety hazards. If an extractables study truly represents the worst-case leachables situation, then there is little value in testing a leachate for leachables that have been established are likely to be safe as extractables, based on worst-case extractables data.

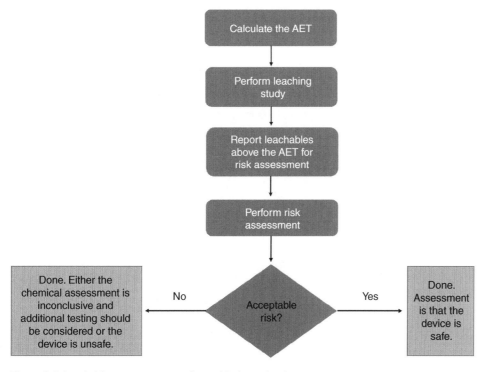

Figure 9.4 Leachables assessment performed independently.

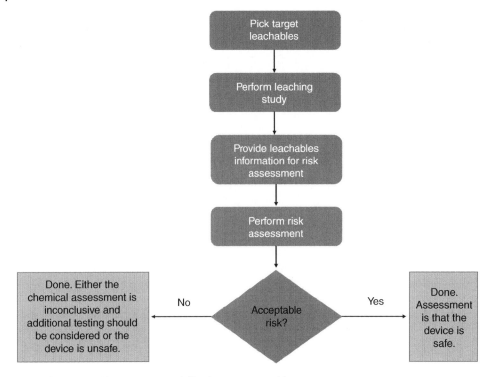

Figure 9.5 Leachables assessment following an extractables assessment.

The process of performing a leachables assessment based on targeting is established in Figure 9.5.

When targeted testing is performed, there is always the nagging doubt concerning the presence of leachables in the leachate that were not previously established as extractables. When it is anticipated that additional leachables that were not revealed as extractables can be present, the leachables study should also include screening for the additional leachables.

When exposure to a medical device's leachables profile can be established as being acceptable (e.g. by comparing the exposure to individually established permissible daily exposures, PDEs), then the chemical characterization process is deemed to have been completed. If the toxicological risk assessment of the leachables data determines that a chemical or chemicals are a risk to the patient using the leachables data, then it is difficult to avoid the evident conclusion that the device is not safe for use in the defined clinical application. It is certainly the case that there is no further chemistry-based testing that would manage the assessment outcome of potentially unsafe leachables. If the leachables characterization does not support a toxicological risk assessment conclusion that all leachables present an acceptable health risk, the chemical characterization process has been completed but cannot be used to support biological evaluation. The need for further assessment (e.g. biological testing) or for mitigation should be evaluated.

Of course, it goes without saying that if the compositional, extractables, and/or leachables assessment does not support a toxicological risk assessment conclusion that all the various substances present an acceptable health risk, then the medical device can either be rejected as unsafe (and never marketed) or re-designed to manage, mitigate, or eliminate the safety hazard(s).

9.6 FDA Recognition of Part 18

There is an expectation among the various users of a Standard such as Part 18 that if medical device sponsors take the actions required by the Standard and if the device meets the specifications contained in the Standard, that these circumstances will contribute to the regulatory approval of the medical device for commercialization. For this expectation to be realized, regulatory agencies must accept the Standard as actionable, relevant, applicable, and appropriate.

The FDA communicates its position with respect to a Standard via a process termed recognition. That is, the FDA communicates to other stakeholders that it "recognizes" the Standard as meeting the requirements of actionable, relevant, applicable, and appropriate. Recognition is the process whereby the FDA identifies Standards to which manufacturers of medical devices may submit a declaration of conformity to demonstrate they have met relevant requirements in the Federal Food, Drug, and Cosmetic Act

(FD&C Act). The FDA may recognize all, part, or none of a Standard established by a national or international Standards development organization (SDO). Any interested party may submit a request for recognition to the FDA.

In general, the FDA actively assesses the impact of new consensus Standards and revisions of existing Standards on the premarket review process and recognizes these Standards, as appropriate. As the FDA determines that new or revised Standards are appropriate for meeting requirements under the FD&C Act for medical devices, it updates the Recognized Consensus Standards Database on the FDA's website. Once the FDA has decided to recognize a Standard, it issues a recognition number and provides a "Supplemental Information Sheet" in the database. To formally recognize such Standards, FDA periodically publishes a recognition list in the Federal Register. The FDA usually performs the activity at least twice annually.

The FDA has published its recognition status for Part 18 [9]. The Extent of Recognition for part 18 was Partial recognition. The following part(s) of the Part 18 standard were not recognized:

- Clause 5.5 second and third sentences.
- Clause 7, second paragraph, phrase "to assist in any toxicological risk assessment."
- Table D.3 in Clause D.5 of Annex D.
- Formula E2 and paragraph preceding Formula E2 and two paragraphs following. Formula E2 in Clause E.3 of Annex E.
- Example C2 in Clause E.4 of Annex E.

Clause 5.5 second and third sentences address the relationship between the AET and a method's LoQ as follows:

> "The AET should preferably be derived from a safety-based threshold (such as the TTC) but if this is not practically achievable, an analytical threshold, such as the Limit of Quantification (LOQ) can be used as the reporting threshold. However, the difference between the AET and the LOQ shall be considered in the toxicological risk assessment and the difference shall be justified."

The FDA's rationale for not recognizing this text was communicated as "Clause 5.5 second and third sentences are in conflict with published literature articles" and provides two literature references. Although it would be an interesting debate to consider the entire body of literature on this subject in an attempt to establish a scientific consensus, ultimately the point to be learned is that situations where the analytical sensitivity of the test methods are insufficient to achieve the AET result in a gap in chemical characterization's ability to establish the safety of a medical device. When such a gap exists in chemical characterization, it must be acknowledged and managed.

Clause 7, second paragraph of Part 18 reads as follows, with the text in question italicized: "As necessary and appropriate, identified substances in the test solutions could be grouped into compound classes, based on structural or functional group similarities, *to assist in any toxicological risk assessment*." Considering this text, the FDA maintains that the phrase "to assist in any toxicological risk assessment" is in conflict with published literature and provides three literature references. With this text, the FDA is communicating its reluctance to endorsed the grouping of structurally and functionally similar compounds for the purpose of toxicologically assessing the compounds as a group (as opposed to assessing the compounds individually).

Table D.3 in Clause D.5 of Annex D is a table that lists "Potential surrogate extraction vehicles for correlating chemical to biological testing." In its rationale for recognition, the FDA notes that "it is not supported by the cited article and in conflict with the published document." Despite the title of the Table noting that the listed solvents are *potential* surrogates, the fact that the Table in question appears in an informative annex, the circumstance that the table is limited to extraction studies for correlating chemical and biological test results and the note that points out "Inclusion of vehicles here does not fully justify their use in chemical-biological comparisons," the FDA is communicating to users of Part 18 that justification of these extraction solvents merely by citing their presence in Part 18 will not be viewed as being adequate justification by the FDA.

In Formula E2 and the paragraph preceding Formula E2 and two paragraphs following, Formula E2 in Clause E.3 of Annex E speaks to the adjustment of the AET with an uncertainty factor, UF. The FDA notes that "Formula E2 and paragraph preceding Formula E2 and two paragraphs following Formula E2 in Clause E.3 of Annex E include technical errors and will be corrected by ISO in the following publication." This equation is in fact incorrectly stated in the current text of Part 18 but has been correctly written in Chapter 3, Section 3.2.3.3 of this book.

Example C2 in Clause E.4 of Annex E illustrates the calculation of the AET for a medical device that is permanently implanted (e.g. a cardiovascular stent) and the circumstance where a single device is used. This calculation is based on a DBT of 120 µg/day. In its rationale for recognition, the FDA notes that "Example C2 in Clause E.4 of Annex E is in conflict with another recognized standard, see clause 5.2 and Table 1 of ISO TS 21726 listed below" and is communicating that they do not agree that 120 µg/day is the proper DBT to be used in this situation.

In terms of timing of its partial recognition, the following text is provided in reference 9. FDA recognition of ANSI/AAMI BE83:2006/(R)2011 [Rec# 2-238] will be superseded by recognition of ISO 10993-18 Second edition 2020-01 [Rec# 2-276]. The FDA will accept declarations of conformity, in support of premarket submissions, to [Rec# 2-238] until 10 July 2022. After this transition period, declarations of conformity to [Rec# 2-238] will not be accepted.

9.7 Medical Device Packaging

One way that a medical device and a pharmaceutical drug product are similar is the fact that they are both packaged; that is, both the drug product and medical device are contained within a packaging system. Since they are both packaged, arguably they both share the same concern with respect to leachables derived from the packaging.

Regulatory guidelines for packaging of drug products were addressed previously in Chapter 7, Sections 7.3.1 and 7.3.3. As these guidance documents passively exclude medical device packaging from their Scope (meaning that instead of specifying that medical device packaging is out of scope they merely fail to mention whether medical device packaging is in scope or not), this exclusion is more territorial than philosophical. Considering the FDA Container Closure guidance specifically, it was written and is enforced by FDA's Center for Drug Evaluation and Research (CDER) and Center for Biologics Evaluation and Research (CBER). Medical devices, on the other hand, are regulated by FDA's Center for Devices and Radiological Health (CDRH). It would be unlikely that a guidance written by CDER or CBER would address topics enforced by CDRH and vice versa, and yet philosophically packaging presents the same challenges for both drug products and medical devices. Surely a packaging system should:

- Protect the packaged item, regardless of whether the item is a dosage form or medical device,
- Be compatible with the packaged item, regardless of whether the item is a dosage form or a medical device,
- Be safe for use with the packaged item, regardless of whether the item is a dosage form or a medical device,
- Function properly.

Furthermore, it surely must be the case that "packaging components that are compatible with an item will not interact sufficiently to cause unacceptable changes in the quality of either the item or the packaging component," regardless of whether the item is a dosage form or a medical device. Therefore, it surely must be the case that, at a high level, requirements for packaging systems for dosage forms are the same requirements for packaging systems for medical devices. Moreover, if the requirements are the same for packaging systems for dosage forms and medical devices at a high level, then the means of qualifying packaging for dosage forms and medical devices should be the same, at least at a high level.

It is in the details where the qualification of packaging for dosage forms and medical devices may differ. For example, consider the requirement that the packaging and the packaged item be compatible. For a drug product, an incompatibility between the drug product and its packaging is manifested as:

- Loss of potency due to absorption or adsorption of the active drug substance;
- Loss of potency due to degradation of the active drug substance induced by a chemical entity leached from a packaging component;
- Reduction in the concentration of an excipient due to absorption, adsorption, or leachable-induced degradation;
- Precipitation;
- Changes in drug product pH;
- Discoloration of either the dosage form or the packaging component, or an increase in brittleness of the packaging component.

Clearly, some of these incompatibilities do not specifically translate well from a dosage form to a medical device. Nevertheless, the intent is clear, packaging for medical devices should neither adversely affect the medical device's ability to perform its clinical functions nor the medical device's shelf-life.

Moreover, consider the safety requirement for drug product packaging: "packaging components should be constructed of materials that will not leach harmful or undesirable amounts of substances to which a patient will be exposed when being treated with the drug product." Surely, this is readily translated to medical device packaging as follows: "packaging components should be constructed of materials that will not leach harmful or undesirable amounts of substances to which a patient will be exposed when being treated with the medical device."

If the expectations for the impact of packaging on a patient are essentially the same whether the packaged item is a dosage form or a medical device, then philosophically the qualification process for medical device packaging should be similar to the qualification process for drug product packaging as it was established previously in Section 7. That is, a medical device's packaging system should be tested for extractables that could possibly be transferred to the packaged medical device during its storage over shelf-life. The conditions under which the extraction is performed should mirror the conditions of contact between the packaging system and the medical device during its packaged shelf-life.

Which brings one to the whole point of this narrative, which is to address the misconception that medical device packaging should be extracted under the conditions specified in ISO 10993 Parts 12 or 18. As was noted previously, the extraction conditions specified especially in Part 12 reflect extraction conditions justified as being appropriate to address the conditions of contact between the medical device and the patient on whom the medical device is used. Except in the most unlikely of circumstances, the conditions of contact between the medical device and its packaging over shelf-life will not be the same conditions of contact as those between the medical device and the patient during clinical use. Given these differences in contact conditions, it is clear and obvious that extractions designed to address device–patient contact (Parts 12 and 18) cannot and should not be applied to studies whose intent is to address device-packaging contact. This means, for example, that it is improper to extract the packaging of a permanently implanted medical device in the same way that the medical device itself is extracted. Specifically, this means:

- No mandatory exhaustive extraction.
- No mandatory three extraction solvents of different polarity.
- No mandatory 50 °C for 72 hours.

This should apply for medical device packaging unless it happens that one or more of these extraction conditions are appropriate to mimic the conditions of contact between the packaging and the packaged medical device.

There is one other important distinction between medical device packaging versus drug product packaging and that concerns the nature of leachables testing. It has been argued that testing drug products for packaging-related leachables should include both target analysis for major extractables as potential leachables and screening analysis for those leachables that were not established as extractables. This latter requirement is predicated on the possibility of so-called "reactive leachables;" that is, substances that are derived from the interaction of extractables with substances in the drug product to produce reaction products. It is these reaction products that are taken as leachables that were not first extractables.

The argument for requiring screening leachables testing is that "one can never know whether reactive leachables can occur or not when you are dealing with drug products, so you better do the screening just to be sure." For many medical devices, the formation of "reactive leachables" can probably be discounted based on the nature of the medical device and thus it is reasonable to propose that screening medical devices for so-called "reactive leachables" should only be required when there is a realistic possibility that such "reactive leachables" could be formed by the interaction of a medical device with packaging-related extractables.

Another factor to consider in terms of screening is environmentally-derived contaminants. Environmentally-derived contaminants, which are substances from the environment that can breach the barrier of the packaging during storage and which become entrained in or on the packaged item, are clearly substances that are not related to extractables. The fact that they are not true leachables is semantically important but relatively unimportant in terms of potential adverse safety effects. If the packaging system for a medical device is not a barrier to such environmentally-derived contaminants, then testing of the medical device for leachables must include screening, not because the environmentally-derived contaminants are leachables but rather because leachables screening is the only place in the medical device's chemical characterization process that these environmentally-derived contaminants can be addressed.

Lastly, it is observed that while many pharmaceutical drug products are liquids, most medical devices are solid items. Thus, extraction methods and circumstances that are relevant for packaging used for liquid drug products are likely to have little or no relevance for packaging used for solid medical devices.

9.8 Case Studies

Due at least in part to the revision of the MDRs, extractables and leachables from medical devices is one of the current hot topics in the E&L community and there are several recent publications on this topic. Recent publications that provide information on the identity and concentrations of extractables from medical devices are discussed as follows.

The use of 3D printed medical devices is a growing industry across the globe, thanks to product customizability, short lead times, cost efficiency, and the ability to create unique devices not possible with traditional manufacturing techniques. Researchers from the University of Auckland, Train Industries, and Pace Analytical Life Sciences note that "there currently are not any publicly available studies investigating the extractable profiles of … printed medical devices" [10]. To fill this gap, these authors report a study where four different printed polymeric materials (polylactic acid, FDA-approved polylactic acid (PLA), polyethylene terephthalate glycol (PETG), and polycarbonate) were extracted using solvents of varying polarity and analyzed for organic and elemental extractables. In this study, rectangular 3D printed coupons were created from each polymer to represent medical devices. These "devices," as well as the filament materials used to create them, were extracted by immersion in three different solvents of varying polarity: water, isopropyl alcohol (IPA), and hexane. Each device was extracted using 50 ml of the pure solvent for greater than 72 hours at 50 °C. Following extraction, all devices were visually similar to pre-extraction observations. The devices showed no signs of swelling, regardless of the solvent used, and their morphology remained unchanged after extraction. The resulting extracts were tested using a suite of analytical techniques targeting volatile, semi-volatile, and low-volatility organics (headspace GC/MS, GC/MS, and LC/ToF-MS) and elemental extractables (ICP/MS).

In summarizing their results, the authors noted that:

- The 3D printing process did have a notable effect on the chemicals extracted from the devices.
- Both the largest amount and number of organic compounds were observed in the IPA extracts.
- The water extracts contained the smallest amounts of observed components.
- The hexane extracts did not contain extractables that were not also contained in the IPA extracts.

More specifically,

- Volatile organic compounds were not observed above the method's detection limit (10 μg/device) in the water extracts (the other extracts were not tested).
- Of the 15 elements targeted by ICP/MS analysis only lead (Pb) was observed in the printed device extracts, with levels ranging from 0.11 ng/g in the device fashioned from PETG to 1.46 ng/g in the device fashioned from PLA. The authors suggest that the Pb was incorporated into the device via the brass printing nozzle.
- Fifteen separate extractables were detected using the GC/MS technique. The polylactic acid precursor, L-lactide, was observed in all PLA extracts, and had the largest measured concentrations of any analyte detected across all materials. One of the tested PLA materials had trace levels of extractable DEHP. The only reported extractables from the PETG-based devices were several unspecified siloxanes, while triphenyl phosphate was the only observed extractable from the PC-based device.
- The main class of organic compounds detected by LC/MS was PLA oligomers derived from the PLA polymers. The extracts of one of the PLA-based devices also contained polypropylene glycol (PPG) oligomers. Although PPG is a co-polymer of PLA, this compound was not listed as a material component by the material's vendor. Extracts of one of the PLA-based devices also included DEHP and the antioxidant Irganox 1010.

- The major non-volatile extractable attributed to the PETG-based device was Latyl Yellow 3G (colorant).
- The major non-volatile extractable attributed to the PC-based device was Bisphenol A bis (diphenyl phosphate), clearly correlated to the PC's monomer (Bisphenol A).

In recent years, several cases of allergic contact dermatitis (ACD) associated with glucose sensors and insulin pumps have been reported. Isobornyl acrylate (IBOA) and *N,N*-dimethylacrylamide (DMAA) have been identified as the major culprit allergens [11–15]. In a recent study, researchers from Lund University and Helsingborg Hospital reviewed 11 cases where diabetes mellitus patients were investigated for ACD reactions to medical devices and specifically Dexcom G6[®], CGM system (Dexcom, Inc., San Diego) [16]. Extracts from the medical devices were analyzed for potential allergens. Specifically, extracts were prepared from separate components of the medical devices (the adhesive patch, the sensor, and the plastic support) by cutting the items into small pieces and extracting the pieces in 5 ml acetone at room temperature for 72 hours. Thereafter the extracts were concentrated to a final volume of 0.2 ml using a rotary evaporator. These concentrated extracts were analyzed by gas chromatography–mass spectrometry (GC/MS) after being diluted 10 times and filtered. The authors report that a majority of the patients had relevant allergies and particularly an allergy to isobornyl acrylate, which was extracted from the glucose sensor system Dexcom G6. Specifically, the extracts made from the adhesive patches contained IBOA in concentrations of 0.1–0.7 μg/patch, the extracts of the sensors had 0.8–1.3 μg IBOA, and no IBOA was detected in the extracts of the plastic support. The GC/MS analyses indicated the presence of methyldehydroabietate in the extract of the adhesive patches.

Bisphenol A (BPA) belongs to a group of chemicals used in the production of polycarbonate, polysulfone, and polyethersulfone, which are used, among other applications, in the manufacture of dialyzers. To assess the potential risk of BPA exposure to dialysis patients through dialyzer use, the BPA levels in extracts from polypropylene- and polycarbonate-containing dialyzers were determined by researchers from Fresenius Medical Care and Exponent, Inc. [17] Both simulated-use and exaggerated extractions were performed. Prior to extraction, each dialyzer was primed (as is done clinically) using a saline solution (0.9% NaCl) according to the device's Instructions for Use (IFU). After priming, the simulated-use extraction was performed by recirculating 1 l of 17.2% ethanol solvent (justified as a blood simulant) through the bloodline of each dialyzer at 37 °C for 24 hours. The dialysate compartment was filled with an aliquot of the extraction solution. After recirculation, the extract solution from the blood compartment was collected for analysis of BPA levels. The extraction solution from the dialysate compartment was not collected, as theoretically BPA in the solvent reached equilibrium between the blood and dialysate compartments in the dialyzer.

An exaggerated extraction was performed by filling the blood and dialysate compartments with an extraction solution consisting of 95% ethanol diluted in saline and storing the filled dialyzer under static conditions for 72 hours at 37 °C.

The BPA concentration in the extracts was determined via UPLC-QTOF-MS with an electrospray ionization (ESI) source operating in negative ionization mode.

The mean BPA concentrations were 3.6 and 108.9 ppb (μg/kg) from simulated-use and exaggerated extractions, respectively, from polycarbonate-containing dialyzers. No BPA was detected from polypropylene-containing dialyzers. This information was toxicologically safety assessed considering that one device is used per hemodialysis session, the mean BPA concentration measured in the samples represents a patient exposure, a patient body weight of 60 kg, and a time-weighted daily average exposure to reflect dialysis sessions occurring three times per week. The authors concluded that the findings suggest that "there is an acceptable level of toxicological risk to dialysis patients exposed to BPA from use of the dialyzers tested in the current study."

Recently, a team of authors from the FDA's Office of Science and Engineering Laboratories, Center for Devices and Radiological Health (CDRH) and Office of Food Safety, and Center for Food Safety and Applied Nutrition have considered the use of mathematical migration modeling to estimate patient exposure to medical device leachables [18]. These authors note that extraction testing is routinely performed under aggressive conditions with respect to the presumed clinical environment to estimate exposure to potential leachable chemicals from a device material. These authors suggest that although this approach is generally considered protective, it can lead to exceedingly conservative exposure estimates. Therefore, they propose that a promising alternative is to base exposure estimates on conservative physics-based models. To examine this proposition, the authors applied conventional extraction techniques to model systems consisting of high-density polyethylene (HDPE) "spiked" with a variety of common antioxidants found in medical device materials. Specifically, HDPE disks were produced without (neat) and with additives, including butylated hydroxy toluene (BHT), Irganox 2246, Irganox 1076, and Irganox 1010 at a level of 1 wt%. Portions of the disks were extracted with either isopropanol or hexanes at 50 °C for 6, 24, or 144 hours with agitation. Extracted substances were measured via ultra-high-pressure liquid-chromatography

(UHPLC) coupled with a time-of-flight mass spectrometer (LC/Q-ToF MS) and a diode array detector (DAD) at 280 nm. Extracted substances were identified via MS and quantified via UV.

Considering the test results, the authors reported that asymptotic extraction was achieved for all intentional additives in both extraction solvents over the time course of the study. The authors also noted that the choice of solvent had a strong and consistent effect as the amount of extractables released was much higher in hexane compared with IPA over the same time period. This phenomenon was attributed to the difference in the extent of swelling of the polymer matrix by the respective solvents; while hexane caused the neat polymer to swell approximately 11 wt%, IPA resulted in less than 1% swelling.

While this book was in proofing, authors from the FDA CDRH published a Perspective that describes the elements of medical device extractables studies and provides an overview of the current practices, identified gaps, and emerging practices that may be adopted on a wider scale in the future". Although the authors note that "the findings and conclusions in this manuscript have not been formally disseminated by the Food and Drug Administration and should not be construed to represent any agency determination or policy", it is clear that the manuscript reflects the opinions and recommendations of these influential CDRH individuals.

In addition to providing an extensive bibliography (252 citations), the manuscript contains the following:

1. The term non-target analysis (NTA) is preferred versus screening.
2. The manuscript notes that knowledge of raw materials and manufacturing improves the analytical chemistry testing.
3. Exhaustive extraction is emphasized, noting, however, that "alternate methods … can be used under certain circumstances".
4. NTA usually includes the use of polar, semi-polar and non-polar solvents, elevated temperatures and longer extraction times.
5. The topic of extraction solvents is addressed in detail.
6. The topic of AET and the use of the UF is considered although no specific values for the UF are advocated.
7. The manuscript speaks to the aspect of method qualification, including system suitability.
8. The manuscript contains a recommendation to use ESI versus APCI as the preferred ionization technique for LC/MS.
9. A discussion of quantitation is provided, emphasizing the use of surrogate standards and pointing out perceived shortcomings in the use of RRF databases.

10. The manuscript considers the challenges of data processing (and notes the many opportunities for, and consequences of, poor data processing practices).
11. Identification strategies, processes and classification is discussed.

As practical constraints prevent a more in-depth discussion of the Perspective, including constructive debate of article's contents, readers are referred to the article to perform their own more extensive review.

References

1 ISO 10993:1(2018). Biological evaluation of medical devices – Part 1: Evaluation and testing within a risk management process. International Organization for Standardization; 2018-10.

2 Guidance for Industry and Food and Drug Administration Staff. Use of International Standard ISO 10993:1, Biological evaluation of medical devices – Part 1: Evaluation and testing within a risk management process. US Department of Health and Human Services: Food and Drug Administration; Center for Devices and Radiological Health, Center for Biologics Evaluation and Research. 4 September 2020.

3 USP <87> Biological Reactivity Tests, In Vitro. USP43-NF38; pp. 6514. Official as of 1 August 2016.

4 USP <88> Biological Reactivity Tests, In Vivo. USP43-NF38; pp. 6516. Official as of 1 August 2013.

5 ISO 10993:12(2012). Biological evaluation of medical devices – Part 12: Sample preparation and reference materials. International Organization for Standardization; 2012-07.

6 ISO 10993:18(2020). Chemical characterization of medical device materials within a risk management process. International Organization for Standardization; 2020-01.

7 ICH Guideline Q3C (R6) on Impurities: Guideline for residual solvents. European Medicines Agency; EMA/CHMP/ICH/82260/2006, Committee for Human Medicinal Products. 9 August -2019.

8 USP <1058>. Analytical instrument qualification. USP43-NF38; pp. 7522. Official as of 1 August 2017.

9 US Food and Drug Administration. Part B: Supplementary Information Sheet (SIS). FR Recognition List Number 054. FR Recognition Number 2-276. Date of Entry 07/06/2020. https://www.accessdata.fda.gov/scripts/cdrh/cfdocs/cfStandards/detail.cfm?standard__identification_no=41050. Accessed 4 January 2021.

10 Rindelaub, J.D., Baird, Z., Lindner, B.A., and Strantz, A.A. (2019). Identifying extractable profiles from 3D printed medical devices. *PLoS One.* https://doi.org/10.1371/journal.pone.0217137 .

11 Herman, A., Aerts, O., Baeck, M. et al. (2017). Allergic contact dermatitis caused by isobornyl acrylate in Freestyle® Libre, a newly introduced glucose sensor. *Contact Dermatitis* 77 (6): 367–373.

12 Raison-Peyron, N., Mowitz, M., Bonardel, N. et al. (2018). Allergic contact dermatitis caused by isobornyl acrylate in OmniPod, an innovative tubeless insulin pump. *Contact Dermatitis* 79 (2): 76–80.

13 Herman, A., Baeck, M., de Montjoye, L. et al. (2019). Allergic contact dermatitis caused by isobornyl acrylate in the Enlite glucose sensor and the Paradigm MiniMed Quick-set insulin infusion set. *Contact Dermatitis* 81 (6): 432–437.

14 Hyry, H.S.I., Liippo, J.P., and Virtanen, H.M. (2019). Allergic contact dermatitis caused by glucose sensors in type 1 diabetes patients. *Contact Dermatitis* 81 (3): 161–166.

15 Mowitz, M., Herman, A., Baeck, M. et al. (2019). *N,N*-dimethylacrylamide – A new sensitizer in the FreeStyle Libre glucose sensor. *Contact Dermatitis* 81 (1): 27–31.

16 Svedman, C., Bruze, M., Antelmi, A. et al. Continuous glucose monitoring systems give contact dermatitis in children and adults despite efforts of providing less "allergy- prone" devices investigation and advice hampered by insufficient material for optimized patch test investigations. *J. Eur. Acad. Dermatol. Venereol.* https://doi.org/10.1111/jdv.16981. 08-Oct-2020.

17 Badding, M.A., Vargas, J.R., Fortney, J. et al. (2020). Toxicological risk assessment of bisphenol a released from dialyzers under simulated-use and exaggerated extraction conditions. *Regul. Toxicol. Pharmacol.* 118: 104787.

18 Turner, P., Elder, R.M., Nahan, K. et al. (2020). Leveraging extraction testing to predict patient exposure to polymeric medical device leachables using physics-based models. *Toxicol. Sci.* 178 (1): 201–211.

19 Sussman, E.M., Oktem, B., Isayeva, I.S., Liu, J., Wickramasekara, S., Chandrasekar, V., Nahan, K., Shin, H.Y., Zheng, J. (2022). Chemical characterization and non-targeted analysis of medical device extracts: A review of current approaches, gaps, and emerging practices. *Biomat. Sci. Eng.* https://doi.org/10.1021/acsbiomaterials.1c01119.

10

General Principles for Risk Assessment of Extractables and Leachables

10.1 Toxicological Safety Risk Assessment of Extractables and Leachables

10.1.1 Introduction

To this point in this book the discussion has been highly focused on the chemical aspects of performing extractables and leachables assessments. While the toxicological dimension of the assessment has certainly been acknowledged, the mechanics of performing a toxicological assessment have largely been neglected.

There are two reasons that this is the case. The first reason is process-driven; that is, the toxicological risk assessment can only occur after the chemical information has been generated. Thus, from a process perspective it is proper to address the chemical data generation process first and to address the toxicological data interpretation process second.

The other reason is simply a matter of expertise. At this point of the book, it is impossible to escape the fact that the data generation aspects of chemical assessment are numerous and complex. This truth in no less valid for the toxicological safety risk assessment process. The effect that chemicals can have on the human body is incompletely understood; thus, toxicological safety risk assessment is a complex process that often relies on considerable training, experience, and expertise to manage shortcomings in the available information and the available science. Simply stated, it is beyond the capabilities of this author to address the topic fully and effectively and it is foolhardy to try and "fake it." Those readers who seek to understand toxicology essentials more completely will have to seek that knowledge elsewhere. Rather than address the science of toxicology, this chapter will focus on a simplified discussion of the toxicological assessment process.

10.1.2 Key Definitions and Concepts

ISO 10993:17, Establishment of allowable limits for leachable substances [1] specifies a method for the determination of allowable limits for substances leachable from medical devices. This Standard is currently under revision and it is possible that the revised Standard will be available for public review by the time this book is actually published. Definitions of the following terms critical to toxicological risk assessment anticipate what might be in the normative text in the revised Standard, fully recognizing that the text that is actually present in the draft Standard when it is released may differ, perhaps significantly, from the text provided as follows.

Appreciable harm to health
An adverse reaction, such as altered morphology, physiology, growth, development, reproduction, or lifespan that (i) impairs function of an organism/system, (ii) reduces capacity to tolerate impaired function, or (iii) increases susceptibility to other influences that impair function.

Cancer risk
The probability of an increase in the occurrence of cancer resulting from exposure to a carcinogen at a specified exposure dose (ED) and duration (e.g. lifetime).

Human carcinogen
Human data demonstrates a causal association between exposure to the constituent and human cancer.

Margin of Safety (MoS)
Ratio of the tolerable intake (TI) and maximum exposure estimate.

Point of departure (POD)
Dose that marks the beginning of a low-dose extrapolation.

Release kinetics
The quantity of a constituent that is released from a medical device as a function of time.

Suspected human carcinogen
Toxicological data indicates a causal relationship between exposure to the constituent and cancer. Suspected human carcinogen applies when human data is inadequate to establish an association between

Extractables and Leachables: Characterization of Drug Products, Packaging, Manufacturing and Delivery Systems, and Medical Devices, First Edition. Dennis Jenke.
© 2022 John Wiley & Sons, Inc. Published 2022 by John Wiley & Sons, Inc.

exposure to the constituent and cancer. Suspected human carcinogens are established by experts in chemical carcinogenesis based on a weight-of-evidence test.

Systemic toxicity

Harm(s) that occur at a different organ/system distal of the site of contact. Systemic toxicity can occur within a few days of a one-time exposure (i.e. acutely) or delayed after repeated exposure (sub-acute, sub-chronic, or chronic) to a toxic dose of a constituent.

Tolerable intake (TI)

Estimate of the daily exposure (DE) of a constituent over a specified time period, on the basis of body mass, that is considered to be without appreciable harm to health. TI is expressed in milligrams (mg) per kilogram (kg) of body weight (bw) per day (i.e. mg/kg bw/day).

Toxic dose

A dose capable of eliciting appreciable harm to health.

Toxicological risk

Probability of an adverse reaction occurring in response to a specified level of exposure.

Toxicological risk assessment

The determination of a DE for a constituent that could elicit appreciable harm to health.

Uncertainty factor(s) (UF)

Numerical value that accounts for uncertainties when extrapolating a POD to individuals who could be exposed to constituent(s) of toxicological concern.

Worst case

Assumption and/or clinical condition that intentionally results in a maximum human DE estimate.

10.1.3 The Toxicological Safety Risk Assessment Process Simplified

Although the science of toxicology may be complex, the mathematical process of performing a toxicological safety risk assessment is deceptively simple. The assessment is based on a concept that is widely attribute to Paracelsus, a Swiss physician, alchemist, lay theologian, and philosopher of the German Renaissance, who is widely regarded as the "father of toxicology" based on his concept of dose-response; that is, that "all substances are poisons; there is none which is not a poison. The right dose differentiates a poison from a remedy." A corollary to this wisdom is the assertion that for many toxicological effects, there exists a threshold that establishes that dose at which a chemical's effect on a person's health becomes adverse. A patient who is exposed to the chemical in an amount lower than the threshold will likely exhibit no appreciable ill-health effects while the health of a person exposed to

the same chemical in quantities above the threshold will be adversely affected.

As is the case with any simplifying generalization, there is much debate in the toxicological community in terms of which toxic effects (toxicological endpoints) truly exhibit classical dose-response behavior. Furthermore, the means by which the dose threshold is calculated is open to discussion and interpretation. However, for those toxic effects that exhibit dose-response behavior and for which a dose threshold, otherwise known as a TI, can be established, the toxicological risk assessment mathematically involves a comparison of the dose threshold, referred to as the TI and the actual dose, DE (see Figure 10.1). The TI is calculated for a particular entity based on the toxicological properties of that entity; the DE is calculated for a particular circumstance based on the level of the entity that is transferred to the patient daily. When the DE is less than the TI, it is likely that the chemical being assessed will have a negligible adverse effect on patient health. When the DE exceeds the TI, it is likely that the chemical being assessed will adversely affect the patient's health.

The relationship between the TI and the DE can be expressed in the MoS, which is the ratio of the TI to the DE (Equation 10.1). Equal values for the DE and the TI produce an MoS of 1, which becomes the de facto threshold. When the DE is less than the TI and the compound is likely to have a negligible adverse effect, the MoS has a numerical value greater than 1. When the DE is greater than the TI and the compound is likely to have a discernible adverse effect, the MoS has a numerical value less than 1.

Margin of Safety (MoS)

$$= \text{Tolerable Intake (TI)} \div \text{Daily Exposure (DE)} \quad (10.1)$$

A literal interpretation of the MoS is ill-advised. Because of the assumptions that are inherent in the calculations of the TI and DE, an MoS value slightly above or below the 1.0 threshold cannot be definitely interpreted. That Is, rather than interpreting the MoS as a solid, well-defined cut-off,

Figure 10.1 Mathematical concept of a toxicological risk assessment. The assessment involves a comparison of the tolerable intake and the daily exposure.

it is more appropriate to consider the MoS to be a somewhat indistinct value that defines a region of reasonable concern. Thus, for example, an MoS of 1.5, which could be interpreted as "no risk" or a MoS of 0.5, which could be interpreted as probable risk, may in fact be "within the margin of error" and both would properly be interpreted as reflecting an uncertain effect on patient health that requires further assessment.

Furthermore, the magnitude of the MoS cannot be interpreted as a relative indicator of the toxicological hazard. That is, a chemical that has an MoS of 10 is not 10 times "safer" than a chemical with an MoS of 1. Rather, the magnitude of the MoS is an indicator of the certainty in the conclusion that can be drawn from the MoS. Thus, the conclusion that a compound with a MoS of 10 is safe is more certain than the conclusion that a compound with an MoS of 2 is safe.

Moreover, the larger the MoS, the more uncertainty in the chemical test results or the toxicological determination of the TI that can be tolerated. For chemical testing, this is relevant to the question of the quantitative nature of an extractable's or leachable's reported concentration. For example, it was noted previously that a concentration estimated using a single surrogate standard-related response factor is somewhat inaccurate due to the variation in response factors across the universe of extractable, leachables, and possible surrogate standards. Considering GC/MS specifically, it is generally accepted that the reported concentration could be off by a factor of 2 or more, either high or low, which on the face of things seems to be a rather large amount. However, if the toxicological safety risk assessment of the compound in question produced an MoS of 10, even if the compound's concentration had been underestimated by a full factor of four, the "correct" MoS would be 2.5 and one would still conclude that the compound was likely to be safe. In this case, the somewhat inaccurate concentration estimate would still be accurate enough to establish the compound's likely toxicological effect. On the other hand, had the original MoS been 3, a factor of 4 reduction in the MoS would change the assessment conclusion from "likely safe" to "possibly unsafe," which would surely trigger the additional analytical action of securing a more accurate and quantitative concentration estimate.

Similarly, the magnitude of the MoS can have a bearing on the toxicology side of the assessment, specifically the determination of the TI. Again, consider the case where the outcome of a toxicological risk assessment is an MoS of 10. Furthermore, imagine that in the calculation of the TI (to be discussed in greater detailed later in this chapter), the toxicological assessor used a factor (F6) to account for the bioavailability of the extractable of interest. Armed with relevant but not necessarily definitive bioavailability data, the assessor decided to assign F6 a value of 2. It is possible that another toxicological assessor could have taken a more conservative approach and assigned F6 a value of 10, reducing the calculated TI by a factor of 5. Although such a change has a discernible effect on the value of the TI, the MoS calculated using the more conservative TI (MoS = 2) still supports the conclusion that the compound of interest is likely to be appropriately safe.

10.1.4 Toxicological Safety Risk Assessment of Pharmaceutical Packaging

It is intuitively obvious why a drug product's packaging is of concern to a toxicologist considering the drug product's safe use. For one thing, at least the primary packaging is in direct contact with the drug product. Moreover, the drug product and packaging likely have been in contact for a long period of time, as many drug products have labeled shelf-lives measured in years. Lastly, the drug product and the packaging are likely to be quite different chemically and thus will be prone to interaction.

Packaging influences drug product safety by adding impurities, termed leachables, to the drug product. Any adverse effect the leachable may have on the health of a patient receiving the drug product as part of a clinical therapy will depend on the substance's inherent toxicity and the patient's exposure to the leachable as dictated by the concentration of the leachable in the drug product, the clinical dosing of the drug product, and the drug product's route of administration.

The purpose of the toxicological safety risk assessment is to establish the likely effect that the leachables will have on patient safety. The assessment can be based on extractables as well, considering the premise that extractables represent probable, worst-case leachables. Regardless of whether the subject of the assessment is an extractable or leachable, completion of the assessment requires the following information about the substance of interest:

- Compound's identity, including
  - Proper scientific (e.g. IUPAC) name
  - Appropriate reference information (for example, CAS registry number)
  - Chemical formula and chemical structure (useful in cases where the assessment requires QSAR analysis or identification of a surrogate)
- Compound's concentration in the drug product (either the direct result of testing a drug product or an estimation from an extraction study)

Ideally, both the identity and concentration are reported with an associated degree of certainty in case the uncertainty must be accounted for in the assessment.

Previously, the process of toxicological safety risk assessment was trivialized by describing it as consisting of four steps:

1. Establishing a TI by review of the compound's available toxicological information and considering the drug product's route of administration,
2. Establishing a patient DE based on the compound's concentration in the drug product and the product's clinical dosing,
3. Comparing the TI and DE via the use of the MoS, and
4. Interpreting the MoS.

There is a wide variety of approaches that have been proposed for the toxicological safety evaluation of extractables and leachables. For illustration purposes, a systematic, comprehensive approach to safety evaluation that involves utilization of generic thresholds, predictive *in silico* (quantitative) structure activity relationship ((Q)SAR) approaches, experimental study results from publicly, available sources, and development of evaluation thresholds and toxicological qualification testing, developed by scientists from the Extractables and Leachables Safety Information Exchange consortium (ELSIE) [2], is illustrated in Figure 10.2.

One proceeds through the Evaluation process as follows:

1. The proper input information is provided by the chemist for all relevant substances (extractables or leachables) above an established reporting threshold (e.g. AET).
2. The patient's DE to a substance is established. In its flow diagram, the ELSIE authors refer to the DE as a PDE (patient daily exposure). The term DE is used in this discussion to avoid confusion between a PDE that refers to permissible daily exposure and a PDE that refers to a patient daily exposure.
3. The resultant DE is compared to the appropriate threshold for mutagens, such as the threshold of toxicological concern (TTC) from ICH M7 [3].
   a. If the DE is less than or equal to the TTC, then the substance of interest is concluded to present no safety alert (that is, the substance will not be likely to adversely impact patient health) and the assessment is completed.
   b. If the DE is greater than the TTC, then the substance presents a safety alert (that is, the substance could possibly adversely impact patient safety). In this case, the safety alert is further investigated.
4. The conclusion that the compound presents a safety alert is predicated on the substance being mutagenic. The substance's mutagenicity is established by literature review and/or appropriate *in silico* QSAR evaluation.
   a. If the substance is established to be mutagenic, then the substance is established to be a likely threat to patient safety. Either the level of the substance in the drug product must be reduced (for example, by replacing the substance's source material in the packaging) or the packaged drug product is concluded to be unsafe and unsuitable for use.
   b. If the substance is not mutagenic, its safety assessment continues with a consideration of other toxicological endpoints such as local effects (sensitization and irritation), repeat dose toxicity, carcinogenicity, reproductive toxicity, etc.
5. Once the available toxicological information has been collected and assessed, the assessor considers whether the information is sufficiently robust that a TI can be established.
   a. If a TI cannot be established, then further assessment is required, as noted in Step 6.
   b. If a TI can be established, then the toxicological assessment is concluded with the comparison of the DE and the TI (e.g. calculation of the MoS).
6. The DE and TI are compared.
   a. If the TI < DE (MoS < 1), then it is concluded that the substance is a safety hazard, in which case either the substance's level in the packaging is reduced, the substance's leaching is reduced, or the packaged drug product is concluded to be unsuited for commercial use.
   b. If TI ≥ DE (MoS ≥ 1), then the substance of interest is concluded to present no safety alert (that is, is not likely to adversely impact patient health) and the assessment has been completed.
7. In the absence of a TI, the ELSIE authors proposed that the DE be compared to a more generic (meaning generalized and broadly applicable) safety threshold such as a Cramer TTC or the Product Quality Research Institute's (PQRI's) SCT or QT. Such an approach is highly conservative, as the process of establishing the SCT or QT (for example) is such that these thresholds are, by design, highly protective of patent safety. As was the case in step 5, the DE is compared to the generic safety threshold in the same manner as noted in steps 3 and 5.

Thus, the toxicological safety risk assessment has three possible outcomes:

1. The packaged drug product as developed has been qualified as likely to be safe for its developed use with respect to leachables,
2. The packaged drug product has been qualified, after its packaging has been modified to mitigate any potentially unsafe leachables, as likely to be safe for its developed use with respect to leachables,
3. The packaged drug product has been established as likely to be unsafe in its developed use.

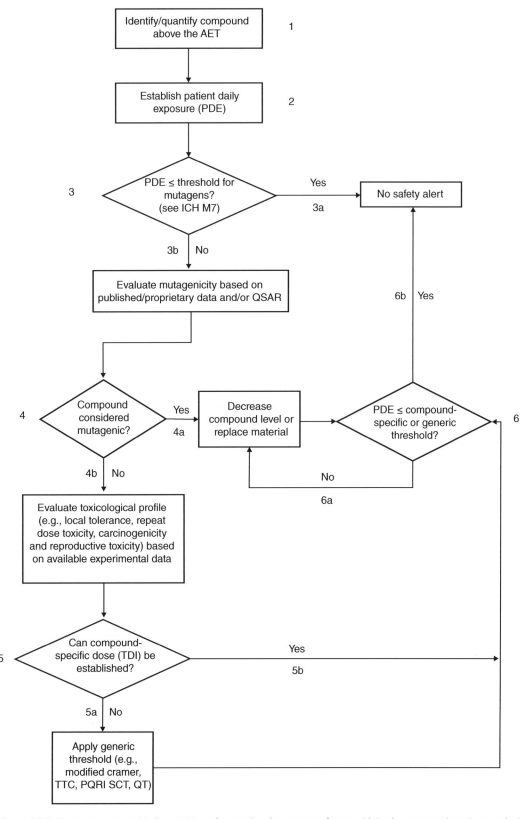

Figure 10.2 Example extractable/leachable safety evaluation process for non-biologic parenteral, oral, or topical drug products. Source: Adapted from reference [2].

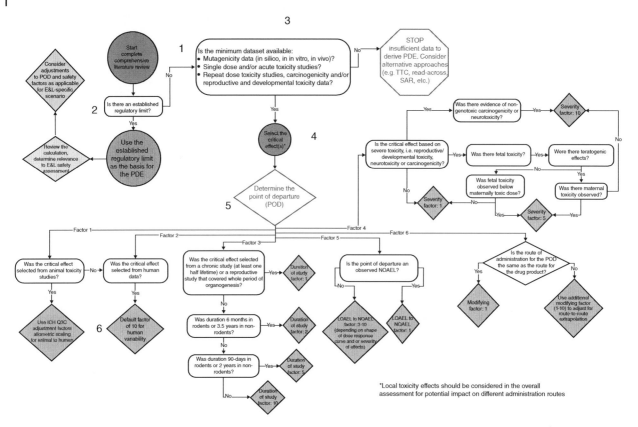

Figure 10.3 Decision tree for deriving a tolerable intake, TI, which here is called a permissible daily exposure (PDE). Source: From reference [4] with permission.

The critical role of the TI in toxicological safety risk assessment is evident; it should not be surprising, then, that much attention is focused on its derivation and justification. In an ideal world where abundant, high-quality, and relevant toxicological data is available, the determination of the TI becomes a standardized exercise of mathematics and expert judgement. In the real world where leachables are all too frequently chemically exotic and poorly studied, the determination of a proper TI is as much an art to the toxicologist as is securing an identification for a never-reported-before leachable to the analytical chemist.

Following up on their manuscript that defined the previously discussed process for performing a toxicological safety risk assessment, the ELSIE toxicology team devised a process map (decision tree) for deriving a substance's TI (which they termed a permissible daily exposure, PDE) [4]. The process, illustrated in Figure 10.3, begins with a comprehensive literature search and review (step 1) to establish whether there are any existing thresholds or limits set by regulatory bodies and/or health authorities that have been scientifically peer-reviewed and appropriately derived for a similar route of administration. Should such a threshold exist, it can typically be used as the TI; however, the threshold should be reviewed to consider whether

adjustments are necessary, for example, to account for route of administration.

If an established regulatory limit has not been developed, the comprehensive literature review is re-visited to determine if there are sufficient data to derive a substance-specific TI (step 2). In general, the minimum dataset should consist of an *in silico* analysis for mutagenicity and repeat dose toxicity studies (minimum 28-day study), ideally administered via a route of exposure relevant to the drug product being evaluated. If there is insufficient information for a substance to derive a TI, then alternate assessment approaches are necessary, including read-across (i.e. the use of a surrogate compound with robust and adequate toxicological data), QSAR, or the use of general thresholds, etc.

If the literature search was able to surface sufficient information to establish a TI, the literature is examined to establish the "critical effect," i.e. the most sensitive and/or relevant indicator of an adverse effect in the non-clinical study observed in the most relevant species/target organ, through the most relevant route of administration (step 4). For all critical effects identified, a NOEL or NOAEL should be established that represents the highest tested dose at which no critical effect is observed. If a NOEL/NOAEL is not available, the LOEL/LOAEL may be used. The lowest dose from non-clinical studies or human clinical data that show a certain response level for the critical effect is selected as the POD, which is the value with which the subsequent calculation of the TI is performed (step 5).

Once the POD has been established, the mechanics of the calculation of the TI begins (step 6). In essence, the calculation of the TI is accomplished by adjusting the POD with various modifying or adjustment factors that address the nature of the study in which the POD was derived and accounts for differences between those study factors and factors relevant to patient exposure to the substance being evaluated. The general equation used for this adjustment has been established in ICH Q3C as follows [5]:

$$\text{TI} \left(\mu g/day \right) = \left(\text{POD} \left(\mu g/kg - day \right) \times \text{BW} \left(kg \right) \right)$$
$$\div \left(F1 \times F2 \times F3 \times F4 \times F5 \times F6 \right) \quad (10.2)$$

where the individual factors and their acceptable values are summarized in Table 10.1.

When the calculation of the TI becomes formularized, it is possible to complete the calculation without considering the shortcomings in the calculated result. For example, consider the magnitude of the adjustment required, that is the product of factors F1 through F6. Surely it is the case that the magnitude of the adjustment required speaks to the validity of the calculated TI. In the case of a smaller adjustment required, one is confident in the calculated TI as it is based on relevant, reliable, and "high quality" data. Alternatively, in the case of larger required adjustments one is less confident in the calculated TI as it is based on tangential, unreliable, and "low-quality" data. Surely there is a point at which the adjustment is so large that the ability to perform the TI calculation is over-ruled by the fact that the compounded uncertainty in the calculated

Table 10.1 Modifying factors used in the calculation of the TI per Equation (10.2).

Factor	Purpose	Acceptable values
F1	A factor to account for extrapolation between species, based on a power relationship between body surface area and body weight for various laboratory species and humans	F1 = 2 for extrapolation from dogs to humans F1 = 2.5 for extrapolation from rabbits to humans F1 = 3 for extrapolation from monkeys to humans F1 = 5 for extrapolation from rats to humans F1 = 10 for extrapolation from other animals to humans F1 = 12 for extrapolation from mice to humans
F2	A factor to account for accounts for human variability in pharmacokinetics and pharmacodynamics among the general population	F2 = 10
F3	A factor used for the extrapolation from the duration of the study used for the POD to chronic/lifetime exposure.	F3 = 1 for studies that last at least one half-lifetime (1 year for rodents or rabbits; 7 years for cats, dogs, and monkeys) F3 = 1 for reproductive studies in which the whole period of organogenesis is covered F3 = 2 for a 6-month study in rodents, or a 3.5-year study in non-rodents F3 = 5 for a 3-month study in rodents, or a 2-year study in non-rodents F3 = 10 for studies of a shorter duration.
F4	A factor that may be applied in cases of severe toxicity (e.g. non-genotoxic carcinogenicity, neurotoxicity, or teratogenicity).	F4 = 1 for fetal toxicity associated with maternal toxicity F4 = 5 for fetal toxicity without maternal toxicity F4 = 5 for a teratogenic effect with maternal toxicity F4 = 10 for a teratogenic effect without maternal toxicity F4 = 10 if the critical effect is most severe (e.g. non-genotoxic carcinogenicity or teratogenicity observed at a sub-maternally toxic dose)
F5	A factor that may be applied if the NOEL was not established	F5 = 1 to 10, depending on dose selection and shape of the slope of the dose–response curve F5 = 3, based on an average LOAEL:NOAEL ratio determined from analysis of a large database of toxicology studies
F6	A factor that may be applied to account for route-to-route absorption/bioavailability	F6 = 1 if the oral bioavailability is ≥90% F6 = 2 if the oral bioavailability is ≥50% but <90% F6 = 10 if the oral bioavailability is ≥1% but <50% F6 = 100 if the oral bioavailability is <1%

TI makes it largely unreliable, irrelevant, and unusable. That is, the more a NOEL/NOAEL needs to be adjusted to be applicable to a human exposure situation, the less likely it is that the calculated TI effectively accounts for the safety risk posed by the substance being evaluated. For example, Parris et al. performed TI calculations for four well-known extractables, bisphenol A, Irganox 1010, Irgafos 168, and butylated hydroxy toluene (BHT) [4]. The combined adjustment factors required to produce their TI values ranged from 100 for BHT to 60 000 for BPA, both of which are relatively small compared to the largest possible combined adjustment factor of 1.2×10^6 if all the individual adjustment factors are set at their largest possible value. Using a similar process to calculate PDE's of organic extractables relevant to biotechnology products, Li et al. reported several PDE values that were derived using combined adjustment factors of 1×10^6 [6].

It is reasonable to question the propriety of using TIs calculated with excessively large combined adjustment factors in a toxicological safety risk assessment. As noted by Sussman et al., "when a composite factor approaches values that large, the total amount of uncertainty is considerable and the risk assessor should consider whether it is unreasonably large and whether another risk assessment strategy should be used" [7]. Although the impropriety is widely acknowledged in the toxicology community, there is a decided lack of clarity and consensus in terms of what the "acceptability" threshold is for the combined adjustment factors. The US EPA recommends a maximum composite adjustment factor value of 3000 for traditional factors (i.e. F1–F5, and thus exclusive of route-to-route bioavailability extrapolations) and suggests avoiding the derivation of an exposure limit that involves application of a 10-fold assessment factor in four or more areas of extrapolation [8]. In the medical device standards, ICH 10993-17:2002(E) [9] states that "Any situation that results in a modifying factor greater than 10 000 is indicative of a high degree of imprecision in the analysis and consideration should be given, in such cases, to the urgent need for additional data."

In any event, the discussion of calculating the TI ends with the following commentary from Sussman et al. [7]:

"… Each adjustment factor should be considered individually, but they are not necessarily independent and a potential exists for compounding effects leading to ADEs (their terms for TI) that are lower than necessary. This should be taken into account when reviewing the overall composite factor for the assessment. The interdependence of the factors should be identified and adjusted, as appropriate, and the maximum composite value should be set at a 'reasonable' magnitude. Scientific judgement is needed to understand the interdependence of these factors and to quantify the overall adjustment …."

10.1.5 Toxicological Safety Risk Assessment of Medical Devices

These days it is an unavoidable circumstance that any book on the topic of extractables and leachables is out of date the minute it is published, as so much of the E&L space is undergoing revision and modification. This is especially true in the case of the toxicological safety risk assessment of medical devices, whose international standard is ISO 10993:17. As this book is being written, the current official version of this standard is 10993:17(2002) (R2012) [1], which was reaffirmed most recently in 2016. However, Part 17 is under active revision and it may be possible that a new draft standard will be in its early stages of public review and adoption by the time this book is actually published. Thus, it is useful to try to anticipate the contents of the new version of the standard, as it is the new standard that readers will need to work with and navigate moving forward. However, the reader must understand that the consequence of doing so is the possibility that the standard's text differs, perhaps significantly, from the following text.

As was the case with all standards, understanding the standard is facilitated by understanding the terminology. Terms that might be useful in considering the revised text of Part 17 were defined in Section 10.1.2 and additional terms are as follows:

Exposure dose (ED)
  Quantity of a chemical constituent that contacts the body. If the ED is estimated, it is termed an estimated exposure dose (EED). When the EED represents a worst-case (maximum exposure) estimate, it is abbreviated as EED_{max}.
Tolerable contact level (TCl)
  Level of surface-contact exposure to a chemical constituent that is without appreciable harm to health, normally expressed in $mg/cm^2/day$ of body surface.
Margin of safety (MOS)
  Ratio of the chemical's tolerable contact level, TI, or TTC (if applicable), and its worst-case estimated exposure dose (EED_{max}).
Toxicological screening limit
  Total cumulative DE dose to a leachable, over a specified time period, that will be without appreciable harm to health.

Although the mechanics of deriving quantities such as the TCL, TI, and EED could differ significantly between pharmaceutical packaging and medical devices, the actual use of these quantities in the toxicological risk assessment is not significantly different for these two items. Thus, the risk assessment process developed for medical devices and illustrated in Figure 10.4 is largely applicable to both pharmaceutical packaging and medical devices and consists of four essential steps:

1. Derivation of the TCL or TI,
2. Derivation of the ED,
3. Calculation of the MoS,
4. Interpretation of the MoS in the content of probable harm to health.

Often, one is challenged to manage confounding factors that arise during the toxicological risk assessment that are likely to be relevant to packaging assessment as well. For example, it is not uncommon that a risk assessor is faced with leachables whose available toxicity information is insufficiently robust to allow for the calculation of a credible TCL or TI. In extreme cases, such as for less frequently encountered and/or more chemically "exotic" leachables, little or no toxicological information may be available for the specific leachable of interest. When toxicological information is absent for an identified chemical constituent, it is possible that a POD and toxicological data can be based on an analog compound, defined as a "substance(s) with similar molecular, physical, chemical and toxicological properties." The approach of basing the toxicological risk assessment of one compound based on an analog has been

termed "read across." The selection of an analog shall be based on similarity of the following chemical properties:

- Molecular structure (e.g. the type and arrangement, including bonding, of elements in a specified order),
- Physical (e.g. molecular weight, boiling point, vapor pressure, density, solubility), and
- Chemical (e.g. toxicity, reactivity, stability, acidity/alkalinity).

In such a circumstance, the selection of a structural analog must be justified and documented.

Another approach to the circumstance of insufficient toxicological information to establish a TCL or TI is to adopt a value such as the TTC as the TCL/TI. As noted by Broschard et al. [2], "Conservative generic or default safety thresholds are helpful tools in cases where little or no data exists on the chemical compounds of interest, and a compound specific PDE cannot be developed." While such an approach is simple enough in concept, a significant shortcoming lies in the conservative and general nature of the TTC. As TTCs are conservatively set at low values, it is likely to be the case that a toxicological risk assessment based on the TTC will overestimate the hazard associated with a leachable, raising the possibility that a safe leachable is assessed as having a significant likelihood for harm.

Another situation that can confound toxicological risk assessment is the opposite situation to the above in which there is an excess of relevant information; for example,

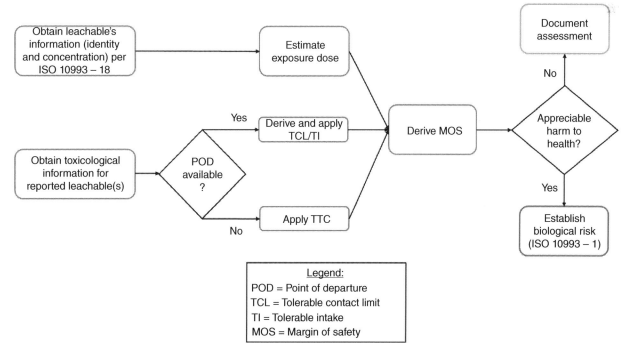

Figure 10.4 Toxicological risk assessment process.

multiple NOAELs have been reported for the same leachable or the leachable is capable of producing multiple adverse health effects. In the case of multiple NOAELs, it is reasonable to propose that the highest NOAEL be used, where the objective for selecting the highest NOAEL is preventing over estimation of the toxicological risk. In the case of multiple adverse health effects, that is, when a substance produces more than one type of harm under the same clinically relevant exposure, the lowest POD should be used, where the objective in selecting the lowest POD is to prevent other clinically relevant harm(s) that occur at higher dose(s).

It is beyond the scope of this book to predict what detailed information will likely be contained in the revision of Part 17 with regard to the conduct of a toxicological risk assessment. For a more complete understanding of the toxicological risk assessment process applied to medical devices, the reader is referred directly to the standard, when it is made available to the public

- The medical device manufacturer is responsible for assessing whether an ED of a chemical constituent is without appreciable harm to health.
- The toxicological risk assessment process includes the:
  o Determination of leachables exposure with information obtained according to ISO 10993:18.
  o Comparison of the quantity of each leachable present in, on, or released from the medical device to its respective relevant toxicological threshold.
- A chemical can be concluded to present a negligible risk of having an adverse effect on patient health and safety when both of the following apply:
  o MoS exceeds 1.
  o The chemical is not a human carcinogen, suspected human carcinogen, or its cancer risk is less than or equal to 1 in 10^{-5}.
- When the MoS is equivocal and does not clearly establish safe use, the toxicological risk should be addressed according to ISO 10993:1.
- A MoS below 1 generally indicates the EED_{max} is, or could be, a harmful dose; however, an MOS below or near 1 is likely to be equivocal.
- MoS values of two or more chemical constituents are to be combined if:
  o Exposure to each chemical occurs simultaneously (e.g. chemical constituents are present in, on, or released from, the same medical device) and
  o The chemical constituents produce the same harm at the same target organ or system and the same mode of action.
- Toxicological risk is addressed according to ISO 10993:1 when any of the following apply:
  o The MOS is below 1.

o The cancer risk of a human carcinogen, or suspected human carcinogen, exceeds 1 in 10^{-5}.

Lastly, it is noted that the proper reference for the application of TTCs to medical devices is ISO/TS 21726:2019 [9], and not ICH M7, which applies specifically to drug products.

A final dimension to be considered in the toxicological risk assessment of medical device leachables is the leachable's rate of release. It is the need to address release kinetics that most definitively differentiates risk assessment of devices from assessment for packaging.

When a patient is exposed to packaging-related leachables, the patient is exposed to whatever leachables are present in the packaged drug product at its time of use. For packaging, there is no opportunity for the patient exposure to leachables to change during the course of clinical use.

This same situation is not the case for implanted medical devices. Although the device will have an established amount of leachables, the impact that the leachables could have on patient health will depend on the rate at which those leachables are leached from the medical device. It is surely the case that the patient health impact of leachables will be quite different if the leachables are released essentially all at once (within one day of implantation) versus if the leachables are released at a constant rate over an extended period of time.

Leachables or extractables studies performed on implantable medical devices are not typically designed to produce unambiguous kinetic release information. For example, the multi-step exhaustive extraction required for permanent implants is typically designed for efficiency (achieving exhaustion in the fewest number of extraction steps) and not to produce relevant kinetic release information; that is, the exhaustive extraction is designed to efficiently establish the total pool of leachables and not to establish the rate of release of those leachables.

If the design of the exhaustive extraction study does not provide extraction kinetics, then either:

1. The experimental design of the extraction study must be altered,
2. The exhaustive extraction study must be augmented with a kinetic study,
3. The result for individual extraction steps should be considered to be kinetic data,
4. Release rate kinetics must be assumed (subject to justification and review of that justification), or
5. Release rates must be estimated by mathematical means, for example, see references [10] and [11].

10.1.6 Toxicological Safety Risk Assessment of Manufacturing Components

At a high level, the toxicological risk assessment of a leachable derived from a drug product's manufacturing system proceeds no differently from that of the assessment of a packaging-related leachable or a leachable from a medical device. That is, a TI is established for the leachable, the patient's ED is calculated, and both items are used to calculate a MoS. Furthermore, establishing a TI is a function of the leachable and not its source, and thus there is no difference in how a TI is calculated, leachables from manufacturing versus a packaging-derived leachable. The challenge in the toxicological risk assessment of manufacturing system-related leachables is related to establishing patient exposure. If manufacturing-related leachables were generally delineated by testing the manufactured product, then calculation of the ED would be straightforward and no different from the calculation for packaging-related leachables. However, it is generally the case that manufacturing-related leachables are inferred from extractables testing of manufacturing components. In this context, it is the complexity of the manufacturing process that complicates the determination of the ED.

To illustrate this point, consider a manufacturing process for a biopharmaceutical that consists of three steps: harvest, purification, and sterilization. During the harvest operation, cell growth occurs and the biopharmaceutical agent is produced in large plastic biocontainers. During purification, processes such as ultrafiltration are used to segregate the target protein from the other components that make up the process stream. Sterilization of the drug product is accomplished by sterile filtration of the process stream during the final fill. The process solution is transported between steps via plastic tubing.

Each one of these components (biocontainer, ultrafiltration membrane, sterilizing filter, and tubing) can be profiled for extractables and each one of the extractables in the component's profile is a potential leachable in the manufactured drug product. If the toxicological assessment is to be performed on the extractable's data, then the patient ED to each extractable (as a potential leachable) must be determined.

It is this determination of ED based on extractables that is most challenging as it is typically based on worst-case assumptions. For example, consider the case of an extractable that is found in all four manufacturing components. The most conservative approach to estimating the ED is to assume that the extractables levels are additive; that is, the total patient exposure to the shared extractable is the sum of the patient exposures calculated for each individual component. However, in many circumstances,

the extractable's concentration in the process stream is not additive and is limited by solubility constraints or other effects. Thus, the actual ED will be less than that obtained with the additive approach and the additive approach will produce an exaggerated estimated exposure.

Moreover, it is generally well-established that certain manufacturing processes are able to remove extractables from the process stream, for example, diafiltration. Thus, an ED based on the summation approach would be overestimated, as extractables from the biocontainer and the tubing connecting the biocontainer to the diafiltration component would likely be cleared, in all or in part, from the process stream by the diafiltration process and not become leachables in the manufactured drug product.

Lastly, the circumstance that a manufacturing batch produces multiple doses of drug product must be taken into account in the assessment. For packaging-related leachables, exposure is readily calculated by the number of containers of drug product that are needed to achieve the proper clinical dose. For manufacturing extractables, the distribution of the manufacturing extractables into all the manufactured units must be considered. For example, consider the case of the sterilizing filter. An entire batch of manufactured products is sterilized by passing the process stream through the filter. Extractables from the filter are distributed in the filtered drug product, which ultimately is packaged to produce individual drug product units. The complication is establishing the distribution of the extractable in the individual drug product units.

The simplest possible distribution of the extractable is an even distribution. For example, if the filter produces 1 l of drug product and each unit package contains 10 ml of the drug product, then the extractables amount in each of the one-hundred packaged unit is 1% of the total amount of the extractable extracted from the filter. One could imagine that this would be the case if the filtered drug product were pooled after filtration but prior to final fill.

However, imagine that sterilization and filling are continuous processes; that is, the effluent from the filter immediately flows into a filling machine which dispenses the drug product into its unit packages. In this case it is possible that extraction of the filter during use is uneven and that the extractables are extracted from the filter at higher levels at the beginning of use and at lower levels at the end of use as the filter becomes depleted in extractables. In this case the extractables will not be evenly distributed among all the drug units, complicating the calculation of ED.

The bottom line here is that establishing the correct ED requires an in-depth understanding of the manufacturing process and the physicochemical factors that affect the extent of leaching in multiple drug-component contact situations, such as those that occur during manufacturing.

10.1.7 Toxicological Safety Risk Assessment of Incompletely Identified Compounds

Previously, the confounding circumstance of having insufficient toxicological information for a leachable requiring assessment was discussed. An equally unfortunate circumstance is the situation where a leachable is reported without a fully elucidated identity. For example, a leachable could be partially identified by providing its empirical chemical formula (e.g. "C9H10O2"), or providing a generic structural descriptor such as that it is "a phthalate" or "related to BHT," or "an aliphatic hydrocarbon," or "an isomer of xylene," or "a rubber oligomer."

Clearly, a partial identity is not the most desirable launch point for a toxicological risk assessment as it does not provide focus in terms of which compound's toxicological data is relevant. Nevertheless, a partial identity can be a proper basis for a meaningful assessment. The key to the partial identity's utility to enable the assessment is the ability to use the partial identity to link the compound of interest to an appropriate analog.

From the perspective of toxicological risk assessment, the example of partial identities provided in the previous paragraph are not equally enabling. Because the leachable has not been identified, it can only be toxicologically assessed by establishing an analog. Thus, the partial identity must be sufficiently information-rich that it facilitates the selection and justification of an analog.

With this caveat in mind, one can re-examine the partial identity examples noted previously. Clearly, an empirical formula alone is not an adequate identification to support an assessment as the empirical formula is insufficiently descriptive and precise to serve as the basis for analog selection. This is the case as the empirical formula provides no evidence concerning the structure of the compound of interest. Considering the partial identities that speak to the structural nature of the compound of interest, the specifics of the description will dictate the effectiveness with which a relevant analog can be established. For example, a generic description that describes the compound's class, such as "it is phthalate," is useful to the extent that it limits potential analogs to compounds in that class (e.g. phthalates). However, it is clear that the toxicology of phthalates varies somewhat as a function of the phthalate's structural characteristics and thus the partial identity of "it is a phthalate" provides only limited benefit in terms of selecting an analog. Given the conservative nature of toxicological risk assessment, it is likely that the chosen analog will be that phthalate with the worst toxicological profile (e.g. lowest NOAEL). This same outcome is obtained with other structure-based partial identities including that it is "an aliphatic hydrocarbon" or "a rubber oligomer."

A partial identification that links one compound to another related substance would tend to focus the selection of an analog toward the related substance. For example, the partial identification "BHT-related" would generally suggest that BHT would be the appropriate analog. Such a choice of an analog is valid only to the extent that there is a meaningful structural relationship between the compound of interest and its related substance. If the functionalities shared by the two compounds are the same functionalities that dictate the compound's toxicity, then use of the surrogate for assessment will likely produce a robust assessment. Alternatively, if the functionalities that are common are not toxicologically significant, then using the related substance as an analog likely will produce an incorrect assessment.

Considering the example of partial identities provided previously, the identification "it is an isomer of xylene" is the most powerful from the perspective of selection of an analog, as it is clear that the proper analog can only come from a very small number of possibilities. Presumably one would choose the analog of xylene that has the least favorable toxicity profile or the isomer of xylene that has the most robust toxicological dataset.

QSAR analysis of partial identities is a useful means of establishing whether the partially identified compound is likely to possess alerting structures; that is, structures that are correlated with certain toxicity endpoints. Thus, for example, the generic structure of "a phthalate" can be accessed via QSAR to establish the presence of alerting structures. The presence of an alerting structure might provide the assessor with an indication of what type of POD would be proper. For example, if the compound of interest's structure produced an alert for mutagenicity, then a TTC would arguably be the proper POD for the assessment.

Furthermore, QSAR analysis of a compound and its proposed analog would be one means of justifying the analog. That is to say that the QSAR analysis of the compound and its proper analog should result in the same structural alerts. Clearly an analog that lacks the structural alerts for the compound it is representing is a poorly chosen analog.

In the extreme situation, a leachable that has been established to be present at a level that requires that it be toxicologically assessed cannot be even partially identified (the dreaded "unknown"). In this case, the only option that is seemingly available to the toxicological risk assessor is to use the TTC value as the POD in the risk assessment. However, in practice this is a solution that solves nothing. It is likely that the means by which the leachable was established as requiring assessment involved the AET; that is, the leachable required toxicological assessment because its level in the medical device was above the AET. If the AET itself was derived from the TTC, or a quantity related to the TTC, then the circumstance that the leachable is above the AET translates to the circumstance that the leachable is above the TTC as well. Now this would be the case for an unadjusted AET, that is an AET that is not adjusted to take into account response factor variation. For an AET that has been adjusted for response factor variation, the MoS that results from application of the TTC is affected by the adjustment or uncertainty factor used on the AET.

For example, consider the case of a medical device whose AET was established from a TTC of 120 μg/day based on the patient exposure to one device per day (AET = 120 μg/day × 1 day/1 device = 120 μg/device). If this AET is adjusted by an uncertainty factor of 4 (for example, for GC/MS data), then the adjusted AET becomes 30 μg/device. In this example, the unfortunate toxicologist has four unidentified leachables that require assessment because they are present in the device at levels at or above the AET; specifically, one each at a level of 30 μg/device, 60 μg/device, 120 μg/device, and 180 μg/device. Calculating the MoS based on the TTC produces the following results: MoS for the 30 μg/device leachable = 4, MoS for the 60 μg/device leachable = 2, MoS for the 120 μg/device leachable = 1, and MoS for the 180 μg/day device = 0.7.

Thus, the effect of the AET adjustment for leachables identified as unknowns is that such unknowns will be assessed as likely to be safe if their DE falls in the range between the adjusted AET and the unadjusted AET. Otherwise, the unidentified leachables will be assessed as likely to be harmful based on the TTC approach.

10.2 The Special Case of Biologics

A "biological product," or biologic, is "a virus, therapeutic serum, toxin, antitoxin, vaccine, blood, blood component or derivative, allergenic product, protein (except any chemically synthesized polypeptide), or analogous product … applicable to the prevention, treatment, or cure of a disease or condition of human beings" [12]. Biological drug substances are derived from living cells or organisms, possess a distinctive three-dimensional structure, are typically of high molecular weight, are often unstable at room temperature, and are prone to degradation via aggregation, deamidation, oxidation, and formation of clipped variants.

Leachables present in biologics can compromise patient health as a result of their direct or inherent toxicity. Moreover, leachables can interact with proteinaceous biological drug substances and cause or accelerate their degradation. For example, the abundance of both hydrophilic and hydrophobic sites on the surface of a protein and the protein's extensive surface area provide ample opportunity for leachable–protein interactions. As a result of such an interaction, the protein may change conformation, reducing its biological effectiveness or causing it to become immunogenic. Biologic products pose a unique set of safety risks due to the biologic's ability to become unstable, or to adopt multiple conformations or alterations in its primary structure. Thus, the safety assessment of leachables present in biologics must consider both primary and secondary safety effects.

Examples of these secondary effects have been well-documented in the literature. For example, one of the most widely documented instances of an unanticipated incompatibility between a container closure system and a protein drug product is that of EPREX® (epoetinum alfa) and its pre-filled syringe packaging system [13–15]. Reformulation of this product (substitution of polysorbate 80 for human serum albumin as a stabilizer) triggered an increase in the incidence of antibody-mediated pure red cell aplasia (PRCA) with EPREX use by chronic renal failure patients. The cause of PRCA was directly linked to the formation of neutralizing antibodies to both recombinant and endogenous erythropoietin in patients administered EPREX. One potential root cause of this phenomenon involved leached substances. Leaching studies determined that the polysorbate 80 extracted low levels of vulcanizing agents (and related substances) from the syringe's uncoated rubber components. This leaching issue was addressed by replacing the rubber components with components coated with a fluoropolymer, which served as an effective migration barrier and greatly reduced the leaching of the rubber's components. After the conversion to the coated components, the incidence of PRCA returned to the baseline rate seen for all marketed epoetin products. This strong circumstantial evidence that leaching of the vulcanizing agent was the root cause of the observed effect was further strengthened when the adjuvant effect of the leached agent (and related substances) was confirmed in animal models. Nevertheless, the link between the extracted vulcanizing agent and the adverse patient effect has not been conclusively established and there are alternate proposals as to the root cause of this phenomenon.

Another example of this phenomenon involved the interaction between a specific protein, IgG2, and leached acrylic acid [16]. Researchers at Amgen discovered that acrylic acid, a component of the acrylic adhesive that can be used to attach the needle to the glass barrel of a pre-filled syringe, could be extracted from the syringe at levels approaching 5 mg/l and that this leachable could interact with proteins at three specific sites, producing a small amount of protein modification (0.02–0.3%). Furthermore, the researchers postulated that the modified protein could be harmful (due to the toxicity of acrylic acid) and suggested that the properties of the protein could be changed due to charge and hydrophobicity variations.

The presence and impact of tungsten in biologics stored in pre-filled syringes have been widely reported. During the manufacture of glass syringe barrels, a tungsten pin is used to form the inner needle channel. Under the high-temperature conditions of contact, tungsten can oxidize in the presence of air and interact with the glass to form residuals. As these residuals may not be removed from the syringe during subsequent washing, they contact the drug formulation, providing opportunities for the tungsten residuals and the drug product to interact [17].

These tungsten residuals have been reported to have a marked effect on drug products stored in pre-filled syringes. Lee et al. report that concentrations of tungsten as low as 1 ppm in the drug product can produce measurable protein aggregation [17]. Similarly, Osterberg reported that tungsten leached from a syringe caused the oxidation of the contained drug product, resulting in drug degradation and aggregation [18]. Rosenberg and Worobec also report the tungsten-induced aggregation of a protein arising from oxidation mediated by leached tungsten [19]. Markovic also documents a case in which leached tungsten oxide triggered protein oxidation followed by aggregation [20]. Wen et al. report a case in which protein aggregation results from small particles of tungsten that were present in the drug product as a result of its storage in a pre-filled syringe [21]. Similarly, Seidl et al. report the tungsten-induced denaturation and aggregation of Epoetin Alfa and suggest this as the potential root cause for increased immunogenicity reported during pre-marketing clinical trials [22]. Mensch and Davis have reported that protein-induced protein aggregation can be inhibited by the addition of a cationic surfactant to the drug product formulation [23]. Based on an extensive characterization of pre-filled syringes and protein aggregates formed when the syringes were filled with protein-containing drug products, a team of Amgen researchers were able to establish that leached tungsten was the root cause of the observed aggregation [24].

Although much of the published literature focuses on tungsten, there is no reason that aggregation should be limited to this particular metal. In fact, researchers at Novo Nordisk have reported that both aluminum and zinc have the ability to affect the physical stability and secondary structure of a model protein, whereas potassium and magnesium had no effect on the protein's stability profile [25].

Silicon oil is a lubricant used in many container closure applications including pre-filled syringes and stoppered vials and its potential impact on the viability of protein drug molecules is well documented. Considering pre-filled syringes specifically, Sharma notes that silicone oil has been implicated in the induction of protein aggregation, although evidence of large changes in the protein in the presence of the lubricant is limited [26]. Jones et al. report a significant induction of protein aggregation in four proteins of various molecular weights in the presence of silicone oil and suggest that the most likely explanation for aggregation is that the silicon oil directly affects intermolecular protein interactions or exerts an effect via a solvent [27]. Markovic documents an instance in which the drug product developed thread-like, gelatinous particles when stored in syringes for only a short period of time (>1 hour) due to product interaction with the silicone oil [20]. Mechanisms of oil–protein interactions, the effect of excipients on oil-induced protein agglomeration, and means to reduce the level of extractable silicone oils have been reported [28–30]. Lastly, Bee et al. have considered the effect of leachables on the stability of biopharmaceuticals, noting many of the situations cited previously (EPREX, tungsten, silicone oil, etc.) [31].

It is often said that hindsight is 20/20 and that is certainly the case with the safety effect of leachables in biologics. Given what is known today in the literature, the finding of leachable tungsten from a packaging component to be used with a biologic would surely raise the red flag that this is likely not to be an optimum situation. However, in the context of the moment, and using tungsten as an example, the finding of tungsten as a leachable from glass syringes was unexpected and the effect that tungsten could have or the biologic was unpredictable. Thus, the assessment of secondary safety effects of leachables in biologics cannot be handled in the same way as assessment of primary effects can be assessed. Although the approach of "find the leachables, identify the leachable, quantify the leachables and infer the leachable's likely safety" impact works well for assessing primary safety effects, it does not work well for assessing secondary effects for the simple reason that there is no way to infer the leachable's secondary effect. For example, even if the developers of a biologic were alerted to the fact that tungsten was an extractable from the biologics packaging, this information would not lead to an assessment of any secondary safety impacts of the leachable for the simple reason that the effect of tungsten on a particular protein in a particular biologic cannot readily be ascertained. Thus, the effect of the leachable on the biologic's safety cannot be inferred by extractables or leachables information alone.

It is for this reason that it is critical in the development of a biologic that its compatibility with its packaging is established by direct experimentation (e.g. stability studies). If the phenomenon of protein alteration by leachables on patient health cannot be inferred, then it must be experimentally measured.

10.3 Biological Safety Risk Assessment

10.3.1 Key Definitions and Concepts

Biocompatibility is, by definition, a measurement of how compatible a device is with a biological system. The purpose of performing biocompatibility testing is to determine the fitness of a device for human use, and to see whether use of the device can have any potentially harmful physiological effects. The primary purpose of a device biocompatibility assessment is to protect patient and user safety.

The biocompatibility of a device depends on several factors, including:

- The chemical and physical nature of its materials of construction
- The types of patient tissue that will be exposed to the device
- The duration of that exposure

Unlike chemical characterization, where there is very little specific guidance that prescribes tests and test methodologies, biocompatibility testing requirements are clearly enumerated and standardized test methods have been established and are universally adopted. Considering medical devices, biocompatibility testing requirements and methods are communicated in the ISO 10993 guideline, Biological Evaluation of Medical Devices. Part 1 of the standard addresses evaluation and testing within a risk management process, Part 2 covers animal welfare requirements, and Parts 3 through 20 are guidelines for specific test procedures or other testing-related issues. For packaging and manufacturing systems, biocompatibility test requirements are specified in USP <1031>, The Biocompatibility of Materials used in Drug Containers, Medical Devices, and Implants [32], and test methods are specified in <87> Biological Reactivity Tests, *in vitro* [33]and <88> Biological Reactivity Tests, *in vivo* [34].

Whether biological testing requirements are established using ISO 10993:1 or USP <1031>, the selection of the suggested tests is a two-step process. Focusing on medical devices specifically, medical devices are divided and subdivided according to the nature and extent of their contact with the body (for example, see Table 10.2). Each sub-category is then further subdivided according to the duration of the contact between the device and the body; the duration of contact categories includes limited (less than 24 hours), prolonged (24 hours to 30 days), and permanent (more than 30 days). The biological testing that is required for each of the established sub-categories is then provided in both the ISO and USP standards. In ISO 10993:1, the testing requirements for the various sub-categories are listed in its Table 1 while in USP <1031>

the testing requirements are listed in its Tables 3–5, Test Selection Matrices for the various device categories. It is to be expected that the longer and more "intimate" the contact is between the body and the device, the more extensive are the testing requirements.

Both ISO 10993:1 and the USP <1031> emphasize that the testing matrices they provide are best interpreted as being recommendations and that the proper strategy for establishing the biocompatibility of a test item is most appropriately customized on a case-by-case basis. Thus, the testing requirements in these standards are not reproduced in this book, as so doing fosters the "turn off your brain and check the box" approach that is the hallmark of inefficient and ineffective biocompatibility assessment. Thus, the reader is directed toward these documents not only to find the requirements tables but also to review the proper use of the tables in establishing a proper, effective, meaningful, and robust biocompatibility testing strategy.

Demonstrating biocompatibility for a medical device should proceed in three steps:

- First, a Biological Evaluation Plan (BEP) is created to review device materials, identify potential risks, and suggest possible evaluations and testing to address the risks identified based on the nature of patient contact of the device. This serves as an initial risk assessment outlined in ISO 10993:1 and provides good internal documentation of the approach used to address biocompatibility.
- Second, the device is evaluated and tested using a variety of methods to address the potential risks identified in the BEP. These risks (and associated tests) include those listed in Annex A of ISO 10993:1. Often, this is accomplished using a combination of the following:
  - o Traditional *in vivo* or *in vitro* biological tests.
  - o Chemistry tests followed by toxicological risk assessment.
  - o Written assessment based on scientific literature information and clinical use of the materials.
- Third, the results of all tests and written evaluations should be summarized in a Biological Evaluation Report (BER) that is submittable in regulatory filings.

10.3.2 Biological Tests and Test Methods

As noted previously, one of the shortcomings in having a clear direction in terms of test requirements is that people recognize the tests by name but know very little about the purpose, nature, and context of the underlying tests. Thus, the tests most commonly performed in biocompatibility assessment are described in general as follows, with the caveat that this author has limited familiarity with biochemical test methods.

Table 10.2 Categorization of medical devices.

Device type	Contact medium	Definition	Examples
Surface device	Intact skin	Devices that contact intact skin surfaces only	Electrodes, external prostheses, fixation tapes, compression bandages, and monitors of various types
	Mucous membrane	Devices communicating with intact mucosal membranes	o Contact lenses, urinary catheters, intravaginal and intraintestinal devices (stomach tubes, sigmoidoscopes, colonoscopes, gastroscopes), endotracheal tubes, bronchoscopes, dental prostheses, orthodontic devices, and IUDs
	Breached or compromised surfaces	Devices that contact breached or otherwise compromised external body surfaces	Ulcer, burn, and granulation tissue dressings or healing devices and occlusive patches
External communicating device	Blood path – indirect	Devices that contact the blood path at one point and serve as a conduit for entry into the indirect vascular system	Solution administration sets, extension sets, transfer sets, and blood administration sets. **This is also the likely classification for pharmaceutical packaging.**
	Tissue/bone/dentin	Devices communicating with tissue, bone, and pulp/dentin system	Laparoscopes, arthroscopes, draining systems, dental cements, dental filling materials, and skin staples. This category also includes devices that contact internal tissues (rather than blood contact devices). Examples include many surgical instruments and accessories.
	Circulating blood	Devices that contact circulating blood	Intravascular catheters, temporary pacemaker electrodes, oxygenators, extracorporeal oxygenator tubing and accessories, hemoadsorbents, and immunoabsorbents
Implant device	Tissue/bone	Devices principally contacting bone	Orthopedic pins, plates, replacement joints, bone prostheses, cements and intraosseous devices
		Devices principally contacting tissue and tissues fluid	Pacemakers, drug supply devices, neuromuscular sensors and stimulators, replacement tendons, breast implants, artificial larynxes, subperiosteal implants and ligation clips
	Blood	Devices principally contacting blood	Pacemaker electrodes, artificial arteriovenous fistulae, heart valves, vascular grafts and stents, internal drug delivery catheters, and ventricular assist devices

Cytotoxicity (Cell Culture). Cell culture assays are used to assess the biocompatibility of a material or extract through the use of isolated cells *in vitro* and are useful in evaluating the toxicity or irritancy potential of materials and chemicals. Commonly used cytotoxicity tests include:

- In the Direct Contact procedure, a piece of test material is placed directly on to cells growing on a culture medium. During incubation, leachables from the test material can diffuse into the culture medium and contact the cell layer. A cytotoxic response is indicated by malformation, degeneration, and lysis of cells around the test material.
- In the Agar Diffusion procedure, a thin layer of nutrient-supplemented agar is placed over the cultured cells. The test material (or an extract of the test material dried on filter paper) is placed on top of the agar layer and the cells are incubated. A zone of malformed, degenerative, or lysed cells under and around the test material indicates cytotoxicity.
- In the MEM Elution assay the test material is extracted according to actual use conditions or to exaggerate those conditions. The resulting extracts are transferred on to a layer of cells and incubated. Following incubation, the cells are examined microscopically for malformation, degeneration, and lysis of the cells.

Sensitization Assays. Sensitization studies establish whether a material contains leachables that cause adverse local or systemic allergic or hypersensitivity effects after repeated or prolonged exposure. Sensitization assays, typically performed on extracts of the test material, include:

- In the Guinea Pig Maximization Test (Magnusson-Kligman Method), the test material is mixed with complete Freund adjuvant to enhance an immunological response. The test animals are exposed intradermally to the test extract, along with an adjuvant to enhance the immune reaction of the test animal. A short time later, the test animal is exposed to a lower concentration of the test material, and its allergic reaction, if any, is measured. As an alternative, the local lymph node assay (LLNA) *in vivo* test can be used.
- The Murine LLNA determines the quantitative increase in lymphocytes in response to a sensitizer via an animal model. A skin sensitizer will induce the epidermal Langherhans cells to transport the allergen to the draining lymph nodes, which in turn causes T-lymphocytes to proliferate and differentiate.

Irritation Tests or Intracutaneous Reactivity tests estimate the local irritation potential of test materials using sites such as skin or mucous membranes, usually in an animal model.

- In the Intracutaneous Test, test material extracts are injected intradermally and the injection sites are scored for erythema and edema (redness and swelling).
- In the Primary Skin Irritation test, an extract is applied directly to intact and abraded sites on the skin of a rabbit. After a 24-hour exposure, the material is removed and the sites are scored for erythema and edema.
- In Mucous Membrane Irritation Tests, extracts are applied to various mucous membranes (vaginal, rectal, penile, and hamster cheek pouch) and the sites are scored for irritation.

Acute systemic toxicity tests establish whether substances extracted from the test article produce systemic (as opposed to local) toxic effects. The test material extracts and negative control blanks are injected into mice, which are observed for toxic signs just after injection and at four other time points.

Material Mediated Pyrogen tests evaluate the potential of a material to cause a pyrogenic response, or fever, when an extract of the material is introduced into the blood of the test subject (rabbit).

Sub-acute/Sub-chronic Toxicity tests are used to determine the potentially harmful effects of longer-term or multiple exposures to test materials and/or extracts in an animal model. These tests are performed for a duration up to 10% of the total lifespan of the test animal. Animal models and dosing methods for sub-chronic toxicity testing are selected based on the actual use conditions of a medical device.

Implantation Tests are used to determine the biocompatibility of medical devices or biomaterials that directly contact living tissue other than skin (e.g. sutures, surgical ligating clips, implantable devices, intraocular lenses, etc.). In these tests, the test material is implanted into the body of the animal to assess its localized adverse reactions. The test is based on grading acute and sub-acute tissue reactions at 7 and 28 days, respectively, following implantation in rats.

Genotoxicity evaluations use a set of *in vitro* and *in vivo* tests to detect mutagens, substances that can directly or indirectly induce genetic damage directly through a variety of mechanisms.

The most common test for mutagenicity is the Ames test, which detects point mutations by employing strains of the bacteria that have been selected for their sensitivity to mutagens (*Salmonella typhimurium* and *Escherichia coli*). The Mouse Lymphoma assay is another common procedure that uses mammalian cells to detect point mutations and it can also detect clastogenic lesions in genes (chromosome damage). *In vitro* and *in vivo* methods, including the Chromosomal Aberration and the Mouse Micronucleus assays, are cytogenetic assays that allow direct observation of chromosome damage.

Hemocompatibility tests are required for all blood-contacting medical devices as all materials are to some degree incompatible with blood because they can either disrupt the blood cells (hemolysis) or activate the coagulation pathways (thrombogenicity) and/or the complement system.

- The hemolysis assay measures the damage to red blood cells when they are exposed to materials or their extracts, and compares it to positive and negative controls.
- Coagulation assays, including the Prothrombin Time Assay (PT) and the Partial Thromboplastin Time Assay (PTT), measure the effect of the test article on human blood coagulation time.
- The *in vitro* Complement activation assay measures complement activation in human plasma as a result of exposure of the plasma to the test article or an extract. The measure of complement actuation indicates whether a test article is capable of inducing a complement-induced inflammatory immune response in humans.

10.4 The Use of Auxiliary Information in Safety Risk Assessment

It is possible that the outcome of the biocompatibility assessment, with or without the toxicological risk assessment of extractables or leachables, may be equivocal and it may not be possible to definitively establish whether a medical device, packaging system, or manufacturing component is safe for its intended use. When reasonable testing has been concluded and produces an equivocal safety risk assessment, it can be challenging to find a reasonable answer to the question "what more can I do to establish that my item is safe to use?"

In a court of law, innocence or guilt cannot always be definitively established and it is up to the judge and jury to weigh the preponderance of evidence to reach a verdict. The same concept can be applied, to some extent, in the context of establishing that an item is safe to use in medical applications. For example, when safe use cannot be definitively established, the weight of the preponderance of evidence can be used to "tip the scales" toward one conclusion or the other.

Auxiliary information is information, other than the toxicological risk assessment (and its supporting extractable or leachables data) and biocompatibility testing, that speaks to the safe use of medical items and their materials of construction. Types of auxiliary information that speak to safe use include:

- Results of tests conducted in accordance with the relevant Pharmacopeia (e.g. USP <661>, Pharm Eur 3.1, 3.2).
- Conformity to relevant food contact regulations (e.g. EU 10/2011, 21 CFR 177).
- Compliance with compositional regulations and standards (e.g. REACH, Prop 65, CONEG).

The use of auxiliary information to establish safe for use is based on the following logic:

- A material that has been tested by, and meets the requirements of, a relevant compendial monograph is likely to be safe for use in medical applications. This logic is an important part of regulatory guidelines for packaging such as the FDA container closure guidance [35] and the EMEA guidelines for immediate packaging [36].
- A material that has been tested by, and meets the requirements for, use as food packaging is likely to be safe for use in medical applications in general and packaging in particular.
- A material that meets compositional regulations and standards that limit the use of substances of concern is likely to be safe for its intended use in medical applications.

Individual pieces of auxiliary information are rarely sufficient to support a safe use conclusion, except in circumstances where the conditions of use of the medical item are so benign or mild that interaction between the item and a patient is so limited that the safety risk can be established to trivial or negligible on the basis of nature of contact alone. However, as the individual pieces of auxiliary information are orthogonal and complementary, collectively they can be a compelling indication of likely safe use, especially when taken together with the a toxicological and biocompatibility risk assessment.

10.5 Are Organic Extractables and Leachables Inherently Unsafe?

The essential assumption of the safety assessment process employed to qualify medical items as being safe in their intended use is that the medical items are unsafe until they are proven to be safe. As the safety of medical items is predominantly dictated by the biological effect of substances that are leached from the medical item during its intended clinical use, then the supposition that a medical item is unsafe until proven otherwise is a supposition that leachables are unsafe until they are proven otherwise.

It is this supposition of "guilty until proven innocent" that has led to the proliferation of extractables and leachables testing in the pharmaceutical and medical industry. Although the vast experience in extractables and leachables toxicological safety assessment is that it is rare that an unsafe leachable or extractable is encountered in a commercially viable medical product, the proponents of even more rigorous extractables and leachables testing are emboldened by the relatively rare cases of commercial products exhibiting undesirable and largely unexpected adverse health effects due to leachables or substances incorrectly described as leachables.

It is reasonable and appropriate to at least consider the possibility that the proliferation of extractables and leachables testing is an over-reaction to a perceived risk and that the steps taken to address safety risks are too extreme in the content of the actual safety risk itself. In this context, then, it is reasonable to consider the question of the inherent toxicity of extractables and leachables from materials commonly used in pharmaceutical packaging drug product manufacturing systems and medical devices.

Truly it is beyond the scope of this book and the expertise of its author to provide a definitive answer to this question. Nevertheless, there is relevant information in the public domain that speaks to this question.

Case 1: In their proposal of a holistic extractables and leachables program for biotechnology products, Li and associates from Amgen and the FDA compiled a dataset of 151 organic extractables obtained from devices, packaging and manufacturing components consisting of polymers including PE, PP, PC, PTFE, PVDC, EVOH, EVA, PA, PET, PS, PDMS, and bromobutyl and chlorobutyl elastomers [6]. PDE values could be confidently derived for 103 of the

Figure 10.5 Distribution of proposed permissible daily exposure (pPDE) values for 103 organic extractables associated with items related to biotechnology products. Source: Data from Li et al. [6].

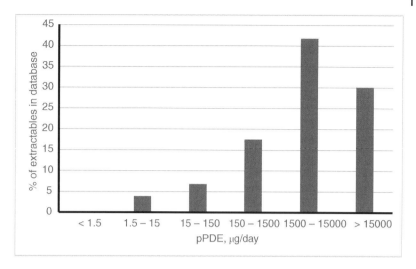

organic extractables using relevant toxicological data, the calculation method described in ICH Q3C (including F1 through F5 plus F6 to account for bioavailability) and based on a body weight of 60 kg. Using the calculated PDE values, these 103 organic extractables were assigned to Cramer Classes. It is noted that this assignment is based on the PDE value and not on the compound's structure. Furthermore, the Cramer Class assignments for the remaining 49 organic extractables were obtained by *in-silico* assessment.

A majority of the organic extractables (73%) were classified as Cramer Class I, low toxicity, TTC = 1800 μg/day. An additional 17% of the organic extractables were classified as Cramer Class II, intermediate toxicity, TTC = 540 μg/day, and 10% of the organic extractables were classified as Cramer Class III, high toxicity, TTC = 90 μg/day.

Considering the extractables with calculated PDE values, the distribution of the PDEs is illustrated in Figure 10.5. None of the extractables had a PDE of less than 1.5 μg/day, the SCT established by the PQRI for parenteral drug products; the lowest calculated PDE was 3 μg/day. Approximately 10% of the organic extractables had PDEs lower than the chronic TTC for 120 μg/day from ICH M7. These authors concluded that "based on the risk analysis, it became apparent that high concern compounds were only a small subset of the dataset."

Case 2: The PQRI Parenteral and Ophthalmic Drug Product (PODP) Leachable and Extractable team proposed a Classification Strategy based on assessment of a database of 606 organic extractables using *in silico* software (Toxtree) based on chemical classification approaches developed by Cramer and later refined by Munro. As these approaches do not account for a carcinogenic endpoint, the PODP team also evaluated the 606 chemical databases using Deductive Estimation of Risk from Existing Knowledge software (DEREK, Lhasa, Ltd).

The results of the PQRI classification of the organic extractables are shown in Figure 10.6 [37]. Slightly over 53% of the organic extractables were classified as Cramer Class I (low toxicity) while approximately 38% were classified as Class III (high toxicity). Approximately 11% of the compounds were determined to be potential mutagens via the DEREK and SARAH analysis. Approximately 83% of the DEREK/SARAH Class IV mutagenic compounds were also Cramer Class 3 chemicals (most toxic).

Case 3: In a 2014 publication, Jenke and Carlson collected toxicological information for over 540 extractables identified in laboratory testing of polymeric materials used

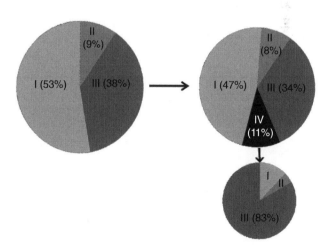

Figure 10.6 Distribution of classifications for 606 organic extractables. The entire database was grouped using the Cramer classification, producing the distribution shown on the left. To address mutagenicity, compounds were also grouped based on DEREK and SARAH, producing the distribution on the right. The "validity" of the DEREK and SARAH Class IV designations was addressed by considering the Cramer classification of the Group IV compounds; 83% compounds with a mutagenicity alert were also classified as Cramer Class III. Source: Data from Paskiet [37]. (*See insert for colour representation of the figure.*)

in pharmaceutical applications [38]. Relevant toxicological endpoints, such as NOELs (no observed effects), NOAELs (no adverse effects), TDLOs (lowest published toxic dose), and others were collated for these extractables or their structurally similar surrogates and were systematically assessed to produce a risk index (RI), which represents a daily intake value for life-long intravenous administration, similar to a PDE but lacking the rigor with which a PDE is established. This systematic approach used four uncertainty factors (UFs), each assigned a maximum value of 10, which considered the quality and relevance of the data, differences in route of administration, non-human species to human extrapolations, and interindividual variation among humans. Specifically, the available toxicology data (NOAEL, NOEL, LD50, TD_{LO}) were adjusted with appropriate and relevant UFs, consistent with well-recognized toxicological practices, as follows:

$$RI = \left[\left(\text{toxicological value, mg/kg} - \text{day} \right) \right. \\ \left. \times 70 \, \text{kg} \right] / \left[(UF1 \times UF2 \times UF3 \times UF4) \right] \, (10.3)$$

where:

UF1 = 10 was used to account for the interindividual variation among humans.

UF = 10 was used to take into account the inherent differences between man and other species.

UFs 3 and 4 were used to account for the quality and relevance of the data. A UF (T3) of 10 was used to account for routes of administration other than intravenous. Additionally, a UF4 of 10 was used in all cases where a NOEL, NOAEL, or comparable no-effect level was not available.

The individual RIs were obtained for the extractables ways:

1. For extractables with sufficiently rigorous toxicological data, the RI was calculated using that data.

2. For extractables without sufficiently rigorous toxicological data, the RI was calculated using a structural surrogate for that extractable. Whenever possible, the structural surrogate used was an extractable whose RI had been calculated via Equation 10.3 above.

3. For extractables without sufficiently rigorous toxicological data and whose surrogate was not an extractable, the RI was calculated using that a non-extractable structural surrogate for that extractable.

In addition to the RI values, all extractables and most of their surrogates were classified for structural safety alerts using Cramer rules and for mutagenicity alerts using an *in-silico* approach (Benigni/Bossa rule base for mutagenicity via Toxtree). Lastly, *in vitro* mutagenicity data (Ames *Salmonella typimurium* and Mouse Lymphoma tests) were collected from available databases (Chemical Carcinogenesis Research Information, CCRIS, and Carcinogenic Potency Database, CPDB). The CPDB is a widely used international resource containing the results of 6540 chronic, long-term animal cancer tests on 1547 chemicals. The CCRIS is a database sponsored by the National Cancer Institute containing primarily *in vitro* carcinogenicity and mutagenicity test results for over 8000 chemicals.

The results of the analysis of the resulting information were as follows. The frequency distributions of the resulting data were established; in general RI values were normally distributed around a band ranging from 5 to 20 mg/day (Figure 10.7). The RI associated with the 95% level of the cumulative distribution plot was approximately 0.1 mg/day (Figure 10.8). Thirteen extractables in the dataset had individual RI values less than 0.1 mg/day. Additionally, approximately 50% of the extractables were classified in Cramer Class 1 (low risk of toxicity) and approximately 35% were in Cramer Class 3 (high risk of toxicity). Lastly, roughly 20% of the extractables triggered

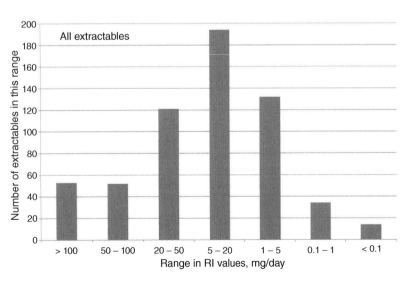

Figure 10.7 Distribution plots, RI values for all extractables. In general, the extractables' RI values fell within the range of 0.05 to 20 mg/day, although there were individual extractables whose RI values fell outside this range. The frequency distribution is more or less normal, centered on a band of SCTs between 5 and 20 mg/day. Thirteen individual extractables had RI values that were less than 0.1 mg/day. Source: From reference [38], with permission.

Figure 10.8 Cumulative distribution plots, RI values for all extractables. The critical RI value corresponding to the 95% level on the cumulative distribution plot was approximately 0.1 mg/day. Thus 95% of the extractables had an RI value greater than 0.1 mg/day. Source: From reference [38], with permission.

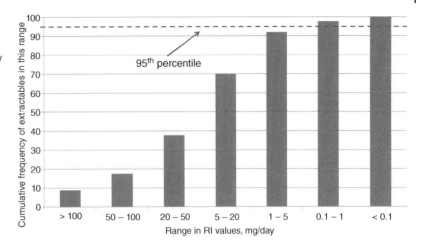

Figure 10.9 Distribution plots, Cramer class versus RI values, all extractables. In general, the distribution of the RI values is similar across the three Cramer classes, although the distribution is shifted slightly toward the lower RI values for the Cramer Class 2 and 3 extractables. Source: From reference [38], with permission. (*See insert for colour representation of the figure.*)

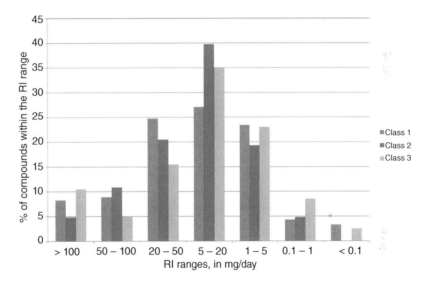

either an *in vitro* or *in silico* alert for mutagenicity. When Cramer classifications and the mutagenicity alerts were compared to the risk indices, extractables with safety alerts generally had lower RI values, although the differences in the RI data distributions, extractables with or without alerts, were small and subtle (Figure 10.9).

It is important to emphasize a point made earlier, which is that the RI values are not equivalent to PDE values, as they are not assessed as rigorously as are PDEs. Nevertheless, it is instructive to compare the RI values with toxicological thresholds that can be applied to extractables and leachables. Relevant thresholds for comparison include the 120 μg/day acute TTC per M7, the 5 μg/day threshold QT proposed by the FDA to reflect general toxicity, and the 1.5 μg/day chronic TTC per M7 and the chronic SCT per PQRI PDP. As noted in Table 10.3, approximately 3% of the extractables considered in this study (13 extractables) had RI values below the chronic TTC of

120 μg/day and only one extractable had an RI below the QT of 5 μg/day and the SCT of 1.5 μg/day.

Case 4: In a 2015 publication, Jenke developed and justified a Risk Evaluation Matrix for estimating the safety risk associated with extractables from plastic materials used in pharmaceutical applications and to apply that matrix to approximately 510 organic extractables to assess the risk that they would accumulate in drug products at levels sufficiently high to affect patient safety [39]. This Matrix considered the two dimensions of risk, severity (hazard) and probability, as follows: the hazard (as established by the inherent toxicity of the extractable) and the probability of occurrence (as established by the frequency with which extractables are present in polymeric materials, the amounts at which the extractables are present in the materials, and the propensity of the extractables to accumulate in the drug products as leachables).

Individual organic extractables were assigned a Risk Score based on three dimensions (Figure 10.10):

Table 10.3 Extractables with RI values less than 120 μg/day.

Extractable	CAS RN	RI, μg/day	Cramer class	Mutagenicity alert
1-Methyl benzene	98-82-8	70	1	None
Acetophenone	98-86-2	70[a]	1	None
Pyridine	110-86-1	70	3	None
Benzaldehyde	100-52-7	63	1	Yes[b]
Toluene	108-88-3	56[a]	1	None
Carbon disulfide	75-15-0	53	3	None
Pentanal	110-62-3	40	1	Yes[b]
Oleamide	301-02-0	35	1	None
Tetramethyl-butylphenol	140-66-9	35	1	None
Succinic acid	110-15-6	11[a]	1	None
5-Quinolinol	578-67-6	6	3	None
Oleic acid	112-80-1	6[a]	1	None
Acrylonitrile	107-13-1	1	3	Yes[c]

a) These compounds had numerous toxicological endpoints from which the RI could be calculated. The reported RI represents the lowest value obtained.
b) Mutagenicity alerts both *in vitro* and *in vivo*.
c) Mutagenicity alert *in vitro*.

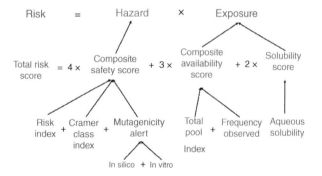

Figure 10.10 Process used to calculate the total risk score (TRS). Source: From reference [39], with permission.

1. The safety profile of the extractable (reflected in the Composite Safety Score),
2. The "impact" of the extractable, based on the frequency with which the extractable is encountered in extractables studies and its total pool in tested materials (reflected in the Composite Availability Score), and
3. The ability of the extractable to be leached from materials, based on the extractable's aqueous solubility (reflected in the Solubility Score).

Factors that were considered in establishing the Composite Safety Score included the extractable's RI per reference [38], structure–activity analysis of the extractable (Cramer classification), and reported *in vitro* or *in silico* mutagenicity Alerts. Factors that were included in the Composite Availability Score included the frequency with which the extractable was reported in extraction studies and the total pool of the extractable in the materials for which it was identified as an extractable.

Ultimately, a Total Risk Score (TRS) was calculated for each individual extractable using a mathematical combination of the three individual risk dimension scores as shown in Figure 10.10. This mathematical process produced a range of possible TRS values of 0 (lowest possible risk) to 55 (highest possible risk). The total number of combinations that could produce a TRS score were somewhat subjectively considered to group the TRS into four categories: lowest risk (TRS ≤ 13), moderate risk (14 < TRS < 22), intermediate risk (23 < TRS < 35), and highest risk (TRS ≥ 36).

The frequency distribution of TRS scores for all 505 organic extractables in illustrated in Figure 10.11. In general, approximately 15% of the extractables were classified as lowest risk and 3% were classified as the highest risk. The lowest risk extractables were generally insoluble, infrequently encountered in extractables profiles, and had a low toxicity. Alternatively, the highest risk extractables were generally highly soluble, frequently encountered in extractables profiles, and had higher intrinsic toxicity.

Ultimately, the TRS can be used for selecting a material for use in a polymeric packaging system, medical device, or manufacturing component used with an aqueous contacting medium, if potential extractables associated with the item are known. For example, a material that contains only lower risk extractables would be a good candidate for use in these medical items while a material that contains higher risk extractables would be a poor candidate for use. Furthermore, the TRS can be used to facilitate the development and justification of analytical screening methods used to characterize extracts for extractables. It is well known that even the most rigorous analytical screening approaches for organic extractables are not universal and that some extractables elude detection. Clearly, the safety risk associated with "missed" organic extractables is higher if the compounds that are missed have the higher TRS scores. Thus, the analytical approach must be optimized so that the higher risk compounds are not missed.

Also recently, researchers from W.L. Gore & Associates considered the issue of the proper POD for the AET linked to medical devices, specifically whether a POD of 120 μg/day was appropriate for short- and long-term medical devices [40]. These authors were concerned about this topic as they recognized a "lower value often requiring sensitivities beyond analytical capabilities."

Figure 10.11 Distribution of total risk scores (TRS) for the 505 organic extractables. The TRS is normally distributed around a median TRS of 20 (moderate risk). Source: From reference [39], with permission.

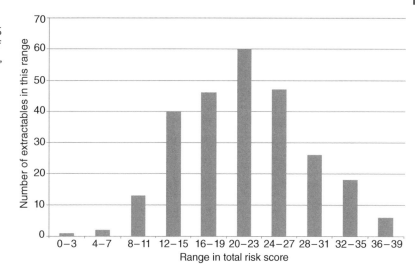

To address this issue, the authors reviewed the CPDB, using the TD_{50} values for non-Cohort of Concern substances to extrapolate the doses to an excess cancer risk of 10^{-5} and to calculate the total doses. These authors report that of the 199 substances evaluated, only two posed an excess risk at an AET calculated as 1.5 μg/day and only seven more posed at excess risk at an AET calculated as with 120 μg/day. Furthermore, over 95% of non-Cohort substances would not pose an excess cancer risk using an AET calculated with 120 μg/day. Although these authors used this analysis to justify 120 μg/day as the proper POD, their finding suggests that extractables are, by and large, relatively safe substances.

It is unlikely that the five cases just discussed are sufficient to answer the question "are organic leachables and extractables inherently safe (or at least safer than the general population of known mutagens and carcinogens that are used to establish thresholds such as the TTC and the SCT)?" To be sure, the question as stated is untenable in its simplicity and it is likely that the most proper answer is "it depends on the circumstances." However, perhaps the presented information is sufficient to initiate constructive and scientific debate around topics such as:

- Can known and frequently encountered extractables be classified into three groups, somewhat empirically described as generally recognized as safe, potentially unsafe, and potential bad actors? If such a classification were possible (and it was for smaller populations of residual solvents and elemental impurities), then testing and qualification strategies in particular circumstances could be based on this information. For example, substances established to be potential bad actors could be addressed via validated target analytical methods whereas substances generally recognized as safe could be addressed with qualified limit tests.

- Can thresholds that are the current basis of the AET be derived using datasets that contain only extractables or leachables? So doing would mean that the AET is directly linked to the universe of organic extractables and leachables, as opposed to the AET being ultimately based on compound populations that likely focus on high-risk impurity compounds regardless of whether they are encountered as extractables or leachables.

Lastly, the ELSIE consortium brought together subject matter expert toxicologists to derive TTC values for organic, non-mutagenic E&L substances administered parenterally. A total of 488 E&L compounds from the ELSIE database were analyzed and parenteral point of departure (PPOD) estimates were derived for 252 of these compounds. The PPODs estimates were adjusted to extrapolate to subacute, subchronic, and chronic durations of nonclinical exposure, the lower 5th percentiles were calculated and an additional 100-fold adjustment factor was used to account for nonclinical species and human variability to derive the parenteral TTC values for E&Ls. The resulting parenteral TTC values are 35, 110, and 180 μg/day for human exposures of >10 years to lifetime, >1–10 years, and ≤1 year, respectively [41].

10.6 Assessing Risks Other Than Patient Safety

It is an incontrovertible fact that risk assessment of extractables and leachables is dominated by a single adverse effect, patient safety. This is true for two reasons: first, because of the criticality of preserving a patient's health and, second, because the safety assessment can be accomplished. Despite the numerous challenges that have been enumerated in actually performing a toxicological safety risk assessment, the fact of the matter remains that the calculational aspects of the assessment process themself are simple and straightforward. The leachable is identified and quantified and the toxicologist uses this information to establish exposure and TI. Thus, the effect can be inferred based on leachables testing and test results.

This ability to test and infer outcomes can also be harnessed in considering other adverse effects of leachables. For example, the effect of leaching low molecular weight organic acids from gamma irradiated plastics on drug product pH can largely be predicted based on acid/base chemistry using the identity of the leachable and the amount that has been leached.

However, there are other adverse effects of leachables that cannot be inferred simply by knowing the identity of the leachable and its leached concentration. For example, the degradation of proteins by tungsten, leached from glass pre-filled syringes, was largely unanticipated before the phenomenon had been observed and reported. That is to say, even if tungsten had been reported as an extractable from a syringe, it would have been unlikely that its effect on protein stability would have been inferred based on this knowledge only.

Another example of this situation, this time concerning manufacturing systems, is the case of bDTbPP, bis(2,4-di-tert-butylphenyl)-phosphate. In 2013, the world of single-use manufacturing systems was startled by the report of reduced cell viability in cell cultures caused by the use of single-use bioprocess bags. First reported in the literature by researchers from Amgen [42], this leachable, a radiation-induced secondary degradation product of the antioxidant Irgafos 168, was established to be detrimental to cell proliferation at concentrations of 0.1 mg/l or lower. When the effect was first documented and before the link between this specific extractable and the effect was established, this phenomenon was quite confounding. Different cell lines were affected in different amounts, the effect was not caused by all bioprocessing bags, and the effect was variable between batches of bioprocessing bags. The connection between this phenomenon and a single bag-related leachable was difficult to make.

Ultimately, of course, the phenomenon was studied and understood, the root cause established and corrective actions taken. However, the point that is relevant to the discussion at hand is the simple fact that this effect would not have been inferred based on extractables or leachables studies alone. That is, had bDTbPP been reported in an extraction study, it is highly unlikely that this compound would have raised a flag of concern. The levels at which this compound is extracted are relatively low, the compound itself is somewhat unstable, and the compound's toxicity is largely unremarkable. It is likely the case that this compound would be present in extracts at levels below

the AET and thus that it would not even be reported for toxicological risk assessment. Even if it was above the AET and toxicologically risk assessed, it is likely that the assessment would have concluded that the likelihood of it having an adverse effect on patient health would be negligible. Furthermore, prior to 2013, there would have been no reason to suspect that this compound would materially affect drug product manufacturing operations. The bDTbPP compound would have been just another of those many leachables that are reported at low levels that ultimately are judged to be safe and thus of little importance.

The point that is being made here is that leachables can have effects that cannot be inferred from the results of an extraction or leaching study; that is, the effect cannot be inferred knowing the leachable's identity and concentration. Thus, an assessment for these effects cannot be made or inferred on the basis of extractables and leachables data, as they can only be observed during compatibility studies.

Ultimately, the registration stability study performed on a packaged drug product is the definitive compatibility assessment for a packaged drug product. Similarly, the production of the first commercial product batch is the definitive compatibility assessment of a manufacturing component. Clinical safety studies performed on medical devices serve to establish the compatibility of the device and the patient. If there are incompatibilities between the packaged and the packaged drug, between the manufacturing component and the process stream, or between the device and the patient, these studies will surely reveal them. However, the commercial consequences of discovering incompatibilities in these stages of commercialization are catastrophic and it is poor practice to wait until these activities to address the issue of potential incompatibilities.

Thus, confirmation of compatibility early in the drug development process is an essential aspect of extractables/leachables impact assessment. An extractables/leachables program that relies only on extractables or leachables profiling and that does not include compatibility confirmation is a deficient E&L program.

Ultimately, it is highly desirable that compatibility confirmation be performed in a manner that is more efficient than "put them together and wait for something bad to happen." For example, there is always the possibility that extracts or drug products could be screened for potential incompatibilities. Considering bDTbPP-type incompatibilities specifically, several cell lines that are particularly sensitive to this compound have been proposed as screening tools for this effect [43, 44]. In a similar vein, a simple assay with glutathione as the surrogate compound has been proposed as a means of screening extracts for protein-reactive rubber oligomers [45].

Lastly is the possibility that certain functionalities can be linked to adverse effects, in much the same way that QSAR analysis links functionality to certain toxicological endpoints. For example, Li et al. suggest that organic leachables that are either Michael acceptors, Schiff base formers, acylating agents, or nucleophilic substitution agents are likely to be protein-reactive [6].

References

1 ISO 10993:17(2002) (R2012). Biological evaluation of medical devices – Part 17: Establishment of allowable limits for leachable substances.

2 Broschard, T.H., Glowienke, S., Bruen, U.S. et al. (2016). Assessing safety of extractables from materials and leachables in pharmaceuticals and biologics – current challenges and approaches. *Regul. Toxicol. Pharmacol.* 81: 201–211.

3 Guidance for Industry. M7(R1) Assessment and Control of DNA Reactive (Mutagenic) Impurities in Pharmaceuticals to Limit Potential Carcinogenic Risk. US Department of Health and Human Services: Food and Drug Administration; Center for Drug Evaluation and Research (CDER), Center for Biologics Evaluation and Research (CBER). March, 2018.

4 Parris, P., Martin, E.L., Stanard, B. et al. (2020). Considerations when deriving compound-specific limits for extractables and leachables from pharmaceutical products: four case studies. *Regul. Toxicol. Pharmacol.* 118: 104802.

5 Guidance for Industry. Q3C Impurities: Residual Solvents. US Department of Health and Human Services: Food and Drug Administration; Center for Drug Evaluation and Research (CDER), Center for Biologics Evaluation and Research (CBER). December, 1997.

6 Li, K., Rogers, G., Nashed-Samuel, Y. et al. (2015). Creating a holistic extractables and leachables (E&L) program for biotechnology products. *PDA J. Pharm. Sci. Technol.* 69 (5): 590–619.

7 Sussman, R.G., Naumann, B.D., Pfister, T. et al. (2016). A harmonization effort of acceptable daily exposure derivation – Considerations for application of adjustment factors. *Regul. Toxicol. Pharmacol.* 79: S57–S66.

8 US Environmental Protection Agency. 2002. A review of the Reference Dose and Reference Concentration Processes. EPA/630/P-02/002F.

9 ISO/TS 21726(2019). Biological evaluation of medical devices – Application of the threshold of toxicological concern (TTC) for assessing biocompatibility of medical device constituents.

10 Chandrasekar, V., Janes, D.W., Saylor, D.M. et al. (2018). Conservative exposure predications for rapid risk assessment of phase-separated additives in medical device polymers. *Anal. Biomed Eng.* 46 (1): 14–24.

11 Turner, P., Elder, R.E., Nahan, K. et al. (2020). Leveraging extraction testing to predict patient exposure to polymeric medical device leachables using physics-based models. *Toxicol. Sci.* 178 (1): 201–211.

12 US Code, 42, US Code § 262. Regulation of biological products.

13 Sharma, B., Bader, F., Templeman, T. et al. (2005). Technical investigations into the cause of the increased incidence of antibody-mediated pure red cell aplasia associated with Eprex®. *Eur. J. Hosp. Pharm.* 5: 86–91.

14 Boven, K., Knight, J., Bader, F. et al. (2005). Epoetin-associated pure red cell aplasia in patients with chronic kidney disease: solving the mystery. *Nephrol. Dial. Transplant.* 20 (Suppl. 3): ii30–ii40.

15 Pang, J., Blanc, T., Brown, J. et al. (2007). Recognition and identification of leachables in Eprex® pre-filled syringes: an unexpected occurrence at a formulation-component interface. *PDA J. Pharm. Sci. Technol.* 61 (6): 423–432.

16 Liu, D., Nashed-Samuel, Y., Bondarenko, P.V. et al. (2012). Interactions between therapeutic proteins and acrylic acid leachable. *PDA J. Pharm. Sci. Technol.* 66 (1): 12–19.

17 Lee, H., Nashed-Samuel, Y., and Fujimori, K. Tungsten leaching from pre-filled syringes and impact on protein aggregation, poster. Presented at the PDA Extractables/Leachables Forum, Confronting Extractables and Leachables Issues in an Evolving Industry; 6–8 November 2007; Bethesda, MD, USA.

18 Osterberg, R.E. (2005). Potential toxicity of extractables and leachables in drug products. *Am. Pharm. Rev.* 8 (2): 64–67.

19 Rosenberg, A.S. and Worobec, A.S. (2004). A risk-based approach to immunogenicity concerns of therapeutic protein products, Part 2: Considering host-specific and product specific factors impacting immunogenicity. *BioPharm. Int.* 2004: 34–40.

20 Markovic, I. (2006). Challenges associated with extractable and/or leachable substances in therapeutic biologic protein products. *Am. Pharm. Rev.* 9 (6): 20–27.

21 Wen, Z.Q., Torraca, G., Yee, C., and Li, G. (2007). Investigation of contaminants in protein pharmaceuticals in pre-filled syringes by multiple micro-spectroscopies. *Am. Pharm. Rev.* 10 (5): 101–107.

22 Seidl, A., Hainzl, O., Richter, M. et al. (2012). Tungsten-induced denaturation and aggregation of

epoetin alfa during primary packaging as a cause of immunogenicity. *Pharm. Res.* 29: 1454–1467.

23 Mensch, C.D. and Davis, H.B. (2012). Inhibition of tungsten-induced protein aggregation by cetyl trimethyl ammonium bromide. *PDA J. Pharm. Sci. Technol.* 66 (1): 2–11.

24 Liu, W., Swift, R., Torraca, G. et al. (2010). Root cause analysis of tungsten-induced protein aggregation in pre-filled syringes. *PDA J. Pharm. Sci. Technol.* 64 (10): 11–19.

25 Christensen, S., Moeller, E.H., Bonde, C., and Lilleoere, A.M. (2007). Preliminary studies of the physical stability of a glucagon-like peptide-1 derivate in the presence of metal ions. *Eur. J. Pharm. Biopharm.* 66: 366–371.

26 Sharma, B. (2007). Immunogenicity of therapeutic proteins. Part 2. Impact of container closures. *Biotechnol. Adv.* 25 (3): 318–324.

27 Jones, L.S., Kaufmann, A., and Middaugh, C.R. (2004). Silicone oil induced aggregation of proteins. *J. Pharm. Sci.* 94 (4): 918–927.

28 Felsovalyi, F., Janvier, S., Jouffray, S. et al. (2012). Silicone-oil-based subvisible particles: Their detection, interactions, and regulation in prefilled container closure systems for biopharmaceuticals. *J. Pharm. Sci.* 101 (12): 4569–4583.

29 Ludwig, D.B., Carpenter, J.F., Hamel, J.B., and Randolph, T.W. (2010). Protein adsorption and excipient effects on kinetic stability of silicone oil emulsions. *J. Pharm. Sci.* 99 (4): 1721–1733.

30 Chan, E., Hubbard, A., Sane, S., and Maa, Y.F. (2012). Syringe siliconization process investigation and optimization. *PDA J. Pharm. Sci. Technol.* 66 (2): 136–150.

31 Bee, J.S., Randolph, T.W., Carpenter, J.F. et al. (2011). Effects of surfaces and leachables on the stability of biopharmaceuticals. *J. Pharm. Sci.* 100 (10): 4158–4170.

32 USP <1031>. The Biocompatibility of Materials Used in Drug Containers, Medical Devices, and Implants. USP43-NF38, pp. 7313. Official as of 1 December 2020.

33 USP <87>. Biological Reactivity Tests, *in vitro*. USP43-NF38, pp. 6514. Official as of 1 August 2016.

34 USP <88>. Biological Reactivity Tests, *in vivo*. USP43-NF38, pp. 6516. Official as of 1 August 2013.

35 Guidance for Industry. Container Closure Systems for Packaging Human Drugs and Biologics: Chemistry, Manufacturing, and Controls Documentation. US Department of Health and Human Services: Food and Drug Administration; Center for Drug Evaluation and Research (CDER), Center for Biologics Evaluation and Research (CBER). May 1999.

36 Guideline on Plastic Immediate Packaging Materials. European Medicines Agency Inspections; Committee for Medicinal Products for Human Use (CHMP), Committee for Medicinal Products for Veterinary Use (CVMP). CPMP/QWP/4359/03. EMEA/CVMP/205/04. London, 19 May 2005.

37 Paskiet, D. Overview of Thresholds and Best Practices for Extractable and Leachables (L&E); Thresholds and Best Practices for Extractable and Leachables. 3rd PQRI/FDA Conference on Advancing Product Quality. Washington, DC, 22 March 2017.

38 Jenke, D. and Carlson, T. (2014). A compilation of safety impact information for extractables associated with materials used in pharmaceutical packaging, delivery, administration and manufacturing systems. *PDA J. Pharm. Sci. Technol.* 68 (5): 407–455.

39 Jenke, D. (2015). Safety risk categorization of organic extractables associated with polymers used in packaging, delivery and manufacturing systems for parenteral drug products. *Pharm. Res.* 32 (3): 1105–1127.

40 Kennedy, T.A. and Spinti, M.J. (2021). How sensitive does chemical characterization of medical devices need to be? Calibration of analytical evaluation thresholds with the carcinogenic database. *Regul. Toxicol. Pharmacol.* 122: 104899.

41 Masuda-Herrera, M., Bercul, J.P., Broschard, T.H., et al. Development of duration-based non-mutagenic thresholds of toxicological concern (TTC) relevant to parenteral extractables and leachables (E&Ls). *PDA J Pharm Sci Technol.* 2022. Jan 14;pdajpst.2021.012693. doi: 10.5731/pdajpst.2021.012693.

42 Hammond, M., Marghitoiu, L., Lee, H. et al. (2014). A cytotoxic leachable compound from single-use bioprocess equipment that causes poor cell growth performance. *Biotechnol. Prog.* 30: 332–337.

43 Kelly, P.S., Dorvial-Gracia, N., Pare, S. et al. Improvements in single-use bioreactor film material composition leads to robust and reliable Chinese hamster ovary cell performance. *Biotechnol. Prog.* 35 (4): e2824. https://doi.org/10.1002/btpr.2824. Epub 6 May 2019.

44 Eibl, R., Steiger, B.N., Fritz, C. et al. (2014). *Standardized Cell Culture Test for Early Identification of Critical Films for CHO Cell Lines in Chemically Defined Culture Media.* Dechema. ISBN: 978-3-89746-149-9.

45 Haep, R., Stratmann, A., Watt, S.A., and Martens, L. (2018). Protein reactive extractables – A screening assay which adds additional value to the traditional E/L workflow. *Pharm. Ind.* 80 (1): 104–114.

11

Focus on Emerging Concepts

11.1 A Dose of Reality

It has been hard work creating (for me) and reading (for you) the text in the proceeding 10 chapters, which generally focused on issues of science, technology, and practice. Now, as we close this book, we can indulge ourselves a little by humoring our more philosophical selves by considering the current and future state of affairs with respect to the enterprise we call "extractables and leachables" or "E&L."

To put us in the right frame of mind, I am reminded of an old E&L story. After a long day at a well-attended E&L conference, three E&L experts meet at the bar. Eventually, as almost always happens, the three experts start talking shop. One of the experts starts talking about the problems that were encountered in profiling a particular packaging system component used for a drug product with a large daily dose volume and a chronic clinical indication (i.e. extremely low analytical evaluation threshold, AET). "Yeah," says the expert, "we burned a lot of lab time trying to get down to the AET and then identifying all the little peaks that were barely above the baseline." The other two experts shook their heads sadly in the "been there, done that" sort of way, because, after all, we have all been there and done that. Feeling the need for a little more sympathy, the first expert goes on; "yeah, there was one extractable that we identified as a weird anti-oxidant-related degradation product, that took us forever to figure out."

Now instead of providing the desired sympathy, all of a sudden both of the other experts start to get agitated. Both blurt out, almost simultaneously, "That sounds like something we encountered recently as well!" Well, now Pandora's box has been opened for sure and the experts start sharing notes. Without betraying confidences, the three are able to ascertain that indeed they all have been testing the same item. Again, practicing due diligence, they shared and compared the extractables profiles. And then the conversation goes south. "What?" exclaimed expert 1, "how in the world did you get three dozen peaks in the APCI + ion mode chromatogram? I'll bet my new ICP/MS that this material has no more than 10 non-volatile extractables!" To counter, expert 2 stares the other two down and shouts "how in the world did you get that ID for the GC peak at 11.5 minutes? It is not the 2,4-isomer, you moron, it is the 2,6-isomer." Expert 3, ever the peacemaker, does his best to calm the others down until his eye is drawn to the reported concentration for the D4 siloxane that all three measured via LC/MS. "Are you freaking kidding me?" he bellows to the other two. Because you see, not only do the results obtained by the first two experts differ by a factor of 10, but expert 3's result is a factor of three different in the other direction!

You know, maybe the story isn't very funny, because E&L experts tend to be a rather dry group. However, consider the implications of this alleged situation by asking yourselves, "What is wrong with this picture?" Well, one thing for sure is that three E&L experts are reporting different extractables profiles for the same item. Is it not logical, from a simplistic perspective, that one item has one extractables profile?

However, there is more that is wrong with this picture than the obvious. Point 2 is, here we have three E&L efforts, likely taking hundreds of manhours and easily costing tens of thousands of dollars each, all spent on the same item. Again, is it not logical, from a simplistic perspective, that an item's extractables profile need be generated only once?

Dig deeper and there is yet a third point. It turns out that this particular item is widely used in pharma in very similar applications. In fact, the first packaged drug product that used the item was commercialized three years ago. So, the item is safe, everyone knows that it is safe and yet here we are, still testing it (and reporting the results as if it is the first time the reviewing regulatory agency has seen this data)!

I am absolutely convinced that if an alien E&L expert were to land on planet Earth and demanded to review our E&L protocols and practices, that they would, after reviewing what we have, report back to their leaders that they encountered "an evolving culture where practitioners have fully grasped the importance of this issue but have yet to develop an effective and efficient means of addressing the

Extractables and Leachables: Characterization of Drug Products, Packaging, Manufacturing and Delivery Systems, and Medical Devices, First Edition. Dennis Jenke.
© 2022 John Wiley & Sons, Inc. Published 2022 by John Wiley & Sons, Inc.

issue. Although we cannot intervene because of the Prime Directive, it would be interesting to revisit this planet in 15 years and see how the science, practices, and policies have evolved."

If there is any truth to the above characterization of E&L as it is practiced in 2022, it is fair to ask three questions:

1. How did we get here?
2. What should things look like?
3. How can we transition from where we are to where we would like to be?

I do not intend to address point 1 in more than a general way. We have got where we are today for four reasons:

1. We needed to act before we had the wisdom and the technology to act (and thus we have been making it up and playing catch-up as we go along).
2. We talk about risk-based approaches but in fact what we mean are "no-risk approaches."
3. We have a hard time balancing self-interest with our responsibility to "the common good."
4. It is easier to "go with the flow" than to "make waves."

At this point, let me be clear. None of the statements above are directed toward, or an indictment of, an individual, a company, or an organization, either individually or collectively. There is no hidden "conspiracy theory" or collusion that has got us to, nor dooms as to stay at, the place that has been described previously. However, it is our collective inaction and inability to craft a consensus that preserves the status quo and prevents us from achieving a more sustainable and effective future state.

Considering the second point, a recent Commentary on study design considerations for E&L screening studies [1] noted that:

> "… it is an unfortunate reality that an E/L screening study at multiple labs will, more often than not, yield datasets that are unaligned to at least some degree. Experience suggests that these differences can be attributed to at least four factors: how material samples are extracted (when performing an extractables assessment), how samples are prepared for analysis, the analytical method used for the analysis, and how the data is evaluated."

It is certain that one can envisage an E&L future where extractables datasets will more closely align because the above-mentioned aspects have been addressed.

Focusing further on the future state of the E part of E&L, that future state is characterized by five features:

1. Extractables profiles that reflect the test article's worst-possible (as opposed to worst-imaginable) clinical conditions of use (i.e. extractables = leachables),

2. Extractables profiles that are reproducible,
3. Extractables profiles that are complete,
4. Extractables profiles that are accurate, and
5. Extractables profiling that is time- and cost-efficient (reducing unnecessary redundancy).

It is this author's hypothesis that these features are attained by:

1. Proper design of the extraction study,
2. Proper analytical testing, including standardization of methods and mitigation of testing errors,
3. Information sharing, and
4. Elimination of unnecessary and/or redundant testing.

Each of these is considered in greater details as follows.

11.2 Designing and Implementing Better Extractions

An extraction study is designed to accomplish its purpose. As stated previously, the future of E&L is best served when the purpose of the extractables profiling activity is to produce an extractables profile that matches the leachables profile; that is, extractables = leachables. To wit, an extractables profile of a packaging system should mirror the leachables profile of the packaged drug product. Similarly, an extractables profile for a manufacturing component should mirror the component's profile of PERLs. Lastly, the extractables profile of a medical device should reveal those substances that the patient using the medical device is exposed to, and the amount of exposure, during the device's clinical use.

An unspoken part of the above dialog is that the extractables profile should reflect the worst-case leachables profile; that is, all likely leachables at their highest likely concentration. My more conservative colleagues would propose that, in fact, an appropriate extractables profile would, in all three cases, slightly overexaggerate the leachables profile to provide a "margin for error." This is a reasonable point if it is not taken to an extreme. For example, consider a packaging system used for a family of aqueous water-soluble drug products. It can be proposed, and it has been required, that the extractables characterization of the packaging system include an extraction performed with a non-polar solvent, as use of a non-polar extraction solvent is the best way to ensure that non-polar compounds present in the packaging system can be accounted for. Although this is a true statement, it is irrelevant in terms of the objective of the extractables study being the generation of a reasonably exaggerated leachables profile. This is the case because most (if not all) non-polar substances will be either poorly soluble or insoluble in the aqueous drug product and thus cannot and will not be leachables in meaningful quantities. Thus, while a hexane extract of the packaging system will certainly reveal any non-polar extractables, this is meaningless information because the same extractables will not be leachables. In this case, the extractables profile does not mimic the leachables profile and it does not exaggerate the leachables profile. Rather, it distorts the leachables profile, which serves no useful purpose.

There is a train of thought that maintains that the purpose of an extractables study is to reveal all possible extractables in their highest possible extractable amount. This is the train of thought that says "use three solvents of varying polarity (polar, semi-polar, and non-polar) and perform an exhaustive extraction" in every situation. Arguably, the required testing accomplishes the objective, with the caveat that using a polar extraction solvent, typically water, without considering the effect of pH on leaching, ensures that the resulting extractables profile will not be the comprehensive profile that is desired. In any event, this strategy will reveal all manners of extractable substances and will establish the total amount of these substances that can be extracted from the test article in virtually every imaginable (and likely unimaginable) circumstance.

Ignoring the considerable expense in implementing and completing an extraction study that requires three solvents and likely at least three sequential extraction steps, the question becomes, "what do I do with this data now that I have it?" Well, the typical response would be "toxicologically safety risk assess each extractable using a daily exposure linked to the total amount extracted (meaning it all comes out in one day)."

Now there may be clinical applications where the conditions of this experiment match the clinical conditions of use. That is, the item in question contacts all manners of fluids (or other contact media such as a tissue) during use and is completely extracted (or dissolves) in one day of clinical use. However, logically these applications are somewhat rare and it is more commonly the case that contact is limited to a small number of well-defined contact fluids (or media such as a tissue) and that complete extraction is not accomplished in a single day. Thus, in a majority of cases, the "multiple polarity solvents, exhaustive extraction" approach greatly exaggerates and arguably distorts the patient safety risk. Thus:

1. Assessments are (unnecessarily) performed on an (inappropriately) large number of substances at (unattainably) high levels,
2. The resulting assessments are more likely to identify potentially risky substances (bad actors),

3. Perfectly safe items are unnecessarily rejected or re-engineered to mitigate the bad actors,
4. Leachables studies are unnecessarily performed to establish the true clinical behavior of the extractables (bad actors and not-so-bad actors alike).

From the limited perspective of patient safety, these four points are irrelevant; that is, patient safety is not compromised and possibly is enhanced by the assessments that are based on the worst possible scenario. However, there are other factors to consider. If the intent of medical product development also includes the concepts of "quick to market" and "cost-effective therapies," then the four points are clearly sub-optimal. The cost of re-engineering, increased assessment and testing and other related items surely are reflected in the ultimate cost of a commercialized therapy. Delays in product launch translate to situations where patient access to life-enhancing or even life-saving therapies is also delayed, leading to unnecessary suffering.

The need to perform extraction studies, which by their design provide a reasonable "margin for error" to properly balance all risks associated with medical therapies, is clearly established; however, establishing the need is quite different from accomplishing the objective. It is obvious, and it goes without saying, that the most efficient and effective extraction study is the one that is designed to fit a specific set of circumstances. However, there are two significant tail winds that prevent the adoption of "custom extraction studies" as acceptable regulatory practice. The first of these tailwinds is practical; given the great diversity of drug products (and thus drug product packaging), the great diversity in medical devices and their clinical use and the great diversity of manufacturing operations (and thus their components of construction and conditions of use), one could literally envision thousands of "custom" extraction study designs, each of which would have to be judged as being acceptable and appropriate on their own merits. Furthermore, it is likely that two custom extraction studies would produce different extractables profiles for the same test article.

The situation created by thousands of custom extractables study designs leads us to the second tailwind, which is the lack of a consensus on the essential science of extraction. To be blunt, the success of securing regulatory approval for an extraction study rests on the sponsor's ability to convince the reviewer that the study was properly designed and implemented to produce the necessary "margin for error." If the sponsor and the regulator do not agree on the essential science of extraction, then it is that much more likely that the sponsor and the reviewer will disagree on the adequacy of an extraction. Moreover, consider the situation of a reviewer being presented with files from sponsors of similar products who themselves do not agree on the fundamental science of extraction. The reviewer is likely to receive two completely different extraction studies, both claiming to reflect "the best available science."

No, the "custom" approach to designing an extraction is not a viable option. Furthermore, the "extract the heck out of it until there is nothing left" approach is not a universally viable option. Thus, a compromise is in order. An appropriate compromise between these two extremes is obtained when:

1. The universe of medical products and medical applications is broken up into small enough groups that a standardized approach is possible for each group,
2. Areas of agreement on fundamental principles can be leveraged to produce the standardized approach.

At first glance it may seem that accomplishing this task is an unsurmountable challenge; however, there is a success story that can be cited to serve as an indication that "it can be done." This success was achieved in the area of components used in pharmaceutical manufacturing, which was previously considered in Chapter 8. It is useful to consider this example in greater detail, focusing on factors that enabled success (see Table 11.1), as so-doing might demonstrate the means by which the success can be repeated in other circumstances and situations.

The first success factor in the development of standardized extractions for manufacturing components is the aspect of finding a champion with clout and deep pockets. It is one thing for an individual author like myself to go on and on in his book about "standardization is the path to salvation." As one might expect, the net result is barely a whisper and few converts. However, when users band together and in one voice proclaim "we need and will have a standard extraction" (BioPhorum) or when an organization with standards-setting authority concludes "the industry needs a standard approach" (USP), then the whisper becomes a roar. Furthermore, when the users and the standards-setting organization collaborate, then the roar becomes a boom that reverberates throughout the entire industry. Finding the champion with deep pockets is important in the beginning of the development of the standard extraction for the simple reason of resources. It is going to take time, patience, lab work (and more patience) to achieve the objective of an accepted and universally applied standard extraction.

We will come back to the "with clout" aspect of this success factor later.

The second success factor is to gain a profound and fundamental understanding of the conditions of use, as it is this profound and fundamental understanding of the conditions of use that defines and justifies what is an acceptable standard practice. One gains a profound and fundamental understanding of the conditions by focusing on what is being used and how is it being used. To accomplish this objective, BioPhorum surveyed its membership to establish what constituted the general practice in pharmaceutical manufacturing, specifically focusing on unit operations, what items are used in those unit operations, what is the nature (e.g. composition) of those items, and what are the operating conditions under which the items are used. Clearly, this is critical information, as designing (and justifying) an extraction that reflects conditions of use surely requires that the conditions of use be known.

It is at the point that process information has been obtained that the information can be interpreted in the context of designing appropriate extractions. Perhaps it is intuitively obvious that, for example, a sterilizing filter (for example) is not composed of the same materials of construction and is not used in the same manner as a media bag (as another example), but it is the collection and interpretation of process intelligence that replaces intuition with fact.

Table 11.1 Success factors in designing and implementing a standardized extraction.

Success factor	Description
1	Identify and secure a powerful champion with deep pockets.
2	Gain a profound and fundamental understanding of the conditions of use.
3	Subdivide the total population of relevant test items into classes for which a standard extraction can be developed.
4	Develop the philosophy and objectives of the standardized extraction.
5	Couple the profound and fundamental understanding gained in factor 2 with proper scientific principles to design a standard extraction that is consistent with the philosophy and accomplishes the objectives enumerated in factor 4.
6	Pressure-test the standard extraction, both experimentally and through peer review, to find unanticipated difficulties in implementation. Revise and adjust the standard extraction appropriately.
7	Address exceptions to the standard extraction.
8	Communicate, adopt, and enforce the standard extraction.

The first accomplishment to be secured by the interpretation of process intelligence is the ability to "cut the meat into chewable pieces," that is, divide the universe of manufacturing components into its various classes (success factor 3), with the intent of developing a standard extraction for each class. Thus, if it is impossible to devise and justify a single extraction protocol for all manufacturing components, it may be possible to devise (and justify) individual extraction protocols for individual classes of manufacturing components that share certain common characteristics, such as function, composition, and conditions of use. In the case of manufacturing components, it was relatively simple to realize that the proper classification was the type (function) of the manufacturing component. Thus, tubing (used to transport fluids) is separated from bags (used to store fluids), which is separated from filters (used to filter fluids).

Once the entire population has been compartmentalized into manageable classes, the task of using the process intelligence to establish the extraction conditions can occur. It is at this point that science meets philosophy (success factors 4 and 5), as the science is applied to support the philosophy. Armed with the process intelligence, the BioPhorum working team adopted the strategy that the extraction process should reflect the worst-case conditions of use among its members, supposing that so doing would mean that "everyone is covered." It is this supposition, augmented by the process knowledge, that shaped the BioPhorum's original extraction protocol, consisting of six extraction solvents, two extraction temperatures, and multiple extraction durations at the elevated extraction temperature.

Working collaboratively with BioPhorum but with an independent perspective, the USP used essentially equivalent process knowledge but a different testing philosophy to come up with its own standard extraction protocol. Where the BioPhorum philosophy was "test at worst case and everybody is covered," the USP approach was "aim for the middle" and test consistently with the more commonly encountered conditions of use so that almost everybody is covered. Thus, the USP standard extraction protocol had fewer extraction solvents, less extreme extraction solvents (generally), and only one extraction temperature and one extraction duration.

Although, at first glance, the BioPhorum approach appears to be more advantageous, as "everybody is covered," this so-called "universal coverage" comes with a price, both literally and figuratively. Literally, accomplishing an extraction study with six solvents and multiple timepoints is an expensive proposition, with "full BPOG testing" typically requiring several hundreds of thousands of dollars to complete. Speaking more figuratively, in many cases the BPOG extraction is quite exaggerated when compared to many conditions of use scenarios. For example, the pH of BioPhorum's 0.5 M NaOH extraction solution is approximately 13.6. While there are certainly manufacturing operations where the process stream pH is this high (for example, pH adjustment and flushing), a majority of process solutions rarely have a pH higher than 10. Being nearly four magnitudes more caustic than a pH 10 extraction solvent, the 0.5 M NaOH solution has the potential of altering a material's extractables profile by using base hydrolysis, for example. Thus, a polycarbonate component will essentially depolymerize at a pH of 13.6, resulting in the high quantities of extracted bisphenol A (BPA). In this way, the "price penalty" that is paid by performing the BioPhorum's extraction is an altered or distorted extractables profile.

It is clear that the outcome of BioPhorum and USP advocating different extraction protocols is not optimum for the various stakeholders. Thus, a reconciliation and compromise based on both science and practicality was in order (success factor 6). To wit:

- The BioPhorum 0.1% PS-80 extraction solution was eliminated based on the practical difficulty associated with screening this solution and the scientific reality that the extractables profiles of 50% ethanol and 0.1% PS-80 are similar (making the 0.1% PSA extraction solution redundant).
- The acidic pH extraction solutions (0.1% phosphoric acid, pH 2 for BioPhorum and a pH 3 salt/acid mixture for USP) were established experimentally to have approximately the same "leaching power" and thus were established to be interchangeable.
- The circumstance of the basic extraction solutions (0.5 M NaOH, pH 13.5 for BioPhorum and pH 10 buffer for USP) producing different extraction profiles was experimentally verified and thus these extraction solvents were established as being non-equivalent. To manage the difference, the USP high pH extraction's instructions were modified to the extent that although the pH 10 buffer was the recommended extraction solvent, an extraction solvent at higher pH, possibly even the 0.5 N NaOH required by BioPhorum, could be used when the pH of the process solution was greater than 10.
- The ambient temperature, 30-minute extraction required in the original BioPhorum protocol was eliminated as being unnecessary.

Once the standard extraction has been revised and optimized, it is important to realize that invariably there will be exceptions. It could turn out that a material of construction is incompatible with the extraction, turning the extraction into a dissolution. It may be that the surface area to solution volume ratio called for in the standard extraction cannot be achieved in a particular circumstance. It may be that an item is used for a longer time or at a higher temperature than that specified in the standard extraction. Or it could be something else.

In any event, potential users of the standard extraction must be given some guidance in terms of what to do if the "one size fits all" standard extraction truly does not fit their circumstances or else these potential users will become challengers to the standard extraction because, after all, no one likes to be left up the creek without a paddle.

In the case of the USP's standard extraction protocol, such guidance is reflected in the text of USP <665> that describes alternate extractions.

To a certain extent, the generation of the pressure-tested, peer-reviewed standard extraction can be considered to be the end goal. However, the practical fact of the matter is the standard extraction is just a scientific curiosity until it is adopted and universally implemented. Although we might want to believe in a world where the standard extraction would be adopted because "it is the right thing to do," the reality is that the standard extraction, because it is likely that it is not the most efficient extraction in every circumstance (that is the penalty of standardization), will only be adopted if there is a tangible incentive to do so. This is why success factor 1 required that the extraction's champion needs to have both deep pockets (to be able to procure the resources required to develop the standard extraction) and "clout," as it is the clout that provides the incentive for adoption.

In the case of the standard extraction for manufacturing components, the two champion organizations had different means of applying clout. BioPhorum, as a community primarily of users, was able to apply financial clout, essentially suggesting to vendors that the BioPhorum members would buy only those components that were supported by BioPhorum testing. In this way, the vendors were incentivized to perform BioPhorum testing and supply BioPhorum test results (in a standardized format no less). USP, as a standards-setting organization and specifically, at least in theory, the standards-setting counterpart to the standards-enforcing regulatory authority, are able to incentivize drug product sponsors to perform <665> testing by noting that such testing is a pre-requisite to obtaining regulatory approval to market the drug product.

The success of standardizing extraction protocols for manufacturing components has not been matched for pharmaceutical drug products or medical devices. It is time to change that circumstance.

There is one remaining point to make before the discussion of standardization moves on to the analytical aspect of E&L profiling. This remaining point is to observe that even implementation of a standard extraction will not achieve the objective of consistently producing a compositionally equivalent extract unless the standard extraction is purposely designed to do so. It is intuitively obvious that an extraction that achieves equilibrium can be more easily and consistently reproduced than an extraction where diffusion is still occurring when the extraction is completed. Moreover, extractions that are stopped when the speed of diffusion is high will be less reproducible than extractions that persist until the speed of diffusion has slowed. This is the case as extractions are rarely conducted under the exact conditions with which they are designed. For example, consider the guidance in ISO 10993:12 for the generation of an extract for biological characterization, where one of the recommended extraction cycles is $(70 \pm 2)\,°C$ for (24 ± 2) hours. If lab A runs the extraction at $68\,°C$ for 22 hours and lab B runs the extraction at $72\,°C$ for 26 hours, both labs can rightly claim that the extraction was accomplished per ISO 10993:12. However, unless equilibrium is achieved at a temperature lower than $68\,°C$ and at an extraction duration less than 22 hours, the two labs will have produced extracts that are inherently and unavoidably different in their compositions.

Thus, whenever the purpose of the extraction study and the clinical use conditions allow, standardized extraction should achieve equilibrium by design and intent.

11.3 Achieving Better Analytical Outcomes

11.3.1 Re-stating the Case for Standardized Methods

It is largely luck, and not an assured outcome, that two labs, screening the same extract for extractables or drug product with leachables, will report the same results, including:

1. The same number of reported compounds,
2. The same identities for reported compounds,
3. The same concentrations for reported compounds.

This statement is not a commentary on the qualifications of any labs involved in E&L testing (commercial, academic, or internal) but rather the direct and logical outcome when the two labs use different analytical approaches and methods to accomplish the same objective. If the desired outcome is comparable results, then the test methods must be standardized. Furthermore, since standardized methods are produced by consensus, they are more likely to reflect the true state of the scientific art versus a method developed by an individual testing laboratory.

In the absence of a true breakthrough in analytical technology, analytical methods for screening extracts for extractables or drug products and medical devices for leachables are mature and well-established. That is, virtually every practicing E&L laboratory screens samples for organic extractables or leachables using a combination of GC/MS with headspace sampling, "direct injection" GC/MS, and "direct injection" LC/MS, where the term "direct injection" is in parentheses because preparation steps may be necessary to make extracts or drug products suitable for "direct injection." Arguably, the individual methods used by individual testing laboratories are largely interchangeable in terms of their capabilities, as competitive pressures force the universal adoption of truly superior methods. Methods that are superior in one aspect or for one class of compounds are generally deficient in another aspect or for another class of compounds because all methods must balance the same operational challenges. Thus, even if an individual laboratory's methods were superior, that superiority would be fleeting. Even if the individual laboratory were able to maintain its own intellectual property, eventually the capabilities of that technology would become expected practice and other laboratories, supported by appropriate budgets and staffed by dedicated and competent analysts, would be forced to meet the challenges of the expected practice by "discovering" either the superior method or a suitable alternative.

The same argument can be made for the means of preparing samples for instrumental analysis and the processing of analytical data.

It is naïve to propose the standardization of methods without recognizing the practical headwinds that thwart the development of the standardized practices. Asking individual organizations to adopt standard practices is a burden given the organization's investment in its own existing practices. Replacing a perceived competitive advantage with a standard approach has business and marketplace implications. Suggesting that practices be standardized presumes that the industry has the ability to come together and collectively establish the standard and, furthermore, that the industry has the ability to enforce the standard.

Realistically, the science- and regulatory-driven pressure to standardize will be meet by the equal and opposite pressure of the vested interest in maintaining the status quo to preserve perceived "competitive advantages." At best one anticipates standardization will occur only in the case of low hanging fruit; for example, standardizing the internal standard used in GC/MS.

The challenge of standardizing analytical screening methods is best illustrated by the "low hanging fruit" example of GC/MS. GC/MS screening for organic extractables or leachables has gained the status of the "lowest hanging fruit" primarily on the basis of standardization that has already occurred. Thus, for example, the GC/MS

analytical instrumentation is largely standardized in terms of the most impactful operating conditions (e.g. ionization voltage of 70 eV). Most laboratories use the same type of column to produce the chromatographic separation, most temperature programs are comparable in terms of their starting temperature, ending temperature, and temperature program, and most labs use comparable mass ranges. Although there is certainly some lab-to-lab variation in the choice of internal standards, it is likely to be the case that the universe of routinely employed internal standards is small.

Given these circumstances, one can be emboldened by the possibility that standardization of the GC/MS method could be achieved without undue sacrifice on the part of the analytical community. That is, until one understands the practical reality that even if these parameters become standardized there are many more aspects of GC/MS analysis that contribute to variation in results. Even if the column is standardized, the column's condition at the time of use is not. Not every lab cleans or replaces the injector liner with the same frequency. Does one perform a spilt or a split-less injection? If split-less, what is the split ratio? And so on and so forth.

In the event that it was possible to standardize the instrumentation and the instrumental operating parameters, the issue of sample preparation practices remains. It is a practical reality of analysis by GC/MS that many commonly employed extraction solutions cannot be directly injected and thus they must be solvent-exchanged to be analytically compatible with the instrumentation. It is also a practical reality that when faced with low AETs or complicated samples, extracts are manipulated (before or after solvent switching) to minimize analytical interferences and maximize analytical sensitivity. In a recent review of the analytical challenges and recent advances in the identification and quantitation of extractables and leachables, researchers from SGS Health Sciences and Nexus Pharmaceuticals identified means of preparing aqueous extracts for GC analysis [2]:

1. Traditional liquid–liquid extraction (LLE)
2. Traditional solid-phase extraction (SPE)
3. Solid-phase microextraction (SPME)
4. Dispersive liquid–liquid microextraction (DLLME)

Considering the relative merits of these techniques, these authors note that "no one technique can be said to be suitable in every situation" and it is logical to observe that no one technique can claim to be better than the others in terms of its ability to handle the diverse chemistries exhibited by polar and non-polar, volatile and non-volatile organic extractables from an expanding universe of chemically compositionally diverse materials used in pharmaceutical and medical applications. While one methodology may be superior for one class of compounds, it would be likely to be deficient for another set of compounds. Furthermore, there are practical considerations in terms of sample preparation method selected. Lastly, even if a standard method were to be selected, what about the operating conditions? LLE, for example, can be accomplished with various solvents, with various phases ratios, with different number of repetitive extractables, with and without pH adjustment, etc. It seems impossible, then, that standardization of GC/MS is possible. However, perhaps a history lesson is in order. When faced with the issue that "GC is likely missing extractables and leachables," the industry migrated to adopt the now standard practice of screening by both GC and LC. Perhaps a similar approach would be possible to establish standardized sample preparation practices; that is, rather than developing a single standardized sample preparation method that everybody agrees is "the best we can do" (even knowing full well it is not good enough), perhaps two orthogonal but complementary standardized sample processing methods can be established that together truly minimize coverage gaps.

11.3.2 Reducing, Mitigating, and Eliminating Errors and Variation in E&L Testing

If the challenges of standardizing extractables and leachables screening methods appear to be insurmountable, then perhaps the benefits of standardization can be achieved if the methods' output (analytical data) is considered. Rather than standardizing the process by which an output is generated (that is, standardizing the analytical method), one standardizes the output by establishing a minimum quality standard for the output. Methods whose output is able to meet the minimum quality standards are said to be standardized and are likely to produce correct and comparable lab-to-lab data, regardless of any methodological differences between labs.

Relevant minimum quality standards can be applied to the three major objectives of screening, each of which is addressed as follows:

1. Detecting and reporting a compound
2. Identifying a compound
3. Quantifying a compound

11.3.2.1 Detecting and Reporting a Compound

Before a compound can be identified and quantified, it must be detected. Once a compound has been detected, the decision whether to report the compound, potentially triggering the actions of identification and quantitation, must be made and justified.

The "detection power" of an analytical method is measured in two dimensions, the number and chemical diversity of compounds that produce a response (breadth and coverage) and the magnitude of the response produced (sensitivity).

Breadth and Coverage of the Method An analytical method's breadth or scope is defined as the number of compounds that produce a recognizable response and the responding compounds' key physicochemical characteristics. Chromatographic screening methods applied to extractables and leachables are often described as being broad-scope or having a considerable breadth, as it is expected that the methods will produce adequate responses for a large number of diverse compounds.

The term "coverage" refers to the ability of the chromatographic method to respond to all possible organic extractables/leachables and thus is related to breadth. However, whereas breadth is related to the number of compounds producing a viable analytical response, coverage is related to the percent of compounds in the entire population of potential extractables/leachables that produces a viable analytical response. Logically, a method with a greater breadth (greater number of compounds producing a useful response) will have a greater coverage.

One measure of a method's breadth is the method's list of detectable compounds. As the method is used to support studies, the method's list of detected compounds grows and the method's breadth is established (i.e. "good for C1–C10 alkanes, not so good for branched fatty acids"). Typically, the list grows as the method is applied to specific projects and thus it is likely that the contents of the list will vary somewhat from lab to lab depending on the products and projects supported by that lab. For example, even if the same method is used by two labs, one specializing in medical device testing and the other specializing in packaged drug product testing, the method's breadths defined by each lab are likely to consist of slightly different specific compounds but similar compound classes. As the list grows, it is likely that the method's breadth expands somewhat and that breadth can be more precisely defined.

As an alternative to a "listing" approach to defining breadth, breadth can be established by a method's ability to respond to intentionally chosen marker compounds. For example, compounds that elute early and late (so-called anchor compounds) can be used to establish the "boundaries" of the method while marker compounds of a particular functionality (or of known significance) can establish the ability of the method to respond to necessary compounds or compound classes. For example, markers chosen to reflect functionalities might include (as appropriate), acids, alcohols, aldehydes, ketones, siloxanes, etc., while markers chosen to reflect significance might include (but are not limited to) so-called "bad actors," such as di-(2-ethylhexyl) phthalate (DEHP) and BPA as compounds of known toxicological concern and bis(2,4-di-tert-butylphenyl)phosphate (bDtBPP) as a compound known to reduce yield in biomanufacturing.

Comparing a method's experimentally determined breadth, checked by testing a set of marker compounds or a list of "must have" compounds, between labs serves as a means of standardizing output. Methods from various labs that produce an acceptable response for each member of the set of critical compounds are said to be standardized with respect to breadth (that is, the methods have the same breadth).

Sensitivity of the Method The AET is that concentration of an organic extractable or leachable at which the compound is likely enough to be unsafe that it requires formal safety assessment. Compounds at levels at or above the AET are thus reported for safety assessment. It is intuitively obvious that the sensitivity of a chromatographic screening method must be sufficient that extractables can be detected at least at the level of the AET, if not lower, and thus LoD ≤ AET is an appropriate requirement for testing laboratories.

If all labs calculate the AET in the same manner (for example, they all use the same uncertainty factor (UF)), use the same value for the AET and meet the requirement that LoD ≤ AET, then the number of compounds reported by each lab should be highly similar because all labs are able to meet the same reporting threshold. However, as the labs may have different LoDs, it is reasonable to expect that they may have detected a different number of compounds below the AET. If the labs choose to report these compounds, then there is a clear opportunity for the two labs to have two different lists of extractables.

This circumstance is readily and easily managed by asking labs to report all compounds at or above the AET in a different manner (or in a different place in the report) than compounds below the AET but above the LoD. In so doing, the lab-to-lab extractables profiles will be more similar at or above the AET, dispelling the perception that extractables profile differences lab-to-lab are rampant in the industry. Although there may be greater lab-to-lab differences in the compounds that are below the AET, such differences are largely academic from the perspective of adversely affecting the rigor and adequacy of the safety assessment, as compounds below the AET are not safety-assessed (because, by definition of the AET, such compounds are safe).

In a certain way, the aspect of sensitivity for the screening method is a refinement of the aspect of breadth. Whereas breadth reflects the ability to produce a detectable response, sensitivity requires that the magnitude of the produced response be above a fixed value. Given this similarity, the approach taken to establish sensitivity is essentially the same as the approach taken to assess breadth and in fact it is most efficient if breadth and sensitivity are assessed by same data produced by the same experiment. However, breadth is an invariant characteristic of an analytical method while sensitivity is a situational characteristic of an analytical method as the AET can vary from situation to situation. Thus, it is possible that two labs could have methods with equal breadths but an unequal ability to meet the sensitivity requirement.

The above discussion works well in the situation where the AET is achievable by all testing laboratories. However, one can envisage situations where the AET is so low that it is an insurmountable challenge for one or more (or perhaps all) labs to meet the expectation that LoD ≤ AET. Invariably and unavoidably, this differing ability to "go low" will lead to lab-to-lab variations in the reported extractables/leachables profiles.

11.3.2.2 Mathematical Toxicological Safety Assessment Simplified

The science and practice of toxicological safety risk assessment has been considered previously (Chapter 10, Section 10.1). Here, a high-level description of the mathematical process of safety risk assessment is relevant when considering how the risk assessment can account for analytical variations in compound identification and quantitation.

Greatly oversimplified, the toxicological assessment process is the mathematical comparison of patient's daily exposure to a leachable (DE) to an exposure level that is considered to be safe, such as the tolerable daily intake (TDI). A margin of safety (MoS) can be calculated as the ratio of TDI to DE:

$$\text{TDI}/\text{DE} \tag{11.1}$$

If the DE is less than or equal to the TDI (MoS ≥ 1), the compound of interest is assessed as having a negligible adverse effect on patient health and safety.

Considering the terms in Equation (11.1), DE is calculated from the level of a leachable in a drug product (CL) and the drug product's maximum daily dose volume (MDDV). Thus, the DE can be adjusted to account for the uncertainty in CL by adding a UF to the DE calculation, e.g.:

$$\text{DE (μg/day)} = [\text{CL (μg/ml)} \times \text{MDDV (ml/day)}] \div \text{UF} \tag{11.2}$$

The TDI is established based on the identity of the compound of interest, as identity is the link between the compound and its toxicological data. Thus, the TDI can be adjusted to account for uncertainty in the leachable's identity by adding a UF to the TDI calculation:

$$\text{TDI (final, μg/day)} = \text{TDI (initial, μg/day)} \times \text{UF} \tag{11.3}$$

The net result of the UF adjustment to either the DE or TDI (or both) is a smaller value for MoS and therefore a more conservative estimation of a compound's potential toxic effect.

11.3.2.3 Identifying a Compound

Considering identification at a high level, it is likely that "everybody follows the same strategic process for identifying extractables" as identifications are typically secured in one of three ways:

1. Matching compound-related information (mass spectrum, retention time) to libraries and/or databases of compiled information,
2. Interpreting analytical data (structure elucidation from an interpreted mass spectrum) by an expert,
3. Considering related information (composition, similarities between extractables).

However, it is in the tactics where individual practices differ. Considering matching, for example, individual practitioners may have different criteria as to what constitutes an acceptable match. Moreover, expert interpretation is characteristically variable, depending on the skills, training, and experience of the expert.

Thus, it can be a difficult task to standardize practices for establishing the identity of a compound that produces a chromatographic peak (presumably using mass spectrometric detection).

Alternatively, standardization can be driven by requiring that identities reported for safety assessment meet a certain standard of science. To accomplish this purpose, it is first noted that identifications have already been classified with respect to their scientific rigor; see, for example, Table 11.2. Thus, one notes that a confident identity, supported by two corroborating pieces of evidence (for example, spectral matching and structure elucidation), is more scientifically rigorous than a tentative identity, supported by a single piece of evidence (e.g. spectral matching or structure elucidation). It follows, then, that a safety assessment performed on a confidently identified compound is likely to be more correct in establishing the compound's patient safety impact than an assessment performed on a tentative identity.

As Identification Categories have already been created and are routinely reported, the challenge here is simply establishing that level of identification that is sufficiently rigorous to serve as the basis for safety assessment. To this end, it is the opinion of this author that all safety assessments should be based on compounds with a confirmed identity and, when that is not possible, on confidently identified compounds.

While it is fine to have this opinion, it is naïve to expect that confirmed or confident identities can be secured for each and every extractable (or leachable) in each and every circumstance. This means that saying "all IDs must be confirmed or confident" without allowing for exceptions ignores the practical realities of E&L analysis and thus is not an actionable recommendation. It seems clear to this author that rather than forbidding the use of lower confidence identities in a safety assessment, a more actionable approach would be to require that there be a consequence for using lower confident identities. That is, a lower confidence identity could be used as the basis for a safety assessment but only when reasonable options for obtaining a more confident ID have been exhausted. Furthermore, if a lower confidence identity is used in the safety assessment, one must adjust the DE to a lower value to account for the additional uncertainty when such lower confidence identities are used. Specifically, when one uses a partial or tentative identity, an uncertainty factor of $UF = 2$ must be added to the calculation of the DE.

Now it could be suggested that a UF should be higher for a partial identity than for a tentative identity as the partial identity is more uncertain. However, this true statement is typically addressed during the toxicological safety assessment. That is, a partially identified compound is safety assessed by either read-across to a surrogate or use of a Threshold of Toxicological Concern (TTC), for example the mutagenicity TTC per ICH M7, as the TDI. In the case of read-across, a UF is usually added to the TDI

Table 11.2 Identification categories. The shaded category is the minimum reporting requirement.

Available identifying information	Identification category (per reference [3])	Supporting information
Insufficient information to propose a specific identity	Unidentified	N/A
	Partial	One item of equivocal evidence
Sufficient information to propose a specific identity	Tentative	One item of definitive evidence
	Confident	Two corroborating items of definitive evidence
	Confirmed	Three (or more) corroborating items of definitive evidence

Table 11.3 Accuracy scale for quantitation. The shaded category is the minimum reporting requirement.

Quantitation category (per reference [3])	Accuracy range
Estimated	50% < accuracy or accuracy > 200%
Semi-quantitative	50% ≤ accuracy ≤ 200%
Quantitative	80% ≤ accuracy ≤ 120%

determination. In the case of use of a TTC as the TDI, it is noted that, by its nature, the TTC is highly conservative and thus a TDI derived from a TTC is highly conservative. In these ways, the safety assessment of a partially identified compound is already sufficiently conservative that using a larger UF factor to account for the increased uncertainty of a partial identity is, in a certain respect, akin to accounting for the same aspect twice.

11.3.2.4 Quantifying a Compound

The aspect of quantifying a leachable can be similarly addressed in terms of accounting for uncertainty in the leachable's calculated concentration in the safety assessment. To wit, quantitation categories have already been developed and are in common use (see Table 11.3) and thus the challenge becomes simply establishing accuracy requirements for each category and establishing categories that are a suitable basis for safety risk assessment.

To accomplish this purpose, Table 11.3 contains not only the quantitation categories but also accuracy ranges to fit each category. For example, a quantitative concentration is one that is generated by an approach that has an accuracy of $100 \pm 20\%$, which is reasonable for trace level analysis in often difficult test matrices. More importantly, this author suggests that a semi-quantitative determination of concentration, typified by an accuracy of "plus or minus a factor of 2" are adequate for a safety risk assessment.

As was the case with identification, it is naïve to require that all safety assessments be based on semi-quantitative data (or better) as there are likely to be circumstances where an estimated concentration is as good as one can obtain. Thus, this author recommendss that a UF = 2 be applied to the determination of the PDE if the PDE is based on an estimated concentration (see Equation (11.2)). As was the case for identity previously, use of the UF in quantitation produces a smaller MoS and thus a more conservative safety risk assessment.

It is reasonable to consider which of the quantitation methods used in E&L screening can meet the accuracy criteria, not for individual compounds but for what is likely to be a suite of compounds, each of whose concentrations must be determined. It is almost a certainty

that the use of individual standard curves (based on an authentic reference standard) for each analyte will meet the requirement for a quantitative method. It is possible that the use of analyte-specific relative response factors (also based on an authentic reference standard and collected in a database), which are essentially one-point calibration curves, will meet the requirement for a quantitative method and is highly likely to meet the requirement for a semi-quantitative measurement. It is this author's opinion that the matching of an analyte to a surrogate standard, whether by retention time or structure, will not result in a quantitative result. It is highly unlikely that use of an internal standard as a single surrogate standard will produce a quantitative result. Whether the result is semi-quantitative or estimated depends on the variation in the response across analytes and the laboratory's skill at choosing an acceptable surrogate. For example, consider the case of GC/MS. Several authors have suggested that response factor variation in GC/MS is generally "a factor of 2 either high or low" [1, 4–6], when the internal standard is chosen at the mean or median response of a response factor database. In this case, the accuracy requirement of 50–200% can be met and the method can be established to be semi-quantitative. However, improper choice of the surrogate (i.e. picking an internal standard above or below the median) will produce accuracies that are either biased high or low and thus will fall outside of the range for semi-quantitative meaning so that the method produces only an estimated result.

The situation is much less promising for LC/MS, where response factor variation is much greater than that for GC/MS [1, 5, 7]. Thus, LC/MS concentrations based on an internal standard are inherently estimates and therefore require a UF = 2 if they are used as the basis of a safety risk assessment.

11.3.2.5 Conclusion

Thus, I have proposed that a certain degree of standardization can be achieved by establishing requirements and specifications for the screening method's output (that is, the reported extractables or leachables profile). Supporting this proposition, the following performance expectations have been established:

To address a method's Detectability:

- Breadth is established by verifying that the method produces an acceptable response to all members of a selected and justified text mixture containing known extractables or leachables.
- Coverage is established by the laboratories' determination and use of the AET. The requirement with respect to the method providing acceptable coverage is that the AET should be linked to an internal standard

whose response factor is equal to the median of the response factors for a relevant population and that an AET so linked to the proper internal standard should be adjusted via a justified UF, which is generally between 2 and 5. The combination of the linkage and adjustment produces a final AET.

- Sensitivity is established by noting that the method's LoD for all compounds in the justified test mixture (see Breadth) should be less than or equal to the final AET (see Coverage).

To address a method's ability to support compound identification, the requirement is that any and all identities reported for the purpose of patient safety risk assessment must achieve the Confident level, meaning that the identity is supported by two independent but corroborating items of evidence. If tentative or partial identities are used in a toxicological risk assessment, a UF of 2 is applied to the TDI value.

To address a method's quantitative ability, the requirement for screening chromatographic methods is that all concentrations reported for the purpose of safety risk assessment must be semi-quantitative, herein defined as a method that has been demonstrated to produce acceptable accuracy (e.g. 50–200%) for all members of a selected and justified test mixture. If this criterion is not met and an estimated concentration is used to establish the PDE, a UF of 2 is to be applied.

Although it is highly desirable that standardization is achieved by meeting expectations for output, it is recognized that there may be circumstances when meeting the specifications are either not practically possible or unnecessary. Considering the practically possible, it is unrealistic and frankly unfair to require semi-quantitation and confident identities with no recourse. There will be circumstances where the level of effort required to elevate an estimated concentration to semi-quantitation and/or a tentative identification to confident is beyond reasonable expectations. Rather than leaving the unlucky sponsor in this situation where the impossible is expected, the sponsor is advised to reduce the DE or increase the TDI via application of a UF to account for uncertainty in the measurement.

Even if it is likely that an identification, quantitation, or both can be elevated with a reasonable effort, it may still be the case that the effort is unnecessary in the context of a successful risk assessment. For example, consider a situation where a tentative identity and an estimated concentration are the points of departure for a toxicological safety risk assessment, producing an MoS. If the calculated MoS of the tentatively identified, estimated concentration extractable is large, for example 100, application of UFs for both identification and quantitation would reduce the calculated MoS to 25, which for all practical purposes still establishes that the extractable in question is likely to be safe. In this situation, a definitive conclusion about a safety impact can be drawn when the UF correction is used and thus it may be easier to use the UF correction than it is to secure the better identity and concentration.

The concept of a test mixture has been mentioned several times but the test mixture has not been specified. Failing to do so is a simple matter of competence and collaboration. It is likely to be the case that the least qualified person to specify the test mixture is someone, such as myself, who has not sat in front of a chromatographic instrument as an operator for almost a decade. It is also likely that there will be some degree of professional debate as individual experts and/or individual organizations have their own opinions in terms of which compounds (and how many compounds) make up an achievable but challenging test mixture. No, establishing the test mixture is a team, not an individual, sport and I hope that this section of my book is a catalyst for the industry to embrace this challenge and get to work.

In closing, I note that standardization is the only viable solution to the issue of inter-lab and intra-lab variation in reported extractables and leachables profile. If it is to be expected that extractables and leachables profiles should be reproducible with a lab and between labs, then standardization is the unavoidable pathway to reproducibility. This pathway can either be entered at the beginning, by adopting standard methods, or at the exit by requiring standardized performance.

11.3.3 The Final Word on Setting the Uncertainty Factor, UF, for AET Adjustment

In Chapter 3, Section 3.2.3.3, values for the UF, used to adjust the AET lower to account for response variation across extractables and leachables, were considered. Based on an analysis of response factors for approximately 220 compounds, UF values of 2, 6, and 3 were noted for GC/MS, LC/MS, and the combination of GC with multiple detectors (MS, FID) and LC with multiple detectors (MS, UV, CAD) [5].

Recently, I have been able to revisit this topic using a database of RRF values provided by Nelson Labs. As a portion of a much larger database of chromatographic data for organic extractables and leachables, Nelson Labs shared a dataset of 1196 compounds whose RRFs have been addressed for three techniques commonly employed in the chromatographic screening of extracts for extractables and drug products for leachables; GC/MS, LC/MS-APCI, and LC/MS-ESI. This dataset allows us to revisit the concept of the proper value for the UF, considering the individual methods and the combination of the methods.

To begin this discussion, it is proper to revisit the terms coverage and % coverage. In the context of the AET, coverage is defined as the number of compounds that are flagged to be at or above the AET when they are present in a sample at the AET level. Clearly, compounds with a response equal to or greater than an internal standard used to define the AET will be flagged as being above the AET without the AET being adjusted by the UF. However, compounds with responses less than the internal standard that sets the AET will not be flagged as being above the AET. These compounds that are not flagged will not be reported for safety risk assessment, even though it is proper that they would be reported and safety assessed. This situation is rectified by UF adjustment of the AET to lower values; the larger the UF, the smaller the adjusted AET and the greater the number of compounds that are properly flagged as having a concentration equal to or greater than that of the internal standard.

Before considering the use of the Nelson dataset to set a value for UF, it is important to understand the dataset. The dataset is comprised of compounds whose response in three screening methods, GC/MS, LC/MS-APCI, and LC/MS-ESI, has been established. The dataset is such that all 1196 compounds produced a response in at least one of the targeted techniques, GC/MS, LC/MS-APCI, and LC/MS-ESI. Thus, the combination of the three techniques could provide 100% coverage of the dataset if the AET were set at the response of the poorest responding analyte in the combined RRF database. However, the individual techniques did not necessarily produce a response for each compound in the dataset; while a majority of the compounds (91.5%) produced a response in GC/MS, many fewer compounds produced responses in LC/MS-APCI (47.3%) and LC/MS-ESI (30.5%). This means that individually none of the methods is able to supply 100% coverage, no matter how large the UF (as the UF cannot adjust a non-response).

It is important to understand the basis of why the combination of the methods is more effective in covering the dataset than the individual methods. In order for a single method to provide coverage, that single method must produce an acceptable response for all compounds. However, in the case of multiple methods, only one of the methods must produce an acceptable response for each analyte. Considering a set of three methods, a compound is covered if the response in one method is adequate, even if the compound does not even produce a response in the other two methods.

Given the distribution of RRF values, it is possible to establish the % coverage for the individual methods as well as the combination of the methods as a function of UF. Considering the combining of the methods, it is noted that there are several ways that this can be accomplished. Here, two approaches were taken. In the first approach, the RRF that is assigned to a compound is the largest RRF for that compound by any technique; that is, each compound in the dataset is assigned the largest RRF obtained by either of the three methods. In the second approach, the RRF that is assigned to a compound is that RRF that has a value closest to 1 (that is, a situation where the analyte and the internal standard produce the same response when they are present in a sample at the same concentration).

For example, consider an analyte whose GC/MS, LC/MS-APCI, and LC/MS-ESI RRFs are 0.2, 0.9, and 1.3. The value assigned to this analyte under the maximum response rule is the LC/MS-ESI response of 1.3; however, under the "closest to 1" rule, the response by LC/MS-APCI (0.9) is assigned.

This % coverage versus UF plots for the individual and combined methods are shown in Figures 11.1 and 11.2, where Figure 11.2 is an expanded view of Figure 11.1, more effectively illustrating the lower UF region of the curves. Reflecting the concept of diminishing returns, one notes that the greatest gains in % coverage come at the lower UF

Figure 11.1 Percent coverage of the adjusted AET as a function of the value of the uncertainty factor (UF) applied to the AET; all compounds. The arrows illustrate the point of diminishing return, where further increases in UF do not significantly increase the % coverage. The individual methods do not achieve 100% coverage as there are analytes in the Nelson dataset that do not produce responses by those methods. As the UF increases, the incremental increase in % coverage gets smaller, eventually reaching a point of diminishing returns where large increases in UF produce only small increases in % coverage. (*See insert for colour representation of the figure.*)

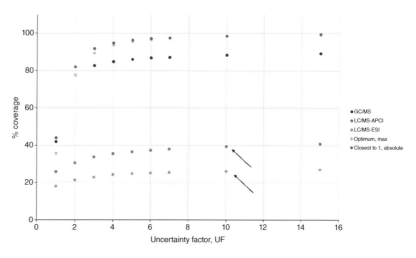

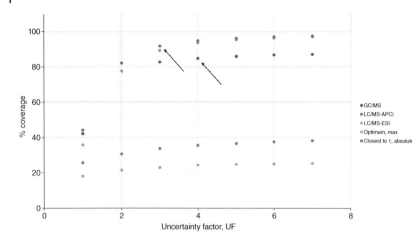

Figure 11.2 Percent coverage of the adjusted AET as a function of the value of the uncertainty factor (UF) applied to the AET; an expanded UF axis versus Figure 11.1 for all compounds. The arrows illustrate the point of diminishing return, where further increases in UF do not significantly increase the % coverage. UF values to be generally used in setting the AET are those that are at the point of diminishing returns. (*See insert for colour representation of the figure.*)

values and that eventually large increases in UF produce small gains in % coverage.

Based on the concept of diminishing returns, UF values are set at 4 and 10 for the GC and LC methods respectively, and a value of 3 for the analytical approach that uses the methods collaboratively [8]. Considering the combined approach of all three methods, a UF of 3 results in approximately 90% coverage, whereas a UF of 4 to 5 is necessary to achieve a coverage of 95% or better.

As was noted earlier, a UF of 4 for the GC/MS method and 10 for the LC/MS methods produce low % coverage values for the entire dataset as many compounds in that dataset produce no response for the individual methods. However, if one considers only those compounds that produce a detectable response, a UF = 4 for GC/MS produces a coverage of approximately 93% while a UF of 10 applied to either LC/MS approach produces a coverage of approximately 85%. The lower percentage for the LC/MS methods reflects the fact that these methods have many compounds with a low response, in addition to the numerous compounds that do not produce any response.

Lastly, it is noted that this data highlights the complementary nature of the GC/MS and LC/MS methods. That is,

the reason that the combination of the methods is able to produce significantly better coverage than GC/MS alone is that the LC/MS methods "fill in" the response gaps in the GC/MS method (and vice versa).

In a way, the previous discussion is less a discussion on managing the AET and more about managing expectations on the coverage of E&L screening in general. This is the case because the AET question is directed toward detected peaks (as in "is this detected peak above the AET?") and not to the larger question of the ability to detect peak, which was addressed in the previous discussion. Although focusing on the narrower topic of detected peaks will have no effect on the combined method statistics (as all peaks were detected in the combined methods), focusing on detected peaks will have a material effect on the statistics for the individual methods in general and the LC/MS methods in particular since they had many non-responding compounds.

The % coverage plot for the various analytical approaches using only those compounds that produce a measurable response is shown in Figure 11.3. Here we see the method's ability to effectively support the purpose of the AET, which is to distinguish between potentially meaningful and irrelevant peaks (compounds) based on their concentration

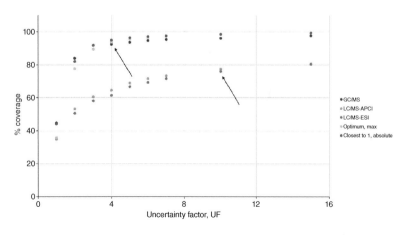

Figure 11.3 Percent coverage of the adjusted AET as a function of the value of the uncertainty factor (UF) applied to the AET; responding compounds only. The arrows illustrate the point of diminishing return, where further increases in UF do not significantly increase the % coverage. The LC/MC methods do not achieve 100% coverage as there are analytes in the Nelson dataset that do not produce small responses by those techniques. As the UF increases, the incremental increase in % coverage gets smaller, eventually reaching a point of diminishing returns where large increases in UF produce only small increases in % coverage. (*See insert for colour representation of the figure.*)

relative to a threshold. In Figure 11.3 we see that the method's ability to deal with responding compounds in the context of the AET is much improved versus the method's ability to intrinsically produce responses (comparing % coverage in Figures 11.1 and 11.3). Considering the GC/MS method alone, the % coverage is approximately 90% with a UF of 3. The LC/MS methods individually do not achieve 90% coverage (the maximum coverage with a large UF is approximately 80%), due to the many compounds that produce only very small responses by these methods.

Employing the concept of diminishing returns to Figure 11.3 results in the conclusion that a UF = 4 for GC/MS and a UF of 10 for LC/MS (both APCI and ESI) are the appropriate values for the UF that balance the % coverage with the analytical challenges of meeting an AET adjusted with the UF. For the GC/MS method, a UF of 4 provides 92% coverage of all the GC/MS-responding compounds in the Nelson dataset.

It is noted that the Nelson dataset itself is strongly influenced by poorly responding compounds due to its generous definition of what constitutes a measurable response. To wit, the lowest RRF present in the dataset is 0.001; in other words, an analyte whose response is 1/1000th the response of an internal standard. It is generally the case that such low responses, which can only be secured via analysis of a standard in which the analyte is highly concentrated versus the internal standard, are recorded as no response by other laboratories.

11.4 Information Sharing for the Common Good

Consider the following story as an introduction to the topic of information sharing in E&L. A wise man discovered the one and only true secret of the stock market, applicable to any individual and any circumstance. Realizing the value of this discovery and being an entrepreneur at heart, the wise man started a financial advising business. For each client, the wise man would create a financial plan, customized for the individual and his or her circumstances, but essentially based on the same concept and the same secret.

Clients receiving the plans were dismayed at the cost but recognized the value of what they had been given. Because the clients desired to preserve a competitive advantage over the other people playing in the market, they never revealed the secret and never discussed the analysis.

Ultimately the individual who profited the most from this scenario was the wise man, who had essentially sold the same secret to numerous individuals. This is contrasted versus the much smaller profit the wise man might have made

if he had revealed the secret in its entirety to the general public.

So, what does this have to do with E&L? Well, in E&L the secret is an extractables profile of a particular item or the leachables profile of a particular drug product or medical device. When was the last time a colleague walked up to you and said "hey, want to see my extractables profile of the rubber stopper from company XYZ?" Unless you are talking about components used in pharmaceutical manufacturing, chances are it never happened. Nobody shares extractables (or leachables) profiles because they are a big secret.

It is a misconception that the extractables profile itself is a big secret. In fact, it is likely that everyone who uses the test item (and there are likely to be a lot of users) also has performed an extraction study and thus possesses an extractables profile for the item. Thus, it is not the profile itself that is a secret; rather the secret is how many people out there actually have redundantly and repeatedly generated what arguably should be the same extractables profile for the same item. So, the big secret is not the extractables profile; rather, the big secret is that the extractables profile is not a secret. Individuals and organizations just think that the profiles are a secret and they act accordingly.

There are two reasons why this code of silence is most unfortunate. The first has to do with economics. That is, a not-insignificant portion of money spent on E&L is creating redundant information. Consider a stopper, where one user spent $50K for an extractables profile. Other users of the same stopper have their own extractables profiles for the same stopper and have the $50K expense to prove it. One could argue that the money spent in this redundant testing was not well spent.

Another issue with this redundant testing is that it propagates the perception of lab-to-lab variation. In the absence of a standard extraction protocol, it is likely that each time the stopper was tested it was extracted somewhat differently; for example, at lab 1 the study might have been performed at 70 °C for 24 hours while lab 2 might have performed the extraction at 50 °C for a week. Two labs, same stopper, different extractables profile. No wonder there is concern about lab-to-lab variation!

The second unfortunate reason has to do with setting science-based and practical standards and fully understanding the risk posed by extractables and leachables. Setting science-based standards can only be accomplished if the science is known by everybody, as setting standards is a team sport. If everybody is hording their own information, science is not known, at least in the sense that the same science is known to all the scientists. If standards setting organizations have access to only a small part of the science, and if the part of the science that they have is incomplete, then how is anyone ever going to establish a consensus, which is the first step to setting a standard? Rather than using the big picture evidence in a shared body of science to drive consensus, individuals argue about the small picture details of what did or did not happen in their particular study, often without sharing the underlying data. Rather than sharing a common foundation of all the science, the individuals within the organizations have pieces of information that may be biased somewhat, not because the data is of poor quality but rather because the data reflects a limited experience. The outcome of such a circumstance is predictable; either one ends up with no standards because there is no consensus or one ends up with unfortunate standards generated from either an incomplete or distorted understanding of the science.

To be clear and to reiterate a point made at the beginning of this chapter, no one is playing the blame game here. No one individual, organization, agency, consortium, and/or association is responsible for the status quo. In fact, we are all to blame, present company included, because, as a community of practice, we settled for second-class science and an inefficient business model.

11.4.1   Reduction of Redundant Testing

I was surprised to see that the word redundant has several meanings. According to the Merriam-Webster dictionary, redundant has the following meanings:

- Exceeding what is necessary or normal: superfluous
- Characterized by or containing an excess
- Characterized by similarity or repetition

When I refer to redundant testing, I am speaking to the latter point, where redundant implies unnecessary replication. Applied to E&L testing, redundant testing occurs when two (or more) organizations, both interested in using the same item, independently generate the item's extractables profile. In this case, the first time the profile is generated is significant, while all other times the profile is generated are redundant. Similarly, redundant testing occurs when two drug products, packaged in the same container closure system, and with similar compositions and shelf-lives, are both required to be profiled (screened and targeted) for leachables. In this case, the generation of the leachables profile for the first drug product is essential, while generation of the leachables profile for the second (and third and fourth) drug product is redundant.

The "everybody has to have their own extractables profile" business model is not a fundamental law of nature that cannot be broken; in fact, we have discussed previously a noteworthy exception, components used in pharmaceutical manufacturing. However, because this business model is the status quo, there is significant momentum that thwarts its being replaced with another paradigm. Reinforcing the momentum is the absence of two change-enabling circumstances, the carrot and the stick.

First the carrot. Redundancy in extractables profiling persists in large part as extractables profiling is perceived as being user-centric. The thinking here is that since it is the user's responsibility to possess extractables data it must be the user's responsibility to generate the data. Since the users typically do not collaborate (for example, "you do the study for the stopper, I'll do the study for the vial and then we will share the results"), they must perform redundant testing.

Furthermore, redundancy in extractables and leachables profiles persists, at least in part, due to the perception that "the profiles are not really redundant because they represent different circumstances." Considering the use of the same part by two users, for example, user 1 could argue that "I need a profile different from user 2 because my conditions of use are different." To me this is not an adequate justification for two extraction studies but rather is the justification for one study that is able to address both uses.

It is obvious that there are two solutions to the issue of redundant, user-driven testing. The first solution is vendor testing. That is, the vendor is responsible for performing the testing and is obligated to share the test results with all users. Historically, this solution has enjoyed widespread user support but has secured lesser support from vendors.

There are a number of credible reasons why vendors might resist vendor testing as being the solution to the redundancy problem and there is nothing fundamentally improper in the vendor's reluctance to accept this responsibility. Many of the credible reasons are beyond the capabilities of this author to address. However, one issue is the lack of a standardized extraction protocol. That is, while a vendor might agree to doing one standardized extractables study on their item and share this information with all users, it is understandable that the vendor would be reluctant to perform a custom extraction study for every potential user of the item. So, the carrot here is the development and adoption of an efficient standard extraction protocol. Presented with an efficient standardized testing process, the vendors' reluctance to perform testing (and share test results) might be mitigated somewhat.

Therefore, it up to the community of practice to come up with and agree to an efficient standard extraction protocol and to offer this protocol to the users as a carrot. Furthermore, it is necessary that the users adopt the approach of "One test, one time and we, as users, will never bother you again (until you make a change)."

The standard extraction protocol solves the problem of the "how do we test?" but opens up (perhaps) the question "what do we test?" Here one has to exhibit a sense of shared responsibility and fair play. Surely it is logical that the vendor's responsibility is fulfilled if they test the item as they provide it. Furthermore, it is logical that the user would perform any testing that addresses any action the users perform on the item prior to their using the item. For example, if a stopper is provided irradiated, then the vendor should test the irradiated stopper. However, if the stopper is provided unsterile and is irradiated by the user, then surely it is the vendor's responsibility to test the non-irradiated stopper and the user's responsibility to account for the effect of irradiation.

Therefore, one makes the vendor's job of supplying useful extractables profiles to all users easier by providing the vendors with an efficient standard extraction protocol. That is the carrot. So now for the stick. The stick is creating the circumstance where nothing can happen until and unless the vendor supplies users with the extractables profile generated by the standard extraction protocol.

Two versions of the stick are readily envisaged. Version 1 involves the power of economics and in this case it is the users who hold the stick. Version 1 is realized when the users join together and declare "we will not buy any item that is not supported by a study performed via the standard extraction protocol." Arguably, the rapid adoption of the BioPhorum extractables protocol for manufacturing components is an example of this successful use of this stick. Version 2 involves the power of regulation and in this case it is the regulators who hold the stick. Version 2 is realized when the regulators say "we will not approve any product that does not have an extractables profile that was generated by application of the standard extraction protocol."

If one supplies an effective means of compliance (the carrot) and then provides the proper incentive for compliance (the stick), then surely compliance is assured.

An alternative to "get the vendor to do it" is information sharing. That is, one user has generated an extractables profile for a stopper. Another user has generated an extractables profile for a vial. Somehow the two users find out that they intend to use the same stopper and vial. Information sharing is accomplished when the users agree that "I'll give you my stopper study if you give me your vial study" as opposed to both users performing both studies. In another manifestation of the information sharing scenario, a third party becomes the clearinghouse for extractables information. Perhaps there is an organization whose members are willing to contribute extraction studies (ideally performed via a standard extraction protocol) to the organization. Contributed extraction studies are then made available to other members for use. Perhaps the organization secures funding and actually initiates extraction studies itself. Then the results of those studies are available to all the organization's members.

Realizing such a desirous situation requires only that a place for housing the shared information be established and an organization willing to perform and manage the housing be identified (and funded). Everything else is just details, right? How hard can that be?

Of course, the answer to the somewhat facetious question is "pretty hard" or the situation would have been realized already and I would not be open to the accusation of "merely blowing off some steam."

11.4.2 Elimination of Unnecessary Testing

Extractables and leachables testing serves a purpose; that is, extractables and leachables data is used to establish that the item that has been tested is safe for human clinical use. It follows, then, that extractables and leachables testing is unnecessary if it is already known and accepted that the item is safe.

It is a fundamental toxicological tenet that every chemical is potentially unsafe, depending upon exposure. Is it not then equally valid that certain chemicals can be declared to be safe when used in certain clinical applications that limit clinical exposure to safe levels? Moreover, is it not also valid that certain items, known to contain safe chemicals, can themselves be declared to be safe for use in certain applications?

For example, let us consider the case of a plastic that contains additive X. Let us further suppose that the plastic containing additive X is commonly used in pharmaceutical packaging, meaning it is likely that the plastic, and the packaging system containing it, have been extensively profiled to establish the extraction and leaching behavior of additive X. Furthermore, if the packaging is used in currently marketed drug products, then it must be the case that the extractables and/or leachables data has been toxicologically safety assessed and additive X has been established to be safe for use. Finally, if the packaging is used for more than one drug product and/or by more than one sponsor, then it is likely that the conclusion that additive X is safe has been reached time and time again.

It seems reasonable that under these circumstances, additive X and the plastic that uses it could be accepted and designated as GRAS (generally recognized as safe), sparing current and future users the considerable burden of performing unnecessary testing whose outcome is as certain as the rising sun, death, and taxes.

If we only had a mechanism, the mandate, and the conviction to make this happen! Although some traction has been gained in this area via the concept of a "medical grade plastic" [9], regulatory acceptance of such a designation in lieu of testing is uncommon.

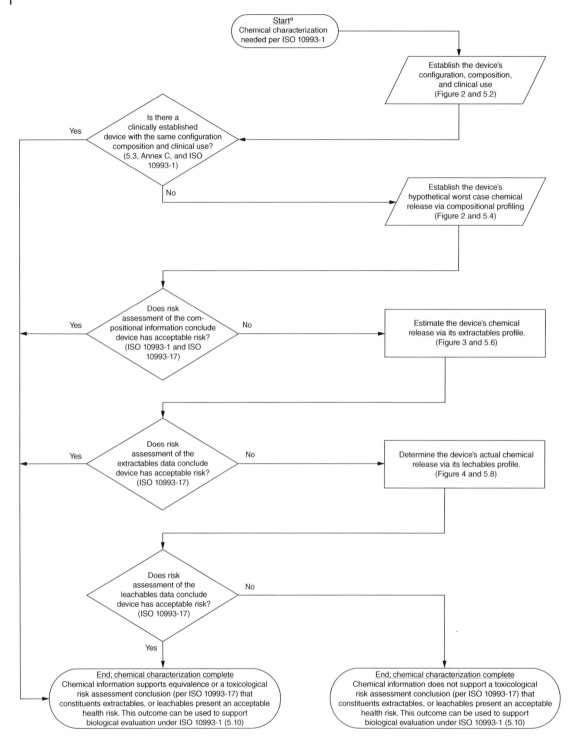

Figure 11.4 General chemical characterization process per Figure 1 of ISO 10993:18(2020).

In the scenario just described, it was the entire extractables study itself that was unnecessary. Another possibility is that certain aspects of the study are unnecessary because the data resulting from those aspects serve no useful purpose, a circumstance that I will refer to as over-testing.

Perhaps the most obvious example of this over-testing is reflected in the difference between risk management and risk avoidance. In risk management, one accepts the possibility that there is some level of risk that is acceptable and a process is managed to achieve the acceptable level of risk.

In risk avoidance, one rejects the possibility that there is a level of acceptable risk (i.e. that no risk is acceptable) and a process is managed so that no risk is ever present. Surely, we can all agree that when there are no consequences and when it can be achieved, risk avoidance is a noble goal and desirable outcome. However, the pursuit of perfection is often the anthesis of progress.

At the risk of seeming unfair by singling out (or picking on) medical devices, device extractables testing will be used as an example of risk avoidance versus mitigation in E&L. If one wants to know all the substances that could possibly be extracted from the medical device by any solvent (making the leap that extraction by a solvent mimics the interaction of the medical device and the human body) and the maximum amount that could be extracted by those solvents, one performs an exhaustive multi-step extraction of the device with solvents of multiple chemical properties (e.g. polar, semi-polar, and non-polar). The resulting extractables profile, listing all possible extractables and establishing their total amount in the device, surely can be used to establish the absolute worst-case total patient exposure to devise leachables (ignoring for a moment kinetics). If the toxicological assessment of the extractables data is based on the patient exposure assumption that the patient is exposed to these extractables on the first day of the device's use, then surely this is the worst-case toxicological assessment.

The combination of worst-case chemistry and worst-case toxicology reflects the risk avoidance strategy. If, after the worst-case extraction study and the worst-case toxicological assessment, it can be concluded that the device possesses a negligible risk of an adverse health effect, then this conclusion is valid for any way the device might be used, and all safety risks have been avoided.

For medical devices whose conditions of use reflect the worst-case situation described above (i.e. the medical device is fully leached in its first day of use) and for all cases where the toxicological safety assessment does in fact conclude "no risk," then the "worst-case chemistry + worst-case toxicology" approach has accomplished its objective and everyone is happy. It is only when this approach produces a negative outcome (that is, one or more extractables are potentially unsafe) that one understands that applying the worst-case combination of extraction plus toxicology to all medical devices, or to medical devices where the underlying assumption of "it all comes out in one day," is clearly not the case and is not sound risk management but rather the poster child example of generating (and assessing) unnecessary and irrelevant information.

The ISO standard for the chemical characterization of medical device, ISO 10993:18(2020) [10], emphasizes this point in its systematic approach to the chemical characterization of medical devices, illustrated in its

Figure 1 – General chemical characterization process (reproduced here as Figure 11.4) and described in Section 5 of the standard. Figure 11.4 clearly positions worst-case (compositional profiling), estimated case (extractables profiling), and actual case (leachables profiling) as appropriate parts of the chemical characterization process, while the text of Section 5 clearly advocates the use for the flow chart, coupled with a profound understanding of the nature of the device's contact with the human body during its clinical use, in order to establish that testing is most likely to reflect the clinical situation and advises investigators to perform the most relevant testing. This approach is manifested clearly in Table 2 of the standard (reproduced here as Table 11.4), which advises investigators about the proper use of various extraction strategies.

I want to close this section with a confession. Every time I look at Table 2 from ISO 10993:18(2020), and especially the footnotes, I feel an overwhelming sense of doom and failure as the approach of exhaustive extraction with three solvents of varying polarity for every medical device represents the ultimate triumph of empiricism versus science, irrationality versus logic, and avoidance versus management.

Table 11.4 Recommended extraction conditions for medical devices per Table 2 of ISO 10993:18(2020).

Contact category	Recommended extraction conditions	Credible alternatives
Limited contact devices	Simulated use conditions[a]	Exaggerated conditions
Prolonged contact devices	Exhaustive conditions	Exaggerated conditions[b],[c]
Long-term contact devices	Exhaustive conditions	Exaggerated conditions[b],[c],[d]

a) Note that some legal authorities (e.g. US FDA) can request exaggerated extraction, unless otherwise justified.

b) Examples of instances where exhaustive extraction would not typically be required include:
   - Single-use devices used for less than 24 hours, where repeat use of a new device each day would result in categorization as prolonged or long-term contact;
   - Single-use devices used for several days, where repeat use of new devices would result in categorization as prolonged or long-term contact;
   - Reusable devices, where a patient may be exposed to repeated use of the same device, resulting in categorization as prolonged or long-term contact; when an exaggerated extraction is used for a reusable device, the extraction should properly account for the duration of each individual use.

c) Exaggerated conditions can be appropriate for external communicating or non-absorbable surface contact devices, with justification.

d) An example is a device comprised entirely of non-absorbable metal (e.g. a vascular stent), because migration of constituents from within the material is not possible and the constituents of interest are related to the surface only and exaggerated extraction can be adequate to generate a complete extractables profile.

11.5 Replacing the Lab with a Computer (Mathematical Modeling)

The promise and challenges in using mathematical modeling as a means of numerically (as opposed to experimentally) generating extractables or leachables data has been discussed previously, for example, in Chapter 2, Section 2.2.5. The mathematics of extraction or leaching, both of which are diffusive processes, are well-known and well-established; however, necessary input data for the relevant equations, such as a compound's plastic/liquid partition coefficient ($K_{p/l}$) or coefficients, diffusion coefficient (D), total pool in the test article (m_T), and the slope and intercept values for the van Hoff equation (necessary for temperature projections), are often unavailable and/or difficult to procure. Moreover, while the mathematics are straightforward for a simple test article (e.g. a uniformly shaped item consisting of a single homogeneous material), they become much more formidable for the compositionally and dimensionally more complex test articles that are packaging components, manufacturing components, and medical devices.

Complicating factors notwithstanding, when the necessary input information is either available or obtained, mathematical calculations effectively reproduce experimental extraction profiles. For example, as noted in Figure 11.5, researchers from Sartorius Stedim Biotech GmbH are mathematically able to mimic the effect of temperature, duration, extraction solvent, and extraction stoichiometry (surface area to solution volume ratio) on the extractables profile of a multi-layer film used in pharmaceutical manufacturing components such as bioreactors and storage bags [11]. Additionally, researchers

at the FDA are making progress in terms of developing mathematical models that mimic the leaching of chemical substances from medical devices [12].

Lastly, several research teams have recently published texts that provide the sound fundamental science which enables the mathematical modeling of leaching processes [13–16].

11.6 Consider the Impact of a Result Before Attempting to Improve It

I am not sure why it should be this way but E&L is practiced as if it should be a one-pass activity. That is to say, you do one extraction, you do one set of testing (and at one time it was that you did one test), and you process the data once and somehow this is supposed to produce all the answers with great certainty and high confidence, where every relevant entity is discovered, and is adequately identified and quantified. Notice the use of the word adequately as opposed to other works such as accurately or correctly, as its use is intentional, not accidental.

In many cases, such a first pass may not produce "gold standard" data for every analyte. For example, if a detected compound does not produce a hit in a response factor–MS–retention time database, then this compound will not have a quantitative concentration and confirmed ID in the first pass, in which case one is faced with the dilemma of possibly not having regulatorily acceptable data. Faced with this circumstance, one takes one or two actions to improve the (perceived) quality of the data. One possible action is to improve the process. For example, instead of quantitation with an internal standard, analytes are matched to surrogates by retention time, structure, or some other characteristic. The second possible action is to do a second pass, where the second pass focuses on doing a better job for the compounds that were not in the database.

If one gets too serious about demands such as "all reported concentrations must be quantitative" and "all identities must be confirmed," they are missing the point that E&L is not all about quantitation and identification but rather is all about the interpretation of concentration and identity. One cannot establish, based on a leachable's measured concentration and reported identity whether it is safe; rather, this leachable's information needs to be processed, evaluated, and interpreted before a safety determination can be made.

It is a possible, then, that one can accomplish the objective of safety assessment to everyone's satisfaction without having a concentration generated by a five-point calibration curve based on an authentic reference standard(s) or that one does not have to synthesize and purify an exotic leachable just so that a confident identity becomes

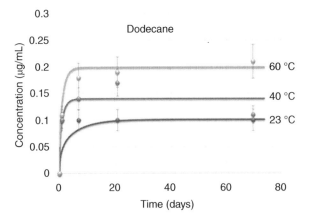

Figure 11.5 Extraction of dodecane from a multi-layered polyolefin film into ethanol as a function of storage time and temperature. The circle datapoints reflect actual experimental data (with ±15% error bars) while the lines reflect the mathematically-derived extraction profiles. The mathematical profiles provide an excellent fit to the experimental data. From reference [12] with permission.

a confirmed one. If the conclusion of safe-for-use based on a semi-quantitative concentration (or, heaven forbid, a concentration estimate) is sufficiently compelling, then it is possible that the greater precision and certainty of assessment that one would get with a quantitative concentration is unnecessary, as it would produce no different outcome.

What is being proposed here is that there is actually a step between the first and second pass, which is to preliminarily assess the possible safety effect of the compound based on first pass information. If such an assessment produces an overwhelmingly definitive positive outcome (that is, the compound is indisputably safe), then it is proposed that there is no need to get better data and the second pass is eschewed. Only if the outcome of the preliminary assessment is inconclusive or negative (the compound is likely to be unsafe), is the second pass analysis warranted.

Previously, the mathematics of toxicological safety risk assessment was discussed (Chapter 10, Section 10.1.3), simplifying the assessment by focusing solely on the MoS. If one accepts the MoS as an adequate reflection of safety, then it can be proposed that a preliminary MoS can be the guide to determining whether a second pass is required or not. For example, let us presume that a compound discovered by GC/MS is above the AET and is not in an organization's internal E&L database. Thus, the organization resorted to an internal standard to generate a concentration estimate and to a spectral match to an external database (e.g. NIST/Wiley) to obtain a tentative identity. Before the organization invests whatever resources are necessary to secure a more certain concentration and identity, it is proposed that one should first calculate the MoS based on the information that is available and then act according to the MoS result. For instance, consider the situation of a definitively positive MoS, say MoS = 50, which supports the conclusion that the compound is safe. However, someone might suggest that this outcome is bogus because it was based on an estimated concentration and a tentative identity, following up on the previous instruction to "go back and do better." Well, I would submit to you that so doing has no impact on the outcome of the assessment and thus that "going back and do better" serves no useful purpose. For example, since the concentration was based on GC/MS one could suggest, quite conservatively, that its concentration could have been underestimated by a full factor of 10. Although this is unlikely, as an uncertainty of a factor of 4 is more appropriate for GC/MS, consider the effect that this would have on the calculated MoS. If one presumes that the MoS was based on a concentration an order of magnitude too low, then the MoS based on the actual concentration would be 5. As an MoS of 5 is arguably interpreted as "safe to use," then securing better data has no impact on the outcome. This would suggest, then, that there is no compelling reason to pursue the better data.

On the other hand, if the initial calculation of the MoS produces a value of 5 and one suspects that the concentration on which the MoS is based might be underestimated by a factor of 10, then this analysis of the MoS confirms that for this analyte it is unavoidably necessary to go back and produce better data at a second pass.

In essence, the proposal here is not to reject "low quality" data as being inadequate for safety assessment until a preliminary safety assessment has confirmed the need for higher quality data. A significant amount of time and effort can be spent (perhaps wasted is a better verb) securing higher quality data that in the end has no material effect on the outcome of the impact (safety) assessment.

11.7 Recent Publications

As I am sure I mentioned previously, it is next to impossible to prepare a book that captures all the current developments in E&L as key events will certainly occur between the time the book is closed to new content and the book is read. However, I will use this section of Chapter 11 to capture, to the extent possible, recent noteworthy E&L-related publications. At this late stage in the book generation process, it is not possible to place these manuscripts in their appropriate chapters within the book.

It is obvious that extractables and leachables are, or are derived from, substances that are present in the plastic materials of construction. Thus, information sources that address such substances and speak to their safe use are quite valuable to E&L assessment. In this regard, two important manuscripts have recently been published. Groh et al. have compiled a database of Chemicals associated with Plastic Packaging (CPPdb), which includes chemicals used during manufacturing and/or present in final packaging articles [17]. The CPPdb lists 906 chemicals likely to be associated with plastic packaging and 3377 substances that are possibly associated. Of the 906 chemicals likely to be associated with plastic packaging, 63 rank highest for human health hazards and 68 for environmental hazards according to the harmonized hazard classifications assigned by the European Chemicals Agency within the Classification, Labeling, and Packaging (CLP) regulation implementing the United Nations' Globally Harmonized System (GHS). Further, 7 of the 906 substances are classified in the European Union as persistent, bio-accumulative, and toxic. In considering the most hazardous chemicals likely to be associated with plastic packaging, these authors note the following [17]:

- Chemicals used as monomers and intermediates in plastics and substances likely to be used as solvents were among the most hazardous plastic packaging-associated substances identified.

- When judged based on monomers, some polymers used in packaging, including PS, PVC, PC, and PU (the latter used often in adhesives), are regarded as highly hazardous, while polyolefins and PLA are considered to be of lower hazard.
- The majority of plastic packaging-associated substances identified as the most hazardous for environmental and human health were in fact plastic additives representing diverse chemical groups used for a variety of functions.
- A particularly prominent group of hazardous additives consisted of substances containing metals, including cadmium, chromium, lead, mercury, cobalt, tin, and zinc.
- Surfactants and their degradation products form another large group of highly hazardous substances likely to be associated with plastic packaging.
- Substances possessing endocrine disrupting, PBT (persistent, bio-accumulative. and toxic), or vPvB (very persistent, very bio-accumulative) properties are also considered hazardous and therefore are of immediate concern.
- Other groups of plastic packaging-associated chemicals (of concern) include preservatives, UV stabilizers, and antioxidants.

A key aspect of this publication is its supplementary files, which contain the accumulated data for the compounds addressed by these authors, including:

- Supplementary file 1. A list of all categories extracted from the US EPA's CPCat database for this study.
- Supplementary file 2. Hazard grade scores assigned to different CLP classifications in this study.
- Supplementary file 3. Snapshot of the database of Chemicals associated with Plastic Packaging (CPPdb, Lists A and B) at the time of manuscript submission.
- Supplementary file 4. The lists of the most hazardous chemicals likely to be associated with plastic packaging (that is, from the CPPdb_ListA) identified in this study.

Along similar lines, and expanding the focus to include plastics in general (and not just plastic packaging), researchers from the Institute of Environmental Engineering, ETH Zürich, systematically investigated plastic monomers, additives, and processing aids on the global market based on a review of 63 industrial, scientific, and regulatory data sources [18]. In total, these researchers identified more than 10 000 relevant substances and categorized them based on substance types, use patterns, and hazard classifications wherever possible. In so doing, over 2400 substances were identified as substances of potential concern as they met one or more of the persistence, bioaccumulation, and toxicity criteria in the European Union.

Key findings of this research group included:

- Overall, 55% of the substances identified are categorized as plastics additives, 39% as processing aids, and 24% as monomers, with significant overlaps between these three categories. Due to a lack of information, 30% of the substances were uncategorizable regarding their functions.
- A total of 2263 compounds were reported for use in the following applications with high exposure potential: 2109 in food-contact applications, 522 in toys, and 247 in medical items including masks.
- Reported hazard classifications were available for about 6400 substances (61%), whereas about 4100 substances (39%) lacked any reported hazard classifications in the considered regulatory databases.
- Of the 6400 compounds that could be classified, 3950 substances (37%) did not meet any of the considered hazard criteria and therefore were designated as substances of a low level of concern. Another 2486 substances (24%) met one or more of the hazard criteria considered and are identified as substances of potential concern; among them, 1254 substances (12%) are also high production volume chemicals (HPVCs) and thus of a high level of concern. The remaining 1232 substances (12%) were classified as having a medium level of concern.

As was the case with the previously discussed reference, a key aspect of this publications is its supplementary information, listed in Table 11.5, which contains the accumulated data for the compounds addressed by these authors.

Two recent publications deal with identifying and justifying simulated extraction solvents. Based on Abraham general solvation models, and considering that water saturated n-butanol (called wet n-butanol) is chemically corresponding to blood, Jianwei Li concluded that "ethanol/water (60/40, V/V) and ethanol/water (50/50, V/V) … can be used as blood simulating solvents in chemical characterization study of medical device" [19]. Strobel et al addressed the leaching properties of polysorbate 80-containing solutions, as this surfactant is a common solubility enhancer in drug product formulations [20]. Using a linear solvation energy relationship (LSER) approach, these authors studied the partitioning of neutral chemicals from polysorbate 80 (PS 80) micelles to water. Predicted partition coefficients were converted to a concentration-dependent solubilization strength of aqueous PS 80 solutions. This solubilization strength represents a key parameter to project equilibrium levels of leaching from pharmaceutical plastic materials. For example, Figure 11.6 illustrates the effect of increasing the polysorbate 80 content on the solubility of three compounds on varying polarity.

Table 11.5 Contents of supporting information supplements to the *Wiesinger Deep Dive into Plastic Monomers, Additives and Processing Aids* manuscript.

Sheet	Overview	Detailed description
S1	Plastic-relevant sources	Description, information content, data preparation, and basic overview over plastics-relevant sources collected for this study.
S2	Sources for further information	Description, information content, data preparation, and basic overview over sources containing further information used for this study.
S3	Keywords for polymer categorization	Categories, keywords, regular expression implementation, and reliability of keywords used to identify the polymer type of a substance.
S4	Keywords for function categorization	Categories, keywords, regular expression implementation, and reliability of keywords used to identify the function of a substance.
S5	Keywords for industrial sector categorization	Categories, keywords, regular expression implementation, and reliability of keywords used to identify the industrial sector of use of a substance.
S6	Keywords for substance type categorization	Regex-Implementation of keywords used to identify substance types.
S7	Manual error checks for preliminary categorizations	Manual error checks of a random selection of preliminarily assigned uses (including compatibly polymer type, function, and industrial sector of use). Based on the identified errors, keywords in Sheet S3-S5 were improved.
S8	Substances potentially used as plastic monomers, additives, and processing aids (Plastic MAP)	Substances potentially used as plastic monomers, additives, and processing aids. Chemical identifiers, substance type information, confidence assignment, use-related information, and tonnage are shown for each substance.
S9	Substances of potential concern (SopC) used in plastics	Substances of potential concern used as plastic monomers, additives, or processing aids. Chemical identifiers, substance type information, confidence assignment, hazard classification, regulatory information, use-related information, and tonnage are shown for each substance.

Figure 11.6 Effect of polysorbate 80 content on the water solubility of organic compounds of varying polarity. Source: Data from Strobel et al. [20].

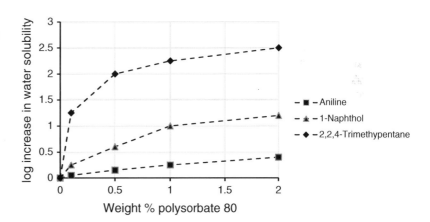

Several authors published the results of extraction and/or leaching studies. Researchers at Amgen continued their "holistic" evaluation of extractables from manufacturing components and container closure systems used with biotechnology products, publishing data on pre-filled syringes [21] and stoppered vials [22]. Considering pre-filled syringes, the "holistic" approach was successful in the qualification of five syringe systems, made from different materials, including glass and plastics, for use with 12 different biologic drug products of various modalities and formulations. Considering the stoppered vials, a total of 12 drug products filled in different size glass vials capped with laminated and non-laminated rubber stoppers made from three different rubber formulations were investigated for extractables and leachables.

As Amgen's "holistic" approach has been applied in multiple situations, it is instructive to review the approach's details. Extractions are accomplished with multiple extraction solvents (see Table 11.6) and the extractions are performed at a temperature and for a duration that is projected to accelerate a one-year shelf-life at 5 °C, 70 °C for 4 days, but not degrade the test articles. Container-closure systems are tested intact (not as components) by filling to some fraction of their nominal volume (20–100%). When

Table 11.6 Model extraction solvents used in the Amgen "holistic" approach.

Solvent description
Water
20% Acetonitrile, 20% ethanol, 60% high-pressure liquid chromatography water
20% Acetonitrile, 20% ethanol, 60% 10 mM ammonium formate, pH 3.0
20% Acetonitrile, 20% ethanol, 60% 10 mM ammonium formate, pH 7.0
20% Acetonitrile, 20% ethanol, 60% 10 mM ammonium formate, pH 9.0
20% Acetonitrile, 20% ethanol, 60% 1 mM sodium phosphate monobasic, 1 mM NaCl, pH 3.0
20% Acetonitrile, 20% ethanol, 60% 1 mM sodium phosphate monobasic, 1 mM NaCl, pH 9.0

orientation is important (e.g. stoppered vials), filled test articles are extracted for two days each in an upright and inverted configuration.

Screening of method-compatible extracts for organic and inorganic extractables is accomplished by a suite of analytical methods including:

- GC/MS
- Headspace GC/MS
- LC/MS/MS
- LC/UV
- LC/ELSD
- ICP/MS

Additional analytical details can be found in the cited manuscripts.

The resulting extractables profiles were toxicologically assessed to establish target leachables. For organic extractables, literature searches were conducted for toxicology data and supplemented with *in silico* analyses as necessary, using Toxtree software. For data-rich compounds, Permitted Daily Exposure (PDE) values were derived following the principles and methods of ICH Q3C(R7), with an additional modifying factor to account for bioavailability. For data-poor compounds, their structures were analyzed using Toxtree for alerts for mutagenicity and systemic toxicity. The ICH M7 TTC was used to screen for DNA reactive (mutagenic) potential. Extracted elements were assessed versus relevant ICH Q3D PDEs, where available, and versus derived PDEs when the element was not addressed in Q3D.

Leachables testing was performed in the specified drug products over shelf-life. Leachables testing included both target analysis using validated methods and screening. Leachables were toxicologically assessed versus concentration limits derived from PDE values.

The extractables and leachables profiles thus established are tabulated in the respective cited manuscripts. It is noteworthy that both studies conclude that profiles of detectable leachables are predicted by the extractables profiles, as the leachable profiles were a subset of the extractable profiles [21, 22].

Moving beyond their holistic approach, Amgen researchers have documented the circumstance of leachables in a drug product that are derived from a non-direct component of its container/closure–delivery system [23]. Specifically, these researchers performed E&L assessments on a battery-powered delivery device for storing and delivering therapeutic protein products. During the development and qualification of the device, these authors reported that a novel compound 2-hydroxy-2-methylpropiophenone (HMPP) and related compounds were observed in both extractables and leachables studies and were ultimately ascribed to the non-product contact device batteries, in which an HMPP photo-initiator was used as a curing agent in the battery sealant to prevent leakage of the battery electrolytes. To resolve this circumstance, seven battery candidates from different suppliers were screened and one new battery was successfully qualified for the delivery device.

Gamma irradiation is an established sterilization technology applicable to packaging and manufacturing components and medical devices that may be restricted in the future by the availability of ^{60}Co as the irradiation source and irradiation capacities. X-ray technology is considered an alternative type of radiation for sterilizing so-called single-use (SU) biotechnology product manufacturing equipment. Prior to its widespread adoption, however, extractables profiles of the materials subjected to X-ray irradiation need to be compared to profiles established for gamma irradiated materials to estimate the patient safety and product quality risks associated with this new technology. Researchers at Sartorius Stedim Biotech GmbH addressed this need by generating extractables profiles for materials sterilized by these two techniques [24]. Extraction studies were performed on two articles, the first of which were bags manufactured from a multi-layered film that was specifically developed for biopharmaceutical applications and used in bioreactors, storage and shipping bags, or mixing bags. These bags have a structure of polyethylene/ethylene vinyl alcohol/polyethylene (PE/EVOH/PE) and whose additive package (of the polyolefin PE layer) includes common plastic additives such as antioxidants and releasing agents. The second test article was the Eastman Tritan™ MX731 copolyester (PCTG), consisting of three monomers: terephthalic acid, which is the reactant in the dimethyl terephthalate polymerization process, and the two diols, 1,4-cyclohexanedimethanol and 2,2,4,4-tetramethyl-1,3-cyclobutanediol. Although this test article contains no additional stabilizers, it does contain processing aids such as releasing agents, e.g. stearates or palmitates.

Extractions were performed at 40 °C for either 21 or 70 days using water, ethanol, 1 M NaOH (not used with the PCTG material due to compatibility issues), and 1 N HCl. The resulting extracts were characterized for organic and elemental extractables using a typical suite of screening analytical methodologies including HS-GC/MS, GC/MS, LC/UV, LC/MS, IC, ICP, as well as general test methods such as pH, TOC, and conductivity, as appropriate.

These authors report the full extractables profile obtained for both test articles. Ultimately, these researchers established that the extractables profiles of the tested polymers were not affected by the type of irradiation used for sterilization.

Researchers from Lonza AG and Goethe University Frankfurt have reported the leaching of 4-hydroxynonenal (HNE) from administration sets used in the delivery of biopharmaceutical drug products in potentially toxic quantities, linking leachables to the sets of PVC components [25]. These researchers note that clinical administration sets should, like manufacturing materials and container closure systems, be in the focus of routine leachables studies and that manufacturers of clinical administration sets are responsible for avoiding safety-concerning chemicals such as HNE.

D. Budde (Sartorius Stedim Biotech GmbH) and associates have published several manuscripts describing the effect of extractables or leachables on biological systems or biopharmaceutical drug products. In one recent study, the effect of a set of leachable model compounds (LMCs) on a Chinese hamster ovary DG44 cell line producing an IgG antibody was investigated considering cell growth, cell cycle distribution, and productivity [26]. LMCs studied as a group included bis(2,4-di-tert-butylphenyl) phosphate (bDtBPP), a known radiolysis product of the antioxidant Irgafos 168, butylated hydroxytoluene (BHT), a commonly used antioxidant, acetophenone, a plastics manufacturing residual, and tris(2-ethylhexyl) trimellitate (TOTM), a

plasticizer. Based on their investigation, these researchers concluded that although they observed growth-inhibiting effects of the LMC mix in both the dose–response experiments and a fed-batch run, the cell specific productivity was not impaired. Furthermore, since worst-case approaches were applied for defining the LMC concentrations used in this study, the authors anticipated that under normal process conditions the impact of leachables on CHO cell cultures would be even lower. The authors suggested that especially during a cultivation carried out in perfusion mode, the impact of leachables can be expected to be negligible due to lower leachable concentrations, as the cultivation medium is exchanged several times during a process. Since no effect on cell cycle, apoptosis, or cell viability was observed, long-term effects of LMCs on CHO cell cultivation were not anticipated by the authors and they expect that the impact of the LMCs will likely decrease over time with growing cell numbers. The authors report that the inhibiting effect of a LMC mix on a CHO cell culture can be lower than the expected cumulative effect of the individual LMCs and that the inhibiting effect is greatly reduced in the combination of BHT and TOTM.

In a second manuscript, Budde and Jurkiewicz studied the effect of single-use (SU) component-related leachables on cell and gene therapy using a CAR-T process [27]. These authors identified SU tubing materials as having the highest negative effect on process productivity and reported two cases of growth inhibition:

- A TPE tubing caused reduced cell growth of a virus producing HEK293T/17 SF cell line, even at relatively low S/V, potentially linked to the presence of bDtBPP.
- Cell growth issues for T-cells were detected with PVC extracts.

There has been much discussion on the circumstance of response factor variation across organic extractables and leachables, and the impact of that variation on the reporting and accurate quantitation of these substances. Considering the reporting of extractables/leachables, the response factor variation is important as it complicates the application of the AET, specifically causing analytes with low response factors to be incorrectly identified as having a concentration below the AET level when in fact their concentration in a tested sample is above the AET. This error is critical, as it means that poorly responding, potentially unsafe compounds are not being reported for toxicological safety risk assessment. When this error occurs, the AET becomes less protective in the sense that it no longer serves its purpose of protecting patients from hazardous extractables or leachables.

To ensure protection, an uncertainty factor can be used to adjust the AET downward, thereby addressing response variation. However, Jordi and Heise note that in some cases the response variation is so great and the resulting uncertainty factor so large that the revised AET is below the achievable sensitivity specifications of even state-of-the-art, expertly operated instrumental technologies [28]. These authors advocate an analytical strategy based on methods with multiple complementary and orthogonal detection techniques (a multi-detector approach) to mitigate the problem of response factor variation and thus eliminate the need for large uncertainty factors, noting the complementary nature of the multiple techniques and the circumstance that all analytes need only to be detected by at least one of the combination of detectors for the AET to be protective.

Using a database of response factors for over 200 known and commonly encountered extractables or leachables, these authors suggest that an uncertainty factor of 3, applied to an analytical approach based on GC/MS-FID and LC/UV-MS-CAD, will ensure that 97% of the compounds in the database will be properly flagged as being at or above the AET when the compounds are present in a sample at the AET level.

11.8 USP Biological Reactivity Testing

Biological Reactivity or Biocompatibility testing, as specified in ISO and USP documents, was addressed previously in Chapter 10, Section 10.3. As the focus of this book is chemical characterization and the interpretation of chemical data to infer safe use, the previous discussion of this topic was brief and largely high-level. A further, somewhat detailed discussion of the USP biological reactivity chapters, USP <87>, <88>, and <1031> is warranted as these chapters are undergoing major revision.

Chemical characterization and biological reactivity testing are similar as they serve the same purpose, establishing the safe use of packaging systems, manufacturing systems, and medical devices. A further similarity is that the test methods rely on generating extracts of the test item in question and then testing of the resulting extracts. However, the two approaches differ considerably in terms of the testing of the extracts and inferring patient safety impacts from the test results. In chemical characterization, the extracted substances are identified and quantified and the safety impact of resulting extractables is inferred by toxicological assessment of the existing data. In biological reactivity testing, the extracts are not chemically characterized but rather are applied to biological systems, with the effect of the application of the extract on the test system being noted. Presumably an adverse impact on the test system would translate into the test article having an adverse effect on patient safety.

To a very real extent, chemical characterization followed by toxicological safety risk assessment and biological reactivity tests are complementary means of assessing safe use. Furthermore, to a certain extent, the two approaches can be considered to be confirmatory and interchangeable when they both can address the same aspect of safe use. For example, if the mutagenic properties of extractables are established via safety risk assessment, it is reasonable that the outcome of the assessment would be the same as if mutagenicity were established biologically.

It is this complementary and/or confirmatory aspect of biological reactivity testing that makes it an appropriate topic for consideration in a book on chemical characterization. Furthermore, a discussion of biological reactivity testing is appropriate in this book as the USP approach to biological reactivity testing is evolving.

Considering medical devices, biocompatibility testing of medical devices per ISO 10993 was discussed previously in Chapter 10, Section 10.3. There it was noted that ISO 10993:1 provided the framework for testing requirements while other chapters in the 10993 series provided test methods. The same approach is used by the USP, where chapter <1031> establishes the testing requirements for pharmaceutical packaging and manufacturing components and chapters <87> and <88> provide the test methods.

In 2005, the USP commissioned an Expert Panel to revise these biological reactivity chapters. Expected outcomes for the revisions included:

- A reduction in the amount of redundant testing of existing plastic and elastomeric materials,
- The elimination of unnecessary animal testing for new materials,
- A refinement of the type of testing performed to align with the potential risk,
- The replacement of outdated tests with new tests,
- The alignment of USP tests and specifications with relevant ISO 10993 standards, where possible.

At the time this book is being written, USP chapters <87>, <88>, and <1031> have already undergone two cycles of revision and the currently revised version has been published in the *Pharmacopeial Forum* (*PF*), Volume 47(4), triggering additional stakeholder review and commenting.

Considering changes to tests and testing methodology, chapter <87>, *in vitro* testing, was changed to replace obsolete tests with alternate, more relevant tests. Specifically, considering Cytotoxicity Tests, the existing Agar Diffusion Test was replaced by the Neutral Red Uptake (NRU) Test, aligning <87> with ISO 10993:5(2009), Tests for *In Vitro* Cytotoxicity. Considering Genotoxicity Tests, four genotoxicity tests were added to <87>, aligning <87> with ISO 10993:3(2014), Tests for Genotoxicity. These added tests included:

- Bacterial Reverse Mutation (Ames) Assay
- *In Vitro* Mammalian Cell Chromosomal Aberration Test
- *In Vitro* Mammalian Cell Gene Mutation Test
    o Human lymphoblastoid TK6 cells
    o Mouse lymphoma L5178Y cells
- *In Vitro* Mammalian Cell Micronucleus (MNvit) Test

Furthermore, Chapter <88>, *in vivo* testing, was revised to:

- Replace the Plastic Classification system with a single class referred to as Pharmaceutical Grade Polymeric Materials
- Delete the Implantation Test requirement for packaging/delivery systems
- Delete the Biological Safety Test
- Add Sensitization Tests

In its current form, <87> classifies packaging and manufacturing components into six categories (I through VI) based on the required testing. Presumably the classification, reflecting the risk of a test item being incompatible with living systems, was a guide to appropriate testing with greater risk (and higher-class designation) being linked to greater and more rigorous testing. However, practical application of <88> has established that the sorting of plastic materials into six categories no longer serves the current purpose and that in practice only Class VI is currently used by vendors and end users. Thus, the existing Class VI Classification has been replaced by the term Pharmaceutical Grade Polymeric Materials. Classification of a material as a Pharmaceutical Grade requires the same test procedures as those currently linked to Class VI classification of plastic and elastomeric materials for packaging/delivery systems.

Removing the Implantation test from <88> is obvious as packaging and manufacturing system materials are not implanted into patients under the material's conditions of use. The Safety Test – Biologicals, currently in <88>, was removed because the associated FDA regulation, FDA Code of Federal Regulations, Section 610.11, was revoked on August 3, 2015. Relevant Sensitization Tests from USP <1184>, including the

- Magnusson & Kligman Guinea Pig Maximization Test (GPMT)

- Standard Buehler Test (SBT)
- Local Lymph Node Test (LLNT)

were moved into <88> to align it with ISO 10993:10(2121), Tests for skin sensitization, and allow USP <1884> to be deleted.

USP informational chapter <1081> was significantly modified to serve the new purpose of supporting and explaining the revised USP chapters <87> and <88>. Cosmetically, the title of this chapter was changed to *The Biocompatibility of Pharmaceutical Packaging/Delivery Systems and Their Materials of Construction*. Contextually, the scope of the chapter was changed so that it is specifically applicable to plastic materials of construction and plastic and elastomeric components for pharmaceutical packaging/delivery systems and packaging of combination products, and is no longer applicable to medical devices and implants. Significant changes to the content of <1031> include:

- The addition of sections covering:
    o Overview of Biocompatibility Evaluation
    o Pharmaceutical Grade Polymeric Materials
    o Regulatory Expectations
    o Risk-Based Approach to Biocompatibility Evaluation
    o Biological Reactivity Test Considerations
    o Investigating a Biological Reactivity Test Failure
    o Chemical Assessment
    o Overall Biocompatibility Evaluation
    o Glossary
    o References
- The development of a Risk-Based Approach to Biocompatibility Evaluation, which is summarized in seven steps, including:
    o Step 1. Gather Information
    o Step 2. Risk Analysis
    o Step 3. Informational Gap Analysis and Testing
    o Step 4. Risk Evaluation
    o Step 5. Risk Control
    o Step 6. Document Risk Evaluation Results
    o Step 7. Periodic Monitoring/Lifecycle Management
- The generation of a Biological Reactivity Test Selection section, which addresses:
    o Test Article Selection and Sample Preparation Considerations
    o *In Vitro* Test Selection
    o *In Vivo* Test Selection
    o Development of Acceptance Criteria for *In Vivo* and *In Vitro* Tests.

Considering this last point, for example, Table 11.7, appearing in the current revision of <1081>, establishes *In Vivo* and *In Vitro* Test Acceptance Criteria.

Table 11.7 *In Vitro* and *In Vivo* Test Acceptance Criteria per the revised USP <1031>.

Route of administration	Cytotoxicity	Irritation	Sensitization	Genotoxicity (*in vitro*)	Acute systemic toxicity
Dermal	Mild reactivity (Grade 2)	Non- to slight irritant	Non-sensitizer	Non-mutagenic	Meets systemic injection test criteria in <88>
Parenteral	Mild reactivity (Grade 2)	Non- to slight irritant	Non-sensitizer	Non-mutagenic	Meets systemic injection test criteria in <88>
Inhalation	Mild reactivity (Grade 2)	Non-irritant	Non-sensitizer	Non-mutagenic	Meets systemic injection test criteria in <88>
Ophthalmic	No reactivity (Grade 0)	Non-irritant	Non-sensitizer	Non-mutagenic	Meets systemic injection test criteria in <88>

11.9 I'll Pass on That

Previously I have spent considerable text discussing response factor variation and its ramifications on both the AET and the quantitation of extractables and leachables. I have also described first pass data processing as the first, and not necessarily the last, step in the analysis of E&L data. The purpose of this section is to note that the use of the UF and the adjusted AET is neither appropriate nor necessary for a class of leachables, specifically those leachables whose identities are confirmed and whose concentrations are accurately established via reference standard information collected in an analytical database. Furthermore, it will be established that expanding such an analytical database to include toxicological information, such as a compound's Tolerable Intake (TI), makes the issue of "is it above or below the AET" a moot point.

Considering this first point, I note that chromatographic data obtained by applying screening analytical methods to drug products for the purpose of leachables screening can initially be processed (i.e. "first pass" processing) by one of two approaches:

1. Establishing whether the chromatographic peaks are above or below the AET (largely ignoring the identity of the compounds responsible for the peaks).
2. Establishing the identity of the compounds that are responsible for the peaks (largely ignoring whether they are above or below the AET).

At first glance, the first approach seems to offer greater efficiency, as it potentially eliminates the necessity to secure identities for peaks that are deemed to be below the AET, which is especially valuable when those identities are difficult to procure. However, inherent in the first approach is the determination and use of the UF to establish the adjusted AET, with all its inherent issues.

Considering the second approach, such an exercise can reveal leachables whose identities are confirmed and whose

concentrations are accurately established. Confirmed identities and accurate concentrations are obtained when the "first-pass" processing is supported by a database of collected identifying information (e.g. retention times and mass spectra) and quantifying information (e.g. response factors). In essence, such a database is generated via the analysis of authentic reference standards and subsequent recording of the relevant information. The properties of chromatographic peaks obtained during leachables screening are compared to the properties of compounds present in the database and if there is a match, the compound is identified to a confirmed level and quantified to at least a semi-quantitative level.

Leachables whose concentrations have been semi-quantitatively established in this manner can be compared to the AET with virtually 100% confidence that the comparison is valid. Adjustment of the AET with a UF, and establishing the proper value for the UF, is unnecessary.

At second glance, and considering the previous point, the second approach is more efficient, as it insures an AET assessment that is 100% accurate (at least for those leachables present in the database).

It is clear that the relative merits of approaches 1 and 2 depends on the proportion of leachables that can be confidentially identified. If no leachables can be confidently identified, then approach 2 has no value and approach 1 must be pursued. If all leachables can be confidently identified, then approach 2 alone accomplishes the objective of the AET comparison and approach 1 is not necessary.

It is likely that the most commonly encountered circumstance is that the leachables discovered by screening will fall into both identity groups (confident and not confident), and thus a hybrid approach to AET adjustment and interpretation is proposed as follows:

1. Establish those peaks in the chromatogram whose compounds can be identified to a confirmed level. For these compounds, establish the compound's concentration in the drug product via the compound's response factor

and compare that concentration to the AET to establish whether the compound is present in the drug product at a level above or below the AET.

2. For those peaks in the chromatogram whose compounds cannot be confidently identified, calculate a UF, apply the UF to adjust the AET, and compare the peak's response to the adjusted AET to establish whether the compound is present in the drug product at a level above or below the adjusted AET.

In either case 1 or 2, compounds established to be present in the drug product at levels that are at or above the AET undergo a formal toxicological safety risk assessment.

It is noted in passing that such a hybrid approach is efficient as (1) obtaining confirmed IDs (and accurate quantitations) is no more complicated than comparing experimental chromatographic data with tabulated reference standard data contained in the database (a process that can be easily automated) and (2) it produces a higher proportion of AET comparisons that are certain to be correct.

To this point, the proposed approach is enabled by an analytical database. It is an interesting exercise in what perhaps could be termed "wishful thinking" to consider the possibility of a combined analytical and toxicology database. For example, what if a Tolerable Intake (TI) were to be established for each of the compounds in the database by application of rigorous and appropriate toxicological processes of assessment? If that were the case, then the issue of "is the compound above the AET?" is largely irrelevant because the available information is sufficient to address the bigger question, which is "is the compound safe?" To wit, one envisions the following "first-pass" data evaluation process:

1. Establish those peaks in the chromatogram whose compounds can be identified to a confirmed level. For these compounds, establish the compound's concentration in the drug product via the compound's response factor.
2. For confidently identified compounds, use the compound's concentration in the drug product, together with drug product dosing information, to establish the patient's Daily Exposure (DE).
3. Compare the DE to the compound's TI (collated in the database) and calculate the Margin of Safety as the ratio of the TI to the DE (i.e. MoS = TI/DE).
4. Establish the safe use implications of the compound via the MoS.
5. For those peaks in the chromatogram whose compounds cannot be confidently identified, calculate a UF, apply the UF to adjust the AET and compare the peak's response to the adjusted AET to establish whether the compound is present in the drug product at a level above or below the adjusted AET.

6. For those compounds from step 5 that are at or above the AET, perform a rigorous toxicological safety risk assessment.

Now there are those individuals (likely toxicologists) who might suggest that reducing toxicological safety risk assessment to the calculations and interpretation of an MoS (as suggested in step 3) is a gross oversimplification of the art and science of a toxicological safety risk assessment. To those individuals I would offer the following rebuttal:

1. This author readily acknowledges the art and science that goes into rigorously establishing a proper value for a TI,
2. The author readily acknowledges the art and science that goes into the interpretation of so-called borderline MoS values (that is, MoS values that are close to 1, meaning that the MoS itself may not be definitive),
3. The author notes that although toxicological safety risk assessment cannot be trivialized into a mathematical exercise, at some point the assessment process is facilitated and completed by the application of mathematics.

In closing, a "first-pass" strategy, enabled by a database of collected analytical and toxicological data, is the most efficient and effective means of ensuring that necessary leachables are definitively assessed for patient safety impact. This author challenges E&L practitioners, both analytical and toxicological, to accelerate their generation, collection, and collation of relevant data (such as retention time, MS spectrum, response factor and TI) for reference compounds and to use such collated data to make leachables assessment more efficient and more accurate.

11.10 The End of E&L?

It is certain that fruition of some of the items noted previously in this chapter would have a profound impact on E&L in the future. Standardization of extraction protocols and standardization of screening analytical testing methods will change screening studies from "investigational studies … (that are) developed and implemented on a lab-by-lab basis using the individual expertise and philosophies of the scientists at each organization" [1] to more-or-less commodity testing performed per universally accepted and applied protocols, specifications, and standard operating procedures. Widespread implementation of the standardized methods, which will include exacting time-of-use method qualification (system suitability testing) will mitigate lab-to-lab variation in E&L reported profiles. Reductions in redundant testing, secured by either information sharing or GRAS designation of substances, materials, components, systems, and/or devices, would certainly lead to a contraction in the testing capacity required to support drug product and medical device development, registration, and commercialization. Acceptance of migration modeling as a viable alternative to laboratory testing would further contract the necessary testing capacity.

The development and continued population of informational databases, based on the analysis of authentic reference materials or standards, will result in fewer compounds going unreported, in more accurate reported concentrations, and in more correct reported identities for a greater number of the most commonly encountered extractables and leachables. Moreover, well-populated databases will provide the data from which critical generalizations can be developed, for example, by establishing the proper value of the UF or defining proper surrogate standards for semi-quantitation of compounds for which analytical standards are not available.

Furthermore, if one adopts the strategy of "eliminating extractables studies when one can screen the drug product for leachables down in the AET level (or lower)," then the logical consequence will be fewer extractables studies being performed.

For these and many other reasons, one expects that in the future there will be fewer extractables and leachables studies that will be performed and that those studies that are performed will leverage standardized approaches to be more efficient, effective, and definitive. Furthermore, the studies that are preformed will be necessitated by true product innovation to provide necessary and meaningful suitability for use information and not be required to support either the status quo or incremental existing product extensions to placate a "check the box" qualification mentality. Although leaching studies will be preferred (as they reflect the reality of the product and its suitability for use), there will remain circumstances where simulation by extraction is necessitated by practical considerations. Furthermore, extraction studies will still be appropriate for purposes other than leachables estimation and suitability for use evaluation (for example, control of incoming materials).

As long as there is a need for chemical information to be the basis of suitability for use assessment, E&L will remain a necessary part of pharmaceutical drug product and medical device development, registration, and commercialization. However, in order to remain a credible source of such chemical information, or for chemical information to remain as a proper foundation for suitability for use assessment, E&L in the future must be practiced strategically and tactically in a very different way than it is practiced today.

References

1 Zdravkovic, S. (2021). https://doi.org/10.5731/pdajpst .2020.012624, 2021). A commentary on several study design considerations pertaining to the screening of substances extracted and/or leached from pharmaceutical contact materials. *PDA J. Pharm. Sci. Technol.* 76 (6): 536–552.

2 Singh, G., Lu, D., Liu, C., and Hower, D. (2021). Analytical challenges and recent advances in the identification and quantitation of extractables and leachables in pharmaceutical and medical products. *Trends Anal. Chem.* 141: 116286.

3 Jenke, D. (2020). Identification and quantitation classifications for extractables and leachables. *PDA J. Pharm. Sci. Technol.* 74 (2): 275–285.

4 Jenke, D. and Odufu, A. (2012). Utilization of internal standard response factors to estimate the concentration of organic compounds leached from pharmaceutical packaging systems and application of such estimated concentrations to safety assessment. *J. Chromatogr. Sci.* 50: 206–212.

5 Jordi, M.A., Rowland, K., Liu, W. et al. (2020). Reducing relative response factor variation using a multidetector system for extractables and leachables (E&L) analysis to mitigate the need for uncertainty factors. *J. Pharm. Biomed. Anal.* 186: 1–14.

6 Mullis, J.O., Granger, A., Quin, C., Norwood, D.L. The analytical evaluation threshold (AET) concept, sensitivity and analytical uncertainty. In *Conference Proceedings, Leachables and Extractables.* Smithers Rapra, Dublin, Ireland, March, 2008.

7 Jenke, D. and Liu, N. (2016). Chromatographic considerations in the standardization of liquid chromatographic methods used for extractables screening. *J. Liq. Chromatogr. Relat. Technol.* 39 (13): 613–619.

8 Jenke, D., Christiaens, P., Beusen, J.M. et al. (2021). A practical derivation of the uncertainty factor applied to adjust the extractables/leachables Analytical Evaluation Threshold (AET) for response factor variation. *PDA J. Pharm. Sci. Technol.*, pdajpst.2021.012692. https://doi .org/10.5731/pdajpst.2021.012692.

9 Medical Grade Plastics (MGP). VDI 2017. Engl. VDI-Gesellschaft Materials Engineering. 2019-07. 10. ISO 10993:18(2020). Chemical characterization of medical device materials within a risk management

process. International Organization for Standardization; 2020-01.

10 ISO 10993:18(2020). Chemical characterization of medical device materials within a risk management process. International Organization for Standardization; 2020-01.

11 Hauk, A., Pahl, I., Dorey, S., and Menzel, R. (2021). Using extractables data from single-use components for extrapolation to process equipment-related leachables: The toolbox and justifications. *Eur. J. Pharm. Sci.* 163: 105841.

12 Turner, P., Elder, R.M., Nahan, K. et al. (2020). Leveraging extraction testing to predict patient exposure to polymeric medical device leachables using physics-based models. *Toxicol. Sci.* 178 (1): 201–211.

13 Cartledge, B., Carmichael, J., Haag, N. et al. (2020). Predicting extractables and leachables from container stopper data form migration kinetics studies can be used to develop models that predict the levels of extractables and leachables at different temperatures and timepoints. *BioPharm Int.* 33(8): 40–44.

14 Zhu, T., Chen, W., Jafvert, C.T. et al. (2021). Development of novel experimental and modeled low density polyethylene (LDPE)-water partition coefficients for a range of hydrophobic organic compounds. *Environ Pollut.* 2021 Dec 15;291:118223. doi: 10.1016/j.envpol.2021.118223. Epub 2021 Sep 22.

15 Egert, T. and Langowski, H.C. (2022). Linear solvation energy relationships (LSERs) for accurate prediction of partition coefficients between low density polyethylene and water – Part I: Experimental partition coefficients and model calibration. *Eur. J. Pharm. Sci.* 172, 106137. https://doi.org/10.1016/j.ejps.2022.106137.

16 Egert, T. and Langowski, H.C. (2022). Linear solvation energy relationships (LSERs) for accurate prediction of partition coefficients between low density polyethylene and water – Part II: Model evaluation and benchmarking. *Eur. J. Pharm. Sci.* 172, 106138 https://doi.org/10.1016/j.ejps.2022.106138.

17 Groh, K.J., Backhaus, T., Carney-Almroth, B. et al. (2019). Overview of known plastic packaging-associated chemicals and their hazards. *Sci. Total Environ.* 651: 3253–3268.

18 Wiesinger, H., Wang, Z., and Hellweg, S. (2021). Deep dive into plastic monomers, additives, and processing aids. *Environ. Sci. Technol.* 55: 9339–9351.

19 Li, J. (2021). Evaluation of blood simulating solvents in extractables and leachables testing for chemical characterization of medical devices based on Abraham general solvation model. *J. Mol. Liq.* https://doi.org/10.1016/j.molliq.2021.1116995.

20 Strobel, A.B., Egert, T., and Langguth, P. (2021). Predicting leachables solubilization in Polysorbate 80 solutions by a linear solvation energy relationship (LSER). *Pharm. Res.* https://doi.org/10.1007/s11095-021-03096-8.

21 Ronk, M., Liu, J., Gallegos, A. et al. (2020). Holistic extractables and leachables program: evaluations of pre-filled syringe systems for biotechnology products. *PDA J. Pharm. Sci. Technol.* 74 (6): 627–643.

22 Qi, L., Liu, J., Ronk, M. et al. (2021). A holistic approach of extractables and leachables assessment of rubber stoppered glass vial systems for biotechnology products. *J. Pharm. Sci.* 110: 3580–3593.

23 Liu, J., Ronk, M., Luo, Z. et al. (2021). Observation and mitigation of leachables from non-product contact materials in electromechanical delivery devices for biotechnology products. *J. Pharm. Sci.* https://doi.org/10.1016/j.xphs.2021.08.007.

24 Menzel, R., Dorey, S., Maier, T. et al. (2021). X-ray sterilization of biopharmaceutical manufacturing equipment – Extractables profile of a film material and copolyester Tritan™ compared to gamma irradiation. *Biotechnol. Prog.* e3214. https://doi.org/10.1002/btpr.3214.

25 Schroder, A., Mahler, H.C., Sayed, N.B. et al. (2021). 4-Hydroxynonenal – A toxic leachable from clinically used administration sets. *J. Pharm. Sci.* 110 (9): 3268–3275. https://doi.org/10.1016/j.xphs.2021.05.014.

26 Budde, D., Albano, G.L., Noll, T., and Jurkiewicz, E. (2021). Interaction of leachable model compounds and their impact on Chinese hamster ovary cell cultivation. *Biotechnol. Prog.* 37: e3150. https://doi.org/10.1002/btpr.3150.

27 Budde, D. and Jurkiewicz, E. (2021). Risk analysis of leachables in cell and gene therapy using a CAR-T model process. *Int. J. Pharm.* 607 (25): 121015. https://doi.org/10.1016/j.ijpharm.2021.121015.

28 Jordi, M. and Heise, T. (2021). An analytical strategy based on multiple complementary and orthogonal chromatographic and detection methods (multidetector approach) to effectively manage the analytical evaluation threshold (AET). *PDA J. Pharm. Sci. Technol.* 75 (3): 289–301. https://doi.org/10.5731/pdajpst.2020.012047.

A

Principles for Identifying Organic Extractables and Leachables

A.1 Identification Classes, Processes, and Practices

A.1.1 Introduction

It is generally accepted good practice that organic extractables and leachables are discovered by complementary and orthogonal chromatographic techniques whose separation method provides the necessary selectivity and resolution and whose detection method provides the necessary information from which an identity can be inferred and a concentration can be estimated. Mass spectrometry is the chosen method for identification, as the mass spectral properties of the discovered extractables or leachables provide multiple avenues for identification. Mass spectrometry is also suitable for quantitation, with certain caveats. Thus, the typical orthogonal and complementary chromatographic techniques used in the screening for organic extractables or leachables are Headspace GC/MS (volatile organic compounds), GC/MS (semi-volatile organic compounds), and LC/MS (non-volatile organic compounds).

Securing a substance's accurate concentration and correct identity is essential in establishing the substance's potential impact as a leachable. This Appendix focus on identification, specifically addressing the process by which mass spectral data and other supporting evidences are used to secure, judge, and justify complete and correct identities for extractables or leachables.

A.1.2 Identification

Once a chromatogram is obtained by analyzing a drug product (for leachables) or an extract (for extractables), the chromatogram is reviewed to establish those chromatographic peaks that (i) are absent from the associated blank (and thus are legitimate analytes) and (ii) whose responses are greater than the response produced by a standard compound present in a sample at the level equal to an established threshold (e.g. the Analytical Evaluation Threshold, AET, for patient safety assessment).

Once a peak has been established as having a response greater than the AET, the compound responsible for the peak must be identified and quantified, enabling the extractable's potential patient impact assessment. Typically, a mass spectrum, secured for each peak in the chromatogram, is the basis for identification.

By definition, a compound has been identified if it can be assigned a proper chemical name, an appropriate identifying number (for example, a CAS registry number), and a chemical structure with an acceptably high degree of confidence. A compound can be described (or partially identified) if it can be assigned to a class of compounds that share a common functional characteristic. For example, an analyte can be identified as a diethyl phthalate or can be described as "a phthalate."

A meaningful, rigorous, proper, and accurate risk assessment of an extractable or a leachable can only be obtained when the extractable (or leachable) has been correctly identified with a high degree of confidence. If a compound remains unidentified, if the compound has been mis-identified, or if there is little confidence that the proposed identity is correct, then the impact assessment, if it can even be completed, will likely be flawed as:

a. An unidentified compound's impact cannot be established,
b. A mis-identified compound's assessment will be erroneous (as it is based on the properties of the wrong compound), and
c. A low-confidence proposed identity is likely to be incorrect – triggering an erroneous assessment.

An identification error is a fatal error as an error in identification can only be fixed by securing the correct identity. Such a fatal error unavoidably undermines the overall impact assessment of a medical device or pharmaceutical container/closure component or system.

The likelihood that an inferred impact is, in fact, the actual impact depends, in large part, on the certainty of the information upon which the impact assessment is based. If there is great uncertainty (and therefore lack of confidence)

Extractables and Leachables: Characterization of Drug Products, Packaging, Manufacturing and Delivery Systems, and Medical Devices, First Edition. Dennis Jenke.
© 2022 John Wiley & Sons, Inc. Published 2022 by John Wiley & Sons, Inc.

in the identity of a leachable, then there is corresponding great uncertainty (and therefore lack of confidence) in the subsequent assessment. It is for this reason that:

- Identities must be secured with the highest possible certainty.
- Identities must be reported with an indication of their uncertainty.

A.1.3 Identification Processes

"Identification" is not a direct product or outcome of screening as the test methods used for screening do not produce the identity as a direct outcome of the test; rather, identification involves processing and analysis of generated test data. Various means for processing and analyzing mass spectral and supporting data to secure a compound's identity are discussed as follows.

A.1.3.1 Mass Spectral Matching

Once a peak's mass spectrum has been obtained, identification of the compound that produced the mass spectrum can be proceed via a process termed mass spectral matching. Mass spectral matching is based on the premise that a library of reference mass spectra for relevant organic compounds exists and involves the comparison of library spectra with the experimental mass spectrum of the compound of interest. Matches between library spectra and the experimental mass spectrum of interest are ranked in terms of the degree of agreement. The better the agreement, the higher the ranking and the greater the likelihood that the compound responsible for the library spectrum is also responsible for the test spectrum.

The following are crucial questions to consider when using mass spectral matching with an external library as the sole means of identifying a compound:

- How comprehensive and relevant is the library?
- What constitutes a good and acceptable similarity or match between an experimental and a reference mass spectrum?
- If several "good" matches exist, which match is the "best" and thus establishes the compound's identity?

Answers to these questions, and the concepts behind them, are contained in Section A.2.

A.1.3.2 Manual Mass Spectral Interpretation – Structural Elucidation

"Manual" mass spectral interpretation (structure elucidation) can be pursued to secure the compound's identity. In structural elucidation, a qualified mass spectroscopist uses established principles of mass spectrometry to interpret the mass spectrum, translating its key features into the compound's structural features. The types of information that could assist a qualified mass spectrometrist in securing an identification include (but are not limited to):

- The molecular formula (e.g. accurate mass measurements),
- The presence of specific elements (isotopic data),
- Substructural evidence (fragment interpretation),
- "De novo" structural elucidation or a provable mass spectrum similar to a compound that is identified as confirmed, confident, or tentative.

As postulating a chemical structure solely based upon mass spectral information and interpretation is not an easy task, the mass spectrometrist should be aware of the consequences of postulating an identity, as the identity is the basis for a subsequent toxicological evaluation of the compound. Cases where the wrong identity for the compound is elucidated will inevitably jeopardize the overall safety evaluation of the material, device, or container/closure component or system.

Mass spectral interpretation (MSI) is considered in greater detail in Section A.3.

A.1.3.3 Additional Evidences – Securing the Identity with the Highest Confidence

Additional evidences that can support a matched or elucidated identity include (but are not limited to):

- A provable similarity between the identified compound and a second compound with a confirmed identity,
- A provable relationship between the identified compound and the known composition of the test article,
- A good match between the identified compound's retention index and a database of retention indices, and
- Interpretation of a compound's accurate mass spectrum or MS/MS interpretation.

Section A.4 will further elaborate on the various identification evidences that may assist in establishing and augmenting the level of identification of the compound of interest.

A.1.4 Practical Considerations in Identification

It is reasonable to consider what level of identification should be achieved to support a rigorous toxicological safety assessment. Although an impact assessment can be based on a tentative identification, as the tentative identification provides the essential and necessary information (name, structure, and identifier such as the CAS number), if there is a lack of confidence in the tentative identity, then there will also be a lack of confidence in the certainty of the impact assessment. While a lower level of confidence might be acceptable when the outcome of the impact assessment is an emphatic "no impact," such a low level of confidence might not be acceptable in cases where the potential impact is "too close to tell." When the impact assessment produces a "too close to tell" outcome, the higher the level of confidence in the identification, the more willing the impact assessor will be to accept the outcome of the assessment.

For example, consider the case of a toxicological safety risk assessment where the assessor is faced with the challenge of establishing the adverse effect that a leachable might have on a patient's health. An assessment measure that is commonly used in such a toxicological safety risk assessment is the margin of safety (MoS), where the MoS is the ratio of a leachable's permissible daily exposure (PDE), which is linked to the leachable's identity, versus its total daily intake (TDI), which is based on the leachable's concentration and the drug product's prescribed dosing. When the PDE > TDI (MoS > 1), the patient intake is less than the maximum permitted exposure and the leachable poses a negligible risk of having an adverse patient safety effect. When the PDE < TDI (MoS < 1), the patient intake is greater than the maximum permitted exposure and the leachable could potentially adversely affect patient safety.

In the case of a large MoS (MoS > 10 as an example of an assessment whose outcome is an emphatic "no safety risk exists"), a PDE based on a tentative identification might be acceptable as there is a sufficiently large "margin of error." That is, the tentative identity is likely to be correct enough that even if the identity is wrong in terms of the specific compound, it is likely right in terms of the compound class. Having a "margin of error" of an order of magnitude should provide a sufficient "safety cushion" that the tentative identity, even if it is not completely correct, is adequately protective of a patient's safety.

Alternatively, consider the case where the resultant MoS = 2, where the "margin of error" is not adequate and it is obvious that basing the PDE on a tentative identity is not adequately protective of the patient. If the tentative identity turns out to be incorrect, it is possible that the correctly identified compound would have a PDE that is sufficiently lower than the PDE based on the incorrect identity that patient safety could be compromised.

It is for this reason that it is generally accepted that a rigorous and acceptable toxicological safety risk assessment should be based on either a confident or, preferably, a confirmed identity.

A.2 Identification via Mass Spectral Matching

A.2.1 Introduction to Mass Spectral Matching

The process of mass spectral matching is self-explanatory. Screening chromatographic analysis has produced chromatographic peaks associated with extractables or leachables, each of which has an associated mass spectrum. This test spectrum can be compared to a compiled library of reference mass spectra and an acceptable feature-by-feature match between test and reference spectra would suggest that the analyte that produced the test spectrum and the reference compound that produced the reference spectrum are one and the same.

The primary outcome of library matching is a list of "hits," those library spectra that have some degree of similarity to the test spectrum. In most commercially available libraries, "hits" are further delineated by providing critical information about the compound that is the "hit," such as names, chemical structures, CAS numbers, and possibility other information. Each "hit" is typically accompanied with a numerical value (e.g. match factors, probability scores, etc.) that is obtained by an established matching algorithm and that indicates the degree to which the hit's mass spectrum corresponds to the test spectrum. The simplest interpretation of the "hit list" is that the hit with the highest match score provides the identity of the compound of interest. Even if the top hit is not taken as the compound's identity, it is generally "assumed" that the proper identity for the compound of interest is among the highest ranked "hits."

Although the concept of spectral matching is intuitive and straightforward, it is not without its challenges. It is evident that matching can only be successful if the mass spectrum of the compound of interest is present in the library.

If the compound's mass spectrum is present in the library, the best-case outcome of matching is that the compound with the highest "hit score" will be the compound of interest. However, there is absolutely no guarantee that this will always be the case in all circumstances as the link between the library and the compound of interest depends upon the selected mass spectral match criteria and the similarity between the analytical conditions used to produce the test and reference spectra, among other factors.

Spectral matching is a well-established strategy that has long been applied in the reporting and identification of detected (non-target) analytes in environmental analysis [1]. Relying solely on mass spectral matching to secure an identity, however, is an inappropriate interpretation practice as more incorrect results are probably reported because of the sole reliance on mass spectral library search programs than have been reported due to all the other types of errors that can occur in mass spectral data [2]. A high correlation (*high match index or probability of match*) between an unknown spectrum and a library spectrum does not necessarily mean that the unknown has been identified unequivocally. Additionally, the match with the highest match score is not always the correct identity. As there are no universally applied and accepted mass spectral match "quality criteria" that unequivocally establish that a matched identity is, in fact, the true identity of a compound, it is recommended "good practice" to have all match-based identities reviewed by a mass spectrometrist to substantiate the tentative identity, regardless of the mass spectral match quality. The poorer the mass spectral match between the mass spectrum of the compound to be identified and the reference mass spectrum, the greater is the need for a visual inspection and MSI by a mass spectrometrist to substantiate the matched identity. Additionally, the identification hypothesis should be corroborated by additional information such as retention behavior, etc. [3].

A.2.2 Relevance of External Mass Spectral Libraries for Identification

An ideal mass spectral library for identification would:

1. Contain as many of the potential extractables and leachables as possible (increasing the likelihood of securing a match),
2. Contain no substances that are not extractables or leachables (decreasing the possibility of an incorrect match or false positive),
3. Be well controlled and maintained,
4. Be constantly and routinely updated (but in a controlled manner),
5. Be peer-reviewed,
6. Contain spectra that are secured under standardized analytical conditions,
7. Have scientifically validated search algorithms and scientifically vetted means of establishing the match factor,
8. Have a user-friendly output that supports review, interpretation, and assessment,
9. Be compatible with all data platforms used in all generally available commercial instrumentation, and
10. Be universally available, accessible, and employed.

Considering point 1, extractables and leachables will be included in the most commonly used commercial spectral libraries (e.g. NIST, Wiley) if they are commonly encountered organic compounds such as residual solvents, monomers, processing aids, and commonly used additives (anti-oxidants, plasticizers, slip agents, acid scavengers, nucleating agents, curing agents, and others). However, the mass spectra of many other extractables and leachables will not likely be included in the most commonly available commercial mass spectral libraries if these compounds are infrequently encountered outside of extractables/leachables studies. Although a number of these more exotic extractables/leachables may be known to the industry – and their mass spectra may be represented in internal, closely-held mass spectral libraries – they are less likely to have been reported and incorporated into the commercial databases.

The two most commonly used chromatographic methods for the screening of organic extractables or leachables

are gas chromatography (GC) and liquid chromatography (LC). Mass spectral matching is a particularly powerful tool for securing identities in GC/MS for the simple reason that the operating conditions for the mass spectrometer in GC/MS have been standardized (e.g. electron impact (EI) mass spectra that were recorded at an ionization energy of 70 eV). Because of the highly standardized mass spectrometric data acquisition parameters, EI mass spectra are very reproducible across different GC/MS platforms and reproducible across test systems and commercial libraries. Well-known and well-controlled large commercial reference libraries, such as the NIST/EPA/NIH Mass Spectral Library and the Wiley Registry of Mass Spectra, are widely used tools to facilitate spectral matching of GC/MS data. Such databases fulfill most of the ideal requirements listed previously with the notable exception that the databases contain many more compounds that are not extractables or leachables than they contain extractables and leachables, thus increasing the likelihood that false and generally unlikely identifications are secured.

Mass spectral matching to external libraries is less successful when applied to LC/MS. The fact that either in-source fragmentation spectra or multi-stage MS (MS^n) fragmentation spectra are necessary for mass spectral matching in LC/MS, adds to the complexity of the identification process. As there are no standard ionization and fragmentation settings for LC/MS detection and because the ionization can be strongly influenced by the chromatographic conditions, no commercial libraries are available that can readily and reliably be used to perform a useful first-pass mass spectral matching for each and every LC/MS instrument platform. Although NIST and some instrument vendors have started to create collision-induced dissociation (CID)-based MS^n spectral libraries that can serve as a resource to scientists who seek to establish a compound's identity in LC/MS, these mass spectra were acquired with a certain and specific combination of instrumental detector settings. Wrong selection of the precursor ion or deviations between the experimental conditions or instrument type used to collect the test and library spectrum may produce aberrations between the experimental mass spectrum and the library mass spectrum, complicating the identification process and leading to lower confidence in its outcome. As a consequence, LC/MS matches should be considered as supporting information for identifications made by other methods, for example, structure elucidation by experienced mass spectrometrists.

A.2.3 Detection and Discrimination of Analyte Signals (Spectra) for Identification

It is intuitive that the best identifications are secured when the chromatographic peaks are pure and the mass spectra are uncompromised. However, peak co-elution cannot always be avoided and it can be challenging to resolve peaks sufficiently so that useful, uncompromised responses can be obtained for the co-eluting analytes. When the determined mass spectra of the detected analytes are compromised due to spectral contamination, poorer identifications are produced. Typical sources of spectral contamination include ion signals from co-eluting compounds, column bleed, solvent tailing, or even electronic noise.

As visual inspection of complex chromatograms is an ineffective means of resolving chromatographic peaks and their associated mass spectra, data processing techniques are often applied. Minimally, background subtraction should be used to establish the mass spectrum that will serve as the basis for identification. To support a higher quality of identification, application of a deconvolution-based approach is preferable as it delivers better quality mass spectra (i.e. free of interferences) and thus reduces the risk of misidentification.

Deconvolution is the process of computationally extracting analyte signals from a complex mass chromatogram, resulting in the elimination of background noise and the spectral separation of co-eluting compounds. The deconvolution process involves multiple steps such as noise analysis, peak shape analysis (ions belonging to the same peak should have the same apex and peak shape), and the assembly of a deconvoluted spectrum [4]. As deconvolution algorithms and related parameters can differ among various software platforms, it is possible to generate slightly different deconvoluted spectra for the same raw data. To facilitate identification, most instrument software platforms enable the combination of deconvolution with a mass spectral library search.

A.2.4 Evaluation of Mass Spectral Matching Results

When an identification is based solely on a spectral match, the identification may be judged solely on the calculated similarity values (i.e. the highest match factor wins) without subsequent critical review of the spectra. However, uni-laterally choosing the highest ranked hit as the reported identity is problematic at best and has been established to generate false positives in many cases [4].

Match factors (MFs) are calculated values that indicate the similarity of two spectra by comparing individual m/z values and their corresponding intensities. In practice, however, there is no "universal" MF threshold value that exclusively establishes that the corresponding match-based identity represents the true identity of a compound. The underlying reason for this is the varying degree of spectral uniqueness among the universe of chemical compounds. Certain compounds may have a rather unique spectrum and are thus more likely to be correctly identified, while others may have a spectrum that very closely resembles the spectra of many other compounds. Nevertheless, the "goodness" of a mass spectral match factor can be correlated with the probability that the match factor has suggested the right compound. The lower a mass spectral match factor, the lower the quality of the fit and the more MSI efforts are necessary to justify an identification decision that was based solely on mass spectral matching. As a general rule of thumb, hits with an MF below 700 are generally associated with a very low probability that the identification is correct; nevertheless, practitioners do report identifications based on an MF below 700, particularly when an MF below 700 is the highest ranked or the only match.

Various approaches can be used to review mass spectral matching results; such approaches are addressed in reference [5].

A.2.5 Examples of Identification by Mass Spectral Matching

A.2.5.1 Example 1: Correct Identification for Best Hit (MF > 900); GC/MS

The top-five hit list for a compound of interest is presented in Table A.1 with the respective MS spectra being displayed in Figure A.1. All five hits have relatively high match factors. The highest ranked candidate has a high match factor of 946 and visual review of the mass spectra

Table A.1 Top five ranked hit list for a mass spectral matching example with an excellent (> 900) match score and a high probability of securing a correct tentative identity for a compound, matched using NIST MS Search v2.3. Experimental RI = 1865.

Rank	Candidate	Match	Reverse match	Probability (%)	Retention index (RI)
1	4-(1-Methyl-1-pheylethyl)-phenol	946	947	78.1	1859
2	4-(1-Methyl-1-phenylethyl)-phenol acetate	892	893	15.3	1947
3	4-Cumylphenol, 2-methylpropionate	858	859	3.92	2098
4	o-(α,α-Dimethylbenzyl)-phenol	826	840	1.04	–
5	2-Methyl-2-[4-(1-methyl-1-phenylethyl)phenoxy]propanoic acid	816	816	0.74	2304

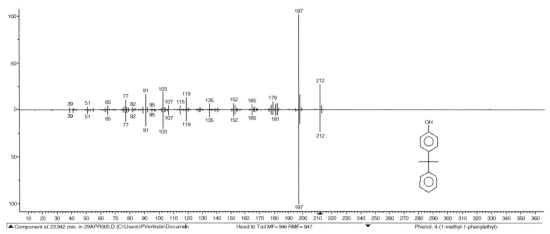

A: Rank 1 candidate spectrum

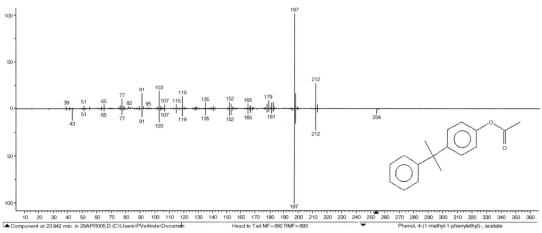

B: Rank 2 candidate spectrum

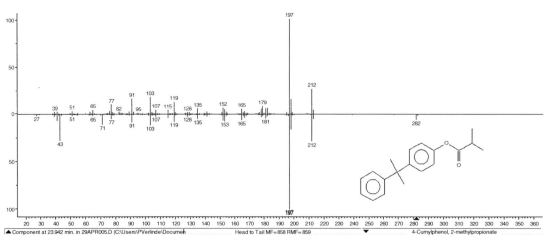

C: Rank 3 candidate spectrum

Figure A.1 Mirror plots of an unknown compound's mass spectrum (top) versus the matched spectra of the top five ranked identification candidates (bottom) where the Rank 1 candidate represents the correct identification (Table A.1). (A) Rank 1 candidate spectrum. (B) Rank 2 candidate spectrum. (C) Rank 3 candidate spectrum. (D) Rank 4 candidate spectrum. (E) Rank 5 candidate spectrum.

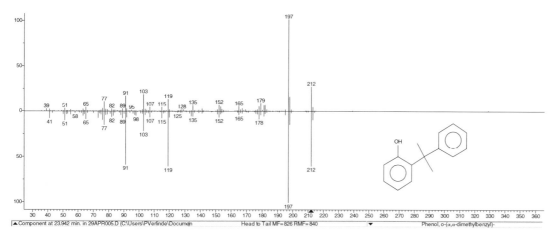

D: Rank 4 candidate spectrum

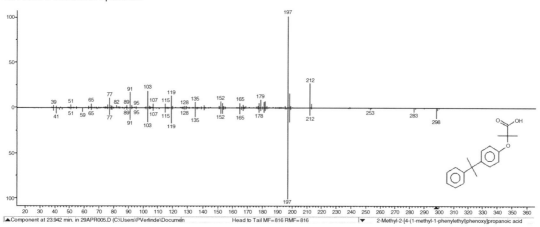

E: Rank 5 candidate spectrum

Figure A.1 (*Continued*)

shows an almost perfect mirror image match, without missing characteristic ions, between the experimental or library mass spectrum (spectrum A in Figure A.1) and the test spectrum. This is in contrast to the lower ranked hits, where the image match is not nearly as perfect. The second ranked spectrum contains an additional peak at m/z 254, which is not detected in the experimental spectrum (spectrum B in Figure A.1). The third ranked spectrum contains additional peaks at m/z 71 and 282 compared to the experimental spectrum (spectrum C in Figure A.1). These observations lead to the conclusion that top hit, 4-(1-methyl-1-pheylethyl)-phenol, has a high probability of being the correct tentative identity for the compound. In addition, comparison of the unknown's retention index (1865) to the experimental retention indices present in the library further corroborates the identification and typically would support upgrading the identification to the confident level.

A.2.5.2 Example 2: Correct Identification for Best Hit (800 < MF < 900); HS-GC/MS

This example shows a top-five hit list for a second compound of interest (Table A.2), where only the best hit has a match factor above 800. Visual inspection of the spectra (Figure A.2) shows a good mirror plot for the best ranked hit; the only marked difference is the relative intensity of the peak clusters at m/z 325 and m/z 341. The lower ranked hits, on the contrary, show clearly deviating features such as additional or missing m/z values and very different relative intensities. In addition, Table A.2 shows that the probability score of the best hit is very high and markedly different from the score of the lower ranked hits. Therefore, the compound can be tentatively identified as dodecamethylcyclohexasiloxane with an appropriate degree of confidence. Never-the-less, this tentative identification should be further reviewed to increase the likelihood that it is the compound's true identity.

Table A.2 Top five ranked hit list for a mass spectral matching example with a good (800–900) match score and a high probability of securing a correct tentative identity for a compound, matched using NIST MS Search v2.3. Experimental RI = 1320.

Rank	Candidate	Match	Reverse match	Probability (%)	Retention index (RI)
1	Dodecamethylcyclohexasiloxane	870	876	96.5	1335
2	5-Amino-1-methyl-1H-pyrazole-4-carboxamide, 3 TMS derivative	634	776	9.96	1837
3	2,4-Diemthoxy-2-(tert-butyldimethylsilyl)oxychalcone	608	723	0.28	–
4	4,6-Diemthoxy-2-(tert-butyldimethylsilyl)oxychalcone	607	725	0.27	–
5	4,6-Diemthoxy-2-(tert-butyldimethylsilyl)oxychalcone (isomer 2)	602	715	0.22	–

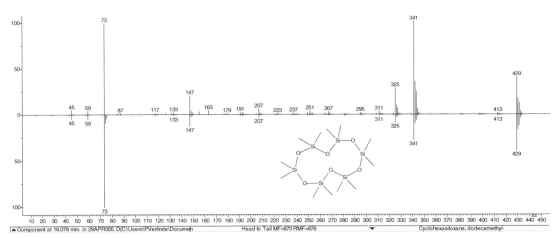

A: Rank 1 candidate spectrum

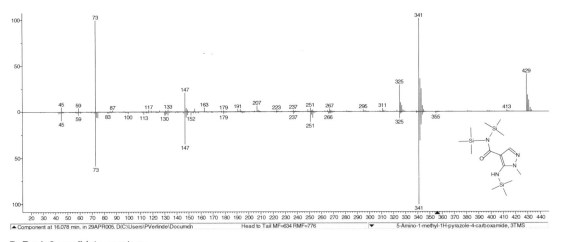

B: Rank 2 candidate spectrum

Figure A.2 Mirror plots of an unknown compound's mass spectrum (top) versus the matched spectra of the top five ranked identification candidates (bottom) where the Rank 1 candidate represents the correct identification (Table A.2). (A) Rank 1 candidate spectrum.
(B) Rank 2 candidate spectrum. (C) Rank 3 candidate spectrum. (D) Rank 4 candidate spectrum. (E) Rank 5 candidate spectrum.

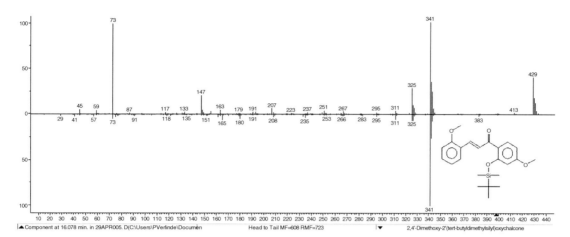

C: Rank 3 candidate spectrum

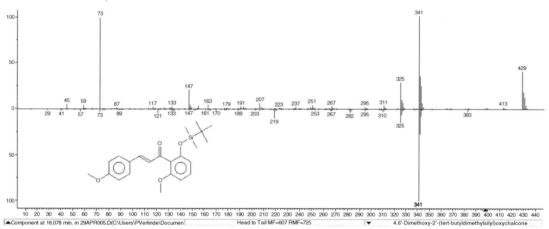

D: Rank 4 candidate spectrum

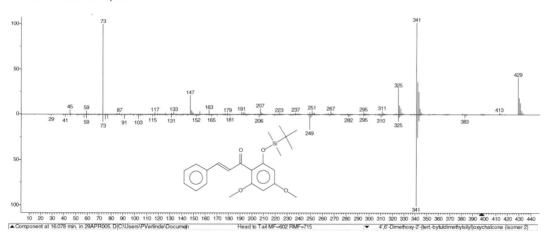

E: Rank 5 candidate spectrum

Figure A.2 (*Continued*)

A.2.5.3 Example 3: Incorrect Identification for Best Hit (800 < MF < 900); GC/MS

Table A.3 shows another example for a third compound of interest where the highest MF value is approximately 800. Upon cursory examination, this top hit seems to provide a good and acceptable match. Mirror plots for the five best hits (Figure A.3), however, show that none of the library spectra are very good matches to the experimental spectrum. Either there are a number of characteristic ions that are present in the experimental spectrum and not in the library spectrum, or vice versa. For instance, peaks at m/z 49, 97, and 112 in the best hit library spectrum (spectrum A in Figure A.3) are missing in the experimental spectrum, which indicates that this identity is incorrect. The same decision can be made for the other hits (spectra B-E in Figure A.3) because they either lack

Table A.3 Top five ranked hit list for a mass spectral matching example with moderate and low match scores and a low probability of securing a correct tentative identity for a compound, matched using NIST MS Search v2.3.

Rank	Candidate	Match	Reverse match	Probability (%)	Retention index (RI)
1	2-Methyl-2-heptanol	808	813	29.1	895
2	(S)-2,5-Dimethyl-2-hexanol	806	809	26.9	857
3	2-Methyl-2-propanol	781	850	8.20	491
4	2-Methyl-2(1-methylethoxy)-propane	761	822	3.73	651
5	1-(1,1-Dimethylethoxy)-2-methyl-propane	756	764	3.01	763

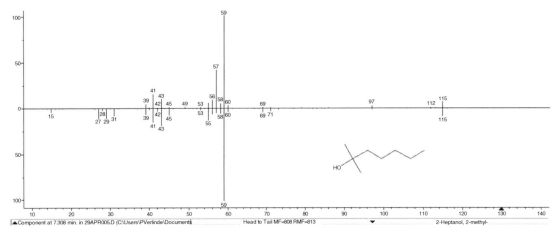

A: Rank 1 candidate spectrum

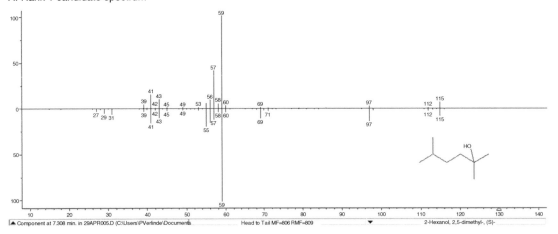

B: Rank 2 candidate spectrum

Figure A.3 Mirror plots of an unknown compound's mass spectrum (top) versus the matched spectra of the top five ranked identification candidates (bottom) where none of the identification candidates is correct (Table A.3). (A) Rank 1 candidate spectrum. (B) Rank 2 candidate spectrum. (C) Rank 3 candidate spectrum. (D) Rank 4 candidate spectrum. (E) Rank 5 candidate spectrum.

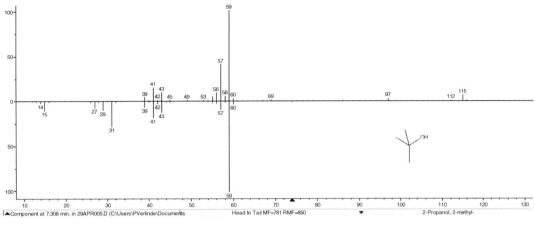

C: Rank 3 candidate spectrum

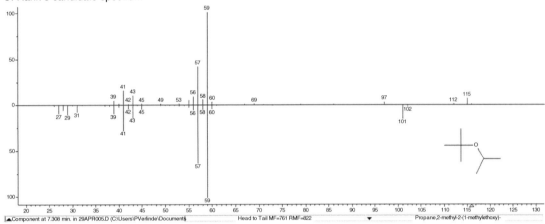

D: Rank 4 candidate spectrum

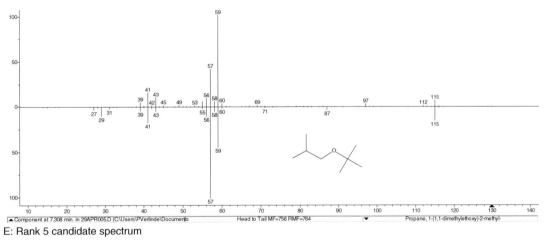

E: Rank 5 candidate spectrum

Figure A.3 (*Continued*)

characteristic ions in the test spectrum or have deviating relative intensities.

Moreover, the unknown's retention index does not correspond to any retention index in the hit list. Consequently, further MSI efforts by an expert are necessary to identify the compound of interest (whose correct identity is 2,4,4-trimethyl-2-pentanol).

A.2.5.4 Example 4: False Positive Identification for Best Hit (700 < MF < 800); GC/MS

Identification based on mass spectral matching becomes even more difficult when the quality of the match factors deteriorates further, as reflected in even lower MF values. An example of this is shown in Table A.4 and the associated Figure A.4, where the MF values are between 750

Table A.4 Top five ranked hit list for a mass spectral matching example with moderate and low match scores and a low probability of securing a correct tentative identity for a compound, matched using NIST MS Search v2.3. Experimental RI = 1178.

Rank	Candidate	Match	Reverse match	Probability (%)	Retention index (RI)
1	2-Tert-butoxytetrahydrofuran	727	750	16.0	–
2	1, (1,1-Dimethylethoxy)-2,2-dimethyl-propane	725	729	14.8	–
3	4-Butoxy-2-butanone	725	726	14.8	–
4	4-Iso-butoxy-2-butanone	701	702	4.96	–
5	4-Sec-butoxy-2-butanone	701	701	4.96	–

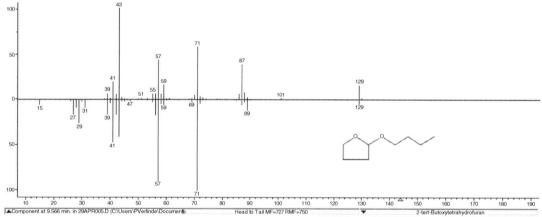

A: Rank 1 candidate spectrum

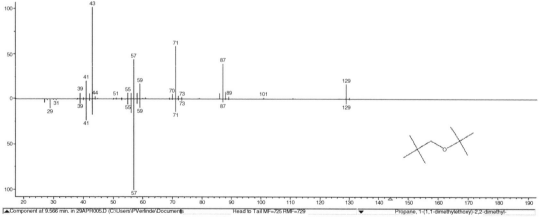

B: Rank 2 candidate spectrum

Figure A.4 Mirror plots of an unknown compound's mass spectrum (top) versus the matched spectra of the top five ranked identification candidates (bottom) where none are correct (Table A.4). (A) Rank 1 candidate spectrum. (B) Rank 2 candidate spectrum. (C) Rank 3 candidate spectrum. (D) Rank 4 candidate spectrum. (E) Rank 5 candidate spectrum.

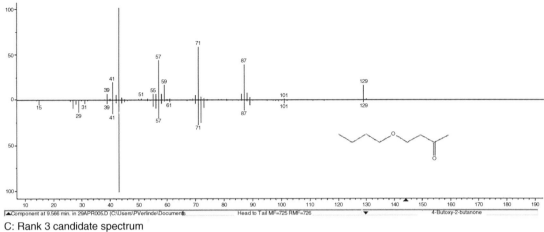

C: Rank 3 candidate spectrum

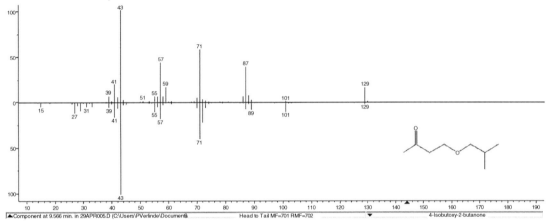

D: Rank 4 candidate spectrum

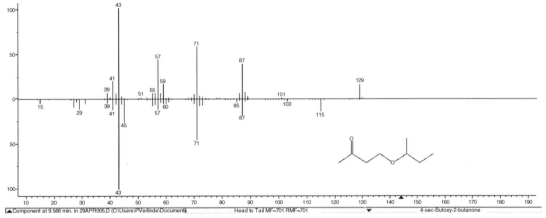

E: Rank 5 candidate spectrum

Figure A.4 (*Continued*)

and 700 for the five best ranked hits. A visual inspection, performed by an experienced mass spectrometrist, would reveal that none of the candidate library spectra fit the experimental spectrum of the compound of interest. Thus, "identification" of the unknown as the highest ranked hit would result in a false identity. Consequently, additional efforts in MSI are essential to secure the correct identity of this compound.

A.2.6 Closing Comments

Spectral matching is not an exact science and the following recommendations are made to assist in securing the right identity:

- There is no value for the mass spectral match factor (MF) that unequivocally guarantees that the correct identity of a compound has been determined, based upon the MF alone. It is clear that the exclusive reliance on mass spectral match factors without any expert review cannot robustly and routinely lead to correct identifications. Reporting the highest ranked hit as an analyte's tentative identity can lead to false positive identifications.
- An expert should *always* visually evaluate the spectra of match candidates in the mass spectral matching's hit list, regardless of the quality of the MF.
- It is good and necessary practice that tentative identifications based on mass spectral matching are always substantiated by comparative spectrum plots such as mirror plots, as such comparative data allows one to visually confirm the quality of, and increase the confidence in, the fit of the matched spectra.
- The lower the MF, the more intense the MSI exercise will need to be to secure the identity of the compound based solely upon the merits of its mass spectrum.
- While MFs above 80% (or 800 depending on the scoring scale) may lead, after a careful mass spectral evaluation, to a correct identification, the probability of securing the right identity via mass spectral matching decreases quickly below these values. When the MFs are below 70% (or 700), the probability of correctly identifying a compound based solely upon the MF is extremely low and the compound remains unidentified, although it is possible that the match is sufficient to substantiate and support a partial identification.
- Identifications based solely on mass spectral matching are tentative identifications, as the identification is a "one"-dimensional identification, where the one piece of evidence is its mass spectrum. Such an identification can be augmented by analyzing an authentic standard (mass spectrum and retention time: confirmed identity) or by obtaining additional supporting documentation.

A.3 Identification by Mass Spectral Interpretation

MSI is the process of securing a compound's identity solely by expert interpretation of the information that is available in the compound's mass spectrum.

A.3.1 Introduction to Mass Spectral Interpretation

Although it is not the intent of this document to provide a comprehensive and detailed discussion of all the fundamentals of MSI, essential principles and practices are discussed and illustrated. In general, the MSI identification consists of three consecutive steps:

1. Determining which peak in the mass spectrum corresponds to the molecular weight of the molecule. When the spectrum is acquired with an accurate mass high-resolution instrument, the mass-to-charge ratio (m/z) can be used to generate a candidate molecular formula.
2. Establishing whether the compound of interest contains certain elements, such as chlorine or bromine, which have specific isotope patterns that translate into recognizable spectral features, namely specific relative abundances of monoisotopic masses.
3. Performing de novo structural elucidation. All peaks in a mass spectrum with an m/z value below that of the molecular ion are formed during the ionization (or MS/MS fragmentation) of the compound of interest and relate in one way or another to substructures or functional groups of that substance. An expert in fragmentation chemistry can relate this information to the molecular ion and propose a tentative structure via a process that is generally referred to as "*de novo* structural elucidation." Although algorithms have been developed to assist in certain aspects of such an elucidation, structural elucidation is invariably a subjective interpretation performed by an expert mass spectrometrist. Therefore, any identity secured by structural elucidation is classified as being a tentative identification until additional collaborating data allows for an upgrade of the identification level.

A.3.2 Identification of the Molecular Formula

Establishing a compound's elemental composition (molecular formula) from its mass spectrum typically starts with determining which ion peak in the mass spectrum represents the molecular ion, the m/z value of the intact, ionized molecule.

The ability and strategies to ascertain the (pseudo) molecular ion depend heavily on the type of ionization technique used. For example, GC/MS spectra are generally acquired with electron ionization (EI) which produces a radical molecular ion $M^{+\bullet}$ with highly variable intensities. The intensity of the molecular ion depends on its propensity to decompose into several smaller fragments, which in turn is dictated by the stability of the ion under the applied ionization conditions. For instance, the molecular ion is usually very intense for compounds that are highly stable under these ionization conditions, such as (*poly*)aromatics, while it is rarely detected in spectra of largely unstable compounds such as aliphatic alcohols, highly branched compounds, and polyether glycols. Therefore, an independent assignment of the molecular ion for EI spectra, while not impossible, can be prone to the error of incorrectly picking an *m/z* value, which is, in fact, associated with a fragment of the molecular ion.

More intense molecular ions are usually produced by "soft" ionization techniques, such as atmospheric pressure chemical ionization (APCI) or electrospray ionization (ESI) for LC/MS methods and chemical ionization (CI) for GC/MS methods. Ionization of the molecule can result in protonated $[M + H]^+$ or deprotonated $[M - H]^-$ ions depending on the polarity of ionization. In addition, adducts may be formed during ionization by reaction or clustering of the molecule with chemical entities present in the sample or mobile phase or due to reaction with a reagent gas. For instance, adducts with alkali or ammonium salts (e.g. Na^+, K^+, NH_4^+ in the positive mode and Cl^- in the negative mode) are frequently observed in APCI or ESI spectra. In addition to adducts, soft ionization may also be associated with in-source fragmentation depending on the ionization conditions, the stability of the (pseudo) molecular ion and, in case of CI, on the proton affinity of the molecule. In general, a thorough evaluation of adducts and in-source fragments is necessary to confirm the molecular ion. Additionally, dimeric ions or even higher clusters can also be formed in the case of APCI or ESI.

For both hard and soft ionization technologies, identifying the molecular ion in a mass spectrum is a subjective interpretation performed by a mass spectrometry expert and thus there is a degree of uncertainty in the interpretation. Unfortunately, the degree of uncertainty cannot easily be expressed as a mathematical number such as the probability score used in mass spectral matching.

Although establishing the (pseudo) molecular ion with unit mass resolution is a significant step in compound identification, such information in itself is rarely adequate to secure even a tentative identification. However, if the molecular weight of the ion can be established with a high degree of resolution, the exact (or accurate) mass so secured could be used to generate a short list of candidate compounds whose molecular formulas have molecular weights equal to the determined accurate mass. This exact mass information can be obtained with a high resolution–accurate mass spectrometer (HRAMS) such as time-of-flight, orbitrap, or ion cyclotron resonance mass spectrometers. On such instrumentation, ions that only slightly differ in *m/z* value would be detected as isobaric signals. For example, diethyl fumarate and 2-fluorobiphenyl have the same unit mass of 172 Da and thus would be indistinguishable on this basis alone. However, their accurate masses (172.0730 Da and 172.0683 Da, respectively) are sufficiently different that they would be readily distinguished on the basis of the elemental compositions obtained using HRAMS, since such mass measurements enable the determination of the ion's elemental composition by considering the sum of the exact masses of various nuclides ($C_8H_{12}O_4$ and $C_{12}H_9F$, respectively).

The generation of molecular formulas from accurate mass information is usually assisted by software algorithms using user-defined search constraints. Search criteria include the species and quantity of allowed elements, allowed mass accuracy (depends on the resolution of the mass spectrometer), the charge state (e.g. singly or multiply charged), and the allowed electron state, which refers to the number of electrons (even or odd) and depends on the ionization technique. Soft ionization techniques normally produce ions with an even electron state. EI spectra, on the other hand, generate a radical molecular ion with an odd number of electrons while fragments of the molecular ion can either have an odd or even number of electrons.

In addition to establishing molecular formulas, accurate mass information can even give structural information based on the number of ring and double-bond equivalents rule, which is a conventional measure of the degree of the unsaturation of an organic molecule corresponding with the lowest formal valence state of the elements present in its elemental formula.

Depending on the mass resolution and mass accuracy of an HRAMS measurement and the structure of the compound of interest, it may or may not be possible to distinguish the correct elemental formula from among the multiple candidate elemental formulas that are a reasonable match to the accurate mass established via the *m/z* peak of the (pseudo) molecular ion. In certain cases, the correct molecular formula can be established by evaluation of the isotopic data.

A.3.3 Interpretation of Isotopic Data

Most elements appear naturally as a mixture of stable isotopes; for example, natural carbon is a mixture of 98.9% of isotope ^{12}C and 1.1% of isotope ^{13}C. The natural isotopic composition of a molecule is reflected in the mass

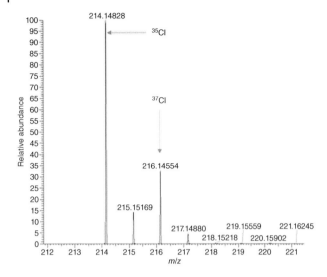

Figure A.5 Example of a diagnostic bromine isotope pattern for elemental formula $C_{13}H_{23}Cl$ rubber oligomer. The 3 : 1 size of the two chromatographic peaks represent the natural abundances of these isotopes (75.77% ^{35}Cl and 24.23% ^{37}Cl).

spectrum by the presence of isotopic clusters. Such a cluster is composed of distinct monoisotopic masses with relative abundances that reflect their distinct isotopic compositions. By consensus, the first peak in the cluster of peaks corresponding to the most abundant isotopes of a given ion is designated as X. The isotopic regions in a mass spectrum that corresponds to 1 or 2 (or more) mass units further away from X are designated as X + 1, X + 2, etc., regions. The common elements such as C, H, N, and O – which have a diagnostic isotopic pattern with relatively low abundance of their first isotope at X + 1 – do not exhibit very intense and obvious isotopic compositions when measured as unit mass. On the other hand, certain elements such as Cl, Br, S, K, and Si have very characteristic and intense isotopic distributions up to the X + 2 region. For instance, chlorine isotopes have a natural composition of 75.77% ^{35}Cl and 24.23% ^{37}Cl. Therefore, a molecular formula of $C_{13}H_{23}Cl$ has an average molecular weight of 259.231, which in the mass spectrum will be observed as distinct peaks at 214.148 and 216.146 Da with almost equal intensities, as shown in Figure A.5.

These diagnostic isotopic clusters are readily recognizable by a mass spectrometry expert and can be used to reveal the presence of specific elements. In addition, algorithms have been developed to predict the presence of certain elements such as chlorine and bromine (e.g. NIST/EPA/NIH MS Search software). Accurate mass data are not required to infer the presence of these elements from their isotope patterns, although such data would certainly reinforce the isotopic evidence for the presence of certain elements in other cases.

A.3.4 Interpretation of Mass Fragments: *De Novo* Structural Elucidation

The ion peaks present in a mass spectrum can be interpreted to establish the presence of functional groups or substructures in the compound of interest, to place the compound of interest into certain structure-based classes (e.g. alcohols or phthalates), or even to propose a tentative molecular structure, as the ions formed during ionization, represent either the ionized molecule or ionized fragments thereof. Fragmentation of a molecule principally occurs in a predictable and reproducible way within the boundaries of the applied instrumental parameters. The general mechanisms for such fragmentation reactions have been extensively described in authoritative reference works on MSI [1, 2, 6, 7]. For example, cleavages resulting in the loss of neutral molecules (water, carbon monoxide, methanol, etc.) are usually produced by structural rearrangements or proton shifts of the ionized molecule.

Another type of cleavage involves the loss of a radical fragment (e.g. methyl radical), which is almost exclusively observed in EI spectra. A cleavage generates two types of informative spectral values, namely the mass value of the formed fragment and mass differences of that fragment with a heavier fragment or the molecular ion, commonly referred to as "losses."

An experienced mass spectrometrist who has an in-depth knowledge of fragmentation rules can relate observed mass values and losses to specific fragment structures that are linked together to establish a logical fragmentation pathway. Moreover, an extensive chemistry background is imperative to assess whether or not a proposed structure is viable, that is, that it is thermodynamically stable and whether it is likely to be detected with the applied technique.

The ultimate goal of structural elucidation is to elucidate as many fragments as possible and to link the fragments together via a rational pathway, as so doing limits the number of possible structures to the smallest number of candidates. In general, the more fragments that can be fitted into a defendable fragmentation pattern for a proposed structure, the greater the likelihood that the identity established by elucidation is, in fact, the correct identity.

In the initial stages of elucidation, it often happens that numerous structure candidates can be proposed that fit the generated molecular formula or observed fragments to varying degrees. As a general rule, the relative percentage of peaks that can be rationalized by a fragmentation pathway for a given structure is directly related to the likelihood that the spectrum indeed corresponds to that structure; that is, the higher the percentage of rationalized peaks, the greater the likelihood that the elucidated identity is the correct identity. A complicating factor in structural elucidation, however, is that not all structures have unique mass spectra. This is often the case for compounds with very similar structures. For instance, the degree and position of branching of hydrocarbon chains or the exact stereochemistry of a molecule often cannot be inferred from a mass spectrum. Therefore, the confidence level of identifications that are solely based on structural elucidation is limited to tentative identification at best. This is the case even for compelling elucidations, as the identification is still based only on one dimension of information. A higher level of confidence can be achieved by gathering additional data, such as retention time, MS/MS spectra, or spectra recorded with a different type of ionization. In addition, other corroborating data, such as the result of an identification found in another orthogonal and complementary technique (such as GC/MS identification results for LC/MS compound identifications), disclosed compositional data of the material of construction, or other analytical techniques that can assist in the confirmation of the elucidated structure (e.g. NMR on the isolated compound), can assist in upgrading the elucidated structure from a tentative to a confident or confirmed identity. It goes without saying that the highest level of identification is obtained by confirming the mass spectrum (and the associated retention time) of the tentatively identified compound with its authentic standard and/or providing sufficient corroborating information so that the chances of an incorrect identification are small.

A.3.5   Securing Partial Identities

Occasionally, partial or tentative identities can be deduced for an extractable based on similarities between the extractable's mass spectrum and similar mass spectra for other extractables that have been accurately identified themselves. The underlying principle is that spectra from molecules with very similar structures also have similar spectral features. This is particularly relevant for sample spectra that are not present in a library of reference spectra. The similarity between spectra is not limited to the circumstance that the sample and reference spectra contain equal mass values or relative abundances, but may also include equal losses.

Because a structure cannot be established, identifications secured in this manner are classified as partial identifications. Some examples of these partial identities include:

- EI spectra of phthalate esters contain an intense m/z 149 ion, which is often the only major peak. The molecular ion is often not detected.
- EI spectra of aliphatic hydrocarbons are characterized by a typical pattern of m/z 43, 57, 71, 85, etc.

- The presence of ions m/z 77 and 91 in an EI spectra is diagnostic of the presence of phenyl and benzyl substructures, respectively.
- CI spectra that contain a mass difference of 18 Da indicates the loss of water, which is typically observed in alcohols or acids but not in ketones.

Another example of a partial identification for a siloxane-type compound is shown in Figure A.6.

A.4   Additional Evidences Supporting Higher Level Identifications

An identification secured by either mass spectral matching or MSI is a tentative identification as it is based on one dimension of identifying information. Additional information about the compound of interest should be pursued to either corroborate (or refute) the tentative identification. Depending on the quantity and nature of the corroborating data, tentative identities can be substantiated and "elevated" to either confident or confirmed identifications.

Tentative identifications can be "elevated" to at least confident status using the mass spectral information itself. For example, if the same tentative identity is secured by mass spectral matching and MSI, then these two independent corroborating outcomes "elevate" the tentative identification secured with both processes to a confident identification.

Additionally, identifications can be also be substantiated by accumulating independent evidences or evidences. It is logical that the more additional evidences that are gathered, the more certain the identification becomes. Various means of "elevating" identities are considered and examples are provided; however, it is outside the scope of this document to provide an exhaustive list of additional evidences.

A.4.1   Retention Time/Index Matching

As a result of the chromatographic process, compounds of interest are separated in terms of the time it takes the compounds to emerge (elute) from the chromatographic column and be detected by the mass spectrometer. This elution, or retention, time will depend on the chemical and physical nature of the compounds of interest and is therefore diagnostic for specific compounds. However, even with the excellent separation efficiencies (resolution) achievable by modern chromatographic methods applied to extractables/leachables screening, a specific retention time is not necessarily unique to a single specific organic compound

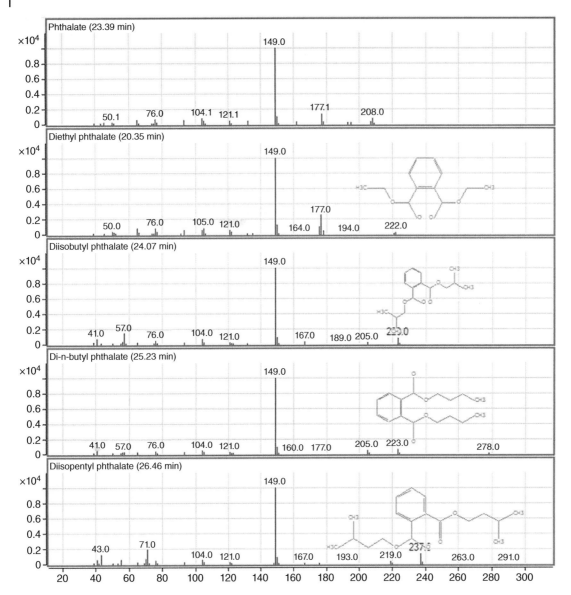

Figure A.6 Example of a partial identification in which the mass spectrum of the compound of interest (top spectrum) contains typical fragments with *m/z* values 149, 177, 76, 104, and 121. The mass spectra of phthalates, confirmed with authentic standards, are shown for comparison and have the same characteristic pattern. This justifies a partial identification for the compound of interest (it is a phthalate). However, because neither the spectrum nor the retention time of the peak associated with the compound of interest is an acceptable match with the same data for any of the confirmed phthalates, the identification cannot be elevated to tentative status.

(that is, it is not uncommon that several compositionally dissimilar compounds have comparable retention times). Thus, the retention time itself is not a sufficiently diagnostic property of a compound that it alone can be used to secure a tentative identity. Rather, retention time is corroborating information for identities secured by another means, such as mass spectral matching or interpretation.

Use of the retention time to support an identification is similar to the use of a spectral matching to secure a tentative identity. Similar to a mass spectrum, the retention time can be matched to a laboratory-generated database of retention times, producing a hit list of potential candidates.

Presumably, a test compound whose mass spectrum and retention matches the mass spectrum and retention time of a reference compound in a database has been confidently established to be the reference compound.

Retention time matching is essentially a "free" piece of information as the retention time is obtained via the same analytical activity as the mass spectrum. Securing retention time as a corroborating piece of data does not require re-analysis of the sample to secure additional information, which is required, for example, if corroborating information is obtained via a different analytical technique (for example, NMR).

For retention time to be useful as an identification tool, it is imperative that the retention time must be accurate and reproducible over time and across different instruments. However, shifts in retention time occur frequently. Routine maintenance procedures such as column trimming alter retention times. In a multi-instrument laboratory running the same method on multiple instruments, the retention times for each instrument are likely to differ from each other, even when care is taken to ensure that all instruments are operated using identical conditions. These differences in retention times confound efforts to use retention time as a means of identification. This is especially true for LC-based separations.

To a certain extent, retention time differences can be managed by two methodologies, Retention Time Locking (RTL) and Relative Retention Times (RRT, alternatively referred to as the Retention Index, RI). RTL is the ability to very closely match retention times on one system to those in another system by adjusting the chromatographic conditions and is more typically applicable to GC. In GC, for example, adjusting the inlet carrier gas pressure will change retention times in an even and predictable manner. Thus, retention times on a given system can be closely matched to those on another system by altering the inlet carrier gas pressure in one or both of the systems. A specific compound (usually the Internal Standard for Injection) is used for both developing the locking calibration and locking all applicable systems.

Alternatively, expressing retention behavior on a relative scale using retention indices (RIs) or linear retention indices (LRIs) can be used as corroborating information. The advantage of the RI is that retention indices do not depend on the exact column dimensions, flows, or temperature-programming. However, they do depend on the type of stationary phase (for example non-polar versus polar phases).

The use of RLT, RRF, and/or RI to facilitate retention-based matching is discussed in greater detail in references [8, 9].

Identification corroboration via retention matching is most effective when the reference retention data are obtained through analysis of reference materials under standardized chromatographic and MS conditions that are identical to the conditions applied in routine laboratory operations.

A.4.2 Tandem Mass Spectrometry

The interpretation of MS/MS (or more generally MS^n) spectra can either lead to the proposal of a tentative structure or add confidence to a tentative structure that has already been proposed. The most common type of MS/MS

analysis is the acquisition of product ion scans, which is achieved by isolating a certain precursor ion followed by fragmentation of that ion into products ions. (Depending on MS technology and instrument vendor, such MS/MS analyses can either be set up manually in a separate run or be performed along with the acquisition of screening data, for instance through selection of the top most intense ions for isolation and fragmentation.) A good choice of precursor ion selection would be to pick the molecular ion as it results in a spectrum of product ions that have an unequivocal relationship with the molecular structure. MS/MS analyses are particularly useful to obtain fragmentation data when the ionization method yields very few structurally informative fragments (e.g. APCI spectra, which only contain the molecular ion). Furthermore, an MS/MS spectrum has a higher level of selectivity compared to MS^1 scan data as the in-source fragmentation in MS^1 could be obscured by other ions generated from the matrix in the course of the ionization process or by co-elution with other compounds present in the sample.

In the case of co-elution, mass spectral deconvolution is a powerful tool to resolve spectra from co-eluting compounds that is effective with a vast majority of acquired spectra. However, complete resolution of complex mass chromatograms by deconvolution will not be possible in all cases.

A.4.3 Additional Evidences from Orthogonal Techniques

Compounds can be detected by multiple analytical techniques and tentatively identified by independent assessment of the evidence from each technique. When this occurs, the independent assessments (which produce the same identities) are mutually corroborative and the identification, supported by two-dimensional data, is "elevated" to confident.

For example, take the relatively simple and common case where an extractable produces a response in both GC/MS and LC/MS. In this case, and without any additional testing, two identical tentative identities secured by both techniques independently corroborate one another, resulting in an elevated confident identity. Alternatively, a tentative identity secured by one method can be used to tentatively identify a peak that is unidentifiable by the second method.

Another means by which information from an orthogonal technique can facilitate procuring the correct identity is when compounds with the same m/z co-elute. This may, for instance, be the case for caprolactam and 2-methyl-1-pyrrolidinone in an LC/MS analysis. While these compounds, both with the elemental formula $C_6H_{11}ON$, may co-elute in the LC/MS chromatogram, they do not co-elute in GC/MS. Therefore, the identity of the detected compound associated with the LC/MS peak at the proper retention time and with a detected *m/z* of 114.091 can be uniquely attributed to either caprolactam or 2-methyl-1-pyrrolidinone depending on which compound is reported in the GC/MS data.

Another manifestation of the orthogonal technique approach is the use of a non-chromatographic method, such as NMR, to independently secure an unknown's identity. This identification strategy is also described in USP <1663> [10] as follows: "Although these identification categories are based upon mass spectrometry, it is possible to use data from other analytical techniques to assist in the extractables identification. Such techniques include GC/FTIR (Fourier Transform Infrared Spectroscopy) and LC/NMR (Nuclear Magnetic Resonance Spectroscopy)."

While the power of NMR as an identification method is well known, use of this technique in E&L laboratories is limited by certain practical realities. One such practical reality is access to NMR technology. Although access to NMR technology may be straightforward for larger pharmaceutical companies, access may be problematic for E&L labs in a contract research environment. A second practical reality is that NMR can only come to relevant conclusions if the neat (pure) "unknown compound" can be isolated and investigated. Furthermore, the sample amount requirements to perform an NMR experiment on this neat chemical compound – often a few milligrams of the purified "unknown compound" at least – often requires

intensive sample preparation steps, such as isolation of the compound through fraction collection.

A.4.4 Derivatization

Derivatization is the chemical treatment of an extract to convert an extractable (or extractables) into a more analytically expedient form. Derivatization is performed to increase the sensitivity, selectivity, or thermal stability of a compound for a certain technique. Trimethylsilylation and methylation, for example, are common techniques used in GC/MS to increase the volatility and hence the sensitivity of polar molecules. Derivatization using halogenated acyl groups is another example and is used to increase the sensitivity for detection with an electron capture detector (ECD) or a mass spectrometer with electron capture negative chemical ionization.

Additionally, the selectivity of the derivatization reaction can also be exploited to identify the presence of certain functional groups. Trimethylsilylation, for instance, will derivatize all functional groups with active H atoms (e.g. acids, alcohols, amines) such that each active H atom is replaced by a trimethylsilyl (TMS) group. These changes will also be reflected in the mass spectrum by an increase in molecular weight of 72 Da for each TMS group. Comparison of MS chromatograms associated with non-derivatized and derivatized extract indicates whether or not the extract contains analytes whose structures include derivatizable groups and, if there is an analyte with derivatized groups, how many derivatized groups the analyte possesses (e.g. Figure A.7).

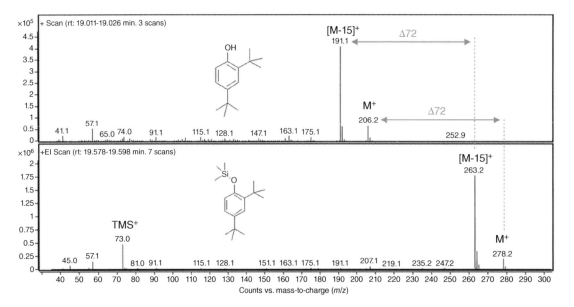

Figure A.7 Comparison of the EI mass spectrum of 2,4-di-tert-butylphenol with the mass spectrum of its trimethylsilyl (TMS) derivative. A mass difference of 72 Da is observed for the molecular ion (*m/z* 206 versus *m/z* 278) and demonstrates that the molecule contains one derivatizable group (in this case a hydroxyl group). The same mass difference is also observed for the abundant [M-15]⁺ peak (*m/z* 191 versus *m/z* 263). Furthermore, ion *m/z* 73 is also diagnostic for the trimethylsilyl group.

A.4.5 Indirect Inferences

In some cases, it is possible to support an identification with an indirect inference; that is, secondary information is used to infer whether a proposed identity is likely or not. For instance, knowledge of a test article's composition can facilitate the identification of its associated extractables, as it is likely that the extractables include the ingredients themselves or reaction products of these ingredients. Thus, the decision between two possible identities can be made on the basis of only one of the structures being related to a known test article ingredient.

As an example, consider the hypothetical list of ingredients for a polyolefin material given in Table A.5. Each ingredient serves a specific purpose, either to protect the polymer from oxidation (Irganox 1076 as a primary antioxidant protecting the polymer during use, Irgafos 168 as a secondary antioxidant protecting the polymer during its manufacturing, and calcium stearate as an acid scavenger) or to enhance the functionality of the polymer (monostearin as a lubricant).

This polyolefin material, with a known composition described in Table A.5, is then subjected to an extraction with an organic solvent, followed by an extract analysis via GC/MS. The resulting chromatogram (Figure A.8) contains six peaks whose associated compounds can be confidently identified as follows.

First, the mass spectra for compounds 1 and 2 can be readily matched with a high match score to library spectra for Irgafos 168 and Irganox 1076, which is corroborated by an expert review of the mass spectral matches and which therefore produces a tentative identification. However, knowing that these compounds are intentionally present in the extracted material makes it all the more likely that these tentative identities are in fact the correct identities and thus the composition information is sufficiently corroborative that the tentative identities can be "elevated" to confident identities, based on this two-dimensional corroboration.

Furthermore, compound 3 in Figure 3.8 was also tentatively identified as tris(2,4-di-tert-butylphenyl) phosphate, the well-known and well-characterized oxidized form of Irgafos 168, via expert-reviewed mass spectral matching. Given the presence of Irgafos 168 in the test material, it is very likely that the oxidized form of Irgafos 168 will also be present in the material, as it is by its sacrificial oxidation that Irgafos 168 protects the polyolefin. Thus, compound 3 is confidently identified as the oxidized form of Irgafos 168 based on corroborating information of a mass spectral match and logical inference of the presence of this compound in the test article.

A similar logic can be applied to peak 4, which was also tentatively established to be 2,4-di-tert-butylphenol via an expert-verified mass spectral match. As it is well established in the chemical literature (for example, reference [11]) that 2,4-di-tert-butylphenol is a degradation product of the material's ingredient Irgafos 168, the combination of compositional information and the scientific literature corroborates the tentative identification, allowing it to be elevated to confident status.

Table A.5 Table with a hypothetical list of ingredients for a material of construction (in this case, a polyolefin).

Compound	Synonym	Function
Polyolefin	–	Polymer
Tris-(2,4-di-tert-butylphenyl) phosphite	Irgafos 168	Antioxidant
Octadecyl 3-(3,5-di-tert-butyl-4-hydroxyphenyl)-propionate	Irganox 1076	Antioxidant
1,2,3-Propanetriol-1-octadecanoate	Monostearin	Lubricant
Calcium dioctadecanoate	Ca-stearate	Acid scavenger

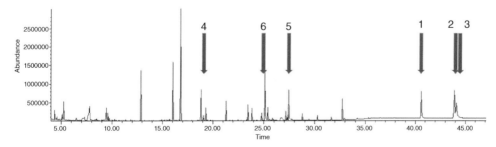

Figure A.8 GC/MS chromatogram of an organic extract of the material of Construct with a fictious composition as described in Table A.5. Based upon the provided information in the ingredients list, six (at least) identifications can be upgraded from a tentative identification to a higher class of identification (e.g. confident) using this information.

Peak 5 is also tentatively identified by expert-reviewed mass spectral matching, in this case as stearic acid. One notes that stearic acid is not listed as an intentional ingredient and at first glance the tentative identification does not appear to be corroborated by composition. However, closer examination of the ingredient list reveals that calcium stearate was added as an acid scavenger to the polyolefin. While this is not a one-on-one correlation (indeed, calcium stearate is not the same exact molecule as stearic acid), once it is understood how the "acid scavenger" mechanism works (illustrated here with acetic acid as the acid being scavenged), it becomes obvious that the acid scavenger's action results in the formation of stearic acid:

$$Ca(stearate)_2 + 2\,HCOOH$$
$$\rightarrow 2\,Stearic\ Acid + Ca(HCOO)_2$$

Once again, composition corroborates a tentative identity, elevating the identity to confident status.

Lastly, peak 6 is tentatively identified as palmitic acid, again based on expert-reviewed mass spectral matching. Because this compound is not a listed ingredient, at first pass this identification does not appear to be corroborated by composition. However, calcium stearate additives are generally natural products that rarely are as pure as analytical-grade reagents. In fact, the calcium stearate additive is likely to be a mixture of both stearate, palmitate, and even lower molecular weight fatty acid salts. Thus, the calcium stearate is a logical source of palmitic acid and once again compositional information corroborates a tentative mass spectral match identity to elevate its status to confident.

Thus, on the basis of tentative identities secured by expert-reviewed mass spectral matching and corroborated by compositional knowledge, all six extractables noted in Figure A.8 have been confidently identified.

Information from a partially elucidated extractables profile can either facilitate an identification or be used as collaborating information to elevate an identification. For example, consider the case where a homologous series of compounds with a certain functionality (for example, a homologous series of 2-methyl-2-alkanols) was detected and the identities of a number of those homologous compounds were confirmed via the analysis of authentic standards (Figure A.9). It is very clear that the major peaks in the chromatogram are all part of a homologous series of extractables, differing in mass 28. Via available authentic reference standards, the peaks at 10.23 and 23.53 minutes are confirmed to be alkanols of increasing ring size. However, the next compound in the series (peak at 14.29 minutes) can only be tentatively identified, via mass spectral matching due to a lack of an available reference standard. However, the fact that the compound is so clearly the "next step up" in the homologous series surely

supports the proposition that the tentative identity can be elevated to at least confident status.

A.4.6 The Use of a Database to Capture the Identification Efforts

The practice of using corroborative data to augment and support higher level identifications, as well as the efforts to secure the identity of the compounds through mass spectral matching or MSI, can be quite time-consuming, labor-intensive, and expensive, requiring expert scientific, process, and material knowledge and advanced analytical capabilities.

It is evident that once a compound has been identified and has been assigned an elevated identification class, the supporting analytical data (such as mass spectral fragmentation or retention time) is fixed, as long as the analytical methods and instrumental settings remain unchanged. This circumstance supports the generalization that "once a compound has been identified to a certain class, it remains identified to that class until the analytical method is changed."

Thus, there is significant value in capturing completed identifications, as it makes little sense to perform the identification exercise all over again, for each analytical event. An appropriate means of capturing identities, and documenting the identification process, is via an internal database.

The concept of "once identified always identified" is a powerful means of making identification efficient and reproducible, but is only enabled if the identification information is captured in an accessible database.

A.4.7 Conclusion

Securing the correct identity of an extractable or leachable is essential, as it is the correct identity that enables a compound's impact assessment. If a compound cannot be unequivocally identified, the overall impact assessment of the compound will be compromised.

Nevertheless, not all extractables and leachables can be unequivocally identified, even with the best available analytical data, the most complete material and process information, and the highest level of scientific expertise. To ensure that users of an identity understand the relative certainty that the identity is correct and to provide scientists with an aid for judging the value of the data they have collected, a hierarchy or classification of identities has been established:

- *Partial.* No full identity of the compound can be determined, but certain general functionalities can be ascertained.

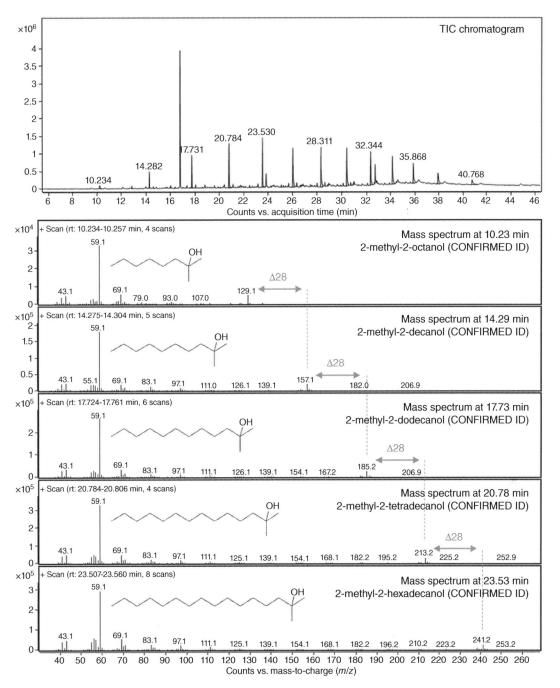

Figure A.9 Identification of a homolog series of 2-methyl-2-alkanols extractables. The chromatogram contains multiple peaks at regular retention times. Additionally, the mass difference for the [M-15]⁺ peak between each homolog (i.e. 129, 157, 185, 213, and 241) amounts to 28 Da, which corresponds to an ethylene unit. The peaks at 10.23 and 23.53 minutes are identified as confirmed identifications based on authentic reference standards. Based on these patterns the other peaks can be identified as 2-methyl-2-alkanols with different chain lengths.

- *Tentative.* One-dimensional identification, only based upon one piece of information
- *Confident.* A two-dimensional identification, based upon at least two independent pieces of corroborating data.
- *Confirmed.* A three-dimensional identification, based upon three or more independent and complimentary pieces of corroborating data.

Of course, the fifth identification class, unknown, is reserved for those compounds that cannot be identified to any degree.

Clearly, the ultimate objective of the identification process is to secure a confirmed identity. When the available information is insufficient to support this level of certainty (for example, a reference standard is not available to secure the confirmation), other classes have been established to communicate the certainty in the identity, based on the amount and rigor of the supporting information.

The most likely identification class secured through the typically employed identification processes (mass spectral matching and MSI) is tentative. Although a tentative identification is the minimum appropriate for impact assessment, one understands that there is a possibility that the tentative identity is incorrect, leading to a flawed impact assessment. Therefore, the goal of the identification process is to secure as high an identification level as data and insight will support.

Means of "elevating" a tentative identification via corroborating information include:

- Chromatography and associated retention time considerations (e.g. "Retention Index" Matching for GC/MS),
- Tandem mass spectrometry,
- Additional evidences from orthogonal techniques,
- Derivatization, and
- Indirect inferences

Lastly, collation of identities in a database prevents one from re-inventing the wheel by re-identifying the compound each time it is encountered across studies.

A.5   Final Thoughts

The use of mass spectral detectors in the screening process to account for all organic extractables and leachables that are present in a material, component, or device – or compounds that may have leached out and proceed to

have direct (for medical devices) or indirect (for drug products) patient exposure – has been widely accepted and implemented.

It is generally accepted that expert-reviewed mass spectral matching is a reliable means of tentatively identifying organic extractables and leachables via GC/MS. However, there are many circumstances where mass spectral matching will not produce an unequivocal identification. In such cases, more information about the compound's identity can be obtained via MSI. It should be noted that for mass spectra generated via LC/MS, MSI is often the only way to increase the level of identification for a detected compound, as there are no universal mass spectral libraries available that are optimized for LC/MS.

It is also noted that high confidence toxicological safety risk assessments can only be obtained from highly confident (or confirmed) identities. Thus, every effort should be made to elevate tentative identifications to confident or confirmed identities via corroborating information.

Lastly, it is emphasized that postulating a chemical structure, solely based upon mass spectral information and interpretation, is not an easy task. The mass spectrometrist performing this task should be aware of the importance and consequences of postulating a defined chemical structure, as this information will be used to link the chemical compounds to its toxicological information and will be the basis for a subsequent toxicological evaluation of the compound. Cases where the wrong identity for the compound is postulated will inevitably bias the overall safety evaluation of the material, device, or container/closure component or system.

References

1 Kind, T. and Fiehn, O. (2007). Seven Golden Rules for heuristic filtering of molecular formulas obtained by accurate mass spectrometry. *BMC Bioinform.* 8: 105.

2 Watson, J. and Sparkman, O. (2007). *Introduction to Mass Spectrometry: Instrumentation, Applications and Strategies for Data Interpretation*, 4e. Chichester: Wiley.

3 Smith, R. (2004). *Understanding Mass Spectra: A Basic Approach*, 2e. Hoboken, NJ: Wiley.

4 Stein, S. (1995). Chemical substructure identification by mass spectral library searching. *J. Am. Soc. Mass Spectrom.* 6: 644–655.

5 Hites, R. and Jobs, K. (2018). Is nontargeted screening reproducible? *Environ. Sci. Technol.* 52 (21): 11975–11976.

6 McLafferty, F. and Turecek, F. (1993). *Interpretation of Mass Spectra*, 4e. Sausalito: University Science Books.

7 Harrison, A.G. (1992). *Chemical Ionization Mass Spectrometry*, 2e. CRC Press.

8 Hoffmann, T.G.M. (2019). The next level in chemical space navigation: Going far beyond enumerable compound libraries. *Drug Discov. Today* 24 (5): 1148–1156.

9 Jenke, D., Christiaens, P., Baeten, J. et al. (2021). Good identification practices for organic extractables and leachables via mass spectrometry. *Am. Pharmaceut. Rev.* ISBN 978-0-578-95605-3.

10 USP <1663> Assessment of extractables associated with pharmaceutical packaging/delivery systems. USP 43/NF 38, pp. 8422. Effective 1 Dec 2020.

11 Jenke, D., Swanson, S., Edgcomb, E. et al. (2005). Strategy for assessing the leachables impact on material change in a container/closure system. *PDA J. Pharm. Sci. Technol.* 59 (6): 360–379.

Glossary and Abbreviations

Terms and abbreviations noted here have been used, sometimes frequently, throughout the book. Abbreviations and terms used infrequently, such as those used in an individual equation, are defined and described in the text discussing the equation. Abbreviations applied to specific chemical compounds are typically explained at the point at which the chemical is mentioned. This list of abbreviations contains chemicals that are mentioned frequently.

A

AAF — Accelerated Aging Factor, a term used to establish acceleration extraction and leaching conditions

AAMI — Association for the Advancement of Medical Instrumentation. AAMI is an organization for advancing the development and safe and effective use of medical technology through the generation of standards.

AAT — Analytical Action Threshold. The AAT reflects the lowest concentration of an extractable in an extract or a leachable in a drug product at which an analytical method is able to reliably perform the functions of detection, identification, and quantitation.

AAT — Accelerated Aging Time, a term used to establish acceleration extraction and leaching conditions

ACD — Allergic Contact Dermatitis. ACD is a form of dermatitis/eczema caused by an allergic reaction to a material, called an allergen, in contact with the skin.

AET — Analytical Evaluation Threshold. The AET is that concentration of an extractable in an extract or a leachable in a drug product at or above which the extractable or leachable must be assessed for its product safety impact. Such an assessment typically requires that the extractable or leachable be detected, identified, and quantified. Extractables or leachables that are present at levels below the AET are deemed to be safe and do not require assessment. The AET is derived from the SCT or an otherwise applicable quantity.

ANDA — Abbreviated New Drug Application. An ANDA contains data that, when submitted to FDA's Center for Drug Evaluation and Research, Office of Generic Drugs, provides for the review and ultimate approval of a generic drug product. Generic drug applications are called "abbreviated" because they are generally not required to include pre-clinical (animal) and clinical (human) data to establish safety and effectiveness.

ANSI — American National Standards Institute. ANSI facilitates the development of American National Standards (ANS) by accrediting the procedures of standards developing organizations (SDOs) and approving their documents as American National Standards (ANS).

APCI — Atmospheric Pressure Chemical Ionization (MS). A form of MS ionization used in tandem with LC separations.

API — Active Pharmaceutical Ingredient. An API is any substance or mixture of substances intended to be used in the manufacture of a drug product and that, when used in the production of a drug, becomes an active ingredient in the drug product. Such substances are intended to furnish pharmacological activity or other direct effect in the diagnosis, cure, mitigation, treatment, or prevention of disease or to affect the structure and function of the body. Analogous to DS.

ASTM — ASTM International, formerly known as American Society for Testing and Materials, is an international standards organization that develops and publishes voluntary consensus technical standards for a wide range of materials, products, systems, and services.

B

bDtBPP	bis(2,4-di-tert-butylphenyl) phosphate, an antioxidant-related extractable with a known ability to reduce yields in biopharmaceutical manufacturing by affecting cell growth and viability
BEDS	BioPhorum Extractables Data Summary
BEP	Biological Evaluation Plan
BHT	Butylated hydroxytoluene
BLA	Biologics License Application. A BLA is a submission to the FDA to secure approval for marketing a biologic product.
BPA	Bisphenol A, an extractable of higher safety concern and also the monomer of polycarbonate plastics
BPC	Base Peak Chromatogram
BPOG	BioPhorum Operations Group. BPOG mission is a cross-industry collaboration between biopharmaceutical developers, manufacturers, and suppliers whose purpose is to accelerate the advancement of biologics manufacturing in terms of product and process quality.
BPSA	Bio-Process Systems Alliance. BPSA is an industry-led international industry association dedicated to encouraging and accelerating the adoption of single-use manufacturing technologies used in the production of biopharmaceuticals and vaccines.
BW	Body Weight

C

C	Centigrade
C18	Chromatography stationary phase based on an octadecyl carbon chain (C18)-bonded to a support (typically silica) used in HPLC
C_x	Concentration of compound x
CAS RN	Chemical Abstract Services Registry Number. The CAS RN is a unique numerical identifier assigned by the Chemical Abstracts Service (CAS) to every chemical substance described in the open scientific literature, including organic and inorganic compounds, minerals, isotopes, alloys, and non-structurable

materials (UVCBs, substances of unknown or variable composition, complex reaction products, or of biological origin).

CBER	FDA'S Center for Biological Evaluation and Research
CCRIS	Chemical Carcinogenesis Research Information System. The CCRIS database contains chemical records with carcinogenicity, mutagenicity, tumor promotion, and tumor inhibition test results.
CCS	Container-closure System
CDER	FDA's Center for Drug Evaluation and Research
CDRH	FDA's Center for Devices and Radiological Health
CET	Compound-specific Evaluation Threshold. The CET is a concentration threshold that is specific for a specific extractable or leachable. If an extractable's or leachable's measured concentration is less that the CET, then the adverse patient safety impact of the extractable or leachable is deemed to be negligible. If the extractable's or leachable's measured concentration is greater than the CET, then the extractable or leachable has the potential to adversely impact patient safety. The CET is derived from the TDI.
CFR	Code of Federal Regulations. The CFR is the codification of the general and permanent rules and regulations (sometimes called administrative law) published in the Federal Register by the executive departments and agencies of the federal government of the United States.
CGT	Cell and Gene Therapy drug products
CHO	Chinese Hamster Ovary cells. CHO cells are an epithelial cell line derived from the ovary of the Chinese hamster, often used in biological and medical research.
CI	Chemical Ionization (MS)
CID	Collusion-Induced Dissociation
cm	Centimeter
CMC	Chemistry and Manufacturing Controls
COC	Cyclic Olefin Copolymer
CONEG	Coalition of Northeast Governors. The abbreviation is typically applied to legislation, passed by the Coalition, to create consistent requirements for the reduction/elimination of four specific toxics in packaging sold or distributed in the Northeast US.

CP	Comparability Protocol, a comprehensive, prospectively written, plan for assessing change control
CPDB	Carcinogenic Potency Database. The CPDB reports analyses of animal cancer tests on 1547 chemicals.
CRO	Contract Research Organization, typically an organization that performs E&L testing
CT	Charge transfer

D

D	Diffusion coefficient
δ	Thickness in an appropriate unit of measure
DAS	Distance Along the production Stream, e.g. upstream or downstream
DB-5MS	A non-polar, 5%-phenyl-methylpolysiloxane stationary phase used in GC
DBT	Dose-Based Threshold, such as the TTC or SCT
DCM	Dichloromethane
DDA	Data Dependent Acquisition
DE	Daily Exposure, that actual amount of a leachable a patient is exposed to in a day
DEHP	Di-(2-ethylhexyl phthalate). DEHP is a plasticizer for plastics used in pharmaceutical applications such as poly (vinyl chloride), PVC.
DEREK	Derek Nexus is expert, knowledge-based toxicology software that predicts whether a compound will likely trigger an alert for a variety of toxicity endpoints.
DP	Drug Product
DPI	Dry Powder Inhaler
DR	Dilution Ratio
DS	Drug Substance. Analogous to API

E

E	Extractable, a substance or chemical entity, extracted from a test article by an extraction vehicle that is present in that vehicle because it contacted the test article under specified laboratory test conditions, including temperature, duration, process (e.g. refluxing, sonication, etc.) and dimensions of contact (e.g. ratio of test article weight to extraction vehicle volume).
E&L, E/L	Extractables and Leachables, an inexact and largely inappropriate shorthand of chemical characterization

E_b	Equilibrium Binding Constant, conceptually equivalent to a partition coefficient
ED	Exposure (contact) Duration
EDX	Energy Dispersive X-ray spectroscopy
EED	Estimated Exposure Dose (from ISO 10993:17); EED_{max} is the maximum estimated exposure dose
EI	Electron (impact) Ionization (MS)
ELSIE	Extractables and Leachables Safety Information Exchange. ELSIE is a consortium comprised of pharmaceutical, biotechnology, and medical device companies whose mission is to advance the understanding of extractables and leachables and their impact on product quality and process performance.
EMEA	European Medicines Agency. The EMEA is a decentralized agency of the European Union (EU) responsible for the scientific evaluation, supervision, and safety monitoring of medicines in the EU.
EPA	Environmental Protection Agency
EPDM	Ethylene propylene diene monomer rubber
EPO	Erythropoietin. EPO is a glycoprotein cytokine secreted mainly by the kidney in response to cellular hypoxia; it stimulates red blood cell production in the bone marrow.
ESI	Electrospray Ionization (MS). A form of MS ionization used in tandem with LC separations.
ET	Exposure (contact) Temperature
ETOH	Ethanol
EVA	Ethylene vinyl acetate (polymer)
EVOH	(poly)Ethylene vinyl alcohol copolymer
E/W	A mixture of Ethanol and Water

F

FDA	Food and Drug Administration. The FDA is responsible for protecting and promoting public health through the control and supervision of food safety, tobacco products, dietary supplements, prescription and over-the-counter pharmaceutical drugs (medications), vaccines, biopharmaceuticals, blood transfusions, medical devices, electromagnetic radiation emitting devices, cosmetics, animal foods and feed and veterinary products.
FTIR	Fourier Transform Infrared spectroscopy

G

g	Gram

GC	Gas Chromatography
GC/FID	Gas Chromatography with Flame Ionization Detection
GC/IR	Gas Chromatography with Infrared spectrometric detection
GC/MS	Gas Chromatography with Mass Spectrometric detection
GFAAS	Graphite Furnace Atomic Absorption Spectroscopy
GMP	Good Manufacturing Practices

H

HILIC	Hydrophilic Interaction Liquid Chromatography
HS	Headspace, a gas generation/sampling system as the "front-end" of a GC system
HPLC	High Performance Liquid Chromatography. Analogous to the term LC
HPVP	High Production Volume Plastic
hr	Hour
HRAMS	High-Resolution Accurate Mass Spectrometry

I

IC	Ion Chromatography, a screening method that targets inorganic anions and low molecular weight organic acids
ICH	International Council for Harmonization of Technical Requirements for Pharmaceuticals for Human Use. ICH is an initiative that brings together regulatory authorities and the pharmaceutical industry to discuss scientific and technical aspects of pharmaceutical product development and registration. The mission of the ICH is to promote public health by achieving greater harmonization through the development of technical Guidelines and requirements for pharmaceutical product registration.
IC/MS	Ion Chromatography with Mass Spectrometric detection
ICP/AES	Inductively Coupled Plasma Atomic Emission Spectroscopy. Analogous to the term ICP/OES. A test method for quantifying the levels of elements in a sample.
ID	Identity
ICP/MS	Inductively Coupled Plasma Mass Spectroscopy. A test method for quantifying the levels of elements in a sample. ICP/MS is generally more sensitive than ICP-AES.

ICP/OES	Inductively Coupled Plasma Optical Emission Spectroscopy. Analogous to the term ICP/AES
IFU	Instructions For Use, typically packaged with the drug product or medical device.
IND	Investigational New Drug. The IND is the means through which the sponsor technically obtains an exemption from the FDA to transport or distribute a DP across state lines (for example, to support clinical trials).
IPA	Isopropyl Alcohol
IS	Internal Standard, an analyte added to a sample at a known concentration to facilitate the use of the AET or to provide a means of assessing analyte loss during sample testing.
ISO	International Organization for Standardization. ISO is an independent, non-governmental international organization for developing voluntary, consensus-based, market-relevant International Standards that support innovation and provide solutions to global challenges.
IV	Intravenous or "through the vein"

J

JP	Japanese Pharmacopeia

K

$K_{P/L}$	Plastic/Liquid Partition Coefficient, the ratio of the concentration of a substance in contacting P and L phases at equilibrium
kGy	Kilograys, a dose of radiation

L

l	Liter
L	Leachable, a substance or chemical entity, leached from a packaging system, a manufacturing component, or a medical device by a pharmaceutical product, process stream or a body fluid/tissue, that is present in the pharmaceutical product, process stream or body fluid/tissue because these objects contacted the system, the component, or the device under the manufacturing, distribution, storage, or clinical use conditions.
L & E	Leachables and Extractables, typically a reference to chemical characterization, analogous to E&L
LC	Liquid Chromatography, analogous to the term HPLC

LC/CAD	Liquid Chromatography with Charged Aerosol Detection
LC/ELSD	Liquid Chromatography with Evaporative Light Scattering Detection
LC/MS	Liquid Chromatography with Mass Spectrometric detection
LC/PDA	Liquid Chromatography with UV absorbance detection using a Photodiode Array UV spectrometer
LC/UV	Liquid Chromatography with UV absorbance detection. Analogous to the term LC/PDA but generally applied to a measurement made at a single detection wavelength
LD_{50}	LD_{50} is the amount of a material, given all at once, that causes the death of 50% (one half) of a group of test animals
LLDPE	Linear Low-Density Polyethylene (polymer)
LLE	Liquid-liquid Extraction
LoD	Limit of Detection. The LoD is the lowest quantity or concentration of a component that can be reliably detected with a given analytical method.
LOAEL	Lowest dose at which there was an observed toxic or Adverse Effect
LOEL	Lowest dose at which there was an Observed Effect
LoQ	Limit of Quantitation. The LoQ is the lowest analyte concentration that can be quantitatively determined with a stated accuracy and precision.
LRI	Linear Retention Index. A retention time expressed relative to the retention time of a standard compound.
LSER	Linear Solvation Energy Relationship
LVP	Large Volume Parenteral, a parenteral dosage form packaged in a unit dose of greater than 100 ml.

M

M	Molar Concentration (moles/l)
M_a	Mass of substance a
MBT	2-Mercapto-benzothiazole, a "bad actor," "special case," or "cohort of concern" compound that is exceptionally toxic and for which the AET does not apply.
MDDV	Maximum Daily Dose Volume
MD	Medical Device
MDE	Maximum Daily Exposure. The MDE is the maximum amount of a specific extractable or leachable that the patient would be exposed to on a daily basis. The MDE is obtained as the product of the extractable's or leachable's concentration and the daily dose volume of the drug product.
MDI	Metered Dose Inhaler
MDR	Medical Device Regulations. The European MDR is a new set of regulations that governs the production and distribution of medical devices in Europe, and compliance with the regulation is mandatory for medical device companies that want to sell their products in the European marketplace.
MEHP	Mono-(2-ethyl hexyl) phthalate
MF	Match factor, applicable to compound identification by mass spectral matching (MSM)
mg	Milligram
µg	Microgram
ml	Milliliter
min	Minute
MoS	Margin of Safety
MS	Mass Spectrometry
MS^n	Multi-stage mass spectrometry where the value of n is the number of stages
MSDS	Material Safety Data Sheet. An MSDS is a safety document required by the Occupational Safety and Health Administration (OSHA) that contains data about the physical, chemical, and toxicological properties of a particular substance.
MSI	Mass Spectral Interpretation
MSM	Mass spectral matching
MW	Molecular weight
m/z	Mass to Charge ratio

N

N/A	Not Applicable
ND	Not Detected or Not Discernible
NDA	New Drug Application. An NDA is the documentation of the sponsor of a new drug submitted to the FDA to secure marketing approval.
NaOH	Sodium hydroxide
ng	Nanogram
NIBRT	National Institute for Bioprocessing Research and Training. NIBRT is a global center of excellence for training and research in bioprocessing, including chemical characterization.
NIH	National Institute of Health
NIST	National Institute of Standards and Technology. NIST's mission is to promote US

innovation and industrial competitiveness by advancing measurement science, standards, and technology in ways that enhance economic security and improve our quality of life.

nm	Nanometer
NMPA	National Medical Products Administration. The NMPA is the Chinese government's administrative body responsible for regulating pharmaceuticals, medical devices, and cosmetics in China.
NMR	Nuclear Magnetic Resonance spectroscopy or spectrometer
NOAEL	Highest dose at which there was No Observed Toxic or Adverse Effect
NOEL	Highest dose at which there was No Observed Effect
NTA	Non-Targeted Analysis, similar to the term "screening"
NVR	Non-Volatile Residue. NMR is a measure of the total amount of non-volatile substances present in a solution.

O

ODP	Ophthalmic Drug Product
OINDP	Orally Inhaled and Nasal Drug Product
OOS	Out-of-Specification. An OOS result is a test result that falls outside the test specifications.

P

P	Plastic
PA	Polyamide polymer
PAH	Polycyclic Aromatic Hydrocarbon, a "bad actor," "special case," or "cohort of concern" class of compounds that are exceptionally toxic and for which the AET does not apply.
PBS	Phosphate Buffered Saline. PBS is a buffer solution commonly used in biological research. It is a water-based salt solution containing disodium hydrogen phosphate, sodium chloride and, in some formulations, potassium chloride and potassium dihydrogen phosphate. The buffer helps to maintain a constant pH of approximately 7.4. The osmolarity and ion concentrations of the solution matches that of the human body (isotonic).
PBT	Very persistant, very bio-accumulative; applied to a compound or element

PDE	Permissible Daily Exposure, analogous to a TI or TDI
PDP	Parenteral Drug Product
PDMS	Polydimethylsiloxane polymer
PE	Polyethylene polymer
PERL	Process Equipment-Related Leachable. A PERL is a substance that has been extracted from a manufacturing component and which resides in a process stream.
PES	Polyethersulfone polymer
PET	Polyethylene terephthalate polymer
PETG	Polyethylene terephthalate glycol polymer
PFA	Perfluoroalkoxy polymer
PFI	Process Fluid Interaction, as in an interaction between a process fluid and the manufacturing component
PFS	Pre-filled Syringe
Pharm Eur	European Pharmacopeia
PI	Polyisoprene rubber
pK_a	Acid dissociation constant
PLA	Polylactic acid polymer
PNA	Polynuclear Aromatics, a "bad actor," "special case," or "cohort of concern" class of compounds that are exceptionally toxic and for which the AET does not apply; typically linked to rubber items
PO	Polyolefin polymer
PoD	Point of Departure. The PoD is the dose that is used as the beginning point of a low-dose extrapolation or calculation; for example, the SCT used to calculate an AET.
POD	Point of Departure, the starting point of the calculation of the TI and MoS
PODP	Parenteral and Ophthalmic Drug Product, consisting of parenteral drug products (DBP) and ophthalmic drug products (ODP)
$P_{o/w}$	Octanol/water partition coefficient
PP	Polypropylene polymer
ppb	Parts per billion. The ppb is a unit of measure corresponding to 1 ng/g (or equivalent units) on a weight basis and 1 ng/l (or equivalent units) on a volume basis.
ppm	Parts per million. The ppm is a unit of measure corresponding to 1 µg/g (or equivalent units) on a weight basis and 1 µg/l (or equivalent units) on a volume basis.
PQRI	Product Quality Research Institute. PQRI is a non-profit consortium of organizations working together to generate and share timely, relevant, and impactful information that advances global drug product quality, manufacturing, and regulation.

PRCA   Pure Red Cell Aplasia, an affliction linked to the EPREX® E&L case

Prop 65   Proposition 65. Prop 65 is legislation that requires businesses to provide warnings to Californians about significant exposures to chemicals that cause cancer, birth defects, or other reproductive harm. Proposition 65 requires California to publish a list of chemicals known to cause cancer, birth defects, or other reproductive harm.

PS   Polystyrene polymer

PSU   Polysulfone polymer

PTFE   Poly-(tertrafluoroethylene) polymer

PU   Polyurethane polymer

PVC   Poly (vinyl chloride) polymer. PVC, after it has been plasticized, is employed in medical items such as bags and tubing.

PVDF   Poly(vinylidene difluoride) polymer

Q

Q_{10} Aging factor, a term used to establish acceleration extraction and leaching conditions

QbD   Quality by Design; the concept that quality should be built into a product and not established by testing of the product.

QSAR   Quantitative Structure Activity Relationship. QSAR models summarize the relationship between the chemical structural features of a compound and its physicochemical properties or biological activities.

QT   Qualification Threshold. The threshold below which a given non-carcinogenic leachable is not considered for safety qualification (toxicological assessment) unless the leachable presents structure-activity (SAR) concerns.

Q-TOF   Quadrupole Time-of-Flight mass spectrometer

R

REACH   Registration, Evaluation, Authorization, and Restriction of Chemicals. REACH is a regulation of the European Union, adopted to improve the protection of human health and the environment from the risks that can be posed by chemicals, while enhancing the competitiveness of the EU chemicals industry.

RF   Response Factor, the ratio of an analyte's concentration versus the magnitude of analytical response to the analyte at that concentration

RI   Risk Index. The RI is a quantity, somewhat akin to a TI, that establishes the level at or above which a compound has a higher risk of producing an undesirable outcome.

RRF   Relative Response Factor, the ratio of the response factor of one compound versus another (for example, an analyte versus an internal standard)

RRT   Relative Retention Time, the retention time of an analyte in reference to the retention time of a standard anchor compound

Rs   Response produced on an analytical instrument by an analyte of interest s.

RSD   Relative Standard Deviation

RT   Retention Time

RTL   Retention Time Locking, is a means of producing the same retention time obtained on one chromatographic system (typically GC) with another similar system equipped with the same nominal column.

S

S_e Conditional Solubility of compound e

S_o Intrinsic Solubility

SA_n Surface Area of item *n*

SARAH   Sarah Nexus is a statistical software tool that provides accurate mutagenicity predictions based on a compound's structure.

SCT   Safety Concern Threshold. The SCT is the maximum amount (dose) of any leachable (or extractable as a potential leachable) that a patient can receive on a daily basis which presents a negligible safety risk to the patient. The SCT is generally derived by a toxicological consideration of large groups of substances, which may or may not be extractables or leachables, and is based on an acceptable occurrence of an adverse safety outcome (e.g., 1 in 100 000 chances of developing cancer due to daily exposure to the compound dosed to a patient at the SCT amount).

SDO   Standards Developing Organization

SEM   Scanning Electron Microscopy

SEP   Standard Extraction Protocol for manufacturing components per USP <665>

SOP   Standard Operating Procedure

SS   Surrogate Standard. A substance used to facilitate the quantitation of the analyte. The

	SS response is applied to all analytes to provide for their quantitation.
SST	System Suitability Test, a test used to establish that a method is functioning properly at its time of use and through that use.
STD	Standard Deviation
SUS	Single-Use (SU) System, a manufacturing system that is used once and then discarded
SUT	Single-Use Technology, see SUS.
SA/V	Surface Area of an item versus the Volume of solution in contact with that item, typically applied to an extraction
SVP	Small Volume Parenteral, a parenteral dosage form packaged in a unit dose of 100 ml or less

T

t	Time, measured in appropriate units
t_p	Thickness of an item p
T	Temperature, measured in appropriate units
TCI	Tolerable Contact Level, the level of a substance leached from a medical device that does not cause contact-related adverse effects such as irritation
TD	Thermal Desorption, a sample introduction method for GC analysis
TD_{LO}	The lowest dose of a substance introduced by any route, other than inhalation, over any given period of time, and reported to produce any toxic effect in humans or to produce tumorigenic or reproductive effects in animals.
TDI	Tolerable Daily Intake. The TDI is the maximum amount (dose) of a leachable (or an extractable as a probable leachable) that a patient can receive on a daily basis without an adverse safety impact. The TDI is determined for individual extractables or leachables based on an assessment of the available toxicological safety data. Analogous to a PDE.
TGA	Thermo-Gravimetric Analysis
TI	Tolerable Intake. The TI is the maximum amount (dose) of a leachable (or an extractable as a probable leachable) that a patient can receive without an adverse safety impact. The TI should be annotated to include the duration of exposure.

TIC	Total Ion Current (chromatogram)
TLC	Thin-Layer Chromatography
TMS	Trimethylsilylation, a derivatization process designed to make organic compounds easier to chromatograph
TOC	Total Organic Carbon, a measurement of the total amount of carbon present in a sample
TOF	Time of Flight (mass spectrometer)
TPE	Thermoplastic Elastomer
TRS	Total Risk Score
TSE	Transmissible Spongiform Encephalopathies
TSL	Toxicological Safety Limit
TTC	Threshold of Toxicological Concern. The TTC is a level of human intake or exposure that is considered to be of negligible risk, despite the absence of chemical-specific toxicity data

U

UF	Uncertainty Factor, typically used to adjust a quantity, such as the AET, for the uncertainty involved in establishing the quantity
ULDPE	Ultra-Low-Density Polyethylene (polymer)
UPLC	Ultra-high Pressure Liquid Chromatography
USP	United States Pharmacopeia or United States Pharmacopeial Convention. The United States Pharmacopeia (USP) is a compendium of drug information for the United States. The United States Pharmacopeial Convention establishes written (documentary) and physical (reference) standards for medicines, food ingredients, dietary supplement products, and ingredients.

V

V_L	Volume of liquid L
vPvB	Very persistent, very bio-accumulative. Applied to compounds to communicate their toxic nature

W

WFI	Water For Injection

Index